Pests of Landscape Trees and Shrubs

An Integrated Pest Management Guide
THIRD EDITION

UC Statewide Integrated Pest Management Program

Steve H. Dreistadt
Writer

Jack Kelly Clark
Principal Photographer

Tunyalee A. Martin
Content Supervisor

Mary Louise Flint
Technical Editor

Oakland, California • 2016
Publication 3359

University *of* **California**
Agriculture and Natural Resources

To order or obtain ANR publications and other products, visit the ANR Communication Services online catalog at http://anrcatalog.ucanr.edu, or phone 1-800-994-8849. Direct inquiries to

University of California
Agriculture and Natural Resources
Communication Services
2801 Second Street
Davis, CA 95618
Telephone 1-800-994-8849
Email: anrcatalog@ucanr.edu

Third edition, 2016
© 1994, 2004, 2016 The Regents of the University of California
Agriculture and Natural Resources
Reprint December 2019

Publication 3359
ISBN-13: 978-1-60107-864-3

Library of Congress Cataloging-in-Publication Data

Names: Dreistadt, Steve H., author. | Clark, Jack Kelly, photographer. | University of California Integrated Pest Management Program, issuing body.
Title: Pests of landscape trees and shrubs : an integrated pest management guide / UC Statewide Integrated Pest Management Program ; Steve H. Dreistadt, writer ; Jack Kelly Clark, principal photographer ; Tunyalee A. Martin, content supervisor ; Mary Louise Flint, technical editor.
Other titles: Publication (University of California (System). Division of Agriculture and Natural Resources) ; 3359.
Description: Third edition. | Oakland, California : University of California, Agriculture and Natural Resources, 2016. | Series: Publication ; 3359 | Includes bibliographical references and index.
Identifiers: LCCN 2015044249 | ISBN 9781601078643
Subjects: LCSH: Ornamental trees--Diseases and pests--Integrated control--California. | Ornamental shrubs--Diseases and pests--Integrated control--California. | Ornamental trees--Diseases and pests--California. | Ornamental shrubs--Diseases and pests--California.
Classification: LCC SB763.C2 D74 2016 | DDC 635.9/7709794--dc23
LC record available at http://lccn.loc.gov/2015044249

Project Management: Stephen Barnett. Design: Celeste Rusconi. Illustrations: A. Child, David Kidd, Celeste Rusconi, W. Suckow, Robin Walton, and Valerie Winemiller. Proofreading and indexing: Hazel White. Print Coordination: Ann Senuta. Archivist: Evett Kilmartin. Production assistance: Sueanne Johnson.

Photo credits are given in the captions. J. K. Clark: pp. 1, 3, 17, 65, 125; L. R. Costello, p. 43; S. Paisley, p. 253; J. D. Becker, p. 289. Cover design: Celeste Rusconi, photos: E. Kilmartin (top), Karey Windbiel-Rojas (middle), ANR (bottom).

UC PEER REVIEWED This publication has been anonymously peer reviewed for technical accuracy by University of California scientists and other qualified professionals. This review process was managed by ANR Associate Editor for Urban Pest Management Mary Louise Flint.

Printed in Canada on recycled paper.
3m-rep-12/19-SB/CR/WS

PRECAUTIONS FOR USING PESTICIDES

Pesticides are poisonous and must be used with caution. READ THE LABEL CAREFULLY BEFORE PURCHASING A PESTICIDE AND READ THE LABEL AGAIN BEFORE OPENING A PESTICIDE CONTAINER. Follow all label precautions and directions, including requirements for protective equipment. Use a pesticide only on the plants or site specified on the label or in published University of California recommendations. Apply pesticides at the rates specified on the label or at lower rates if suggested in this publication. In California, all agricultural uses of pesticides must be reported, including use in many nonfarm situations, including cemeteries, golf courses, parks, roadsides, and commercial nurseries. Contact your county agricultural commissioner for further details. Laws, regulations, and information concerning pesticides change periodically, so be sure the publication you are using is up-to-date.

Legal Responsibility. The user is legally responsible for any damage due to misuse of pesticides. Responsibility extends to effects caused by drift, runoff, or residues.

Transportation. Do not ship or carry pesticides together with food or feed in a way that allows contamination of the edible items. Never transport pesticides in a closed passenger vehicle or in a closed cab.

Storage. Keep pesticides in original containers until used. Store them in a locked cabinet, building, or fenced area where they are not accessible to children, unauthorized persons, pets, or livestock. DO NOT store pesticides with foods, feed, fertilizers, or other materials that may become contaminated by the pesticides.

Container Disposal. Dispose of empty containers carefully. Never reuse them. Make sure empty containers are not accessible to children or animals. Never dispose of containers where they may contaminate water supplies or natural waterways. Home-use pesticide containers can be thrown in the trash if they are completely empty and they were used by the homeowner on their own property. Professional applicators and anyone who performs pest control for others or applies pesticides as part of their job should contact the local county agricultural commissioner to learn the proper methods for handling and disposing of empty pesticide containers.

Protection of Nonpest Animals and Plants. Many pesticides are toxic to useful or desirable animals, including honey bees, natural enemies, fish, domestic animals, and birds. Certain rodenticides may pose a special hazard to animals that eat poisoned rodents. Desirable plants may be damaged by misapplied pesticides. Take precautions to protect nonpest species from direct exposure to pesticides and from contamination due to drift, runoff, or residues.

Permit Requirements. Certain pesticides require a permit from the county agricultural commissioner before possession or use.

Plant Injury. Certain chemicals may cause injury to desirable plants (phytotoxicity) under certain conditions. Always consult the label for limitations. Before applying any pesticide, take into account the stage of plant development, the soil type and condition, the temperature, moisture, and wind. Injury may also result from the use of incompatible materials.

Personal Safety. Follow label directions carefully. Avoid splashing, spilling, leaks, spray drift, and contamination of clothing. NEVER eat, smoke, drink, or chew while using pesticides. Provide for emergency medical care IN ADVANCE as required by regulation.

Contributors and Acknowledgments

This book was produced under the auspices of the University of California, Agriculture and Natural Resources, Statewide Integrated Pest Management (IPM) Program.

Steve H. Dreistadt, Writer

Jack Kelly Clark, Principal Photographer

Tunyalee A. Martin and Petr Kosina, Content Supervisors

Joyce Fox Strand, Associate Director for Communications

Kassim Al-Khatib, Director

Mary Louise Flint, UC Agriculture and Natural Resources Associate Editor

TECHNICAL COORDINATORS FOR THE THIRD EDITION

Mary Louise Flint, UC IPM Program and Entomology, UC Davis

Pamela M. Geisel, UC Statewide Master Gardener Program

Deborah Mathews, Plant Pathology, UC Riverside

Tedmund J. Swiecki, Phytosphere Research

Cheryl A. Wilen, UC IPM Program and UC Cooperative Extension, Orange, Los Angeles, and San Diego Counties

CONTRIBUTORS TO THE THIRD EDITION

J. Ole Becker, Nematology, UC Riverside

Lisa A. Blecker, UC Statewide IPM Program

Lawrence R. Costello, UC Cooperative Extension, San Mateo and San Francisco Counties

A. James Downer, UC Cooperative Extension, Ventura County

Richard Y. Evans, Plant Sciences, UC Davis

Janet Hartin, UC Cooperative Extension, San Bernardino County

Janine K. Hasey, UC Cooperative Extension, Sutter and Yuba Counties

Darren L. Haver, UC Cooperative Extension, Orange County

Mark Hoddle, Entomology, UC Riverside

Donald R. Hodel, UC Cooperative Extension, Los Angeles County

Chuck A. Ingels, UC Cooperative Extension, Sacramento County

John N. Kabashima, UC Cooperative Extension, Orange and Los Angeles Counties

John F. Karlik, UC Cooperative Extension, Kern County

Vincent F. Lazaneo, UC Cooperative Extension, San Diego County

Lorence R. Oki, Plant Sciences, UC Davis

Dennis R. Pittenger, UC Cooperative Extension, Central Coast, South Region, and Los Angeles County and Botany and Plant Sciences, UC Riverside

Karrie Reid, UC Cooperative Extension, San Joaquin County

John A. Roncoroni, UC Cooperative Extension, Napa County

Andrew M. Sutherland, UC IPM Program and UC Cooperative Extension, Alameda, Contra Costa, San Francisco, San Mateo, and Santa Clara Counties

Steven V. Swain, UC Cooperative Extension, Marin County

Steven A. Tjosvold, UC Cooperative Extension, Santa Cruz and Monterey Counties

TECHNICAL COORDINATORS FOR SECOND EDITION

Laurence R. Costello, UC Cooperative Extension, San Mateo and San Francisco Counties

A. James Downer, UC Cooperative Extension, Ventura County

Clyde L. Elmore, Weed Science Program, UC Davis

Donald R. Hodel, UC Cooperative Extension, Los Angeles County

John N. Kabashima, UC Cooperative Extension, Orange County

Edward J. Perry, UC Cooperative Extension, Stanislaus County

Robert D. Raabe, Division of Environmental Science, Policy, and Management, UC Berkeley

Pavel Svihra, UC Cooperative Extension, Marin County

Cheryl A. Wilen, UC IPM Program and UC Cooperative Extension, Orange, Los Angeles, and San Diego Counties

TECHNICAL COORDINATORS FOR FIRST EDITION

Carlton S. Koehler, Extension Entomologist Emeritus, UC Berkeley

Arthur H. McCain, Extension Plant Pathologist Emeritus, UC Berkeley

Robert D. Raabe, Plant Pathologist Emeritus, UC Berkeley

CONTRIBUTORS TO THE PREVIOUS EDITIONS

David H. Adams, California Department of Forestry and Fire Protection

Michael Baefsky, Baefsky & Associates

Bethallyn Black, UC Cooperative Extension, Contra Costa County

Pamela S. Bone, UC Cooperative Extension, Sacramento County

Heather Costa, Entomology, UC Riverside

Laurence R. Costello, UC Cooperative Extension, San Mateo and San Francisco Counties

Richard S. Cowles, Connecticut Agricultural Experiment Station

Donald L. Dahlsten, Environmental Science, Policy, and Management, UC Berkeley

Dean R. Donaldson, UC Cooperative
Extension, Napa County

A. James Downer, UC Cooperative
Extension, Ventura County

Lester E. Ehler, Entomology, UC Davis

Clyde L. Elmore, Weed Science Program,
UC Davis

Richard W. Harris, Environmental
Horticulture, UC Davis

Donald R. Hodel, UC Cooperative
Extension, Los Angeles County

Chuck A. Ingels, UC Cooperative
Extension, Sacramento County

John N. Kabashima, UC Cooperative
Extension, Orange County

John F. Karlik, UC Cooperative Extension,
Kern County

Carlton S. Koehler, Entomology,
UC Berkeley

John Lichter, Tree Associates

Richard S. Melnicoe, Toxicology, UC Davis

Arthur H. McCain, Plant Pathology,
UC Berkeley

Timothy D. Paine, Entomology,
UC Riverside

Edward J. Perry, UC Cooperative
Extension, Stanislaus County

Dennis R. Pittenger, UC Cooperative
Extension and Botany and Plant
Sciences, UC Riverside

Robert D. Raabe, Environmental Science,
Policy, and Management, UC Berkeley

Pavel Svihra, UC Cooperative Extension,
Marin County

Cheryl A. Wilen, UC IPM Program and UC
Cooperative Extension, Orange, Los
Angeles, and San Diego Counties

Ellen M. Zagory, Davis Arboretum,
UC Davis

SPECIAL THANKS

The following have generously provided information, offered suggestions, reviewed draft manuscripts, identified pests, or helped obtain photographs:

Maria Alfaro, Bob Allen, Edith B. Allen, Vonny M. Barlow, Walter J. Bentley, Alison M. Berry, James A. Bethke, Karen Beverlin, Matthew Blua, Patrick Brown, Margaret A. Brush, Robert L. Bugg, David W. Burger, Linda Farrar Bybee, Kathleen Campbell, Bob Cordrey, Donald A. Cooksey, Cheryl Covert, Don Cox, David W. Cudney, Kent M. Daane, Jerry Davidson, John Debenedictis, Paul De Ley, Joseph DiTomaso, Linda L. Dodge, Dean R. Donaldson, Thomas D. Eichlin, Lynn Epstein, Donald M. Ferrin, Debbie Flower, Gordon W. Frankie, Mach T. Fukada, Matteo Garbelotto, Rosser W. Garrison, Sal Genito, Raymond J. Gill, Deborah D. Giraud, Thomas R. Gordon, Patricia Gouveia, David A. Grantz, Marcella E. Grebus, Walter D. Gubler, Bruce Hagen, Kenneth S. Hagen, Susan E. Halbert, Richard W. Harris, David R. Haviland, Michael J. Henry, Gary W. Hickman, Carole Hinkle, Raymond L. Hix, Christine Joshel, Andrea Joyce, Greg Kareofalas, Harry K. Kaya, David Kellum, Ann I. King, Shawn King, Bruce C. Kirkpatrick, Steven T. Koike, John LaFlour, W. Thomas Lanini, Vernard R. Lewis, Robert F. Luck, James D. MacDonald, Armand R. Maggenti, Jennifer Manson-Hing, Nelda Matheny, Doug McCreary, Michael V. McKenry, John McKnight, Gregory E. McPherson, John A. Menge, Laura Merrill, Roland D. Meyer, Jocelyn G. Millar, Richard H. Molinar, Joseph G. Morse, Eric C. Mussen, Eric T. Natwick, Julie P. Newman, Robert F. Norris, Kenneth Nunes, Patrick J. O'Connor-Marer, Barbara L. P. Ohlendorf, Loren R. Oki, Michael P. Parrella, Gale Perez, Mitch Poole, Dan Pratt, Alexander H. Purcell, Richard A. Redak, Cheryl A. Reynolds, Steve Ries, David M. Rizzo, Karen L. Robb, Phillip A. Roberts, William Roltsch, Robin L. Rosetta, Celeste Rusconi, Kay Ryugo, Heather Scheck, Steven J. Seybold, Dave A. Shaw, Dave Shelter, Andrew J. Storer, Larry L. Strand, Glen Struckman, Mac Takeda, Beth L. Teviotdale, Timothy Tidwell, Lucy Tolmach, Diane E. Ullman, Lucia Varela, Baldo Villegas, Heather Yaffee, Doug E. Walker, Edward Weber, Becky B. Westerdahl, Karen Wikler, Karey Windbiel-Rojas, Frank P. Wong, David L. Wood, Stuart Wooley, Laosheng Wu, Frank G. Zalom, Roger T. Zerillo, Robert L. Zuparko

Contents

CHAPTER SEVEN
WEEDS 253

Chapter 1   What's in This Book

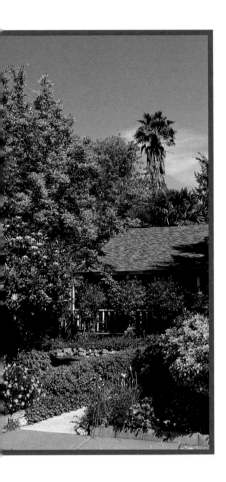

This book is for landscape professionals, home gardeners, and pest managers. Its purpose is to encourage maintenance of healthy trees and shrubs through integrated pest management (IPM).

Chapter 2 describes how to develop an IPM program. Methods include selecting plants that are well adapted to the environment and resistant to pests and using an appropriate combination of cultural practices and biological, mechanical, and physical controls. Pesticides are also used in many integrated pest management programs, but this book generally does not make specific recommendations because availability and appropriate uses of pesticides periodically change. Where pesticides are mentioned, less-toxic materials are emphasized because they generally are more compatible with IPM programs. For more information, including specific pesticide recommendations, consult the online guides and *UC IPM Pest Notes* at ipm.ucanr.edu. For more help, see the "Suggested Reading" at the back of this book or contact the University of California (UC) Cooperative Extension office and UC Master Gardener Program in your county.

Landscape design, planting, and cultural care activities (e.g., proper irrigation) that prevent and minimize damage to trees and shrubs are discussed in Chapter 3. Abiotic disorders are discussed in Chapter 4. These include problems caused by adverse environmental conditions, inadequate cultural care, and other nonliving factors. Subsequent chapters cover pest identification, biology, monitoring, and management for plant pathogens (Chapter 5); insects, mites, and snails (Chapter 6); weeds (Chapter 7); and nematodes (Chapter 8). Vertebrate pests are covered in *Wildlife Pest Control Around Gardens and Homes* and the online *Pest Notes*.

If you are uncertain of the cause of unhealthy plants, consult the two tables at the back of the book (Chapter 9). The "Problem-Solving Guide" briefly summarizes damage symptoms that can occur on many species of trees and shrubs. The "Tree and Shrub Pest Tables" are organized alphabetically according to host plant genus, or scientific name, and list the common problems of over 200 genera or species of trees and shrubs occurring in California landscapes. Both tables direct you to sections of the book that picture and discuss these common causes of problems.

For identification and biology of pests affecting herbaceous ornamentals and flowering annuals, consult publications such as *Integrated Pest Management for Floriculture and Nurseries* or (for commercial growers) the online *Floriculture and Ornamental Nurseries: UC IPM Pest Management Guidelines*. Home fruit and vegetable garden pest management is discussed in *The Home Orchard: Growing Your Own Deciduous Fruit and Nut Trees, Pests of the Garden and Small Farm*, and the *Pest Notes* and online guides.

A list of suggested reading, a glossary, and an index are provided at the back of the book.

Chapter 2 Designing an IPM Program

Integrated pest management (IPM) is an ecological management strategy that prevents or reduces pest damage with minimum harm to human health, the environment, and nontarget organisms. To apply IPM, take preventive measures, correctly diagnose and identify the cause of problems, and monitor plants and pests to determine whether and when to take action. Use knowledge of plant and pest biology to take the appropriate control actions, which may be a combination of biological, chemical, cultural, mechanical, and physical management practices. Use pesticides only if monitoring reveals that they are the best course of action. Choose and apply them in ways that avoid disrupting other IPM management tactics.

WHICH ORGANISMS ARE PESTS?

Many types of organisms can damage trees and shrubs or otherwise be undesirable in landscapes. Landscapes can also be damaged by abiotic disorders caused by adverse environmental conditions and inappropriate cultural practices. Abiotic factors and pest organisms often work in combination to damage or kill plants.

Pests include certain fungi and other disease-causing organisms, insects, mites, mollusks (snails and slugs), nematodes, vertebrates, and weeds. However, in each of these groups there are many species that are beneficial or do not harm plants; in fact, the great majority of organisms in the landscape are desirable components of the ecosystem.

The mere presence of organisms with the potential to become pests may not be cause for concern. For example, many fungi and other microorganisms that can cause disease are continually present in the environment; they usually become damaging only when conditions favor disease development or when environmental stress (e.g., cold soil) or poor cultural practices (e.g., improper irrigation) weaken a plant. Insects, mites, and nematodes that can cause damage when they are abundant can be harmless or even beneficial when their numbers are low; the presence of a few individuals of these plant-feeders provides food to maintain natural enemies that help to prevent outbreaks.

Organisms become pests when they compete with, feed on, or infect desirable plants or other organisms to the extent that they cause undesirable effects. Some organisms are pests because of their excrement, such as aphids' sticky honeydew. Pests reduce landscape quality and function, and they range in severity from problems that are merely annoying or unattractive to organisms that threaten the survival of desirable plants.

The extent to which insects, fungi, weeds, and other organisms are pests depends mostly on how much they interfere with the specific purposes for which plants are grown. Location, plant value and vigor, the species of plant-feeding organisms present, and the attitude and knowledge of people using the landscape also influence whether certain organisms are pest problems.

Your choice from among the many plant species and cultivars will minimize or favor pest problems. For example, rose cultivars resistant to black spot and powdery mildew are good choices when planting roses in landscapes. *Photo:* J. K. Clark

1. Prevention through good cultural practices   2. Pest and symptom identification   3. Regular monitoring

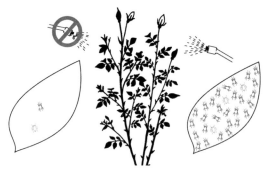 

4. Action thresholds: Treat only when necessary   5. Integration of appropriate management methods

Figure 2-1.

The major components of integrated pest management.
1. Prevention, such as not irrigating established plants around the root crown, which promotes root decay pathogens.
2. Correct pest identification and diagnosis of the cause of plant damage symptoms.
3. Regular surveying of valued plants for damage, pests, and conditions that contribute to problems, such as inappropriate cultural practices.
4. Control action guidelines, which entail tolerating some level of certain pests and damage if they do not threaten plants' health and survival.
5. Appropriate management, including a combination of biological, chemical, mechanical, and physical controls where needed, such as hoeing weeds and applying mulch.

IPM PROGRAM COMPONENTS

Effective, environmentally sound pest management requires considerable forethought, knowledge, and observation. If you wait until a tree or shrub becomes unhealthy or heavily damaged by pests, the only effective option might be to apply a fast-acting pesticide, or it may be too late for any method to be effective except plant replacement.

Focus on keeping plants healthy by using good landscape maintenance practices and taking actions to prevent problems in landscapes. Examine valued plants regularly for pests, damage, and inappropriate cultural practices; keep records of any problems you encounter. Learn the normal healthy appearance of your plants so you can recognize when a plant appears abnormal or when pest abundance or damage is approaching levels that warrant management action. Select control methods that are effective under your growing conditions and unlikely to harm people or the environment. Often, combining several methods provides the most reliable control. Five components are necessary for successful IPM programs (Figure 2-1):

- Prevent problems whenever possible.
- Diagnose accurately the cause of problems and correctly identify pests and natural enemies.
- Monitor regularly to gather information on which to base IPM decisions.
- Develop and use site-specific action guidelines and thresholds to help you decide whether a pesticide application or other pest management action is warranted.
- Manage pests using a combination of appropriate methods.

PREVENTION

Prevention is the most important component of IPM. The best way to avoid most problems is to select the appropriate plant for each site and provide what it requires for healthy growth. To avoid most landscape pest problems

- Design landscapes carefully.
- Prepare sites correctly before planting (e.g., use good sanitation, control weeds, and loosen soil).
- Select healthy plants and choose pest-resistant cultivars and species that are well adapted to local conditions.
- Plant properly.
- Provide appropriate cultural care.

Irrigation, fertilization, pruning, soil aeration and moisture, other environmental factors, and cultural practices are directly linked to many abiotic disorders, insect outbreaks, and pathogenic diseases. Read Chapters 3 and 4 to learn the best practices essential for maintaining healthy plants. Review the "Tree and Shrub Pest Tables" at

Prevent weeds and improve plant growth by applying and maintaining a layer of coarse organic mulch that is 3 to 4 inches deep. *Photo:* J. K. Clark

the back of this book to learn the specific pests to which your plants are susceptible, then read those sections in this book and take the recommended steps to avoid those problems. Talk to University of California (UC) Cooperative Extension Advisors and Master Gardeners, certified nurserypersons, and landscape professionals to learn of their experiences with your particular plant species under local growing conditions. For additional resources consult the "Suggested Reading" at the back of this book.

Exclude Foreign Pests. Many of our worst pests were introduced from other states or countries. Argentine ant, Dutch elm disease, sudden oak death, various bark- and wood-boring beetles, weeds such as Scotch broom and yellow starthistle, and many others have been inadvertently brought into California on contaminated plants, seeds, soil, or wood. To prevent new pest introductions during planting and travel

- Do not bring fruit, plants, seeds, wood products, or soil into California unless you know they were certified as being pest-free or inspected by agricultural officials. This includes

some online purchases that may not go through the required inspection process.

- Buy only pest-free plants from reputable local nurseries.
- Before purchasing plants from outside the local area or moving plants across county lines, learn whether any quarantine prohibits movement of certain plants.
- Purchase firewood near where you will burn it and leave any unused wood on site rather than moving it or bringing it home.
- Clean up, dispose of, or remove diseased or infested plant material in ways that ensure the pathogen or other pest is not moved to new locations (e.g., by properly composting the material in situations where composting can be effective).
- Take any unfamiliar pests to your county agricultural commissioner or UC Cooperative Extension office for identification or telephone 1-800-491-1899. Be sure you put pests in a sealed bag or bottle before transporting them.

Many introduced pests that are in California have not yet spread throughout the entire state. To exclude foreign pests, avoid bringing potentially contaminated or infested equipment, plants and plant products, and soil into your site.

PEST IDENTIFICATION AND SYMPTOM DIAGNOSIS

Many pests or the damage they cause look similar, especially to the untrained eye. Some pests can be easily confused with beneficial or innocuous organisms. Often when plants have an unhealthy appearance there is no obvious cause, for example, when the pest's feeding location is hidden (e.g., under bark) or the pest is microscopic (most fungi).

The damage may not be caused by a pest. Symptoms caused by adverse soil conditions, overwatering or underwatering plants, pesticide toxicity, and other abiotic disorders can be incorrectly blamed on insects, mites, or pathogens. For example, causes of spotted leaves (see Table 5-1) include abiotic disorders, disease-causing microorganisms, and insects.

Plants also are frequently subject to more than one disorder or pest problem at a time. Diagnosing the specific causes that produce certain symptoms can be a challenge. Proper pest identification and correct diagnosis of the cause of unhealthy plants are essential for choosing the right control actions. Even closely related pests often require different management tactics, and some species require no action at all.

Learn the cultural and environmental conditions required by each plant and check that these are being adequately

Proper pest and natural enemy identification are essential. Some people may mistake this large hover fly (*Scaeva pyrastri*) larva for a caterpillar, but this beneficial insect eats aphids, not plants. *Photo:* J. K. Clark

provided. Look for sometimes subtle differences between the appearance of unhealthy and healthy plants of the same species. Patterns in the symptoms may provide clues to the cause. Obtain information about the recent history of affected plants, environmental conditions, the site, and cultural practices. Use appropriate tools, such as binoculars, hand lens, pocket knife, shovel, soil sampling tube, and reference materials like this book.

Use the descriptions and photographs in this book to help you recognize common pests of trees and shrubs in California, the types of damage those pests cause, and the methods for distinguishing them. Because of the broad scope of this book and because new plant and pest species are often introduced from elsewhere, some of the pests you may encounter may not be pictured or described here. The cause of some problems can be reliably diagnosed only by experienced professionals or by submitting proper samples to a diagnostic laboratory for testing.

Consult the online guides and *Pest Notes* at ipm.ucanr.edu for current information. Especially useful printed publications include *Abiotic Disorders of Landscape Plants: A Diagnostic Guide, The Biology and Management of Landscape Palms, California Master Gardener Handbook, A Field Guide to Insects and Diseases of Oaks, Oaks in Urban Landscapes,* and *Pests of the Native California Conifers.*

REGULAR MONITORING FOR PESTS

Go out into the landscape on a regular basis and systematically check for pests, damage symptoms, and site conditions and management practices that may damage plants. Frequency and method of inspection will vary with the season, potential problems, plant value, plant growth stage, and resources. Weekly or even daily inspections may be needed for certain plants during times of the year when they are susceptible to problems that can develop quickly. For example, rainy weather favors development of leaf spot

and rust diseases, and certain problems develop only when plants are producing new growth. Time invested in monitoring can help avoid plant damage and reduce the extent of necessary management actions. If problems are not detected until they become obvious, your management options may be limited to pesticide use or plant replacement.

Develop a routine that is adequate and efficient for the areas under your management and examine plants in a systematic manner. Use a predetermined pattern of inspection to collect information in the same manner each time, allowing you to compare results among inspection dates. For example, start with buds or flowers, then inspect succulent new growth, younger leaves, older leaves, main stems, the trunk, and the basal root crown. Be sure to examine both the upper and lower surfaces of leaves. In addition to close-up inspection, examine plants from a distance, looking for subtle changes in canopy density and foliage color in comparison with surrounding plants and using your knowledge of how healthy plants should appear. Examine plants in locations with different environmental conditions, such as both sunny and shady sites.

If problems such as root and crown rot or vascular wilt disease are suspected, temporarily remove soil to inspect the basal trunk and consider shaving off a thin slice of bark to inspect a portion of the cambial layer and wood just beneath cambium. Check soil compaction and moisture conditions, for example, by using a soil probe or tube. See Chapters 4 to 8 for more problem-specific monitoring recommendations.

Keep written or electronic monitoring records, such as using the landscape map method (see Figure 7-2). Some professional landscape managers enter their records into a computer database and summarize and analyze them using graphical display or other software (Figure 2-2). Professional managers of large numbers of plants can evaluate and compare the effectiveness of management practices in their situation by conducting field trials, as discussed in *IPM in Practice.*

Examine valued plants regularly for pests, damage, and inappropriate cultural practices. Keep records of any problems that you encounter. Examine plants in a systematic manner, inspect all plant parts that may be infested or show symptoms, and view plants both close up and from afar. *Photo:* J. K. Clark

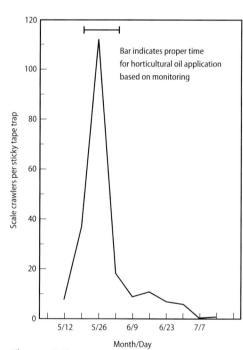

Figure 2-2.
Scale crawler abundance monitored using sticky tape traps to determine the most effective time to make a foliar application of horticultural oil, as discussed in Chapter 6. Some professional landscape managers enter their monitoring records into a computer database and summarize and analyze them using a graphical display such as this.

Compare monitoring results from different dates to determine whether problems are increasing or decreasing, whether control action is needed, and how effective were previous management activities. Record the date, specific location, host plant, presence or absence of pests or actual counts of pest abundance (and perhaps also natural enemies), description of procedures, and name of the sampling individual. Note pest management activities, such as any pesticide applications. Record other actions and weather events that may influence pests. For example, the reproductive and feeding rates of most insects and mites increase with increasing temperature; tracking temperature and time in units called degree-days assists with treatment decisions when managing certain insect pests, as discussed in Chapter 6.

ACTION GUIDELINES AND THRESHOLDS

For many pest species a certain number of pest individuals and some amount of damage can usually be tolerated. The difficulty may be in determining the action threshold—the point at which control action must be taken to prevent unacceptable damage. Since agricultural crops are grown for profit, action thresholds are based on economic criteria; action is warranted when it will improve crop quality or yield and provide increased revenue that exceeds the extra cost of management.

There are few formal action thresholds or guidelines for pests in landscapes, partly because of the difficulty in defining what level of pests or damage is intolerable. Many pests are annoying in landscapes or make plants unsightly but do not seriously damage or kill the plant. Even with pests that can threaten plant health or survival, aesthetic damage is often noted at levels well below those that threaten the plant.

The pest numbers or damage level when action must be taken to deter undesirable damage, the aesthetic threshold or tolerance level, varies with the attitude and knowledge of people using the landscape as well as the value of the landscape. For

example, there may be no reason to control aphids infesting a tree in the middle of a lawn, but aphids can be very annoying when they infest limbs overhanging a driveway or patio, fouling those surfaces with dripping honeydew.

You will find recommendations throughout this book to help you determine whether management may be needed and the best time to take action to avoid or reduce specific pest problems. Many plants are more vulnerable to pest damage at certain times in their development; for instance, during the first year or two after establishment or during certain seasons. Differences in susceptibility mean that action thresholds also differ over the growing season and as the plant develops. Other conditions may affect a plant's ability to tolerate pest damage; for example, plants weakened by water deficit, weed competition, root disease, adverse soil conditions, or injury must be more carefully protected because they are less tolerant of additional stresses or more pests.

Timing of actions is often critical for effective management. For example, once symptoms become obvious, it is often too late to control many plant diseases effectively. Sometimes the pest life stage that damages plants is not the life stage

susceptible to control action. Many times the appropriate action may not be to apply pesticides but rather to use cultural practices such as proper irrigation and pruning. If you are limited to management methods that take several days or months to provide control or that only kill a small proportion of the pests, you will have to allow for more time than you would with faster-acting measures.

How to Establish Control Action Guidelines or Thresholds. Establish control action guidelines or thresholds for highly valued or problem-prone plants by systematically monitoring landscapes, keeping good records, and judging the health and quality of plants in comparison with pest scouting and control records. If using thresholds, they should be quantitative or numerical to be useful.

Suggested numerical thresholds are provided for a few pests in Chapter 6, as in "Aphids" and "Elm Leaf Beetle." For example, thresholds could be based on the percentage of plants or leaves found to be damaged or infested during visual inspection or the number of pests dislodged per branch beat sample. Guidelines other than plant damage or pest counts might include consideration of weather conditions, history of problems in the landscape, stage of

Aesthetic tolerance for pest damage varies between individuals. The yellow leaf blotches on this Chinese lantern are caused by *Abutilon mosaic virus*. This virus causes no apparent harm to plants, and propagators deliberately select infected plants because many people like these variegated abutilon cultivars. *Photo: J. K. Clark*

plant or weed development, when weeds reach a certain height, the presence of signs such as rodent burrows or droppings, or after you get a certain number of calls or complaints about a specific problem.

Action guidelines or thresholds are helpful only when used with accurate pest identification and careful monitoring. Keep records of pests, determining factors for treatments you made, and the results of any management activities. These records will help you to develop and refine action guidelines that work best for your situation in the future.

Experiment over time to develop action guidelines appropriate for your situation. Be flexible in adjusting guidelines and thresholds and adapting monitoring techniques and management methods as appropriate.

MANAGEMENT METHODS

The primary pest management methods are biological, chemical, cultural, mechanical, and physical. Before applying these methods, determine whether action is needed and is likely to be effective. If it is too late for control to be effective or if the problem is minor or does not threaten plant health, consider taking no action or applying other methods. When action is needed, combine two or more methods whenever possible to provide more effective control. The methods summarized below are detailed later in the pest-specific chapters.

Cultural Control. Cultural controls are modifications of planting and maintenance activities that reduce or prevent pest problems. Properly design landscapes (e.g., use hydrozones, grouping plants with similar irrigation needs). Select plants that are adapted to local site conditions and, if available, resistant to pests. Plant properly and irrigate, prune, and provide other maintenance, as discussed in Chapter 3, to minimize plants' susceptibility to pests and to increase their ability to tolerate pest damage. Improve soil aeration and drainage to prevent abiotic disorders and pathogenic diseases favored by adverse soil conditions and unhealthy roots.

Mechanical Control. Mechanical controls use labor, materials not considered to be pesticides, and machinery to reduce pest abundance directly. For example, hand-pull, hoe, flame, mow, and use mulch or a string trimmer to control weeds (see Chapter 7). Install copper bands around trunks and planting areas to exclude snails and slugs. Apply sticky material around trunks to prevent canopies from being infested by ants, weevils, and other flightless insects. Clip and dispose of foliage infested with insects that feed in groups and are limited to small portions of a plant, such as tentmaking caterpillars. Hand-pick snails and large insects. Prune out or rake up foliage and twigs infected with certain pathogens to prevent their propagules from spreading and infecting healthy plant tissue.

Physical Control. Physical controls (also called environmental controls) suppress pest populations and prevent pest damage by altering temperature, light, and humidity. Control black scale, cottony cushion scale, and some other scale species by thinning plant canopies in hot areas of California to increase scale exposure to heat mortality and parasites. Control certain foliar diseases by thinning the plant canopy or cutting back nearby plants to increase light exposure and air circulation and reduce humidity.

To reduce light exposure and sunburn or sunscald to the trunks of young or heavily pruned woody plants, apply white interior (not exterior) latex paint diluted with an equal amount of water. Interior water-based paints are safer to trees than are oil- and water-based exterior paint. Whitewash on bark also helps to avoid attack by wood-boring insects that are attracted to injured trunks.

Biological Control. Biological control uses competitors (antagonists), pathogens, parasites, and predators to control

A raspberry horntail larva is tunneling inside this wilted rose shoot. This pest is easily managed with mechanical control, the pruning off of wilted shoots below any noticeable damage. *Photo:* J. K. Clark

If problems such as root crown rot or vascular wilt disease are suspected, remove soil from around the root crown and shave off a thin slice of bark. Inspect a portion of the cambium (inner bark) and wood for decayed or discolored tissue. *Photo:* J. K. Clark

Biological control has been used most successfully to control pest insects and mites. This bigeyed bug nymph (*Geocoris* sp.) is feeding on a moth egg, preventing a plant-chewing caterpillar from hatching. *Photo: J. K. Clark*

pests, especially pest insects and mites. For example, certain flies, beetles, and wasps feed on and parasitize aphids and mealybugs. To enhance the effectiveness of natural enemies, control ants that protect honeydew-producing insects from parasites and predators, reduce dust, and avoid pesticides that kill natural enemies and disrupt biological control. Plant a diversity of flowering and nonflowering species to provide habitat and food for beneficial predators and parasites.

Periodically releasing commercially available natural enemies may control target pests in a few specific situations, but conserving resident natural enemies is usually a more economical and effective strategy. As with other methods, biological control is most effective when integrated with other strategies, such as applying selective and nonpersistent pesticides instead of broad-spectrum pesticides with long-lasting residues that kill natural enemies. See Chapter 6 for more information.

Chemical Control (Pesticides). Pesticides are chemicals that kill, prevent, or repel pests and reduce pest damage. You can quickly obtain temporary control of certain pests or prevent their damage if you choose the correct pesticide and apply it at the right time in an appropriate manner. Consider alternatives before using a pesticide; cultural practices and other alternatives often provide more long-lasting control. If you use a pesticide, combine or follow up its use with nonchemical control methods where feasible.

Follow all label directions—if you use an incorrect pesticide, the wrong rate, or improper application method, you can do more harm than good. Incorrect pesticide use can damage plants or natural enemies. For example, many weed-and-feed lawn care products contain broadleaf herbicides (e.g., dicamba, 2,4-D) that can damage or kill nearby trees and shrubs with roots growing under treated turfgrass. Carbamate (e.g., carbaryl, or Sevin), organophosphate (acephate, malathion), and pyrethroid (bifenthrin, cyfluthrin, permethrin) insecticides leave residues that remain toxic to honey bees, other pollinators, and beneficial parasites and predators for days or even weeks after application. This persistent toxicity can prevent biological control by harming natural enemies that migrate in after application. Systemic insecticides (e.g., dinotefuran, imidacloprid, and other neonicotinoids) translocate within plants to contaminate flowers and may harm bees and adult natural enemies that feed on pollen and nectar.

Many pest insects and mites that feed while exposed can be controlled by a thorough spray of insecticidal soap, oil, or pyrethrin (pyrethrum). These insecticides are less harmful to natural enemies since they do not have persistent residues and natural enemies will repopulate the site quickly after an application. Spot-spraying only the most-infested plant parts can provide adequate short-term control and may provide better control over the long term because spot-spraying also reduces harm to natural enemies.

Be aware that pesticides differ in their effectiveness against pests. For example, the systemic insecticide imidacloprid controls most soft scales and many leaf- and root-chewing beetles, but it does not control cottony cushion scale and most armored scales and caterpillars (moth larvae). Before using a pesticide, understand its relative toxicity, mode of action, and persistence, and be aware of its safe and legal uses. Read the label carefully

Broad-spectrum pesticides can sometimes be used selectively. This ant bait station enclosure prevents most nontarget species from being exposed to pesticide. Ants are attracted to the insecticide and sugar water, which they carry back to their nest, poisoning the queen and other ants that live underground, where they are protected from pesticide sprays. *Photo: C. R. Reynolds*

before you buy the pesticide and read the label again before you use it so you will understand associated hazards. Be certain it is registered and appropriate for use on the plants or site where it will be applied. For more information, see "Pesticides" in Chapters 5 to 7, *Pesticides: Safe and Effective Use in the Home and Landscape Pest Notes* and the active ingredients database online at ipm.ucanr.edu, and books such as *Lawn and Residential Pest Control, Landscape Maintenance Pest Control,* and *The Safe and Effective Use of Pesticides.*

Types of Pesticides. Pesticides are categorized according to the type of organism controlled. Bactericides control bacteria. Insecticides control insects. Fungicides control fungi. Herbicides kill weeds or plants. Miticides or acaricides control mites. Molluscicides control snails and slugs. Rodenticides control mice and other rodents.

Other ways of characterizing pesticides include

- mode of action, or site of action (see "Pesticide Resistance," below, for more information)
- chemical class (e.g., carbamate or organophosphate)
- source of the material (e.g., botanicals are extracted from plants; inorganics like copper and sulfur are refined

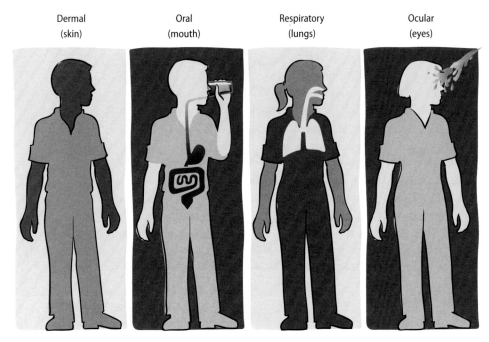

| Dermal (skin) | Oral (mouth) | Respiratory (lungs) | Ocular (eyes) |

Figure 2-3.
The common ways that people are exposed to pesticides are through the skin (dermal), mouth (oral), lungs (respiratory), and eyes (ocular).

from minerals; synthetics such as organophosphates and pyrethroids are manufactured from petroleum)

The chapters in this book concentrate on the more IPM-compatible and less-toxic pesticides. For more specific pesticide information, consult the active ingredients database and *Pest Notes* at ipm.ucanr.edu.

Pesticides Are Toxic. All pesticides are toxic (poisonous) in some way. The common ways for people to be exposed to pesticides (Figure 2-3) are through the

- skin (dermal)
- mouth (oral)
- lungs (respiratory)
- eyes (ocular)

The signal words DANGER, WARNING, or CAUTION on a pesticide label indicate the product's potential of immediate or acute injury if a person is exposed to it. The CAUTION pesticides are least acutely toxic. DANGER (sometimes including a skull and crossbones and also labeled POISON) indicates the most toxic pesticides. Most home-use pesticides are CAUTION products; most WARNING and DANGER pesticides can only be used by trained professionals who may need to

be certified or licensed by the California Department of Pesticide Regulation (DPR).

Some pesticides are suspected of causing long-term health effects, but this information is not provided on the label. A Safety Data Sheet (SDS), formerly called Material Safety Data Sheet (MSDS), detailing potential hazards is available for each pesticide from the pesticide manufacturer or from the National Pesticide Information Center at npic.orst.edu/gen.htm, telephone 800-858-7378.

Pesticide Selectivity. Selective pesticides are toxic only to the target organism and related species. Broad-spectrum (nonselective) pesticides kill many different species. For example, certain strains of the selective insecticide *Bacillus thuringiensis* (Bt) kill only moth and butterfly larvae, while the broad-spectrum carbamate, organophosphate, and pyrethroid insecticides kill both caterpillars, their natural enemies, and many other insects.

In addition to describing the range of organisms susceptible to a pesticide, selectivity also refers to the manner of use. Broad-spectrum pesticides sometimes can be used selectively by modifying applica-

tion timing, equipment, and method. For example, dormant-season application of narrow-range oil to kill scale insects may reduce the impact on natural enemies in comparison with a spring, in-season spray because natural enemies tend to be inactive during winter or are not present in the treatment area. Pesticides for ant or rodent control can be mixed with bait and enclosed in a container that prevents most nontarget organisms from being exposed to the pesticides. Spot treatments, such as insecticide bark banding instead of spraying the whole plant canopy, help control elm leaf beetle larvae migrating down the trunk without killing predators and parasites that occur on leaves. Selective pesticides and application methods for insects, plant pathogens, and weeds are discussed in Chapters 5 to 7.

Pesticide Toxicity to Natural Enemies. Pesticides can severely disrupt biological control. Natural enemies often are more susceptible to pesticides than are pests due to many ecological and physiological factors. In comparison with most pests, natural enemies are more active searchers, resulting in increased contact with treated surfaces. Even if beneficial organisms survive an application, low levels of pesticide residues can have adverse sublethal effects on natural enemy longevity, fecundity (reproduction), and ability to locate and kill pests. Natural enemies require pests as hosts, so few predators and parasites will be present after spraying reduces pest numbers; pests therefore get a head start in reproducing, and natural enemy populations will lag behind until sufficient numbers of pests develop to attract natural enemies and support their reproduction. When possible, rely on selective insecticides and those with little or no persistent (residual) toxicity to natural enemies, as listed in Table 6-3.

Pesticides Can Damage Plants. Phytotoxicity is the ability of chemicals to injure plants. Pesticide injury to desirable plants usually occurs because pesticides have been used carelessly or in a manner contrary to the label. Common mistakes are applying excess amounts, allowing spray to drift, failing to obey label precautions, treating sites or desirable plants not listed on the label, or using a sprayer contaminated with herbicides to apply other materials. Environmental stress, such as drought, heat, or wind, or sensitivity of particular plant cultivars can influence whether and how severely phytotoxicity develops from exposure to pesticides. Because herbicides are made specifically to kill plants, they pose the greatest risk of unintended damage to desirable plant species, as discussed in "Pesticides and Phytotoxicity" in Chapter 4.

Pesticides Can Contaminate Water. Creeks and rivers are being contaminated with pesticides. These pesticides are harming aquatic life and can also affect the quality of our drinking water.

Pesticides reach surface waters through household drains and storm drains (Figure 2-4). When you apply a pesticide, some of the material may move to other locations in air, water, and eroded soil. Rain and irrigation runoff from landscapes flow down the streets through gutters into storm drains. The storm drain runoff flows

Runoff into storm drains usually flows directly to surface waters without any treatment to remove pesticides and other contaminants. *Photo:* J. K. Clark

through pipes directly into our creeks, lakes, rivers, or the ocean without any treatment process. Sewers run from drains within the home and carry wastewater from sinks and toilets to treatment plants. Wastewater treatment plants remove organic solids and disinfect pathogens before discharging water into rivers or the ocean, but wastewater treatment plants do not always detoxify pesticides, thus potentially releasing them into waterways. Be aware that California surface water regulations prohibit certain pesticide applications near or onto pavement and other hard surfaces and this use limitation may not be listed on the product label; contact the county agricultural commissioner for more information.

Use alternatives to pesticides whenever possible. If you use pesticides, follow all instructions on the product label and in the sidebar "Keep Pesticides Out of Our Waterways."

Pesticide Resistance. Many insects, mites, and pathogens and some weeds have developed resistance to certain pesticides. Resistance occurs when a pest population is no longer effectively controlled by a pesticide that previously provided control (Figure 2-5). A different type of pesticide or some other control measure must then be substituted to manage that pest. For example, some populations of annual (rigid) ryegrass and mare's tail (horseweed) are resistant to glyphosate herbicide. Resistance is a serious problem in crops and nursery production. Pests that develop resistance in plant nurseries can be introduced into landscapes on new plantings.

Take as many steps as possible to avoid creating resistant pest populations (Table 2-1). Delay the development of pesticide resistance by using biological, cultural, mechanical, and physical controls whenever possible. When pesticides are applied, choose selective materials, use them in combination with alternatives, and make spot applications whenever possible. Avoid repeatedly applying pesticides in the same chemical class or those with the same mode of action; alternate (rotate) applications among several pesticides with different

modes of action. To identify pesticide modes of action, see Table 5-5 (example fungicides), *Fungicides Sorted by Modes of Action* (online at frac.info), *Classification of Herbicides According to Mode of Action* (hracglobal.com), and (for insecticides) *IRAC Mode of Action Classification* (irac-online.org).

Deciding to Use a Pesticide. Before using any pesticide, be sure you need it.

- Is a pest really causing your problem? Something else (e.g., improper irrigation) may be the cause of the observed damage.
- Is a pesticide application warranted? Learn whether the particular pest seriously threatens plant health or survival. For temporary (e.g., seasonal) problems unlikely to worsen, perhaps patience and tolerance (no control actions) are acceptable.
- Can a pesticide application control the pest now or prevent damage in the future? For many pest problems, there are no effective pesticides or an effective application or product may be used only by a licensed pesticide applicator. Also, the time when a pesticide application can be effective (e.g., early in the growing season) may be different from the time when damage or pests are apparent.
- Can nonpesticide tactics contribute to control? For certain problems, nonchemical methods are more effective, longer-lasting, or provide the most satisfactory results when used with a pesticide. For example, using drip irrigation instead of sprinklers will help reduce the incidence of leaf spot and rust diseases.

Hire a Licensed Pesticide Applicator. If you do not have the time or ability to research your pest problem and effectively and safely apply the appropriate material to control it, you may want to hire a licensed pesticide applicator or pest control business to do the job. Before agreeing to a service, ask the company what materials will be applied, and how and whether least-toxic pesticides can be effective and used when

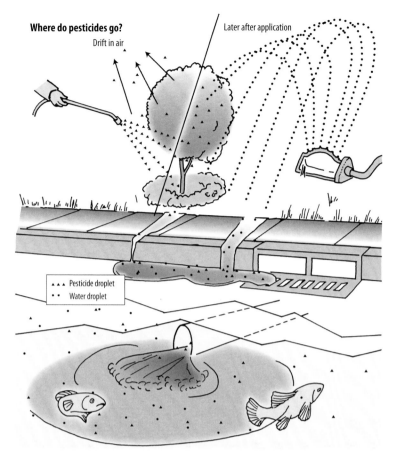

Where do pesticides go?

Drift in air

Later after application

▲▲▲ Pesticide droplet
• • Water droplet

Figure 2-4.
Creeks and rivers are being contaminated with pesticides, which are damaging aquatic life and can affect the quality of our drinking water. Do not allow soil or water to move into storm drains from areas recently treated with pesticides. Do not dump unwanted pesticides down the sink or into gutters or storm drains. Do not dispose of pesticide containers in the garbage unless they are completely empty, home-use-product containers.

appropriate. See *Hiring a Pest Control Company Pest Notes* for how to select a contractor.

Licensed pesticide applicators have access to certain pesticides and equipment not available in retail stores. Effectively managing certain problems, such as termite infestations or pests infesting large trees, can require pesticides, equipment, or technical training not available to home gardeners. Although professional services may be expensive, the investment may be worth it to solve a serious problem.

Choose the Correct Pesticide. Make sure you have correctly identified the pest and have considered nonchemical alternatives before you purchase or apply a pesticide. Read the label carefully before deciding which pesticide to purchase and apply. Do not use a pesticide unless the host plant or location to be sprayed is listed on the label. Do not use pesticides labeled for use only

on ornamental plants on plants that will be eaten, such as herbs and fruit trees that are also used for landscaping.

Read the precautionary statement on the label to be sure desired plants won't be

injured and to be aware of and avoid any associated hazards to humans, pets, beneficial organisms, or the environment. Choose the least-toxic pesticide that will be effective. Purchase only the amount of pesticide you expect to use within a few months; proper disposal of unused pesticide may be difficult.

Store Pesticides Safely. Store pesticides only in the original container and with the

Table 2-1.

Take Steps to Avoid Pesticide Resistance.

- Monitor landscapes or plants before treating and identify specifically which pest species are present.
- Evaluate whether pesticide application is truly warranted (see "Deciding to Use a Pesticide").
- Minimize the amount and frequency of applications and the extent of area treated. For example, spot-spray only locations and plant parts where target pests occur.
- Use biological, cultural, mechanical, and physical control alternatives whenever possible.
- If pesticides are applied frequently, rotate (alternate) applications among several pesticides with different modes of action.
- If control results are unsatisfactory, reexamine the effectiveness of your techniques, evaluate whether your expectations are realistic, and reconsider using alternative methods. Consider sending samples to a laboratory that can test for pesticide resistance if resistance is suspected.

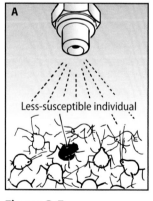

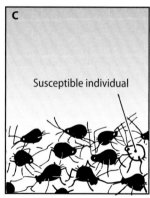

Figure 2-5.
Resistance to pesticides develops through genetic selection in populations of pests, including insects, mites, pathogens, and weeds. A. Certain individuals in a pest population are naturally less susceptible to a pesticide than other individuals. B. These less-susceptible pest biotypes are more likely to survive an application and to produce offspring that are also less susceptible. C. After repeated applications over several generations, the pest population becomes resistant to that pesticide because it consists mostly of less-susceptible individuals. Applying the same pesticide, or other chemicals with the same mode of action, is no longer effective.

appropriate label. Never store pesticides in soft-drink bottles or other food or drink containers. Store pesticides in a locked and labeled area or cabinet that is out of reach of children and pets. Do not store pesticides near beverages, clothing, feed, food, or rags. Protect stored pesticides from moisture and extreme heat or cold. Be sure the storage area is well ventilated to prevent the accumulation of toxic fumes. Check pesticide containers periodically for leakage or corrosion. Store pesticide containers in a plastic tub to facilitate cleanup if there is a leak. Properly dispose of damaged or leaking containers and unwanted pesticides if the products cannot be applied to sites for which they are labeled or you no longer need them.

Apply Pesticides Safely. Most pesticide products for use in the home and garden have relatively low toxicity to humans and are therefore unlikely to injure people if label directions are carefully followed. Unfortunately, many users fail to completely read or follow all the product directions. Read the label instructions before you mix and apply a pesticide, and follow the label exactly. It is illegal and may be dangerous to disregard pesticide label instructions. See *Pesticides: Safe and Effective Use in the Home and Landscape Pest Notes* for more information.

All people who mix, apply, or otherwise handle pesticides at work or as part of their job must be properly trained in advance on each pesticide they use, and employers must maintain written evidence of this training. People who use pesticides as part of their job may need to be licensed or certified by DPR, or have their work supervised by a certified or licensed person. Anyone who applies pesticides as part of a landscape maintenance business must be DPR certified, even if they only occasionally apply pesticides. Consult *Lawn and Residential Pest Control, Landscape Maintenance Pest Control, The Safe and Effective Use of Pesticides,* and the DPR website, cdpr. ca.gov, for more information.

Appropriate Application and Equipment. Be sure you have the proper equipment for applying the pesticide safely and to protect yourself from exposure. Even when applying the least-toxic pesticides, the minimum you should wear includes chemical-resistant gloves, eye protection that also covers the brow and temples, a long-sleeved shirt, long pants, and closed shoes (Figure 2-6). Unless the label states otherwise, avoid using cotton gloves or lightweight dust masks that may absorb pesticide and prolong its contact with your skin. Read the pesticide label carefully for any additional protective requirements.

Pesticide application equipment ranges from a simple hand pump spray bottle to power-driven machines (Tables 2-2, 2-3, and 2-4). What equipment to use varies according to the application site, your choice of pesticide, and your willingness to work with more complicated application devices. A ready-to-use product in a trigger pump sprayer eliminates the need to dilute and mix pesticides or purchase special equipment and is excellent for spot treatments on small plants. A compressed air sprayer requires careful maintenance and operation as well as accurate measuring of chemicals.

Figure 2-6.
Minimize your exposure to pesticides. Anyone who applies or mixes pesticides in California is required to wear eye protection and chemical-resistant gloves, even if the label does not say so. Unless otherwise stated on the label, also wear at least long pants, a long-sleeved shirt, and closed shoes. If spraying overhead, wear a waterproof hood or waterproof brimmed hat that can be washed after each use.

Keep Pesticides Out of Our Waterways

Pesticides applied in landscapes can move off target by drifting in the air or washing off plants and other surfaces into storm drains and waterways. Take steps to keep pesticides out of our water.

Be aware of weather and do not apply pesticides when it is windy or if rainfall is expected within 24 to 48 hours. Exceptions include certain preemergence herbicides that need about $1/2$ inch of rainfall (or irrigation) soon after the application to move herbicide into soil.

Do not apply pesticides to hard surfaces such as driveways, gutters, sidewalks, or storm drains, as they can easily be washed off.

Do not broadcast (widely disperse) pyrethroids or other broad-spectrum, residual insecticides onto pavement or soil; instead, make spot applications where appropriate. Be aware that surface water regulations restrict certain pyrethroid uses, and these use limitations may not be listed on product labels. Contact the county agricultural commissioner for more information.

If fertilizer or pesticide granules land on pavement or other hard surfaces, blow or sweep them onto the application site within the landscape.

Check pesticide labels and follow any warnings regarding use near bodies of water such as creeks, lakes, and rivers.

Never dispose of pesticide concentrates, leftover spray solution, or pesticide-contaminated rinse water in gutters, storm drains, sinks, or toilets.

Do not clean pesticide mixing and application equipment in a location where rinse water could flow into gutters, sewers, storm drains, or open waterways.

Never apply more than the amount (rate) or number of applications listed on a pesticide label.

Learn which pesticides are more easily carried in surface runoff and have greater potential to move off-site during irrigation or rain. To find the leaching and runoff risks of specific pesticides, see Pesticide Selection to Reduce Impacts on Water Quality and the UC IPM WaterTox database at ipm. ucanr.edu/TOX/simplewatertox.html.

If you use a concentrated pesticide, it must be diluted to the proper concentration before application. During mixing, it is often helpful to pour pesticides and fill the sprayer over a plastic tub to contain possible spills. Pour any spills or splashes caught by the tub into the sprayer before you finish filling the spray tank with water.

Keep a set of measuring spoons or cups for use only with pesticides; write "PESTICIDE ONLY" on them to distinguish them from kitchen utensils and store them with your pesticides and away from food preparation areas. If you apply herbicides, use a sprayer specifically for that purpose and label it "WEEDS ONLY." Otherwise, herbicide residue in the sprayer may injure plants if you use the same sprayer later for applying other pesticides or fertilizers.

Check Equipment Before Use. Fill the sprayer with clean water and operate it before adding pesticides. Look for leaking connections, bulging or cracked hoses, cracked or leaking tanks, and worn or plugged nozzles. Repair or replace faulty equipment before adding any pesticide or making an application.

Properly measuring and diluting concentrated pesticides is essential for their effective and safe use. Never use more than what the label directions say to use. The pest will not be killed any faster by using larger amounts, and you will be wasting the pesticide, your time, and money while potentially causing injury to desirable plants and contaminating the environment. Mix only as much as you need immediately. Place a label or tag on the sprayer and include the product name, signal word from the pesticide label, and your name and telephone number. Unless you apply all the material on a site that the label says the product can be used on, the only legal way to dispose of pesticides is to take them to a hazardous waste disposal facility or hazardous waste drop-off location. Do not store leftover pesticide solutions as the product effectiveness or application qualities may change.

Avoid widespread applications of the pesticide throughout your landscape; use spot treatments if appropriate. Properly apply the pesticide to the target plant or site and in ways that do not contaminate other plants or areas. Never smoke, drink, eat, or use the bathroom after a pesticide application without washing your hands and face first. Take a shower as soon after application as possible. Wear coveralls or change your clothes before leaving the application area, as pesticides will be on your clothes and can transfer to your car seat or to people or pets that come in contact with you. Wash clothing used during a pesticide application separately from other laundry.

Injecting or Implanting Systemic Pesticides. When applying a systemic pesticide, spray bark or foliage or make an application to soil as directed on product labels whenever possible instead of injecting or implanting trees (see Figure 6-9). If permitted on the product label, root-absorbed pesticide can be applied into soil using a powered or hand-pumped soil injector (Table 2-2), drenched onto soil immediately adjacent to the trunk, or applied to nearby bare soil, lawn, or planting beds where absorbing tree roots occur.

It is difficult to consistently inject or implant pesticides at the proper depth into trunks or roots. Additionally, these application methods can injure trees, create wounds, and provide entry sites for plant pathogens or other pests. Especially avoid methods that create large wounds, such as implants placed in holes drilled in trunks. For some injectable devices, the applicator must remain at the site until the pesticide is entirely in the tree. Do not implant or inject roots or trunks more than once per year.

Avoid application methods that use the same device, such as drills or needles, to contact internal parts of more than one tree unless the tools are scrubbed clean and sterilized before moving to the next plant. Contaminated tools can mechanically spread certain pathogens from one tree to another, including bacteria (such as slime flux or wetwood), fungi (canker stain of sycamore, Dutch elm disease, and Fusarium wilt), and certain viruses. Whenever working on plants known or suspected of being susceptible to mechanically transmissible pathogens, clean and disinfect tools before working on each new plant to reduce the chance of spreading pathogens. Before chemical disinfection, remove all plant material and scrub any plant sap from tools or equipment that penetrate bark. Bleach, and alcohols and certain other materials to a lesser extent, can be effective disinfectants if applied to debris-free tools, as discussed in "Disinfectants" in Chapter 5. Depending on the disinfectant, at least 1 to 2 minutes of disinfectant contact time between uses may be required. Consider rotating work among several tools, using a freshly disinfected tool while the most recently used tools are being soaked in disinfectant.

Dispose of Containers Properly. With containers that held concentrated liquid pesticide, rinse the empty container three times (triple rinse) immediately after emptying it and before you finish filling the spray tank. First drain the empty container into the spray tank for at least 30 seconds. Next, fill the container about one-quarter full with clean water, close the container, and gently shake or roll it to rinse all interior surfaces. Drain the rinse material into the spray tank and continue to let the material drain for at least 30 seconds after the container is mostly empty and has begun to drip. Repeat the rinsing procedure two more times, then fill the spray tank to the proper level. This diluted solution can then be applied to any labeled site.

Punch holes in the empty container so it cannot inadvertently be reused for other purposes. Do not attempt to rinse or puncture sealed, pressured containers.

Empty containers of home-use pesticides may be disposed of in the trash if they were used by the homeowner on their own property. Professional applicators and anyone who performs pest control for others or applies pesticides as part of their job should contact the local county agricultural commissioner to learn the proper methods for disposing of empty pesticide containers.

Table 2-2.

Selection Guide for Nonpowered and Hand-Operated Application Equipment for Liquid Pesticides.

	TYPE	USES	SUITABLE FORMULATIONS	COMMENTS
	Aerosol can	Insect control on small plants and limited areas.	Liquids must dissolve in solvent; some dusts are available.	Very convenient. Good for spot applications. High cost per unit of active ingredient.
	Hose-end sprayer	Home garden and small landscaped areas. Used for insect, pathogen, and weed control, where water pressure is sufficient.	All formulations. Wettable powders and emulsifiable concentrates require frequent shaking.	Convenient and low-cost way of applying pesticides to small outdoor areas. Cannot spray straight up. Install an anti-siphon device on the hose-end connector to prevent pesticide from being sucked into the water line if water pressure drops. Disadvantages include poor spray coverage, nozzle clogging, and inaccurate metering of the active ingredient. Low purchase price can be negated by the waste and environmental contamination from excessively high spray volume.
	Trigger pump sprayer	Insect, pathogen, and weed control on relatively small areas.	Liquid-soluble formulations best.	Low cost and easy to use. Good for spot applications.
	Trombone sprayer or slide sprayer	For treating shrubs and medium-sized trees up to about 25 feet tall.	All formulations. Wettable powders and emulsifiable concentrates require frequent mixing or stirring.	Applies a uniform concentration of active ingredient if properly mixed. Relatively easy to clean and maintain. Moderately priced. Effort is required to maintain pressure by continually operating a push-pull slide handle. Requires practice to obtain uniform spray coverage. Drift or dripping onto applicator is possible when using to spray tree canopies.
	Compressed air sprayers	Many commercial and homeowner applications. Can develop fairly high pressures. Used for insect, weed, and pathogen control.	All formulations. Wettable powders and emulsifiable concentrates require frequent shaking.	Good overall sprayer for many types of applications. Relatively inexpensive. Needs thorough cleaning and regular servicing to keep sprayer in good working condition and to prevent clogging and corrosion of parts. Difficult to calibrate.
	Backpack sprayers	Same uses as compressed air sprayers.	All formulations. Wettable powders and emulsifiable concentrates require frequent shaking.	Durable and easy to use. Requires periodic maintenance. May be heavy for long periods of use. Needs frequent pumping to maintain pressure, otherwise output is not consistent. Some brands have a pressure regulator to help compensate for this.
	Hand-pump soil injector	For applying root-absorbed pesticides.	Liquid formulations, most commonly systemic insecticides.	Avoids wounds caused by injecting roots or trunks. Various equipment is available; this example has most of the benefits and drawbacks listed for backpack sprayers.

Once used with herbicides, avoid using the sprayer for other pesticides. Adapted from O'Connor-Marer 2000.

Table 2-3.

Selection Guide for Powered Liquid Pesticide Application Equipment.

	TYPE	USES	SUITABLE FORMULATIONS	COMMENTS
	Powered backpack sprayer	Landscape, right-of-way, aquatic, forest, and agricultural applications.	All. Some may require agitation.	Applies a uniform concentration of active ingredient. May be heavy for long periods of use. Requires frequent maintenance. Generally too expensive for individual home garden use.
	Controlled droplet applicator	Used for application of herbicides (e.g., systemics) and some insecticides. Some are hand-held, while others are mounted on spray boom. Produces uniform droplet sizes.	Usually water-soluble formulations.	Plastic parts may break unless handled carefully.
	Low-pressure sprayer	Used in commercial applications for insect, pathogen, and weed control. Used with mounted spray booms or hand-held equipment.	All. Equipment may include agitator.	Useful for larger areas. Powered by own motor or external power source. Frequent cleaning and servicing is required. Expensive.
	High-pressure hydraulic sprayer	Landscape, right-of-way, and agricultural applications. Use on dense foliage and large trees and shrubs.	All. Equipment may include agitator.	Useful for larger areas. Important to clean and service equipment frequently. Requires own motor or external power source. Abrasive pesticides may cause rapid wear of pumps and nozzles. Expensive.

Once used with herbicides, avoid using the sprayer for other pesticides. Adapted from O'Connor-Marer 2000.

Table 2-4.

Selection Guide for Dust and Granule Application Equipment.

	TYPE	USES	SUITABLE FORMULATIONS	COMMENTS
	Mechanical dust applicator	For small areas.	Dusts.	May have bellows to disperse dust. Requires care to avoid drift. Do not breathe dust.
	Hand-operated rotary spreader	Landscape, aquatic, and some agricultural areas.	Granules or pellets.	Suitable for small areas. Easy to use. Inexpensive. Can be difficult to uniformly apply an accurately metered amount of active ingredient. Because they throw material outwards, rotary spreaders may apply granules to nontarget areas including driveways, sidewalks, and streets.
	Mechanically driven drop spreader	Turf and other landscape areas. Also commonly used in agricultural areas.	Granules or pellets.	Requires accurate calibration.
	Powered granule spreader	Large landscape applications (e.g., golf courses).	Granules or pellets.	Some units may have blowers to disperse granules. Others may distribute granules along a boom. Frequent servicing and cleaning is required.

Adapted from O'Connor-Marer 2000.

Chapter 3   Growing Healthy Trees and Shrubs

Select an appropriate plant for each location and provide for its basic growth requirements. If plants are well adapted to local conditions, the environment is conducive to good growth, and proper care is provided, plants will be healthier and more aesthetically pleasing. Healthy trees and shrubs are less frequently attacked by bark beetles and wood borers and typically are more tolerant of other pest presence.

This chapter summarizes how to care for woody plants to minimize disorders and pest damage and grow healthy, sustainable landscapes. For more information, consult *Abiotic Disorders of Landscape Plants: A Diagnostic Guide; Arboriculture: Integrated Management of Landscape Trees, Shrubs, and Vines; California Master Gardener Handbook; Plant Health Care for Woody Ornamentals;* and other publications listed in "Suggested Reading" at the back of this book.

GROWTH REQUIREMENTS

Plants are living organisms that require carbon dioxide, energy, oxygen, and water. Most plants use sunlight to produce their own food (stored chemical energy). This food-producing process (photosynthesis) occurs in green tissue, primarily leaves. Plants need essential elements, carbon dioxide, and appropriate light, temperature, and water to carry out photosynthesis.

As people modify the landscape, they affect the availability of resources that plants need. Water and oxygen availability to roots is affected by aeration, drainage, irrigation, and changes in the composition, density, grade, structure, and texture of soil. For example, plants often become unhealthy when people irrigate too frequently, which restricts oxygen availability to roots. Nutrient availability to plants may be increased by fertilizing, applying amendments and mulches that add organic matter, or by allowing fallen leaves to remain and decompose. However, nutrient deficiency symptoms in plants often occur because of adverse soil conditions, altered soil pH, injured roots, or restricted root growth, which prevent roots from absorbing the available nutrients in soil.

Temperature and light vary naturally according to the weather, but they can be manipulated by irrigating, pruning, adding or removing plants, modifying structures or pavement, and by planting at a suitable location. Depending on their type and location, mulches or ground covers can increase or decrease soil temperatures or light around plants.

All the basic requirements for growth must be properly maintained for plants to have maximum resistance to damage from insects and pathogens. Failing to provide an appropriate environment and adequate care increases the likelihood that disorders or pests will injure or kill plants.

PLANT DEVELOPMENT AND SEASONAL GROWTH

Learn to recognize the normal changes in plants so you do not confuse their normal appearance with disorders or pest problems. Deciduous woody perennials typically drop leaves in the fall before entering winter dormancy and regrow foliage in the spring. However, some species, like the California buckeye, drop leaves early during hot, dry summer weather. Evergreen trees and shrubs retain some foliage year-round but still exhibit

Plants are living organisms that require carbon dioxide, oxygen, nutrients, sunlight, water, and adequate soil and space for growth. Providing plants with a good growing environment and proper cultural care is the most critical aspect of pest management. *Photo:* J. K. Clark

seasonal changes in growth, flowering, foliage production, and leaf fall. For example, evergreen conifers typically drop their oldest needles in the fall while retaining their youngest needles, those produced during the last few years. Most broadleaf evergreens drop older foliage during spring as they relocate nutrients to new growth (Figure 3-1). Changes in temperature, moisture, and especially in the amount and length of daylight induce seasonal changes in plant growth and appearance, such as flowering and leaf flush and growth.

Although deciduous trees may be without leaves for several months each year, tissues beneath bark and in roots are alive all year. Improper watering, excessive light, extreme temperatures, drying winds, and other adverse conditions can damage plants even when they are dormant.

Cultural activities that improve plant health and actions to prevent and manage pests must be properly timed to be effective. Pest abundance and damage are also linked to the seasonal cycle of plant growth. For example, the time of year when plants are pruned affects many important diseases and pest insects, as discussed later in "Pruning and Pest Management."

DESIGN A PEST-TOLERANT LANDSCAPE

Effective pest management begins before the landscape is planted. Design landscapes to provide an optimal living environment for plants. Select species and cultivars that resist common invertebrate and pathogen problems and that are well adapted to local conditions.

Group together (cluster) plants that have similar cultural and environmental requirements. Separate drought-tolerant species from those that are adapted to summer rainfall and will require regular irrigation in most urban areas of California. Keep irrigated ground covers, turf, and weeds back from the trunks of young trees and shrubs. Plants near trunks of young trees and shrubs dramatically retard growth, and irrigating near trunks can cause drought-adapted plants to die from excess moisture or root and crown diseases. Although after establishment many trees adapted to summer rainfall grow well in turf, they may not get enough water and nutrients and may be damaged by mowers, string trimmers, and certain pests

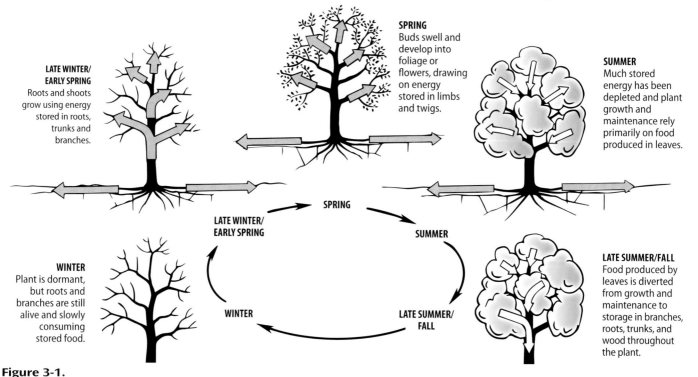

Figure 3-1.
The seasonal growth cycle of a typical deciduous tree. Wide arrows indicate the direction of major energy flow in the form of carbohydrates.

Good design minimizes pest problems. For example, this close spacing of plants shades out weeds.
Photo: J. K. Clark

if turf and ground covers grow too near to their trunks.

Determine the Expectations. Before preparing and planting the site, decide what aesthetics and functions (e.g., shading and visual screening) are desired and how much time, money, and other resources will be budgeted. If the desires are for high visual appeal, very few pests, and almost no plant damage, more effort and inputs will be required to develop and maintain the landscape so that it meets these expectations. Be aware that inadequate site preparation, improper planting, selecting pest-prone cultivars, planting species not adapted to local conditions, or using inappropriate cultural practices can cause landscapes to perform poorly regardless of pest control efforts. Strive for aesthetically desirable and functional landscapes that improve and protect the quality of our air, soil, and water, as discussed below in "Sustainable Landscapes."

SELECT THE RIGHT TYPE OF PLANTING FOR THE SITE

Consider surface and internal soil drainage, soil characteristics, water quality and availability, and other site conditions before selecting the plant species to grow there. Assess how much light and heat occur based on climate, exposure, and the influence of nearby structures, pavement, and existing plants. Determine proximity to pavement, structures, overhead lines, and underground utilities that may be damaged by growing limbs or roots.

Select species suited to the site's conditions by matching the plant to the location. Examine the space available for growth and learn about the mature size of candidate plants. Give limbs and roots plenty of room to grow and use only plants that will fit at maturity. Most smaller trees should be placed at least 6 feet from structures and at least 3 feet from any paved area; larger trees should be placed even farther away.

Look for overhead obstacles. Do not plant tall-growing species beneath utility lines. Utility companies are required to prune trees that grow into overhead lines; this can severely disfigure trees and promote decay, structural failure, and insect attacks, and increase utility costs. For more information, consult resources such as *SelecTree: A Tree Selection Guide* (online at selectree.calpoly.edu) and *Trees Under Power Lines.*

PROVIDE FOR ROOTS

Healthy roots are vital to plant survival. Nutrients, oxygen, and water are absorbed by root tips and their mycorrhizae (see below). Roots produce compounds essential to the plant, store food, and support the aboveground plant structure.

Damage appearing on aboveground parts often occurs because roots have been crushed, cut, exposed to toxic chemicals or misapplied pesticides, overwatered or underwatered, smothered, or otherwise improperly cared for. Excessively wet conditions and soil compaction are probably the most common landscape problems. Insects and pathogens that attack foliage, limbs, or trunks also can cause more serious damage if roots are unhealthy.

Roots are often neglected because they grow underground and are not seen. Provide roots with proper soil conditions and adequate space. Determine whether the site has suitable soil conditions (discussed below) and look for barriers to root growth before planting. After the first few years of growth in suitable conditions, lateral (horizontally growing) roots of healthy plants often extend well beyond the canopy or drip line. Woody plants may also have structural roots that grow downward and help anchor the tree, as well as relatively shallow absorbing roots with concentrations of root hairs that take up water from the soil. Often about 90% of woody plant roots grow in the top 3 feet of soil, most in the top 1 foot (Figure 3-2). Actual root systems can vary greatly, depending in part on cultural practices, plant species, and soil conditions. Ensuring that there is an adequate volume of good quality soil for future root development and properly preparing the planting area are important for the survival and satisfactory performance of trees and shrubs.

Determine whether the location is adequate for good plant growth. If the tree in this photo is retained, the site should probably be modified to provide more space for root and trunk growth as the plant matures.
Photo: J. K. Clark

Figure 3-2.
Healthy roots are vital to plant survival. Woody plants have several different types of roots. Under good root growth conditions, typically about 90% of roots grow in the upper 3 feet of soil. Up to about 70% of roots are in the top 1 foot. Because roots need air, even more of a tree's roots are near the surface if soils are compacted or waterlogged. Conversely, drought-adapted species and trees in deep or well-drained soil that receive only rainfall or infrequent deep irrigation have more of their roots growing deeper below ground than shown here. Roots typically extend beyond the tree canopy drip line. Actual root systems can vary greatly, depending in part on cultural practices, soil conditions, and the species of plant. For example, unlike the central taproot of many broadleaf trees and shrubs, plants such as bamboo, palm, and yucca have threadlike or ropelike roots that spread laterally to form a fibrous mat near the soil surface.

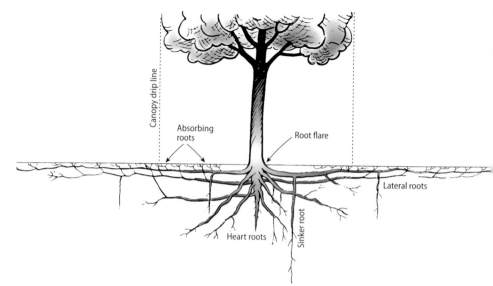

CONSIDER MYCORRHIZAE

Most healthy trees have beneficial fungi growing in their absorbing roots; the mutually beneficial association between such a fungus and the plant is called a mycorrhiza (plural, mycorrhizae; adjective, mycorrhizal). Mycorrhizal root tips are a primary location where plants absorb nutrients (especially phosphorus) and water. Mycorrhizae can increase growth rates and improve the drought tolerance of plants and may also help protect plants from soilborne pathogens.

Mycorrhizal fungi are common in soils where the plant species with which they associate have grown previously, and these beneficial fungi naturally colonize the roots of host trees planted there. Applying commercial mycorrhizal inoculants does not benefit trees in most urban landscapes. Exceptions might be obligately mycorrhizal plants, such as conifers and oaks, that are planted in constructed or highly disturbed soils and will be managed as naturalized areas without receiving fertilization and irrigation.

Use cultural practices that preserve native topsoils where mycorrhizal fungi are naturally abundant. Avoid soil compaction, overirrigation, and excess fertilization, particularly with phosphorus or quick-release synthetic formulations. Apply coarse organic mulch, as discussed in Chapter 7. Avoid fumigating soils, applying fungicide drenches, or contaminating soils with toxic materials that kill beneficial soil organisms.

CHOOSE THE RIGHT TREE OR SHRUB

Proper plant selection is one of the best ways to avoid pest problems. Each plant species or cultivar grows best under a specific range of soil, temperature, sunlight, and water conditions. Some plants tolerate a wide range of conditions, while other species survive only within a narrow range. Many plants that thrive in the eastern states with summer rainfall do well in California only when they are irrigated regularly. Likewise, a plant that thrives along California's coast may grow poorly in the warmer, drier interior valleys and be more likely to die due to environmental stress and pests.

Learn which species or cultivars are adapted to local conditions. For example, look in nearby arboretums, parks, and University of California (UC) Master Gardener demonstration gardens and choose from among plants performing well there. Many local park agencies and public utilities will provide a list of trees recommended for planting in that community. Seek advice from local experts, such as UC Cooperative Extension Advisors (ucanr.edu/County_Offices) and Master Gardeners (mg.ucanr.edu), certified arborists, or certified nurserypersons. Consult publications such as *California Master Gardener Handbook*, *The New Sunset Western Garden Book*, the UC Davis *Arboretum All-Stars* (online at arboretum. ucdavis.edu), *SelecTree* (selectree. calpoly.edu), and the Local Government Commission's *Tree Guidelines* for coastal southern California, the Inland Empire, and the San Joaquin Valley (lgc.org).

Choose plants by looking to see what species and cultivars are doing well under conditions similar to those of your planting site, such as in your neighborhood, nearby arboretums, parks, and UC Cooperative Extension demonstration gardens. This blue-flowering ceanothus and yellow-flowered flannel bush are well adapted to dry areas of California. Flannel bush especially is easily killed by irrigating it during summer. Other attractive, drought-tolerant plants for our Mediterranean climate include the *Arboretum All-Stars* pictured online at arboretum. ucdavis.edu. *Photo:* J. K. Clark

Climate. Most landscape plants are adapted to either summer drought or summer rainfall. Species adapted to summer rainfall are generally those native to the eastern United States, northern Europe, or eastern Asia, where summer rainfall occurs. Most of California has a Mediterranean climate. Winters are cool and wet, summers are warm and dry, and much of the state receives little or no precipitation from late spring through early fall.

Consider planting native California species or ornamentals from other parts of the world that also have a Mediterranean climate (Figure 3-3); these species should perform better with less irrigation and other maintenance than species adapted to summer rain. Be aware that some drought-adapted plants are dormant or do not have lush foliage during summer, so expectations for their appearance should differ in comparison with species that are adapted to summer rainfall.

California encompasses many different climate zones. Even native plants must be matched to local site conditions and pro-vided with the cultural care to which they are adapted. For example, Monterey pine and Monterey cypress from the coast and giant sequoia from the Sierra do poorly in hot, dry, interior areas of the state regardless of how much water they are given. In its native range, Monterey pine commonly lives 80 to 100 years; in interior areas, it typically dies within 20 to 30 years.

Site Environment. Consider the local site environmental conditions. Within each climate zone the light, temperature, and wind can vary dramatically over distances ranging from several miles, for example due to hills and valleys, and between locations only a few feet apart due to the influence of buildings, pavement, and surrounding vegetation.

Environmental Stresses. Wind dehydrates and tatters leaves and breaks limbs. Too much or too little sunlight causes foliage of susceptible species to discolor, die, and drop. Excess cold, heat, light, or water causes cracked and sunken bark. These wounds promote wood-boring insects, bark cankers, and decay fungi.

Determine the direct and reflected light conditions and range of temperatures and windiness at the site and choose species that tolerate those site conditions. If you plant species especially susceptible to mechanical or temperature injury or moisture stress, locate them where they will be sheltered from those conditions. For more discussion see "Sunburn," "High and Low Light," and "Wind" in Chapter 4.

Soil. Determine the key chemical and physical properties of the soil where you plan to plant, as discussed in "Prepare the Site." Many California soils are alkaline, compacted, and poorly drained, especially in urban areas. Before developing large commercial landscapes or sites suspected of having soil conditions that are highly adverse to good plant growth, consider having soils tested by a laboratory for nutrient levels, pH, salinity, and texture. Learn which plants tolerate local soil conditions and choose from among those species and cultivars. If necessary for that site, aerate, deep-till the soil, change the grade, provide for surface drainage, or install

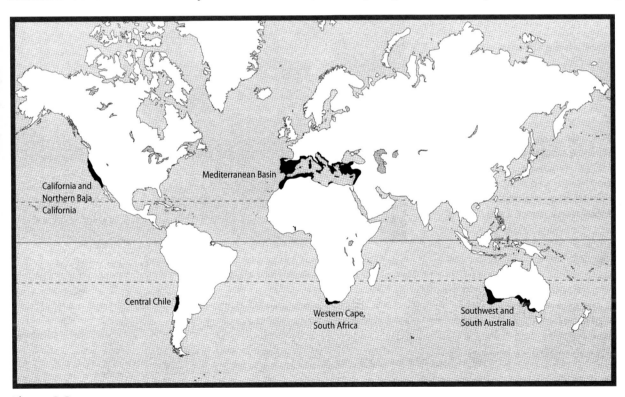

Figure 3-3.
Mediterranean climate regions of the world. Much of California has a Mediterranean climate: cool, moist winters and hot, dry summers. California native plants and species from other Mediterranean regions labeled on the map are generally better adapted to California climates. *Source:* Southern California Research Learning Center 2012.

drain pipe or drain tile before planting. Seek professional advice on which measures are appropriate for your situation.

Water. Choose plants that will perform to your expectations with the quantity and quality of water available at the site. Most species need some irrigation during plant establishment, and species that are adapted to summer rainfall will need regular irrigation throughout their life. Even established, drought-adapted landscape tree and shrub species often perform better when provided occasional summer irrigation. However, if drought-adapted species are planted in areas with frequent summer irrigation, they can be damaged or killed. Group plants with similar irrigation needs so that plants adapted to summer rainfall can easily be irrigated differently from drought-adapted species.

Consider water quality in addition to irrigation frequency and quantity. For example, certain plants grow poorly if water mineral content is high, as it is in some irrigation water, as discussed in "Salinity" and "pH" in Chapter 4.

Select Healthy Plants. Choose good-quality nursery stock (Table 3-1). Investing in better-quality plants can pay great dividends in lower maintenance costs and better performance. Look for a trunk that is tapered (wider at the bottom), with branches placed so that only about one-half of the foliage is on branches originating in the upper one-third of the trunk. For most tree species, choose plants that have a dominant (tallest, thickest) central leader, lateral branches that are well spaced both vertically and horizontally around the trunk, and main branches with wide angles of attachment to the trunk. Preferably there should be extra or "temporary" branches on the lower two-thirds of a tree's trunk so that some branches can be cut off as the young tree grows, as discussed below in "Pruning."

Avoid improperly pruned trees, such as those that had their central leader clipped; this practice is done in some nurseries to produce more compact lateral growth that appears attractive when trees are young.

Such improper pruning of young trees can lead to serious structural problems once the plants mature.

If possible, select trees that are not staked, as they will have sturdier trunks. Untie any stakes and avoid plants that bend over sharply near the soil line. For more on what to look for and do, consult resources such as the video *Training Young Trees for Structure and Form.*

Check Plants' Roots. Trees and shrubs in containers should be well rooted in the soil mix. When the trunk is carefully lifted, both the trunk and root ball should move as one; if the trunk can be raised 1 or 2 inches before the container moves, roots may be poorly developed or extensively circling. Feel below the soil surface and, if permitted, use a hose to wash away topsoil close to the trunk (this soil can be replaced) or examine smaller plants by temporarily removing them from the container. Avoid plants with major roots

that are kinked or circling near the trunk; these will eventually become girdled by their own root system and grow poorly, break off, or die. Smaller roots circling the container periphery can be spread or cut before planting, but if larger roots or roots near the trunk are kinked, reject the plant.

Choose plants with healthy roots. Small roots should be firm and white or pale-colored. Diseased or decayed roots (pictured in Chapters 5 and 8) commonly are soft and dark, and their outer portion strips away readily when pulled. Avoid plants infested with insects or diseases that may cause serious problems in landscapes. See Table 3-1 for a nursery plant selection checklist. Consult *Arboriculture: Integrated Management of Landscape Trees, Shrubs, and Vines* and *Guideline Specifications for Nursery Tree Quality* for more suggestions.

Pest Resistance. In some cases, pest-resistant cultivars or species can be selected that otherwise perform and

Table 3-1.

Nursery Tree and Shrub Selection Checklist.

LOOK FOR
- Species or cultivars well adapted to heat, light, soil, water, wind, and other environmental conditions where they will be planted.
- A plant that at maturity will fit into the space provided for roots and branches.
- Pest-resistant species or cultivars, where available.
- Roots and crown area free of galls, insects, rots, and wounds.
- Large roots, root crown (collar), and trunk that are not kinked and main roots that do not circle the trunk.
- Roots that are not a solid mass or are too small in comparison with aboveground parts.
- A plant that is well rooted in the container so the root ball will not fall apart at planting.
- Good overall appearance, color, leaf size, and vigor.
- Branches distributed radially around and vertically along the trunk.
- A tree with a tapered trunk and a single, relatively straight central leader.
- A trunk that is without wounds and that can stand without being staked.

AVOID
- Species or cultivars poorly adapted to local environmental conditions.
- Plants that at maturity will be too large for the available space.
- Species or cultivars prone to local pest problems.
- Injured, distorted, diseased, or girdled trunks, roots, or crown area.
- Encircling or kinked roots, a root mass too small in comparison with aboveground plant parts.
- Large roots growing out of the container into the native soil.
- Discolored, distorted, or undersized foliage.
- Trees without tapered trunks and that lack a single, relatively straight central leader.
- Trees with large branches close together on the trunk.
- Tree trunks that can't stand without being staked.

Adapted from Burger et al. 2002; Harris, Clark, and Matheny 1999.

Select new plants that have desirable root and trunk structure and are free of important pests. Reject plants with roots that are soft and dark or have outer portions that strip away readily from their core. See the pictures of healthy and unhealthy roots near the beginning and end of Chapter 5. *Photo:* J. K. Clark

Reject container-grown plants with poor structure, like this kinked, circling, major root in the crown area. This defect is likely to slowly girdle and kill the tree as it grows. *Photo:* J. K. Clark

look similar to susceptible plants. Avoid planting species or cultivars known to be prone to serious problems in your area. Do not replant in locations where plants have been killed or severely damaged by disease unless you select a species or cultivar highly resistant to that disease. Do not plant species highly susceptible to root and crown diseases in poorly drained, compacted soils. Improve drainage or plant high, such as on a mound or soil berm.

Consult Table 3-2 for resistant species or cultivars before selecting plants. Tables of species resistant or susceptible to *Phytophthora* and *Verticillium* serve as a guide for selecting plants to avoid these diseases. Resistance is not the same as immunity. Plants may become affected by problems to which they are resistant if they are stressed because of poor cultural care or other factors. New plant cultivars and better information are constantly being developed; consult a knowledgeable UC Cooperative Extension Advisor or Master Gardener, cer-

Table 3-2.

Pest-Resistant Alternative Species or Cultivars for Common Problems on Woody Landscape Plants.

HOST PLANT	PEST	RESISTANT OR LESS-SUSCEPTIBLE ALTERNATIVES
many species	broadleaf mistletoes	page 284
many species	crown gall	page 108
many species	Phytophthora root rot	Table 5-13, page 123
many species	Verticillium wilt	Table 5-12, page 118
acacia	acacia psyllid	Table 6-6, page 171
alder	flatheaded alder borer	black alder, page 229
ash	anthracnose	Moraine, Raywood, Shamel, Table 5-6, page 82
birch	bronze birch borer	non-white-barked birch, e.g., *Betula alleghaniensis*, *B. lenta*, or *B. nigra*, page 227
box elder	boxelder bug	male box elder, page 206
ceanothus	stem gall moth	Table 6-10, page 213
crape myrtle	powdery mildew	Table 5-7, page 91
cypress	cypress canker	Table 5-9, page 99
cypress	cypress tip miner	Table 6-11, page 217
dogwood	anthracnose	Table 5-6, page 82
elm, Chinese	anthracnose	Brea and Drake cultivars, page 82
elm	Dutch elm disease	resistant elms, Table 5-11, page 114
elm	elm leaf beetle	resistant elms, Table 5-11, page 114
elm	European elm scale	hackberry, zelkova, page 197
eucalyptus	longhorned borers	Table 6-7, page 173
eucalyptus	tortoise beetles	Table 6-7, page 173
eucalyptus	redgum lerp psyllid	Table 6-7, page 173
euonymus	euonymus scale	*Euonymus alata*, page 190
euonymus	powdery mildew	variegated cultivars
fuchsia	fuchsia gall mite	Table 6-18, page 250
Indian laurel fig	Cuban laurel thrips	*Ficus microcarpa* 'Green Gem'
juniper	cypress tip miner	Table 6-11, page 217
palm, California fan	diamond scale	Mexican fan, non-California fan palms, page 87
pear, ornamental	fire blight	Capital and Chanticleer cultivars, page 84
pepper tree	peppertree psyllid	page 174
pine	Nantucket pine tip moth	Table 6-12, page 219
pine	sequoia pitch moth	Table 6-17, page 240
poplar	Cytospora canker	Easter, Nor, Mighty Mo, and Platte poplar hybrids, page 100
rhododendron	powdery mildew	Table 5-7, page 91
rhododendron	root weevils	Table 6-5, page 161
rose	certain petal chewing beetles, e.g., hoplia beetle and rose curculio	red roses and other cultivars with darker-colored petals, pages 163, 165
rose	powdery mildew	Table 5-7, page 91
sycamore and London plane	anthracnose	Bloodgood, Columbia, and Liberty cultivars, Table 5-6, page 82
sycamore and London plane	powdery mildew	Columbia, Liberty, and Yarwood cultivars, Table 5-7, page 91

Resistance is not the same as immunity. Plants may become affected by problems to which they are resistant if plants are stressed because of poor cultural care or other factors.

Shrubs and trees have different irrigation requirements than turf. Group together plants with similar water use rates (use hydrozoning) so they can be irrigated the same way. Use sidewalks, driveways, and edging or headers to separate plants with different cultural needs. *Photo: J. K. Clark*

tified arborist, or certified nurseryperson for assistance in selecting pest-resistant plants.

Plant Compatibility. Group together plants having compatible growth characteristics, similar water use rates (hydrozoning), and other cultural care needs. For example, many trees native to the eastern United States can grow well in turfgrass but should not be mixed with most California natives and drought-adapted plants. Also, avoid planting aggressive, rapid-growing ground covers and turf species close to or among shrubs and young trees. Locate incompatible species apart from each other or separate them with structures or pavement, or with wood, plastic, metal, or concrete barriers, edging, or headers extending well below ground (see Figure 7-4).

Edible Landscapes. Landscapes increasingly incorporate attractive plants that also provide food. Citrus are California's most popular landscape fruit trees. Attractive and nutritious alternatives include fruiting olives, Fuyu persimmons,

Tiger figs, Santa Rosa weeping flowering plum, Stella cherry, and numerous varieties of grapes, pomegranates, and southern highbush (low-chill) blueberries.

Many of the disorders and host-specific pests of fruit and nut trees and shrubs are not discussed in this book. For information on edible landscapes and their management, see the *California Master Gardener Handbook, The Home Orchard: Growing Your Own Deciduous Fruit and Nut Trees, Pests of the Garden and Small Farm,* and the Fruit & Nut Research and Information Center (online at fruitsandnuts.ucdavis.edu).

Sustainable Landscapes. Plantings are "sustainable" when they are designed and maintained to minimize the need for inputs and do not deplete or permanently damage natural resources. Sustainable landscapes improve and protect the quality of our air, soil, and water by using noninvasive ornamentals that thrive under local site conditions and need minimal irrigation, fertilization, and pesticide application. To make landscapes sustainable

- Protect water quality and quantity.
  * Irrigate appropriately and avoid runoff.
  * Divert downspouts and other runoff into landscapes to replenish soil moisture and reduce stormwater runoff into drains and gutters.
  * Use rain barrels and recycled water for irrigation where feasible.
  * Use permeable surface paving to increase water infiltration into the soil instead of using concrete or other impermeable surfaces.
- Grow a diversity of plant species and manage the landscape to create wildlife habitat (e.g., for bees, birds, and butterflies) and reduce the risk of wildfire.
- Recycle materials and keep greenwaste on-site for mulch and compost to improve soils; conserve energy, topsoil, and water; and reduce discards to landfills.
- Nurture the soil by mulching, protecting against compaction, and reducing fertilizer use.

- Select trees and shrubs known to be resistant to key local pests and use IPM to manage problems, as discussed throughout this book.
- Conserve energy.
  * Reduce water use to reduce energy consumption to move water.
  * Plant deciduous trees to provide shade on the west and south sides of homes and other buildings.

See *Sustainable Landscaping in California* for more information.

SITE PREPARATION AND PLANTING

Properly prepare the soil and control weeds before planting, as discussed in Chapter 7. Perennial weeds especially are easier to control effectively if you take action before planting.

After controlling weeds, loosen compacted soils, break up any hardpan, and (if warranted) amend soil, such as by adding organic matter. Properly grade soils or create berms or raised beds for plantings so that water drains well, especially around roots, which helps to prevent root disease. Next, install the infrastructure (e.g., fencing, headers and edging, irrigation systems, outdoor lighting, and paving); especially install anything that involves excavating soil or using heavy equipment that might damage plant roots or aboveground parts. Finally, choose species that are well adapted to conditions at that site and properly plant the landscape.

PREPARE THE SITE

Many urban soils have been disturbed during construction (e.g., compacted, graded, and filled with debris) and differ greatly from the native topsoil. Urban soils often exhibit slow infiltration and percolation of water (slow drainage) because they are compacted; contain impervious layers (hardpan) or layers of different texture; or have high clay content. To help learn about local soil and its possible effects on root health, examine nearby plants, observing their species and maturity. Note how well they are growing and the frequency and type of irrigation they receive.

To assess soil and drainage at a site, dig under the surface, extract soil cores, or conduct a percolation test, as discussed below. If poor drainage or other adverse soil conditions are found, determine the cause and identify appropriate remedies. Planting species more tolerant of soggy soils is one option. However, especially where slow drainage is the problem, drainage usually must be improved if young trees and shrubs are to grow well.

Before planting it may be necessary to loosen compacted topsoil or to break up or penetrate hardpan if impervious layers occur within about $1\frac{1}{2}$ to 2 feet of the surface; loosening soils to a depth of 3 feet or more can improve plant growth in some species. Use a backhoe, pickax, powered auger, trencher, or deep ripping to penetrate hardpan and layered soil. Loosen compacted soil by deep ripping or plowing and possibly by rototilling or using a digging fork or shovel. Mix different layers well to provide a relatively uniform soil texture and to minimize distinct boundaries between layers.

Planting on a broad soil berm or raised bed 1 to 2 feet tall may provide adequate aeration and drainage for good plant growth, but subsurface layers may still need to be loosened before planting so roots can penetrate the soil beneath mounds to anchor the trees well and avoid tipping trunks. Where mechanical methods are inappropriate (e.g., when existing tree roots are present) soil may gradually loosen (over months) if you

- Apply 1 or 2 inches of well-composted organic matter to the soil surface.
- Cover that with another inch or two of coarse organic mulch.
- Water as needed to keep the organic material slightly moist.
- Reapply mulch as it decomposes.

Assess Drainage. One method to assess infiltration rate (drainage) of topsoil is to perform a percolation test (Figure 3-4). Dig a 12-inch-deep hole the width of a spade. Roughen the bottom and sides to eliminate any glazed, hard-packed soil. Fill the hole to the top with water at least once and soak the surrounding surface with a sprinkler or hose so that the soil around the hole becomes saturated with water. Wait 24 hours, refill the hole with water, and observe how long it takes for all the water to drain. If any water remains after 24 hours, drainage is probably too slow. A desirable rate of soil drainage for many landscapes is about 1 to 2 inches per hour (roughly 6 to 12 hours to drain a 12-inch hole).

It is usually desirable to know the drainage and water-holding characteristics of soil deeper than 1 foot, except possibly when growing turfgrass or short, shallow-rooted plants. Sources of this information include the soil survey maps available online from the Natural Resources Conservation Service (NRCS) (websoilsurvey.nrcs.usda.gov), UC Soils to Go (soilstogo.uckare.org), and smartphone apps such as SoilWeb (ucanr.edu/SoilWeb).

Digging, feeling soil, and observing how quickly water drains when poured into a hole are among the techniques for assessing soil conditions. Investigation may reveal that the planting location or method, the species grown, or a combination of factors may need to be modified to allow landscape plants to perform as desired. *Photo: J. K. Clark*

At many developed sites, soils have been moved around or imported, so the soil types and layers found may be atypical for the area. Use a soil sampling tube or dig a 1-foot-deep hole and compare your topsoil with survey maps to reveal whether your soil matches the maps. If the topsoil differs from published surveys (for example, because it has been disturbed), deeper soils may also differ. It may be feasible to investigate soil types by digging down 3 feet or deeper using a spade, although con-

Figure 3-4.
Many urban soils have poor permeability and infiltration of water (slow drainage). Wet soil that drains slowly can damage roots and make plants susceptible to root decay pathogens. One method to assess infiltration rate of topsoil is to conduct a percolation test. Dig a 12-inch-deep hole the width of a spade. Roughen the bottom and sides to eliminate any glazed, packed soil. Fill the hole to the top with water at least once and soak the surrounding surface so that soil around the hole becomes saturated with water. Wait 24 hours, refill the hole with water, and observe how long it takes for all the water to drain. If any water remains after 24 hours, drainage probably is too slow. An optimal rate of soil drainage for many landscapes is about 1 to 2 inches per hour (roughly 6 to 12 hours are required to drain a 12-inch hole).

siderable effort may be required. An auger (either hand or powered) can remove soil for inspection up to 6 feet or more belowground. Often it is desirable to have an expert visit the site and dig or core under the surface to assess the soil conditions and extent of any soil disturbance.

PREPARE THE SOIL

Determine whether the soil has chemical or physical deficiencies before deciding whether to amend it. Adding gypsum (calcium sulfate) then leaching heavily with water low in salts may improve soils that are sodic (high in exchangeable sodium), but adding gypsum is only beneficial if sodium is the problem and a soil test is required to determine this. Species adapted to well-drained, acidic soils will do poorly in highly alkaline, poorly drained soils unless drainage is improved and soil is amended (e.g., adding organic matter and elemental, or soil, sulfur) before planting. For more information on soil problems, see Chapter 4, especially the sections "Iron," "Nutrient Deficiencies," "pH," and "Salinity," and consult related publications in "Suggested Reading."

Mixing organic matter into a small volume of soil before planting a tree is not recommended and can cause trees to become a failure hazard to people and structures. Adding organic amendments can create topsoil that has a different water-holding capacity than the sublayers, causing shallow, inadequate root development. Amending the entire potential root zone of trees is generally not practical, and amending around established plants will damage roots. However, in some situations adding peat or well-composted organic matter may improve soil structure and workability (tilth), increase water-holding capacity in sandy soils, and help improve drainage of clay soils. If organic matter is added to soil, it should be well decomposed or well composted and should constitute no more than about 20% of the soil volume in the upper 12 inches of soil or the anticipated rooting zone of the mature plant.

Thoroughly mix the organic matter into the topsoil. If mixing in organic matter, evaluate whether adding a modest amount of nitrogen fertilizer is appropriate to compensate for a temporary reduction in nitrogen availability to plants that can occur as soil microorganisms further decompose the organic matter. Because organic matter will gradually decompose and cause plants and soil to settle, planting on a berm or a mound is an especially good idea in soil amended with organic matter.

Provide for necessary irrigation. Anticipate plants' immediate and long-term irrigation needs so you can avoid system modifications later that can damage plant roots. Properly designed, operated, and maintained low-volume systems (e.g., drip emitters) conserve water, reduce weed growth, and help to avoid certain disease problems.

PLANT PROPERLY

Site preparation and the correct timing of planting are essential for helping new plants become well established. Late fall (especially) through early spring are generally good times to plant in most of California. Avoid planting during hot summer weather, except palms, which can establish well with summer planting. Depending on conditions and preparations needed at that site, weed control and other work may need to begin well before planting.

Planting too deeply or not deeply enough are common problems. Planting too deeply favors root and crown diseases to which young plants are especially susceptible. Planting trees and shrubs too shallowly can make them unstable and can lead to root damage from exposure and excessive drying.

Mark out a planting space that is at least two to three times the diameter of the root ball. Preparing a much larger planting space can greatly benefit trees. If needed, loosen the soil within this area. However, in the center of the hole where the plant will be placed, loosen soil only to the depth somewhat less than the root ball so the root crown can be planted slightly high with the root ball on firm or settled soil.

In most situations plant the root crown

about 2 inches higher than soil level for a 5-gallon or larger container, and 1 to ½ inch or less above soil level for a 1-gallon container. Planting "high" is especially important if the soil is compacted, drains poorly, has been loosened deeper than the root ball, or if the soil is highly amended such that the plant is likely to settle as organic matter decays. Avoid planting in a depression or low-lying area (see Figure 5-3).

Place the plant in the hole and orient the main stem to a vertical position. If there is a "crook" or indentation at the bud union (where the roots and scion were grafted), orient this toward the north to avoid sunburn. Cut any wires or rope around the root ball and pull them away. Remove any burlap, other root wrapping, or container before planting. Exceptions include bougainvillea, which does not tolerate root disturbance and can be planted after cutting out (removing only) the bottom of its container. If the root wrapping is biodegradable (e.g., authentic burlap and not a synthetic lookalike) it may be acceptable to plant with the wrapping left on the root ball to avoid disturbing roots. Check for roots that circle the container and gently spread or cut them before planting. Cut any broken or encircling roots that are too large to spread. Do not use the plant if it is extensively root-bound or has major roots kinked or encircling the trunk or root ball; such plants will perform poorly and may be blown over by the wind.

Backfill the hole with native soil after properly positioning the plant and preparing the roots. Do not cover container soil with field soil, as the difference in texture can prevent water penetration into the container soil, where all the roots are located. Loosening the periphery of the container soil and mixing it with native soil to minimize a distinct layer between the two types of soil might be beneficial in some cases, but it can damage fine feeder roots growing on the periphery of the root ball. Settle soil near the root ball after planting, such as by watering thoroughly to eliminate air holes or introducing a hose into the bottom of the planting hole and

Planting too shallowly or allowing drainage water or irrigation to wash away soil leads to root damage from exposure and excessive drying. Conversely, planting too deeply (pictured later) can lead to root and crown diseases. *Photo:* J. K. Clark

watering; add soil as needed. Do not tamp or press down on the soil, especially after watering, or you may compact the soil. See "Care for Young Trees and Shrubs," below, for watering recommendations.

Keep an area 2 feet in diameter or larger free of turf or other vegetation around the trunk of young woody plants to reduce trunk damage from lawn mowers and weed whips and to minimize competition that stunts tree growth. Apply 3 to 4 inches of coarse organic mulch over the entire prepared area, except keep mulch at least 6 inches away from the stem or trunk to avoid problems such as excess moisture where the trunk and roots meet (which can promote disease).

Remove any labels, tree tags, trunk wrapping, or protective tape that came with the new plant as these can restrict trunk growth and seriously injure young plants. If the trunk will not be shaded by its branches or nearby features and sunburn is a concern, consider applying white interior latex paint diluted with an equal amount of water to the trunk to help prevent sunburn. Interior water-based paints are safer to trees than are oil- and water-based exterior paints. After planting, remove any nursery stakes and loosely re-stake the tree if needed. Consult publications such as *Planting Landscape Trees* for more details.

STAKING

Trees should be staked only during the first year or so after planting and only if needed to protect or support the trunk or anchor the root ball (e.g., at windy sites). After planting, remove the nursery stake that came with the container and loosely re-stake the plant if it is needed. If leafy branches are relatively abundant on the tree, thin the canopy slightly (shorten or remove a few branches). If thinning allows the tree to remain upright without support, staking is not necessary. If staking is necessary, do not fasten trunks firmly to a stake; trunks must be allowed to flex some with the wind in order to develop strength.

Use two stakes in most situations and tie the trunk at just one level so the trunk is free to flex below the tie as well as above it and the treetop is kept upright. Three stakes may be appropriate in very windy locations. Some innovative single staking systems that provide support and allow trunks to flex are also suitable. Where sunburn may be a problem, consider locating one of the stakes southwest or west of the

trunk to help shade the bark, unless stakes should be located otherwise because of prevailing winds. At windy locations when using two stakes, orient the stakes so they and the trunk form a line perpendicular to the direction of prevailing winds, so the trunk is not blown directly toward either stake. Stakes should be tall enough to be easily seen and located near the edge of the root ball. After tying the trunk, cut the stakes off just above the ties so that stakes do not rub against limbs or the trunk.

To determine the proper staking height of trunks that cannot stand upright without support

- Locate the lowest level at which the trunk can be held upright.
- Place the tie(s) about 6 inches above this point.

To avoid deforming growth, do not tie trunks within 2 feet of the growing point of the leader.

Trunks must be allowed to flex with the wind in order to develop stem strength. Too many ties (shown here) and not removing the nursery stake that came with the container are common problems of newly planted trees. *Photo:* J. K. Clark

Ties were left on this young tree too long, causing wood to grow around the top tie. The branch at the right has a canker from rubbing a stake that is too tall and too close (see the close-up photo in Chapter 4). Remove any stakes and ties after a year or so. If the trunk is then unable to stand alone, determine the cause and if possible remedy the problem before properly restaking the trunk loosely and at only one level. *Photo:* J. K. Clark

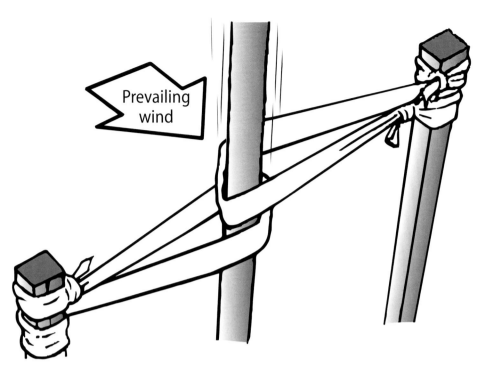

Figure 3-5.
Stake trees during the first year or so after planting only if needed to protect or support the trunk or anchor the root ball. Use two stakes in most situations and tie the trunk at just one level so the trunk is free to flex above and below the tie. Orient the stakes and the trunk so they form an imaginary line perpendicular (90 degrees) to the direction of prevailing winds, so the trunk is not blown directly toward either stake. Cut stakes off above the support ties.

Ties should be belt-like straps with a broad surface that form a loose loop around the trunk, contacting the trunk without cutting into bark. Ties should be made of a flexible or elastic material such as polyethylene tape or rubber tubing (old bicycle tubes can be cut and split open lengthwise). The preferred staking method is to use two ties, each about 18 inches long, attached to opposite posts. Circle each tie around the trunk and attach both ends to the same stake (Figure 3-5) or cross or overlap the ends to form a figure eight. Remove any stakes after a year or so; if the trunk is unable to stand alone, determine the cause and if possible remedy the problem before re-staking. Consult *Planting Landscape Trees* or *Arboriculture: Integrated Management of Landscape Trees, Shrubs, and Vines* for more details.

CARE FOR YOUNG TREES AND SHRUBS

Learn the cultural requirements of plants under local conditions. Water, fertilize (only if appropriate), and prune young plants correctly. Proper cultural care is critical to keeping plants healthy and minimizing pest damage. If a pest or damage does appear, reevaluate cultural practices to determine whether improper care has contributed to the problem. Proper cultural care alone may provide the solution. Other activities, such as pesticide applications, may be of little benefit to plants if cultural care is improper.

Keep soil moist but not soggy in the root ball and for several feet outward around newly planted trees and shrubs. Proper irrigation frequency depends on the plant, soil, and weather. For about the first 2 months, spring or summer plantings may need to be irrigated almost daily during hot, dry weather because their roots are confined to the small volume of soil

that was the old soil ball in the container. Water the root ball directly during early establishment of new plants because water often does not move easily from surrounding soil into the root ball. Place irrigation emitters so that they wet the root ball and nearby soil; avoid the common mistake of locating emitters too far away from new plants. However, do not allow water to puddle around the base of the trunk (the root crown or collar) and minimize direct wetting of the trunk.

Until you learn how irrigation practices affect subsurface soil moisture in your specific situation, periodically dig a shallow hole or use a soil tube or moisture indicator probe to check subsurface moisture levels. As roots grow into the surrounding native soil, apply larger amounts of water less frequently and at increasing distances away from the trunk, but keep the root crown area dry. It may take several months for this root growth to occur, depending on temperature and vigor of the transplant. After the initial establishment period, gradually increase the interval between irrigations and encourage good root growth through infrequent, thorough soakings at least out to the canopy drip line (see Figure 5-2). Allow the surface and upper topsoil to dry between waterings, but make sure soil several inches below the surface remains moist, especially in the original root ball. Avoid frequent sprinkling that wets only the surface; this may drought-stress the plant and encourage undesirable shallow root growth.

Fertilize woody plants sparingly or not at all during their first growing season in the landscape. Exceptions include mixing a modest amount of slow-release nitrogen into soil before planting if soil is amended with organic matter. Nitrogen is usually the only nutrient to which woody plants respond in most soils. Exceptions may include sandy soils that are especially low in nutrients and palms, as discussed later. If young plants will be fertilized, avoid applying too much and incorporate only slow-release, nonmanure fertilizers into the planting hole.

After planting, prune only to remove damaged or diseased branches and to remove those that have serious structural problems (e.g., extremely narrow crotch angles) or interfere with (rub or cross) more desirable branches. During the next 4 or 5 years of growth, prune young woody plants if needed to encourage good structure; establishing a central leader or dominant main terminal is especially important on young trees.

Hire a certified arborist or other professional to prune trees or consult resources such as the video *Training Young Trees for Structure and Form*. Avoid excessive pruning, which can ruin tree structure and retard overall growth by removing food-producing foliage. Leave some temporary short branches along the trunk or main stem during the first few years; these protect tender bark from injuries and sunburn and greatly improve trunk growth and strength.

Prevent weeds, turf, and ground covers from growing near the trunk of young trees and shrubs, as their roots compete for resources and can seriously retard young plant growth. Apply and maintain 3 to 4 inches of coarse mulch over a 2-foot

Applying white, interior latex paint diluted with an equal portion of water to trunks helps to prevent sunburn of bark. If you plant trees in lawn, choose species tolerant of summer irrigation and keep a 2-foot-diameter or larger area around the trunk free of turf and other vegetation. Plant the root ball a little above soil grade and do not make a watering basin. *Photo:* J. K. Clark

radius or larger area beneath new trees and shrubs. Benefits of mulching include reduced weed competition, retained soil moisture, reduced soil splashing, increased soil microbial activity, improved soil structure, and moderated root-zone temperatures.

WATER MANAGEMENT

Improper water management is probably the biggest cause of poor performance and survival of trees and shrubs. Monitor plants' appearance, growth rate, and root zone soil moisture to determine whether irrigation amount and frequency is adequate, inadequate, or excessive and whether the system layout (e.g., location of emitters) is appropriate. When appropriate, dig up and examine small roots to become familiar with their appearance, as pictured in Chapters 5 and 8.

WATER AND PEST PROBLEMS

Too much or too little water damages or kills plants (Tables 3-3 and 3-4). Overwatering (watering too frequently or applying excessive amounts) is especially common in established landscapes. Underwatering of the root balls is common in new plantings, which are shallowly rooted and dry out quickly between waterings.

Insufficient water causes leaves to droop, yellow, wilt, or drop. Water deficiency promotes certain pests, sunburn, shoot and branch dieback, bark cracking, cankers, and overall sparse growth. For example, the *Botryosphaeria* fungus commonly causes cankers and branch dieback on drought-stressed giant sequoia planted outside its native range. Mites, some leaf-sucking insects, and most wood-boring bark beetles, flatheaded and longhorned beetles, and clearwing moths cause more damage to plants stressed from

drought; once trees become severely infested by borers, they decline and usually die.

Overwatering and excess soil moisture excludes from soil the oxygen that roots need to survive. Water dripping, ponding, or spraying near the basal trunk and soggy soil around the root collar are primary causes of root and crown diseases. Pathogens such as *Dematophora* and *Phytophthora* spp. are present in many soils, yet usually become damaging only when wet conditions favor them. Excessive water uptake can cause trunk splitting in palms and bark splitting in certain trees (e.g., × *Chitalpa tashkentensis*).

Wetting areas where there are no desirable plants encourages germination of weed seeds. Splashing water spreads fungal spores and wets foliage, promoting foliar and fruit diseases such as anthracnose, brown rot, leaf spots, and rusts. Minimize or prevent weed growth and many foliar diseases by using low-volume drip irrigation or minisprinklers instead of overhead sprinkling.

The seasonal timing of irrigation also is important in disease development. For example, oak root fungus (*Armillaria mellea*) is present on dead or living roots in many soils and becomes active when soils are warm and moist. Native oaks are susceptible to the disease but usually are not damaged in California, where soils normally dry out as they warm during summer. However, when people water native oaks frequently during the summer or alter soils, moist roots and warm soils coincide, predisposing oaks and many other species to infection and death by *Armillaria*.

Drought-adapted plants, such as certain eucalyptus, may benefit from deep, supplemental water at 1- or 2-month intervals during the summer, especially during years of abnormally low rainfall. Supplemental irrigation may also be appropriate if drought-adapted trees have been injured (e.g., by cutting roots) or soil moisture has been altered from historical levels (e.g., changes in drainage) and after consecutive years of abnormally low precipitation. For

Prolonged wetting of the basal trunk promotes root and crown rot diseases. If irrigation basins are used, instead of a hole as shown here, plant on a central mound so the root crown is kept dry. *Photo:* J. K. Clark

Table 3-3.

Common Problems Associated with Underwatering.

DAMAGE SYMPTOMS	CAUSE	MANAGEMENT
bark or branch cankers	abiotic disorder	pages 44, 57, 58, 60–63
	canker disease fungi	page 97
bark weeping or resin exudation	abiotic disorder	pages 44, 57
	bacteria	page 110
	Botryosphaeria canker	page 98
	wood-boring insects	pages 220–244
bark with holes or sawdust	wood-boring insects	pages 220–244
leaves bleached or stippled	mites	page 245
	sucking insects	pages 170–212
leaves drop prematurely	abiotic disorder	page 44
leaves spotted	many causes, Table 5-1	page 68
shoot or branch dieback	abiotic disorder	page 44
	Botryosphaeria canker	page 98
	wood-boring insects	pages 220–244

Abiotic disorders are noninfectious plant diseases induced by adverse environmental conditions, as discussed in Chapter 4. Some of these symptoms can also be due to other causes not listed here.

Table 3-4.

Common Problems Associated with Overwatering or Poor Water Placement.

DAMAGE SYMPTOMS	CAUSE	MANAGEMENT
bark or trunk splitting	abiotic disorder	avoid applying excessive water
branch cankers	anthracnose diseases	page 81
branches die back	abiotic disorder	pages 44, 57, 58, 60–63
	root and crown diseases	pages 119–124
fruit spotted or discolored	fungal diseases	pages 81, 86
leaves yellow, wilt, or drop prematurely	abiotic disorder	pages 44, 50–57
	anthracnose diseases	page 81
	root and crown diseases	pages 119–124
	vascular wilt diseases	pages 112–118
leaves spotted or discolored	bacterial and fungal pathogens	page 86

Poor water placement includes wetting leaves, trunks, or root collars and allowing water to pond around trunks. Some of these symptoms can also be due to other causes not listed here.

native oaks and other species adapted to summer drought, irrigation should simulate natural patterns and be applied mostly during the normal rainy season (late fall to early spring), tapering off during the dry season. Exceptions where summer watering may be warranted include planted oaks, which may be adapted to summer irrigation from the start, and in certain situations where soils have been disturbed. Be aware that certain eucalyptus and non-native oaks are not adapted to summer drought and require more frequent irrigation when planted in California.

IRRIGATION

Irrigation is required to maintain most urban landscapes in California, where rainless weather prevails throughout much of the growing season. Early morning or just before dawn is generally the best time to irrigate. Irrigating around dawn reduces water loss from soil evaporation and minimizes the length of time when foliage is wet, thereby discouraging the development of certain foliar diseases. Predawn irrigation improves sprinkler efficiency and the uniformity of water distribution because there is generally less wind and more water pressure. Irrigating during late evening or night can minimize evaporation, but avoid overhead sprinkling if foliar diseases are a problem because leaves will remain wet longer than with irrigating around dawn. After plants are established, apply the water around and beyond the drip line, not near the trunk where the root collar should remain dry (see Figure 5-2).

The appropriate irrigation frequency and the volume of water to apply during each watering vary greatly. Considerations include moisture demand according to plant species and season, microclimate, root depth, drainage patterns, irrigation system type and efficiency, and soil texture, structure, and depth. When irrigating overhead, make sure that enough water is applied at each irrigation to penetrate mulch and wet the underlying soil to the average depth of the root zone. Periodically check the placement of drip emitters

Basin irrigation is often used in dry locations that do not have an installed irrigation system. However, it may be best to avoid basins around trunks, except for about the first year after planting. Continually enlarge any basin to encourage lateral root growth as the tree matures. Break berms down during the rainy season to prevent waterlogging. *Photo: J. K. Clark*

This drip system, which will be covered with coarse organic mulch, helps conserve water. Low-volume irrigation also reduces soil compaction and weed growth problems associated with sprinklers. After the tree is established, move drip emitters farther away from the crown root. *Photo: J. K. Clark*

and the direction of spray heads to be sure they are applying water where it is desired and to a depth of about 12 to 18 inches in the entire root zone of trees and shrubs.

Rooting depth varies according to plant age, type, species, soil, and moisture conditions. Plants generally have shallower roots when they are young or receive relatively frequent and light irrigations or are growing in compacted or poorly drained soils. Roots are usually deeper in older plants, as well as in plants growing in well-aerated soils with good drainage and less-frequent but deep irrigation. However, about 90% of tree and shrub roots are usually in the top 3 feet of soil (Figure 3-2).

Water demand or loss depends on the environment and plant species. Water depleted from the soil through a combination of evaporation from the soil surface and transpiration by plants is called evapotranspiration (ET). A plant's demand for water generally increases when weather is sunny, hot, and windy and when humidity is low.

Soil Properties. Soil characteristics affect the air and water available to plant roots. *Texture* is the relative proportion of different sizes of soil particles, including sand (large soil particles), silt (intermediate sizes), and clay (the smallest particles). *Structure* is the arrangement of these soil particles into larger aggregates. *Pore spaces* are the voids between the solid particles, which fill with air or water or both. *Field capacity* is the amount of water that can be held in pore spaces after excess water has drained. As plants use water, the remaining water becomes more tightly held in

soil. Because of the attraction between soil and water, plants can extract only a portion of the water in soil—the available water. The wilting point occurs when plants have extracted all the available water. Loam soils (those with roughly equal proportions of clay, sand, and silt) that are not compacted generally provide the best combination of available water (Table 3-5) and adequate oxygen for roots (Figure 3-6) because they provide a wide range of pore sizes, with the smaller pores holding moisture while the larger pores drain and allow air to enter the

soil. For more information on soil and irrigation, consult publications such as *Water Management.*

Table 3-5.

Approximate Amounts of Available Water When Soils Are at Field Capacity.

SOIL TEXTURE	INCHES OF AVAILABLE WATER PER FOOT OF SOIL
sand	0.5–1.0
sandy loam	1.0–1.5
clay loam	1.5–2.0
clay	1.5–2.5

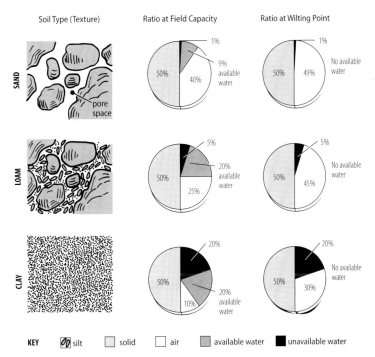

Figure 3-6.
The amount of air and water available to plant roots depends on soil texture (particle sizes) and structure (particle arrangement). Sandy soils have large pore spaces that contain large amounts of air; large spaces allow water to drain quickly, so the soil dries relatively fast. Clay soils have many small pore spaces that retain water, but much of this water is too tightly held to be available to roots. Small pore spaces drain poorly (slowly) and often provide insufficient space for oxygen needed by roots. Silt particles are intermediate in size between clay and sand. Loam soils have roughly equal proportions of sand, silt, and clay. Loam soils that are not compacted generally provide the best balance between water-holding ability and aeration.

Estimating Irrigation Needs. Schedule irrigation by observing plants and monitoring evapotranspiration or soil moisture in the root zone. These techniques assume that plants are correctly planted, well rooted, and have been growing well. Combine more than one method for the best results.

Observe Plants. Examine plants regularly for symptoms of water stress. Early drought-stress symptoms exhibited by broadleaf plants include wilting of leaves and normally shiny green foliage that becomes faded, dull, or grayish. Growing tips may wilt in the afternoon and recover by the next morning. As water deficit becomes more severe, plants may not recover from wilt. As symptoms progress, leaf margins or interiors turn yellow or brown, foliage dies and drops, and twigs, branches, and eventually the entire plant may die. Certain plants may exhibit symptoms first because

Early symptoms of drought stress (water deficit) include temporary wilting of foliage during the day, as with these pittosporum terminals. *Photo:* J. K. Clark

of topography, drier soil (e.g., due to inefficient irrigation), or they are less-drought-tolerant species. Inspect these plants more frequently and use them as indicators of drought stress and irrigation need.

Monitor Soil Moisture. Schedule irrigation by monitoring soil moisture. The frequency of monitoring varies greatly, depending on the factors discussed above. Soil around young plants during hot weather may need to be monitored daily; every few weeks may be adequate when monitoring around mature trees during more favorable weather. Sample soil from the root zone in several different areas of the landscape to assess overall irrigation needs and determine whether water is being applied deeply and uniformly enough.

Monitor soil moisture by digging a shallow hole with a trowel or other small digging tool that minimizes root injury. Alternatively, use an auger, soil probe, or soil sampling tube, such as those pictured in Chapter 8. Examine soil moisture in the rooting zone to a depth of about 1 foot.

Soil lightens in color when it is dry. To estimate soil moisture by how soil feels and molds in your hand, consult the table in *Water Management.* For example, medium- and fine-textured soils such as loam and clay can be molded, rolled, or squeezed into a ball when wet. If soil molds into a ball but does not crumble when rubbed, it is too wet. If the soil can be molded and crumbles when rubbed, the moisture content is probably suitable, except that sandy soil crumbles even when moist. If soil does not mold, it is too dry.

You can use tensiometers to monitor soil moisture if you properly install and maintain them. These devices are buried so their bottom contacts soil particles in the root zone, perhaps about 1 foot deep. Some tensiometers can be wired into irrigation system controls to assure watering occurs only when soil reaches a preset dryness threshold. Be sure to locate tensiometers in areas representative of plants' irri-

gated root zone and use the measurement in the driest part of the landscaped area to trigger the need for irrigation.

Irrigation can also be scheduled using "smart" irrigation controllers with sensors that measure moisture-dependent electrical properties of soil, including capacitance sensors, electrical-resistance or soil-moisture blocks (gypsum blocks), and granular matrix sensors. For discussion of these, see the UC Land, Air and Water Resources website, lawr.ucdavis.edu, and publications such as *Arboriculture: Integrated Management of Landscape Trees, Shrubs, and Vines.*

Monitor Evapotranspiration (ET). Irrigation can be scheduled by monitoring ET, the combination of evaporation of water from soil surfaces and transpiration from plants, commonly expressed in inches per day. Reference ET (ET_o) provides a standard estimate of a plant's water demand based

Monitor soil moisture at regular intervals to help determine the frequency and amount of irrigation. This Oakfield soil tube has just been withdrawn from the ground after collecting soil up to 1 foot deep. The tube's side is cut away for inspection and removal of a soil core for examination to determine the depths where soil is moist. *Photo:* J. K. Clark

on climatic conditions over a given period of time, such as a day or month. It is the amount of water used by well-watered cool-season turfgrass, which typically requires more water than trees and shrubs. After they are established, most woody landscape plants perform well by irrigating them with about 50 to 60% of ET_o.

A standard recommendation is to irrigate established woody landscapes when evapotranspiration monitoring indicates that about 50% of available soil water has been used. For example, assume most woody plant roots are in the upper 2 feet of a loam soil. At field capacity (after irrigation and initial drainage), uncompacted loam soil contains about 1.5 inches of available water per foot (Table 3-5), or a total of 3 inches of available water in the upper 2 feet. Thus, irrigation would occur when ET accumulates to 1.5 inch (50% of available water). If a typical summer ET_o is 0.2 inch per day, many woody plants would need 0.1 inch of water per day, and use up the 1.5 inches of available water in 15 days (1.5 inches of water/0.1 inch of water per day = 15 days). Therefore, 1.5 inches of water should be applied after 15 days. This method of irrigation scheduling is called a water budget. Irrigation and rainfall represent deposits and daily ET rates represent withdrawals.

Some irrigation controllers can access historical ET_o data or current reference ET information gathered on-site or online and use these to schedule irrigation automatically and adjust for seasonal changes. Historical and current (real-time) reference ET values for many locations at sites throughout California are available online from the California Irrigation Management Information System (CIMIS), cimis.water.ca.gov, and UC IPM, ipm.ucanr.edu. Some irrigation controllers use private, commercial sources of ET_o data.

Evaporation outdoors can be monitored on-site by regularly measuring water loss from a shallow pan or an atmometer (a porous, water-filled tube from which water evaporates at a rate similar to plants). Automated evaporation pans and atmome-

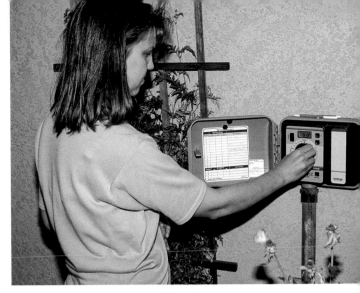

A properly operated irrigation controller can help you avoid over- or underwatering plants. With this type, you must periodically adjust the settings manually as plants' need for irrigation changes with the weather and seasons. Some "smart" irrigation controllers use on-site soil moisture sensors or weather-dependent evapotranspiration (ET) data to automatically adjust the watering schedule. Weather-based controllers can use historical monthly averages of ET, on-site ET sensors, or remote ET data that are broadcast, such as via radio signals. *Photo: S. E. Lock*

ters use a sensor to monitor water level and send that information to a data logger or automated irrigation control system.

For more information, consult the *ET_o Zones Map* on the CIMIS website and publications such as *Determining Daily Reference Evapotranspiration* and *Evapotranspiration and Irrigation Water Requirements*.

IRRIGATION METHODS

Basin, sprinkler, and low-volume soaker or drip irrigation systems are common in landscapes. A basin is formed by creating a berm of soil several inches high that encompasses the drip line of the young tree or shrub bed. Water is provided within the berm by installing an irrigation head, using a hose or tank truck, or (after plant establishment) by relying on runoff or precipitation. Do not irrigate so frequently that topsoil within a basin is constantly wet. Either plant on a central mound or slope soil within any basin to keep water away from the plant's root collar. Break down berms during prolonged rainy weather to prevent water from ponding around the trunk (see Figure 5-2). If basins are used beyond the first year or two after planting, gradually move the berm farther from trunks to increase the area of irrigated soil.

Sprinklers irrigate the soil and also wash dust from plants and increase humidity in landscapes. However, sprinklers may distribute water unevenly and waste water, especially in windy conditions. They can also compact the surface of bare soil,

increase weed germination and growth, and promote certain foliar diseases by splashing fungal spores and wetting foliage. Reclaimed water, if it has high salinity and is applied with sprinklers that wet leaves, can injure foliage. In comparison with low-volume irrigation systems, high-volume sprinkler irrigation of trees and shrubs is less efficient because water is dispersed widely, making it suitable only for relatively large plantings with uniform water needs.

Low-volume systems emit water directly on or below the soil surface using drippers or emitter nozzles to wet only plants' root zone or a limited area of soil. They waste comparatively little water and reduce or avoid compaction, certain pathogen problems, and extensive weed growth associated with high-volume sprinkler irrigation. Low-volume systems can be more expensive to install than sprinklers and may require more maintenance and skill to use, especially to develop appropriate irrigation schedules and manage salinity. Be aware that it can be difficult to monitor and maintain systems installed belowground or beneath mulch. When using a low-volume system, shrubs in dry areas may need to be occasionally washed of dust to keep them healthy.

For more information, consult publications such as *Drip Irrigation in the Home Landscape, Water Conservation Tips for the Home Lawn and Garden,* and *Water Management.*

Planting species that require little irrigation, as shown here, is one method of conserving water. Even when woody species requiring more water are grown in California landscapes, increasing the interval between irrigations often improves plant health and wastes less water. *Photo:* K. Reid

CONSERVE WATER IN LANDSCAPES

Consider using drought-tolerant ornamentals and save the water for fruit trees and vegetable gardens. Some, but not all, California natives and plants from other areas of the world with a Mediterranean climate perform acceptably with limited irrigation once they become established. Group together plants with similar water requirements (use hydrozones) so that they are irrigated the same, appropriate amount.

Install an efficient drip or low-output sprinkler irrigation system. Maintain and operate irrigation equipment properly and regularly inspect systems to ensure they are applying water to the desired area. Irrigate only when needed; monitor ET, plants, or soil moisture, as discussed above, to help you decide the appropriate amounts to apply and the proper interval between irrigations. Modify irrigation throughout the year to match plants' changes in irrigation needs.

Avoid runoff by improving soil permeability and infiltration (drainage) and modifying irrigation. For example, to avoid runoff and allow water to soak into soil,

cycle irrigation systems on and off in several short periods that add up to the total run time instead of irrigating continuously for one long period.

FERTILIZING WOODY PLANTS

Seventeen elements are required for plant growth. Carbon, hydrogen, and oxygen are supplied by air and water. The remaining fourteen elements are minerals provided by the soil. The minerals plants require in the greatest amounts are the six macronutrients, commonly separated into three primary and three secondary macronutrients (Table 3-6). Plants need only tiny amounts of eight micronutrients, including iron and zinc.

Micronutrient deficiency symptoms in landscape plants are rarely due to a deficiency of micronutrients in soil. For example, symptoms of manganese deficiency in palms and iron and zinc deficiency in many plant species are relatively common but usually are the result of adverse soil conditions and anything that injures roots or reduces root growth, inhibiting plants' ability to absorb nutrients. True deficiencies (insufficient amounts in the soil) of

the macronutrients nitrogen, magnesium, and potassium are somewhat common in containers or planter boxes, fruit and nut trees, palms, and in certain atypical soil types.

Although there are many fertilizer recommendations for landscapes, adding fertilizer to established woody plants is not necessary in most situations. Common causes of deficiency symptoms include insufficient nutrients in soil (mostly macronutrients), cool soil temperatures, high pH, inappropriate irrigation (usually irrigating too frequently), mechanical (physical) injury to roots, poor drainage, and root decay pathogens. With a few exceptions, do not fertilize established woody plants unless the cause of unhealthy plants has been definitely diagnosed as insufficient soil nutrient levels.

Nitrogen is the most commonly applied fertilizer and the element usually most limiting to plant growth. However, excess fertilization with nitrogen can increase maintenance and cause pest problems, so routinely applying nitrogen is not recommended in most situations. Exceptions include fruit and nut trees and plants growing in very sandy or highly leached soils. It may be desirable to add nitrogen to young trees and shrubs so they fill the site more quickly. For most palms, regular application of a slow-release fertilizer containing a 2:1:3:1 ratio of N:P:K:Mg (nitrogen:phosphorus:potassium:magnesium) is recommended.

Be aware that adding nutrients will not improve the appearance of foliage damaged by other causes. Improper or excessive fertilization damages plants, is detrimental to soil chemistry and microorganisms, and can pollute groundwater and surface water. Consult Chapter 4 for descriptions and pictures of nutrient deficiency symptoms and the recommended remedies. For fertilization recommendations, consult *The Biology and Management of Landscape Palms, Fertilizing Landscape Trees,* and *The Home Orchard: Growing Your Own Deciduous Fruit and Nut Trees.*

Table 3-6.

Mineral Nutrients Essential for Plant Growth.

MACRONUTRIENTS, PRIMARY
nitrogen (N)
potassium (K)
phosphorus (P)
MACRONUTRIENTS, SECONDARY
calcium (Ca)
magnesium (Mg)
sulfur (S)
MICRONUTRIENTS
iron (Fe)
chlorine (Cl)
manganese (Mn)
zinc (Zn)
boron (B)
copper (Cu)
molybdenum (Mo)
nickel (Ni)

Nutrients are listed in decreasing order of their abundance as commonly found in dry-weight plant tissue. Not listed here are carbon, hydrogen, and oxygen, which are provided by air and water. Consult Chapter 4 for descriptions and pictures of nutrient deficiency symptoms and recommended remedies. See *Fertilizing Landscape Trees* and *Soil and Fertilizer Management* for more information.

This Catalina ironwood (*Lyonothamnus floribundus*) has chlorotic and necrotic leaves symptomatic of severe iron and nitrogen deficiencies. However, highly alkaline soil and unhealthy roots are the actual causes of this damage. Fertilization is unlikely to improve foliage appearance, and the plant will not perform well unless soil conditions are improved. *Photo:* J. K. Clark

FERTILIZATION AND PESTS

Avoid overfertilization, especially with high-nitrogen fertilizers, as excess nitrogen can cause plants to grow undesirably, increasing the need for pruning that creates wounds susceptible to wood decay fungi. Excessive fertilizer may kill or injure roots, increasing plants' susceptibility to root-feeding nematodes and decay fungi. Overfertilization promotes excess foliage that shades the inner canopy and understory plants, increasing certain foliage diseases such as powdery mildews and rusts and causing shaded parts to die back. Fertilization results in the need for more pruning and irrigation, and can shorten a plant's life by causing it to outgrow available space. Application of nitrogen late in the growing season may delay dormancy in deciduous plants, increasing the likelihood of cold weather damage to plants, which favors certain foliar blight pathogens.

Too much fertilizer also can increase populations of pests that prefer succulent new growth, such as aphids and psyllids. Cypress bark moth larvae in natural situations feed primarily on Monterey cypress cones. In landscapes, they often infest trunks and limbs because landscape cypresses are fertilized and watered to promote rapid growth, resulting in thin bark that is susceptible to bark moth attack. Fertilizing oaks may promote distorted terminals; abnormal lateral bud growths called witches' brooms are caused by powdery mildew fungus infection of succulent new growth produced during the dry season when oaks are fertilized and excessively irrigated. Do not fertilize pines that exhibit cankers or ornamental rosaceous plants infected with fire blight as fertilization increases susceptibility to these diseases.

NUTRIENT DEFICIENCIES

Deficiencies reduce shoot growth and leaf size and cause foliage to discolor, fade, distort, or become spotted, sometimes in a characteristic pattern that can be used to identify the cause. Fewer leaves, flowers, and fruit may be produced, and more severely deficient plants become stunted, exhibit dieback, and are prone to other maladies. Learn to recognize symptoms of nutrient deficiency (see Table 4-2 for a summary); nitrogen and iron deficiency symptoms are the most common, but other deficiencies can occur. Deficiency symptoms can occur temporarily when plants produce new foliage during spring and soils are too cool and wet for roots to absorb sufficient nutrients; unless there are other problems, the plants' healthy appearance will return once the topsoil dries and warms.

Diagnose the actual cause of symptoms by investigating whether soil conditions are adverse (e.g., from overirrigation) or roots are unhealthy. Consider laboratory testing of soil and symptomatic foliage to learn whether the situation is among those relatively few instances where nutrients (usually macronutrients) may truly be deficient, as discussed in Chapter 4. See *Abiotic Disorders of Landscape Plants: A Diagnostic Guide* and *Fertilizing Landscape Trees* for more information.

PRUNING

Branches are commonly cut off to direct plant growth, open the canopy structure, and improve performance. However, improper pruning damages plants and causes pest problems. The International Society of Arboriculture, National Arborists Association, and other tree care organizations publish pruning standards to help ensure that trees are kept attractive, healthy, and safe. Only qualified arborists should prune large trees or trees near power lines.

Large trees should be pruned only by a competent professional. Seek quality advice and tree work from an arborist who is certified by the International Society of Arboriculture (ISA) or registered with the American Society of Consulting Arborists (ASCA). Interview candidate arborists to ascertain their expertise and to explain clearly the work desired. *Photo:* J. K. Clark

REASONS FOR PRUNING

The primary reasons for pruning are to improve plant appearance, health, safety, and structure. Prune landscape plants to remove damaged or diseased wood, to induce strong structure in young plants, and to maintain mature plant health and structure. Enhancing flowering or fruiting, improving appearance or form, and controlling size are other common reasons to prune. In landscape situations, many trees must be pruned to promote the development of strong branches that are well spaced both vertically and radially around the trunk. Pruning young plants properly during their first few years of growth is vital to minimize structural problems at maturity. Thinning (removing some) branches also provides more light to the plants below and reduces wind resistance, avoiding deformities or breakage. Obstruction of views or interference with utility lines is also remedied by pruning plants that grow too large for the space provided.

To minimize pruning requirements, select species that mature to a size appropriate for the location and avoid excess fertilization and overirrigation. Where plants are too large, consider replacing them with smaller species that have been well cared for and correctly pruned in the nursery, as summarized in Table 3-1.

PRUNING AND PEST MANAGEMENT

Proper pruning can control or prevent certain pests. Excessive or unnecessary pruning makes plants prone to certain pest problems and can induce limb growth with poor structural strength. Trees do not "heal" wounds the way people do; trees seal injuries. Although smaller cuts can eventually be closed by new growth, the wound is forever contained (compartmentalized) within the tree. Avoid unnecessary pruning because wounds are entry sites for decay and disease organisms and termites, which can remain active inside trees even after wounds have closed.

Correctly prune plants when they are young to minimize the need to remove large limbs later, thereby avoiding large pruning wounds that commonly become colonized by wood decay fungi. Remove damaged or diseased limbs and consider pruning out pests confined to a small portion of the plant. Where appropriate, prune to increase air circulation within the canopy, which reduces humidity and the incidence of certain foliar diseases.

Prune off declining or dead limbs and remove dying trees that are infested with bark- and wood-boring pests. Properly handle wood so the developing insects cannot emerge and attack other parts of the plant or nearby plants. For example, remove bark or split the wood to promote drying or solarize logs beneath clear plastic tarps, as discussed in Chapter 6 in "Bark Beetles."

Time the pruning to avoid the flight season of pest insects because the adults are attracted to fresh wounds where they feed or lay eggs or introduce plant pathogens. For example, prune eucalyptus during December or January (in southern California) or from November through March (in northern California) to avoid attracting eucalyptus longhorned borers, which fly during the other times of the year. Prune elms only during the late fall and winter, when the bark beetles that spread Dutch elm disease are not active. Avoid pruning plants susceptible to powdery mildew during the late spring through fall. Pruning stimulates succulent new growth, which is susceptible to powdery mildew fungi under moderate temperatures (60° to 80°F) and shady conditions (e.g., when foliage is present) during the dry season. See Tables 5-2 and 6-13 for the recommended pruning times to control certain diseases and insects affecting specific plant species.

Keep pruning cuts as small as possible to promote more rapid wound closure. Make pruning cuts correctly, just outside the branch bark ridge and branch collar. Do not prune too much at once or soon after plants begin flushing new foliage. Plants store energy in their trunk and limbs (in addition to roots), and sufficient photosynthetic surface

This California black walnut was topped, encouraging growth of branches weakly attached below the cut on both limbs. A major limb was also removed below the first main branch crotch, leaving a large wound, which closes slowly and has developed decay, as pictured close-up in "Wood Decay" in Chapter 5. *Photo:* J. K. Clark

(primarily foliage) is required to manufacture enough food. Topping trees or otherwise removing excess wood in one season stimulates production of vigorous, dense growth, which is susceptible to breakage and shades interior branches. Removing too much foliage exposes previously shaded bark, which can result in sunburn and bark cankers or attack by fungi and wood-boring insects.

Keep pruning tools clean. Contaminated tools may spread bacterial gall of oleander and olive, fire blight, vascular wilt fungi, and various other plant pathogens. To help avoid spreading pathogens, scrub tools clean with detergent and water before moving to use them on another plant. In situations where pathogens are known to be easily spread during pruning, consider sterilizing tools (e.g., scrubbing them clean then soaking them in bleach) before each cut, as discussed in "Disinfectants" in Chapter 5. After sanitizing tools, promptly rinse, dry, and lubricate the metal with oil to avoid corrosion and rust.

WHEN TO PRUNE

The best time to prune depends on the age and species of the plant, condition of the host, purpose of pruning, time of year (Figure 3-7), and whether there are pests to be avoided (see Tables 5-2 and 6-13). Remove hazardous branches whenever they appear. Remove damaged branches and branches with included bark (where two limbs meet at a narrow angle, causing embedded bark and weakly attached limbs). Prune to improve structure and shape, especially when plants are young. In most cases, encourage development of a single dominant leader and do not prune its tip. Suppress (subdue) branches that compete with the main leader and that have a branch diameter that is about one-half or more than that of the trunk, such as by thinning or heading it back to a smaller branch, called drop crotch pruning.

If many branches should be removed, reduce excessive shoot growth and dwarfing (retarded growth) by spreading pruning over several years. Young trees are

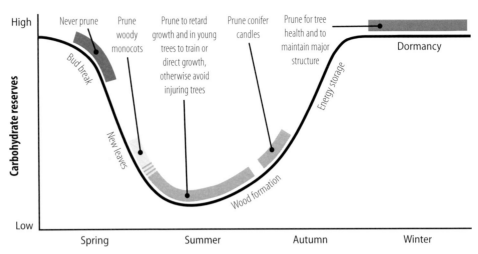

Figure 3-7.

Carbohydrate reserves (energy storage) in plants vary with the season and the extent to which stored carbohydrate is lost in plant parts that are removed. To promote plant health and direct woody growth as desired, pruning should usually be timed according to the annual variation in stored carbohydrates, as shown here. For example, winter dormancy is often the optimal pruning time for deciduous plants because they better tolerate the energy lost in pruned parts when carbohydrate reserves in remaining plant parts are at their maximum. For broadleaf evergreens, pruning during late winter to early spring, just before their normal growth season, minimizes dwarfing (reduced growth). See Tables 5-2 and 6-13 for the recommended pruning times to control certain pests affecting specific plant species.

often pruned during midsummer to direct development of good structure because regrowth then is often more predictable and less vigorous. To minimize dwarfing, avoid pruning deciduous plants in spring or early summer shortly after leaves have flushed. Prune evergreen species just before their normal growth season, usually in the spring. If flowers and fruits are important, time pruning to avoid removing flower- and fruit-bearing wood. For example, for plants that bloom on last year's wood (e.g., cherries), prune later after spring flowering. For plants that bloom on current year's growth (peaches), prune earlier, before new growth begins.

HOW TO PRUNE

Heading and thinning are the two primary types of pruning cuts. Each promotes a different plant response, and a mixture of heading and thinning cuts on the same plant can be desirable.

Heading removes part of a branch terminal back to a bud or a smaller branch. Pinching, tip pruning, shearing, stubbing, and topping are all types of heading cuts. Heading stimulates new growth from buds just below the cut. This cutting is used to form hedges and can be desirable for

inducing branches on vigorous shoots of young trees. However, the resulting foliage and shoots are often dense and on old branches and weakly attached, so they may break off easily. In some cases, a headed branch dies or produces only weak sprouts.

A thinning cut removes a branch at its point of attachment or origin (Figure 3-8). Compared with a heading

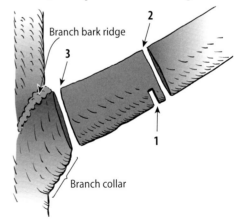

Figure 3-8.

Remove a branch by making the pruning cut just outside the branch bark ridge and branch collar, as indicated by the number 3. When removing a limb larger than about 2 inches in diameter, make three cuts in the order indicated. Make the first cut from below, about one-fourth of the way through the limb and 1 or 2 feet from the trunk. Make the second cut about 2 inches beyond the first cut, cutting from above until the limb drops. Make a final cut at number 3. To prevent the weight of the stump from stripping off bark, support the stump if needed until the cut is complete.

cut, growth near the pruning site is less vigorous after a thinning cut; thinning cuts promote more evenly distributed growth throughout the plant and a more open structure. Thinning cuts may be more selective and time-consuming, but they result in stronger structure and retain more of the plant's natural shape.

The location of the pruning cut in relation to the branch attachment influences the size of the wound, extent of callusing, wound closure time, and potential decay. Make most pruning cuts just outside the branch collar and branch bark ridge (Figure 3-8). Do not cut flush with the main limb or trunk and do not leave stubs. Cutting just outside the branch bark ridge reduces the wound size, reduces the exposure of trunk tissue to infection, and preserves the attachment zone, which is most resistant to decay and contains tissue best able to close over the wound after a cut.

Pruning shears or loppers can be used for small-diameter cuts. A scissors-type or bypass pruner makes cleaner cuts than do anvil shears. Do not use hedge shears to prune trees. Use a pruning saw when cutting branches greater than about ½ to 1 inch in diameter. On branches larger than about 2 inches in diameter, avoid tearing bark or splitting the wood by making three cuts, as illustrated in Figure 3-8.

When pruning young trees during their first few years of growth, leave temporary branches along the trunk to nourish, strengthen, shade, and protect the lower trunk from injury; keep these temporary branches less than about 12 inches long. During deciduous species' leafless season, remove the largest temporary branches each year during the first 4 or 5 years of a tree's growth. For evergreen species, prune off temporary branches just before they flush new foliage, usually in spring. Use the remaining branches to develop the maturing tree's structure.

Applying tree seals, paints, or wound dressings to pruning cuts or other injuries is not beneficial and may be detrimental to tree health. However, wounds on the south and west sides of trunks that become

The lower left pine was topped to prevent damage to power lines. Terminals are dying back and bark beetles are attacking the tree, which died 3 months after this photograph was taken. Do not plant tall-growing species beneath overhead utility lines. *Photo:* J. K. Clark

newly exposed to the sun may be painted with white interior latex paint diluted with an equal amount of water to help prevent sunburn.

The International Society of Arboriculture and the National Arborists Association provide pruning guidelines for professional tree care. Other information sources include the demonstration video *Training Young Trees for Structure and Form* and the publications *Plant Health Care for Woody Ornamentals* and *Woody Landscape Plants*. For large trees, trees near power lines, and if you are not certain about proper pruning techniques, hire a certified or registered arborist.

AVOID TOPPING TREES

Topping (also called dehorning, coat or hat racking, or stubbing) is the drastic heading of the central leader or large branches in mature trees. Main limbs are often sheared as with a hedge, leaving stubs. Topping is a poor pruning practice sometimes used to shorten tall trees, remove hazardous or diseased limbs, or to prevent interference with overhead utility lines.

Topping broadleaves and conifers ruins their structure, and topping palms will kill them. Drastic pruning is rarely justified simply because trees are believed to be too tall. Removing extensive canopy may not leave enough foliage to manufacture sufficient food and may cause roots to die and the tree to decline. The large wounds left by topping often fail to close and are susceptible to internal decay and attack by wood-boring insects. Topping encourages growth of branches weakly attached below the cut, which become susceptible to wind breakage.

When it is necessary to reduce the size of a tree's canopy, selectively remove upper limbs back to lower lateral branches (drop crotch pruning). This proper method is more time-consuming and expensive, but it avoids future expense from improper pruning and provides a more attractive, healthier, and safer tree. Where trees are repeatedly pruned because they grow too tall, replace them with lower-growing species. An exception is pollarding, where young trees are trained to develop a specific structure and new shoots on permanent branches are removed each year throughout the life of tree species that tolerate this severe pruning, such as mulberry and sycamore.

INJURIES, HAZARDS, AND PROTECTING LANDSCAPES

Prevent injuries to trees and shrubs. Wounds attract boring insects, serve as entry sites for disease-causing organisms,

Wounds, such as these made to implant insecticide, directly injure trees and provide entry sites for bacterial wetwood and wood decay fungi. Unless tools (such as drills or needles) are scrubbed and washed clean and sterilized between uses, plant pathogens may be mechanically spread by contaminated equipment. *Photo:* J. K. Clark

Protect trees from likely injury, such as by installing barriers to keep vehicles away from trunks and root zones. *Photo:* J. K. Clark

and can lead to limb, trunk, or root failure and tree death. Plant injury occurs when people cut or crush roots, improperly stake young trees, inject or implant trunks, make pruning wounds, and operate equipment or vehicles that strike bark or compact soil over roots. Deer, gophers, rabbits, mice, and other animals chew bark and wood. Drought, frost, hail, ice, lightning, and snow are among the environmental conditions that can injure plants and cause damage symptoms, as described in Chapter 4.

Make pruning cuts properly, just outside the branch bark ridge and branch collar (Figure 3-8). Keep weed trimmers and lawn mowers away from trunks. Choose plants that are well adapted to local environmental conditions so they are less likely to be damaged by moisture or temperature extremes. Provide proper cultural care so plants are

less likely to be injured and better able to tolerate damage. Protect plants during construction, as discussed below.

RECOGNIZE HAZARDOUS TREES

Damaged or unhealthy trees that may fail (drop limbs or fall over) are hazardous if located where they are likely to injure people and damage property. Conversely, in more natural settings dead or declining trees provide benefits such as cavities for wildlife habitat and recycled nutrients.

Some tree hazards can be avoided by preventing injuries to roots and aboveground parts. Examine trees regularly to see that they are receiving proper

cultural care and are not hazardous. Table 3-7 lists some signs to look for that can indicate that trees may be hazardous. Be aware that some hazards are difficult to detect, such as internal decay or unhealthy roots. If trees are located where their failure could cause injury or damage property, have them regularly inspected by a competent expert, such as an arborist who is certified by the International Society of Arboriculture (ISA) or registered with the American Society of Consulting Arborists (ASCA). Also see publications such as *Evaluation of Hazard Trees in Urban Areas* and *Recognizing Tree Hazards.*

PROTECT TREES DURING CONSTRUCTION

Protect trees during construction or they may decline, become hazardous, or die as a result of construction-related damage. Forests, oak woodlands, and urban lots with large trees are often developed because mature trees make these sites desirable to people. However, in constructing homes and roads and while installing amenities, trees are often killed outright or their lives are greatly shortened. Stress from construction-related damage can increase a tree's susceptibility to many pests. Negligent or

Table 3-7.

Some Warning Signs That Trees May Be Hazardous.

- Brackets, conks, mushrooms, or other fungal decay fruiting bodies growing on bark or out of the tree base.
- Cankers or wounds in bark or wood.
- Cavities on the main trunk or at the tree base.
- Cracks in the main trunk or at crotches (where limbs fork).
- Dead or dying limbs.
- Fissures in soil near the base of trees.
- Trunks that lean or tilt instead of growing upright.
- Some hazards are difficult to detect. Have a competent plant care professional inspect trees regularly for potential hazards if trees are located where their failure could cause injury or damage property.

Damaged bark with the bracketlike fruiting bodies of the common split gill fungus (*Schizophyllum commune*) indicate that this tree is dying and is hazardous because it may break or drop limbs. *Photo:* J. K. Clark

At construction sites, fence off trees around the drip line or beyond to create a protection zone and prevent equipment and activities from damaging roots or trunks. *Photo:* J. K. Clark

thoughtless activities may wound limbs, roots, and trunks. Injuring roots by crushing or cutting them or by compacting soil are very serious problems that construction often causes. Wet soil is especially susceptible to compaction by heavy equipment. Compaction or changes in soil grade or drainage can deprive roots of water and oxygen. Changes belowground promote root and crown diseases and predispose trees to attack by bark beetles and other wood-boring insects. Adverse effects may not become apparent until several years after the injury or stress.

Consider tree preservation during the planning stage of development projects. Consult a certified arborist or other tree care professional for help in protecting trees. Check county or city ordinances before working around mature trees as they may be city-owned or protected by ordinances.

How to Protect Trees. Fence off individual trees or groups of trees around the drip line or beyond to provide protection zones that prevent equipment and activities from damaging roots and trunks and compacting the soil around the trees. Most roots are near the soil surface, and many extend much farther than the tree canopy spread (Figure 3-2). If soil must be driven on or temporarily used to store heavy materials, apply and maintain a 6-inch thick layer of course mulch to reduce soil compaction. Minimize changes in soil grade and drainage; compaction and changes in soil contour alter surface and subsurface water flow on which established vegetation may depend. If roots have been crushed or cut, monitor the root zone in the undisturbed area and ensure that the soil is adequately irrigated.

In areas that will include both landscape and paving (e.g., commercial parking lots where shade trees will be

planted), consider using special structural soil mixes. These specially engineered soil mixes can be compacted to the legal density required to ensure pavement integrity while still providing the air- and water-holding properties vital for root growth.

Do not place fill around trunks. If the grade must be elevated, construct a stone or concrete well around each trunk. Before placing the fill, installing a drainage system on top of the existing soil may help provide oxygen and appropriate water to established roots once the original soil level is covered. If grade must be lowered, construct retaining walls to preserve as much of the original grade, roots, and soil as possible, at least within the drip line of established trees. If soil is undisturbed on one side and adequate cultural care is provided, a tree may survive if soil on the other side is removed no closer to the trunk than about halfway between the trunk and drip line. If grade must be lowered near trunks, consider removing these plants, as they are likely to become diseased and hazardous, and eventually die.

Be realistic in assessing which trees are worth saving and are likely to survive construction. When many large roots are cut, the tree may be more likely to be blown over by wind, especially if the soil is wet. Remove declining or severely injured trees and trees that are likely to die rather than leaving them to become a hazardous and expensive problem after sites are occupied; trees can be left to protect other trees during construction, then removed soon afterward. Consider removing trees if the drainage, root zone, or soil grade will be seriously disturbed or if species with incompatible cultural needs will be planted nearby (e.g., frequently irrigated turfgrass near native oaks).

Minimize trenching near trees and tunnel under roots instead of cutting them. Because roots grow outward from the trunk, where trenching is necessary, dig radially from trunks (as shown here) to minimize root cutting instead of digging perpendicular (crosswise) to the direction of major root growth. *Photo:* J. K. Clark

Avoid crushing or cutting roots, especially those larger than about 2 inches in diameter. Trench for utilities away from roots, combine utilities in a single trench, and tunnel beneath roots instead of cutting them. Instead of digging around roots, use hydroexcavation or pneumatic-excavation techniques that remove soil but leave roots intact. Locate septic systems away from trees; chemicals used in these systems may leach and damage roots or roots may grow into the systems and cause damage.

Use partially permeable materials (e.g., bricks instead of concrete) if extensive areas near trunks must be paved. Use caution when applying wood preservatives; they may kill nearby vegetation through direct contact or by leaching. Be aware of fire hazards from natural or planted vegeta-

Many areas of California are at risk from wildland fires. Appropriate plant selection, placement, and management dramatically reduce the likelihood that vegetation fires will burn structures. *Photo:* J. K. Clark

tion around buildings. Use good judgment by planting species with compatible water requirements in an existing landscape. For more information, consult *Compatible Plants Under and Around Oaks, Living Among the Oaks, Protecting Trees When Building on Forested Land, and Trees and Development.*

MINIMIZE FIRE HAZARDS

Wildfires are a serious hazard throughout much of California where hot, dry weather often prevails. All plants will burn if conditions are suitable, but proper maintenance, placement of plants, and avoidance of fire-prone species dramatically reduce the likelihood that vegetation fires will burn structures. Structure type, building materials, topography, and other factors are also important in minimizing fire hazards, but only landscape plants are discussed here.

Large plants provide more fuel than small plants, and they present a greater fire hazard the closer they are to structures. In fire-prone areas, keep larger plants farther away from structures, and prevent large plants from contacting structures or overhanging roofs. In some areas prone

to wildfires, laws require a vegetation-free area around structures; consult local fire officials.

Provide plants with proper cultural care, especially appropriate irrigation, to improve plant resistance to fire. Unwatered landscapes generally increase the risk of fire. Some drought-adapted species may benefit from watering every 1 to 2 months during the dry season, and the increased moisture may decrease their flammability. Prune lower limbs to provide a fuel break between the ground and tree canopies. Prune out dead branches and remove dead or dying plants. Thin crowns by pruning limbs that form bridges between tall plants. Avoid grouping together a progression of shorter to taller plants; these provide a fuel "ladder," allowing a ground fire to reach the tree canopies, where fire can readily spread from one tree to the next.

In fire-prone areas, minimize the build-up of litter (leaves, bark, etc.) around trees and structures. Keep flammable mulches well away from homes and other structures. For more information, consult *A Property Owner's Guide to Reducing Wildfire Threat* and *Home Landscaping for Fire.*

Chapter 4   Abiotic Disorders

Abiotic disorders are plant problems caused by adverse environmental conditions and other nonliving factors. Many disorders develop because the species planted is not well adapted to conditions at that location, the site was not well prepared before landscaping, or because of improper planting or cultural care. Causes of abiotic disorders in approximate order of their importance include water deficit and excess, aeration deficit (e.g., from soil compaction or poor drainage), nutrient deficiencies (which are often due to adverse soil conditions, such as high pH), salinity and specific ions (e.g., boron), problems related to temperature and light, herbicide damage and other chemical phytotoxicity, and mechanical injuries.

The first step in remedying an abiotic disorder is to identify the species of plant and the growing conditions and cultural practices to which it is adapted. Abiotic disorders can develop if a plant's needs are not being met, as discussed in Chapter 3. Many abiotic disorders can be recognized by their characteristic damage symptoms (e.g., distorted, discolored, or dying foliage). However, diagnosing the cause of disorders can be difficult. Different causes can produce the same symptoms, and more than one cause often adversely affects plants at the same time.

Distinguish abiotic disorders (also called noninfectious diseases) from similar damage caused by pests such as insects, mites, nematodes, pathogens, and vertebrates. In addition to directly damaging plants, abiotic disorders can predispose trees and shrubs to attack by insects and pathogens. Disorders and pests often act in combination, such as when drought stress or root injuries make trees susceptible to bark beetles and root decay pathogens. Consult "Problem-Solving Guide" in the back of this book to help you diagnose the cause of abiotic disorders based on symptoms. Field and laboratory tests can help you assess soil-related causes of disorders (Table 4-1). Consult *Abiotic Disorders of Landscape Plants* and other publications in "Suggested Reading" at the back of this book for more information.

Table 4-1.

Tests to Assess Soil-Related Disorders.

SUSPECTED DISORDER	ANALYSES	SEE PAGES
salt injury, salinity	total dissolved solids (TDS) in water or electrical conductivity (EC) of soil	50–53
sodic soil, high sodium availability	sodium adsorption ratio (SAR) or exchangeable sodium percentage (ESP) in soil	50–54
pH problems	hydrogen ion concentration (pH) and calcium carbonate (percent lime) in soil	53
micronutrient deficiency	soil pH; confirm by foliage tests or chelate application	45–54
macronutrient deficiency	soil testing may be useful, foliage tests may be helpful for nitrogen and phosphorus	45–50
specific ion toxicity	concentrations of ammonium, boron, chloride, and sodium in soil and irrigation water	52
aeration and water deficits caused by soil compaction, restrictive subsurface layers, low water-holding capacity, or poor infiltration or soil permeability (slow drainage)	depth to hardpan (auger or dig into soil); soil bulk density, moisture and organic matter content, percolation (a field test), and texture (laboratory test)	25–26, 44–45

Source: Abiotic Disorders of Landscape Plants.

Branch dieback on ash caused by chronic (prolonged) soil water deficit. Irrigating too frequently also causes dieback. You must correctly diagnose the cause of damage to determine the proper management actions.
Photo: J. K. Clark

WATER DEFICIT AND EXCESS

Inappropriate irrigation is probably the most common cause of landscape plant damage and can adversely affect most any aspect of plant growth and development. Inadequate water causes foliage to wilt, fade in color, die along margins and tips, and drop prematurely. Prolonged but mild moisture deficit results in smaller leaves, slower growth, fewer flowers and fruit, dieback, and increased susceptibility to wood-boring insects and other pests, which eventually can kill plants. Severe lack of soil moisture can result in wilting, leaf discoloration, and defoliation.

Overirrigation (e.g., irrigating too frequently or applying excessive amounts of water) is also common in landscapes. Overirrigation in combination with poor drainage (slow infiltration and percolation) is especially damaging to plants. Excess soil moisture deprives roots of oxygen, killing roots and increasing their susceptibility to root decay pathogens (especially *Phytophthora* spp.) and nematode damage. As roots die, foliage may discolor and die. Foliar symptoms resulting from overirrigation can resemble the symptoms from underirrigation because prolonged soil saturation causes roots to die and plants become unable to take up sufficient water. For more information see "Water Management" in Chapter 3.

AERATION DEFICIT

Aeration is the process of air passage though pore spaces in soil to provide an adequate supply of oxygen. Because roots require oxygen for growth and the uptake of nutrients and water, insufficient oxygen in the root zone is a serious, often life-threatening problem.

A short-term (acute) aeration deficit (for hours to days) commonly causes wilting and premature leaf drop. Sometimes the entire plant dies. Chronic aeration deficit (for weeks to months) kills roots, stunts growth, causes abnormally short shoots and small leaves, and may lead to the gradual decline and death of plants. Leaves can suddenly collapse and turn brown and branches may die back from the tips, although only a few branches or one limb may die. In some plants, chronic aeration deficit causes stem or bark cankers, resinous or gummy exudate on bark, adventitious (atypical, not normally present) roots near the basal stem, or unusually abundant and large lenticels (corky pores, or rows of narrow openings) in bark. Certain species tolerate very low soil aeration or tolerate it during dormancy.

Aeration deficit, sometimes called waterlogging, is especially common in irrigated landscapes. Excess soil moisture prevents adequate aeration by impeding oxygen movement through soil and leaving few water-free pore spaces for oxygen to occupy. Irrigating too frequently or applying excess amounts of water, especially when drainage is poor (slow), is the most common cause of aeration deficit. Compaction, flooding, hardpan (an impermeable soil layer), insufficient soil volume for root growth, poor soil structure and texture, a shallow water table, and surface barriers such as pavement are other common causes of insufficient oxygen for roots. Planting in too deep of a hole, especially in poorly drained soil when the hole is backfilled with amended soil, can cause aeration deficit. Insufficient soil oxygen, excess soil moisture, and root decay pathogens often act in combination to damage or kill plants.

To diagnose aeration deficit, examine soil and roots and suspect aeration deficit if

- Soil or roots smell like rotten eggs.
- Soil is bluish gray or black and roots appear discolored, rotted, or water-soaked.
- Water ponds (accumulates) on the soil surface or drains slowly (e.g., topsoil remains soggy).

To determine the cause of aeration deficit, assess drainage, check soil moisture, determine whether the grade has been changed, and especially evaluate irrigation practices. Send soil samples to a diagnostic laboratory to determine soil texture and whether excess sodium is present (movement of air and water are restricted in sodic soils; see "Salinity," below). Consider measuring soil bulk density, a soil's weight per volume (e.g., g/cm^3), by core sampling or carefully excavating specific volumes of soil, then thoroughly oven-drying and weighing them. Bulk density higher than critical levels for that soil texture as determined from published tables indicates that soil is compacted and lacks sufficient porosity for good plant growth. A percolation test, as described in "Prepare the Site" in Chapter 3, measures soil drainage (infiltration rate), which is one indicator of aeration.

Prevent aeration deficit through good site preparation, appropriate planting, and proper cultural practices, as discussed in "Site Preparation and Planting" and "Irrigation" in Chapter 3. If periodic soil saturation is unavoidable, consider replacing flood-sensitive plants with species more tolerant of wet soils. For more information, consult publications such as *Abiotic Disorders of Landscape Plants; Arboriculture: Integrated Management of Landscape Trees, Shrubs, and Vines;* and *Soil and Fertilizer Management.*

Slow growth, canopy thinning, and yellowing in Victorian box (*Pittosporum undulatum*) caused by aeration deficit. Where water is puddling (lower left) rhaphiolepis formerly grew and some of the remaining shrubs are dying back because excess irrigation and poor drainage result in insufficient oxygen in soil, which kills roots. *Photo:* L. R. Costello

NUTRIENT DEFICIENCIES

Plants require certain mineral elements for healthy growth. Deficiencies reduce shoot growth and leaf size and cause foliage to discolor, fade, and distort, sometimes in a characteristic pattern that can help you identify the cause. Fewer leaves, flowers, and fruit may be produced, and more severely deficient plants become stunted, exhibit dieback, and are predisposed to other maladies.

Sufficient micronutrients (e.g., iron, manganese, and zinc) are often present in landscape soils, but they may be deficient in sandy soils. Deficient macronutrients (nitrogen, phosphorus, and potassium) are more common in containers or planter boxes and when growing fruit and nut trees and palms. Most nutrient deficiency symptoms are not due to a deficiency of nutrients in soil; they usually result from other causes, especially adverse soil conditions and anything that injures roots or restricts root growth thereby limiting nutrient uptake. Common causes of deficiency symptoms include high soil pH (especially with plants adapted to acidic soil), inappropriate irriga-

tion, mechanical (physical) injury to roots, poor drainage, and root decay pathogens.

With certain exceptions (e.g., fruit and nut trees, palms, and perhaps roses and certain other heavily flowering shrubs), fertilization of established woody plants is not recommended unless specific knowledge of the local soil or a soil analysis indicates soil nutrients are insufficient. Applying nitrogen to young trees and shrubs can help plants reach a desirable size sooner. Adding nutrients will not improve the appearance of foliage already damaged by other causes and may divert attention from solving the true cause of unhealthy plants. Symptoms of common deficiencies are summarized in Table 4-2 and are discussed below. Consult publications such as *Abiotic Disorders of Landscape Plants* and *Fertilizing Landscape Trees* for more information on prevention, diagnosis, testing, and remediation of nutrient deficiencies.

Table 4-2.

Common Mineral Deficiency Symptoms.

SYMPTOMS	DEFICIENCY
New broadleaf foliage is undersized and yellow to whitish, except for green along veins. Brown dead spots may develop at margins, tips, and between veins. Conifers' upper canopy and palms' newer fronds are yellow; lower foliage appears normal. Common in plants growing in sandy soils and in species adapted to acidic soils that are grown in high-pH soils.	iron or manganese*
Older leaves or needles are uniformly pale green to yellowish. New growth is sparse and undersized but usually is green. Plants grow slowly and may drop foliage prematurely. Palms develop color gradation, with the oldest fronds most chlorotic, sometimes completely yellow or whitish. Many California soils have low organic matter content and a low supply of nitrogen.	nitrogen
Broadleaves' young foliage is dark green. Leaves overall or veins only may be purplish, especially on leaf undersides. Conifers' older needles dull or gray green and drop prematurely. Rare in landscapes.	phosphorus
Foliage growth is sparse. Older broadleaf foliage is yellowish or brown at tips and margins and between veins. Leaf edges crinkle or curl. Conifer needles dark blue-green, yellow, or brown and undersized, with dead tips. Older palm fronds have yellow, orange, or brown in flecks or spots or along margins, tips, and midribs. Rare in California landscapes except in fruit and nut trees grown in sandy soils and in palms.	potassium
Broadleaf leaves are uniformly yellowish or pale between veins, especially new growth. Spring leaf flush and blossoming may be delayed. New leaves may be small, narrow, and grow in tufts. Older leaves may drop prematurely. Conifer needles yellowish, undersized, and may drop prematurely. Sandy soils may lack zinc.	zinc*

Micronutrient deficiency (*) symptoms especially are usually not due to insufficient nutrients in soil; for all nutrients, deficiency symptoms commonly result from adverse soil conditions or anything that injures roots or restricts root growth, thereby inhibiting nutrient uptake.

NITROGEN

Slow growth and uniform yellowing of older leaves are usually the first symptoms of nitrogen (N) deficiency. Nitrogen-deficient plants produce smaller than normal fruit, leaves, and shoots, and these can develop later than normal. Broadleaf foliage in fall may be more reddish than normal and may drop prematurely. Nitrogen-deficient conifers may develop few or no side branches, and lower canopy needles may be short, close together, and yellowish, while the upper canopy appears normal. Nitrogen-deficient palms develop a color gradation, with the oldest leaves being most chlorotic and completely yellow or whitish in severe cases.

In trees and shrubs, insufficient nitrogen in soil is more likely a problem for fruit and nut trees, palms, and plants growing in soil that is sandy, highly leached, or in containers and planter boxes. Unless a modest amount of nitrogen fertilizer is added to soils amended with large amounts of undecomposed organic matter, nitrogen deficiency can occur because decomposer microorganisms consume the available nitrogen.

When established woody plants exhibit nitrogen deficiency symptoms, the cause is often unhealthy roots or adverse soil conditions, including aeration deficit, compaction, cool soil temperature, mechanical injury to roots, poor drainage, overirrigation, root-feeding insects or nematodes, or root decay pathogens. Adding nitrogen will not remedy these other causes. For example, plants may exhibit nitrogen deficiency in early spring when soils are too cold or wet to allow roots to absorb sufficient nutrients for new growth. Once topsoil drains and warms, foliage develops its healthy appearance.

Visual symptoms (e.g., uniform yellowing of older foliage) can strongly suggest nitrogen deficiency, especially if you can rule out other causes that produce similar symptoms. To help confirm nitrogen deficiency, laboratory analysis of properly collected soil samples can be useful but difficult to interpret. Trees have a relatively large rooting volume, and nitrogen availability varies with soil conditions, which can change over relatively short time periods. Leaf analysis can be obtained, but there are no guidelines for appropriate nitrogen levels in most ornamentals, although conifer and broadleaf foliage typically has 1 to 3% total nitrogen. If testing foliage, you may want to conduct separate leaf analyses on current-season foliage of unhealthy and healthy plants of same species growing nearby and compare the tests' results. If testing soil, you may want to take separate samples from problem and healthy areas.

If you determine that nitrogen is needed, use the correct type, rate, and method of fertilization for that situation. Inappropriate fertilization can damage plants and cause other problems, as discussed below in "Excess Nitrogen."

Nitrogen is commonly provided to plants as organic matter (e.g., organic mulch that slowly decomposes), or as inorganic compounds (e.g., ammonium nitrate or ammonium sulfate). Commercial slow-release fertilizers (e.g., sulfur- or polymer-coated urea) provide the easy handling of synthetic fertilizers but also slow-release characteristics. Although more expensive than other preparations, these can be a good choice for adding nitrogen to deficient soils.

PHOSPHORUS

Symptoms vary greatly when plants are deficient in phosphorus (P). In broadleaf plants, young leaves may be dark green and have purplish veins, especially on the underside of leaves. Older leaves can develop an overall purplish tint and tip dieback. Leaves may be smaller than normal, curled or distorted, and drop prematurely in the fall.

In phosphorus-deficient conifers, foliage on older trees is discolored gray-green or dull blue-green. With severe deficiency, few or no new needles are produced, and needles die prematurely, starting with lower needles and progressing upward. Seedling needles can turn purple, starting at their tips and progressing inward and upward through the canopy.

Phosphorus generally occurs in soil in adequate amounts for most trees and shrubs. Exceptions are where topsoil has been removed during site development and in soils derived from serpentine minerals, such as in the Sierra Nevada foothills and parts of the Coastal Range. A plant may be phosphorus deficient if tests of current-season foliage of conifers or woody broadleaves find less than 0.1% phosphorus, or if comparison testing reveals substantially less phosphorus in foliage of symptomatic plants versus nearby healthy plants of the same species.

If phosphorus is deficient in leaves, the likely causes are the soil or root problems described above for nitrogen. Certain herbicides also cause leaf distortion and curling that can resemble phosphorus deficiency symptoms. In unusual cases when soil is phosphorus deficient, add a fertilizer such as ammonium phosphate. Relatively

Relatively uniform yellowing of leaves indicates nitrogen deficiency in this rhododendron. The somewhat greener veins in newer growth may indicate iron or zinc deficiency, which commonly is due to high-pH soil. Most deficiency symptoms are caused by adverse soil conditions and poor root health that prevent nutrient uptake. Learn the site conditions so you can determine the cause and effective remedy. *Photo: J. K. Clark*

large amounts of chicken manure also provide phosphorus but can be detrimental because manure often has a high salt content. Regardless of the form, adding phosphorus to most soils in California is rarely beneficial to landscape plants.

POTASSIUM

In California, potassium (K) deficiency is common in palms, but otherwise it usually occurs only in fruit and nut trees grown in sandy soils. Potassium deficiency in broadleaves causes leaves to turn yellow and then brown at the tips and margins and between veins. Older leaves are affected first and can entirely discolor, crinkle, curl or roll along edges, and die. In potassium-deficient conifers, older foliage turns dark blue-green, progressing to yellow then reddish brown. Needles are often undersized, with brown, dead tips.

All palm species are susceptible to potassium deficiency. Symptoms always appear first in older leaves and eventually progress into younger leaves, but symptoms otherwise vary among palm species. The most common symptom is yellow or orangish flecks or spots on older leaves, which appear translucent when discoloring is observed from below. In some species, yellowing begins at the leaf margins or tips, and leaves gradually become entirely yellow, then brown and withered. Leaf midribs may be yellow instead of their normal green.

Except for palms, potassium deficiency symptoms are usually due to poor soil conditions and unhealthy roots, as discussed above in "Nutrient Deficiencies." Some sucking insects, foliar pathogens (e.g., diamond scale fungus of palms), and preemergence herbicides cause similar damage symptoms.

Diagnose potassium deficiency based on visual symptoms, plant susceptibility (palms and fruit and nut trees), and the presence of soil conditions that favor deficiency (high leaching, sandy soil, sparse topsoil).

To correct a deficiency, spread organic mulch beneath plants and apply potassium fertilizer, preferably slow-release forms (e.g., potassium silicate or sulfur- or polymer-coated potassium products). Potassium sulfate may be used, and potassium will be held by organic matter and clay particles. Especially on sandy soils, avoid readily leached materials (e.g., potassium nitrate and potassium sulfate). Potassium nitrate may also cause an excess of nitrogen unless nitrogen deficiency is also a problem. Do not use potassium chloride where chlorine or salt toxicity are problems. Because a high potassium concentration reduces magnesium availability (and excess magnesium makes potassium unavailable), it may be best to add both potassium and magnesium in combination.

A slow-release fertilizer containing magnesium and a 3-1-3 ratio of nitrogen, phosphorus, and potassium is good for palms. Symptomatic palm foliage will not recover, and you must wait for new growth. To avoid aggravating potassium deficiency, do not remove symptomatic leaves until they have turned entirely brown.

NITROGEN, PHOSPHORUS, AND POTASSIUM (NPK)

So-called complete fertilizers contain nitrogen (N), phosphorus (P), and potassium (K), listed as NPK on the fertilizer label. Except when growing palms or at very sandy or highly leached sites, soil around landscape trees and shrubs is rarely deficient in all three elements. Adding sufficient complete fertilizer to provide the deficient element can result in an excess of other nutrients and may contribute to salinity problems and pollute water.

Established woody plants should be fertilized in response to specific needs. Complete fertilizers are generally not recommended for woody landscape plants, except for palms and possibly other woody monocots (plants with a single seed leaf). Also avoid products containing both fertilizer and pesticide.

IRON

Iron deficiency symptoms are common in California landscapes even though sufficient iron is present in most soils. Azaleas,

Phosphorus deficiency commonly causes older leaves to curl, distort, and be smaller than normal (as with this pear tree) and develop a purplish tinge or atypical dark green color. Phosphorus deficiency symptoms are typically due to poor soil conditions or unhealthy roots. Except where topsoil has been removed and in serpentine soils, adding phosphorus is rarely beneficial to woody landscape plants. *Photo:* J. K. Clark

Potassium deficiency is common in palms, where symptoms always appear first in older leaves. Symptoms vary by species, but orange or yellow mottling and spotting of leaves is the most common symptom, as in this queen palm. *Photo:* J. K. Clark

Yellowish leaves with curled edges, especially around the leaf tip, are symptomatic of insufficient potassium in broadleaves. Potassium deficiency is rare in woody broadleaves, except in fruit and nut trees growing in sandy soil, as with this citrus. *Photo:* D. Rosen

practices and the soil environment. To improve aeration and reduce waterlogging, increase the interval between irrigations to the maximum extent that still provides adequate moisture to maintain good plant growth. If plants are small, consider digging them up and replanting them on a broad mound raised several inches to improve drainage. Amend soils to improve drainage, lower pH, and increase organic matter before planting or replanting species adapted to acidic soils.

If soil is alkaline, about 6 months before planting add 1 to 4 pounds of elemental sulfur per 100 square feet of soil surface, mix or rototill it in the top 6 inches, and irrigate. Use the lower amount in sandier soils and the higher amount in soils high in clay. Around established plants, use a soil probe or an auger to create holes around the drip line, then fill the holes with soil mixed with 2 to 3 teaspoons of elemental sulfur. Bacteria slowly convert sulfur to sulfuric acid, which lowers soil pH and increases iron availability slowly over months when soil is moist, warm, and well aerated. Acidification is not effective if soil is high in calcium carbonate, as discussed below in "pH" and "Buffering Capacity."

Alternatively, before planting, add acidic sphagnum peat or organic matter that has been well composted and mix it into the top 1 to 2 feet of soil at a rate not exceeding 20% of soil volume. Be aware that amended soils will settle as organic matter decomposes, causing new plants to settle in the planting hole and become subject to root and crown diseases. When planting in amended soil, compost organic matter well before use, form soil into a broad mound, place the root ball on solid

citrus, gardenias, rhododendrons, and other plants that are adapted to acidic soil are especially prone to iron deficiency when soil pH is above about 7.5 (alkaline). Iron deficiency is also common when soils are high in calcium, poorly drained, too cool, or waterlogged and when root health is impaired by pathogens or other causes.

Iron deficiency in broadleaves causes young foliage to be bleached, chlorotic, or pale between distinct narrow green veins. Fading appears first around leaf margins, then spreads inward until only the veins are green on younger foliage. When severe, young leaves remain undersized, turn almost white, and develop black or brown spots, margins, and tips. Twigs may die back and leaves may drop prematurely.

In iron-deficient conifers, the upper canopy becomes chlorotic because new needles are undersized and yellow, while older, lower canopy foliage remains green. In iron-deficient palms, new leaves typically are uniformly chlorotic.

Tissue analysis may not be reliable for diagnosing iron deficiency for plants in the landscape. Diagnose this malady based on visual symptoms, soil tests (for example, showing high pH or low organic matter), and knowledge of existing cultural practices and soil conditions and whether iron deficiency is common in that species. Recognize that certain preemergence herbicides and manganese deficiency cause similar damage symptoms. Manganese and iron deficiency have mostly the same causes and remedies.

Management. Remedy iron deficiency primarily by improving cultural

Yellow new growth with distinctly green veins indicates an iron deficiency in this toyon. Sufficient iron is present in many soils, although sandy soils are commonly deficient. Iron deficiency symptoms are usually caused by adverse soil conditions such as high pH and poor drainage. *Photo:* J. K. Clark

soil, and plant about ½ to 2 inches or more above the native soil line, as discussed in "Plant Properly" in Chapter 3.

One remedy is to apply iron chelate according to the product label, such as to foliage. If used, apply chelate in combination with measures to improve the plant's culture, environment, and soil conditions. Iron is also applied by trunk injections, but injections injure trunks and may mechanically spread plant pathogens. If applying inorganic fertilizers, switch from nitrate- to ammonium-based compounds, such as ammonium nitrate, which gradually lowers soil pH.

Regularly place composted organic matter as mulch on top of the roots of established plants to eventually (slowly) remedy iron deficiency as organic matter decays and soil becomes more acidic. Mulch provides many benefits in addition to increasing nutrient availability.

MANGANESE

Manganese (Mn) deficiency occurs primarily in palms and plants growing in adverse soils (e.g., alkaline, cool, or poorly drained). Manganese deficiency in broadleaves causes new leaves to be chlorotic, with wide, green areas along the veins. On severely affected leaves, brown dead spots develop between veins; leaf margins may become crinkled, curled, or wavy; and shoot growth is reduced.

In conifers, manganese deficiency symptoms closely resemble iron deficiency. New needles are stunted and chlorotic, while older, lower-canopy foliage remains green.

In palms, new leaves are uniformly chlorotic with necrotic streaks, and younger leaves remain smaller than older fronds. As manganese deficiency worsens, emerging new leaves and older leaves are necrotic, distorted, and withered, giving palm canopies a frizzled, scorched, undersized appearance.

Preemergence herbicides cause similar symptoms but primarily damage old leaves first, while manganese deficiency occurs on new leaves. Manganese deficiency symptoms closely resemble iron deficiency, and any differences are too variable and unreliable to visually distinguish these maladies.

Diagnose and remedy this malady as described for iron deficiency. Lower soil pH, increase soil organic matter, and otherwise improve the plant's cultural practices and growing environment. Manganese chelates can be applied as labeled to new foliage as a quick, temporary remedy.

ZINC

Zinc (Zn) deficiency, sometimes called little-leaf disease, most often occurs in fruit and nut trees and woody species in soils that are sandy or abnormally high in organic matter. Zinc deficiency is rare in palms.

In broadleaves mildly deficient in zinc, leaves are uniformly yellowish or pale between the veins and may develop dead spots. Symptoms are usually most apparent on new foliage in the spring. Severely deficient plants bloom and leaf out late, sometimes several weeks later than normal. When buds open, leaves are small, narrow, pointed, and yellowish, and internodes are shortened, resulting in tufts of leaves (rosettes, or witches' brooms). Older leaves may drop prematurely. Zinc deficiency symptoms often resemble iron or manganese deficiency or phytotoxicity from glyphosate or preemergence herbicides.

Zinc-deficient conifers have undersized, yellowish needles that may drop prematurely. When plants are severely affected, branches are undersized and may die back.

Zinc is usually present in soils in adequate amounts, except where development or grading removed topsoil. More commonly, plants cannot adequately absorb the zinc that is present because of high soil pH, unhealthy roots, and inappropriate cultural practices and adverse environmental conditions, as discussed for iron and nitrogen deficiencies.

Diagnose zinc deficiency based on characteristic foliar symptoms and assessment of conditions that affect zinc availability.

Yellow elemental sulfur has been spread on the ground beneath this tree. Once irrigated in, sulfur helps to gradually remedy nutrient deficiencies caused by high soil pH. Sulfur is phytotoxic if applied to foliage. It is also relatively insoluble in water, so it is much more effective to mix sulfur with native soil and backfill the mix into the planting area. Alternatively, mix sulfur and soil and apply it in holes made using a soil probe, which requires many holes. The benefits may not be apparent for months. *Photo:* J. K. Clark

Manganese deficiency in broadleaves causes new foliage to become yellow, except for green along the veins, as in this plum. Iron deficiency symptoms look very similar. Although the green around veins often forms broader bands when manganese is deficient, the differences are too variable and unreliable to visually distinguish these maladies. *Photo:* J. K. Clark

For example, laboratory analysis of soil can reveal high pH or high phosphorus concentrations, which reduce zinc availability.

Remedy zinc deficiency by improving cultural practices and soil conditions to

Zinc-deficient broadleaf trees can develop small, chlorotic leaves in tufts, like this almond. However, some zinc-deficient plants do not develop tufted foliage. Damage resembling this is also caused by glyphosate herbicide. *Photo:* J. K. Clark

facilitate zinc uptake by roots, as described for management of iron deficiency. If appropriate, zinc chelate can be applied according to the product instructions. Be aware that foliar application of zinc chelate can be phytotoxic and the proper timing of application is important.

MAGNESIUM

Magnesium (Mg) deficiency is extremely rare in broadleaves and conifers in landscape, but it is common in palms, especially date palms. In palms, leaf tips and terminals turn bright yellow, while leaf bases and along the midrib remain green. Lower (older) fronds may die prematurely.

This deficiency can be remedied by fertilizing soil with magnesium sulfate. Be aware that adding magnesium can reduce potassium availability; conversely, excess potassium makes magnesium unavailable. It may be best to add both potassium and magnesium in combination, such as by using fertilizers especially for palms. To avoid aggravating this deficiency, do not remove symptomatic leaves until they have turned entirely brown. Symptomatic leaves do not recover and must be replaced by new growth.

NITROGEN EXCESS

Excess nitrogen may kill small roots and increase plants' susceptibility to root pathogens and nematode damage. Overfertilization may cause leaves to turn pale or dark green, gray, or brown at margins and tips. Foliage may wilt temporarily or die prematurely.

Excess nitrogen can cause plants to grow excessively and develop succulent tissue, which promotes outbreaks of certain sucking insects and mites. Excessive nitrogen may also cause fruiting plants to produce relatively more foliage and reduce their fruit production. Nitrogen fertilization commonly increases the need for irrigation and pruning and can cause plants to outgrow the available space and die prematurely. Applying nitrogen can undesirably alter soil pH, contribute to excess soil salinity, pollute water, and, when applied late in the growing season, increase plants' susceptibility to freeze damage.

For established woody species, nitrogen fertilization usually is warranted only for fruit and nut trees, palms, roses, and certain other profusely blossoming shrubs, and for plants growing in soils that are very sandy, highly leached, amended with large amounts of undecomposed organic matter, or in containers or planter boxes. For more information, see the nitrogen deficiency discussion above and "Fertilizing Woody Plants" in Chapter 3.

SALINITY

Salts are compounds that separate into positively and negatively charged molecules or elements in water or moist soil. These charged particles are called ions. Anions are negatively charged compounds or elements, such as nitrate, chloride, and sulfate. Cations are positively charged, such as ammonium, iron, calcium, magnesium, and sodium. Plants obtain most nutrients from dissolved ions and tolerate many types of salts in low concentrations. However, roots and foliage can be injured by exposure to high concentrations of almost any type of salt, including those in certain soils, fertilizers, low-quality irrigation water, ocean spray, and road deicing salt.

Foliage exposed directly to excess sodium from irrigation water or sea spray discolors, beginning terminally and (in broadleaves) marginally, and may drop prematurely. Foliar salt exposure typically produces a distinct pattern of damage, such as only on lower foliage wetted by low-quality irrigation water or on the windward side of plants facing an ocean breeze. Symptoms are more severe in sensitive plant species and as salt concen-

Magnesium deficiency causes palm leaf tips and terminals to turn bright yellow, while leaf bases and along the midrib remain green, as in this date palm. Magnesium deficiency is common in palms but rare in landscape broadleaves and conifers. *Photo:* D. R. Hodel

Chlorotic and necrotic leaf margins and tips in citrus caused by excess nitrogen from the overapplication of urea fertilizer. *Photo:* R. G. Platt

trations increase. Buds, twigs, and entire branches may be killed, leading to witches' broom growth the next season. Severely affected plants may die.

Root exposure to high sodium concentrations causes wilted foliage and stunted plant growth because excessive salts in soil impede plants' uptake of water and cause plants to desiccate; water moves out of roots into soil because water is relatively less concentrated in the soil. If salt is high, but not extremely high, plants may grow slowly but not show other obvious symptoms.

In broadleaves, excess salts carried with water into the plant concentrate at leaf margins and tips, which turn yellow, then brown, usually beginning with older foliage, which dies and drops prematurely. In evergreen broadleaves, foliage damage due to salts may be more pronounced on the south side of plants. Conifer needles turn yellow, then brown from the tip inward, then drop prematurely.

Soils high in exchangeable (readily available) sodium relative to calcium and magnesium are called sodic soils, and their soil pH usually exceeds 8.5. Sodic soils are sometimes evident because of a white or dark crust on the soil surface and very slow water penetration. Sodium may damage roots through direct toxicity and kill sensitive plants. High levels of sodium can destroy the aggregate structure of fine- and medium-textured soils, decreasing porosity and preventing soil from holding air and water, so that plants cannot grow.

Salt toxicity is most common in certain soils, such as on the western side of the

Central Valley of California, northern California's Livermore Valley, parts of the Mojave Desert, some coastal areas in southern California, and filled soils around the San Francisco Bay shoreline. Toxicity also occurs from irrigation with water high in salts, such as from shallow wells, surface water that passes through arid areas, treated (recycled) municipal wastewater, and in cold regions where salt is applied to deice pavement. Frequent irrigation of poorly drained soils and excessive fertilization also can result in salt damage. Even relatively low salt concentrations involving certain specific ions (e.g., boron) can damage plants. Table 4-3 summarizes conditions and locations where salt damage to plants is most likely to occur.

Diagnosis and Management.
Because similar damage can be caused by water deficit and other maladies, test soil and irrigation water to definitely diagnose whether salinity is the cause of symptoms. If irrigating with domestic water, contact the water agency to obtain its test results for salinity and specific ions.

To determine whether excess sodium is present, have a laboratory measure the soil's sodium adsorption ratio (SAR), the ratio of sodium to calcium plus magnesium. Total salt (salinity) in soil is usually measured by a laboratory test of electrical conductivity of an extract (EC$_e$) of a saturated paste (soggy soil) reported in units of millimhos per centimeter (mmhos/cm), which is the same as decisiemens per meter (dS/m), since

This evergreen clematis exhibits terminal leaf necrosis because it is irrigated with salty water. *Photo:* J. K. Clark

Reddish pine needles due to root injury from road deicing salts applied during winter. With root exposure, salt toxicity appears first in older (inner and lower canopy) foliage. The oldest needles have dropped prematurely, giving the canopy a sparse appearance. *Photo:* USDA Forest Service, Rocky Mountain Region, Bugwood.org

1 mmho/cm = 1 dS/m. Plants vary greatly in their tolerance for salinity. Depending on the species, sensitive plants may be damaged if soil EC$_e$ exceeds 2 to 4 mmhos/cm, while tolerant species are not damaged unless soil EC$_e$ exceeds about 8 to 10 mmhos/cm.

Salinity in irrigation water is measured as electrical conductivity (EC$_w$), or more

Table 4-3.

Conditions and Locations Favoring Salt Damage to Plants.

> - Arid regions with salts accumulated near the surface due to deposition or rock weathering and little leaching by rain. There may be a dark or white crust on soil.
> - Coastal sites where plants are exposed to ocean spray.
> - Fertilizers applied excessively, especially products with a high salt hazard or salt index.
> - Irrigation with poor-quality, salty water, such as recycled water (treated municipal wastewater) or from shallow wells or surface water that flowed through arid agricultural areas.
> - Low-lying area with salty water table near the surface.
> - Manures or other salty amendments or amendments of unknown salinity were applied in excess amounts.
> - Snowy locations near pavement where deicing salts were applied.

Adapted from *Abiotic Disorders of Landscape Plants*.

commonly as total dissolved solids (TDS) reported as milligrams per liter (mg/L) or parts per million (ppm). For landscape irrigation, water should generally be below about 1,000 mg/L TDS or below 1 mmho/cm EC_w.

Saline soils cannot be remedied with chemical amendments or fertilizers. They can be remedied by leaching, but first you may need to improve soil drainage, as discussed in Chapter 3, such as by adding subsurface drain pipes or tiles. To leach, apply water that is low in soluble salts to move salts deeper into soil. During each irrigation apply a greater volume of low-salt water than needed to wet the root zone. The extra water is called the leaching fraction. For example, water for a longer amount of time and, if warranted, increase the interval between irrigations to avoid overwatering.

Sodic soils also need leaching, but first apply gypsum so calcium displaces sodium as water moves it through soil. If soils are high in free calcium (lime), sulfur may be applied before leaching. Before leaching sodic soil, have soil tested to determine whether it is beneficial to apply gypsum, sulfur, or other amendments, and if so how much to apply.

To help prevent topsoil salinity, apply mulch around plants to reduce evaporation, since evaporation concentrates minerals near the soil surface where most roots occur. Minimize fertilizer applications where salinity is a problem because fertilizers add salts to the soil. If applying fertilizers, use products with a low salt

hazard (salt index). Obtain an analysis or test the salinity of animal manures, composts, and sewage sludge before deciding to apply them. Instead of using rock salt to deice pavement, consider using alternatives such as calcium magnesium acetate (CMA) or sand. When irrigating, provide enough extra water to move salts below the root zone.

Where salinity may be a problem, avoid planting species with low salt tolerance and instead grow salt-tolerant species. Especially when growing more salt-tolerant species, recycled water such as treated municipal wastewater can be acceptable for use in drip or flood irrigation that does not apply water to foliage. Foliage is more sensitive to salts than roots. Avoid using sprinklers if irrigating with salty water. If foliage is exposed to salts, rinse leaves with good-quality water if possible.

For more information on salinity testing and remediation and lists of plants' salt tolerance, consult *Abiotic Disorders of Landscape Plants* and *Water Quality: Its Effects on Ornamental Plants*.

BORON AND OTHER SPECIFIC IONS

Boron (B) and certain other specific ions (most commonly chloride and sodium) can be toxic to plants even when overall salt measures (EC and TDS, as discussed in "Salinity," above) are within acceptable limits. Toxicity symptoms of specific ions are difficult to distinguish from each other and from overall high salinity and water deficit

symptoms. Typical symptoms are yellowing of leaf margins and tips and sometimes between veins, then necrosis and premature drop of foliage.

For boron, tiny amounts are essential for plant growth, but there is a narrow acceptable range between sufficient and excess concentrations. In many plants, boron accumulates in older foliage, causing the margins or tips of leaves and needles to turn yellow, then brown or blackish, and leaves may drop prematurely. In certain plants (e.g., cotoneaster, gardenia, privet, pyracantha, rhaphiolepis, and syringa) boron accumulates mostly in new terminal growth and symptoms of toxicity appear in young, expanding foliage. In apple, pear, and stone fruits and other *Prunus* spp., boron toxicity symptoms appear in young, expanding foliage and as misshapen fruit and as cankering and dieback of petioles and young twigs. In any plant, bark can crack or become corky, and severely affected plants can die.

High concentrations of boron, chloride, and sodium occur naturally in soils and

Salinity is common in arid regions where salts accumula[te] near the surface, as evidenced by the dark and white crust on soil at the base of this stairway. Certain salt-tolerant species can be planted in saline soil to create a[n] interesting, low-maintenance landscape. *Photo*: J. K. Cla[rk]

water in the situations described in "Salinity," above. Ammonium toxicity can occur after excess application of ammonium fertilizers or incorporation of soil amendments high in ammonium. Diagnose specific ion toxicity based on the visual symptoms in plants, knowledge of locations where those ion concentrations are high, and ion-specific laboratory tests of soil or water. For example, with most landscape plant species, boron toxicity occurs when soil contains about 1 ppm or more of boron or irrigation water exceeds about 0.5 ppm boron.

General salinity tests (measures of total EC and TDS) do not tell you the concentrations of specific ions. Depending on the suspected causes of problems, more than one test may be needed (e.g., for EC, SAR, and each specific ion), and you may need to separately collect more than one type of sample (irrigation water, older plant leaves, and soil). When testing foliage, collect separate samples from nearby plants with apparently healthy foliage for comparison testing.

Manage specific ion toxicity with the methods described for salinity. Modify irrigation to prevent sprinkling of foliage. Apply only high-quality irrigation water and, where soil is high in boron or other salts, improve drainage to gradually leach ions below the root zone. Grow tolerant plant species.

pH PROBLEMS

Soil pH (hydrogen ion concentration) affects the form and availability of nutrients and the ability of roots to absorb nutrients and water. Nutrient deficiency or toxicity symptoms commonly develop when plants grow in soil with adverse pH. The activity of soil microorganisms (e.g., nitrifying bacteria) also depends partly on pH.

A scale from 0 to 14 is used to express pH. A pH of 7.0 is neutral. Lower numbers represent increasingly acidic conditions; higher numbers indicate increasingly alkaline (also called basic) conditions. Because a logarithmic scale is used, ten times as many positively charged hydrogen ions are

Mineral toxicity is causing these ginkgo, or maidenhair tree, leaf margins to turn yellow, then brown and die. Laboratory testing of soil and water may be needed to diagnose the cause of damage such as this. The low-quality irrigation water used here contains about 1,000 ppm total dissolved solids (salinity) and over 0.5 ppm of boron. *Photo:* J. K. Clark

available at a pH 6 than at pH 7. For optimal nutrient availability for plant growth, a good range of soil pH for most landscapes is 6.0 to 7.5.

The primary symptoms of adverse soil pH are similar to those that may occur from nutrient deficiencies or excesses (toxicity). High pH causes interveinal chlorosis and bleaching, pale mottling, and blotchy or marginal necrosis of new growth. Damage primarily is due to reduced availability of iron, manganese, and zinc, so any of the symptoms described earlier for those deficiencies may occur in high-pH soils.

If soil pH is below about 5.5, new foliage becomes chlorotic, distorted, and possibly necrotic, and plant growth slows. In severe cases, affected roots can become discolored, short, and stubby. Symptoms result primarily from aluminum toxicity and deficiencies of calcium and magnesium. Copper and manganese toxicity and phosphorus deficiency symptoms may also occur. Acidic soils occur mostly under conifer forests, in regions with high average rainfall, and in exposed subsoils in the Coastal Range westward to the Pacific Ocean, mostly due to soil cuts and grading during development.

Diagnose adverse pH using knowledge of local soil conditions and species' susceptibility. Especially test the pH of soil from the root zone. If sodic soils are suspected, obtain a value for sodium adsorption ratio (SAR) and a separate test of calcium carbonate (percent lime), as discussed below in "Buffering Capacity."

Grow species tolerant of the pH at that site. To learn what species grow in acidic soils (e.g., in California's North Coast area) versus alkaline soils (many desert locations) consult resources such as *Abiotic Disorders of Landscape Plants* and *The New Sunset Western Garden Book.*

To raise pH in the few landscape locations where pH is too low, mix appropriate amounts of finely ground limestone into soil before planting (soil tests can provide specific rate recommendations) or fertilize with calcium nitrate. To lower pH, mix elemental sulfur, iron sulfate, and organic matter into topsoil, apply mulch and acidifying fertilizers such as ammonium sulfate, and follow other management practices described earlier in "Iron."

Buffering Capacity. The ability of soil to resist change in pH is called its buffering capacity. Soils' response to acidifying amendments is usually estimated by measuring calcium carbonate (percent free lime). Measuring calcium carbonate indicates the amount of acidifying amendments or fertilizers needed to lower the pH. When calcium carbonate is low, the pH of soil more readily changes after the application of certain fertilizers or other amendments (e.g., sulfur) that make the soil more acidic. If calcium carbonate is high, the pH tends to stay high. If both pH and calcium carbonate are high in soil, it is difficult to lower soil pH sufficiently to grow certain plants. Only species more tolerant of high pH may grow there.

These two melaleuca trees have a sparse canopy, extensive limb dieback, and (when viewed close up) foliar symptoms of extreme iron deficiency. The cause is alkaline (high-pH) irrigation water and soil, which inhibit adequate nutrient uptake. *Photo:* J. K. Clark

Acidic soils with high buffering capacity can resist the effects of adding lime or limestone to raise pH. Raising pH is not usually needed for California soils, but may be required for plants where soil pH is naturally low (e.g., many forest soils) or has been lowered by agricultural practices.

HERBICIDE PHYTOTOXICITY

Phytotoxicity is plant injury caused by chemicals, including air pollutants, fertilizers, and pesticides. Because they are designed specifically to kill plants, herbicides, more so than other pesticide types, pose the greatest risk of phytotoxicity when desirable species are exposed to them. Each type of herbicide or chemical family produces characteristic damage symptoms, including

- malformed, distorted leaves and shoots
- stunted shoots and roots
- chlorotic, necrotic, or spotted leaves

Phytotoxicity is most likely when label directions are not followed, such as applying excessive rates or in ways that expose desirable plants that are sensitive to the herbicide. Landscapes can be injured as a result of herbicide application and movement (e.g., drift) from nearby properties, rights-of-way, or turf and when tree roots grow in treated soil. Spray equipment contaminated with minute quantities of herbicides (phenoxy and other types) can damage plants when used later to apply fertilizers, fungicides, or insecticides.

Preemergence Herbicides. With preemergence herbicide phytotoxicity, aboveground symptoms first appear on foliage. Often, damage is observed on leaves as yellow or brown spots where herbicide granules landed and were not washed off. In plants with leaves in whorls, herbicide may accumulate in the whorls and damage will not be observed until new leaves emerge. After root exposure to preemergence herbicide, leaves may develop yellow or white veins, or veins remain green and tissue between them becomes chlorotic or pale. Sometimes chlorosis progresses to necrosis (blotchy, interveinal, or marginal) and foliage dies.

Many preemergence herbicides are tolerated by established woody plants, especially plants with healthy, well-developed root systems. Damage sometimes occurs if preemergence herbicide is applied at high rates, incorporated too deeply into the root zone, is not labeled for the site, or is applied near poorly rooted young plants or shallow-rooted plants. Some preemergence herbicides can persist for months, and if roots grow into treated soil, such as nearby fence lines or roadsides, damage can occur long after the application. In some cases plants will be stunted; in the worst cases, desirable plants will die.

Postemergence Herbicides. Some herbicides are nonselective (broad-spectrum); they can kill both weeds and desirable plants that are sprayed. Selective herbicides kill only certain types of plants, such as only broadleaves or only grasses, as discussed in "Herbicide Types" in Chapter 7. Postemergence herbicides that are nonselective or are selective for (designed to kill) broadleaves are most likely to damage landscapes; avoid using them in established landscapes or use them only when you can avoid exposing desirable plants.

Some of these herbicides move systemically (translocate) within plants. Treating cut stumps with systemic herbicide to prevent resprouting can damage nearby plants because natural root grafts are common between adjacent trees of the same species and herbicide translocates from treated stumps through roots to nearby trees.

Broadleaf Herbicides. The post-emergence broadleaf herbicides include dicamba, triclopyr, and 2,4-D. They can be absorbed through roots and translocate within plants, resulting in twisted shoots and leaf petioles and dwarfed, distorted, and discolored foliage. These herbicides can severely damage or kill broadleaf trees and shrubs if their roots grow under treated lawns, if droplets drift during application or after application when warm weather causes herbicide to vaporize and move in air, or when sprayers contaminated with their difficult-to-remove residues are used to apply insecticides or fungicides. Dicamba and 2,4-D are contained in some lawn "weed and feed" products, and dicamba in particular can injure woody broadleaves exposed to even very low amounts.

Nonselective Herbicides. Postemergence nonselective herbicides include glyphosate, glufosinate, and pelargonic acid. Glyphosate contamination on broadleaves' basal buds, foliage, or thin or green bark causes leaves to turn yellow or mottled green and sometimes die. Plants exposed to this nonselective herbicide in the fall or winter may not exhibit symptoms until months later when new growth appears in the spring. Plant growth is then distorted, and stunted and leaves appear undersized, puckered, and needlelike. Roses in the landscape are quite sensitive to glyphosate absorbed through green stems.

Diagnosing and Managing Phytotoxicity. Investigate whether other causes of similar symptoms are responsible for the damage. Adverse soil conditions (compaction, pH, salinity), leaf spot diseases, natural leaf variegation, nutritional disorders,

This defoliated California bay was severely damaged by systemic herbicide applied to the cut stump (foreground) to prevent resprouting. Because natural root grafts are common between adjacent trees of the same species, trees can be damaged when herbicide translocates from stumps through roots to nearby trees. *Photo:* T. J. Swiecki, Phytosphere Research

Chlorotic veins in almond leaves caused by root exposure to excess preemergence herbicide. Older (larger) leaves are more severely affected than is young foliage, as is typical with preemergence herbicide phytotoxicity. *Photo:* J. K. Clark

In comparison with normal foliage, sycamore leaves on the left are twisted and cupped from exposure to 2,4-D. Broadleaf herbicides such as this are contained in some lawn "weed and feed" products and can severely damage trees and shrubs growing in or near treated lawns. *Photo:* J. K. Clark

When deciduous trees are exposed to glyphosate in the fall, symptoms in new growth the following spring include small, puckered, needlelike leaves. Glyphosate injury can resemble symptoms of severe zinc deficiency. *Photo:* J. K. Clark

root-feeding nematodes, viral diseases, and water deficit can cause symptoms resembling herbicide damage.

Learn what herbicides have been used on-site or nearby, when they were used, at what rate they were used, conditions at time of application, what plants are susceptible to them, and the type of damage caused by them. Herbicide damage is usually most severe on plants nearest to where it was applied. Unlike with diseases and insects, when herbicide is the cause more than one plant species is damaged. Necrotic spots mostly on leaves of similar age may indicate drift onto foliage of a postemergence herbicide. Chlorosis or necrosis caused by a soil-applied preemergence herbicide usually is most prominent in older foliage, where time has allowed the herbicide to accumulate; symptoms are often most prominent in new growth when inappropriate soil conditions or poor root health are the cause of damage. Inspect surrounding plants, including weeds, to observe where herbicide was applied and to help you determine whether injury resulted from root absorption or aerial drift.

You may be able to diagnose phytotoxicity by having a laboratory test soil for preemergence herbicides or test foliage for systemic pesticides or spray residue. Samples should be tested soon after plant exposure. Such tests are expensive, and laboratories require you to identify the specific herbicides for which you want tests. Be aware that some soil-active herbicides can affect plants at concentrations that are below the minimum detection limit of the laboratory.

If preemergence herbicides are suspected, bioassays may be useful as the first step to determine whether herbicide residue is causing the problem. Collect soil from the upper 2 inches and separately from one or more deeper areas in the root zone. Separately collect uncontaminated soil that is otherwise similar. Plant seeds into small containers of these soils. Use species known to be susceptible to the suspect herbicides, such as target weeds listed on the label or sensitive desirable species the label warns you not to spray. Provide good growing conditions and appropriate cultural care and compare the emergence or growth of plants from the different soils. Consult *Abiotic Disorders of Landscape Plants* and the Herbicide Symptoms website, herbicidesymptoms.ipm.ucanr.edu, for help in diagnosing phytotoxicity.

Avoid phytotoxicity by applying herbicides and other pesticides carefully, as directed on the label, and using nonchemical weed management methods when feasible. If phytotoxicity has occurred, be diligent about providing plants with proper cultural care, especially appropriate irrigation. Incorporating activated charcoal, compost, manure, or organic mulch into topsoil and keeping soil moist when temperatures are warm can help reduce the concentration of certain preemergence herbicides through microbial degradation or because the herbicides bind to organic particles. It takes time for herbicide residues to completely degrade and new, uninjured growth to appear.

After being repeatedly sprayed with horticultural oil during hot weather, these rose leaves became chlorotic and necrotic along margins where oil collected. Narrow-range oil can be safely applied to most foliage, but treating during hot weather or repeated applications of most any insecticide over relatively short periods can injure foliage. *Photo:* J. K. Clark

OTHER CHEMICAL PHYTOTOXICITY

Fertilizers, fungicides, insecticides, and plant growth regulators used to slow plant growth or prevent nuisance fruiting can cause phytotoxicity if they are misapplied or used when environmental conditions or cultural practices predispose plants to phytotoxicity. Phytotoxicity symptoms include
- yellow to brown leaf spots
- chlorosis or necrosis of leaf margins, interveinal areas, or entire leaves
- leaf curling and stunting
- premature leaf drop

Damage is possible from applications during or just before hot weather (about 90°F) or if plants are stressed from water deficit. Excess rates, conditions that do not favor rapid drying of spray, treating during certain growth stages (e.g., young plants or during flowering), or not following all label directions increase the risk of phytotoxicity. Treating plants not on the product label and using spray equipment that previously applied herbicides also can cause phytotoxicity. For example, horticultural (narrow-range) oil can remove the desirable bluish cast on foliage of certain spruces. Oil application is not recommended on a few genera of sensitive plants, on any species that is drought stressed, during hot weather, or in mixes with chlorothalonil, sulfur, and certain other fungicides, as discussed in "Oils" in Chapter 6.

If phytotoxicity is suspected, learn what chemicals were used, how and when they were applied, and their rates. Apply the diagnostic methods discussed above in "Herbicide Phytotoxicity." If phytotoxicity occurs, provide damaged plants with good care, especially appropriate irrigation. Modify cultural methods and the plant growing environment to minimize the need for pesticide use. Use nonpesticide alternatives whenever feasible, such as for foliage-feeding insects, as discussed in Chapter 6. Many types of broad-spectrum insecticides (especially carbamates, organophosphates, and pyrethroids) applied during warm weather can induce outbreaks of mites, causing foliage to appear burned and possibly drop prematurely.

CHILLING INJURY

Damage to plants from cold temperatures above freezing (32°F) is called chilling injury. Tropical and subtropical plants (e.g., avocado, banana, and mango) are most susceptible to chilling injury, but many plant species can be injured by a rapid, substantial drop in temperatures. Chilling commonly kills flowers, injures fruit, and causes leaves and shoots to wilt or discolor purplish or red.

FREEZING AND FROST

Damage that occurs from temperatures at or below 32°F is called freeze injury or frost injury. Frost and freezing produce the same damage but occur under different conditions, and some of their management strategies differ. Freezing occurs when air temperatures are below 32°F. Frost occurs when air is warmer than 32°F but plant tissues drop to 32°F or below because plants radiate (lose) heat into the atmosphere, especially during cool, clear nights.

Cold temperature damage causes buds, flowers, and shoots to curl, turn brown or black, and die. Foliage appears scorched because low temperatures severely dehydrate plant tissue. Bark and wood can crack or split, and whole branches or entire plants may be killed if temperatures are below those tolerated by the plant. Cold injury to roots causes these same aboveground symptoms, but soil insulation usually prevents cold injury to roots, except in containers, planter boxes, and raised beds. Often injury is not apparent until days after a freeze and when temperatures rise. Symptoms resembling low-temperature injury are also caused by anthracnose and other leaf and shoot diseases, gas or mechanical injury to roots, phytotoxicity, and water deficit.

Cold injury is most likely to occur during autumn and spring, the coldest times during winter, and when temperatures decline rapidly after warm weather. The time of year, minimum temperature, duration of cold, the rate at which temperatures drop, and plant characteristics (e.g., age, hydration, part affected, whether it has acclimated) influence the severity of damage. Plants that are gradually exposed to increasingly cool weather during the fall become acclimated (hardened) and tolerate more cold than during spring and summer. Broadleaf evergreens and plants not in dormancy generally are the most sensitive to cold. Buds, flowers, younger leaves and shoots, and especially succulent tissues in the spring are the parts most susceptible to freezing and frost injury. However, by the summer, new growth often replaces tissue damaged by cold during spring. Plants adapted to the local environment usually are not permanently harmed.

Prevention and Management. Choose species well adapted to the climate and seasonal temperatures for the location. For information on plants' tolerance to

Freezing weather scorched and killed this lemon foliage. Unless trees pose a hazard, such as falling limbs, do not prune freeze-damaged plants until after you are certain what tissues are dead, preferably by waiting until after new growth would normally develop the subsequent spring. *Photo:* J. K. Clark

Yellow and brown sunburned leaf areas developed on this rarely irrigated euonymus growing in California's hot, arid Central Valley. *Photo:* J. K. Clark

cold, consult the *California Master Gardener Handbook* and *Sunset Western Garden Book.* Consult the *USDA Plant Hardiness Zone Map* to learn the historical winter minimum temperatures at your location.

To increase a soil's ability to absorb heat and warm plants, control weeds and during winter rake away mulch and keep soil bare. When frost or freezing are expected, irrigate dry topsoil at least 3 days before freezing weather to increase the soil's ability to retain heat. When frost is expected, cover sensitive plants overnight with cloth or similar material other than plastic to reduce heat loss to the atmosphere, but leave covers open at their bottom so heat from soil can help warm plants. Remove covers during the day.

During freezing, covering plants is of little help unless a heat source is provided. Placing incandescent lights designed for outdoor use in the canopy may generate enough heat to prevent plants from freezing if plants are also covered. Be sure not to create electrical shock or fire hazards.

Operating misters or sprinklers to wet foliage continuously and flooding beneath plants during freezing are used in some commercial situations, but generally not in landscapes. With sprinklers, the irrigation system and plants must be able to tolerate being covered with ice, and flooding soil can damage plants. Do not combine the use of outdoor lights with flooding or sprinkling.

Do not prune freeze-damaged plants until after you are certain what tissues are dead, preferably by waiting until spring or summer after new growth begins. An exception is limbs or trunks that are hazards and may fall. Replant with more cold-tolerant species.

SUNBURN

Sunburn is damage to bark, foliage, fruit, and other aboveground plant parts caused by excessive exposure to solar radiation. Usually injury is most severe, or present only, on the south and west sides of plants and on the upper side of horizontal branches that are not adequately shaded. Sunscald, certain canker disease pathogens, water deficit, and certain other disorders can cause bark damage that resembles sunburn.

Sunburned bark may discolor, and if the injury is recent it may ooze sap. As affected tissue dries, it becomes cracked or sunken, bark may peel away, and the wood may be attacked by boring insects and decay fungi. Sunburned trunks and limbs can become cankered or girdled and killed. Sunburned broadleaf foliage may appear glazed (abnormally shiny), silvery, or reddish brown and may progress to necrosis beginning at leaf tips, margins, and between veins. Sunburned conifer needles turn black or brown or drop prematurely.

Sunburn often occurs on the stems or trunks of young woody plants. Their bark is thin, and they may not tolerate being exposed to direct sun in landscapes, especially if they were grown close together in nurseries where their trunks were shaded.

Sunburn cracked and cankered the bark and sapwood on the southwest side of toyon. The sunburn developed after overstory limbs that shaded the trunk were pruned out, increasing bark's exposure to solar radiation. *Photo:* J. K. Clark

Apply white, interior latex paint diluted with an equal portion of water to trunks where sunburn may occur. White paint reflects light and reduces bark heating, thereby helping to avoid sunburn. *Photo: J. K. Clark*

Older trees can be damaged when bark is newly exposed to the sun because of pruning or premature leaf drop. Removing structures or trees that provided shade or adding pavement or structures that reflect light or radiate heat around established plants can also lead to sunburn.

Sunburn is usually associated with warm weather and often with insufficient soil moisture availability. Sunburn is common in new plantings that lack a well-developed root system. Restricted soil volumes, inappropriate soil moisture, or anything that makes roots unhealthy or prevents plants from absorbing adequate water may contribute to sunburn. Even in soil that is saturated with water, sunburn may occur.

Management. Plant where roots will have adequate soil volume and sufficient growing space as they mature. Choose plants that are well adapted to the local environment, plant properly, and provide appropriate cultural care. Avoid anything that damages roots or prevents them from absorbing sufficient nutrients or water, including irrigating too frequently.

Encourage desired branch structure by properly pruning and training plants while they are young. Retain some temporary lower branches that help shade the trunk, avoid pruning during summer, and avoid pruning off more than about 20% of the plant canopy during any one year. Apply and maintain appropriate mulch to conserve soil moisture and reduce soil temperatures during summer. Minimize changes to a plant's environment unless deliberately done to improve conditions. Apply white interior (not exterior) latex paint, diluted with an equal portion of water, to the trunks of young trees and to older bark newly exposed to the sun if it is susceptible to sunburn. Interior water-based paints are safer to trees than oil- and water-based exterior paint.

Where appropriate, modify the site to provide partial shade and prevent sunburn. If leaves have not already been killed, sunburn injury to foliage can often be remedied by adequate irrigation, adding shade or shelter, and improving soil conditions. Sunburn injury to bark contributes to tree decline and premature death.

Sunscald

Sunscald is winter injury to bark caused by rapid temperature fluctuations. It occasionally occurs in mountainous areas and the high valleys of eastern California. Bark warms from exposure to sunlight and cools as the sun drops below the horizon. Young trees and trees with thin bark are most susceptible to injury.

Sunscald commonly discolors bark reddish brown, usually on the south and west sides of the trunk and limbs and on the upper side of poorly shaded horizontal branches. Bark then separates from wood, shrinks, appears sunken, and cracks and peels off in patches, exposing wood. Injured branches and trunks develop callus tissue, appear gnarled or rough, and may become girdled and die.

Sunscald damage resembles sunburn of bark. Superficially similar damage can also be caused by boring insects, canker fungi,

freeze damage, and mechanical injuries, such as trunks rubbed by tree stakes or hit by high-pressure sprinkler water.

To prevent sunscald, choose plants well adapted to the climate and retain lower branches for several years after planting. Provide good cultural care (especially adequate irrigation), apply mulch to moderate soil temperature and retain moisture, and whitewash trunks and limbs, as discussed in "Sunburn," above. Tree wrap can be applied in cold climates to insulate bark from heating by sunlight.

High and Low Light

Each plant is adapted to certain amounts of light, depending on its species and previous growing environment. Plant growth can be retarded by either too much or too little light. Excess light can cause foliage to become chlorotic even when temperatures

Sunscald cracked and cankered the bark on this mountain ash. Where freezing temperatures occur and bark is directly exposed to sunlight, young and recently planted trees are most susceptible to sunscald. *Photo: W. Jacobi, Colorado State University, Bugwood.org*

are cool; unlike sunburn, excess light usually does not cause necrosis (tissue death). Aeration deficit, nutritional deficiency, and water deficit also cause symptoms resembling high light damage.

Deficient light often causes elongated, spindly shoots so plants become taller than normal but have thin stems. Foliage can become pale or dark, depending on the plant species. Leaves may become larger and thinner than normal and drop prematurely, beginning with the most shaded foliage. Aeration deficit in the root zone and certain plant growth regulators can cause similar symptoms.

Typical foliage color returns after plants receive appropriate light, but a prolonged light imbalance causes plants to become susceptible to other problems and possibly die. For example, artificial lights at night can alter certain plants' response to seasonal changes in natural light, increasing their susceptibility to frost damage.

Diagnose inappropriate light based on knowledge of the site's conditions in comparison with the plant's light adaptations. Plant only species that are well adapted to the amount of light at the location. At sites receiving full sun, do not plant light-sensitive (shade-adapted) species such as aucuba, camellia, cast iron plant, Japanese aralia (*Fatsia japonica*), *Sarcococca* spp., and vine maple (*Acer circinatum*). Avoid changing the environment in any way that significantly alters the amount of light received by established plants unless this is purposely done to remedy inappropriate light conditions. For example, planting overstory species can provide shade for plants sensitive to direct light. For a list of shade-tolerant species and directions on how to measure whether plants are receiving adequate light, consult publications such as *Abiotic Disorders of Landscape Plants*.

High light yellowed this foliage on camellia, a species that is adapted to partial shade. Overstory plantings to provide shade can help to restore the normal color of leaves damaged by high light. *Photo: R. D. Raabe*

THERMAL INJURY, OR HIGH TEMPERATURES

Plants are injured if temperatures become too high in the canopy or root zone, dehydrating tissues and killing plant cells. Damage is more likely the longer trees are exposed to heat and if they are moisture stressed, otherwise unhealthy, young, or more heat-sensitive species. For example, Monterey pine and other species adapted to cool growing conditions become damaged or killed by heat when planted where summers are warm, as in most inland valleys of California. Extreme heat events can sometimes cause acute thermal damage to species that normally tolerate local conditions.

Thermal injury causes scorched, crisp, brown foliage. Similar damage is caused by cold temperatures, gas or mechanical injury to roots, herbicide phytotoxicity, root disease pathogens, severe fire blight disease, and acute water deficit.

Fire is a common cause of thermal injury and can be diagnosed because it blackens bark in addition to scorching foliage. Usually the fire event itself was observed and reported. Often it is apparent that fire was the cause based on surrounding damage: one-sided injury along edges and distinct boundaries to the burn zone. All plants are susceptible to fire, although species with higher volatile hydrocarbon concentrations in leaves, young woody plants, and plants suffering water deficit tend to ignite more readily. See "Minimizing Fire Hazards" in Chapter 3 for more discussion.

Steam or heat released by equipment, vents, and blacktop paving sometimes causes thermal injury to foliage. Root injury from heat is common in plant containers exposed to direct sun, but it is uncommon in landscapes, except in unshaded raised beds or when roots are near buried heat or steam pipes, usually when pipes leak.

Thermal injury causes brown, crisp foliage, with the most severe damage nearest the heat source, as in this landscape bordering where a wildfire was extinguished. Similar one-sided damage patterns occur when heat from road paving scorches only foliage nearest the paving or exhaust heat or steam kills leaves near vents. *Photo: J. K. Clark*

Rose blossoms sometimes develop abnormal, green, leaflike structures. This malady, called phyllody, results from a plant hormone imbalance in roses exposed to high temperatures or drought stress during flower bud formation. *Photo:* B. S. Ferguson

These scabby blisters (edema) on eucalyptus leaves are apparently harmless to plants. No methods have been proven to prevent this malady in landscapes, but improving soil drainage and avoiding excess irrigation may help to avoid edema. *Photo:* J. K. Clark

Avoid planting near heat sources and choose species well adapted to local conditions. To avoid high-temperature injury to roots, plant properly, provide sufficient soil volume for root growth as plants mature, irrigate adequately, apply organic mulch, and shade the sides of containers and planting boxes.

Organic matter decomposition by microorganisms can generate substantial heat, which in new plantings with overly amended soil, can kill roots. Before mixing organic matter into soil, compost it well. Do not plant in soil that has recently been amended with a large percentage of undecomposed organic matter. If topsoil is amended with undecomposed organic matter at more than about 20% of topsoil volume, do not plant in it until after soil has been moist and warm for 2 months or longer to provide time for decomposition.

EDEMA

Edema is the development of raised, scabby areas on leaves. Although affected tissue is often brown and blisterlike, it is not necrotic. Damage is aesthetic and does not threaten plant health. The specific cause is unknown, but edema often develops in the presence of light when soil is cool and wet and the air is relatively warm. These conditions apparently cause excess moisture to accumulate in leaves, damaging tissue and causing leaf blisters.

Injured tissue cannot be restored, and no methods are known to prevent edema in landscapes. Excess soil moisture may promote edema, so avoid irrigating too frequently and improve drainage where warranted. Avoiding prolonged excess soil moisture will at least help to prevent certain other common maladies.

ROSE PHYLLODY

Rose flowers sometimes develop green, leaflike structures. This uncommon abnormality, called phyllody, apparently is due to a plant hormone imbalance that in roses is usually caused by hot weather or drought stress when flower buds are forming. Certain rose varieties (e.g., floribundas) more often develop phyllody. Phytoplasmas and viruses can also disrupt plant hormones and cause phyllody in many plant species but not commonly in roses. If phyllody occurs on roses with stunted growth or foliage that is yellow or otherwise unhealthy, a phytoplasma or virus may be the cause. When roses have both normal and abnormal flowers but foliage growth and color look healthy, the cause is probably abiotic and when mild weather returns the subsequent flowers should develop normally.

Phyllody does not seriously harm plants. There is no management other than pruning out affected blooms and providing plants with good growing conditions and proper cultural care.

MECHANICAL INJURY

Mechanical injury occurs when plant parts are crushed, cut, punctured, rubbed, or struck due to accidental or deliberate physical actions. Mechanical injury to aboveground parts often produces obvious wounds, such as broken limbs and stripped bark, and causes a wide range of decline symptoms such as discolored foliage, limb dieback, premature leaf drop, slow growth, wilting, and even plant death. Causes of mechanical injury include

- changing soil grade in the root zone
- chewing, rubbing, or scratching by animals
- cutting roots or trunks during excavation or trenching
- cutting trunks or root crowns with mowers or string trimmers
- girdling from stakes or ties
- impact from vehicle bumpers
- operating heavy equipment or laying pavement over roots
- pruning
- vandalism

Major roots were cut and the soil grade lowered around this trunk to install pavement. Cutting and crushing major roots often causes trees to decline, increases their susceptibility to various pests, and reduces anchorage and stability, creating a tree failure (tipping trunk) hazard. *Photo:* J. K. Clark

Other causes of symptoms resembling mechanical injury include canker diseases, chemical phytotoxicity, extreme winds, inappropriate irrigation, poor soil conditions, root decay, vascular wilt pathogens, vertebrate pest chewing, and wood-boring insects.

Sometimes the cause of injury is evident, such as ties or wires embedded in bark are due to improper staking or a mature tree that lacks basal trunk flaring, indicating that the soil grade has been raised. Historical knowledge can help you diagnose the cause, such as knowing the local history of construction or development or whether potentially phytotoxic chemicals were applied. Mechanical injury to roots can be especially difficult to diagnose because wounds are hidden underground and aboveground symptoms may not become obvious until months or years after root injury. Use a pneumatic excavator (e.g., Air Knife or Air Spade) to remove soil around roots or carefully excavate or wash soil away and inspect roots to help determine whether they have been injured.

Prevent wounds to trunks and roots through appropriate site design and planting and good landscape maintenance. Protect trees by installing protective barriers and screens and keeping other vegetation back from trunks. Properly stake young trees and promptly remove support when no longer needed. Avoid injecting root crowns and trunks. Minimize soil disturbance, at least within the drip line of canopies. For more information, consult "Injuries, Hazards, and Protecting Landscapes" in Chapter 3.

HAIL

Hailstones usually tatter, tear, and pit leaves and scar fruit. Sometimes hail causes elliptical wounds in bark or breaks twigs. Serious hail damage can girdle and kill branches.

Hail damage is rarely seen in California landscapes and can be easy to diagnose when the storm event is reported in the news. When hail is the cause of damage, all exposed species at a site can show at least some injury, and the wounds occur on the upper side of plants, especially on parts facing the direction from which storms come. Severe fire blight outbreaks can occur in susceptible species shortly after hail damage causes wounds through which pathogens enter hosts. Canker diseases or excess wind can cause similar symptoms on bark or leaves. Shot hole fungus causes fruit scars resembling hail damage. However, pathogens usually affect only certain species at a site; for example, shot hole infects only *Prunus* spp.

Prune out wood that is seriously damaged. Provide injured plants with proper cultural care, especially appropriate irrigation.

Cankers due to mechanical injury from limbs rubbing against planting stakes. The stakes should have been removed within about 1 year after planting. Newly planted trees sometimes warrant temporary staking, but these stakes were placed too close to the trunk, tied too tightly, and left in place too long. *Photo:* J. K. Clark

Hail made many small scars on this citrus twig. Unlike pathogens that are typically host-specific, hail can damage all exposed species at a site. The location of injury aids in diagnosis, as these wounds all occur on the upper side of plants facing the direction from which storms came. *Photo:* J. K. Clark

High wind tattered and tore these leaves and caused their margins to dry out and die. *Photo:* J. K. Clark

WIND

Windborne sand in desert locations and near beaches can abrade tissue, causing bark and leaf surfaces to appear sandblasted. High wind can break flowers, foliage, and limbs and tear and shred leaves, sometimes called tatters. Cold, hot, or high winds desiccate foliage and bark. If water loss is severe, leaves become necrotic along their margins and tips and drop prematurely. Wind drying of foliage resembles damage from hail, herbicide phytotoxicity, excess salinity or specific ions, and water deficit.

Plants growing at especially windy sites often have smaller-than-normal leaves and are stunted overall. Early abscission (leaf and branch drop) and wind breakage can cause plants to develop a highly "sculptured" structure, such as California's distinctive coastal Monterey cypress with their one-sided foliage and limb growth pointing away from the ocean, as pictured at the beginning of this chapter.

Provide plants with proper cultural care, especially appropriate irrigation, to reduce the adverse effects of wind. If needed during their first year of growth, stake plants properly to allow young trunks to flex and develop strength, as discussed in Chapter 3. Plant susceptible species in more sheltered locations, and choose species carefully for windy sites. Plants that grow fast, become tall, and have broad, thin leaves are usually less tolerant of wind. Smaller plants and those with narrow leaves with a thicker cuticle better tolerate wind. For lists of wind-sensitive and -tolerant plants, consult publications such as *Abiotic Disorders of Landscape Plants.*

LIGHTNING

Lightning strikes can break off a tree's upper limbs or trunk, kill treetops or entire trees, and cause plants to burn or explode. Sometimes lightning kills bark and wood in a long, vertical streak extending from the point of the strike to the ground. This wound can be visible where bark is loosened or blown from the trunk. Lightning can also seriously damage roots or internal tissues of trunks and limbs, even though damage is not visible externally or appears to be minor. Wood beneath bark may grow abnormally and form galls, and roots may grow from unexpected places on the trunk (adventitious roots). After lightning injury, trees often become attacked by wood-boring insects and decay pathogens and die prematurely.

Lightning mostly strikes exposed, isolated, or tall trees at higher elevations in eastern and northern California. Unless the event was observed, plants were inspected soon after lightning struck, or plants show the distinctive damage symptoms from lightning, suspect more common causes that produce similar damage symptoms, including adverse soil conditions, mechanical injuries, and root decay pathogens.

Where lightning is prevalent, lightning rods wired to the ground can be installed at the top of especially valuable, tall specimen trees. The International Society of Arboriculture, isa.org, National Fire Protection Association, nfpa.org, and Tree Care Industry Association, tcia.org, publish standards for lightning rod systems.

Immediately after a lighting strike, have a certified or registered arborist inspect trees for limb, trunk, or root damage that may cause the tree to fail (fall). Immediate repairs include pruning to reduce safety hazards and tacking any loosened bark back into place over wounds, then periodically moistening loosened bark to reduce drying. Provide injured plants with proper cultural care, especially appropriate irrigation. Have trees periodically reinspected to determine whether hazard-reduction pruning, tree removal, or other remedies are appropriate.

When struck by lightning, a continuous strip of bark and sapwood can be blown off a branch and the trunk, leaving a groove of exposed wood from the point of the strike that can extend to the ground. The trunk may split, as shown here, and bark sometimes hangs in shreds around wounds. If the tree survives, exposed wood often becomes attacked by wood-boring insects and decay pathogens. *Photo:* L. R. Costello

GAS INJURY

Landfills, natural gas lines, sewers, and wastewater disposal vents can emit or leak gases that injure or kill foliage or roots. Gas injury symptoms include slow plant growth and foliage that wilts, then turns brown, crispy, and dry. Affected roots may discolor and appear bluish or water-soaked. Plants may only partially leaf out in spring, limbs may die back, and plants can be killed. Gas causes aeration deficit symptoms, but injury may also resemble that from herbicides, root pathogens, or water deficit.

Diagnose gas injury by smelling air and soil near injured plants. Odor of ammonia, natural gas (methane), or hydrogen sulfide (rotten eggs) may be detectable. Soil may be discolored black, bluish, or gray. Look for any pattern to the damage, such as only on the plants near vents or over the gas or sewer line.

Contact the utility company immediately if a leak of natural gas is suspected. With a sewer line break, depending on the situation, the local water agency, property owner, or persons causing damage are responsible for repairing it.

To diagnose whether the cause is gas emissions from garbage decaying at a former landfill, investigate the site's previous land use, test air using special gas monitoring instruments, or send air samples to a laboratory. Grow plant species that are shallow rooted or otherwise more tolerant of landfill emissions, as listed in *Abiotic Disorders of Landscape Plants*.

AIR POLLUTION

Air pollution damage to plants in California is caused mostly by ozone (O_3). Because of the nature of air emissions and pollution control measures, sulfur dioxide (SO_2) and peroxyacetyl nitrate (PAN) rarely damage plants. Ozone and PAN are formed in the lower atmosphere through complex reactions among volatile organic compounds (VOCs) and sunlight. Vehicle emissions and fuel burning are major sources of VOCs, but plants, wildland

Dieback of Canary Island pine caused by a natural gas pipeline leak in the root zone. Other less likely sources of gas injury include leaking from sewer lines and vents and former landfills that have been landscaped. *Photo:* L. R. Costello

fires, and numerous petroleum-derived products (e.g., many pesticides) also emit VOCs. See *Urban Trees and Ozone Formation* for more information on ozone, VOCs, and which plants are high or low producers of biogenic VOCs.

Air pollution reduces crop yields and in landscapes mostly stunts plant growth and may discolor foliage. Damage varies with species, cultivar, and age of plant and with weather and location.

Air pollution damage is difficult to diagnose, in part because symptoms are mostly due to long-term (chronic) exposure to relatively lower pollution levels. Many of its symptoms are similar to and aggravated by those resulting from other stresses, including aeration deficit, foliar pathogens, herbicide phytotoxicity, mite or thrips feeding, nutrient disorders, and water deficit.

This hazy air is caused by aerosols and nitrogen oxide pollutants. Air pollution damage to plants in California is caused mostly by ozone, which is invisible and causes plant damage that is difficult to diagnose. Many causes produce symptoms similar to those of ozone and other causes, and air pollution can work in combination to damage plants. *Photo:* J. K. Clark

Ozone pollution damage can be subtle, such as the yellowish patches in this maple leaf. *Photo:* J. K. Clark

Shortened, chlorotic, and necrotic needles, discolored banding, and tip dieback on pine exposed to excess ozone. Management includes providing good cultural care and planting species less susceptible to ozone. *Photo:* J. K. Clark

OZONE

Ozone naturally occurring in the upper atmosphere (stratosphere) shields plants and animals from harmful solar radiation. At ground level, plant-damaging levels of ozone commonly occur in parts of California during summer and early fall, especially in the Los Angeles basin and the Central Valley. Ozone in the lower atmosphere develops mostly from hydrocarbons and nitrogen oxides from vehicle exhaust and various other sources.

Damage symptoms usually are most apparent on younger plants and older foliage. In broadleaf foliage, ozone causes bleaching, mottling, or discolored patches, flecks, or stippling. Discoloring may be dark, pale, reddish, or a mix of these and other colors between the veins, especially on the upper surface of leaves. Discolored areas can enlarge and extend all the way though leaf tissue, and foliage may drop prematurely. Flowers can become bleached or necrotic and die. Conifer needles may develop yellow bands, flecks, or mottling. Needle tips can turn brown, reddish, or yellow and foliage may drop prematurely.

MANAGEMENT OF AIR POLLUTION INJURY

Provide proper cultural care and improve the growing environment to reduce plant stress. Where air quality is especially poor, plant tolerant species, as listed in *Abiotic Disorders of Landscape Plants*. For large-scale plantings in regions where pollution exceeds air quality standards, consider planting trees that are low emitters of biogenic volatile organic compounds (Table 4-4).

Air pollution is best controlled at its source. Consult the pesticide VOCs emissions calculator, ipm.ucanr.edu/mitigation/reducing_voc.html, and minimize use of emulsifiable concentrate (EC) formulations and other pesticides that produce VOCs. When applying pesticide, minimize the number of applications and the size of treatment areas. Pollution reduction methods include using alternative means of transportation and energy, properly maintaining vehicles and engines, and conserving resources and materials. Grow more plants because certain gaseous and particulate air pollutants are absorbed by bark and foliage.

Table 4-4.

Plants with Low Biogenic Volatile Organic Compounds (VOCs) Emission.

COMMON NAMES	FAMILY
maple	Aceraceae
Brazilian pepper tree, Chinese pistache, smoke tree	Anacardiaceae
catalpa, chitalpa, desert willow, jacaranda	Bignoniaceae
honeysuckle	Caprifoliaceae
cypress, juniper	Cupressaceae
sago palm	Cycadaceae
azalea, heather, manzanita, rhododendron, strawberry tree	Ericaceae
pecan, walnut	Juglandaceae
magnolia	Magnoliaceae
ash, jasmine, lilac, olive	Oleaceae
buckthorn, ceanothus	Rhamnaceae
hawthorn, pome fruit, pyracantha, rose, serviceberry, stone fruits, toyon	Rosaceae
coast redwood	Taxodiaceae
elm	Ulmaceae

For more information on plants with low emission volatile organic compounds (VOCs) and which plants are high emitters of VOCs, see *Urban Trees and Ozone Formation: A Consideration for Large-Scale Plantings.*

Chapter 5   Diseases

Disease is an abnormal condition that damages plants. Pathogenic diseases are caused by infectious organisms. Pathogens and abiotic agents (discussed in Chapter 4, "Abiotic Disorders") often work in combination to damage plants and can cause symptoms on any plant part.

The severity of pathogenic diseases depends on the virulence of the pathogen, the susceptibility and growth stage of the host, and environmental conditions, especially moisture and temperature. The interaction of these factors is called the pest or disease triangle (Figure 5-1). If conditions are poor for pathogen development and plants are otherwise healthy and growing well, many pathogens have little or no effect on their host. The same pathogen can be devastating when conditions are favorable for the pathogen or when host plants are stressed and predisposed to infection or disease development.

Because a pathogen's effect on a plant and the location of symptoms also depend on the site of infection, this chapter groups diseases by affected plant parts.

TYPES OF PATHOGENS

Pathogens include mistletoes (discussed in Chapter 7), nematodes (Chapter 8), and the bacteria, fungi, phytoplasmas, and viruses discussed in this chapter. Only a few species of microorganisms cause disease in plants, and most microorganisms have little or no effect on plants. Some species are beneficial, such as mycorrhizal fungi and nitrogen-fixing bacteria that colonize roots of many plants. Other microorganisms suppress or kill pests or help to decompose dead plants, releasing nutrients that become available for plant growth.

Fungi and Oomycetes. Fungi and funguslike oomycetes are the most common plant pathogens. Diseases they cause include flower blights, heart rots, leaf spots, trunk cankers, root decay, and vascular wilts.

Fungi and oomycetes are usually composed of fine, threadlike structures (hyphae) that form a network or mass (mycelium) growing on or through their host. Fungi and

Figure 5-1.
The pest or disease triangle. Damage results from interactions among a susceptible host plant, environmental conditions, and pest organisms (e.g., the leaf-infecting fungus producing spores illustrated here under magnification). Pests can be present but unable to cause damage if trees are healthy and growing conditions are not favorable for pest development. Abiotic disease requires only a susceptible host and adverse environmental conditions. Time (such as season or duration of disease-favoring conditions) is also a factor. This disease-development interaction is sometimes illustrated as a pyramid with time represented as a fourth axis.

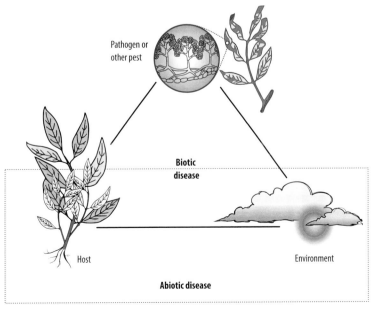

These white basidiocarps, or conks, are the spore-forming structures of a *Ganoderma* sp. fungus and a sign of wood decay in this tree. *Photo:* J. K. Clark

These black, greasy fruiting bodies of *Phaeochoropsis neowashingtoniae* on California fan palm are called diamond scale because they superficially resemble scale insects. This fungus causes severe chlorosis and brown spots on palm fronds, damage that can be confused with that caused by certain noninfectious disorders such as nutrient deficiency. *Photo:* J. K. Clark

oomycetes produce tiny, seedlike spores that spread with moving wind, water, soil, machinery, insects, and animals, including humans. Oomycetes (*Phytophthora* and *Pythium* spp.) also produce zoospores that actively swim through water and wet soil. Some fungi produce other dispersal and infection structures, such as rhizomorphs (rootlike or cordlike masses of hyphae) and sclerotia (compact masses of hyphae). Sclerotia and some types of spores help fungi to survive for long periods in the absence of susceptible hosts, then germinate to infect hosts when conditions (e.g., temperature and moisture) are suitable.

If they are large enough to be seen with the naked eye or a hand lens, mycelia, masses of spores, and spore-forming structures help in identifying fungi. Signs (visible fungi) include mildews and sooty molds (fungal mycelia and spores), rusts (orangish spores of certain fungi), and conks and mushrooms (spore-forming structures). Note that many mushrooms and other fungal structures are produced by beneficial or innocuous species, as discussed under "Mycorrhizae" in Chapter 3 and in the *Mushrooms and Other Nuisance Fungi in Lawns Pest Notes* and other publications listed in "Suggested Reading" at the back of this book.

Bacteria. Bacteria are microscopic, one-celled organisms that grow in or on branches, flowers, foliage, twigs, stems, and roots or other organic matter. Common symptoms of bacterial plant diseases include cankers, leaf spots, galls, scabs, shoot blights, soft rots, and wilts. Bacteria are commonly spread by splashing water, insects, or people moving infested plants. Bacteria must enter plants through openings or wounds. For example, certain leafhoppers and sharpshooters carry (vector) *Xylella fastidiosa* bacteria and infect plants when feeding, causing diseases discussed later in "Bacterial Leaf Scorch, or Oleander Leaf Scorch."

Viruses. Viruses are submicroscopic pathogens that can infect and grow only in living cells. Viral diseases typically deform, discolor, or stunt plants or flowers, fruit, leaves, or shoots. Viruses rarely kill woody plants, and some infected plants exhibit no symptoms. Most plant viruses do not survive for very long outside of living tissue. However, some like *Tobacco mosaic virus* can survive for decades in dried leaf tissue. Many viruses are vectored by aphids, leafhoppers and sharpshooters, or plant-feeding nematodes. Viruses can also be transmitted through infected seeds, plant cuttings, and grafted rootstocks or scions. Many viruses can be spread mechanically, such as on contaminated hands or cutting tools. Once a plant is infected by a virus, it usually remains infected during its entire life. There is no treatment to cure virus-infected plants in landscapes.

Viroids. Viroids cause symptoms similar to those seen in viral infections. Viroids are transmitted by grafting of rootstocks and scions, mechanical means, and by the pollen and seeds of infected plants. There is no cure.

Phytoplasmas. Phytoplasmas are minute organisms smaller than bacteria. They are often spread by leafhoppers, and many are called yellows because leaf chlorosis is a common damage symptom.

Steps to Help You Diagnose Problems

Correctly identify the affected plant species.

Carefully examine symptomatic plants and those nearby, looking for patterns in the distribution of symptoms within and between plants.

Accurately describe the abnormalities and where they occur.

Collect pertinent information, e.g., on cultural practices, site conditions, and weather.

Make a tentative diagnosis by ruling out and including certain possibilities based on available information.

Compare what you see in the landscape with the illustrations and descriptions in this book.

See the "Tree and Shrub Pest Tables" at the end of this book for common damage symptoms and their causes by host plant.

Check plants regularly for stress symptoms, improper cultural care, and disease symptoms and signs. *Photo: J. K. Clark*

Examining plants from a distance helps to reveal problems and any pattern to symptoms. The hackberry trees on the left side of the street have yellow, sparse foliage and dead terminals at the treetop. In comparison, hackberry trees of the same age across the street have a healthy appearance. *Photo: J. K. Clark*

Healthy roots and proper soil conditions are critical for plant growth and performance. Diagnosing the cause of aboveground symptoms may include digging up and examining root crowns or small feeder roots. Healthy roots are typically firm and light colored inside. Disease roots are commonly dark and soft, as pictured later in "Root and Crown Diseases." *Photo: D. Rosen*

MONITORING AND DIAGNOSING DISEASES

To decrease the likelihood of plant damage and improve the effectiveness of your management actions, inspect landscapes and plants regularly for

- adverse growing conditions
- improper cultural care
- disease signs: actual structures of the pathogens, such as fungal mycelia and spore-forming structures
- disease symptoms: unhealthy appearance, such as canopy thinning (abnormally few leaves) and leaf chlorosis or wilting

Time of year influences how frequently to monitor and what signs and symptoms you may see. Bacterial and fungal diseases of fruit, leaves, and shoots cause the most damage during or after wet periods in winter and spring. Root and trunk diseases commonly become more obvious when trees are water stressed during warm weather. Monitor highly susceptible plants at least weekly or even daily during certain times of the year to allow prompt management actions for quickly developing pathogens.

Know the normal appearance of each plant species so you can determine whether plants look unhealthy. To help you accurately diagnose the cause of symptoms, examine as many parts of the affected plant as possible, because some aboveground symptoms like wilt and twig dieback can be caused by root or trunk diseases. Reveal injured tissue and primary symptoms by scraping off small sections of discolored or irregular bark, digging up some feeder roots, or removing soil from the root crown or the main lateral roots extending outward to the edge of the canopy.

Examine several plants if possible with different stages of disease to determine how symptoms change as the disease progresses. Plant parts in the early stage of disease development often show more characteristic symptoms; secondary organisms or other factors that obscure symptoms may later become involved. Do not rely on a single symptom because various causes can produce similar damage (e.g., leaf spots,

Table 5-1). You usually need to observe several different symptoms to identify the cause of disease. Examine plants both close-up and from a distance and look at all affected plant parts, as discussed in Chapter 2 in "Regular Surveying for Pests."

A chisel, hand lens, hand pick, hatchet, pruning shears, sharp knife, and shovel are useful for inspecting diseased plants and their growing site. An ice chest, plastic bags, notebook, pencils, and permanent marker for collecting, labeling, and preserving samples and recording your findings are useful. Some diseases are difficult to identify based only on field symptoms. Professional help and laboratory tests performed on the diseased plants, surrounding soil, and (for comparison) nearby apparently healthy plant tissue are sometimes needed to confidently diagnose the cause(s) of disease.

Compare what you see in the landscape with the illustrations and descriptions in this chapter, which are grouped accord-

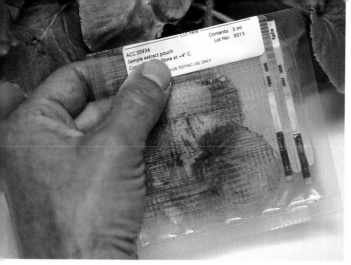

Test kits are available for use in the field to help determine whether certain bacteria, fungi, or viruses are present in plant tissue. With this kit, plant tissue is macerated in a pouch containing liquid. Detector strips are inserted through a slit and certain colors appear if the specific pathogen for which the kit tests is present. Exercise caution when interpreting test kit results, as they can give incorrect results, such as when improper samples are collected. *Photo: J. K. Clark*

ing to the portion of the plant where symptoms most often appear. Consult the "Problem-Solving Guide" and "Tree and Shrub Pest Tables" at the end of this book for lists of common damage symptoms and their causes. Other key information sources include *Abiotic Disorders of Landscape Plants,* the regularly updated resources online at ipm.ucanr.edu, and the University of California (UC) Cooperative Extension office and Master Gardener Program in your county.

Keep records of your observations, including the disease signs and symptoms and the location, date, and plant species. Note the current and past environmental factors that may have contributed to the problem, including humidity, injuries, temperature, pesticide use, soil conditions such as drainage, and the presence of free water (rain, dew, irrigation). Repeated monitoring over time will help you confirm your diagnosis, follow the development of the disease, and make proper management decisions. Revisit problems after taking action to evaluate the effectiveness of your control efforts.

DISEASE MANAGEMENT

It is often too late to provide effective control once disease symptoms appear or become severe. Action may be required before damage becomes apparent. Prevention is the most important method of disease management; for many diseases, prevention is the only effective option.

Choose plants well adapted to the local conditions. Learn the cultural and environmental requirements of your plants and provide them with proper care. Many diseases are more severe on plants that are stressed or growing under adverse conditions. Stress can be caused by soil that is compacted or kept continually too wet or too dry, overfertilization or excess salinity, improper pruning especially at bud break or during early growth flush, and severe insect damage. Physical damage to roots and trunks, changes in soil grade, excessive or misdirected herbicide use, injury from injections or implants, and other damaging factors can predispose plants to disease.

It is more effective and cost efficient to prevent disease by using high-quality, well-adapted, disease-resistant plants growing under good cultural conditions than to manage established diseases. Learn the conditions that promote diseases common to your plants, as discussed below and in Chapters 3 and 4 and take actions to prevent disease.

RESISTANT PLANTS

The species or cultivars planted often determine whether certain diseases are likely to develop or can be avoided. Choose resistant or tolerant cultivars and species, which usually are less likely to become

Table 5-1.

Some Causes of Leaf Spots.

CAUSE	COMMENTS	SEE PAGE
air pollution	Aggravates other causes if air quality is poor.	63
anthracnose	Many hosts; promoted by moisture during new growth.	81
bacterial blight	Dieback, cankers, and oozing twigs may be associated symptoms.	83
chewing, mining, or sucking insects	Insect presence usually helps to identify.	144, 170, 216
Entomosporium leaf spot	Plants in Pomoideae group of the Rosaceae are affected.	87
eriophyid mites	Tiny, elongate mites may barely be seen with a hand lens.	249
leaf blisters	California buckeye and oak leaves affected by *Taphrina* fungi if spring is moist.	
mineral deficiency or toxicity	Often produces characteristic pattern of discoloration helpful in identifying the cause.	45–50
pesticide injury	Commonly herbicides, but other pesticides can be the cause.	54–56
rusts	Orangish or yellowish spore masses, usually on leaf underside or on branches or stems.	94
scabs	Dark, circular, scabby or velvety fungal spots on many hosts.	88
scale insects	Unlike diseased tissue, scales can usually be scraped off.	188
Septoria leaf spot	Spots mostly older leaves; cankers may develop on poplars.	88
shot hole	Almond, apricot, plum, and other *Prunus* spp. are affected.	89
spider mites and red mites	May be webbing, foliage speckling, or tiny mites present.	245
sucking insects	May be dark excrement of thrips, lace bugs, or plant bugs.	170
sunburn	Yellow or brown area beginning between leaf veins, mostly on southwest side of drought-stressed plants.	57
viruses	Streaked, discolored, or distorted foliage.	92
water deficiency	Often preceded by leaf wilting.	44

infected or are not seriously damaged if they do become infected. Because resistance is not the same as immunity, plants may become affected by problems to which they are resistant if they are not well suited to local conditions or are stressed because of adverse site conditions and poor cultural care. Be aware that plants resistant to certain pathogens may be susceptible to other pests.

Landscape plants that are resistant to anthracnose, powdery mildew, root rots, vascular wilts, and certain other pests are listed in Table 3-2. Check with the UC Cooperative Extension office in your county or a certified nurseryperson for the most recent recommendations.

When plants are placed in locations where soil or other environmental conditions are inappropriate for their growth, certain problems develop regardless of how much care plants are given. For example, cypress canker in certain cypress species and Botryosphaeria canker in giant sequoia inevitably develop when these plants are grown in hot interior areas of California. The only effective strategy for some problems is to replace and avoid growing certain plants in inappropriate locations.

QUALITY PLANTING MATERIAL

Many pathogens of trees and shrubs can be transmitted by nursery stock and transplants. Select certified pathogen-free nursery stock when available. Examine young plants for symptoms of root and trunk diseases such as bark wounds and crown gall before purchasing and planting. Expose roots to be sure they are not diseased or excessively kinked or restricted by the planting container. When the container is removed, the root ball (roots and soil together) should remain intact, and when the trunk is carefully lifted both the trunk and root system should move as one. Avoid trees that lack a dominant central leader or have otherwise been improperly pruned, as they are unlikely to develop good structure and will be more prone to breakage

and wood decay. See Table 3-1 in Chapter 3 for a checklist of what to look for when purchasing nursery stock. A relatively small initial investment in higher-quality plants can pay great dividends in improved aesthetic quality, lower maintenance costs, and less disease.

EXCLUDE FOREIGN PESTS

Many of our worst pathogens and other pests were introduced from other states or countries. Dutch elm disease, pitch canker of pines, sudden oak death, and most host-specific diseases of non-native ornamentals were inadvertently brought into California on contaminated plants, seeds, or soil. To prevent new pest introductions during planting and travel

- Do not bring fruit, plants, seeds, or soil into California unless you know they have been certified as being pest-free or

were inspected by agricultural officials.
- Buy only disease-free plants from reputable, local nurseries.
- Before purchasing plants from outside the local area or moving plants across county lines, contact the local county agricultural commissioner to learn whether any quarantines prohibit plant movement.
- Take any unfamiliar pests to your county agricultural commissioner or UC Cooperative Extension office for identification or telephone California's Pest Hotline at 1-800-491-1899.

Many introduced pathogens and other pests that are in California have not yet spread throughout the state. To prevent many pest problems, avoid moving around or bringing into your site potentially contaminated or infested equipment, plants and plant products, and soil.

Mottled yellowing of leaves and odd-shaped fruit are symptoms of huanglongbing (HLB, or citrus greening). Cutting this fruit in half would reveal its asymmetric (lopsided) growth and irregular internal coloration. This exotic, lethal disease and many other pests have been found in California because plants were brought in illegally. Do not bring fruit, plants, seeds, or soil into California unless you know they are certified to be pest-free or were inspected by agricultural officials. *Photo:* S. E. Halbert, University of Florida

Prevent conditions that are stressful to plants. Paving around established trees reduces oxygen and moisture availability to roots and changes temperatures, affecting aboveground plant parts. *Photo:* J. K. Clark

PLANTING SITE AND DESIGN

Plant appropriately and design your landscape to prevent many diseases.

- Select plants that are adapted for your location. Some species require full sun while others do well only in shady areas. Diseases such as powdery mildews and certain rusts are more prevalent in shady areas; sunburn occurs when sensitive plants are planted at bright sites.
- Avoid planting aggressive, rapid-growing ground covers and turf close to shrubs and young trees to avoid the need for frequent edging or trimming, which commonly wounds trunks.
- Improve poorly drained soils before planting to avoid root diseases.
- Group plants according to their water requirements (hydrozones) and do not plant drought-adapted species near plants requiring frequent watering.

- Compatible ground covers and trees can be irrigated with the same system, but turf and many trees have different soil moisture and irrigation needs.
- Around trees and shrubs that are adapted to a Mediterranean climate (summer drought and winter rainfall), use compatible plants that need a similar irrigation, keep the soil bare, or apply mulch.
- Keep mulch 6 inches or more back from the trunk to avoid promoting root crown diseases.
- When replanting after removing diseased trees or shrubs, use species resistant to the disease-producing agent that occurred there if resistant plants are available.
- See Chapters 2 and 3 for more discussion on landscape design and planting to prevent disease.

MULCH

Mulching, discussed in Chapter 7, controls weeds and improves plant growth by conserving soil moisture, improving soil conditions, and inhibiting certain root pathogens. However, certain types of pathogens can be spread in some types of mulch. Depending on the type of pathogen and which plant parts will be used, you can apply the material as mulch around hosts after you chip or grind and properly handle it (e.g., by adequately heating or thoroughly composting).

Certain vascular wilt fungi persist in plant material and soil, as discussed in "Fusarium Wilt of Palms." Chips from plants affected by these pathogens should not be used for mulch or allowed to enter greenwaste. Do not apply fresh material from diseased plants under the same hosts, especially during the rainy season. If you chip and allow mulch to dry out before use, most bacterial and viral pathogens and many fungal diseases that affect the flower, fruit, leaves, twigs, and branches of trees and shrubs will not spread in mulch. Likewise, most wood decay fungi

that affect the upper trunk and branches do not survive in chipped, thoroughly dried wood.

If trees are infected or killed by root pathogens, wood chips from the branches and upper trunk of the trees generally do not pose a risk of infection. However, bark and wood from the lower trunk, root crown, and roots (e.g., from stump grinding), can be contaminated with pathogens such as *Armillaria* and *Phytophthora* spp., and these materials should not be used beyond the already infested area unless they are treated to eliminate the pathogens. Wood from *Armillaria*-killed trees can be used as mulch after it has been chipped well or ground to avoid large (>2-inch) chunks and spread until it completely air-dries. For *Phytophthora* spp. and certain other pathogens, wood chips must be heated sufficiently through proper composting, solarization, steam, or other means to ensure that any spores are killed.

Proper treatment includes finely chopping or grinding the wood then exposing all the material to temperatures of at least 131°F for a period of 3 days or longer, depending on the particular process. See the standards for the various composting methods under "Laws & Regulations" online at calrecycle.ca.gov. After proper composting or other heating methods, store the material on a clean surface and keep it covered to prevent contact with fresh debris, soil, and other untreated material that could contaminate it.

IRRIGATION

Learn the pattern of moisture your plant is adapted to and provide proper irrigation, as discussed in Chapter 3. Overwatering commonly contributes to the development of root diseases and other maladies. Overwatering includes applying too much water and watering too frequently, especially when drainage is poor.

Most plants native to the eastern United States, northern Europe, and eastern Asia require summer irrigation or rainfall. Conversely, except during establishment, avoid frequent summer irrigation of plants adapted to summer drought, such as California oaks and many other native plants. The specific amount and frequency of water needed vary greatly, depending in part on plant species, soil conditions, and the local environment, as discussed in Chapter 3.

Where foliar pathogens are a problem, avoid overhead watering if feasible or water early in the day or around sunrise so foliage can dry quicker. Irrigate established plants when needed near the drip line, not around the trunk (Figure 5-2). Do not let water stand around trunks. If you have irrigation water basins around trunks, break down the soil berm during the rainy season so water can drain away. Provide good soil drainage by gently grading soil surfaces, installing subsurface drains or sumps, and breaking up compacted soil layers before planting. Instead of planting in a low-lying area, plant in raised beds or on a berm, mound, or ridge of soil (Figure 5-3). Prevent soil compaction by applying mulch and avoiding traffic within the drip line of trees, especially when soil is wet.

FERTILIZATION

Many people assume that unhealthy plants will benefit from fertilization, but most established woody landscapes do not require added fertilizer. Inappropriate fertilization can promote fire blight, certain canker pathogens, and injury-related root diseases. Nutrient deficiency symptoms in most woody landscape plants are due to unhealthy roots or adverse soil conditions, which prevent plants from absorbing available nutrients. Avoid routine fertilization with a few exceptions, such as fruit trees, palms, and certain heavily blossoming shrubs. Use laboratory soil and plant tissue tests, where available, to determine the nutrient needs, as discussed in Chapter 4, and determine the actual cause of unhealthy plants before taking management action.

Do not allow water to pond around trunks. Provide good drainage and do not pile soil next to trunks. Many root and crown diseases are caused or aggravated by improper irrigation practices and adverse soil conditions, such as poor drainage (slow infiltration of water).
Photo: J. K. Clark

Figure 5-2.
Do not water established trees and shrubs near the trunk, as this promotes root and crown disease. Water plants when needed around the drip line and beyond. Adjust sprinklers or install deflectors to prevent wetting of trunk bases. Move drip emitters away from the base of the trunk after plants are established.

Figure 5-3.
Avoid planting in a hole or a low-lying area, except when planting in sandy soils. Plant in raised beds or on a ridge or mound of soil several inches high and several feet across in areas where drainage is poor or soil is highly amended and plants will settle as organic matter in the soil decomposes.

Table 5-2.

Pathogens That May Be Managed by Pruning.

COMMON NAME	SCIENTIFIC NAME	HOSTS	PRUNING METHOD AND TIMING
bacterial blight and canker	*Pseudomonas syringae*	*Prunus* spp.	Prune diseased branches back at least 6 inches into healthy-appearing tissue during dry summer weather.
olive knot, oleander gall	*Pseudomonas* spp.	oleander, olive	Prune and dispose of infected tissue during the dry season.
brown rot or blossom and twig blight	*Monilinia* spp.	*Prunus* spp.	Prune infected branches during winter. Thin canopy to increase air circulation.
Chinese elm anthracnose	*Stegophora ulmea*	Chinese elm	Prune infected branches back to the next healthy lateral. For small cankers on the trunk and major limbs, cut entirely around the canker margin about ½ inch into healthy wood.
crown gall	*Agrobacterium tumefaciens*	many hosts	If small, remove gall by cutting into healthy wood during dry season and sterilizing wound edges with brief exposure to flame heat e.g., blowtorch.
cypress canker	*Seiridium cardinale*	see Table 5-9	During hot, dry weather promptly prune dying branches at least 6 inches below any apparent cankers.
Dutch elm disease	*Ophiostoma* spp.	see Table 5-11	If tree is otherwise vigorous and healthy and symptoms are limited to one or a few limbs, promptly prune at least 10 feet back into healthy wood. If damage is more extensive, promptly remove entire tree.
Eutypa canker	*Eutypa lata*	apricot, cherry, grape	Prune at least 1 foot below visible infection. Prune stone fruits during July or August and prune grape late in the dormant season. Immediately flame the cut surface for 5-10 seconds with a propane torch or apply fungicide.
fire blight	*Erwinia amylovora*	apple, crabapple, pear, pyracantha, and other Pomoidea	Entirely prune off diseased branches and make cuts at least 6 to 8 inches into healthy-appearing tissue.[2]
leaf gall	*Exobasidium vaccinii*	azalea	Remove galled tissue during summer.
mistletoe, leafy	*Phoradendron* spp.	many deciduous species	Cut infected limbs at least 1 foot below mistletoe attachment point.
oak twig blight[1]	*Cryptocline cinerescens* and *Discula quercina*	oaks	If limited to a relatively small proportion of the plant canopy, prune by making cuts in healthy wood below infected twigs during dry weather in the summer or fall.
pink rot of palms	*Nalanthamala vermoeseni*	all palms	Prune only during warm, dry weather.
pitch canker	*Fusarium circinatum*	pines	Remove and dispose of infected terminals by cutting well below visibly infected wood.[2]
powdery mildew[1]	*Podosphaera leucotricha*	apple	Prune and dispose of infected shoot tips during winter or early spring.
Septoria leaf spot[1]	*Septoria* spp.	many hosts	On deciduous hosts, prune and dispose of infected wood in the fall after leaves drop.
shot hole	*Wilsonomyces carphophilus*	*Prunus* spp.	Prune and dispose of infected tissue as soon as it appears. After leaf drop, inspect plants carefully and prune varnished-appearing (infected) buds and twigs with lesions.
silver leaf	*Chondrostereum purpureum*	*Prunus* spp.	Prune infected branches during late spring to early fall. Make cuts in healthy wood.[2]
sycamore anthracnose[1]	*Apiognomonia veneta*	see Table 5-6	Prune into previous year's growth to remove and dispose of infected twigs during dry weather.
sycamore canker stain	*Ceratocystis fimbriata* f. sp. *platani*	sycamore and London plane	Prune only during dry weather in December and January. Promptly remove and dispose of infected trees.[2]
western gall rust	*Endocronartium harknessii*	2- and 3-needle pines	Prune and dispose of infected branches during October to January, before spring.
wood decay	various fungi	most trees	Properly prune young trees to promote good structure and avoid the need to remove large limbs from older trees, which commonly leads to decay.

Pruning usually is more effective when combined with other methods, as discussed in the section on that pathogen elsewhere in this book. Dispose of infected material away from healthy trees.

1. Pruning for these diseases and possibly others is practical only on shrubs and small trees.

2. Carefully clean and sterilize tools before reuse (before making another cut) if they contact discolored tissue or resinous wood.

Adapted from Svihra 1994.

PRUNING

Consider pruning off and disposing of localized plant parts with dying or diseased plant tissue to stop or slow the spread of certain pathogens (Table 5-2) and wood-boring insects (see Table 6-13). Make pruning cuts in healthy tissue, well below the infected or infested area, and dispose of prunings away from susceptible plants. Removing some branches, especially in the lower canopy, can reduce the incidence of diseases such as brown rot of tree fruits by improving air flow and reducing the movement of spores from the ground to the canopy.

Pruning off large branches creates wounds that can be invaded by wood decay fungi and other pathogens. Most important wood decay fungi produce infectious spores during early fall to late spring, so avoid making large pruning cuts during this period. The best time to prune for other purposes can depend on plant growth stage and the particular problem. See Chapter 3 for more discussion on proper pruning.

SANITATION

Rake away and dispose of pathogen-infected plant parts that drop, and prune off infected wood, as discussed above, to help control certain pathogens. Work first in pathogen-free areas before working where plants or soil are suspected of being contaminated with pathogens. Clean soil particles and plant parts off shoes and garden tools and wash your hands before moving to another area after working with diseased plants. At least before moving between plants, scrub tools clean with detergent and water to avoid spreading contaminated soil or pathogens from infected plants.

Certain pathogens can be spread mechanically from plant to plant via pruning shears, saw blades, and tools used to inject or implant chemicals into trunks or roots. Pathogens that can be spread this way include bacteria that cause fire blight, galls, and oleander leaf scorch and fungi that cause canker stain of sycamore, Fusarium wilt and pink rot of palms, and pitch canker of pines. When working on hosts of these diseases and other mechanically spread pathogens, consider sterilizing tools with a registered disinfectant after scrubbing tools clean.

Disinfecting Tools. To reduce the likelihood of spreading plant pathogens, depending on the disease and the distribution of the pathogen on and within plants, the recommended frequency of tool disinfection ranges from after each cut to only before moving between plants.

Diluted bleach (sodium hypochlorite) is one of the most reliable disinfectants, but solutions must be made freshly and replaced frequently to maintain sufficient activity. Bleach also corrodes metal, especially tools' cutting surfaces. Undiluted isopropyl alcohol (70% or higher) and certain other materials can be effective (Table 5-3) and may be preferred to avoid tool rust and clothing damage from bleach. Where it is safe to do so, bare metal surfaces can be disinfested by flaming or similarly heating the surface.

Before disinfection, scrub any plant material or sap from blades and tools using detergent and water because contaminants reduce disinfectant effectiveness. Mix a 10 to 20% solution of bleach (e.g., 1 part bleach and 4 to 9 parts of water) and thoroughly squirt it onto debris-free tools or soak tools in the solution. Because it reacts with oxygen in the air, bleach solution becomes ineffective within about 2 hours. Start with freshly mixed bleach and make a new solution every 2 hours or sooner if it becomes dirty.

When using other disinfectants, at least 1 to 2 minutes of disinfectant contact time with clean tools may be required. Consider rotating work among several tools, using a freshly disinfected tool while the most recently used tools soak in disinfectant. Promptly rinse, dry, and oil metal equipment and tools following treatments such as bleach that promote rust. Wear proper eye and skin protective equipment when using disinfectants.

WEED AND INSECT CONTROL

Control weeds, turf, and ground covers near trunks of plants by properly using mulch or maintaining bare soil. Infrequent or light-to-moderate damage by foliage-feeding insects can be tolerated by many landscape plants. Prevent repeated heavy insect damage that occurs more

Table 5-3.

Disinfectant Effectiveness for Preventing Transmission of Fire Blight Bacteria on Cutting Blades.

EFFECTIVE[1]
≥0.5% sodium hypochlorite (≥10% Clorox bleach)
LESS EFFECTIVE[2]
alcohol, isopropyl (undiluted 70–99% rubbing alcohol)
pine oil disinfectant (≥20% Pine-Sol)
soap, phenols, alcohol, and other compounds mixed (≥20% Lysol)[3,4]
LEAST EFFECTIVE[5]
dimethyl benzyl ammonium chloride and dimethyl ethylbenzyl ammonium chloride (e.g., Physan)

Plant care professionals and persons conducting pest management for hire should use only disinfectants that are registered pesticides and must follow the label directions.
1. Spraying a debris-free tool or dipping it in freshly made ≥10% bleach entirely disinfests it of fire blight bacteria. Bleach can damage clothing and severely rust metal tools.
2. Soaking debris-free tools for a least 1 minute is the most effective method of using these disinfectants. Spraying a debris-free tool with this material or quickly dipping the tool in disinfectant is less effective.
3. Lysol concentrated disinfectant, not Lysol cleanser.
4. May cause some rust damage to metal tools.
5. Soaking debris-free tools in these materials for a least 1 minute may disinfect tools, but it is not always effective. Spraying or quickly dipping tools with these materials is less effective.
Adapted from Teviotdale 1991, 1992; Teviotdale, Wiley, and Harper 1991.

than once in a growing season or during consecutive years, especially total loss of foliage, as this weakens plants and can increase their susceptibility to disease.

SOIL SOLARIZATION

To control many weeds and most soil-dwelling insects, nematodes, and pathogens, cover moist, bare soil with clear plastic for at least 4 to 6 weeks during a sunny, warm time of the year as described in "Solarization" in Chapter 7. Solarizing before planting is only effective within several inches of the soil surface, but this is where most root growth occurs when plants are young and most susceptible to pest damage.

BIOLOGICAL CONTROL

Many naturally occurring organisms kill or retard the growth of pathogens. Some bacteria and fungi colonize the surface of aboveground plant parts or roots and suppress infection of these surfaces by pathogens. However, despite their importance, little is known on how to manipulate most of these beneficial microorganisms. For example, suppressive soils contain microorganisms that improve plant growth when added to certain types of pathogen-infested soil. Unfortunately, there are no general recommendations on effectively using suppressive soils in landscapes. A few beneficial microorganisms or their by-products are commercially available for use in preventing disease, as discussed under "Biologicals," below.

BENEFICIAL MICROORGANISMS

Many soil-dwelling microorganisms improve plant growth in ways other than their ability to control pathogens. Many fungi and bacteria break down organic and certain inorganic materials in soil so that nutrients become available for plant growth. Mycorrhizae are beneficial associations between plant roots and fungi. They improve plants' ability to absorb nutrients and apparently increase plants' tolerance to drought and certain pathogens, as discussed in Chapter 3. Soil-dwelling bacteria convert nitrogen-containing materials into nutrient forms that plants can use. Some species of nitrogen-fixing bacteria form nodules on roots, especially roots of plants in the legume family, so determine whether galls on roots are the result of beneficial or harmful organisms before taking control action.

PESTICIDES

With careful cultural management, at least some cultivars of most landscape plants can be grown at a high level of aesthetic quality with little or no pesticide application. Growing species well adapted to local conditions, selecting cultivars resistant to disease, and employing cultural and environmental management practices that prevent diseases are generally the most effective strategies; many disease-producing organisms cannot be effectively controlled once plants become infected and symptoms develop or become severe.

Synthetic (petroleum-derived) and organically acceptable pesticides (e.g., from naturally occurring ingredients) are available to control certain plant pathogens, primarily foliage-infecting fungi. Each pesticide is effective only against certain pathogen species and when applied at the proper time, so accurate identification of the cause of problems and careful application timing are critical to effective pesticide use. Proper mixing (e.g., the correct amount and dilution of the fungicide), effective application technique (e.g., thorough spray coverage), the type of spray equipment to use, and safe pesticide use methods are discussed in Chapter 2. Consider the potential toxicity to humans and the environment before using a specific product, such as by consulting the active ingredients database at ipm.ucanr.edu (Figure 5-4). Follow all label directions and wear at least chemical-resistant gloves, eye protection that also shields the brow and temples (e.g., goggles), and clothing that completely covers arms and legs.

Fungicides and bactericides have either contact or systemic activity. Contact fungicides and bactericides are effective only when susceptible plant parts are thoroughly covered by these materials. Systemic fungicides and bactericides, depending on the material, move short distances within plants (e.g., from the sprayed side of a leaf to the opposite side) or may be transported throughout the plant. Most of the mobile systemics move only upward in plants via the xylem, but a few materials also enter the phloem and can move both upward and downward in plants.

Virtually all contact and most systemic materials act only as protectants; they prevent only the infection of healthy tissue and cannot kill existing infections in hosts. Protectant (preventive) fungicides must be applied before plants become infected or show symptoms. Certain systemic fungicides and (for powdery mildews only) certain oils have curative (or eradicant) activity and can kill an existing infection if they are applied at a sufficient dosage at the earliest stages of infection. However, using fungicides as eradicants increases the chance of selecting for fungicide-resistant pathogens.

For both contact and systemic fungicides, repeated applications may be necessary to protect new growth or during prolonged periods that favor disease development. In many landscape situations, fungicides will not effectively control certain pathogenic diseases, such as root diseases and vascular wilts. Even for diseases that are potentially manageable with pesticides (primarily fungal diseases of blossoms, fruit, and leaves), pesticides are infrequently used on large trees because of the application expense, potential for spray drift, and difficulties of achieving proper application timing and good spray coverage.

Most contact fungicides, including the inorganic fungicides sulfur and copper, are referred to as multisite inhibitors because they affect a variety of biological processes. Their diversity of effects reduces the likelihood that pathogens will become resistant to them but increases the risk of phytotoxicity (plant damage). In contrast, most systemic fungicides affect only a single metabolic site and generally pose a lower risk of phytotoxicity. However, repeated use of single-site inhibitors, especially on foliar pathogens, favors the development of resistance, and certain pathogen populations can no longer be controlled with certain single-site fungicides.

Take steps to minimize resistance development.

- Rely on nonpesticide alternatives where feasible.
- Reduce the number of pesticide applications per season.
- Mix or alternate (rotate) fungicides with different modes of action.

See Tables 5-4 and 5-5, Chapter 2, and the Fungicide Resistance Action Committee website, frac.info, for more information.

Biologicals. Certain beneficial microorganisms or their by-products are commercially available to control plant pathogens. *Bacillus subtilis* can prevent powdery mildew and certain other foliar diseases. Crown gall can be prevented from developing in nurseries by dipping roots in a suspension of *Agrobacterium tumefaciens* 'K-84' or *A. tumefaciens* 'K-1026' (=*A. radiobacter* 'K-1026'). Consult the specific pathogen sections later in this chapter and *Natural Enemies Handbook* for more information.

Biologicals have little or no toxicity to nontarget organisms and people. However, in comparison with conventional synthetic pesticides, they often require more knowledge and careful application to use them effectively.

Botanicals. Plant-derived fungicides include botanical oils such as jojoba and neem oil. Neem oil from the tropical neem tree is primarily used as a curative foliar spray for powdery mildew. Treated plants must be thoroughly sprayed.

Inorganics. Inorganics include potassium bicarbonate (discussed under "Powdery Mildews"), coppers, and sulfur. These are primarily protectants and require thorough plant coverage to provide control. Repeated applications are generally necessary to prevent infection of new growth and to renew deposits removed by rain or irrigation.

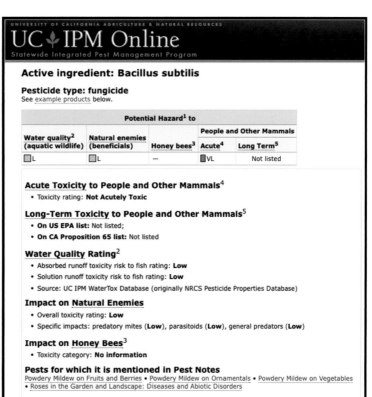

Figure 5-4.
Consider the potential toxicity to humans and the environment before using a specific product. Up-to-date pesticide recommendations and information on their environmental impacts, such as this on the biological fungicide *Bacillus subtilis*, are available at ipm.ucanr.edu.

Bordeaux Mixture. Bordeaux mixture is copper sulfate and lime (calcium hydroxide) mixed in a particular way; it must be used soon after preparation. Bordeaux is highly effective at preventing plant infection by various bacteria and black spot, leaf curl, powdery mildew, and many other fungi. Because it adheres well to plants and persists through extensive rain, when applied in late fall or early winter it provides excellent protection from pathogens during winter and spring. Consult *Bordeaux Mixture Pest Notes* for how to prepare and use Bordeaux.

Copper. Fixed copper fungicides include copper ammonium, copper oxychloride, cupric hydroxide, and tribasic copper sulfate. Fixed coppers slowly dissolve and slowly release the tiny amounts of copper needed to prevent infection by pathogens while reducing the risk of phytotoxicity from overexposing plants to copper. Fixed copper fungicides are easier to prepare than Bordeaux mixture and prevent many of the same foliar bacterial and fungal diseases (Table 5-4). However, copper ammonium may be the only fixed copper readily available to nonprofessionals.

Copper salts of fatty and rosin acids (also called copper octanoate, copper soaps, or organic copper) avoid the sometimes unsightly visible residue left on surfaces and plants by Bordeaux mixture and fixed copper. However, copper salts are not as effective against some diseases as Bordeaux and fixed coppers.

Sulfur. Sulfur can prevent powdery mildews, certain rusts, and scabs. Elemental sulfur is applied as a dust using a dust applicator or by lightly shaking it from a container onto small plants in the morning when they are slightly moist with dew. Sulfur is also available as a finely ground wettable powder or flowable liquid containing a wetting agent and as micronized sulfur (also called flotation or colloidal sulfur). These materials readily mix in water when they are diluted before application.

To avoid inhaling sulfur dust, use a liquid formulation. Micronized and wettable sulfurs are generally easier to apply and provide better coverage than sulfur dust. Sulfur's effectiveness increases with increasing temperature, but avoid its use

Table 5-4.

Selected Biological, Botanical, and Inorganic Pesticides for Pathogen Control.

PESTICIDE (TRADE NAME EXAMPLES)	CURES OR PROTECTS	TYPE	PATHOGENS TARGETED
Agrobacterium tumefaciens (=*A. radiobacter*), nonpathogenic strain (Galltrol-A, Nogall)	P	Bio	crown gall on many hosts, excluding grapes
Bacillus subtilis (Bayer Advanced Natria Disease Control, Plant Guardian Biofungicide, Serenade)	P[1]	Bio	fire blight, powdery mildew, and scab on many hosts
Bordeaux mixture[2]	P	I	anthracnose, bacterial blight and canker, bacterial and fungal leaf spots, and fire blight on many hosts
copper, fixed, including copper hydroxide (Kocide)	P	I	anthracnose, bacterial and fungal leaf spots, fire blight, and others on many hosts
copper, octanoate or soap (Lilly Miller Ready to Use Cueva Copper Soap Fungicide, Ortho Disease B Gon Copper)	P	CS	leaf spots, rusts, and others on many hosts
copper sulfate (Phyton)	P	I	leaf spots, rusts, and others on many hosts
jojoba oil[2] (Eco E-Rase)	C[3]	Bot	powdery mildew on rose and many ornamentals
neem oil (Garden Safe Brand Fungicide 3, Green Light Rose Defense)	P, C[3]	Bot	black spot, powdery mildews, and rusts on many hosts
potassium bicarbonate[2] (Bi-Carb Old Fashioned Fungicide, Eco-Mate Armicarb O, Kaligreen)	P	I	powdery mildew on shrubs such as grape and rose
sulfur[2] (Bonide Sulfur Plant Fungicide, Lilly Miller Sulfur Dust)	P	I	black spot, powdery mildew, rusts, and scabs on fruit trees, roses, and other hosts

KEY

Bio = biological

Bot = botanical

C = curative, or eradicant

CS = copper soaps, copper octanoate or copper salts of fatty and rosin acids

I = inorganic

P = protectant, or preventive

All these products are believed to have a low potential for promoting pathogen resistance. In comparison with synthetic fungicides (Table 5-5), some of these products are less effective if used alone, but they can work well if applied in an integrated program that includes cultural and environmental controls. Often, materials are for use only on certain landscape plants and are effective only against certain pathogen species, such as some species of anthracnose, powdery mildew, or rust, but not others. Certain of these materials may not be registered (legal) for use in your situation. Consult a current label for details on legal pesticide use.

1. Applied as protectant, but may also kill or parasitize pathogenic fungi.

2. In comparison with other materials, is more likely to cause phytotoxicity to foliage under certain conditions.

3. Curative only for powdery mildew.

before and during hot weather because plant damage may result if temperatures exceed about 85°F.

Oils. Certain botanical and petroleum-derived oils can eradicate slight to moderate powdery mildew infections, providing good control. The botanical neem oil is also effective as a preventive for certain other fungi (Table 5-4). Oils are sometimes formulated into products with other active ingredients, but unless otherwise stated on the label they should not be combined with sulfur or applied within 3 weeks of applying sulfur (or longer on certain plants) due to the risk of phytotoxicity. Do not apply oil or other pesticides when plants are drought stressed or when temperatures over 90°F or below freezing are expected within a day. Some botanical oils are more likely to cause phytotoxicity than are petroleum-based, narrow-range (horticultural) oils.

Synthetics. Synthetics include materials with both contact and systemic activity (Table 5-5). To reduce the likelihood that they will become ineffective against certain pathogens, products that have a single site of biological activity especially must be used as indicated in "Pesticide Resistance" in Chapter 2 and as summarized above.

Benzimidazoles. Thiophanate-methyl and thiabendazole are protectant and curative systemics that control many fungi but not oomycetes (downy mildews and *Phytophthora* and *Pythium* spp.). Certain fungi have developed resistance to benzimidazoles, especially some populations of fungi that cause Botrytis blight and powdery mildews.

Chlorothalonil. Chlorothalonil is a contact fungicide with preventive activity against many fungal diseases. As with other protectant fungicides, chlorothalonil must be reapplied to plants as it is washed off by rain and irrigation and to protect new growth that emerges after the previous application. It can be phytotoxic if combined with oil or certain surfactants.

Phosphonates (*Phosphites*). This group includes fosetyl-al and salts of phosphonic acid (e.g., potassium phosphonate). These

Table 5-5.

Selected Synthetic Pesticides for Pathogen Control.

PESTICIDE (TRADE NAME EXAMPLES)	CURES OR PROTECTS	FRAC CODE[1]	SITE[2]	PATHOGENS TARGETED
azoxystrobin (Heritage), trifloxystrobin (Compass)	P[3]	11	single	anthracnose, black spot, downy mildew, leaf spots, powdery mildews, rusts, scabs, and certain others
chlorothalonil (Daconil, Fung-onil, Ortho Disease B Gon Garden Fungicide)	P	M5	multi	black spot, leaf spots, rust, and others
myclobutanil (Spectracide Immunox), triforine (Ortho Rosepride)	C, P	3	single	black spot and other leaf spot fungi, powdery mildews, rusts, and scabs
fosetyl-al (Flanker WDG Ornamental)	P[3]	33	multi	bacterial blight, downy mildew on rose, fire blight, and Pythium and Phytophthora root rots
oil (JMS Stylet Oil, Sunspray)	C	NC	multi	powdery mildew
thiophanate-methyl (Infuse Systemic Disease Control Lawn & Landscape)	P[3]	1	single	black spot, leaf spots, Ovulinia petal blight, powdery mildews, scabs, and others

KEY

C = curative or eradicant

P = protectant or preventive

Often, materials are for use only on certain plants and are effective only against certain pathogen species, such as some species of anthracnose, powdery mildew, or rust, but not others. Certain of these materials may not be registered (legal) for use in your situation, or they are available only to professional applicators. Consult a current label for details on legal pesticide use.

1. FRAC (Fungicide Resistance Action Committee) code refers to the material's mode of action (MOA). Bactericides and fungicides with the same code number have the same chemical mode of action, so pathogens that develop resistance to one of these pesticides typically are resistant to all pesticides with that same MOA (FRAC code number). See frac.info for more information.

2. "Single" materials are active against only one of a pathogen's biochemical sites (a single-site mode of action). "Multi" materials affect two or more biochemical sites in a pathogen. Pathogens are more likely to develop resistance to pesticides with a single-site mode of action, rendering those materials ineffective in certain situations.

3. Can also have some curative efficacy, but are best applied as protectants, before plant infection and significant disease development.

systemic fungicides move both upward in the xylem and downward in the phloem. They have a complex mode of action that includes increasing plant defense reactions to pathogens. Phosphonates primarily are effective against *Phytophthora* and *Pythium* spp.

Sterol Biosynthesis Inhibitors (SBI). These fungicides include several chemical classes. The most widely used group is the DMI (sterol demethylation inhibitor) fungicides, which includes myclobutanil and triforine. They have protective and curative activity against many fungal diseases but are not effective against *Phytophthora* and *Pythium* spp. Because SBI fungicides are site-specific, certain fungi have developed resistance to them.

Strobilurins. Strobilurins (e.g., azoxystrobin, pyraclostrobin), also called QoI

fungicides, are synthetic versions of an antifungal secretion of the *Strobilurus tenacellus* mushroom. They are good protectants against various fungi and oomycetes, and certain products can move upward in xylem and short distances in plant tissue. Some products cause severe phytotoxicity to certain plants. Strobilurins have a single site of activity, so limit how many times they are applied and alternate their use with other fungicides and non-pesticidal controls to minimize the development of pesticide resistance.

Blossom and Fruit Diseases

Many pathogens cause leaves or shoots to discolor, wilt, or die. Some of these diseases, such as sooty mold, leaf spots, and

certain powdery mildews, are aesthetically displeasing but usually do not cause serious, long-term harm to plants. Damage limited to blossoms and fruit interferes with our enjoyment of these but does not threaten woody plants' survival. On the other hand, unhealthy blossoms, leaves, and twigs can be symptoms of certain abiotic disorders, wood-boring insects, or pathogens that damage water- and nutrient-conducting tissues and can kill plants, as discussed later under "Branch and Trunk Diseases" and "Root and Crown Diseases." Most types of symptoms can have many different causes, as summarized for leaf spots in Table 5-1. It is important to correctly diagnose the cause of unhealthy plants so you can know whether control action is warranted and what management methods are likely to be effective.

Ovulinia petal blight fungus caused these water-soaked blotches on azalea. Unless reproductive structures characteristic of the fungus are present, these early symptoms are largely indistinguishable from Botrytis blight damage. However, *Botrytis* infects both green tissue and petals, while *Ovulinia* infects only petals. *Photo:* J. K. Clark

AZALEA PETAL BLIGHT AND RHODODENDRON PETAL BLIGHT

Azaleas and rhododendrons are highly susceptible to *Ovulinia azaleae* petal blight, also called flower blight or Ovulinia petal blight. The fungus infects only petals, causing white to brownish spots that enlarge rapidly. Infected blossoms become droopy, limp, slimy, and sometimes cling to leaves after they die.

Ovulinia petal blight resembles Botrytis blight, discussed below. However, Botrytis blight is a drier rot that also affects dying or inactive green tissue, while Ovulinia petal blight infects only blossoms. *Botrytis* and *Ovulinia* produce similar sclerotia (infectious structures) that are black, flattened, irregular shaped, and about ⅛ to ½ inch long. Sclerotia from fallen flowers produce very small, brownish, inverted mushroomlike or wineglass-shaped apothecia (reproductive structures) about ¹⁄₁₂ inch diameter on stalks about ⅛ to ⅖ inch long.

The biology of Ovulinia petal blight is nearly identical to that of camellia petal blight discussed below. Both fungi infect wet blossoms when temperatures are mild, about 50° to 70°F. One difference is that in addition to sclerotia and apothecia, Ovulinia petal blight also produces colorless conidia (asexual spores) on infected petals. These spores spread in wind and by the movement of flower-visiting insects, espe-

cially bumble bees. This allows the fungus to reproduce and spread between many blossoms within several days of an initial infection.

Sanitation is the primary management method. Remove and dispose of fallen, old, and infected flowers. Gently twist the flower stem and discard it in a covered trash container. Otherwise manage Ovulinia petal blight the same ways discussed in "Camellia Petal Blight." Avoid overhead irrigation and provide good air circulation. Each year, when blossoms are no longer present, apply a fresh layer of uncontaminated organic mulch beneath host plants and maintain about a 4-inch mulch layer to help suppress *Ovulinia* propagules. Application of appropriate fungicides prior to rainy weather can help to reduce infections, but fungicides often provide only partial control unless they are used in combination with the recommended sanitation and cultural practices.

BOTRYTIS BLIGHT, OR GRAY MOLD

Gray mold is named for the gray, brown, or tan fungal spores that develop on infected tissue when conditions are humid or moist, making rotted tissue appear fuzzy when viewed by the naked eye. Disease often starts as tiny, almost translucent spots and then progresses to brown, water-soaked spots or dark decay on fruit, leaves, petals, and succulent stems, which wilt and die.

Various *Botrytis* and *Botryotinia* spp. cause gray mold, especially *Botrytis cinerea*. Succulent tissue of most plants is susceptible. Landscape hosts include azalea, bird of paradise, cacti, coast redwood, fuchsia, giant sequoia, hydrangea, rhododendron, and rose.

Gray mold fungi do not infect woody parts, and healthy, actively growing green plant parts are seldom infected directly. However, once petals or dead or weakened tissues are infected, the fungus can move from infected parts to invade healthy green tissue it contacts.

The fungus can grow on almost any moist or decaying herbaceous vegetation. Weeds and plant debris are common sources of gray mold spores, which are produced in enormous numbers and readily spread in air to nearby plants (Figure 5-5). Spores germinate and produce new infections only when plants are continuously wet or relative humidity is higher than about 90% for 6 or more consecutive hours. The disease can be a problem almost any time of year in coastal areas but usually develops in interior locations only during the late fall through early spring rainy season. Gray mold is particularly troublesome under high humidity and moderate temperatures (70° to 77°F), but it is also active over a broader temperature range.

Provide proper cultural care, use good sanitation practices, and modify environmental conditions where feasible. Remove old blossoms, declining green plant tissue, herbaceous plant debris, and fallen leaves and dispose of them in covered containers or a well-managed compost pile. Avoid wetting foliage and improve air circulation, as described below in "Downy Mildew." Fungicides are only preventative and generally are not very effective in landscapes when conditions (wet plants and susceptible plant material) favor gray mold development. Many gray mold populations are resistant to certain fungicides.

CAMELLIA PETAL BLIGHT

Ciborinia camelliae causes camellia petal blight in all cultivars of *Camellia japonica*. *Camellia sasanqua* is infected less often in California. Camellia petal blight, also called Ciborinia petal blight, initially causes small, brown, irregularly shaped blotches in petals. Spots enlarge rapidly until the entire flower is brown and dead. Except when wet, blighted petals are dry or leathery but do not crumble when handled. Blossoms drop prematurely to the ground, often as intact flowers. Prominent dark brown veins give infected petals a netted appearance. Damage resembling that of camellia petal blight is also caused by Botrytis blight, frost, old age (overmature blossoms), and injury due to chemicals, rough handling, or wind. Symptoms that distinguish camellia petal blight from these other causes include petal veins darker than the surrounding tissue, infections beginning near the central part of the flower (not appearing first near petal margins), and symptoms that occur only on petals.

Ciborinia camelliae produces dark, hard, irregular-shaped sclerotia at the base of infected flowers, where they replace the calyx lobe. Depending on the extent to which nearby sclerotia unite, they typically range in size from $\frac{1}{12}$ to 1 inch. Sclerotia can lie dormant for several years on or near the soil surface. During winter and spring when camellias blossom, sclerotia produce light brown saucer-shaped apothecia (inverted mushroomlike bodies) about $\frac{1}{5}$ to $\frac{3}{4}$ inch in diameter. Apothecia forcibly discharge large numbers of spores that are carried by wind onto emerging blooms, where they germinate and infect flowers when they are wet.

Pathogen development is favored by wet or humid conditions and mild temperatures (about 59° to 70°F) during bloom. Outbreaks are initiated by sclerotia-infested soil received with new plants and by sclerotia persisting beneath established plants that have previously been infected. Sclerotia continue to

Brown, water-soaked rose blossoms infected with *Botrytis*. Changing from overhead watering to drip or flood irrigation would greatly reduce this gray mold problem. *Photo:* J. K. Clark

Botrytis rot is also called gray mold because under humid conditions it develops gray, fuzzy growth. This fungal growth consists of numerous tiny-stalked, spore-forming structures (conidiophores), which can develop on infected young fruit (shown here) and on dead blossoms and leaves. *Photo:* J. K. Clark

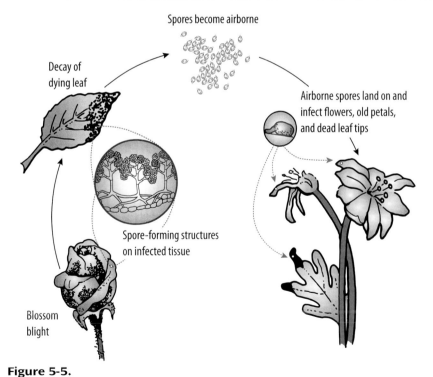

Spores become airborne

Decay of dying leaf

Airborne spores land on and infect flowers, old petals, and dead leaf tips

Spore-forming structures on infected tissue

Blossom blight

Figure 5-5.
Botrytis spores can initiate infections only when they contact tissue that is already injured or declining (left). However, healthy tissue can be infected if it contacts tissue that has already developed Botrytis blight or gray mold disease. Adapted from Agrios 1997.

Prominent brown veins that are darker than the surrounding petal tissue (lower right) distinguish camellia petal blight from similar damage caused by Botrytis blight, frost, old age, and injury due to chemicals, mishandling, or wind. *Photo:* J. K. Clark

produce apothecia for 3 to 5 years after being introduced into soil and the disease cannot be eradicated once present.

Prevention is the best control. Remove the top layer of potting soil when new plants are purchased and replace it with pathogen-free soil. Plant camellias in a well-ventilated location and avoid overhead irrigation. Pull off infected flowers as they appear and collect fallen blossoms and dispose of them in a covered location away from camellias. Do not add camellia petals or leaves to mulch that will be used around camellia, even if it has been composted. It is difficult to expose camellia debris to the 140°F required to kill all of the *Ciborinia* propagules by composting.

Each year, when blossoms are no longer present, apply a fresh layer of pathogen-free organic mulch and maintain a 4-inch layer of organic mulch beneath

and somewhat beyond plants to suppress pathogen spore production. Remove fallen petals and other camellia debris before applying fresh mulch, but otherwise avoid moving or disturbing existing mulch where fungi may be present. Keep mulch thin near the trunk or several inches away from trunk.

Spraying an appropriate fungicide during bloom can help to reduce infections. Depending on the fungicide, reapplication may be needed every 10 to 14 days while conditions remain suitable for the pathogen. Use fungicides only in conjunction with recommended sanitation and cultural practices.

Drippy Nut Disease, or Drippy Acorn

Clear, brownish, or frothy viscous liquid sometimes drips from acorns of coast live oak and interior live oak. Dripping nuts have been an intermittent problem from spring through fall in coastal areas and interior valleys of California, causing a sticky mess on surfaces beneath affected oaks. No methods have been demonstrated to be effective in controlling this bacterial-associated dripping from oaks. The problem does not occur every year and apparently does not threaten tree health.

Drippy nut is caused by *Brenneria* (=*Erwinia*) *quercina*, a bacterium that infects wounds in oak tissue, causing ooze from nuts and the acorn cap, including caps that remain attached after the nut has dropped. Bacteria enter acorns injured by filbert weevils (*Curculio* spp.), filbertworm (*Cydia* =*Melissopus latiferreanus*), and certain other insects.

Where dripping is a problem, wash fouled surfaces regularly while dripping is relatively recent and easier to remove. Pruning branches to reduce canopy overhang in sensitive areas (e.g., patios, driveways) can reduce potential dripping. However, extensive pruning or removing large limbs injures trees and allows entry of decay organisms that can weaken trees, so pruning may not be an appropriate

response to this temporary aesthetic problem.

In addition to dripping from older acorns and caps, profuse dripping or frothy exudate sometimes occurs from very young or barely developed acorns, from distortions where young acorns would be expected to occur, and from leaves or twigs where there are no acorns. This dripping has been attributed to oviposition wounds from several species of oak gall wasps that allow infection by *B. quercina* bacteria. However, neither *B. quercina* or insects or their damage have been definitely associated with some situations where the cause of oak dripping is unknown.

Dripping from oak canopies can also be caused by nectar-producing oak galls produced by certain cynipid wasps (*Andricus*, *Disholcaspis*, and *Dryocosmus* spp.). Aphids, scales, whiteflies, and other honeydew-excreting insects can also cause fine drops of liquid to drip from oak canopies.

Acorns exuding clear to brownish liquid from the cap and nut. This drippy nut disease (drippy acorn) is a poorly understood malady of oaks that occurs after injuries allow bacteria to enter plant tissue. *Photo:* B. Hagen

Leaf and Twig Diseases

ANTHRACNOSE

Anthracnose, often called bud, leaf, shoot, or twig blight, is a group of diseases resulting from various fungi, including *Apiognomonia*, *Colletotrichum*, *Discula*, *Glomerella*, *Gnomonia*, *Marssonina*, and *Stegophora* spp. Damage is most severe when prolonged spring rains occur during or shortly after new growth.

DAMAGE

Anthracnose fungi discolor leaves or kill portions of leaves, entire leaves, small groups of leaves, and twigs scattered throughout the canopy. Small branches can be cankered, girdled, and killed on more susceptible hosts, including some species of ash, dogwood, elm, oak, and sycamore. The resulting regrowth from lateral buds can give trees a gnarled or crooked appearance.

Anthracnose does not seriously harm most hosts. Exceptions are when defoliation occurs repeatedly, and certain species at some locations where branch dieback and cankering are extensive. Chinese elm anthracnose, also called black leaf spot, is a serious disease in coastal areas of California. It produces especially large cankers that can weaken, girdle, or kill limbs and trunks, as discussed later in "Canker Diseases." Dogwood anthracnose can severely canker branches, especially in the lower canopy, causing branch dieback and sometimes death of the entire tree.

IDENTIFICATION AND BIOLOGY

Anthracnose symptoms vary with the host's growth stage and species, weather, and time of year. Small black, brown, or tan spots appear on infected leaves of some hosts, including elm and oak. Dead leaf areas may be more irregular on hosts such as ash. Sycamore anthracnose lesions commonly develop first around the major leaf veins, but scattered shoot terminals and entire leaves and groups of leaves are often killed. Dogwood leaves and flower bracts often develop large brownish lesions with well-defined dark discolored margins that may be grayish or purple.

Anthracnose fungi overwinter primarily in cankers and lesions in infected twigs (Figure 5-6). On evergreen species such as Chinese elm and live oaks, the fungi can occur year-round on leaves as well as woody parts. Spores are produced when infected tissue is wet and spores spread by splashing and windborne rain to infect new twigs and foliage. Spore production and infection continue as long as new growth and wet conditions occur together. Once foliage matures and weather becomes warm and dry, disease development slows or stops.

Terminal dieback and partly killed Modesto ash leaves due to ash anthracnose (*Discula fraxinea*). Avoid this pathogen by planting other species or cultivars because ash anthracnose affects only flowering ash (*Fraxinus ornus*) and Modesto ash (*F. velutina* var. *glabra*). Photo: R. D. Raabe

Sycamore anthracnose (*Apiognomonia veneta*, or *Discula platani*) kills sycamore shoots and leaves. The fungus sometimes causes leaf tissue to die beginning along veins, as with the leaf tip at the center right. Bloodgood, Columbia, and Liberty sycamores are less susceptible or resistant to anthracnose. *Photo:* J. K. Clark

Black leaf spots caused by Chinese elm anthracnose (*Stegophora ulmea*). In coastal California, anthracnose also causes bark cankers and limb dieback that can seriously damage Chinese elm. *Photo:* J. K. Clark

Midvein necrosis of a coast live oak leaf with anthracnose. *Photo:* J. K. Clark

Young, healthy leaves

Spores are released during spring rains

Repeating cycle
(Depends on moist conditions)

Young twigs and new leaves become infect

Canker

Spores are splashed during spring rains

Fungus overwinters in cankers in twigs

Overwintering cycle

MANAGEMENT

Where prolonged spring rains or foggy conditions are common, avoid planting especially susceptible species. Plant resistant or less-susceptible species, such as the ash, dogwood, elm, and sycamore listed in Table 5-6.

There are a number of cultural management techniques. Prevent sprinkler wetting of foliage. Prune and dispose of infected twigs during the fall or winter where practical (e.g., on small trees). For example, Yarwood London plane is highly susceptible to anthracnose, but usually it is undamaged if pollarded (severely pruned) regularly because overwintering infections are pruned off. Rake up and dispose of fallen leaves and do not mulch around hosts using fresh or uncomposted leaves or flowers from plants susceptible to that same anthracnose. Increase sunlight and air circulation (for example, by pruning nearby plants to reduce shade and spacing new plants far enough apart). Provide good cultural care and otherwise minimize stress to improve plants' ability to tolerate anthracnose damage. To stimulate vigorous growth of trees severely affected by anthracnose, you can apply a nitrogen fertilizer after the leaves open and spring rains have stopped.

Certain fungicides help prevent anthracnose on Modesto ash if thoroughly sprayed on all new growth as buds begin

Figure 5-6.

Anthracnose disease cycle illustrated with *Discula fraxinea* infecting Modesto ash. The fungus overwinters in cankers in twigs and sporulates and infects leaves and shoots if new spring growth coincides with rainy weather. By V. Winemiller in Crump 2009.

Table 5-6.

Anthracnose Susceptibility of Some Landscape Trees.

SUSCEPTIBLE	RESISTANT OR LESS SUSCEPTIBLE
Ash (*Fraxinus* spp.)	
Modesto (*F. velutina* 'Modesto')	evergreen, or Shamel (*F. uhdei* 'Shamel')
	Moraine (*F. holotricha* 'Moraine')
	Raywood (*F. oxycarpa* 'Raywood')
Chinese elm (*Ulmus parvifolia*)[1]	
Evergreen	Brea
Sempervirens	Drake
True Green	
Dogwood (*Cornus* spp.)	
Chinese (*C. kousa* 'Chinensis')	bunchberry (*C. canadensis*)
flowering (*C. florida*): many cultivars	Carnelian cherry (*C. mas*)
Pacific (*C. nuttallii*)	Chinese (*C. kousa*): many cultivars
	flowering (*C. florida*): Appalachian Spring, Spring Grove, Sunset
	Japanese cornel (*C. officinalis*)
Oak (*Quercus* spp.)[2]	
white oak (*Q. alba*)	pin oak (*Q. palustris*)
Sycamore (*Platanus* spp.)	
American (*P. occidentalis*)	London plane: Bloodgood,[4] Columbia, Liberty
California (*P. racemosa*)	
London plane (*P. acerifolia*): Yarwood[3, 4]	

1. All cultivars appear to be resistant in warm interior areas of California, where Chinese elm anthracnose is uncommon.

2. Individual oak species and trees vary in susceptibility.

3. Yarwood is usually undamaged by anthracnose if regularly pollarded (severely pruned to remove the previous year's growth).

4. Yarwood is largely resistant to powdery mildew, while Bloodgood is susceptible to powdery mildew.

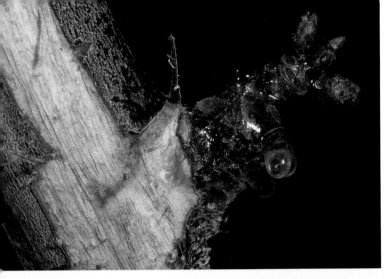

Where *Pseudomonas* bacteria cause brown ooze, cutting shallowly under bark reveals orange to reddish discoloration in infected phloem tissue. Bacterial blight commonly causes shoot dieback as pictured on lilac on the first page of this chapter. *Photo:* J. K. Clark

Bacterial canker causes dark ooze and elliptical wounds on bark. Infections can expand to girdle and kill limbs and occasionally the main trunk. Disease becomes inactive during warm, dry weather, and the next season any active cankers mostly occur at different locations on the tree. *Photo:* J. K. Clark

to open in the spring. If moist weather prevails, additional protective sprays may be needed at intervals of about 2 weeks to protect new growth. Systemic fungicides may be available for professional use. See *Anthracnose Pest Notes* for specific fungicide recommendations.

BLIGHTS

BACTERIAL BLAST, BLIGHT, AND CANKER

Pseudomonas syringae commonly kills blossoms or causes dark lesions on petals and occasionally on fruits. Damage otherwise depends on the host plant and strain of *P. syringae* bacterium and whether infection occurred through blossoms, buds, or wounds in other tissues. Severe damage usually occurs only on lilacs (*Syringa* spp.) and on stone fruits and other *Prunus* spp.

Lilac leaves and sometimes shoots and stems discolor (usually brown or reddish), shrivel, and die and the damage is called bacterial blight. Stone fruits and other *Prunus* spp. develop elongated lesions and brownish ooze on branches and twigs, called bacterial canker. Lesions commonly progress to branch cankers and shoot dieback due to phloem damage, which is visible under bark as brown to reddish discoloration or streaks in the outer wood. Trees can be seriously damaged or killed if major limbs or the trunk become infected.

On apple, citrus, and pear, disease is relatively minor and is called bacterial blast. On apple and pear, usually only flower clusters and a few adjacent leaves are killed. On citrus, small twigs and several leaves in a group are killed scattered throughout mostly the south (wind-exposed) side of the canopy.

On oleander and many herbaceous ornamentals, damage includes blossom and shoot tip dieback, leaf spots and vein blackening, and stem cankers. *Pseudomonas*-infected oleander and olive also develop galls or swellings on leaves and stems, as discussed later in "Olive Knot and Oleander Gall."

Determine whether resistant species or cultivars are available and plant or replace plants using these. Many *Syringa vulgaris* cultivars are highly susceptible to bacterial blight and canker, while *Syringa josikaea, S. komarowii, S. microphylla, S. pekinensis,* and *S. reflexa* are generally less susceptible. Rootstock greatly affects *Prunus* spp. susceptibility to *P. syringae,* and less-susceptible or resistant rootstocks are available for cherry, peach, and plum. Because rootstocks that are less susceptible to certain pathogens are more susceptible to others, consult publications such as *Integrated Pest Management for Stone Fruits* to identify the best cultivars for the common problems in your planting situation.

Prune and dispose of infected twigs and branches during the dry season; new infections are less likely to occur then. Do not wet foliage with overhead irrigation. Provide good cultural care, especially appropriate irrigation. Keep roots healthy, such as by planting on a broad mound of soil where drainage is poor. Bactericide applications have not been found to give reliable control and spraying for *P. syringae* is not recommended.

Fire Blight

Erwinia amylovora causes fire blight bacterial disease in plants in the pome tribe of the rose family. Pear and quince are extremely susceptible. Apple, crabapple, and pyracantha are frequently damaged. Other hosts include cotoneaster, hawthorn, loquat, mountain ash (*Sorbus* spp.), serviceberry, spirea, and toyon.

Fire blight results in a sudden wilting, shriveling, and blackening or browning of shoots, blossoms, and fruit. Dead leaves remain attached when twigs are killed quickly, giving plants a scorched appearance, hence the name "fire blight." Cankers form on twigs and branches, which can become girdled and die. Prolonged serious infections can kill highly susceptible hosts, including Asian, European, and flowering pears.

Identification and Biology

The bacteria overwinter in cracked, sunken cankers in bark. During warm, wet or humid weather, brown droplets containing bacteria ooze from infected cankers and move in splashing water and on insects to infect flowers and new shoots (Figure 5-7). Once flowers are infected, honey bees especially spread the bacteria from one flower to another. Less frequently, new infections are caused after flowering when other insects (e.g., flies) that feed on the ooze move the bacteria to other shoots.

Infections in terminal parts slowly spread downward in the tree, such as from the initial infection points (leaves, shoots, and blossoms) into main limbs and the trunk. The new or spreading infections may not become apparent until the next spring, when terminals die back and bark cankers and oozing appear in new locations as plants begin growing.

Diagnose fire blight by cutting off bark around where healthy tissue meets the edge of an active canker. This should reveal reddish flecking (new, developing infections) in wood adjacent to the canker (dark, discolored wood). As cankers expand, more wood dies, turns brown,

dries out, and becomes sunken. Bark often cracks around the canker edges.

Management

Manage fire blight by growing resistant cultivars, maintaining plants in an appropriate range of vigor, regularly inspecting hosts for infections, and pruning off blighted wood. If fire blight has previously been a problem, consider applying a copper bactericide (e.g., about 0.5% Bordeaux mixture) several times during bloom beginning as soon as blossoms open. A bactericide will not cure existing infections, and even multiple applications are not very effective at preventing new infections when conditions favor the pathogen.

Plant less-susceptible cultivars such as Prairie Fire crabapple, and Capital and Chanticleer ornamental pear. Avoid Aristocrat pear, which is highly susceptible. Avoid excess irrigation and overfertilization because vigorously growing shoots are most severely infected, so cultural practices that favor rapid plant growth can increase the severity of damage.

Remove and dispose of all diseased tissue during summer to early winter when the bacteria are not active. On highly susceptible hosts, remove diseased wood as soon as it appears in spring. When conditions are foggy or wet (favoring pathogen spread), between making each cut, scrub tools clean with detergent and water and consider disinfecting them in a 10 to 20% bleach solution. Dry and oil tools after using bleach to minimize rust.

Making cuts well below diseased tissue, at least 6 to 8 inches into healthy-appearing tissue, is more important than sterilizing tools. To locate the correct cutting site on small branches

1. Find the lower edge of the visible infection in the branch.
2. Trace that infected branch back to its point of attachment.
3. Cut at the next branch juncture down without harming the branch collar.

This will remove the infected branch and the branch to which it is attached.

Fire blight caused this pear blossom and terminal to suddenly wilt, blacken, and die. Many susceptible ornamental species have certain cultivars that are resistant to fire blight. *Photo:* J. K. Clark

Bark removed to reveal how fire blight spreads through limbs in a narrow strip. The main infection canker is blackish to dark brown, encircling the branch and killing the terminal. At the leading edge of this spreading infection (photo center), the pale wood surface is flecked reddish. Cutting off bark to expose healthy (uniformly lighter) wood at the leading edge and all margins of infection can stop fire blight spread in large limbs and the main trunk. With small branches such as this, entirely prune off terminals, cutting at least 8 to 12 inches into healthy tissue. *Photo:* J. K. Clark

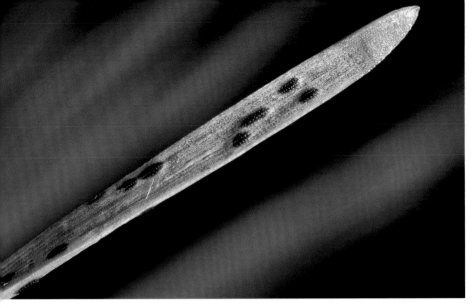

Elytroderma needle cast disease causes reddish brown discoloration and dark, oblong fungal fruiting bodies on pine foliage. Most needle blight and cast fungi infect only current-season foliage and do not seriously harm conifers. However, *Elytroderma* infection becomes systemic, persists in branch tissue, and can cause serious branch dieback. *Photo:* J. K. Clark

If fire blight occurs on the trunk or major limbs, sometimes the infection can be stopped and the wood saved by scraping off the long, narrow strips of bark where infection initially spreads downward in the inner bark. This bark removal is best done in winter when trees are dormant and bacteria are not active:

1. Begin at the main (limb-girdling) canker. Look for long, narrow infection paths of discolored bark, typically up to 1½ inches wide and extending up to 2 to 3 feet beyond the main canker margin.

2. Remove both outer and inner bark (cambial tissue) down to the wood surface in these strips of discolored tissue and also remove 6 to 8 inches more below the infection and on each side (laterally) just past the point of infection.

3. Make sure to remove bark over the leading edge and all margins of infection to expose healthy wood. Under bark, the infection appears as dark brown to blackish wood (the main canker), lighter brown to reddish wood (the infection spread nearest the main canker), red flecking in wood (the leading edge of infection spread), then uniformly light-colored wood (healthy tissue).

For more detailed information including resistant cultivars and fungicide recommendations, consult *Fire Blight Pest Notes*.

NEEDLE BLIGHT AND CAST

Fungi that cause needle blight (dieback) and cast (premature drop) are weak pathogens that mostly infect stressed conifers and only when foliage is immature and wet. Even when environmental, plant, and fungal characteristics coincide to favor needle disease, usually only the foliage produced during a single season is damaged or drops prematurely, so trees rarely suffer severe stress or are killed.

Most needle fungi infect only one or several related conifers and can be identified by their host and characteristic spore-forming structures on dying and dead needles. For example, Douglas-fir needle cast (*Rhabdocline* spp.) affects only Douglas-fir, and mostly those in the North Coast. Elytroderma disease caused by *Elytroderma deformans* affects only pines, commonly those grown alongside lakes and streams. Unlike other needle cast fungi that develop in foliage, *Elytroderma* infection becomes systemic and persists in branch tissue, and it can seriously damage pines.

All conifers are susceptible to brown felt blight, or "snow mold." Needles buried under snow die in dark brown to gray masses matted together with fungal mycelium. Brown felt blight of pines is caused by *Neopeckia* (=*Herpotrichia*) *coulteri*. In other conifers, *Herpotrichia juniperi* (=*H. nigra*) causes brown felt blight.

To minimize needle blights, choose a planting site with good air circulation and avoid planting conifers too close together. Provide appropriate cultural care to encourage vigorous plant growth. When irrigating, use drip irrigation instead of overhead sprinkling and avoid irrigating in the late afternoon or evening. Generally, no other management is practical or necessary.

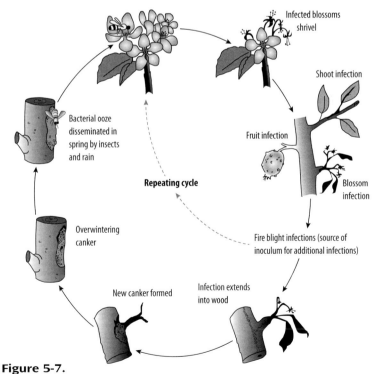

Figure 5-7.

Fire blight disease cycle. The pathogen persists in infected wood and infects new growth mostly through flowers during the spring when the bacteria are spread by insects and splashing water. By V. Winemiller in Teviotdale 2011.

Oak twig blight fungus killed scattered twigs and leaves on this coast live oak. Other pathogens and certain insects cause similar damage, so correctly determine the cause before attempting any control. *Photo:* J. K. Clark

This curled, flattened green stem terminal and bushy foliage of Chinese pistache is called fasciation. *Photo:* P. Svihra

OAK TWIG BLIGHT

Cryptocline cinerescens sporadically causes twig blight in coast live oak, interior live oak, and valley oak. Other anthracnose and leaf spot fungi, including *Apiognomonia errabunda* (=*Discula umbrinella*) and *Septoria quercicola*, can cause similar damage. The fungi kill leaves and twigs from the current-season's growth. The shoots turn white or tan and die and remain on the tree, typically in scattered patches throughout the canopy. These twig and foliar diseases tend to be more severe in years when frequent rains coincide with when new leaves are being produced. *Cryptocline cinerescens* also causes more damage on trees infested with oak pit scale.

Provide infected trees with adequate cultural care, especially appropriate watering. Unless they were raised with irrigation (e.g., planted oaks), avoid irrigating native oaks during the dry season; irrigate during the winter, if needed, when rainfall has been below normal. Avoid applying fertilizer, which can favor excessive shoot growth, leading to a denser, slower-drying canopy more susceptible to these diseases. Pruning may help to control the disease, but this may be feasible only for a few small or specimen trees with limited infections. Prune infected twigs during dry weather in the summer or early fall; make cuts properly (see Figure 3-8) in healthy tissue below infected twigs. Some systemic fungicides provide control if applied within 1 week after pruning. Fungicides alone are not as effective. Dead twigs and leaves remain on the tree for 2 or 3 years, and the tree's appearance does not improve until these drop or are pruned out.

FASCIATION

Fasciation is an abnormal flattening of stems, often appearing as if several adjoining stems have fused. Small stems or leaves growing from distorted stems are abnormally abundant and undersized. Some fasciations are noninfectious, possibly caused by a genetic disorder. The bacterium *Rhodococcus fascians* is a common cause of infectious fasciation, but the cause of many fasciations has not been identified. Fasciation apparently rarely, if ever, threatens the survival of established woody plants.

Fasciation bacteria survive on infected plants and debris, spreading in water to infect through wounds. To help prevent their damage, use good sanitation, avoid injuring plants (especially when plants are wet), and keep the basal trunk as dry as feasible. To control fasciation due to all likely causes (bacterial and genetic), prune off and dispose of distorted tissue and do not propagate or graft symptomatic plants.

LEAF SPOTS

Leaves develop discolored blotches, spots, or holes and can die and drop prematurely due to various fungi and occasionally bacteria. Discolored, dying leaves can also be due to adverse growing conditions, certain insects, inappropriate cultural practices, and virtually anything that makes limbs or roots unhealthy (Table 5-1).

Leaf spot infection and disease mostly occur when new leaf and shoot growth coincide with wet conditions (fog, rain, or overhead irrigation). These diseases include anthracnose, black spot, diamond scale, Entomosporium leaf spot, scab, needle blight and cast, Septoria leaf spot, and shot hole.

BLACK SPOT

Diplocarpon rosae produces black spots on the upper surface of rose leaves and stems. Spots sometimes have feathery margins, and tiny, black, fungal fruiting bodies may be visible in the spots. Infected foliage often yellows and drops prematurely.

Black spot is usually a problem only in foggy or humid coastal areas, but it can occur anywhere leaves commonly remain wet, such as where roses are sprinkler irrigated during the evening or night. For spore production and infection, leaves must remain wet for more than about 7 hours.

Black spot disease on rose leaflets. Infected foliage often turns yellow, except for the black spots. The fungus infects rose foliage when it remains wet for more than about 7 hours. Roses can be entirely defoliated when disease-favoring, moist conditions persist. *Photo:* J. K. Clark

Diamond scale fungus spots the leaves and petioles (rachises) of California fan palm and certain *Washingtonia* hybrids, causing foliage to turn yellow and brown then die prematurely. The pathogen's distinctive, diamond-shaped fruiting bodies are pictured near the beginning of this chapter. *Photo:* D. Hodel

To reduce black spot, irrigate and hose off aphids in the morning instead of the evening or night. Do not plant roses too close together and prune canopies to increase air circulation. Prune off infected stems during the dormant season and dispose of fallen rose leaves and stems away from rose plants.

Avoid planting roses where they will not receive at least 6 to 8 hours of full sun. Choose disease-resistant varieties when planting. Miniature roses are more susceptible than other types, but a few varieties are reliably resistant to all strains of black spot. Where weather favors severe disease, available preventive fungicides include neem oil, potassium bicarbonate, sulfur, and certain synthetic fungicides, as discussed in *Healthy Roses* and *Roses: Diseases and Abiotic Disorders Pest Notes*.

DIAMOND SCALE

Phaeochoropsis neowashingtoniae (=*Sphaerodothis neowashingtoniae*) causes dark, water-soaked leaf spots only in California fan palm (*Washingtonia filifera*) and *W. filifera* hybrids with Mexican fan palm (*W. robusta*). The amount of disease in hybrids is proportionate to their amount of *W. filifera*. Pure Mexican fan palm and other species are not affected.

This fungal disease is named for its shiny, black, somewhat diamond-shaped fruiting bodies that resemble scale insects. These greasy-looking spore-forming bodies are ³⁄₁₀ by ¹⁄₁₀ inch or smaller and form on the upper and lower surfaces of leaf stalks,

blades, and petioles. Heavily infected leaves produce abundant black, dustlike spores that rub off easily, making plants a nuisance. Infected fronds can turn yellow and brown overall, then die prematurely, leaving a sparse, unattractive crown.

Diamond scale by itself typically is not lethal, but it does stress fan palms and increases their susceptibility to other diseases that can lead to plant death. Diamond scale occurs in coastal areas and where interior valleys receive cool, humid ocean air; the disease is rare in the Central Valley and southern deserts of California. Disease is less severe during dry, warm summer and fall weather, when California fan palms grow leaves faster. During the cool, moist winter and spring, palms grow more slowly, the pathogen grows well, and disease advances higher up the crown, causing infected palms to become more unsightly. Older, lower leaves and older fan palms are more severely affected. Young, vigorously growing palms maintain a fuller crown of leaves when affected.

Where marine air favors the disease, grow diamond scale-resistant palms. Good substitute palms include Australian fountain (*Livistona australis*), Chinese fountain (*L. chinensis*), Chinese windmill (*Trachycarpus fortunei*), Guadalupe (*Brahea edulis*), Mexican blue (*B. armata*), pure Mexican fan, and San Jose hesper (*B. brandegeei*) palms.

Keep California fan palms vigorous by irrigating regularly in the summer and possibly during winters with little rainfall.

Avoid soggy soil and provide good drainage. Apply a palm fertilizer (e.g., equal amounts of nitrogen and potassium with half as much magnesium) according to the product directions. Keep ground covers, lawns, shrubs, and weeds at least 2 feet away from the trunk and maintain mulch several inches deep over this area. See *Palm Diseases in the Landscape Pest Notes* and *The Biology and Management of Landscape Palms* for more information.

ENTOMOSPORIUM LEAF SPOT

Entomosporium mespili (named *Diplocarpon mespili* during its sexual stage) causes spotting and premature drop of leaves in most plants in the Pomoideae group of the rose family. Hosts include apple, crabapple, evergreen pear, hawthorn, loquat, photinia, pyracantha, quince, *Rhaphiolepis*, serviceberry, and toyon.

Tiny, reddish spots, sometimes surrounded by a dark red, purple, or yellow halo, appear on infected leaves, and the spots darken and enlarge as the leaves mature. Later, a spore-forming body develops in the center of the spots; these cream-colored specks may appear to be covered with a glossy membrane, beneath which white masses of spores are visible under magnification.

On deciduous plants, the fungus overwinters mainly as spores or mycelia on fallen leaves. On evergreen hosts, the fungus can also remain on leaves on the plant year-round. Raindrops and overhead irrigation splash spores from infected plant

Entomosporium causes reddish spots, sometimes surrounded by a yellow halo; spots darken and enlarge as leaves mature, like these on rhaphiolepis. Switching from overhead sprinkling to drip irrigation greatly reduces this problem in many situations. *Photo:* J. K. Clark

These circular, rough spots on toyon leaves, called scab, are caused by a fungus (*Spilocaea photinicola*). Scab is favored by wet spring weather and little can be done to prevent it when wet weather favors its development. Like most leaf spot diseases, scab apparently does not threaten plant health. *Photo:* J. K. Clark

tissue or contaminated leaf litter to healthy leaves. Fungal infection and disease development can be severe when wet weather or overhead irrigation coincide with new plant growth.

Remove and dispose of spotted leaves on plants and the ground near plants. Use drip, flood, or low-volume sprinklers instead of overhead irrigation. Consider removing ground covers beneath infected shrubs and mulching or maintaining bare soil instead. Where the problem is severe, a copper fungicide or certain other materials can greatly reduce damage if thoroughly sprayed on plants before they are damaged.

SCAB

Fusicladium, Spilocaea, and (mostly) *Venturia* spp. fungi cause leaf and fruit spotting and drop. Hosts include apple, loquat, manzanita, olive, pear, photinia, pyracantha, *Rhamnus,* toyon, and willow. Scab generally is important only on fruit trees grown for harvest, especially apple and pear. The disease is worst when rainfall is high in the spring.

Scab first appears as pale or yellow spots on leaves. As disease progresses, dark, olive-colored spots form on leaves, fruit, and sometimes stems. On the undersurface of leaves, spots may have velvety fungal growth. Affected leaves may twist or pucker, turn yellow, and drop prematurely. Often only a few scattered leaves are affected, but severe disease can affect most of a plant's leaves and fruit.

When scab affects flower stems, flowers may drop. On fruit, spots begin as velvety or sooty gray-black, or greasy lesions that sometimes have a red halo. Lesions become sunken and tan and may have olive-colored spores around their margins. Severely infected fruit distort or crack and drop.

Scab fungi overwinter on fallen leaves and on evergreen hosts on leaves on the plant. Spores on infected tissue are discharged into the air or splashed by water, usually in the spring, and infect host fruit, leaves, and stems if they remain wet for more than about 9 hours. Mild temperatures when plants are continuously wet promotes scab disease; hot, dry weather slows or stops disease development.

When planting apple, crabapple, or pear, choose one of the many scab-resistant species or varieties. For example, Asian pears (*Pyrus pyrifolia*) are less susceptible to scab than European pears (*P. communis*). Remove and dispose of or thoroughly compost fallen leaves. Avoid sprinkling foliage, but if water does hit trees, irrigate early in the day so that foliage dries more quickly. Prune branches to thin canopies, which improves air circulation and reduces humidity.

Various copper, oil, sulfur, and synthetic fungicides may prevent the disease if properly applied from about bud break through a month after petal fall. Spraying is generally not warranted except where the disease is severe on apple or pear fruit. See *Apple and Pear Scab Pest Notes* for more information.

SEPTORIA LEAF SPOT

Several dozen *Septoria* spp. fungi each infect a different group of closely related hosts. *Septoria* spp. (many of which are named *Mycosphaerella* during their conidial stages) cause round or angular, flecked, sunken, or irregular spots on mostly older leaves. Aspen, azalea, cottonwood, hebe, and poplar are commonly infected. *Populus* spp. severely infected by *Septoria populicola* develop both leaf spots and branch cankers (Septoria canker).

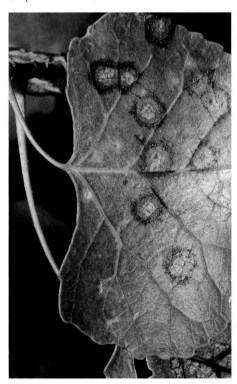

Septoria leaf spot on poplar. Leaf veins limit the spread of the fungus (*Mycosphaerella populorum*), so some of these dead patches have angular edges. *Photo:* J. K. Clark

The biology and management of *Septoria* is similar to that of anthracnose, discussed above. Disease can be common when prolonged, cool, rainy weather occurs during new leaf growth. Where feasible, prune off infected wood in the fall after leaves drop from deciduous hosts and rake up and dispose of fallen leaves away from hosts. Reduce splashing water and humidity within canopies if possible.

SHOT HOLE

Wilsonomyces carpophilus causes shot hole, also called coryneum blight, in *Prunus* spp. Hosts include Catalina and Japanese flowering cherries, ornamental plums, and stone fruits, especially apricot.

Shot hole first appears in the spring as purplish or reddish spots about $\frac{1}{10}$ inch in diameter on new buds, leaves, and shoots. Spots on young leaves commonly have a narrow, light green or yellow margin. The spots expand and their centers turn brown and can drop out, leaving holes. The fungus also causes premature leaf drop, rough and corky or scabby lesions on fruit, mostly on the upper surface, and small cankers on branches. Tiny, dark specks visible with a hand lens form in the brown centers, especially on buds; these dark fungal spores help to distinguish shot hole from other diseases.

The fungus overwinters in infected buds and twig lesions. The spores spread in splashing water, and disease is most severe following warm, foggy or rainy winters and when it rains in the spring during young leaf growth.

Use low-volume sprinklers, drip irrigation, or sprinkler deflectors and prune off lower branches to prevent foliage from getting wet. Prune and dispose of infected plant tissue as soon as it appears. After leaf drop, inspect plants carefully and prune varnished-appearing (infected) buds and twigs with lesions. Diligent sanitation and directing irrigation water away from foliage provide adequate control where the incidence of shot hole is low.

Where disease is severe on high-value

Discolored, scabby spots on fruit and holes in leaves are characteristic of shot hole disease, which affects *Prunus* spp., including flowering cherry, flowering plum, and stone fruit trees. The fungus persists in buds and twigs and disease is favored by wet conditions during winter and spring. *Photo:* J. K. Clark

plants, Bordeaux mixture, fixed copper, or certain synthetic fungicides can be applied after leaf drop. Retreatment in late winter before buds swell or between full bloom and petal fall or both times may be necessary on highly susceptible apricot or if prolonged wet weather occurs in the spring.

MILDEWS AND MOLDS

Distinctly visible mycelia or spore masses resembling a velvety coating on plants are called mildews or molds. These include plant-damaging downy mildews and powdery mildews and the relatively harmless sooty molds.

DOWNY MILDEW

Downy mildew pathogens, including *Peronospora* and *Plasmopara* spp., primarily damage foliage. Hosts include caneberries, rose, various herbaceous ornamentals, and many fruit, grain, and vegetable crops.

Downy mildew causes pale green to yellow areas on the upper surface of leaves, which may drop prematurely, and sometimes discolors buds and stems. Discoloration may turn purplish red to dark brown and necrotic; its shape often depends on leaf veins, which limit the pathogen's spread. Where the lesions occur on the upper surface, fluffy masses of brown, gray, or purplish spores form on the underside of leaves. Because under ideal, cool moist conditions spores persist for only a few days, often no spores are visible even though downy mildew is causing damage.

Downy mildew can be confused with powdery mildew, such as on roses. However, grayish downy mildew spores are almost always limited to the underside of leaves, and viewed under magnification its spores occur in branched stalks like tiny trees. Powdery mildew is common on

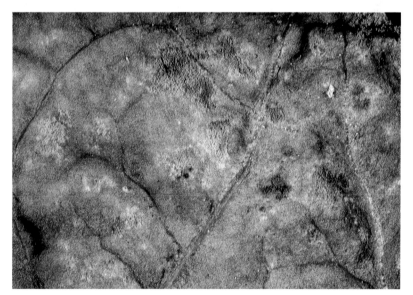

A magnified view of downy mildew sporulation on the underside of a leaf. Unlike powdery mildew, which commonly produces extensive areas of pale, powdery growth on both sides of leaves, the grayish patches caused by downy mildew are almost always limited to the underside of leaves and are less extensive. *Photo:* J. K. Clark

both sides of the leaf and produces spores on chains, and its whitish to gray growth is typically more extensive and prominent than with downy mildew. Powdery mildews thrive at moderate temperatures and low humidity, while downy mildews are favored by low temperatures (40° to 70°F) and require high relative humidity (≥90%) or wet foliage to infect plants and produce spores.

Unlike powdery mildew fungi, downy mildews are oomycetes (not fungi), which partly explains why their biology and management differ. Downy mildew spores are produced only on living plants and spread mostly with air movement. Spores landing on a host germinate and infect within 8 to 12 hours if the plant is wet.

To control downy mildew, provide good air circulation and maintain low humidity; adequately space plants and prune host canopies and nearby and overhead vegetation. Avoid wetting foliage; use drip or low-volume microsprinklers instead of overhead irrigation where feasible. Promptly remove and dispose of infected foliage to reduce pathogen inoculum. Control with pesticides (mostly preventives) is difficult, and modifying environmental conditions can be more effective.

POWDERY MILDEW

Many plants are susceptible to one or more powdery mildew fungi, including *Microsphaera*, *Podosphaera*, *Phyllactinia*, and *Uncinula* spp. Common hosts include apple, crape myrtle, euonymus, European grape, heavenly bamboo, rose, stone fruits, and sycamore.

DAMAGE

Powdery mildew fungi produce pale mycelia and spores on leaves and shoots and (less often) on flowers and fruits. Infected leaves may drop prematurely, and leaves and shoots may become discolored, distorted, and dwarfed, such as oak terminals that form witches' brooms. Weblike russeting may appear on infected fruit. The type and severity of symptoms depend on the species or cultivar of host plant, the age of tissue when infected, environmental conditions, and the specific pathogen involved. Some powdery mildews are unlikely to seriously harm certain plants, while others significantly impact their hosts.

IDENTIFICATION AND BIOLOGY

Some powdery mildew fungi prefer new plant growth, while others are more prevalent on old tissue. Their spores grow on the upper and lower leaf surfaces on strands or in chains, which can be seen with a hand lens. Similar-looking downy mildews grow mostly on the underside of leaves and produce spores on branched stalks that look like tiny trees.

Different powdery mildew fungi cause disease on different plants. These fungi tend to infect either one genus of plants or many species within the same family. For example, hosts of *Sphaerotheca pannosa* include photinia, rose, and certain stone fruits (all Rosaceae), whereas *Microsphaera euonymi-japonici* infects only *Euonymus* spp.

Powdery mildews spread as windblown spores, which do not need free water to germinate and die on host surfaces that are wet for an extended period. Powdery mildews grow best under moderate temperatures (60° to 80°F) and shady conditions and are inhibited by heat and sunlight, especially leaf temperatures above 95°F. Powdery mildew survives only on plant tissue, including as mycelia in dormant buds and sometimes as spherical fruiting bodies (chasmothecia, or cleistothecia) on bark and fallen leaves.

MANAGEMENT

Moderate levels of powdery mildew can be ignored on most plant species, but prevention and immediate control action should be taken to avoid damage to highly susceptible plants, such as certain cultivars of crape myrtle, euonymus, European grape, and rose. To manage powdery mildews

- Grow resistant cultivars and species.
- Reduce shading and increase air circulation within plants.

Powdery mildew is growing on these rose leaves and sepals, but it usually does not grow on the flower petals. *Photo:* J. K. Clark

Some euonymus cultivars are highly susceptible to powdery mildew, which develops in white patches on infected leaves and can cause premature leaf drop. *Photo:* J. K. Clark

This coast live oak shoot has become distorted, dwarfed, and discolored by the powdery mildew fungus *Cystotheca lanestris*, giving it a witches' broom appearance. Eriophyid mites, mistletoe, phytoplasmas, rust fungi, and viruses are among the other organisms that can cause witches' brooms on certain hosts. *Photo: J. K. Clark*

- Provide good cultural care.
- Apply fungicides to highly susceptible hosts when conditions favor disease development.

Choose resistant cultivars and consider replacing problem-prone species. The London plane cultivars Columbia, Liberty, and Yarwood are resistant in comparison with other sycamores. Powdery mildew–resistant crape myrtles include many cultivars with Native American names. Variegated cultivars of euonymus tend to be more resistant to powdery mildew than nonvariegated types, and many roses are reportedly resistant (Tables 3-2 and 5-7). Contact the UC Cooperative Extension office in your county or a certified nurseryperson for current recommendations.

To prevent powdery mildews, plant in a sunny location and adequately space plants. Prune to improve air circulation and increase light penetration. Avoid excessive fertilization and irrigation and prune during the dormant season to avoid promoting susceptible new growth. For example, witches' broom of oak caused by a powdery mildew fungus (*Cystotheca =Sphaerotheca lanestris*) often follows the stimulation of off-season growth resulting from summer irrigation or pruning. Overhead sprinkling plants may reduce powdery mildew infection because spores cannot germinate, and some are killed, when plants are wet. Wet the plants during morning to midday to allow plants to dry before nightfall and reduce the likelihood of favoring leaf spot and rust diseases.

Most fungicides prevent only new infections and may need to be applied repeatedly when conditions favor disease development and to protect subsequent new plant growth. Preventive fungicides include *Bacillus subtilis*, potassium bicarbonate, and sulfur. Horticultural oils (neem and petroleum derived) can eradicate existing infections and have some protectant activity. Certain synthetic fungicides are available to eradicate or prevent powdery mildew and other fungal diseases. Consult *Healthy Roses, Powdery Mildew on Ornamentals Pest Notes*, and *Roses: Diseases and Abiotic Disorders Pest Notes* for more information.

Table 5-7.

Powdery Mildew–Resistant Cultivars.

PLANT	RESISTANT CULTIVARS
crape myrtle	those with Native American–related names, e.g., Catawba, Cherokee, Hopi
euonymus	variegated cultivars more resistant than nonvariegated euonymus
London plane[1]	Columbia, Liberty, Yarwood
rhododendron	*Rhododendron yakushimanum, R. macrophyllum, R.* 'Nova Zembla,' *R.* 'Palestrina'
rose	Meidiland shrub roses, *Rosa rugosa*, many glossy-leafed hybrid teas and grandifloras, those trademarked Carefree, Drift, Knock Out, and Home Run

1. Columbia and Liberty are also resistant to anthracnose. Yarwood is susceptible to anthracnose unless pollarded, severely pruned to remove the previous year's growth.

Powdery mildew (*Microsphaera alni*) caused white discoloration and distortion of these sycamore leaves. Damage occurs on the younger leaves and shoot terminals; most older foliage (early-season growth) is unaffected. *Photo:* J. Karlik

Powdery mildews produce spores on long chains, which you can see by examining whitish growth on leaves with a hand lens. In contrast, downy mildew produces spores on branched stalks that look like tiny trees. *Photo:* J. K. Clark

Blackish sooty mold on oak leaves infested with aphids. Sooty mold is generally harmless, but if intolerable, manage it by controlling the insects that secrete honeydew on which sooty mold grows. See the "Aphids" and "Scales" sections in Chapter 6 for more pictures of this fungal growth. Photo: J. K. Clark

Sooty Mold

Sooty molds are dark, nonpathogenic fungi that grow on plants and other surfaces fouled with insect honeydew. Sooty molds can be distinguished by the fact that their dark mycelial growth can be completely washed off with a forceful stream of water, or moistened and wiped from plant surfaces, revealing a healthy-looking plant surface underneath. Sooty molds, such as *Capnodium, Fumago, Limacinula,* and *Scorias* spp., are generally harmless to plants and can be ignored, except when they are extremely abundant and prevent enough light from reaching leaf surfaces, causing plants to become stressed and prematurely drop fouled leaves.

Even if sooty mold is extensive, do not apply fungicides. Wash or scrape sooty mold from surfaces. If limited to small portions of a plant, prune off fouled and insect-infested parts. Control aphids, mealybugs, and other insects that secrete the honeydew on which sooty mold grows. Selectively control ants, which tend and spread phloem-sucking insects and cause their populations to increase by protecting these pests from parasites and predators. Insecticidal soap controls most exposed-feeding, plant-juice-sucking insects and helps wash away honeydew and sooty mold. For more information, consult *Sooty Mold Pest Notes* and "Ants" and the various sections on sucking insects in Chapter 6.

Symptoms of virus infection in rose can include yellow bands, blotches, lines, oak-leaf patterns, and rings in leaves and vein clearing (yellowing or whitening). Viruses may reduce flowering and plant growth rate, but often only leaves are visibly affected and then only on a portion of the plant. Photo: J. K. Clark

MOSAIC AND MOTTLE VIRUSES

Viruses are submicroscopic pathogens that infect cells, changing some cell functions. Viruses of abutilon, camellia, *Nandina* (heavenly bamboo), rose, and wisteria are discussed here, but many other woody landscape plants may show viral symptoms on occasion.

Viruses can slow plant growth, but most do not seriously harm woody landscape plants. Damage is usually noticeable only in flowers or foliage. Infected blossoms or leaves may become discolored, distorted, spotted, streaked, or stunted. For example, the flowers of infected plants sometimes develop green, leaflike structures, as discussed in "Rose Phyllody" in Chapter 4. The variegation or other appearance changes that viruses cause are sometimes considered to be attractive, such as *Abutilon mosaic virus.*

IDENTIFICATION AND BIOLOGY

Viral pathogens are named after the main or first-recognized host plant and the primary type of damage they cause. For example, *Apple mosaic virus* and *Elm mosaic virus* cause an irregular pattern of discoloration in leaves on apple and elm, respectively. *Hibiscus chlorotic ringspot* and *Prunus necrotic ringspot* cause small yellow or brownish spots or blotches on the leaves of hibiscus or plants in the rose family, respectively.

Viruses can be transmitted by insects feeding on plant sap or mechanically in sap that is spread by hand or grafting tools. Viruses also spread in pollen, seed, or vegetative parts of plants, such as through budding and grafting during propagation. Some viruses have a relatively narrow host range and apparently infect only one genus of plants. Other viruses have a broad host range, such as *Impatiens*

necrotic spot virus and *Tomato spotted wilt virus*, incompletely understood tospoviruses that infect several hundred genera of mostly herbaceous plants, including many flowering ornamentals.

MANAGEMENT

Once a plant becomes infected with virus, it usually remains infected throughout its life. There is no cure or treatment for virus-infected plants in landscapes, and generally none is needed for woody ornamentals.

Provide proper cultural care to improve plant vigor or replace infected plants if their growth is unsatisfactory. Purchase and plant only high-quality, certified virus-free or virus-resistant nursery stock or seeds. Do not graft virus-infected plant parts onto virus-free plants unless you want to introduce the virus. Although certain viruses are spread by aphids and other insects that suck plant juices, controlling insects is generally not a recommended or effective method of preventing virus infection in woody landscapes. It is very difficult to detect or control insects effectively at the low densities that can spread a virus, and to continually provide control throughout the life of perennial plants, especially of insects like the melon and green peach aphids that spread *Cucumber mosaic virus* in nandina and feed on many different plant species.

ABUTILON MOSAIC VIRUS

Abutilon mosaic virus causes vein-limited yellow blotches on leaves of Chinese lantern or Chinese bellflower as pictured in "Action Thresholds and Guidelines" near the beginning of Chapter 2. These leaf blotches are considered attractive, and infected plants are commonly sold as variegated plants. The virus is deliberately spread by propagating with infected stock to produce the bright yellow variegation on leaves. Other viruses also can infect abutilon.

Camellia yellow mottle virus causes irregular yellow areas on foliage and pale blotches on petals (color break). Although discoloring may be aesthetically objectionable, most viruses do not seriously harm woody landscape plants. *Photo:* J. K. Clark

Some virus infections may be considered attractive, like this reddish mottling on new leaves of heavenly bamboo. At least two viruses cause this red discoloration of heavenly bamboo in California. *Photo:* J. K. Clark

CAMELLIA YELLOW MOTTLE VIRUS

Camellia yellow mottle virus causes an irregular yellow mottling of camellia leaves and a mottled whitish pattern in the blossoms. The virus is sometimes deliberately introduced through grafting to produce an attractive leaf or flower variegation. *Camellia yellow mottle virus* is spread by budding, grafting, or rooting cuttings from plants that are infected.

NANDINA MOSAIC VIRUS

In California, at least two viruses occur in heavenly bamboo and cause a mottled red discoloration of the new leaves. *Nandina mosaic virus* spreads mechanically during propagation and pruning and infects only nandina. *Cucumber mosaic virus*, which infects a variety of agricultural and ornamental plants and annual and perennial weeds, can be transmitted mechanically, but in the field it is often spread by aphids, especially the melon and green peach aphids. Aphids acquire the virus while feeding on infected plants and transmit it when they move and feed on other plants.

Wisteria vein mosaic virus caused this yellow mottling and irregular shape in foliage. Although it gives wisteria an unhealthy appearance, the virus does not seriously harm its host. *Photo:* D. Mathews

Bacterial gall is infecting this oleander stem and killed a leaf. This problem is difficult to control and usually is not a serious threat to plant health. *Photo:* J. K. Clark

ROSE MOSAIC VIRUS

Rose mosaic virus, also called *Prunus necrotic ringspot virus,* may be a complex of more than one pathogen. Roses can also be infected by many other viruses. Virus infection causes yellow to brownish lines, bands, rings, vein clearing or yellowing, oak-leaf patterns, or blotches on leaves, sometimes on only a portion of the plant. Virus-infected plants may grow more slowly, produce delayed or fewer flowers, and become more susceptible to frost damage. The severity of damage varies with the host cultivar, and some infected roses exhibit no damage.

Rose mosaic virus and most other rose viruses are not spread by insects or pruning tools. They infect roses through budding, grafting, or rooting cuttings from infected plants. Roses infected during propagation can be symptomless until after they are planted and begin growing in landscapes. Consult *Healthy Roses* for more information.

WISTERIA VEIN MOSAIC VIRUS

Wisteria vein mosaic virus discolors foliage of *Wisteria* spp. shrubs and vines. Common symptoms are vein clearing (yellowing or whitening) and yellow flecking, mottling, or mosaic of irregular size and shape adjacent to or around leaf veins. The virus is mechanically transmitted and

can be introduced on new plants that were infected during their production. Aside from discoloring foliage, the virus does not seriously harm wisteria.

OLIVE KNOT AND OLEANDER GALL

Pseudomonas savastanoi pathovar (pv.) *nerii* causes distorted, swollen, knotlike growths on stems and bark of oleander and olive. On oleander it also galls flower buds and leaves. *Pseudomonas syringae* pv. *savastanoi* infects only olive, galling woody parts, especially the leaf nodes (bud development sites) on branches. Galled parts often die back, but overall plant health is usually not seriously threatened.

The gall-inducing bacteria persist in cracked, distorted, and rough bark. During wet weather in winter and early spring, the bacteria reproduce and are spread by splashing rain and contaminated hands and tools. Spring growth becomes infected through fresh wounds, such as pruning cuts and frost cracks. Additionally, olive can be infected through leaf scars on branches for several days after leaves have dropped.

To prevent olive knot or oleander gall, plant only disease-free nursery stock. If you purchase oleander and olive during late spring through fall you can see whether plants are galled; plants acquired during the rainy season may be infected but have not yet developed galls. Avoid overhead

watering, which spreads the bacteria. It may help to limit pathogen spread if you scrub pruning blades clean and disinfect them in a 10 to 20% bleach solution before each cut, especially if plants must be pruned during the rainy season. However, sterilizing tools is futile if bacteria have already splashed onto plant surfaces that are then cut or wounded during wet conditions.

Regular monitoring and properly pruning off galled parts help to control the disease, but prune only during the dry season, when the bacteria are inactive, and dispose of infected tissue away from hosts. See *Olive Knot Pest Notes* for more information.

RUSTS

Rusts infect many hosts, including birch, cottonwood, fuchsia, hawthorn, juniper, pear, pine, poplar, rhododendron, and rose. These fungi are named for the dry reddish, yellowish, or orange spore masses or pustules that many species form on infected tissue, commonly on the lower leaf surface of broadleaf plants or on the bark or needles of conifers. Infected leaves may turn yellow or brown and drop prematurely.

Rusts infect hosts when plant surfaces are wet and temperatures are mild, mostly during the winter to spring rainy season or almost any time of the year in coastal locations. Rusts are spread primarily by windblown and water-splashed spores and when infected plants (e.g., from a nursery) are moved. In addition to orangish pustules, many species form black overwintering spores on leaves in autumn, which reinfect plants in spring. Each type of rust is specific to certain hosts. Most species of rust fungi have several different-looking life stages and complex life cycles, alternating generations between two host species in different genera or families of plants. Others, such as rose rust (*Phragmidium mucronatum*) and western gall rust (discussed below) are restricted to one host genus and spread, respectively, from rose to rose or pine to pine.

MANAGEMENT OF RUST

Avoid overhead watering. Collect fallen infected leaves and needles and dispose of them away from host plants. Cut off and dispose of diseased leaves, shoots, and branches as soon as they appear, except do not prune woody parts so extensively that plants are seriously damaged. Plant rust-resistant cultivars if available, such as roses in coastal locations where rose rust is a chronic problem. Fungicides applied in the spring can prevent or reduce some rust diseases. The frequent applications required to provide good rust control may not be warranted in many landscape situations. For more information on rose rust, see *Healthy Roses* and *Roses: Diseases and Abiotic Disorders Pest Notes.*

CHRYSANTHEMUM WHITE RUST

Chrysanthemums are susceptible to brown rust (*Puccinia chrysanthemi*) and white rust (*P. horiana*). Chrysanthemum white rust has been the target of eradication and quarantine programs, which may require the destruction of infected *Chrysanthemum* spp.

Chrysanthemum white rust is named for the pale buff or white (sometimes pinkish or brownish) spore-producing pustules it causes on leaf undersides. Both brown and white rusts distort, discolor, and cause premature drop of leaves in chrysanthemums but do not kill hosts. Infection causes pale green, yellow, or white spots on the upper surfaces of leaves, but both of these rusts and other causes can produce similar leaf spotting. Chrysanthemum white rust is identified based on the presence of its pustules, which produce the spores that spread in wind to infect nearby hosts.

Report any suspected white rust infections to the county agricultural commissioner. Plant nonhost species, including pyrethrum (painted daisy, *Chrysanthemum coccineum*) and Shasta daisy (*C. maximum*).

Light-colored white rust pustules on the underside of a chrysanthemum leaf. Report any suspected infections of this quarantine pathogen to your county agricultural commissioner. *Photo:* S. T. Koike

CEDAR, CYPRESS, AND JUNIPER RUSTS

Gymnosporangium spp. cause galls, stunted and bushy branches (e.g., witches' brooms), stem dieback, and orange gelatinous spore masses on conifers. They infect junipers in some urban areas and cedar, cypress, and juniper in the Sierra Nevada. On their rosaceous species alternate hosts (apple, hawthorn, and pear), Gymnosporangium rusts cause nonlethal swellings and colorful spots on fruit, leaves, and twigs.

Prune and dispose of infected twigs in the fall, before spores are produced in the spring. Remove galls by making cuts in healthy wood below swollen, infected tissue. Avoid overhead watering.

Rust, visible as dry, orangish spore masses on the leaf underside, also caused discoloring of the upper side of these rose leaves. *Photo:* J. K. Clark

PINE NEEDLE RUSTS

Coleosporium spp. cause brown, reddish, or yellow spots or bands on pine needles. Needles often turn brown or yellow and develop tiny, pale or dark spore-forming bodies, usually in winter or spring.

Although unsightly, Coleosporium rusts usually do not seriously harm hosts. Most affect only the youngest needles and sporadically so during years when abundant rainfall coincides with new growth.

WESTERN GALL RUST

Endocronartium (=*Peridermium*) *harknessii* causes spherical swellings or galls on branches in two- and three-needle pines, including Aleppo, Bishop, lodgepole, and Monterey pines. In spring, galls develop orange spores, which are spread from pine to pine by wind and infect healthy shoots if shoots are wet. Galled stems may exude sap and sometimes become colonized by other fungi or insects. Beyond the galls, foliage may become stunted and bushy, and the terminal may die back or break off.

Prune and dispose of infected branches during October to January. Large galls on major limbs or the trunk can lead to structural failure; consider removing and replacing trees that may be hazardous.

Western gall rust swellings on pine. In spring, the galls often become covered with orange spores of the rust fungus. *Photo:* J. K. Clark

WHITE PINE BLISTER RUST

Cronartium ribicola infects all white (five-needle) pines and usually kills infected trees. Initial symptoms include tiny, yellow spots and yellow mottling on needles. Twigs turn yellow, then brown and die and remain hanging on the tree. Infections spread to branches, which swell into spindle-shaped galls. Infected bark develops raised blisters and dusty spore masses and becomes yellowish or brown and rough. Diseased tissue increasingly encircles the infected limb or trunk and kills it.

To avoid white pine blister rust, plant resistant or nonhost pine species. Cutting off infected limbs may increase tree longevity. Promptly remove seriously diseased or hazardous trees. Do not move *Cronartium*-infected pines or wood off-site because some parts of California are still free of this introduced pathogen. Removing alternate hosts of the fungus (*Ribes* spp.) within about 500 feet of high-value white pines may help to reduce infections if trees are relatively isolated from other white pines. Fungicides have not been found to be effective in managing this disease.

Pines are susceptible to other less-serious blister and canker rusts. These include stalactiform rust (*Peridermium stalactiforme*), which causes cankers mostly on the lower branches and lower trunk of the yellow (two- or three-needle) pines, including Jeffrey, lodgepole, and ponderosa pines. As with most rusts, stalactiform rust must move between nearby hosts of different species to complete its development. Its alternate hosts are figworts (Scrophulariaceae), especially Indian paintbrush (*Castilleja* spp.).

YELLOWS, OR PHYTOPLASMAS

Phytoplasmas, formerly called mycoplasmas, are minute, single-cell organisms that resemble bacteria. They are often spread during feeding by leafhoppers. Phytoplasmas commonly cause distorted, dwarfed, and yellowish leaves and shoots. Other common symptoms include shortened internodes, abnormal flower and leaf development, and shoot proliferation

White pine blister rust caused this twig to develop rough bark and swell into a spindle-shaped gall. The yellow and brown needles are another indication that the *Cronartium ribicola* fungus is infecting these pines, although unhealthy foliage can also be due to other causes. When the problem is blister rust, pines almost always are killed. *Photo:* USDA Forest Service, Bugwood.org

Phytoplasma diseases are often called yellows because chlorotic foliage is a common damage symptom. Peach yellow leafroll, caused by a phytoplasmalike organism, infects the leaf on the right, which is chlorotic, cupped, and has an enlarged midvein. *Photo: J. K. Clark*

(witches' broom). The flowers of infected plants sometimes develop green, leaflike structures, as discussed in "Rose Phyllody" in Chapter 4.

The symptoms of phytoplasmas can be confused with those caused by viruses. Some phytoplasmas cause severe decline and eventually death of the plant. Many phytoplasmas are poorly known, and their importance as pests may be underestimated.

Peach yellow leafroll and X-disease of cherry and peach are phytoplasma diseases that damage foliage, give fruit a bitter flavor, and cause trees to decline, as discussed in *Integrated Pest Management for Stone Fruits*. Yellows of ash, elm, and lilacs and lethal yellowing of palm are important phytoplasma diseases that so far have been reported only in the eastern United States. Aster yellows complex is the most important phytoplasma infecting herbaceous ornamentals in California, but not woody plants.

No chemicals are effective against phytoplasmas. Control them primarily through proper sanitation, excluding and controlling insect vectors, and using only pathogen-free stock. Remove infected plants that are a source of pathogens, including certain weeds. For example, burclover (*Medicago polymorpha*), clovers (*Melilotus* and *Trifolium* spp.), and dandelion (*Taraxacum officinale*) host X-disease that spreads when certain leafhoppers feed on phytoplasma-infected plants and move to feed on cherry. Cultivars often vary significantly in their susceptibility to phytoplasmas. Seek information on resistant cultivars and plant them where phytoplasmas are a problem.

Limb and Trunk Diseases

Branches and trunks are damaged by many causes, including mechanical injuries, pathogens, weather, and wood-boring insects. Some diseases that affect roots or leaves also damage branches and trunks, as discussed elsewhere in this chapter. Canker diseases, crown gall, pink rot and sudden crown drop of palms, wetwood, and wood decay are discussed below.

CANKER DISEASES

A canker is a localized dead (necrotic) area on branches, trunks, or roots. Cankers vary greatly in appearance but are often a circular or oblong lesion that may be discolored, oozing, or sunken. Cutting under cankered bark usually reveals discolored tissue, which may have a well-defined margin separating it from healthy tissue. Most canker pathogens are fungi (Table 5-8).

Table 5-8.

Some Causes of Cankers.

CAUSE	COMMENTS
Annulohypoxylon or Hypoxylon canker	Many hosts, but in California reported primarily on oaks and tanoak.
anthracnose	Associated with leaf spots and distorted terminals.
bacterial blight	Bark oozes during wet weather; elongated lesions may appear on twigs.
Botryosphaeria canker	Limbs and branches die back on many hosts, wounds ooze on some hosts.
canker rots	Mostly oaks and other broadleaf trees. Surface cankers are associated with wood decay.
canker stain	Infects only sycamore and London plane.
Chinese elm anthracnose	Cankers only Chinese elm (*Ulmus parvifolia*), mostly in coastal areas.
cypress canker	Affects cypress and sometimes arborvitae, *Chamaecyparis*, and juniper.
Cytospora canker	Many hosts, often causes sunken, elliptical lesions.
ficus canker	Kills frequently pruned, drought-stressed Indian laurel fig.
fire blight	Preceded by twig and leaf damage, affects some plants in Rosaceae family.
injuries	Many causes, e.g., equipment or vehicle impact, pruning.
Nectria canker	Many hosts; wilted foliage appearing first in the spring is a common symptom.
oak branch canker and dieback	*Diplodia* dieback affects many oaks, especially coast live oak in southern California.
Phytophthora root and crown rot	Wilting, foliage discoloration, and premature defoliation are common symptoms.
pine rusts	Causes blistered, oozing, or swollen pine bark; may discolor or spot needles.
pitch canker	Pine branches also turn reddish, die back.
pruning wounds	Caused by pruning large limbs or by improperly making cuts.
Raywood ash canker	Botryosphaeria canker when host is drought stressed.
Septoria canker	Affects poplars; leaf spots are present.
sooty canker	Weak, secondary pathogen of many stressed hosts.
sudden oak death	Kills certain true oaks (in red and intermediate oak subfamilies) and tanoak.
sunburn or sunscald	Defoliated, severely pruned, and young trees commonly are affected.
underwatering	Plants are most susceptible to sunburn from heat and light if they lack sufficient water.
walnut thousand cankers	Bark-beetle vectored fungus kills black walnuts.

Bacterial blight and canker, mechanical injuries (e.g., equipment impact and pruning), sunburn, and sunscald also commonly cause cankers.

When cankers entirely circle (girdle) stems or trunks, foliage turns yellow or brown and wilts as the plant dies outward or upward from the canker. Many plant species are susceptible to Botryosphaeria, Cytospora, and Nectria cankers. More host-specific canker diseases include those infecting Chinese elm, cypress, pine, sycamore, and walnut.

To manage canker diseases, do not plant species that are poorly adapted to local conditions. Botryosphaeria canker of giant sequoia, Chinese elm anthracnose canker, and cypress canker of Leyland and Monterey cypress are virtually unavoidable when their hosts are poorly located; planting other species or resistant cultivars is the only practical management strategy. Many canker diseases primarily damage plants that lack proper cultural care. Keep plants vigorous to avoid and limit these diseases. Prune dead and dying branches when they are first observed, making the cuts in healthy wood below any apparent cankers, and use good sanitation to avoid spreading canker fungi on contaminated tools.

ANNULOHYPOXYLON CANKER, OR HYPOXYLON CANKER

Annulohypoxylon (=*Hypoxylon*) spp. fungi are secondary pathogens that can infect hosts without causing obvious symptoms of disease or infection. When hosts become damaged from other pests or injuries, *Annulohypoxylon* rapidly develops and causes white rot in sapwood. For example, after tanoak and true oaks become diseased from *Phytophthora ramorum,* the cause of sudden oak death, *A. thouarsianum* commonly causes sapwood decay that hastens tree death and increases failure potential.

Annulohypoxylon spp. produce hard, black to dark brown spore-forming tissue on the surface of infected bark and wood. With *A. thouarsianum,* these fruiting bodies are black, hemispherical, and often about 1 inch in diameter.

Avoid disease by planting species that are well adapted for conditions at that location. Protect plants from injury, stress, and other pests and provide proper cultural care, especially appropriate irrigation.

BOTRYOSPHAERIA CANKER AND DIEBACK

Several *Botryosphaeria* and *Fusicoccum* spp. fungi cause Botryosphaeria canker and dieback. Giant sequoia is a common host of *B. dothidea,* which also infects alder, coast redwood, incense-cedar, madrone, and many other woody species. Other *Botryosphaeria* (=*Diplodia*) spp. diseases include those discussed below in "Oak Branch Canker and Dieback" and "Raywood Ash Canker and Decline."

The fungi spread primarily in splashing water during rain, and infections may develop slowly for many months before symptoms become visible. Killed branches and sometimes large limbs are scattered throughout the canopy. Branches dying or recently killed from *B. dothidea* are reddish brown and often exude drops of yellowish pitch, while branches with older infections are grayish brown and (on conifers) mostly bare of foliage. On madrone, dead, brownish

Giant sequoia planted in hot areas typically develops Botryosphaeria canker and dieback no matter how good the cultural care provided. *Photo:* J. K. Clark

These dark, hemispherical fruiting bodies on bark indicate the sapwood underneath is decayed by *Annulohypoxylon* fungus. Annulohypoxylon canker develops in injured and severely stressed trees, such as oaks with sudden oak death. *Photo:* J. K. Clark

Dry leaves on a madrone twig killed by *Botryosphaeria dothidea*. At the base of the blackened twig (upper left), there is a well-defined margin between the sunken dead stem and the healthy stem, which is producing swollen callus tissue. *Photo: J. K. Clark*

leaves often remain attached to dead bark that is gray, reddish, or black, depending on how long ago it died.

Botryosphaeria typically infects hosts that are drought stressed or grown away from their native habitat, especially at warmer locations. No matter how good the cultural care provided, species such as giant sequoia are not adapted to heat and develop Botryosphaeria canker where hot weather prevails. Grow species that are well adapted to local conditions and provide them with proper irrigation and other care, as discussed in Chapter 3.

CHINESE ELM ANTHRACNOSE CANKER

Stegophora ulmea causes irregular, black, tarlike spots on leaves, premature leaf drop, and twig dieback in Chinese elm (*Ulmus parvifolia*), as pictured earlier under "Leaf Spots." More serious damage results when anthracnose cankers form in limbs and trunks. Even vigorous Chinese elms may be girdled and killed by this fungal disease in coastal areas of California. Chinese elm anthracnose cankers are rarely a problem in warmer interior areas of California.

Prune infected branches back to the next healthy lateral. Cut out small cankers on the trunk and major limbs before they become large, but do not make large wounds. Around the entire margin of small cankers cut about ½ inch into healthy wood, which should allow wounds to close.

Consider replacing severely infected trees. The Brea and Drake cultivars of Chi-

nese elm are less susceptible to anthracnose. Avoid Evergreen, Sempervirens, and True Green Chinese elms where anthracnose is a problem.

CYPRESS CANKER

Seiridium (=Coryneum) cardinale causes cypress canker, or Coryneum canker, primarily in *Cupressus* spp. Leyland cypress is especially susceptible, as is Monterey cypress when planted away from the coast. The fungus occasionally damages arborvitae, *Chamaecyparis*, and junipers. Resinous lesions form in infected bark and cambium. Infected branches or treetops turn yellow or brown and can become girdled and die. Often the entire plant is gradually killed. The fungus is moved by wind and within plants by splashing water.

Cypress cankers attract cypress bark moths, and their larvae feed and tunnel in cankered bark. These insects are secondary invaders, and their control is generally not warranted as it is the fungus that kills branches and trees, not this insect.

Provide trees with proper care and prune off diseased branches. Plant species

These Chinese elm anthracnose cankers may grow, eventually girdling limbs and causing dieback. Unlike with this tree near the coast, anthracnose cankers are uncommon on Chinese elm growing in drier, warmer interior areas of California. *Photo: J. K. Clark*

Table 5-9.

Cypress Canker Susceptibility of Cypress and Conifers with Similar Foliage When Planted Away from Direct Local Influence of Cool Coastal Climate.

COMMON NAME	SCIENTIFIC NAME
HIGHLY SUSCEPTIBLE	
Leyland cypress[1]	*Cupressocyparis leylandii*
Monterey cypress	*Cupressus macrocarpa*
LESS SUSCEPTIBLE	
arborvitae	*Platycladus orientalis, Thuja occidentalis*
Arizona cypress	*Cupressus arizonica*
Italian cypress	*Cupressus sempervirens*
Mexican cypress	*C. lusitanica* var. *benthamii*
Portuguese cypress	*C. lusitanica*
NOT SUSCEPTIBLE	
incense-cedar	*Calocedrus decurrens*
western red cedar	*Thuja plicata*

1. Reportedly is more susceptible to cypress canker than other hosts even when planted along the coast.

Yellow and brown branches on a Leyland cypress infected with cypress canker. Warm weather stresses this tree, increasing its susceptibility to disease. Even when planted near the relatively cool California coast, Leyland is more susceptible to canker disease than are other cypress. *Photo:* J. K. Clark

Cypress canker causes resinous lesions and discolored bark. Avoid planting susceptible cypress species in hot areas of California, as they will unavoidably become damaged by this pathogen. *Photo:* J. K. Clark

that are well adapted to local conditions and less susceptible to canker (Table 5-9). Instead of cypress, consider planting arborvitae (*Platycladus* and *Thuja* spp.) or (along the coast) incense-cedar (*Calocedrus decurrens*), which resemble cypress but are less susceptible to or not affected by cypress canker. Avoid planting Leyland cypress in California. Do not plant Italian cypress or especially Monterey cypress in inland areas away from the direct local influence of the cool coastal climate.

CYTOSPORA CANKER

Cytospora spp. infect many plants, including aspen, birch, ceanothus, cypress (Italian, Leyland, and Monterey), fir, fruit trees, maple, poplar, redbud, and willow. Cankers on major branches commonly appear as slightly sunken, smooth, roughly elliptical, reddish brown areas and sometimes exude copious resin. On aspen and poplar, *C. chrysosperma* causes sunken lesions that kill many small branches and twigs without forming any definite canker.

On any host, minute pimplelike or warty fungal fruiting bodies may develop in infected bark and produce yellow to red tendrils of spores during moist weather. Infected branches can turn brick-red in the spring, then fade to brown or tan by the fall.

Drought stress and other disorders or pest damage dramatically increase most hosts' susceptibility to Cytospora canker. Provide appropriate soil moisture for species adapted to summer rainfall or riverbank environments if they are planted where summer drought prevails; irrigation should generally be deep and infrequent.

Avoid planting susceptible cypress in warm areas. Grow species that are resistant or not susceptible. Poplar hybrids that show some resistance to *Cytospora* include Easter, Mighty Mo, Nor, and Platte.

Cytospora canker fungus caused this large sunken lesion on the main trunk of corkscrew willow (*Salix matsudana* 'Tortuosa'). *Photo:* J. K. Clark

FICUS CANKER

Indian laurel fig, or Chinese banyan (*Ficus microcarpa* =*F. retusa*), when stressed, is highly susceptible to ficus canker (ficus branch canker, bot canker, or sooty canker) caused by *Neofusicoccum mangiferae* (=*Nattrassia mangiferae*) and a complex of *Botryosphaeria* spp. fungi. The main symptoms are branch dieback, crown thinning, and, if the disease progresses to the trunk, eventual tree death. Disease generally progresses from leaf fading or yellowing, to premature leaf drop, canopy thinning, and then branch death. Apparently healthy branches are often interspersed with diseased or dead branches and new shoots often sprout on the limb or trunk below dead branches. The entire tree can die within 2 or 3 years after the initial symptoms.

The fungi infect Indian laurel fig through mechanical injury or pruning wounds and cause disease when trees are stressed. Advanced age (which reduces tree vigor) and unfavorable growing con-

The early symptoms of ficus canker on Indian laurel fig typically occur on only some of the branches. Symptoms include lack of normal spring growth flush (left and center arrows), where foliage is darker than the normal light green of the healthy new growth visible between the left arrows. Subtly discolored leaves and fewer leaves than normal or crown thinning are also shown (right arrow). *Photo:* D. Hodel

These Indian laurel fig are highly susceptible to ficus canker because they are of advanced age, frequently pruned, and growing alongside pavement and where summer drought prevails. Within 2 or 3 years after unhealthy foliage symptoms first appear, ficus canker can kill the entire tree. *Photo:* D. Hodel

ditions (e.g., the root zone compacted or paved over) make Indian laurel fig highly susceptible to disease. Dying trees also commonly received inadequate irrigation, excessive canopy pruning, or root pruning to repair pavement.

Provide proper cultural care to minimize tree stress. Avoid severe pruning of ficus and conduct needed pruning during dry weather. Prune off cankered or dying limbs at least 6 inches below any cankers. To avoid spreading the pathogens, scrub cutting blades clean and disinfect them between cuts and avoid using chain saws.

FOAMY BARK CANKER

On coast live oak, pink to white frothy exudate on bark is caused by *Geosmithia pallida*, a fungus spread by the western oak bark beetle. Peeling back outer bark reveals brown, discolored phloem tissue and small beetle entry holes, which may ooze reddish sap. This disease causes decline, branch dieback, and tree death. See the sections "Oak Ambrosia and Bark Beetles" and "Walnut Thousand Cankers Disease" for pictures of the beetle vector and damage by a similar fungus. Be alert for new information on this problem.

FOAMY CANKER, OR ALCOHOLIC FLUX

White, frothy material sometimes exudes from cracks or holes in bark, commonly on elm, liquidambar (sweet gum), oak, and Victorian box. Where exudate occurs, the

Alcoholic flux from bark occurs briefly during warm weather, apparently due to microorganisms that infect wounds. This foamy exudate has a pleasant fermentative odor, unlike the rancid smell of wetwood fluids. *Photo:* J. K. Clark

cambium and inner bark (phloem) may be discolored, mushy, or dead. The foamy material appears for a short time during summer and typically on stressed trees. The exudate has a pleasant alcoholic or fermentative odor, unlike the rancid smell of fluids on bark discussed later in "Wetwood, or Slime Flux."

The cause of this malady is unknown, but apparently it is due to various bacteria and yeasts that colonize wounds. Providing plants with proper cultural care and preventing injuries to bark may help to prevent foamy canker.

NECTRIA CANKER

Coral spot fungus (*Nectria cinnabarina*), European canker (*N. galligena*), and other *Nectria* spp. cause leaves and shoots to wilt, often beginning in spring. Infected branches and twigs may die back and cankers (discolored, sunken, often elliptical areas) commonly develop in woody parts.

Nectria infect many woody species through weakened or wounded tissue, such as leaf scars and pruning wounds and when plants are unhealthy or stressed, such as from recent planting. Plants infected when young may die, but *Nectria* spp. rarely, if ever, kill older plants. Callous tissue developing around wounds often limits their spread and prevents girdling of the trunk or limbs. However, cankered limbs are more susceptible to breakage and may be hazardous.

Nectria spp. infection is often overlooked until clusters of the fungal fruiting bodies erupt from bark, usually during spring and summer. Each of these spherical orange or red sporodochia and perithecia (the coral spots) are about $\frac{1}{50}$ to $\frac{1}{16}$ inch in diameter and later turn pale or become covered with white spores. Where these fungal growths occur, cutting away bark reveals a margin separating dead (dark brown, necrotic) and healthy (cream-colored) wood. The green, living cambial layer found just beneath healthy bark is absent where cankers occur.

Proper cultural care and good growing conditions are the most important man-

agement methods. Plant only species well adapted to local conditions and especially provide appropriate irrigation. Avoid wounding or pruning plants, especially if they are growing poorly. Unless pruning must be seasonally timed to avoid other problems listed in Tables 5-2 and 6-13, prune if needed during early summer to reduce the likelihood of *Nectria* infections.

No fungicides cure *Nectria* infections. If plants are heavily infected or susceptible species are newly planted, the frequency of new infections may be reduced by thoroughly spraying a fixed copper or a freshly prepared Bordeaux mixture during early leaf drop, just before the rainy weather. If leaf fall is prolonged by warm fall weather, a second application may be warranted when three-fourths of leaves have dropped.

OAK BRANCH CANKER AND DIEBACK

At least two *Diplodia* spp. fungi cause branch cankers and dieback on oaks in California. *Diplodia quercina* kills small branches (usually those <¾ inch, but up to 4 inches in diameter), causing leaves to turn brown, wilt, die, and remain attached in scattered patches throughout the canopy. Hosts include California black oak, coast live oak, English oak, and valley oak. *Diplodia corticola* (named *Botryosphaeria corticola* during its sexual stage) causes crown thinning (leaves that are pale and smaller and less abundant than normal) and bark bleeding, cracking, and staining on the main trunk of coast live oak in southern California. Trunk symptoms of

D. corticola infection (called bot canker) resemble damage pictured in "Goldspotted Oak Borer" in Chapter 6.

Disease caused by *Diplodia* spp. often can be distinguished by the masses of tiny, roundish, raised, black fruiting bodies (pycnidia) on bark of infected hosts. Peeling off bark around fruiting bodies and cankers reveals dark brown to black dead wood, commonly with a well-defined border between dark diseased and light-colored healthy wood.

Diplodia quercina outbreaks follow years of below-average rainfall. Even drought-adapted oaks may require supplemental irrigation to reduce stress if rainfall has been well below normal. However, irrigation of native oaks should generally be done during the normal rainy season to supplement inadequate natural rainfall. Exceptions include planted oaks adapted

to (raised with) summer irrigation from the start and in certain situations where soils have been disturbed. The specific amount and frequency of irrigation vary greatly, depending on factors such as environmental and soil conditions, as discussed in "Irrigation" in Chapter 3. Be aware that frequent irrigation during the dry season promotes Armillaria root disease and Phytophthora root and crown rot, as discussed later.

Protect trees from other pests and injury and provide proper cultural care. No other specific management is known for *D. corticola*. For *D. quercina*, the disease is not a major problem in most years, and control is usually not needed. For more information, consult *Twig Blight and Branch Dieback of Oaks in California*.

PITCH CANKER OF PINES

Introduced in California's Central and South Coast, *Fusarium circinatum* infects Monterey pine especially, but also Bishop pine, certain other native California pines, and most non-native pines. This fungus also infects, but does not damage, Douglas-fir.

Cutting off a dying limb exposed this oak branch canker, or bot canker, caused by *Diplodia corticola* infecting coast live oak in southern California. This fungus causes yellowing and premature drop of leaves and bleeding, cracking, and staining on limb and trunk bark, symptoms that resemble those of oaks infested with goldspotted oak borer. *Photo:* A. Eskalen

Pitch canker lesions in wood exude copious resin and girdle and kill branches, trunks, and exposed roots. Branch dieback in the upper canopy is often the first obvious symptom from a distance. Multiple branch infections cause extensive dieback. Terminal needles wilt, turn light green, yellow, reddish, then drop prematurely at any time of year. Infected trees are often attacked by engraver beetles and the fungus and beetles can eventually kill pines. Conversely, some severely damaged pines recover naturally and no longer exhibit obvious disease symptoms. Some Monterey pines apparently are resistant to disease even when nearby pines are heavily damaged.

IDENTIFICATION AND BIOLOGY

Pitch canker can usually be diagnosed in the field by its characteristic damage and progression of symptoms. Unlike the natural shed of older (inner branch) needles throughout the tree during late summer and fall, pitch canker discolors and kills all the needles on the ends of branches scattered throughout the tree any time of year.

If pitch canker is suspected, inspect bark for resinous exudate around the junction between discolored and green needles. Cut under pitchy bark longitudinally and look for resin-soaked, amber-colored wood characteristic of pitch canker. A native Diplodia canker and blight (*Sphaeropsis sapinea* =*Diplodia pinea*) also causes shoot dieback and resin-soaked wood resembling pitch canker. However, Diplodia canker usually causes little or no resin flow on bark surfaces, and pitch under bark is usually blackish to dark amber rather than the light amber color of pitch canker resin. Certain insects and other pathogens, often in combination, also cause flagging or other damage resembling that of pitch canker (Table 5-10). For definite identification, properly collect plant samples and submit them to a plant diagnostic laboratory.

Pitch canker fungus is spread by various bark beetles, including cone beetles (*Conophthorus* spp.), engraver beetles (*Ips* spp.), and twig beetles (*Pityophthorus* spp.). The fungus contaminates pruning tools

Where this pine limb was oozing copious resin, scraping off bark revealed discolored, necrotic (dead) cambium and wood caused by the pitch canker fungus. When cankers girdle stems or trunks (entirely circle affected parts), the plant dies outward or upward from the canker. *Photo:* L. D. Dwinell, USDA Forest Service, Bugwood.org

Extensive branch dieback on Monterey pine due to pitch canker. Sanitation and providing pines with good cultural care are the only practical strategies for managing this disease. *Photo:* J. K. Clark

and equipment used on infected trees and spreads on those tools to infect other pines.

MANAGEMENT

To prevent *Fusarium* spread, use or dispose of infected pines locally and do not move soil or pine seed, seedlings, or wood from infested to uninfested areas. Promptly remove and properly dispose of dead trees. Except for hazardous limbs, which should be removed whenever they appear, prune pines only during the late fall and winter when insects that spread the fungus are less active. Pruning off infected limbs temporarily improves tree appearance but does not cure infected trees. After using tools on trees with resinous cankers, or while working on the same tree after

making a cut that contacts resinous wood, carefully scrub the tools clean and soak them in a 10 to 20% bleach solution. Use only manual pruning saws or reciprocating saws with removable blades that can be thoroughly cleaned as disinfecting chain saws is not practical.

Avoid planting Monterey pine in hot areas, because even if not infected with pitch canker, Monterey pine typically declines and dies within about 20 to 30 years no matter how good the cultural care provided. Consider planting only canker-resistant and certified pathogen-free pines. In coastal areas of California, consider planting only non-pine species. For more information, consult *Pitch Canker Disease in California* and *Pitch Canker Pest Notes*.

Table 5-10.

Pine Tree Maladies with Similar Symptoms.

MALADY	OOZING OR STREAMING PITCH	LUMPY, PROTRUDING, OR TUBULAR MASSES	YELLOW TO RED WILTED TIP NEEDLES	YELLOW TO RED UNWILTED TIP NEEDLES	DEAD TIPS, NEEDLE DROP	CONES OR CONELETS ABORT	SWELLING ON BRANCHES	SILK WEBBING ON TIPS
								SYMPTOMS
blight, Aleppo pine	*		**	*	**			
caterpillars					**			**
cone beetles		**				**		
Diplodia canker and blight	*		**	*	**	*		
dwarf mistletoe	*			*	*		**	
injuries, pruning wounds	**	*						
Ips bark beetles		*	*	**	*	*		
pine scales			*	*	**			
pitch canker	**		**	**	**	**		
pitch moths	*	**	*		*		*	
red turpentine beetle		**						
rusts, western gall or white pine blister	*			**	*	*	**	
salt, wind, or drought dieback				**	**			
shade-suppressed branches			*	**	**			
tip moths				**	*			
twig beetles			*	**	**	*		
weevils				*	**			

Other abiotic disorders, such as poor growing conditions and inappropriate cultural practices, as discussed in Chapters 3 and 4, can also cause many of these symptoms. Consult the "Tree and Shrub Pest Tables" at the end of this book for a more complete list of pests affecting pines.
KEY
* Symptom occasionally occurs
** Symptom usually occurs
Adapted from Adams undated; Camilli et al. 2003; Dallara et al. 1995.

RAYWOOD ASH CANKER AND DECLINE

Fraxinus oxycarpa 'Raywood' commonly suffers from dieback of multiple branches throughout the canopy. Although trees usually are not killed, severely affected ash are often removed because of unsightly dieback, reduced shading, and their potential limb drop hazard. *Botryosphaeria stevensii* can usually be isolated from the dead branches and is believed to contribute to the decline. This fungus is a weak (secondary) pathogen, which is aggressive only when trees are stressed, such as by adverse growing conditions.

Raywood ash is apparently less drought tolerant than previously believed and stressful site conditions and especially moisture deficit predispose Raywood ash to *Botryosphaeria* damage. Occasional deep watering during the drought season and

Canopy decline and branch dieback commonly occur when Raywood ash becomes stressed, allowing a secondary (weak) pathogen (*Botryosphaeria stevensii*) to grow and kill branches. This Raywood ash in California's Central Valley receives full (all day) sun and is surrounded by pavement and turf, which is highly competitive for soil moisture. Shown here during late summer, this decline was preceded by severe water deficit. *Photo: J. K. Clark*

pruning to thin canopies and reduce transpiration demand may improve the performance of Raywood ash. Green ash (*Fraxinus pennsylvanica*) appears not to suffer from this problem and is a similar-looking alternative for planting.

SOOTY CANKER

Neofusicoccum (=Nattrassia) mangiferae causes bark cracking and cankers in many hosts, including Indian laurel fig, as discussed above in "Ficus Canker." Damaged bark may peel off or remain tightly attached to dead limbs. Cracked bark can bleed profusely and contain viscous or dried gum. Removing affected bark often reveals black, powdery growth underneath, so this disease is called sooty canker. This dark discoloration occurs under the cambium layer (inner bark) but extends only shallowly into the wood. On infected limbs, leaves suddenly wither, turn brown, dry up, and die, and dead leaves typically remain attached to the twigs.

The fungus is a relatively weak pathogen that usually infects trees through wounds caused by freezing, fertilizer burn, sunburn, or mechanical injury, such as from pruning. Severe damage is more likely on trees lacking good cultural care and a proper growing environment.

SUDDEN OAK DEATH (SOD) AND RAMORUM BLIGHT

Phytophthora ramorum has killed many tanoak (tanbark oak, *Notholithocarpus densiflorus*) and true oaks (*Quercus* spp.) in California forests and woodlands. The cause was unknown for several years and was originally thought to affect only oaks, but many species are susceptible to this funguslike oomycete. Unlike most *Phytophthora* spp., as discussed below in "Root and Crown Diseases," *P. ramorum* primarily affects aboveground plant parts.

Phytophthora ramorum occurs in Europe, Oregon, and at least a dozen counties in California in oak woodlands and forests containing California bay (*Umbellularia californica*) or tanoak and within about 50 miles of the coast. Outside of coastal areas, it can occur in nurseries where irrigation and shading create a cool, moist environment. The pathogen can be spread inadvertently in infected nursery stock or when infested material (e.g., mud, plants, soil, or water) is moved by animals or people. It also spreads aerially in infested stands of tanoak and California bay because these are the main spore-producing hosts.

DAMAGE

Phytophthora ramorum infects the bark of tanoak and certain species of true oaks, causing SOD and trunk cankers that eventually girdle and kill oaks. In most non-oak hosts, *P. ramorum* infects only leaves and in some species small twigs, causing Ramorum blight, which does not seriously damage or kill most non-oak hosts. Over 100 non-oak species are susceptible to Ramorum blight, especially California bay and certain ornamentals: camellia, *Kalmia, Pieris,* rhododendron, and *Viburnum.* Other foliar hosts include big leaf maple, California buckeye, California coffeeberry, California hazelnut, California honeysuckle, coast redwood, Douglas-fir, huckleberry, manzanita, poison oak, salmonberry, and toyon.

Oaks. California black oak, canyon live oak, coast live oak, and Shreve's oak in California forests, woodlands, and urban-wildland interfaces (suburban development in wildlands) have been killed by SOD. True oaks in the white group (a *Quercus* subgenus), including blue oak, Garry oak, and valley oak, do not appear to be susceptible to SOD.

Trunk cankers are most readily detected from the presence of dark brown to reddish ooze or seeping on the lower trunk, usually within 3 to 6 feet of the ground, but sometimes up to 12 feet or higher. Cutting under oozing bark to expose inner bark and outer wood reveals discolored brown tissue separated from healthy reddish or whitish tissue by a distinct black zone line. On some trees, especially canyon live oak and tanoak, little or no bleeding develops. Before oaks die, foliage often becomes pale, then reddish brown and dry. Foliage might appear healthy, then turn brown within 2 to 4 weeks, or the canopy may fade gradually over several months or years, or the tree may fail due to wood decay. Trees typically have been infected for 1 year or more before symptoms become obvious.

Dying oaks commonly develop sapwood decay and dark hemispherical

Under bark, these dark particles that extend shallowly into wood are pycnidia (fruiting bodies) of the sooty canker fungus. Sooty canker disease kills branches of citrus, ficus, mulberry, walnut (shown here), and various other trees. *Photo:* J. K. Clark

fruiting bodies from a secondary pathogen discussed above in "Annulohypoxylon Canker, or Hypoxylon Canker." They also become attacked by the borers discussed in "Oak Ambrosia and Bark Beetles" in Chapter 6, which produce tiny, round holes and red or white boring dust on bark. These secondary pests hasten the decline of oaks dying from SOD.

Tanoak. In tanoak of any age, *P. ramorum* infects branches, leaves and leaf petioles, and twigs, causing shoot dieback and wilting. Infected twigs produce spores that infect the trunk, but tanoak may or may not develop bark bleeding. Leaves may turn brown, and trees appear to die within a few weeks, or tanoak may decline slowly with gradual leaf loss.

Ramorum Blight Hosts. In species that develop only Ramorum blight, infected leaves develop brown or otherwise discolored spots, margins, or tips, sometimes bordered by yellow. Infected leaves may drop prematurely, and small twigs may develop cankers and die back. Ramorum blight usually does not seriously damage hosts, except for madrone saplings and in nurseries on various hosts when conditions favor repeated infections. However, hosts infected with Ramorum blight (especially California bay) can produce rain-splashed and windborne spores that spread to infect nearby susceptible oaks. Some ornamentals infected in the nursery have introduced the pathogen when planted in forests and landscapes.

IDENTIFICATION AND BIOLOGY

SOD occurs mostly in oak woodlands and forests and landscapes within one of the infested counties and near a source of spores (usually California bay or tanoak). For known locations of the disease, see the UC Berkeley Forest Pathology and Mycology Laboratory website, SODmap.org.

When symptoms occur on hosts growing far away from these sporulating hosts or outside of infested forests, the cause is unlikely to be SOD. Other *Phytophthora* spp. (see "Phytophthora Root and Crown Rot") can cause trunk cankers that can

be distinguished only through laboratory tests. Bleeding from the lower trunk can also be due to canker rot fungi and causes discussed in "Armillaria Root Disease," "Bacterial Wetwood," and "Wood Decay" in this chapter and in "Clearwing Moths" in Chapter 6. See the "Tree and Shrub Pest Tables" at the end of this book to help you distinguish among these causes of damage.

Many other maladies cause damage resembling that of Ramorum blight or SOD, and field symptoms are not reliable for diagnosis. Laboratory tests are required to confirm sudden oak death. Contact your local UC Cooperative Extension office or county agricultural commissioner to learn how to collect samples for testing and where to submit them.

The pathogen's biology varies among host species and growing conditions. In true oaks and tanoak, new infections develop when large numbers of spores are deposited on trunks during wet spring weather. The spores spread in splashing rain and longer distances in wind among infected California bay, tanoaks, and other hosts with twig and leaf infections. It does not spread among infected true oaks because trunk cankers do not produce spores. The pathogen can also be transported to new areas through the movement of infected plants (especially nursery stock), soil, and water runoff. During the wet season spores washing into the soil can be spread inadvertently by the activities of bikers, hikers, vehicles, and animals such as deer and horses.

MANAGEMENT

Prevention is the primary means for managing SOD. Provide proper cultural care and protect oaks from injuries. Avoid causing wounds to oak trunks, especially from March through June. Take precautions to avoid introducing infected plant material (nursery stock, firewood, uncomposted mulch) or contaminated soil or water. Stay out of infested sites during wet months, as contaminated mud can easily be spread by boot soles, equipment, and vehicles. Plant only pathogen-free plants, including acorns.

The most reliable early symptom of sudden oak death in Shreve, coast live, and California black oaks is dark sap exuding from the trunk base, as on this coast live oak. However, laboratory testing of a proper sample is the definitive way to diagnose infection by *Phytophthora ramorum*. Photo: P. Svihra

Cutting beneath bark on this California black oak reveals discolored brown inner bark (cambium) with a distinct black zone line separating cankers from healthy, reddish cambium. This discoloration is characteristic of infection by *Phytophthora ramorum*. Photo: P. Svihra

Phytophthora ramorum infection on California bay causes brown or dark, irregular areas in leaves, usually on tips or along edges and often bordered by yellow. On most hosts, *P. ramorum* causes only minor injury to leaves and twigs. California bay are rarely if ever severely harmed by this infection, but they are a major source of pathogen spores that can infect nearby oak trunks. *Photo:* S. A. Tjosvold

Leave diseased wood on-site for firewood or mulch or contact the local county agricultural commissioner to learn the approved methods for disposing of oaks and whether any regulations prohibit the movement of oaks (acorns, foliage, plants, or wood) and other SOD hosts. Even in areas where SOD is common, the distribution of the disease can be patchy, so prevent introduction and minimize movement of potentially infested materials at all locations.

Promptly cut down dead or decayed standing trees where they may be hazardous to people and property if they fail. Large-scale removal of non-oak host plants in landscapes is not a recommended way to prevent disease. Selective removal or pruning of foliar hosts might be helpful in preventing particular oaks from becoming infected, especially if there are few other SOD hosts nearby. Consider pruning off the lower branches of California bay, camellia, rhododendron, and tanoak that are growing within 8 to 10 feet of an oak's trunk; more protection can be provided by pruning foliar hosts to greater distances back from oaks. Remove poison oak vines climbing in oaks.

There is no known cure for SOD. Where oaks are at high risk of infection (in areas of known infestation), preventive pesticides (i.e., phosphonate compounds) can be applied to trunks and lower limbs. Pesticides do not cure infections, but when properly used they help protect trees from disease development. The treatment must be made to healthy trees and may need to be repeated every 1 to 2 years. Because the pathogen's distribution is patchy and somewhat unpredictable, it can be difficult to determine which trees warrant treatment. For more information, see *Nursery Guide for Diseases of* Phytophthora ramorum *on Ornamentals, Sudden Oak Death Pest Notes,* and the California Oak Mortality Task Force website, suddenoakdeath.org.

SYCAMORE CANKER STAIN

Sycamore and London plane trees are susceptible to infection by *Ceratocystis fimbriata* f. sp. *platani.* The resulting malady, also called canker stain, Ceratocystis canker, or sycamore canker, is lethal to *Platanus* spp. Sycamore canker stain occurs in California, at least in Modesto in the northern San Joaquin Valley.

This vascular wilt fungus causes a sparse canopy of small, chlorotic leaves and elongated cankers in large limbs and trunks. The surface of cankers usually has little obvious callus growth along the margins and commonly appears sunken, dark, and flattened, or covered with discolored or flaky bark. Black fruiting bodies may occur on the wound surface. Cutting into cankered cambium, phloem, and sapwood reveals dark discoloration, typically bluish black. In California, infected sycamores and London plane usually die within 1 or 2 years after symptoms first appear. When a dead branch or trunk is cut in cross-section, wood is often stained in a pie or wedge shape with the tip toward the center of the limb, as pictured later in "Verticillium Wilt." Stained wood is not soft or rotted, but secondary pathogens may invade and cause wood decay.

No insect vectors of this fungus are definitely known, and infections at new locations are almost always the result of pruning or other mechanical injuries to trees caused by people. *Ceratocystis* produces sticky spores that remain infective for a month or more and spread readily from one tree to another on tools or equipment, including pruning saws that contact infected trees. *Ceratocystis* spores are able to infect wood only through fresh wounds, but even tiny wounds allow pathogen entry and development. The fungus may spread through natural root grafts from infected sycamores to nearby hosts.

Good sanitation and proper cultural practices are the only management methods. No chemicals effectively control this disease. If symptomatic trees are found in areas where the pathogen has not been reported, send samples of freshly infected wood to a diagnostic laboratory.

Where the fungus is known or suspected to occur, after pruning any *Platanus* spp. or

Chipping away bark revealed dark bluish cambial tissue in a sycamore dying from canker stain fungus. This pathogen spreads on contaminated pruning tools and lawn mower blades that scalp shallow roots. *Photo:* E. J. Perry

using equipment that impacts trees, immediately scrub tools clean with detergent and water and soak them in a 10 to 20% bleach solution. Prevent injuries to bark from lawn mowers, string trimmers, and anything that wounds trunks or shallow roots. Within several feet of trunks, keep soil bare or apply mulch to avoid the common problem of lawn mower blades or string trimmers scraping the root crown or basal trunk. Where turfgrass or ground covers grow under tree canopies, irrigate properly to encourage deeper root growth. Use relatively infrequent, deep watering when needed instead of frequent shallow sprinkling.

Avoid unnecessary pruning of *Platanus* spp. Except to remove hazardous limbs or dead trees whenever they appear, prune only during dry weather in December and January, when the cold inhibits fungal development. Promptly remove infected trees and dispose of the wood away from other *Platanus* spp. Consider trenching at least 2 to 3 feet deep around stumps to eliminate any natural root grafts to nearby *Platanus* spp.

WALNUT THOUSAND CANKERS DISEASE

Walnut trees are being killed by *Geosmithia morbida,* a fungus spread by the walnut twig beetle (*Pityophthorus juglandis*). Each tiny beetle introduces *Geosmithia* spores when it chews into walnut limbs and trunks to feed and reproduce under bark. Immediately bordering each beetle tunnel, the fungus kills a small patch of cambium and phloem (the wood surface and inner bark), causing a small canker. As the beetles reproduce and their offspring reinfest trees, more cankers form and merge into large patches of dead tissue that girdle branches and the trunk and kill walnuts within a few years of the initial infection.

Thousand cankers disease primarily kills California native black walnuts, *Juglans californica* and *J. hindsii.* It also kills the non-native *J. nigra,* which is rare in California. Other *Juglans* spp. occasionally become diseased. Some English walnut

(*J. regia,* the commercial nut species) also have been affected.

From a distance, infected walnuts have yellowish, wilted foliage and branch dieback. Close-up, there are numerous pinhole-sized beetle entrance or emergence holes in bark, often surrounded by dark, wet discoloration, although this can be difficult to observe in dark, heavily furrowed bark. Cutting under cankered bark exposes dark discoloration in inner bark and on the surface of wood, but deeper into wood the tissue remains a healthy pale color.

Depending on the stage of infestation, the irregular tunnels under bark 1 to 2 inches long may contain walnut twig beetles. Adult beetles and mature larvae are about $1/16$ inch long. Larvae are C shaped, legless, and pale with a dark head. Adults are dark brown and cylindrical, and they emerge and fly from about April through September. Walnut twig beetle apparently has two to three generations per year in California.

No pesticides or other methods are known to save trees with thousand cankers disease at the time of this writing. Remove dead and dying walnuts. Burn, bury, chip, or otherwise properly dispose of wood locally to prevent beetles from emerging and to slow the pathogen spread. Do not move walnut firewood or logs from infested areas to locations where the thousand cankers disease is not known to occur. See

the *Field Identification Guide* for walnut twig beetle and thousand cankers disease online at ipm.ucanr.edu for more information.

OTHER LIMB AND TRUNK DISEASES

CROWN GALL

Root crowns infected by *Agrobacterium tumefaciens* develop galls at the soil line or just below the surface. Galls sometimes also develop on limbs, trunks, and roots. Many plants can be infected, especially euonymus, fruit trees, *Prunus* spp., rose, and willow.

Agrobacterium tumefaciens usually does not seriously harm woody plants unless galls occur in the root crown when plants are young. Infected young plants may become stunted and subject to wind breakage and drought stress. If galls are large, water or nutrient transport can be inhibited to the extent that young plants are killed. Crown gall appears to have a relatively minor effect on most older plants.

Crown galls on bark surfaces resemble damage from certain boring insects, woolly aphids, or other pests (Table 6-16). Certain boring insects colonize galled tissue, complicating diagnosis of the original cause. The surface of crown galls and wood underneath is the same color as healthy bark and wood.

Dark stains on bark of California black walnut are symptomatic of thousand cankers disease, caused by a fungus spread by the walnut twig beetle. Several tiny beetle boring entrance and exit holes on bark are visible here. Cutting shallowly under stained bark revealed the discolored, dark cambial tissue killed by the beetle-vectored fungus. *Photo:* L. L. Strand

Trees infected with crown gall commonly have warty tumors on the lower trunk or large roots. A tree as large as this walnut usually tolerates the growths; however, trees infected when they are young may be seriously stressed and grow poorly. *Photo:* J. K. Clark

However, when cut with a knife, crown galls are softer than normal wood and lack the typical pattern of annual growth rings. Galls can be tiny and smooth on young plants but usually are rough and sometimes massive on older trees.

The bacteria reproduce in galled tissue and may slough off and survive in soil for long periods. Bacteria in soil enter plants through wounds, commonly caused during handling in the nursery or when transplanting. The bacteria can also infect newly emerging roots, growth cracks, and wounds caused by sucker removal or string trimmers or other equipment.

Purchase and plant only high-quality nursery stock and avoid injuring trees, especially around the soil line. Dipping seeds or roots in the nursery in a solution of the biological control agents K-84 or K-1026 (nonpathogenic strains of *A. tumefaciens* =*A. radiobacter*) may reduce infections by most strains of pathogenic crown gall bacteria but not the strain commonly affecting grape. Solarization, covering moist, bare soil with clear plastic for 4 to 6 weeks during the sunny, dry season before planting, as detailed in Chapter 7, may reduce crown gall bacteria in the upper soil.

Where crown gall has been a problem, plant resistant species, including birch, cedar, coast redwood, holly, incense-cedar, magnolia, pine, spruce, tulip tree, and zelkova. Galls can be removed by cutting into healthy wood around galls and exposing the tissue to drying. Cut out galls only during the dry season and minimize the removal of healthy tissue. At least on certain fruit and nut trees, gall regrowth is frequently avoided by using a blowtorch to briefly heat and sterilize the edges where galls were removed. Do not burn or char tissue and use caution to avoid fire or injury. If galls encompass much of the crown area, cutting them out causes other problems or tree death. Replace extensively galled trees if their growth is unsatisfactory.

PINK ROT OF PALMS

Nalanthamala vermoeseni causes pink rot in stressed or weakened palms. The disease affects all plant parts, but especially new leaves and the growing tip (apical meristem). Pink rot fungus (previously named *Penicillium vermoeseni* and *Gliocladium vermoeseni*) infects nearly all palm species in California, often causing decline and premature death.

Pink rot causes spots on leaves and petioles (rachises); decays trunks, leaf tips, and bases; and distorts and stunts new leaves. Infected tissue can ooze brownish liquid and commonly is covered with pink spore masses. Leaf spots and rot are most common on California fan palm (*Washingtonia filifera*) and queen palms (*Arecastrum romanzoffianum* and *Syagrus* spp.) growing near the coast. Trunk decay and cavities especially occur in king palm (*Archontophoenix cunninghamiana*).

Slow-growing and wounded tissues are most likely to become infected. Disease is often seasonally cyclic: leaves produced during cool, wet conditions in winter and spring are infected and symptomatic, while summer and fall foliage growth is vigorous and healthy. This often results in a distinctive canopy pattern of a few damaged leaves regularly distributed among mostly healthy leaves.

Select species well adapted for the site and plant and care for them properly because the best management strategy is to prevent palms from becoming stressed and weakened. Instead of California fan palm, which is especially susceptible to pink rot along the coast, plant Mexican fan or other palms resistant to pink rot. Do not plant too deeply or over-irrigate. Provide good drainage and proper nutrition; many palms are often deficient in magnesium and potassium. Protect palms from other pests and injuries. Do not remove green leaves prematurely; remove leaf bases only if they detach from the trunk easily. Prune palms only during warm, dry weather.

Fungicide application may be warrant-

Damaged fronds on this California fan palm were infected by pink rot fungus during cool winter and spring weather when they were small spears and young leaves. When the fronds unfold and mature during summer, this damage becomes apparent, as shown here, and the disease is no longer active. The palm produces disease-free leaves during vigorous summer growth. *Photo:* J. K. Clark

Pink rot spore mass on the trunk and under the rachis base of a stressed California fan palm. Minimize this problem by avoiding injuries and providing palms with appropriate cultural care and good growing conditions. Regardless of their care, each type of palm is adapted to grow well under specific conditions, such as away from cool coastal marine air. *Photo:* J. K. Clark

ed on stressed palms until you correct cultural problems and after heavy pruning to protect fresh wounds from infection. Do not substitute fungicide application alone for proper cultural care.

SUDDEN CROWN DROP OF PALMS

Sudden crown drop is a lethal disease of Canary Island date palm and, less often, other date palms (*Phoenix* spp.). With little or no warning, all the fronds and the upper part of the trunk break off and drop from the top of the remaining trunk. Crowns that dropped have soft, yellowish internal decay that turns blackish or brown as it dries. This decay is not visible from the outside of the trunk as long as sufficient healthy tissue remains to keep leaves appearing normal and attached. Crown drop occurs when internal decay progresses to the point that there is no longer enough healthy tissue for the palm to support itself structurally.

Thielaviopsis paradoxa has been found in palm crowns that suddenly dropped. This decay fungus commonly causes Thielaviopsis trunk rot and crown drop in Florida and elsewhere, but it is not known whether *Thielaviopsis* or other pathogens are the cause of crown drop in California.

Decay usually begins in the upper part of the trunk where wounds allow pathogen infection. Typically there is a sharp transition between decayed and healthy internal tissues. The innermost trunk is healthy and the outer trunk is decayed but contained within (covered and hidden by) intact pseudobark. To help detect this malady, sharply strike the upper trunk with a heavy stick at intervals of 1 foot around and down the upper trunk. Use this sounding method from the base of the "pineapple," the ball-like mass of persistent leaf bases just below the leaves, downward along the trunk for about eight feet. Healthy tissue emits a solid, sharp, tone and the stick bounces back quickly. In contrast, decayed tissue emits a low, dull thud when sharply struck and the stick does not bounce back with much force. To determine the extent of decay that sounding detects, you can probe the trunk with a long, sharp, slender tool. If decay is extensive, the palm should be removed. Use this sounding method once per year on date palms that have been pruned with a chain saw, especially frequently pruned Canary Island date palms located where their failure may injure people or property.

To prevent sudden crown drop provide palms with good cultural care, especially appropriate irrigation and adequate fertilization. Minimize wounds to the upper trunk that facilitate pathogen entry. Avoid the extensive use of chain saws to prune leaves and shape and sculpt pineapples. Avoid using tree climbing spikes and chain saws because they cause wounds and mechanically spread pathogens. When pruning, use straight-edge, manual saws or reciprocating saws with removable blades and scrub clean and disinfect blades frequently, at least when moving between trees.

Prune out and appropriately dispose of infected palm parts, as discussed in "Fusarium Wilt of Palms." Do not recycle or chip diseased or infected palms, such as for mulch, because material can contain and spread pathogens. No fungicides or other methods have been shown to cure or prevent sudden crown drop.

WETWOOD, OR SLIME FLUX

Wetwood is an area of branches or trunks that is discolored, water soaked, and exuding sour or rancid, reddish or brown fluid from bark cracks or wounds. Foamy canker, or alcoholic flux (discussed above) also causes wood

Sudden crown drop causes the leaves and upper trunk, which weigh several tons, to suddenly break off and crash to the ground. The disease affects *Phoenix* spp., especially Canary Island date palm. The cause is uncertain, although decay fungi, excessive wounding, and palm care practices working in combination are the apparent causes. *Photo:* D. R. Hodel

Stained wood exuding fluid around the crotch of this Siberian elm is a result of bacterial wetwood infection. *Photo:* J. K. Clark

to exude fluid, but only for a short time during the summer and that fluid has a pleasant, fermentative odor.

Wetwood is especially common in elm and poplar, but it affects many other plants, including box elder, fruitless mulberry, hemlock, magnolia, maple, and oak. Usually only trees about 10 years of age or older exhibit symptoms. Foliage wilt and branch dieback may occur on severely infected trees, but the disease rarely causes serious harm to trees. Although it can be unsightly, limbs infected with wetwood may be as strong as healthy wood.

Wetwood is caused by several species of bacteria; yeast organisms may also be involved. Wetwood-causing microorganisms are common in soil and water and infect trees through wounds, including sites where pesticides have been injected into trees. Avoid injuries to bark and wood to help prevent wetwood bacteria and yeasts from infecting trees.

WOOD DECAY

Various fungi cause heart rot or sap rot of wood in limbs and trunks. When conditions favor their development, certain fungi (sometimes in combination with bacteria) can decay extensive portions of the wood of living trees in a relatively short time (months to years). Almost all woody species are subject to limb and trunk decay, usually after they are wounded and especially as they become old. Decay often is not visible on the outside of the plant, except where the bark has been cut or injured, when a cavity is present, or when the decay fungi produce reproductive structures.

Heart rot fungi destroy the plants' internal supportive or structural tissues (the central core, or heartwood). Heart rots make trees hazardous because trunks and limbs become increasingly unable to support their own weight and can fail (break), especially when stressed, such as by heavy rain, snow, or wind.

Sap rot fungi decay outer wood (sapwood), which is part of trees' vascular conductive system. Sap rots reduce plants' ability to store food and transport nutrients and water, causing them to decline and become more susceptible to other pests and disorders. Trees can be affected by both heart rot and sap rot fungi; this is common in trees affected by sudden oak death and greatly increases the risk of branch or trunk failure.

Many broadleaf tree species are susceptible to canker rot fungi, which cause heart rot decay that can extend through the sapwood and kill portions of the cambium. This leads to external stem cankers, which may callus over and cause distorted areas of bark, or they may remain open and develop into cavities (tree holes).

IDENTIFICATION AND BIOLOGY

Wood decay fungi are spread by airborne spores that infect trees through injuries and wounds. Often, these injuries are caused by people, such as construction activities, machinery, pruning, and vandalism. Boring insects, extreme temperatures, fire, ice, lightning, and breakage from wind or snow load also cause wounds through which decay fungi infect wood. As discussed in "Armillaria Root Disease" and "Heterobasidion Root Disease," certain fungi principally decay the roots and can spread through natural root grafts to infect nearby hosts.

Wood decay fungi are identified in the field by their host plant, the appearance and location of decay, and differences in the color, texture, and shape of their spore-producing fruiting bodies (structures called basidiocarps, brackets, conks, or mushrooms). Decay fungi are often divided into brown rots, soft rots, and white rots based on differences in the appearance and texture of decayed wood (cellulose, hemicellulose, and lignin). Fruiting bodies are often located around wounds in bark, at branch scars, or around the root crown. They can be annual (e.g., appearing soon after the beginning of seasonal rains), as with the typical fleshy mushrooms

Avoid making large wounds. Fungi have entered where a large limb was pruned from this black walnut trunk, causing internal decay. *Photo:* J. K. Clark

Fruiting bodies of the sulfur fungus, *Laetiporus gilbertsonii =L. sulphureus.* This fungus causes a red-brown heart rot of living trees. The basidiocarp (spore-forming structure) of another wood decay fungus (*Ganoderma applanatum*) is pictured near the introduction to this chapter. *Photo:* E. J. Perry, G. W. Hickman

These perennial conks of *Phellinus robustus* growing on a blue oak are hard, brown, and deeply cracked. This pathogen infects living trees and causes a white rot of heartwood that can spread to sapwood and produce surface cankers. Infected trees pose a hazard of branch and trunk failure. *Photo:* B. Hagan

of *Armillaria mellea*. Or, fruiting bodies can be perennial (grow by adding a new layer each year), as with the leathery half-mushrooms or seashell-shaped brackets of the split gill fungus (*Schizophyllum commune*). Some fungi that decay living trees do not produce fruiting bodies until the wood completely dies or breaks off.

Other common wood decay fungi include *Echinodontium tinctorium*, *Ganoderma applanatum*, *G. lucidum*, *Laetiporus gilbertsonii*, *L. conifericola*, *Phaeolus schweinitzii*, and various *Cerrena*, *Fomes*, *Fomitopsis*, *Inonotus*, *Phanerochaete*, *Phellinus*, *Polyporus*, *Steccherinum*, and *Trametes* spp. For more information and photographs, consult *Diseases of Pacific Coast Conifers*, *A Field Guide to Insects and Diseases of California Oaks*, *Oaks in the Urban Landscape*, and *Wood Decay Fungi in Landscape Trees Pest Notes*.

MANAGEMENT

Wood decay is common in old, large trees, and the disease is very difficult to manage. To help prevent decay, provide proper cultural care and protect trees from injuries, as discussed in Chapter 3. Properly prune young trees to promote good structure and to avoid the need to remove large limbs from older trees, which creates large wounds. Properly prune off dead or diseased limbs; make cuts just outside the branch bark ridge and leave a collar of cambial tissue around cuts to facilitate

wound closure (see Figure 3-8). Make cuts so that rainwater will drain, and avoid leaving stubs. Do not apply wound dressings, as they do not hasten wound closure or prevent decay. Prevent the spread of Heterobasidion root disease by applying a registered borate fungicide to freshly cut conifer stumps.

If trees are located where limb or trunk breakage may injure people or damage property, have them regularly inspected by a qualified arborist for signs of wood decay and other structural weakness. Decayed trees provide important wildlife habitat, but hazardous trees may need to be braced, cabled, trimmed, or removed (see Chapter 3).

VASCULAR WILT DISEASES

Vascular wilts damage the phloem or (more commonly) xylem system that transports nutrients, food, and water among plant parts. These diseases cause foliage to turn pale green, yellow, brown, and wilt, initially in scattered portions of the canopy or on scattered branches. Shoots and branches die, often beginning on one side of the plant, and the entire plant may die. Important vascular wilt diseases include bacterial leaf scorch (oleander leaf scorch), Dutch elm disease, Fusarium wilt, huanglongbing (citrus greening), laurel wilt, and Verticillium wilt.

BACTERIAL LEAF SCORCH, OR OLEANDER LEAF SCORCH

Leaf, branch, and plant death in various hosts is caused by *Xylella fastidiosa*, a xylem-infecting bacterium spread by glassy-winged sharpshooter (*Homalodisca vitripennis*) and certain other leafhoppers. Oleander was the first ornamental recognized as being killed by this bacterium in California, so the disease became known as oleander leaf scorch. Flowering plum, liquidambar, and landscape olives are also being killed by this disease in southern California, and there are many additional hosts, some of which have been confirmed as being damaged by *Xylella* only in the eastern United States. Disease is mostly limited to southern California, because quarantines have limited spread of the glassy-winged sharpshooter, the pathogen's main vector.

Different strains of *X. fastidiosa* cause Pierce's disease in grapes, almond leaf scorch, and *Xylella*-related maladies in other crops. Various strains of *X. fastidiosa* (including the strain damaging oleander) can occur in certain plant species without causing symptoms in those hosts. The relationships among species of leafhopper vectors, host plants, and pathogenic *Xylella* strains are not fully understood.

DAMAGE

Symptoms can appear at any time of the year but commonly appear first or progress most rapidly from late spring through early fall. Initially, foliage yellows overall and droops on one or more branches. Soon the leaf margins or tips turn yellow, then brown, dry out, and eventually die, giving plants a scorched appearance. Whole branches and eventually entire plants die. Adjacent plants of the same species can differ substantially in severity of symptoms. Some infected plants survive only 1 to 2 years after first showing symptoms, while others can survive for several years.

Identification and Biology

Glassy-winged sharpshooter is the primary vector of bacterial leaf scorch. Blue-green sharpshooter (*Graphocephala atropunctata*), smoke tree sharpshooter (*Homalodisca lacerta*), and possibly other leafhopper species, some of which are called sharpshooters, acquire the bacteria when they feed on infected plants and spread it when they move to feed on healthy host plants.

Other causes produce symptoms that resemble bacterial leaf scorch, and diagnosing whether *Xylella* is the cause of damage can be difficult. Diagnostic laboratories that test plant samples for Pierce's disease of grapes can confirm the presence of *Xylella fastidiosa* in hosts. However, current commercial tests do not distinguish the strains of *Xylella* bacteria, such as that causing oleander leaf scorch from that causing Pierce's disease. A plant harboring harmless *Xylella* strains would test positive for *Xylella* even though damage symptoms would actually be due to other causes.

You may need to investigate several potential causes to confirm or rule out bacterial leaf scorch. Except where glassy-winged sharpshooter is established, *Xylella* is less likely to be the cause of symptoms. Plants can be laboratory tested for various vascular wilt pathogens known to affect that host and cause similar symptoms. If drought stress is the cause, plants should recover after irrigation unless drought was too severe or roots are unhealthy. Irrigation will not remedy *Xylella*-infected plants because bacteria are clogging their vascular system. See Table 4-1 for a summary of field and laboratory tests to assess abiotic causes of symptoms. For example, test soil or plant tissue to determine whether leaf yellowing and necrosis are due to excess boron or salinity. Alternatively, irrigate with sufficient quantities of high-quality water to leach minerals to below the root zone, which should allow plants to recover, whereas no improvement will be seen in plants with bacterial leaf scorch.

Yellow, brown, dying leaf margins on oleander are symptoms of bacterial leaf scorch caused by leafhopper-vectored *Xylella fastidiosa*. Laboratory tests can help to distinguish this pathogen from causes of similar damage, including phytotoxicity, salt burn, and severe drought. *Photo:* J. K. Clark

Xylella bacteria can cause leaf scorch, branch dieback, and death of liquidambar, oleander, olive (shown here), purple-leaf plum, and certain other woody species. Because quarantines limit the spread of glassy-winged sharpshooter, the main vector of *Xylella*, damage in landscapes has been limited to southern California as of 2014. *Photo:* J. Kabashima

Management

There is no known cure for *Xylella*-infected plants. Pruning off affected parts can temporarily improve plant appearance but damage will reappear because plants remain systemically infected. Consider gradually replacing extensive plantings of susceptible hosts with a mixture of non-host plants if this disease occurs in your area. Managing leafhoppers has not been shown to stop the spread of bacterial leaf scorch in landscapes.

Comply with any quarantines to prevent the spread of glassy-winged sharpshooter to new locations. In uninfested areas when receiving nursery plants from areas known to be infested, inspect their foliage for sharpshooter eggs, as discussed in "Glassy-Winged Sharpshooter" in Chapter 6, and report any findings to the county agricultural commissioner.

For more information, consult *Oleander Leaf Scorch Pest Notes* and *Glassy-Winged Sharpshooter Pest Notes*.

Dutch Elm Disease

Dutch elm disease has killed several thousand elms in California, mostly in the San Francisco Bay Area and around Sacramento. Millions of elms have been killed in midwestern and eastern states since 1930. The disease is caused by introduced fungi, *Ophiostoma* (=*Ceratocystis*) *ulmi* and *O. novo-ulmi*.

Dutch elm disease initially causes foliage yellowing and wilting, usually first in one portion of the canopy. Leaves then turn brown, curl, and die but remain

Scattered elm branches with unhealthy yellowish or dead brown foliage can be due to Dutch elm disease fungi spread by certain bark beetles. If Dutch elm disease is suspected, promptly report this to the county agricultural commissioner. Elm leaf beetle causes superficially similar damage, but a close-up examination of leaves (such as those dropped on the ground) reveals leaves with holes and scraped surfaces when elm leaf beetle is the cause. *Photo:* W. M. Brown Jr., Bugwood.org

on branches. Cutting off bark reveals brown to blackish streaks in infected wood, which appear as dark concentric rings when infected branches are cut in cross-section.

Do not confuse disease symptoms with elm leaf beetle chewing damage, which also causes leaf browning that viewed from some distance looks superficially similar. Elm leaf beetle chews holes through leaves and skeletonizes leaf surfaces, which can be distinguished by inspecting foliage or dropped leaves close-up.

Dutch elm disease is spread by elm bark beetles (*Scolytus* spp.). Bark beetles emerge from dead or dying trees or from elm logs infected with the fungus and carry spores that infect healthy elms when the adult beetles feed in the crotch of young twigs. People inadvertently spread this pathogen by moving young elms or elm logs infested with beetles or fungi to disease-free areas. Tools used to prune, cut, or inject infected elms spread the fungi to other elms. The fungi also spread from infected elms through natural root grafts to nearby elms.

MANAGEMENT

American and European elms are highly susceptible to Dutch elm disease, which originally is from Asia. Plant elm cultivars or hybrids resistant to Dutch elm disease and elm leaf beetle (Table 5-11), some of which also resist phloem necrosis (elm yellows), a phytoplasma not yet reported in California. Alternatively, plant hackberry (*Celtis* spp.), hornbeam (*Carpinus* spp.), or other species that resemble elms but are not attacked by elm pests.

Most elm species and hybrids are adapted to summer rainfall. Maintain tree vigor by providing adequate summer irrigation in areas with summer drought. Promptly prune off and dispose of dead and dying elm limbs as they commonly are infested with the fungus-spreading elm bark beetles and can pose a limb drop (failure) hazard. Otherwise, avoid unnecessary pruning and properly prune elms, especially when trees are young and preferably during late fall and winter. Bury, chip, or (where permitted) burn freshly cut elm wood. Alternatively, seal elm logs tightly under clear plastic in a sunny location through the warm season and for at least

Table 5-11.

Elm (*Ulmus*) Tree Susceptibility to Dutch Elm Disease (DED) and Elm Leaf Beetle (ELB).

ELM TREE		SUSCEPTIBILITY	
COMMON NAME	SCIENTIFIC NAME	DED	ELB
Emerald Sunshine	*Ulmus propinqua*[3]	R	R
Patriot	(*U. glabra* × *U. carpinifolia* × *U. pumila*) × *U. wilsoniana*	R	R
Prospector	*U. wilsoniana* selection	R	R
Frontier	*U. carpinifolia* × *U. parvifolia*	R	MR
Morton Accolade	*U. japonica* × *U. wilsoniana*[3]	R	MR
American New Horizon	*U. americana* selection	R	S
Homestead	*U. glabra* × *U. carpinifolia* × *U. pumila*	R	S
Morton Glossy Triumph	*U. pumila* × *U. japonica* × *U.?*[3]	R	S
New Horizon	*U. pumila* × *U. japonica*[3]	R	S
Morton Plainsman Vanguard[1]	*U. pumila* × *U. japonica*[3]	R	HS
Morton Red Tip Danada Charm	*U. japonica* × *U. wilsoniana*[3]	R	HS
Morton Stalwart Commendation	*U. carpinifolia* × *U. pumila* × *U.?*[3]	R	HS
Pioneer[1]	*U. glabra* × *U. carpinifolia*	R	HS
American Valley Forge[1]	*U. americana* selection	MR	R
Chinese[2]	*U. parvifolia*	MR	R
zelkova	*Zelkova serrata*	MR	R
Siberian	*U. pumila*	MR	S
Dynasty Chinese[2]	*U. parvifolia* selection	MR	HS
American	*U. americana*	HS	S
English	*U. procera*	HS	HS
Scotch	*U. glabra*	HS	HS

KEY

HS = Highly susceptible

MR = Moderately resistant

R = Resistant

S = Susceptible

? = hybrid cultivar includes some uncertain or unknown elm parentage.

1. Have exhibited poor growth structure and high pruning requirement when young and grown in California.

2. Dynasty is highly susceptible to ELB; most Chinese elms (e.g., Allee, Athena, Drake, Evergreen, and True Green) are resistant to elm leaf beetle. For susceptibility to elm anthracnose, see Table 5-6.

3. Budded onto *U. pumila* rootstock; scientific name is for the scion (upper trunk and canopy) .

Source: McPherson et al. 2009.

7 months, after which they are no longer suitable for beetle breeding, as discussed in "Elm Bark Beetles" in Chapter 6.

Report suspected Dutch elm disease to the county agricultural commissioner. Remove infected elms immediately to eliminate them as a source of the fungi, which otherwise will spread to nearby elms. Digging a trench around infected trees deep enough to cut the roots may prevent the fungi from spreading by root grafts to nearby elms.

Promptly pruning off recently diseased limbs may be an alternative in areas where quarantine regulations do not require the removal of the entire tree. This "therapeutic pruning" can be effective only if done promptly during the first season when disease symptoms appear on a tree. Symptoms must be limited to one or a few limbs, and at least 10 feet of healthy wood (free of visible disease streaking) must separate the infected wood from the pruning point on the main trunk. The trees must be otherwise vigorous and healthy.

No fungicides have been demonstrated to be effective against Dutch elm disease in California. Based on use in the eastern United States, injection of certain systemic fungicides by a competent, professional applicator may prevent Dutch elm disease and help some recently infected trees to survive if properly treated while less than 5 to 10% of their canopy is symptomatic.

Consider fungicide treatment only if Dutch elm disease has been discovered infecting nearby elms. Do not use fungicide alone, as it has limited effectiveness and only if combined with proper pruning and tree care. If the main trunk or many limbs show symptoms, promptly remove the dying tree to reduce infection spread to nearby elms. The prompt detection and removal of dead and dying elms is the most effective method for managing the disease.

FUSARIUM DIEBACK

An introduced *Fusarium* sp. spread by the polyphagous shothole borer (*Euwallacea* sp.) is causing branch dieback, decline, and death of avocado, big leaf maple, box elder,

California sycamore, coast live oak, and liquidambar in southern California. *Albizia julibrissin, Cercidium floridum, Erythrina corallodendron* and other tree species are also attacked by this introduced ambrosia beetle, but some species attacked by the adult beetles are not suitable for reproduction of the beetle or pathogen. This *Fusarium* and ambrosia beetle are different species than discussed below in "Laurel Wilt."

Fusarium dieback symptoms include pale yellowish and wilted leaves, branch dieback, and in some cases death of the tree. Bark develops patches of dark lesions that are dry, water-soaked, or oily-looking, sometimes surrounded by wet discoloration or white crusty or powdery exudate. Within each lesion is a tiny ($\frac{1}{25}$ inch diameter) hole in bark bored by an adult beetle, which is dark brown to black and $\leq \frac{1}{10}$ inch long. Scraping off bark reveals brown, diseased wood surrounding beetle tunnels up to 2 inches deep in wood. Once the beetles and *Fusarium* become abundant in a limb, the branch dies.

Provide trees with proper cultural care and good growing conditions to make them less attractive to boring beetles. Prune off dying limbs and scrub tools clean and disinfect them between cuts. Leave infected wood on-site and chip it for mulch or cut it into logs and properly solarize the logs. Be alert for new information on potential management of this problem. Report this problem (e.g., abundant white sugary exudate from beetle boring holes) to the county agricultural commissioner if found outside of areas known to be infested. See the Center for Invasive Species Research website, cisr.ucr.edu, for more current information and photographs.

FUSARIUM WILT

Fusarium oxysporum affects relatively few woody ornamental species but can kill hosts including albizia, date palm, hebe, and pyracantha. Fusarium wilt symptoms often appear first on one side of a plant and older leaves usually die first, commonly followed by death of the entire plant.

Cutting into infected wood may reveal that vascular tissue has turned brown, often all the way from the shoot to the soil line.

Fusarium spp. are divided into special forms (*forma specialis,* or f. sp.) or subspecies. Each f. sp. of *Fusarium* is specific to certain hosts and does not spread to infect plants in other genera. Most forms of *F. oxysporum* attack only herbaceous plants. Fusarium wilt results from infection through roots by hyphae that germinate from long-lasting survival structures (chlamydospores) in the soil.

There is no cure for Fusarium wilt. It nearly always kills infected hosts, so prevention and exclusion are the only effective strategies. Avoid this problem by replanting at that site using species from different genera than plants previously infected there by *Fusarium*. Obtain new plants from a reliable, high-quality source. Avoid planting in poorly drained soil, provide good cultural care, and especially do not overirrigate as this encourages surface roots that are easily damaged and infected. Avoid using undecayed organic amendments and excessive fertilizer, especially urea, which may promote development of *Fusarium*. Keep other plants away from the base of hosts, especially palms, as nearby plants and their management can injure roots. Promptly remove and dispose of infected trees to reduce pathogen spread to nearby hosts.

FUSARIUM WILT OF PALMS

Fusarium oxysporum f. sp. *canariensis* kills Canary Island date palm (*Phoenix canariensis*). Other forms of *Fusarium* infect palms elsewhere but have not been reported in California.

Symptoms usually first appear in older, lower leaves (fronds) or occasionally first in mid-level leaves, then spread toward the center, upper (newest) leaves. Infected leaves turn yellow, then brown and die and remain hanging on the palm. Leaflets (pinnae) usually discolor and die first at the base, then progressively out toward the tip. Symptoms often occur first on the leaflets on one side of the leaf, but this is not diagnostic as Dothiorella blight, pink rot, and other diseases also cause one-sided death of palm leaves.

Fusarium wilt causes extensive brown to blackish streaks visible externally along the petiole (rachis) connecting the leaf base to the leaf blade and leaflets. Cutting a cross-section through an infected petiole typically reveals reddish brown to pinkish vascular discoloration on the wet (freshly cut), clean surface.

Canary Island date palms die within several months to several years of initial symptoms. Palms growing at hotter, drier interior locations die relatively quickly, while those at cooler, more humid coastal sites decline more slowly and may survive for many years. *Fusarium*-stressed palms become very susceptible to pink rot, a

Extensive external or internal vascular discoloration or streaking along the rachis (petiole or stalk) is characteristic of Fusarium wilt of palms. *Photo:* D. Hodel

mostly secondary disease that hastens palm death, and symptoms of pink rot can mask those of Fusarium wilt.

Fusarium spreads through natural root grafts to nearby palms. Exposed (aboveground) roots can be infected by contact with *Fusarium*-contaminated tools or materials, such as infested mulch or soil. *Fusarium* frequently spreads on pruning tools, especially chain saws, which introduce the fungus into fresh cuts. Contaminated sawdust can drift during pruning to infect nearby palms. The pathogen can survive in soil for at least 25 years, so soil is likely to be infested anywhere *Fusarium*-infected Canary Island date palms previously grew.

Management in Palms. Avoid mulching around Canary Island date palms with municipal greenwaste or anything potentially containing palm pieces. Avoid or minimize pruning and remove only dead leaves. Use excellent sanitation, especially during pruning.

Frequently and extensively pruned palms are more likely to develop Fusarium wilt. Use only manual pruning saws or reciprocating saws with removable blades as disinfesting chain saws is not practical. After working on each palm

- Remove all sawdust and other particles from tools.
- Scrub blades vigorously until they are thoroughly clean.
- Disinfect the equipment by soaking it for 10 minutes in undiluted bleach or a solution of one part pine oil disinfectant diluted in three parts water, or heat saw blades for at least 10 seconds per side with a hand-held butane torch.

For extremely valuable palms, consider using a new saw for each palm and dedicate it for use only on that palm, or afterward use it only on other plant species.

Remove infected palms with as little cutting, digging, and grinding as possible to limit the spread of inoculum. Excavate the root ball and use a crane to remove the entire palm and as much of the attached root ball as possible. Erect barriers to retard the spread of any cutting or grinding

waste. Bag or cover the palm waste and transport it to an incinerator or landfill. Do not allow Canary Island palm to be recycled for compost or mulch. Do not replant that site with Canary Island date palm and consider not planting any palms there.

Protect palms from injury and provide them with optimal growing conditions, especially appropriate water. Ensure proper nutrition by applying special palm fertilizers where needed. See *Palm Diseases in the Landscape Pest Notes* and *The Biology and Management of Landscape Palms* for more recommendations.

HUANGLONGBING (HLB), OR CITRUS GREENING

The bacterium *Candidatus* Liberibacter asiaticus causes huanglongbing, a lethal disease in all citrus varieties. Citrus and orange jasmine, or orange jessamine (*Murraya paniculata*), are the common hosts, but other Rutaceae can be infected, including box orange, calamondin, Indian curry leaf, and wampee (or wampi, *Clausena* spp.).

There is no known cure for huanglongbing. Once a tree is infected, the only method for preventing spread of the disease is to remove and destroy the infected tree. Bacteria infect the phloem vascular system, inhibiting nutrient flow and damaging all parts of the plant. The characteristic early symptoms of this exotic disease are asymmetrical yellowing of leaves (color differs on opposite sides of the main vein) and pale shoots in individual limbs or scattered portions of the canopy, hence its Chinese name, which means "yellow shoot disease." Later, fruit become lopsided and fail to turn a mature color, hence the name "citrus greening." Citrus juice commonly develops a highly acidic or bitter flavor that makes the fruit unpalatable. Leaves and fruit drop, twigs and limbs die back, and trees often die within 3 to 5 years after infection.

Candidatus spp. bacteria are spread by the introduced Asian citrus psyllid, pictured in Chapter 6. This aphidlike insect also damages Rutaceae plants directly by

Mottling and yellowing that cross leaf veins are symptoms of huanglongbing. Where marked by arrows, mottling is asymmetrical: there is a yellow area on one side of the midvein and dark green area opposite. If you suspect that this exotic, tree-killing disease or another new pest is present, promptly report it to the county agricultural commissioner. *Photo:* M. Garnier, INRA, France

Pale yellowish and brown dead foliage on camphor tree with laurel wilt, a fungal disease spread by the redbay ambrosia beetle. Avocado, California bay, and other Lauraceae are also susceptible to these exotic pests that are spreading from the southeastern U.S. *Photo:* C. Bates, Georgia Forestry Commission, Bugwood.org

injecting a toxin during feeding. New shoot growth that is heavily infested by psyllids does not expand or develop normally and may easily break off. Report suspected huanglongbing disease and Asian citrus psyllids to the county agricultural commissioner if they are not known to be established in that area.

LAUREL WILT

Avocado, California bay, camphor, and other Lauraceae family trees in the southeastern United States are being killed by an introduced vascular wilt fungus (*Raffaelea lauricola*) and the redbay ambrosia beetle (*Xyleborus glabratus*). Although not yet reported in California, similar introduced species are damaging avocado and other trees, as discussed in "Fusarium Dieback."

Laurel wilt causes foliage wilting, branch dieback, and discolored reddish and purple leaves that drop prematurely or turn brown and die and remain hanging on the tree for weeks. Affected trees may have pale, stringlike frass (beetle excrement) protruding from bark.

To confirm disease, cut bark from wilted branches to expose sapwood (outer wood). Healthy sapwood is pale colored, while infected wood is discolored black to brown. Cut dying limbs in cross-section to expose beetle tunnels, which extend deeply into wood and are $^1/_{20}$ inch in diameter or less. Tunnels may contain white, legless larvae with a brownish head and dark

cylindrical adults of the ambrosia beetle.

The beetles and fungus are mostly spread long distances by people moving infested firewood and logs. Report suspected laurel wilt disease or stringlike boring beetle excrement on the bark of Lauraceae trees to the county agricultural commissioner. Do not bring plants or wood into California unless you know they have been certified to be pest-free or were inspected by agricultural officials. Buy wood where you burn or use it and do not move firewood or logs between counties.

VERTICILLIUM WILT

Many deciduous trees, shrubs, and herbaceous plants are susceptible to infection by *Verticillium dahliae*. Common hosts include ash, camphor, Chinese pistache, fuchsia, hebe, maple, olive, pepper tree, and rose.

Verticillium wilt causes foliage to turn faded green, yellow, or brown and wilt in scattered portions of the canopy or on scattered branches. Shoots and branches die, often beginning on one side of the plant, and entire plants may die. Small plants may die from Verticillium wilt in a single season, but larger plants usually decline slowly. Mature trees may take many years to die and may suddenly recover if conditions become favorable for plant growth and poor for disease development.

Verticillium persists as microsclerotia in soil. When it is cool, microsclerotia produce hyphae that infect susceptible roots,

spread upward in the current year's growth, and inhibit the plant's nutrient and water transport. In some plants, but not all, peeling off the bark on newly infected branches may reveal dark streaks following the wood grain. Depending on the plant species, the stains are dark black, brownish, gray, or greenish. Infection can occur during the spring but not become apparent until warm weather, when plants are more stressed. A

Brown, dead foliage in scattered patches on one side of a Japanese maple due to Verticillium wilt. *Photo:* J. K. Clark

Dark staining visible [in]
stem and trunk tissue [when it]
is cut in cross-section [is]
characteristic of Vert[icillium]
wilt and certain othe[r]
vascular wilt disease[s.]
Because the disease [begins]
in roots and spreads
upward, the lower tr[unk]
(right) is entirely disc[olored]
but only portions of [the]
upper trunk (left) are
stained, often in a pi[e]
wedge shape with th[e point]
toward the center of [the]
limb. *Photo:* J. K. Cla[rk]

Verticillium wilt damages plants' vascular tissue; peeling back the bark on newly infected branches may reveal dark stains following the grain, as seen on this almond wood. *Photo:* J. K. Clark

laboratory culture from newly infected wood is often required to confirm the presence of *Verticillium*.

Keep plants vigorous by providing for their cultural requirements, especially proper irrigation. Modest amounts of slow-release fertilizer and other appropriate care to promote new growth can increase infected plants' likelihood of survival. If chronic branch dieback develops, prune off any dead wood and have trees regularly inspected for possible hazards; affected trees may need to be removed. Where Verticillium wilt has been a problem, plant only resistant species (Table 5-12) because the pathogen persists in soil.

Mushrooms and Other Spore-Forming Structures in Landscapes

Certain fungi produce relatively large reproductive spore-forming structures. Although the umbrella-shaped (toadstool) fruiting bodies called mushrooms are the

Table 5-12.

Landscape Trees and Shrubs Resistant to Verticillium Wilt.

COMMON NAME	GENUS OR SCIENTIFIC NAME
apple and crabapple[1]	*Malus*
arborvitae	*Thuja*
beech	*Fagus*
birch	*Betula*
box and boxwood	*Buxus*
California bay	*Umbellularia californica*
cedar	*Thuja*
citrus	*Citrus*
dogwood	*Cornus*
eucalyptus	*Eucalyptus*
fig, edible	*Ficus carica*
fir	*Abies*
hawthorn	*Crataegus*
holly	*Ilex*
honey locust	*Gleditsia*
hornbeam	*Carpinus*
katsura tree	*Cercidiphyllum japonicum*
linden	*Tilia*
manzanita	*Arctostaphylos*
mountain ash, European	*Sorbus aucuparia*
mulberry	*Morus*
oak	*Quercus*
oleander	*Nerium oleander*
palms	All genera
pear[1]	*Pyrus*
pine	*Pinus*
pyracantha	*Pyracantha*
spruce	*Picea*
sweet gum	*Liquidambar styraciflua*
sycamore and plane tree	*Platanus*
walnut	*Juglans*
willow	*Salix*

This list provides a guideline only; there is no guarantee that these plants will not be affected. New pathogen strains develop or are introduced. Disease incidence is greatly influenced by cultural care and environmental conditions.
1. Apple, pear, and quince are susceptible to European strains of *Verticillium albo-atrum* not reported in California.
Adapted from Farr et al. 1989; McCain, Raabe, and Wilhelm 1981.

This innocuous slime mold (called dog vomit fungus, *Fuligo septica*) commonly develops in decaying organic matter. If mushrooms or other fungal structures are aesthetically objectionable, cut or rake them away or spray slime molds during their gelatinous (early growth) phase with a forceful stream of water. Reduce irrigation frequency and periodically mix or stir mulch to reduce fungal development. *Photo: R. M. Davis*

most well known, other fungi produce structures called brackets, basidiocarps, or conks. These reproductive structures vary greatly in color, shape, and size as pictured in "Annulohypoxylon Canker, or Hypoxylon Canker," "Armillaria Root Disease," "Nectria Canker," and "Wood Decay." Some fruiting bodies seen in lawns and organic mulch are named for their appearance, including bird's nests, puffballs, and stinkhorns.

Some people become concerned when mushrooms or other fungal reproductive structures appear in lawns, organic mulch, or near or attached to plants. However, few of the fungi that cause plant diseases produce large, obvious fruiting structures. Exceptions include Armillaria root disease, which produces groups of mushrooms around trunks after rainy weather, and various wood decay fungi that produce brackets or conks on declining limbs and trunks, which can indicate trees are hazardous. Many fungi in landscapes are beneficial, because they decompose organic matter (releasing nutrients for plant growth) or form beneficial relationships with roots (mycorrhizae) that improve plant growth. For more information, see the *Mushrooms and Other Nuisance Fungi in Lawns Pests Notes* and *Wood Decay Fungi in Landscape Trees Pests Notes.*

Root and Crown Diseases

Pathogens that infect roots and crowns and can damage or kill trees and shrubs include *Armillaria, Dematophora, Heterobasidion, Phytophthora,* and *Pythium* spp. Because roots and the basal trunk (crown, or root collar) structurally support aboveground parts and transport nutrients and water to the rest of the plant, diseases of the root and crown affect other parts of the plant as well.

Often, the first observed symptom of root disease is chlorotic, discolored, and wilted foliage resembling a nutrient deficiency. As disease progresses, leaves and branches die, and the entire plant can be killed. Similar symptoms can be caused by boring insects and abiotic disorders, including aeration deficit, herbicide phytotoxicity, salinity, and too much or too little water.

Once root disease pathogens become introduced in a location, they are often continuously present there in old roots, stumps, soil, or infected living or dead standing trees. Root diseases commonly develop in nurseries and can be introduced into landscapes on contaminated nursery stock. In many landscape situations, fungicides are not available or effective in controlling these diseases. The most effective management methods are to

- Avoid introducing and moving these pathogens (e.g., use good sanitation and high-quality, pathogen-free nursery stock and soil amendments).
- Grow resistant or tolerant plants (see Table 5-13).
- Plant trees properly and provide good cultural care.
- Prevent conditions that promote disease development.

Plants damaged by root diseases may be helped if the irrigation method is changed from flooding or sprinklers to a drip system and soil is temporarily removed near the trunk to expose and promote drying of the root crown. Protect newly exposed tissue from direct sun or excessive temperature changes. After drying, cover roots to the same level with the same soil, which has been air-dried. Do

Water puddling around this trunk is promoting root and crown disease. Frequent shallow irrigation (such as that applied to the surrounding lawn) is often incompatible with the long-term health and survival of trees, especially in poorly drained soil. *Photo: J. K. Clark*

Some Mushrooms Are Poisonous

Many mushrooms are poisonous, and ONLY an expert can distinguish between edible and poisonous species. There are no simple tests that identify poisonous mushrooms. Do not eat wild mushrooms or other fungal fruiting bodies unless you definitely know the species and that it is not toxic.

Small children and dogs tend to put almost anything in their mouths, including mushrooms. Consider removing (cutting off or raking away) fungal reproductive structures from landscapes frequented by children or pets. You can remove mushrooms growing from buried wood or roots by picking the mushrooms as they appear or by digging out the wood.

not wet or compact the earth. Alternatively, for plants adapted to a Mediterranean climate, cover roots around the crown with pea gravel, which provides good drainage and aeration, and eliminate weeds and ground covers near trunks.

If located where their failure could injure people or damage property, remove and dispose of dying or hazardous trees. Correct any soil or water conditions that promote disease and replant with resistant or tolerant species. Proper planting and appropriate irrigation, as detailed in Chapter 3 and illustrated in Figures 5-2 and 5-3, are especially critical to preventing root diseases.

ARMILLARIA ROOT DISEASE

Armillaria root disease, also known as oak root fungus or shoestring fungus disease, affects bamboo, many deciduous broadleaf and conifer trees, and some common herbaceous ornamentals. *Armillaria* is a native fungus most prevalent where oaks or other native trees once grew.

Armillaria infects and decays cambium of major roots and the basal trunk. This causes leaves to become undersized, discolored, and drop prematurely. As disease progresses, branches begin dying, often first around the tops of deciduous trees or in the lower canopy of conifers. Eventually the entire plant can be killed by the fungus or in combination with bark beetles or other pests. Young plants often die quickly; mature trees may die quickly or slowly, but they can recover, at least temporarily, if conditions become good for tree growth and poor for disease development.

IDENTIFICATION AND BIOLOGY

Several *Armillaria* spp. cause root disease, most commonly *A. mellea*. Between the bark and wood, *Armillaria* forms white, fanlike mycelial plaques that have a mushroomlike odor when fresh and are visible when the bark is removed from infected roots and the lower trunk. Dematophora root rot also causes white growths resembling those of *Armillaria*, but *Dematophora*

When trees are infected by *Armillaria*, groups of short-lived mushrooms sometimes sprout around the trunk after rainfall. *Photo:* J. K. Clark

Rhizomorphs, shown here growing on a large root, are dark stringlike fungal mycelia that resemble small roots. *Armillaria* spreads by rhizomorphs growing short distances through the soil from infected plants, in addition to spreading through direct contact and natural grafts between roots. *Photo:* J. K. Clark

mycelia often are more tan to brown and tend to occur in smaller patches.

During cool, rainy weather, usually in the fall or early winter, clusters of short-lived mushrooms may form at the base of *Armillaria*-infected trees. The mushrooms' caps range from 1 to 6 inches in diameter, always occur in groups and never singly, and have a ring (annulus) on the stalk just under the cap. *Armillaria* mushrooms vary from whitish to honey yellow to brown when fresh but soon shrivel and turn black.

On the surface of infected roots or the root crown *Armillaria* produces rootlike rhizomorphs that can grow short distances to infect roots of adjacent trees. Rhizomorphs have a black to dark reddish brown surface. If you pull them apart, their cottony interior becomes visible. When similar-sized roots are pulled apart for comparison, roots have a more solid, woody interior.

Armillaria develops under warm, moist conditions, for example, when

plants or soil around the roots of California native oaks are irrigated during warm weather. The fungus can survive for many years in dead or living tree roots. Plants become infected when their roots grow and contact infected plants or rhizomorphs or mycelium on dead roots in soil.

MANAGEMENT

Prepare the site well before planting to provide good drainage. Provide appropriate cultural care, especially proper irrigation. Before replanting at sites where *Armillaria* has been a problem, remove from the soil as many roots as possible that are $\frac{1}{2}$ inch in diameter or larger because these can harbor *Armillaria* and avoid planting species known to be susceptible to this pathogen.

Wood from *Armillaria*-killed trees can be chipped and used as mulch with little or no risk of spreading *Armillaria* if properly handled. Before using around host plants grind wood well to avoid large chunks (>2 inches)

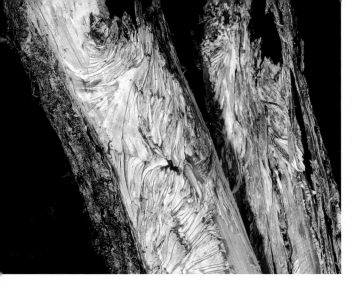

This white fungal mycelia growing in cambial tissue (inner bark) is the most diagnostic sign of Armillaria root disease, revealed here by exposing wood and the underside of bark on the lower trunk. In trees adapted to summer drought, such as native California oaks, summer irrigation favors *Armillaria* fungal development and disease. *Photo:* J. K. Clark

and spread the material thinly during warm weather to completely air-dry or properly compost the material and periodically turn the pile so all parts of it become exposed to hot temperatures for about 2 weeks.

Soil fumigants may be available to professional applicators to treat soil before planting. However, fumigation of field soil usually does not kill the fungus in all roots deep in the soil. Fumigants can be highly toxic to people, and their use is increasingly restricted and are generally not available for use or warranted in landscapes.

DEMATOPHORA ROOT ROT, OR ROSELLINIA ROOT ROT

Rosellinia (=*Dematophora*) *necatrix* causes Dematophora or Rosellinia root rot, also called white root rot. Although relatively uncommon in landscapes, when it occurs it quickly kills hosts, including ceanothus, cotoneaster, fruit trees, holly, poplar, privet, and viburnum. The fungus infects healthy roots when they grow near infected plants, and the pathogen is especially active during moderate temperatures when soils are soggy.

Initially, Dematophora root rot may cause foliage discoloring and wilting throughout the entire plant or in just a portion of the canopy. Branches killed often retain dry foliage. White mycelia may be visible on the lower trunk, in soil over infected roots, growing from infected roots, or beneath bark of the root crown or major roots. These whitish *Rosellinia* growths are much smaller than *Armillaria* mycelia and lack the characteristic mushroomlike odor

produced by fresh *Armillaria*. *Rosellinia* may also produce dark, crustlike mycelia over dead roots or around the root collar.

When *Rosellinia*-infected tissue is sealed in a moist chamber, such as a plastic bag or jar, it produces distinctive, white fluffy mycelia within a few days. However, if Dematophora root rot is suspected, it is best to promptly seek an expert to confirm its presence.

Minimize *Rosellinia*-caused disease by preparing the site well before planting, using high-quality nursery stock that is pathogen-free, and providing appropriate cultural care. Proper irrigation is especially important.

HETEROBASIDION ROOT DISEASE

Heterobasidion spp. fungi can infect almost any native conifer and occasionally certain broadleaves such as madrone. Also called Annosus root and butt rot, the disease kills the basal trunk, root crown, or roots of hosts. Pines and true firs are often killed within several years. Hosts such as Douglas-fir, incense-cedar, hemlock, larch, and spruce usually are not killed outright, but their heartwood and sapwood decay, increasing the likelihood of insect attack and tree failure (e.g., wind breakage).

The fungus spreads by airborne spores that infect trunk wounds and especially freshly cut stumps. It moves through natural root grafts to infect nearby living trees. The fungus can survive for several decades in the roots of killed trees, and groups of trees often die in a gradually expanding

clump as they are infected through root contact with the initial host.

Cutting into *Heterobasidion*-infected heartwood reveals an irregular or blotchy pattern of discolored staining in major roots and the lower trunk or stump. Bark on diseased roots and the basal trunk easily separates from wood, and there may be small white flecks of mycelia on the inner surface of bark and brown resin on the wood surface. *Heterobasidion* conks grow in groups on infected stumps and trunks and are leathery and usually light brown to grayish on top and pale underneath. They vary from $\frac{1}{2}$ inch wide and button shaped to several inches wide and seashell shaped.

Avoid wounding trees and promptly cut down dying and dead conifers. Apply a registered borate fungicide to freshly cut pine and true fir stumps and perhaps to stumps of other conifers to prevent spores from infecting stumps and spreading the fungus through roots to nearby conifers.

White, cobwebby patches of fungus in soil at the base of this apple tree are characteristic of Dematophora root rot. Unlike *Armillaria* mycelia, which occur only under bark, the mycelia of *Dematophora* are less extensive but can occur on top of or under the surface of bark and soil. *Photo:* J. K. Clark

Excess irrigation of nearby turf and compacted, poorly drained soil are promoting the development of *Phytophthora cinnamomi* root and crown rot in this Irish yew, resulting in branch dieback. *Photo:* J. K. Clark

Gumming, a vertical streak, stain, or canker, or a localized cracking of bark are often visible on the trunk of *Phytophthora*-infected trees. However, certain plants often exude liquid when damaged, and this symptom can have other causes. *Photo:* J. K. Clark

PHYTOPHTHORA ROOT AND CROWN ROT

Phytophthora cinnamomi and other *Phytophthora* spp. commonly infect roots and crowns and cause maladies that are also called collar rots, foot rots, and Phytophthora root rot. Phytophthora root and crown rot in California affects many woody species, including acacia, *Agonis,* arborvitae, birch, camphor tree, ceanothus, cedar, *Chamaecyparis,* chestnut, coast redwood, cypress, daphne, dogwood, eucalyptus, fir, *Fremontodendron,* fruit trees, hemlock, holly, juniper, larch, madrone, manzanita, oak, pine, privet, *Prunus,* redbud, *Rhamnus,* rhododendron, tea tree, and walnut. *Phytophthora ramorum* infects aboveground plant parts, as discussed above in "Sudden Oak Death and Ramorum Blight."

Root- and crown-infecting *Phytophthora* cause leaves to wilt, discolor, remain undersized, and drop prematurely. Twigs and branches die back and the entire plant, especially when young, can be killed as roots and vascular tissue die. Plants infected when they are mature grow slowly and may gradually decline.

IDENTIFICATION AND BIOLOGY

Depending on environmental conditions and the species of pathogen and host plant, sap that is black, brown, or reddish may ooze from darkened areas of trunk bark. Cutting away bark from the basal trunk and roots often reveals a brown or reddish streak, stain, or canker in infected wood with a water-soaked margin separating it from the healthy whitish or yellowish wood. Woody roots decaying from *Phytophthora* alone are firm and brittle but eventually soften as secondary decay organisms develop.

Phytophthora spp. can survive in the soil for many years and require high soil moisture to infect host roots. Depending on the species of *Phytophthora,* the pathogen may affect only small feeder roots or rootlets, major roots, or all roots and the crown. *Phytophthora ramorum,* the cause of sudden oak death, infects through host leaves, trunks, and twigs. *Phytophthora* and *Pythium* spp. produce no fruiting bodies visible to the naked eye and are oomycetes, not true fungi. They are related to parasitic brown algae, and the biology and management of *Phytophthora* and *Pythium* spp. differ from that of most true fungi.

Phytophthora spp. are commonly introduced into landscapes through the planting of infested nursery stock. Many nurseries routinely apply pesticides that suppress (but do not eliminate) *Phytophthora,* so infected plants can be symptomless. After planting, disease develops and the introduced *Phytophthora* spreads in the landscape in drainage water and soil movement.

Confirmation of *Phytophthora* and *Pythium* spp. requires sending an appropriate plant sample with viable infection to a diagnostic laboratory. Test kits that employ the serological (antigen-antibody) technique ELISA (enzyme-linked immunosorbent assay) are available for use in the field to confirm the presence of *Phytophthora* with no need for specialized equipment or facilities. However, the use of test kits is limited primarily to nurseries and crops where research-based sampling recommendations have been developed. Caution must be exercised when interpreting test kit results, as several species of *Pythium* that apparently do not cause disease also react with the *Phytophthora* test kits (false positives). Conversely, test kits can fail to detect *Phytophthora* when it is present (false negatives) unless proper samples and a sufficient amount of the pathogen are collected. For more information on using pathogen test kits consult *Easy On-Site Tests for Fungi and Viruses in Nurseries and Greenhouses* or product manufacturers.

MANAGEMENT

Use only nursery stock that has been certified to be free of *Phytophthora* spp. and obtain plants from reputable local suppliers. Carefully select individual plants that are free of symptoms and that come from healthy lots of material. Inspect roots and reject plants that show decayed roots or evidence of crown rot. Learn whether oomycete pesticides were applied in the nursery and consider avoiding such plants, as they may appear to be healthy but can still be infected. Prepare the site well before planting, use excellent sanitation, and provide appropriate cultural care, especially proper irrigation, as discussed in Chapter 3. Where soils are compacted, drain poorly, or are often soggy, improve drainage and plant only species not reported to be susceptible to *P. cinnamomi*, such as those listed in Table 5-13.

Phytophthora in infested soils may be reduced by solarization, steam injection, or by maintaining low moisture levels and growing nonhost plants for extended periods. Certain soil microorganisms suppress *Phytophthora* spp., so adding pathogen-free organic matter to soil before planting to increase microbial activity and applying a 3- to 4-inch layer of chipped wood or other coarse organic mulch on the surface over root zones may be beneficial.

Some pesticides (e.g., phosphonates and mefenoxam) help to control certain *Phytophthora* spp. if combined with practices such as avoiding overirrigation and improving drainage. However, plants larger than medium-sized shrubs can be difficult to effectively treat in landscapes. Many fungicides effective against *Rhizoctonia* and other root decay fungi are not effective against oomycetes (*Phytophthora* and *Pythium* spp.), and vice versa. The genus of pathogen must definitely be known in order to choose an effective pesticide. An ineffective fungicide will not control oomycetes and can kill beneficial soil microorganisms, including those that naturally help to limit certain pathogens.

PYTHIUM ROOT ROT

Pythium spp., and occasionally other soilborne pathogens such as *Rhizoctonia* spp., can cause basal stem decay and root rot, or "damping-off," the death of seedlings that collapse at the soil line under damp conditions. *Pythium* spp. mainly affect fine feeder roots, and most *Pythium* spp. do not cause serious damage to woody landscape

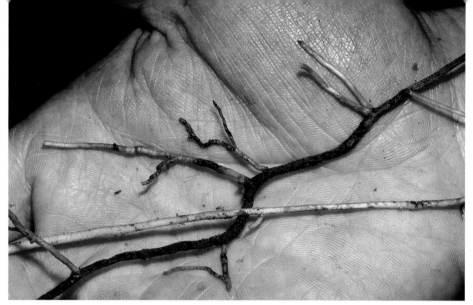

When infected by *Phytophthora*, small roots are commonly dark and soft or mushy. Healthy roots generally are firm and pale colored inside. Other indications of *Phytophthora* infection include roots that are less abundant than normal and that readily separate into inner and outer parts when pulled. *Photo:* D. Rosen

Table 5-13.

Trees and Shrubs Not Reported to Be Susceptible to *Phytophthora cinnamomi*.

COMMON NAME	GENUS NAME
albizia	*Albizia*
alder	*Alnus*
ash[1]	*Fraxinus*
aspen	*Populus*
California buckeye[1]	*Aesculus*
citrus[2]	*Citrus*
cotoneaster[3]	*Cotoneaster*
cottonwood	*Populus*
coyote brush	*Baccharis*
elm	*Ulmus*
euonymus[1]	*Euonymus*
honey locust	*Gleditsia*
linden	*Tilia*
magnolia[1]	*Magnolia*
maidenhair tree	*Ginkgo*
mayten	*Maytenus*
oleander[1]	*Nerium*
podocarpus	*Podocarpus*
poplar	*Populus*
sweet gum	*Liquidambar*
tamarisk	*Tamarix*
toyon[1,4]	*Heteromeles*
tulip tree	*Liriodendron*
zelkova	*Zelkova*

This list provides only a guideline; host vigor and environmental conditions are important in disease development. Some hosts are susceptible to other *Phytophthora* spp.

1. Susceptible to Ramorum blight, *Phytophthora ramorum*.
2. Susceptible to *Phytophthora citrophthora* and *P. parasitica*.
3. Susceptible to *Phytophthora cryptogea* and *P. parasitica*.
4. Susceptible to *Phytophthora cactorum*.

Sources: Farr et al. 1989; Ohr et al. 1980; USDA APHIS 2012.

plants. *Phytophthora* and *Pythium* spp. are closely related, but *Phytophthora* spp. are the more likely cause of root rot in woody landscape plants.

Prevent Pythium root rot and damping-off diseases by properly preparing the site before planting and providing appropriate cultural care, especially proper irrigation and good drainage. Solarizing the soil before planting (see Chapter 7) can temporarily reduce *Pythium* spp. in the upper soil, where most roots occur when plants are young and most susceptible to disease. Some pesticides can be effective against *Pythium* spp., but the genus of pathogen must definitely be known in order to choose an effective product because *Pythium* spp. are oomycetes and are not controlled by certain fungicides.

Chapter 6   Insects, Mites, and Snails

Many insects, mites, and other invertebrates live on and around landscape trees and shrubs. Some are pests because they annoy us or injure ornamentals, but most are innocuous or beneficial and should not be destroyed. Many invertebrates are necessary as food for birds and other wildlife or are valuable parasites or predators that destroy pests. Others break down organic matter so that nutrients are available for plant growth. Insects, including honey bees, are essential for pollinating certain plants so that seeds and fruit are produced.

Some level of most invertebrate species can usually be tolerated without harm to landscape plants. When control is appropriate, selective methods are often available and preferable. This chapter discusses common invertebrate pests and their management. For more information, see the online guides at ipm.ucanr.edu and "Suggested Reading" at the back of this book or contact the University of California (UC) Cooperative Extension office and Master Gardener Program in your county.

DAMAGE

Most invertebrate pests have chewing or sucking mouthparts and damage plants through their feeding. Beetles, caterpillars, snails, and other invertebrates with chewing mouthparts commonly cause identifiable holes in flowers, fruit, leaves, or twigs or cut parts completely from plants. Some chewing insects feed hidden inside trunks and limbs. These boring pests include bark beetles, clearwing moth larvae, and flatheaded and roundheaded borers. Scarab and weevil larvae and some other insects chew on roots. Insects that feed inside plant tissue or chew on roots can cause discolored or wilted leaves and other symptoms of plant decline that may be confused with pathogenic diseases, nutrient deficiencies, or cultural problems.

Sucking pests include aphids, leafhoppers, mites, scales, thrips, and true bugs. Pests with sucking (tubular) mouthparts consume fluids and do not cut off plant parts. Feeding by sucking insects and mites causes buds, fruit, or leaves to discolor, distort, die back, or drop. During feeding, some sucking pests excrete sticky honeydew or spread plant pathogens (e.g., bacteria or viruses).

Exotic Pests. Many of our worst pests were introduced from other states or countries. Argentine ant and many of the mealybug, psyllid, thrips, whitefly, and bark- and wood-boring pest species discussed in this chapter were inadvertently brought into California on contaminated firewood, fruit, plants, or soil. To help prevent pest introductions during planting and travel
- Do not bring any fruit, plants, seeds, or soil into California unless you know they have been inspected by agricultural officials or certified to be pest-free.
- Buy only pest-free plants from reputable, local nurseries.
- Purchase firewood near where you will burn it, leave any unused wood on site rather than moving it or taking it home, and do not move firewood between counties.
- Before purchasing plants from outside the local area or moving plants across county lines, contact the county agricultural commissioner to learn whether any quarantines prohibit plant movement.
- Take any unfamiliar pests to your county agricultural commissioner or UC Cooperative Extension office for identification or telephone the California Department of Food and Agriculture Pest Hotline at 1-800-491-1899.

Insects with chewing mouthparts, like this California oakworm, make distinct holes in fruit, leaves, or stems or entirely clip off plant parts. *Photo:* J. K. Clark

Insects with tubular sucking mouthparts, like this adult stink bug, cause plant parts to discolor, distort, drop, or become fouled with excrement. However, certain sucking insects are beneficial predators that impale their insect prey. *Photo:* J. K. Clark

LIFE CYCLES

Insects, the most common invertebrate pests, have three main body parts (head, thorax, and abdomen) and six legs, at least as adults. The adults of most species have wings, usually two pairs. Most invertebrates begin life as an egg that hatches into an immature form called a larva or nymph.

Immature insects and mites grow by periodically forming a new outer skin or exoskeleton (molting) and shedding their old skin. The series of growth changes from the egg to adult stage is known as metamorphosis. Insects can be divided into two major groups, depending on their type of metamorphosis. Species with complete metamorphosis change greatly in appearance between the immature and adult stages, and much of this change occurs during the nonfeeding pupal stage (Figure 6-1). Immatures of species that undergo complete metamorphosis are called larvae. In groups such as butterflies, moths, and flies, usually only the larval stage causes damage; the adults consume only nectar, pollen, and water. Beetles chew plants during both their adult and larval stages, but the adults often feed on different plant parts than larvae, and one stage typically is more damaging than the other. For example, larvae of certain scarab and weevil species chew roots and can seriously damage plants, while their adults chew leaves, causing cosmetic damage that is harmless to plants.

Insects that undergo gradual or incomplete metamorphosis (e.g., aphids, mealybugs, scales, true bugs, and other Hemiptera) have no pupal stage, and their immatures are called nymphs. Nymphs' appearance differs from adults primarily in size, lack of wings, and coloration (Figure 6-2).

With the most common groups of mites (spider mites and tetranychid predaceous mites), the stage that hatches out of the egg has six legs and is called a larva. Later-stage immatures (nymphs) and adults have eight legs, but otherwise look similar to mite larvae.

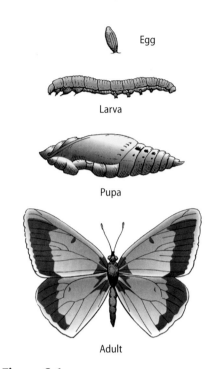

Egg

Larva

Pupa

Adult

Figure 6-1.
Insects with complete metamorphosis develop through four life stages: egg, larva, pupa, and adult. Each stage has a radically different appearance. Most moths and butterflies go through four or five molts during the larval stage. In butterflies and moths, the larva is also called a caterpillar. The pupa of some species (e.g., the one illustrated here) is called a chrysalid or chrysalis; in other species, the pupa occurs within a silk covering called a cocoon.

CONTROL ACTION GUIDELINES, OR THRESHOLDS

With most plant-feeding species the presence of a few individuals does not threaten plant health and usually can be tolerated. Few quantitative thresholds have been established as guidance on what abundance of pests or their damage warrants control action in landscapes, in part because of the lack of research and variation in people's tolerance for pests. Example thresholds are provided for aphid honeydew, defoliating caterpillars, and elm leaf beetle in the sections on those pests. Approximate thresholds can be developed over the long term by regularly monitoring plants and keeping and evaluating records, as discussed in Chapter 2. One strategy is to focus efforts on intolerable problems where reasonable actions can be effective and the cost and effort of control are acceptable. Recognize that in many instances, by the time problems become obvious, there is no effective cure; tolerating damage, waiting (possibly months) until pests develop to a stage vulnerable to control, or in some instances replacing plants may be the only options.

MONITORING AND DIAGNOSING PROBLEMS

Regularly inspect valued plants to determine what invertebrate species are present and their relative abundance. Insect and mite populations can increase rapidly when temperatures are warm; regular monitoring or sampling allows you to recognize developing problems and take action at the proper time. Once pest populations are high or damage becomes extensive, management options become more limited or infeasible.

Proper identification of pest and natural enemy species is essential for successful pest management. One approach is to identify the host plant, then use the "Tree and Shrub Pest Tables" at the end of this book to help diagnose the problem. Alternatively, if you know what kind of pest you have (for example, caterpillars or scale insects), you can go to that section of the book and use the photographs and descriptions to help identify common species. Because of the broad scope of this book and the frequency of new pests being introduced, not all possible landscape pests are included. Take pests or damaged plant parts you cannot identify to a UC Cooperative Extension office, county agricultural commissioner, or a certified nursery professional, or consult resources listed in "Suggested Reading."

After identifying the invertebrates, learn about their biology, potential damage, and management options. You may find that some species, while present near the damage or symptoms, are actually innocuous or beneficial. Even many of the species that cause damage can usually be tolerated at moderate levels without seriously threatening plant health.

Instead of simply monitoring (e.g., visually inspecting or scouting) to detect the presence of pests or damage, managers of high-value or problem-prone plants can benefit from quantitative monitoring called sampling. Select appropriate monitoring methods (Table 6-1) based on knowledge of pest biology, the goals of your monitoring, and available resources.

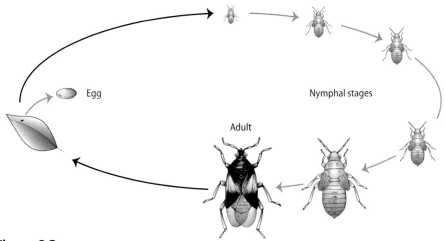

Figure 6-2.

Insects with incomplete metamorphosis develop through three life stages: egg, nymph, and adult. Nymphs resemble adults, except for coloration, size, and the lack of wings. This minute pirate bug (*Orius* sp.) is a beneficial predator of aphids, psyllids, thrips, mites, and insect eggs. Adapted from drawings by Celeste Green in Smith and Hagen 1956.

Table 6-1.

Insect and Mite Monitoring Methods.

METHOD	INVERTEBRATE SPECIES
visual inspection of plant parts	Most exposed-feeding species, including evidence of parasitism and predation. Monitoring tiny pests requires a hand lens.
branch beating	Most exposed, readily dislodged species, especially the adults, including green and brown lacewings, lady beetles, leaf beetles, leafhoppers, mites, nonwebbing caterpillars, psyllids, thrips, true bugs, and weevils.
sticky traps	Adults of many insects, including leafhoppers, parasitoids, psyllids, thrips, whiteflies, and winged aphids.
double-sided sticky tape	Scale and mealybug crawlers.
burlap trunk wraps	Adult weevils.
trap boards	Certain ground-dwelling invertebrates, including snails, slugs, and adult weevils.
pheromone traps	Adults of certain moths and scales, including California red scale, clearwing moths, fruittree leafroller, Nantucket pine tip moth, omnivorous looper, and San Jose scale.
pitfall traps	Adult weevils, ground-dwelling spiders, predaceous ground beetles.
timed counts	Pest individuals that are relatively large and obvious, such as caterpillars, and that occur at relatively low density so they are not observed faster than they can be counted.
honeydew monitoring	Aphids.
frass dropping	Nonwebbing caterpillars.
degree-day monitoring	Species for which researchers have determined development thresholds and rates, including certain caterpillars, elm leaf beetle, and Nantucket pine tip moth.

SAMPLING

Sampling is quantitative monitoring that measures the extent of damage or counts the number (or presence-absence) of pests per plant part (e.g., leaves or 1-foot branch terminals) or monitoring device (e.g., sticky trap card). Counts and comparison of the average number between dates help experienced users to determine

- whether pest abundance is increasing, decreasing, or remaining about the same
- whether management action is needed
- when and where to implement controls
- the effectiveness of management actions

Write down a description of your methods and sample the same way every time to make the results comparable among sample dates and from year to year. Record the date, specific location, host plant and pests sampled, who sampled, the results or counts from your samples, and what management action you took and when you took it. Specialized equipment and techniques for monitoring discussed below include branch beating or shaking, degree-days (temperature monitoring), and sticky traps. Pheromone traps and timed counts are discussed in "Foliage-Feeding Caterpillars." For more information on special tools and how to sample, consult *IPM in Practice* and *Pest Notes* on your particular problem.

Visual Inspection. Inspect a set number of randomly selected plant parts appropriate for that pest, such as leaves or terminals for caterpillars or their eggs. Record the total number of pest individuals found and perhaps also separately record numbers of its natural enemies. Determine the average insects per sample by dividing the total insects by the number of samples inspected.

Foliage may also be inspected for damage, such as by rating each sample from 0 to 10, where 0 equals no damage, 1 equals about 10% damage, 2 is about 20% damage, and so on. Take the average of all samples to estimate overall damage at that location. Because damage indicates past insect activity, make sure that the damaging life stages susceptible to treatment are still abundant before taking control actions.

Presence-Absence Sampling. An alternative to counting insects is to determine the percentage of samples with pests or damage. This presence-absence sampling is quicker than counting each individual, but is generally less precise except where researchers have determined the relationship between the percentage of samples infested and actual average insect densities, as discussed in *Elm Leaf Beetle Pest Notes*.

To presence-absence sample, inspect each sample (e.g., a 1-foot branch terminal) and record whether it is damaged or infested with one or more pests. As soon as you discover damage or pests, move on to inspect the next sample. Calculate and record the percentage of infested or damaged samples:

$$\text{Percentage of samples infested} = (\text{Number of samples infested} \times 100) \div \text{Number of samples inspected}$$

Branch Beating or Shaking. Sweep net or branch beat samples are appropriate for pests that are easily dislodged from foliage (e.g., adults of many species and nonwebbing caterpillars), but not for species that feed in enclosed parts or web themselves in leaves. Hold a light-colored plastic tray, framed cloth, or clipboard with a white sheet of paper beneath foliage and shake or beat the branch a fixed number of times (e.g., once or twice) to dislodge insects onto the collecting surface. Sampling from four locations (e.g., one sample from each cardinal direction) per plant may be sufficient. Alternatively, insert new growth flushes into a sweep net and shake vigorously, then empty the samples onto a clean surface. Count and record the number of pest individuals collected and calculate the overall average (total pests ÷ number of samples, or branches beaten or shaken).

Yellow Sticky Traps. Bright yellow attracts adults of many insects, including adult leafhoppers, parasitoids, psyllids, thrips, whiteflies, and winged aphids. To trap, hang several bright yellow cards cov-

Detect and monitor certain pest and beneficial species by branch beating or shaking to dislodge insects onto a collecting surface, such as a clipboard with a white piece of paper or this sheet placed on the ground. This technique can detect the adults and sometimes immatures of many species, including green and brown lacewings, lady beetles, leaf beetles, leafhoppers, mites, psyllids, thrips, and true bugs. *Photo:* D. Rosen

ered with clear sticky material near plants and inspect trap cards about once a week whenever adults may be present.

Use the same trap type and method throughout the season so that results are comparable among dates. Periodically replace traps so surfaces remain sticky and to prevent them from becoming covered with too many insects (e.g., flies). Preserve old used traps between clear plastic (e.g., sandwich wrap or ziplock bag) and label them with the date and location so you can compare catches between sample dates.

Carefully identify insects in traps, since many of them may be harmless or beneficial. It can be difficult to know the importance of insects trapped, but catching many adults may indicate they are abundant on nearby plants or that they are migrating in and will soon become abundant. Even large numbers of pest species on traps does not necessarily indicate that control action is needed, and there are no specific management guidelines in landscapes based on the number of insects caught in traps. For more information, consult *Sticky Trap Monitoring of Insect Pests* or *Integrated Pest Management for Floriculture and Nurseries*.

Degree-Day Monitoring. The growth and development rate of plants and invertebrates speeds up or slows down depending on the temperature of their environment. The amount of accumulated heat needed for a pest to develop from one point in its life cycle to another (physiological time) can be used to time pest control actions and is more biologically useful than basing management decisions on the calendar date. The unit used to measure physiological time is the degree-day (DD). One degree-day is defined as 1 degree above the threshold temperature maintained for a full day.

The total heat requirement and the lower and upper development thresholds are determined by research studies. A biofix, the date to begin accumulating degree-days, is also needed, and it varies with the species, usually based on a biological event that is important for the pest of interest.

Degree-days for each day are estimated by subtracting the threshold temperature from the average daily temperature for that date. Computerized "sine wave" calculation methods are recommended because they provide more accurate estimates.

Degree-day monitoring tells you when to take action (e.g., when pests will reach life stages susceptible to control actions), but monitoring temperatures does not tell you whether control action is needed; you must still monitor plants to decide whether thresholds are exceeded.

Current temperatures and easy-to-use point-and-click software for calculating degree-days and timing control actions for certain pests can be obtained online at ipm.ucanr.edu. Portable, compact electronic temperature recorders that calculate and display degree-days are also available. Using degree-days for decision making is discussed later in "Elm Leaf Beetle," "Pine Tip Moths," and "Scales."

MANAGEMENT

Integrated pest management (IPM) is a strategy for preventing and minimizing pest damage in ways that reduce risks to human health, beneficial and nontarget organisms, and the environment. IPM employs a combination of biological, cultural, chemical, mechanical, and physical controls as summarized below for invertebrates.

CULTURAL CONTROL

Effective pest management begins when you select plants that are well adapted to their location and properly plant and care for them. For many plants, you can choose similar-looking cultivars or species that are resistant to certain pests. As detailed in Chapter 3, appropriate pruning, watering, and other care keep landscape plants vigorous so they are less likely to be attacked by certain pests and are better able to tolerate pest damage.

MECHANICAL CONTROL

Mechanical controls use labor or non-pesticidal materials to control pests. For

Black vine weevil adults exposed by removing corrugated cardboard wrapped around a trunk. Populations of night-feeding weevil adults, earwigs, and certain pupating caterpillars can be reduced by regularly inspecting trunk wraps and disposing of pests that sought shelter there. *Photo:* P. A. Phillips

example, install copper bands around trunks and planting areas to exclude snails and slugs. Apply sticky barriers on trunks to exclude certain flightless invertebrates. To prevent the spread of bark- and wood-boring insects, chip or grind branches for use as mulch on site or locally. When only a small part of a plant is infested, clip and dispose of infested foliage, such as shoots harboring tentmaking caterpillars or other insects that feed in groups. Hand-pick or trap snails.

Sticky Barriers. Exclude ants, flightless weevils, snails, and certain other invertebrates by encircling trunks with a band of sticky material (e.g., Tanglefoot). Prune branches to eliminate bridges to the ground, structures, and any touching plants that are not also banded. Wrap the trunk with a collar of fabric, heavy paper, or tape, then coat the wrap with sticky material to protect young or sensitive trees from possible bark injury and to facilitate removal. If bark is not smooth, use a pli-

Apply sticky material around trunks to prevent canopies from being infested by ants, flightless weevils, and snails. Apply the sticky barrier over wrapping to protect young or sensitive trees from possible bark injury and to facilitate removal. *Photo:* J. K. Clark

able wrap and wedge it snugly into cracks and crevices. To avoid restricting trunk growth, periodically replace bands and do not wrap trunks tightly. Increase the persistence of sticky material by applying it higher above the ground, which reduces dust and dirt contamination and decreases sprinkler wash-off. Avoid applying sticky material to horizontal surfaces where birds may roost.

A barrier band about 2 to 6 inches wide should be adequate in most situations. Check the sticky material at least every 1 to 2 weeks and stir it with a stick to prevent the surface from becoming clogged with debris or dead insects that allow ants to cross. Periodically remove and relocate any wrap to inspect for and minimize injury to bark.

PHYSICAL CONTROL

Physical or environmental controls suppress pests by altering light, humidity, and temperature. For example, control black scale and certain other scales in hot locations by thinning plant canopies, thereby increasing scale mortality due to heat exposure. Solarize freshly cut wood to prevent bark- and wood-boring insects from emerging and attacking nearby standing trees; pile logs in a sunny location and tightly cover them with high-quality UV-resistant clear plastic for several months. Apply white, interior (not exterior) latex paint, diluted with an equal amount of water, to trunks of young or heavily pruned woody plants to prevent sunburn; interior water-based paints are safer to trees than oil- and water-based exterior paint. This whitewashing of trunks also helps prevent attack by wood-boring insects such as flatheaded and roundheaded borers that are attracted to injured trunks. Spray foliage with a forceful stream of water to dislodge and kill aphids or whiteflies. Wet the underside of spider mite-infested foliage regularly during hot weather to help reduce spider mite populations; the increased humidity and rinsing away of dust improve the reproduction and survival of predatory mites.

BIOLOGICAL CONTROL

Biological control—the action of parasites, pathogens, predators, and competitors to control pests and reduce damage—is very important in the management of invertebrates. Importation (classical biological control), conservation, and augmentation are three tactics for using natural enemies. Become familiar with common beneficial species and how to encourage their activity. For more details, consult the *Biological Control and Natural Enemies Pest Notes* and *Natural Enemies Gallery* online at ipm.ucanr.edu and the *Natural Enemies Handbook*.

Importation, or Classical Biological Control. Classical biological control is the importation, release, and establishment of exotic natural enemies, primarily against pests that have been introduced from elsewhere. Many organisms that are not pests in their native habitat become unusually abundant when they arrive in a new area without their natural controls. Many

Life stages of the vedalia lady beetle, a predator of cottony cushion scale. To become familiar with this and other common beneficial species, see the photographs and descriptions in the *Natural Enemies Gallery* online at ipm.ucanr.edu. *Photos:* J. K. Clark

insects that were formerly widespread pests in California landscapes are now partially or completely controlled by introduced natural enemies (Table 6-2), except where natural enemies have been disrupted, such as by pesticide applications, honeydew-seeking ants, or weather.

Introduction of exotic natural enemies from foreign countries by law must be done only by qualified university or government scientists. Researchers collect natural enemies that kill the pest in their native habitat and study them in an approved quarantine facility to determine whether they will have minimal negative impact in the new country of release. Natural enemies found to be promising are introduced into the new environment so they are reassociated with the pest, after which they may reduce their host's population to a low enough level so that it is no longer considered a pest.

It is important for landscape managers to recognize imported natural enemies and conserve them whenever possible. Because classical biological control can provide long-term benefits over a large area and is conducted by agencies and institutions funded through taxes, public support is critical to the continued success of classical biological control.

Is Biological Control "Safe"? A great benefit of biological control is its relative safety for human health and the environment. Most negative impacts from exotic species have been caused by undesirable organisms contaminating imported goods, by travelers carrying in pest-infested fruit, and from introduced ornamentals that escape cultivation and become weeds. These ill-advised or illegal importations are not part of biological control.

Negative impacts have occurred from poorly conceived, quasi-biological control importations of predaceous vertebrates like frogs, mongooses, and certain fish, often conducted by nonscientists. To avoid these problems, biological control researchers follow government quarantine regulations and work mostly with host-specific natural enemies that pose low risks and can provide great benefits. As a pest comes under good biological control, population densities decline for both the pest and of the biological control agent because host-specific natural enemies cannot prey or reproduce on other species.

Conservation and Enhancement. Conservation is the use of management practices that preserve naturally occurring beneficial organisms. Preserve resident natural enemies by choosing cultural, mechanical, or selective chemical controls that do not interfere with or kill beneficial species. Most pests are attacked by a complex of natural enemies, and their conservation is the primary way to successfully use biological control. Judicious (e.g., selective) pesticide use, ant control, and habitat manipulation (e.g., growing flowering plants that provide nectar or pollen and reducing dustiness that inhibits natural enemy activity) are key conservation strategies.

Pesticide Management. In comparison with their effect on pest species, pesticides often kill a higher proportion of predators and parasites because they are more sensitive to sprays, for example, because their active searching behavior exposes them to a larger dose of pesticide residue than pest species that feed in one spot during their entire lives. In addition to immediately killing natural enemies that are present at the time of spraying, many residual pesticides persist inside or on foliage and kill predators or parasites that migrate in after spraying. Even if beneficial organisms survive an application, low levels of pesticide residues can interfere with natural enemies' abilities to reproduce and to locate and kill pests.

Biological control's importance often becomes apparent when broad-spectrum, residual pesticides (those that persist for days or weeks) cause secondary pest outbreaks or pest resurgence. A secondary outbreak occurs when pesticides applied against a target pest kill natural enemies of other species, causing the formerly innocuous species to become a pest (Figure 6-3). Target pest resurgence occurs when spraying reduces the number of pests but causes an even greater destruction of the pest's natural enemies. The resulting unfavorable ratio of pests to natural enemies permits a rapid increase or resurgence of the primary pest population, sometimes to levels higher than were observed prior to spraying. Carbamate, organophosphate,

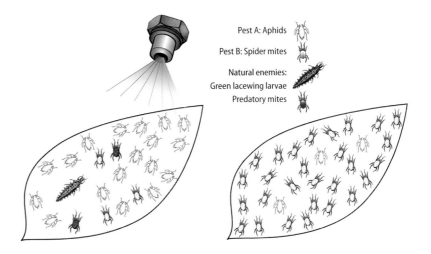

Pest A: Aphids
Pest B: Spider mites
Natural enemies:
Green lacewing larvae
Predatory mites

Figure 6-3.
Killing natural enemies can cause a secondary outbreak of insects and mites. For example, spider mites are often present on plants at low densities but become excessively abundant and cause damage when pesticides applied against other species kill the natural enemies of the spider mites. Here a pesticide applied to kill aphids (Pest A) not only killed aphids but also killed predaceous green lacewing larvae and predatory mites, leading to an outbreak of spider mites (Pest B). Broad-spectrum insecticides applied during hot weather appear to have the greatest effect on mites, sometimes causing dramatic mite outbreaks within a few days after spraying.

Table 6-2.

Woody Landscape Pests Substantially to Completely Controlled by Natural Enemies Introduced for Classical Biological Control.

PEST COMMON NAME	PEST SCIENTIFIC NAME	INTRODUCED NATURAL ENEMIES	SEE PAGE
acacia psyllid	*Acizzia uncatoides*	*Diomus pumilio*,[1] *Anthocoris nemoralis*[2]	138, 170–171
ash whitefly	*Siphoninus phillyreae*	*Encarsia inaron, Clitostethus arcuatus*[1]	185
bayberry whitefly	*Parabemisia myricae*	*Encarsia* spp., *Eretmocerus* spp.	
black scale	*Saissetia oleae*	many species, including *Metaphycus helvolus, Metaphycus bartletii, Scutellista caerulea*	192, 200
bluegum psyllid	*Ctenarytaina eucalypti*	*Psyllaephagus pilosus*	171
brown soft scale	*Coccus hesperidum*	*Metaphycus luteolus, Metaphycus* spp., *Microterys nietneri, Rhizobius lophanthae*,[1] *Chilocorus* spp.[1]	193, 200
California red scale	*Aonidiella aurantii*	*Aphytis melinus, Aphytis* spp., *Comperiella bifasciata, Rhizobius lophanthae*,[1] *Chilocorus* spp.[1]	188, 189, 201
citricola scale	*Coccus pseudomagnoliarum*	*Coccophagus lycimnia, Coccophagus scutellaris, Metaphycus flavus, Metaphycus luteolus*	194, 200
citrophilus mealybug	*Pseudococcus calceolariae*	*Coccophagus gurneyi, Hungariella pretiosa*	
citrus mealybug	*Planococcus citri*	*Leptomastix dactylopii, Leptomastidea abnormis, Cryptolaemus montrouzieri*[1]	186, 188
citrus whitefly	*Dialeurodes citri*	*Encarsia* spp.	
Comstock mealybug	*Pseudococcus comstocki*	*Allotropa convexifrons, Allotropa burrelli, Pseudaphycus malinus*	
cottony cushion scale	*Icerya purchase*	*Rodolia cardinalis*,[1] *Cryptochaetum iceryae*[3]	130, 196–197
dictyospermum scale	*Chrysomphalus dictyospermi*	*Aphytis* spp.	
elm aphid	*Tinocallis platani*	*Trioxys tenuicaudus*	
eucalyptus longhorned borer	*Phoracantha semipunctata*[4]	*Avetianella longoi*	233
eucalyptus redgum lerp psyllid	*Glycaspis brimblecombei*	*Psyllaephagus bliteus*	172
eucalyptus snout beetle	*Gonipterus scutellatus*	*Anaphes nitens*	157
eugenia psyllid	*Trioza eugeniae*	*Tamarixia dahlsteni*	172
European elm scale	*Eriococcus spuria*	*Baryscapus (=Trichomasthus) coeruleus*	197
filbert aphid	*Myzocallis coryli*	*Trioxys pallidus*	
giant whitefly	*Aleurodicus dugesii*	*Encarsia noyesi, Entedononecremnus krauteri, Idioporus affinis*	183
linden aphid	*Eucallipterus tiliae*	*Trioxys curvicaudus*	
longtailed mealybug	*Pseudococcus longispinus*	*Anarhopus sydneyensis, Arhopoideus peregrinus*	
Nantucket pine tip moth	*Rhyacionia frustrana*	*Campoplex frustranae*	218
nigra scale	*Parasaissetia nigra*	*Metaphycus helvolus*	
obscure scale	*Melanaspis obscura*	*Encarsia aurantii*	
olive scale	*Parlatoria oleae*	*Aphytis maculicornis, Coccophagoides utilis*	
peppertree psyllid	*Calophya rubra*	*Tamarixia schina*	174
pink hibiscus mealybug	*Maconellicoccus hirsutus*	*Anagyrus kamali*	
purple scale	*Lepidosaphes beckii*	*Aphytis lepidosaphes, Chilocorus* spp.,[1] *Rhizobius lophanthae*[1]	201
San Jose scale	*Diaspidiotus pernicious*	*Encarsia perniciosi*	
walnut aphid	*Chromaphis juglandicola*	*Trioxys pallidus*	
woolly whitefly	*Aleurothrixus floccosus*	*Amitus spiniferus, Cales noacki*	

Natural enemies for each pest are listed in order of their importance. Biological control is sometimes disrupted, such as by pesticide applications, honeydew-seeking ants, or weather. Certain species are biologically controlled in only portions of California, such as near the coast but not at interior locations, or vice versa.

These natural enemies are parasitic wasps except where noted:

1. lady beetle
2. pirate bug
3. parasitic fly
4. A more recently introduced longhorned borer (*Phoracantha recurva*) is less well controlled.

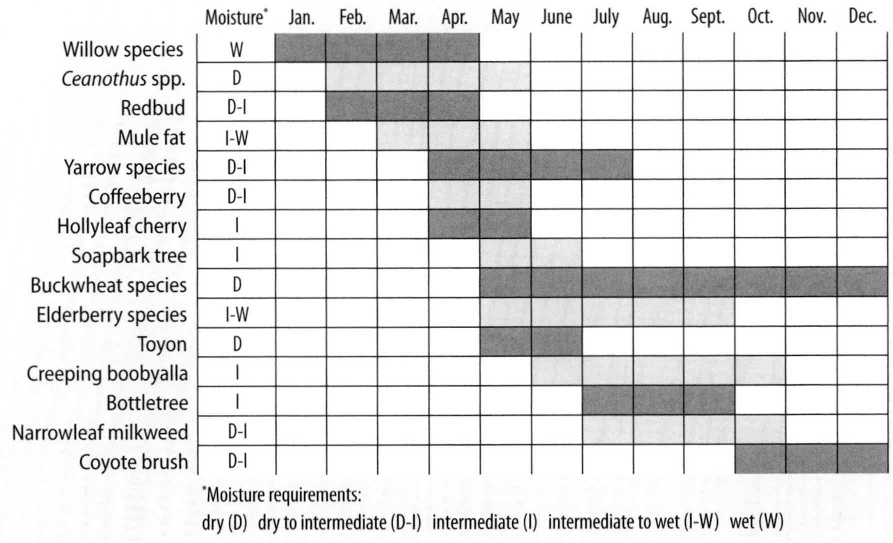

	Moisture*	Jan.	Feb.	Mar.	Apr.	May	June	July	Aug.	Sept.	Oct.	Nov.	Dec.
Willow species	W	■	■	■	■								
Ceanothus spp.	D		■	■	■								
Redbud	D-I		■	■	■								
Mule fat	I-W												
Yarrow species	D-I				■	■	■	■					
Coffeeberry	D-I					■	■						
Hollyleaf cherry	I					■	■						
Soapbark tree	I												
Buckwheat species	D					■	■	■	■	■	■	■	■
Elderberry species	I-W					■	■						
Toyon	D						■						
Creeping boobyalla	I							■	■	■			
Bottletree	I							■	■	■			
Narrowleaf milkweed	D-I							■	■	■			
Coyote brush	D-I										■	■	

*Moisture requirements:
dry (D)   dry to intermediate (D-I)   intermediate (I)   intermediate to wet (I-W)   wet (W)

Figure 6-4.

The flowering periods (the darkened cells) of selected perennial insectary plants. Growing the right combination of flowering plants can provide nectar and pollen for adult natural enemies sequentially throughout the year. For a more detailed list of insectary plants, see *Flower Flies* (Syrphidae) *and Other Biological Control Agents for Aphids in Vegetable Crops* and other publications listed in "Suggested Reading" at the back of the book.

and pyrethroid insecticides are especially toxic to natural enemies. Neonicotinoids (e.g., imidacloprid) and other systemic insecticides that translocate into blossoms can poison adult natural enemies and honey bees that feed on nectar and pollen.

Eliminate or reduce the use of broad-spectrum, residual pesticides whenever possible. Apply pesticides in a selective manner (e.g., spot applications), time applications to minimize impacts on natural enemies (dormant season applications), and choose insecticides that are more specific in the types of invertebrates they kill. Wait until plants are done flowering before treating them with systemic insecticides. Wherever possible, rely on low-persistence insecticides (insecticidal soap and narrow-range oil) or pesticides that are highly selective (*Bacillus thuringiensis*) or somewhat selective (spinosad). See Table 6-3 for a comparison of some common insecticides' relative toxicity to natural enemies.

Ant Control. Ants are beneficial as consumers of weed seeds, predators of many insect pests, soil builders, and nutrient cyclers. Certain species (e.g., fire ants) attack people and pets or are direct pests of plants, feeding on bark or fruit. The Argentine ant and certain others are pests primarily because they feed on honeydew produced by phloem-sucking insects (including aphids, mealybugs, soft scales, and whiteflies). Ants protect these honeydew producers (informally called "Homoptera") from predators and parasites that might otherwise control them. Where natural enemies are present, if ants are controlled, populations of many pests will decline gradually (over several generations) as natural enemies become more abundant. Ant controls include barriers, cultivation, and enclosed insecticide baits, as discussed later in "Ants."

Habitat Manipulation. Good landscape management may enhance natural enemy effectiveness, although there are few research-based recommendations. Dust can interfere with natural enemies and can cause outbreaks of spider mites, so reduce dust by planting ground covers and windbreaks and periodically washing dust from

Table 6-3.

Relative Toxicity to Natural Enemies of Certain Insecticide Groups.

INSECTICIDE	TOXICITY TO PARASITES AND PREDATORS[1]	
	Direct	Residual
microbial (*Bacillus thuringiensis*)	no	no
botanicals (pyrethrins)	yes/no[2]	no
oil (horticultural), soap (potash soap)	yes	no
microbial (spinosad)	yes/no[2]	yes/no[2]
neonicotinoids (imidacloprid)	yes/no[2]	yes
carbamates (carbaryl), organophosphates (malathion), pyrethroids (bifenthrin)	yes	yes

1. Direct contact toxicity is killing within several hours from spraying the beneficial or its habitat. Residual toxicity is killing or sublethal effects (such as reduced reproduction or ability to locate and kill pests) due to residues that persist.

2. Toxicity depends on the specific material and how it is applied and the species and life stage of the natural enemy.

See the active ingredients database at ipm.ucdavis.edu/PMG/menu.pesticides.php for more information about specific pesticides.

Figure 6-5.
Pruning management conserves parasites of certain pests, such as psyllids and whiteflies. Here, clippings from eugenia bushes are left as mulch on the ground for at least 3 weeks. This allows parasites to complete their development and return to the shrubs as adults that parasitize other psyllid nymphs by laying eggs in them. Most psyllids on cut foliage will die. Adapted from illustration by C. M. Dewees from Kabashima et al. 2014.

foliage. Plant a variety of flowering species to serve as insectary plants that provide natural enemies with nectar, pollen, and shelter throughout the growing season (Figure 6-4). Adults of parasites and many predators (e.g., lady beetles, green lacewings, and syrphid flies) feed on pollen and nectar. Even if pests are abundant for the predaceous and parasitic stages, many beneficials will do poorly unless flowering or nectar plants are available to provide food for adults.

Pruning management conserves parasites of pests such as psyllids and whiteflies. For example, eugenia psyllid is partially controlled by a tiny introduced wasp (*Tamarixia dahlsteni*), but parasitic wasps developing within older psyllid nymphs are removed if plants are regularly sheared, as happens with eugenia (Australian brush cherry) managed as topiary plants. Leaving prunings as mulch near plants for at least 3 weeks allows many parasites to complete their development and emerge, while most psyllids on cut foliage will die (Figure 6-5). Alternate pruning (pruning half of the plants or only one-half of a single plant one month and trimming the rest the next month or later) is recommended

for some whitefly-infested plants. Many whiteflies prefer to lay eggs on the succulent new growth that develops after pruning. Alternate pruning provides refuges for whitefly parasites to complete their development and emerge from older growth on untrimmed terminals and attack whiteflies infesting the new growth nearby.

Augmentation. Augmentation is the manipulation of pests or natural enemies to make biological control more effective. When resident natural enemies are insufficient, their populations can be increased (augmented) in certain situations through the purchase and release of commercially available beneficial species. Inoculation and inundation are two tactics for augmenting natural enemies. In inoculative releases, pest populations are low and relatively few natural enemies are released. The offspring of these predators or parasites, not the same individuals released, are expected to eventually provide biological control. The mealybug destroyer lady beetle (*Cryptolaemus montrouzieri*) overwinters poorly in California, and releasing small numbers of mealybug destroyers in the spring to control mealybugs is an example of an inoculative release. Inundative

releases involve large numbers of natural enemies, often released several times over a growing season. The natural enemies released, and possibly their progeny, are expected to provide biological control. Periodically releasing the convergent lady beetle (*Hippodamia convergens*), or "ladybugs," is an example of inundative biological control.

Releasing Natural Enemies Effectively. In most situations, conserving resident natural enemies is more effective than releasing them. Releases are more likely to be effective in the few situations where researchers or pest managers have previously demonstrated success and where some levels of pests and damage can be tolerated. Desperate situations in which pests or damage are already abundant are not good opportunities for augmentation, as pest numbers are too high for the natural enemies to control them quickly. To increase the likelihood that natural enemy releases will be effective

- Accurately identify the pest and its life stages.
- Learn about the biology of the pest and its natural enemies.
- Release the appropriate natural enemy species when the pest is in its vulnerable life stage and at densities where control is feasible.

Unlike insecticides, natural enemy releases cannot be used in a curative manner. Anticipate pest problems and plan releases ahead of time. Begin making releases before pests are too abundant or intolerable damage is imminent. Avoid applying broad-spectrum, residual insecticides (those that persist for days or weeks), or use them as spot sprays against pest hot spots (very abundant, but localized infestations). Remember that natural enemies are living organisms that require food, shelter, and water. Protect them from extreme conditions, such as releasing them at night or early in the day during hot weather.

Effectively releasing natural enemies requires knowledge, practice, and imagination. Releases often fail because information or experience was inadequate, the wrong species was released, the timing was incorrect, pesticides were applied, or the natural enemy species obtained was not effective when released. For example, releases of praying mantids are not recommended. Praying mantids feed indiscriminately on pest and beneficial species, including other mantids and honey bees. Although praying mantids are fascinating creatures, their release is unlikely to provide effective pest control.

TYPES OF NATURAL ENEMIES

The primary natural enemies are parasites, pathogens, and predators. Most parasites and many pathogens and predators attack only one or several related pest species. For example, the convergent lady beetle and syrphid fly larvae feed primarily on aphids; these and other more specialized natural enemies are discussed in the individual pest sections.

PARASITES

A parasite feeds in or on a larger host. Insect parasites (more precisely called parasitoids) are smaller than their host and develop inside or attached to the outside of the host and kill it. Often, only the immature stage of the parasite feeds on the host, and it kills only one host individual during its development (see Figure 6-6), so this type of natural enemy can be effective even when pest densities are low. Adult females of certain parasites (e.g., many wasps that attack scales and whiteflies) feed on the body fluids of their hosts. Pest mortality from this host feeding by adult parasites can be easily overlooked, but is an important source of biological control.

Most parasitic insects are flies (Diptera) or wasps (Hymenoptera). The most common parasitic flies are Tachinidae. Adult tachinids often resemble house flies, and their larvae are maggots that feed inside their host.

Parasitic Hymenoptera occur in over three dozen families, including

- Aphelinidae—about 1,000 known species of tiny wasps that attack aphids, mealybugs, psyllids, scales, and whiteflies.
- Aphidiidae—attack aphids.
- Braconidae and Ichneumonidae—often large wasps, but they do not sting people; about 5,000 species in North America that commonly parasitize beetle, caterpillar, and sawfly larvae and pupae.
- Encyrtidae—over 3,000 species that attack primarily scales and mealybugs, but also beetles, bugs, cockroaches, flies, and moths.
- Eulophidae—attack beetles, caterpillars, flies, psyllids, scales, and thrips.
- Trichogrammatidae—parasitize insect eggs.

A female *Aphidius* laying her egg inside an aphid, where the wasp's larva will feed and eventually kill its host. Before the parasite matures and emerges as an adult, it causes the aphid skin to discolor and become puffy (mummified), as pictured later in "Aphids." *Photo:* J. K. Clark

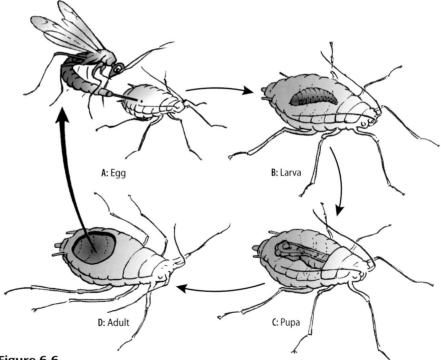

A: Egg

B: Larva

C: Pupa

D: Adult

Figure 6-6.
In many species, only the immature stage of a parasite feeds on the host, as illustrated with this aphid parasite. A. An adult parasite lays an egg inside a live aphid. B. The egg hatches into a parasite larva that grows as it feeds on the aphid's insides. C. The parasite kills the aphid and pupates inside the mummified host. D. The wasp adult chews a hole, emerges, and seeks other aphids to parasitize.

Entomopathogenic nematodes emerging from a diaprepes root weevil they killed. Commercially available beneficial nematodes can be applied to kill certain wood-boring and root-feeding insects where the environment is moist and protected from sunlight. *Photo:* Clayton W. McCoy, University of Florida

PATHOGENS

Pathogens are microorganisms, including bacteria, fungi, certain nematodes, protozoa, and viruses, that infect and kill the host. Populations of aphids, caterpillars, certain mites, and other invertebrates are sometimes drastically reduced by naturally occurring pathogens, although special conditions are often required, such as prolonged high humidity or dense pest populations. Certain pathogens are commercially available, as discussed in "Microbial and Biological Insecticides." These include *Bacillus thuringiensis,* entomopathogenic nematodes (e.g., *Heterorhabditis* and *Steinernema* spp.), fungi (*Beauveria bassiana*), and microorganism by-products (avermectins and spinosyns).

PREDATORS

Predators kill and feed on several to many individual prey during their lifetimes and generally are most effective when prey are abundant. Some predators are specialized

and feed on only one or a few closely related species. Others are general predators that feed opportunistically on a variety of available prey. Some species are predaceous only during their larval or nymphal stages, while others are predaceous as both adults and immatures. Depending on the species, adult predators also (or only) feed on honeydew, nectar, and pollen.

Predators important in the control of insect pests include beetles (order Coleoptera), flies, lacewings (Neuroptera), true bugs (Heteroptera), wasps (Hymenoptera), and spiders (Araneae, or Arachnida). Mites in the family Phytoseiidae are very important in the control of pest mites and certain insects.

BIRDS AND OTHER VERTEBRATES

Insects are important food for many amphibians, birds, mammals, and reptiles. Some bats and birds feed almost exclusively on insects, and many birds that normally feed on seeds or other plant parts rely on

caterpillars and other insects to feed their nestlings. Populations of desirable birds can be increased by growing a mix of trees, shrubs, and ground covers of different size, species, and density and by providing water and perhaps supplemental food. Many insect-eating birds nest in cavities in trees. Although dead and dying trees can rarely be left in urban landscapes because they can be hazardous and may fall, a practical alternative is to provide nesting boxes or birdhouses. Assess the value of dead and dying trees as wildlife habitat along with the hazard when considering tree removal.

ASSASSIN BUGS

Assassin bugs (Reduviidae) are usually brown, black, or reddish, with an elongate head and body and long legs. Both adults and nymphs feed on caterpillars, leafhoppers, and other small- to medium-sized mobile insects, which they ambush or stalk and impale with their needlelike mouthparts.

An adult assassin bug (*Zelus renardii*) eating a lygus bug. Assassin bug nymphs and adults prey on a wide variety of insects. *Photo:* J. K. Clark

Assassin bug eggs are laid on end, glued together in a cluster. These *Zelus* eggs are barrel shaped and dark brown with a white cap, and are found exposed on leaves. *Photo:* J. K. Clark

LACEWINGS AND DUSTYWINGS

Lacewing and dustywing larvae are flattened, tapered at the tail, and have long, curved mandibles for grasping prey. The larvae resemble tiny alligators and feed voraciously on mites and small insects, including aphids, caterpillars, leafhoppers, mealybugs, psyllids, whiteflies, and insect eggs.

Adult green lacewings (*Chrysopa* and *Chrysoperla* spp., Chrysopidae), have slender, green bodies and large, green wings with netlike (lacy) veins. Green lacewings lay oblong green to gray eggs on slender stalks, either singly or in groups. Stalks help protect the eggs from predators and cannibalistic siblings. Depending on the species, adults may feed on insects or only on honeydew, nectar, and pollen.

Adult brown lacewings (Hemerobiidae) resemble green lacewings, except that brown lacewings are typically about half as large and brown. Females lay oblong eggs singly on plants. Larvae resemble those of green lacewings and at maturity pupate in a thin, loosely woven layer of silk.

Dustywings (Coniopterygidae) are less commonly seen than lacewings. Adults resemble a large whitefly or a small lacewing with whitish powder covering the wings. Larvae are commonly gray, white, and black or orangish and pupate in an inconspicuous, flat, white, silken cocoon on the underside of leaves.

LADY BEETLES

Adults and larvae of lady beetles (Coccinellidae), or "ladybugs," are predators of pest mites and most soft-bodied or sessile (immobile) insects. About 200 species of lady beetles occur in California. The convergent lady beetle and many other orangish lady beetle species feed primarily on whatever species of aphids are abundant. Species such as multicolored Asian lady beetle feed on a variety of pests, aphids, psyllids, and scales. The vedalia beetle feeds on only one species, the cottony cushion scale. Other species of lady beetles specialize on mites or certain insects as pictured and discussed in "Aphids," "Mealybugs," "Mites," "Psyllids," and "Scales."

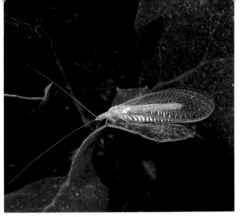

This adult green lacewing (*Chrysoperla* sp.) is named for its green body and wings with netlike veins. Adult brown lacewings look similar but are about half as large and brownish. *Photo:* J. K. Clark

This alligatorlike green lacewing larva (*Chrysoperla rufilabris*) is sucking the body fluids of a rose aphid. Lacewings' varied prey includes caterpillars, leafhoppers, mealybugs, psyllids, whiteflies, and insect eggs. *Photo:* J. K. Clark

Green lacewings lay each oblong egg on a slender stalk. Depending on the species, eggs are laid singly or in groups, as shown here for *Chrysopa nigricornis*. *Photo:* J. K. Clark

A green lacewing pupa, changing from the larval to adult stage within a spherical, silken cocoon. *Photo:* J. K. Clark

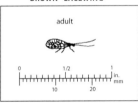

BROWN LACEWING

adult

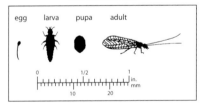

GREEN LACEWING

egg   larva   pupa   adult

Adult, larva (right), and pupa of the introduced multicolored Asian lady beetle (*Harmonia axyridis*). The adult has more than 100 color forms, including orangish or red with no spots or up to 19 large or small spots. Both adults and larvae prey on a variety of pests. *Photo:* J. K. Clark

Syrphid fly eggs are elongate-oval, have a reticulated surface, and are laid singly near aphid colonies. The similar-looking eggs of brown lacewings have a relatively smooth surface and a tiny knob projecting at one end. *Photo:* J. K. Clark

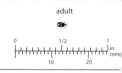

An adult minute pirate bug (*Anthocoris nemoralis*) feeding on an acacia psyllid nymph. These predators consume mites, insect eggs, and various small, soft-bodied pests. *Photo:* J. K. Clark

MINUTE PIRATE BUG

adult

0   1/2   1
mm
10   20

MINUTE PIRATE BUGS

Adults and nymphs of minute pirate bugs (Anthocoridae) commonly feed on aphids, mites, thrips, psyllids, and insect eggs. The adults of many pirate bug species are black or purplish with white markings and commonly have a triangular or X-shaped pattern on the back caused by the folding of their half-dark and half-clear wings. The nymphs are commonly yellowish or reddish brown. *Orius* and *Anthocoris* are two common genera.

PREDACEOUS FLIES

Larvae of many flies prey on mites and soft-bodied insects, such as aphids and mealybugs. Predaceous groups include aphid flies (Chamaemyiidae); aphid midges, also called predatory gall midges (Cecidomyiidae); and syrphid flies (Syrphidae). Larvae are maggotlike and can be brown, green, orangish, yellow, or whitish. Adult syrphids, also known as hover flies or flower flies, resemble honey bees but do not sting. They are often seen feeding on flower nectar. Adult aphid flies are small and chunky. Aphid midge adults are small, slender, and delicate, and larvae are maggotlike, as pictured in "Aphids."

PREDACEOUS GROUND BEETLES

Predaceous ground beetle (Carabidae) adults are commonly black or dark reddish, although some species are brilliantly colored or iridescent. They dwell on the ground, are most active at night, and are fast runners with long legs. Larvae dwell in litter or soil, are elongate, and have a large head with distinct mandibles. Carabids feed on snails, slugs, root-feeding insects, and insect larvae and pupae.

SOLDIER BEETLES

Adult soldier beetles (Cantharidae) are long, narrow beetles, usually red or orange with black, gray, or brown wing covers.

An adult syrphid, also called a flower fly or hover fly, requires pollen to reproduce. It is sometimes mistaken for a honey bee, but syrphid flies cannot sting. Syrphid larvae (pictured in "Aphids") resemble caterpillars. *Photo:* J. K. Clark

SYRPHID

larva   adult

0   1/2   1
in.
mm
10   20

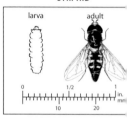

This adult predaceous ground beetle (*Calosoma* sp.) stalks its prey on soil or in litter, mostly during the night. *Photo:* J. K. Clark

PREDACEOUS GROUND BEETLES

Amara adult   Calosoma adult

0   1/2   1
in.
mm
10   20

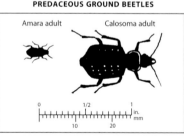

Adult soldier beetles are long and narrow, often with an orangish head and thorax and dark wing covers. They eat aphids and the eggs and larvae of beetles and moths. *Photo:* J. K. Clark

SOLDIER BEETLE

adult

0   1/2   1
in.
mm
10   20

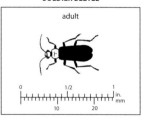

Adults are often observed on flowers, where they feed on nectar and pollen. The dark, flattened larvae are predaceous, as are the adults of many species. They feed on aphids and the eggs and larvae of beetles and moths.

PREDACEOUS MITES

Many species of mites are predators of plant-feeding pest mites and insects. Mites are related to spiders and, unlike insects, do not have antennae, segmented bodies, or wings. Most predaceous mites are long-legged, pear shaped, and shiny. Many are translucent, although after feeding they often take on the color of their prey and may be bright green, red, or yellow. Predaceous mite eggs are colorless and oblong, compared with plant-feeding mite eggs, which are commonly spherical and colored to opaque. One way to distinguish plant-feeding mites from predaceous species is to closely observe them on your plants with a good hand lens. Predaceous species appear more active than plant-feeding species; they stop only to feed. In comparison with pest mites, predaceous mites are often larger and do not occur in large groups. See "Mites" near the end of this chapter for photographs and more information.

An adult predatory mite (*Euseius tularensis*) eating a yellow citrus thrips nymph. Predatory mites are important biological control agents of pest mites. Some predaceous mites also feed on insect eggs and various tiny insects such as scale and whitefly nymphs. *Photo:* J. K. Clark

SPIDERS

Spiders feed mostly on insects, capturing prey in webs or stalking them across the ground or vegetation and pouncing on them. Spiders seek to avoid people, and most are harmless to humans. Spiders have eight legs, two main body parts, and are in the arachnid group along with mites. Common spiders on trees and shrubs include crab spiders (Thomisidae), jumping spiders (Salticidae), and small species of orb-weavers (Araneidae).

To minimize encounters with potentially hazardous widow spiders, reduce clutter near entranceways and around the foundation of buildings and seal cracks and crevices through which invertebrates enter buildings. For more information, see the *Pest Notes: Spiders, Black Widow and Other Widow Spiders, Brown Recluse and Other Recluse Spiders,* and *Hobo Spider.*

This adult funnel weaver spider (*Hololena nedra*, Agelenidae family) is a sit-and-wait predator that captures insects that blunder into its webbing. All spiders are predaceous, usually feeding on insects. *Photo:* J. K. Clark

PESTICIDES

Pesticides are substances applied to kill or repel pests with the intent of reducing pest damage. Some products can quickly reduce the population density of certain pests. Before you buy and use a pesticide, read the label to learn the effective, legal, and safe use of the product.

Understand the relative toxicity, mode of action, persistence, and selectivity of the pesticide before choosing which product to apply. You can then more favorably manage natural enemies in the landscape and avoid other potential problems (e.g., pest resurgence). Insecticide products of low toxicity to humans and pets include botanicals, insecticidal soaps, horticultural (narrow-range) oils, and microbial insecticides. To avoid pest resurgence and outbreaks of secondary pests, use selective pesticides if available, otherwise use less-persistent pesticides because they are less harmful to natural enemies than residual, broad-spectrum insecticides (see Table 6-3).

Some pests develop resistance to pesticides so that spraying becomes ineffective or less effective. Resistance develops because some individuals in a pest population are naturally less susceptible to certain pesticides. These less-susceptible individuals are more likely to survive an application and produce descendants. Repeated applications over several generations eventually result in a pest population composed primarily of less-susceptible or unaffected individuals. To avoid promoting resistance, minimize pesticide applications, make spot applications instead of broadcast spraying, rotate through different classes (modes of action) of insecticides, and use available alternatives, as discussed in "Pesticide Resistance" in Chapter 2.

Pesticides sometimes damage desirable plants (cause phytotoxicity), especially if plants lack proper cultural care, environmental conditions are extreme, or pesticides are used carelessly, as discussed in "Pesticides and Phytotoxicity" in Chapter 4. Sometimes pesticides can increase the reproductive rate of pests (called hormoligosis). Always read and follow the pesticide label directions. For more discussion on using pesticides effectively and minimizing their hazards, see Chapter 2, the pesticide information online at ipm.ucanr.edu, and publications such as *Landscape Maintenance Pest Control* and *Lawn and Residential Pest Control*.

Microbial and Biological Insecticides. Microbial pesticides, including products sometimes called biologicals, are naturally occurring pathogens or their by-products that are commercially produced for pest control. Most microbials affect only a certain group of pests or several related groups. Most have little or no toxicity to humans and have lower toxicity to natural enemies than broad-spectrum, residual pesticides.

Abamectin. Abamectin is a mixture of avermectins, which are derived from the soil bacterium *Streptomyces avermitilis*. Commercial abamectin is a fermentation product of this bacterium. It affects the nervous system of invertebrates and controls leaf beetles, leafminers, mites, and thrips. Abamectin has some translaminar activity (is absorbed short distances into leaves) and typically persists several days or less. Adding horticultural oil to the spray mix may increase the persistence of abamectin within plant tissue. For application to plants, abamectin may be available only to professional pesticide applicators.

Bacillus thuringiensis (Bt). This naturally occurring bacterium is produced commercially by fermentation similar to brewing beer. Different Bt subspecies (ssp.) are available for controlling different pests. *Bacillus thuringiensis* ssp. *aizawai* and Bt ssp. *kurstaki* kill moth and butterfly larvae that eat sprayed foliage. Bt ssp. *israelensis* (Bti) is applied to water to kill black fly and mosquito larvae and to potting soil to kill fungus gnat larvae.

Commercial Bt does not affect humans or most beneficial species. To be effective, Bt must be eaten by the caterpillars (Figure 6-7), so thorough spray coverage is critical. A second application 7 to 10 days after the first is typically recommended

because it degrades quickly, and often not all of the individuals in a pest population are in a Bt-susceptible stage at the same time. *Bacillus thuringiensis* is most effective against early-instar larvae, and for certain species only the younger caterpillars are controlled.

Entomopathogenic Nematodes. Nematodes that kill insects are called entomopathogenic nematodes because hosts are killed by the nematode in combination with associated bacteria that are released into infected hosts, usually within several days of infection (Figure 6-8). Nematodes are microscopic roundworms that also include the pest species discussed in Chapter 8.

Heterorhabditis and *Steinernema* spp. entomopathogenic nematodes are commercially available for control of soil-dwelling insects (including armyworms, cutworms, and root- or soil-dwelling weevils, white grubs, and wireworms) and certain boring pests (carpenterworms and clearwing moth larvae). Application methods include squirting a water solution of nematodes into borer tunnel openings or drenching the soil beneath infested plants when pest larvae or pupae are present. When drenching, the soil temperature should be at least 60°F. Keep soil moist (well irrigated) but not soggy before application and for 2 weeks afterward, as a

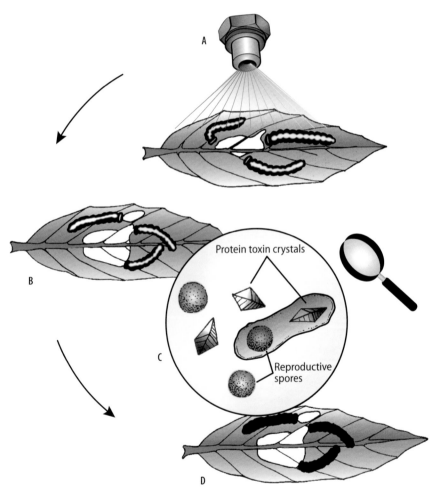

Figure 6-7.
Bacillus thuringiensis controls moth and butterfly larvae. It is most effective against early-instar larvae, and for certain species only the younger caterpillars are controlled. A. Bt must be sprayed during warm, dry conditions to thoroughly cover foliage where young caterpillars are actively feeding. B. Within about 1 day of consuming treated foliage, caterpillars become relatively inactive and stop feeding. C. An enlarged view of Bt, a rod-shaped bacterium containing a reproductive spore and protein toxin crystal (endotoxin); in some commercial Bt formulations, the spores and protein crystals are separate components. D. Within several days of ingesting Bt, caterpillars become dark and die.

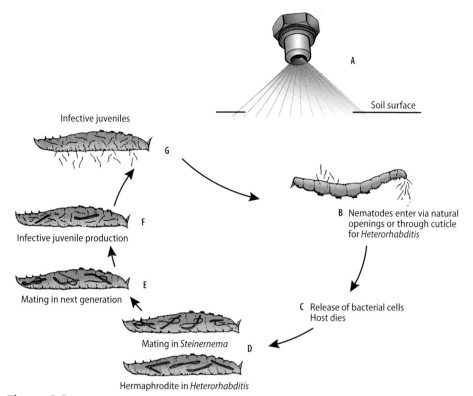

Figure 6-8.
Life cycle of beneficial entomopathogenic nematodes. A. Infective-stage nematodes are applied to soil. B. The nematodes seek a host and enter it. C. The host is killed by nematodes and mutualistic bacteria. D. Nematodes feed, mature, and reproduce in the host, but first-generation *Heterorhabditis* are hermaphrodites. E. Subsequent generations of *Heterorhabditis* and all generations of *Steinernema* produce both males and females. F. Females produce infective-stage juvenile nematodes. G. Infective nematodes exit the dead host and seek new hosts. The entire life cycle from infection of the host to release of the new infective generation takes 7 to 14 days. Adapted from Kaya 1993.

thin film of water on soil particles helps the nematodes disperse as they search for hosts.

Heterorhabditis bacteriophora and *Steinernema carpocapsae* nematodes are sold mostly through mail order. Obtain a fresh product, such as by ordering directly from nematode producers or a primary distributor. Nematodes are perishable, so store them under cool, dark conditions. Nematodes can be mixed and applied in combination with *Bacillus thuringiensis,* but they should not be mixed with potentially toxic materials. Do not mix or store nematodes in water for longer than about 24 hours before application.

Spinosad. Spinosad is a mixture of spinosyns, which are fermentation products from the bacterium *Saccharopolyspora spinosa.* Spinosad quickly kills insects that contact or eat it, including most caterpillars, flies, and thrips and certain species of beetles and wasps. Spinosad has translaminar activity and remains effective for about 1 week or longer after application. Adding horticultural oil to the spray mix and using

water with a pH of 6 to 8 may increase the persistence of spinosad. Spinosad has relatively low toxicity to people but can be toxic to certain natural enemies. It is toxic to bees for several hours after the spray has dried, so do not apply it to plants that are flowering.

Botanicals. Botanical pesticides are derived from plants. They are effective against many exposed-feeding insects and break down rapidly after application, making them relatively safe for the environment and natural enemies. Most have only contact toxicity, provide little or no residual control, and must be sprayed thoroughly where pests are present to be effective. Most botanicals are of low toxicity to humans; however, some can be highly irritating to eyes and skin or are toxic to fish and must be used with care near water.

Limonene (d-limonene) is extracted and refined from citrus fruit peels. Pyrethrum is a mixture of pyrethrins derived from chrysanthemum flowers native to Africa. Pyrethrins are fast-acting in knock-

ing down and paralyzing insects. Because insects often recover after exposure to pyrethrins, to increase effectiveness some products are a mixture that includes pyrethrum and insecticidal soap or piperonyl butoxide (PBO), a petroleum-derived synergist. Azadirachtin is extracted from neem tree seeds. Neem oils can kill exposed, soft-bodied insects and mites on contact and can control powdery mildew and rust fungi.

Inorganics. Insecticides refined from minerals include borate (e.g., boric acid), which is used in ant baits and to control cockroaches. Iron phosphate kills snails and slugs that eat it. Diatomaceous earth upon contact controls certain insects but loses effectiveness when wet and is not recommended for use in landscapes. Sulfur is primarily used to control fungal diseases but also controls mites. Do not treat plants labeled as susceptible to damage by sulfur and do not apply it during very hot or humid weather. Do not mix sulfur with oil or apply it within 3 weeks of an oil application. Sulfur can irritate the skin. Dusts of sulfur, borate (boric acid), and diatomaceous earth are harmful if inhaled, so wear protective equipment and appropriate clothing during mixing and application.

Insecticidal Soaps. Insecticide soaps are salts of fatty acids, such as those from plant oils (coconut, cottonseed, and palm) and potassium-bearing minerals (potash). Insecticidal soap is effective when sprayed onto mites and soft-bodied insects, including aphids, immature scales, leafhoppers, thrips, and whiteflies. Insecticidal soap has low toxicity to humans and wildlife, but it can damage some plants, especially species with hairy leaves. Before treating a plant, consider making a test application to a portion of the foliage and observing it for damage over several days before spraying it further. Do not treat water-stressed plants or when weather is expected to be hot, windy, or humid, because leaf burn (e.g., marginal necrosis) may result. Early morning or evening may be the best application times.

Oils. Narrow-range or horticultural oils

are highly refined petroleum products. Plant-based products include oils from canola, jojoba, and neem. Foliar oil sprays control many types of exposed-feeding invertebrates, apparently by smothering or suffocating insects or disrupting their metabolism. Delayed dormant season sprays (application to deciduous plants after buds have begun to swell in the spring but before leaves flush) control aphid eggs, mites, scales, and other pests overwintering on bark. In comparison with foliar sprays, dormant season or delayed-dormant applications may result in more thorough and effective spray coverage and have less impact on certain natural enemies.

Thorough spray coverage of infested plant parts and timing the application for when susceptible pest stages are present are essential for effective control. Do not apply oil or other pesticides when plants are drought stressed, it is windy, or temperatures are over 90°F or below freezing. Do not spray oil when it is foggy or the relative humidity is expected to be above 90% for 48 hours.

Mixing oil with a small amount of insecticidal soap, pyrethrum, or another insecticide can make the spray more effective. For example, you can apply a mixture of 1% each oil and soap instead of the common rate of 2% when either product is used alone. Check and follow product labels, which may recommend or prohibit specific combinations. Unless the label recommends otherwise, do not mix oil with chlorothalonil, sulfur, and certain other fungicides and do not apply oil within 3 weeks of an application of sulfur-containing compounds (e.g., wettable sulfur).

Use petroleum oils that say "narrow-range," "supreme," or "superior" on the label as these have been refined so they will not damage most plants when applied as directed. Plants in a few genera are sensitive to oils; check the product label for plants identified as susceptible to foliar damage. For example, oil will remove the desirable bluish tinge from blue spruce foliage, although the plants' health is not impaired. For more details on effectively using oils, see *Managing Insects and Mites with Spray Oils.*

Carbamates, Organophosphates, and Pyrethroids. Carbamates (e.g., carbaryl, or Sevin), organophosphates (malathion), and pyrethroids (fluvalinate, permethrin) kill a wide variety of invertebrates that are directly sprayed or that contact or eat treated foliage after applications. They are synthesized from petroleum, and some provide weeks-long control of many pests because they are residual (persistent) insecticides that leave a residue on sprayed surfaces that continues to kill insects after the application.

These pesticides also kill pollinators and natural enemies (see Table 6-3) and can cause spider mite outbreaks, in part by killing predators of spider mites. Some of these insecticides are contaminating urban surface water at levels hazardous to aquatic life, as discussed in Chapter 2 in "Pesticides Can Contaminate Water." Avoid using these broad-spectrum, residual insecticides if alternatives are available.

Neonicotinoids. These are systemic and move within plants (translocate). They provide long-lasting control of many chewing and sucking insects. Depending on the active ingredient and product label, neonicotinoids can be applied as a drench to soil and absorbed by roots, injected into trunks or roots, or sprayed on trunk bark or foliage, then translocate (e.g., from trunks or roots to leaves). Foliar application is not recommended because of drift and negative impacts on natural enemies and pollinators.

Products differ in their persistence and effectiveness. For example, imidacloprid controls European elm scale, most soft scales, and many leaf- and root-chewing beetles but does not control cottony cushion scale and most armored scales and caterpillars (moth larvae). Because neonicotinoids can translocate into nectar and pollen and have adverse effects on natural enemies and pollinators, delay their application where possible until after plants have completed their seasonal flowering and carefully follow the label directions when applying them to fruit trees or other food plants. To avoid injuring trees and mechanically transmitting pathogens, apply them to soil or as a bark-absorbed spray whenever possible, as directed on product labels (Figure 6-9), instead of injecting or implanting trees with them. For more discussion, see "Injecting or Implanting Systemic Pesticides" in Chapter 2.

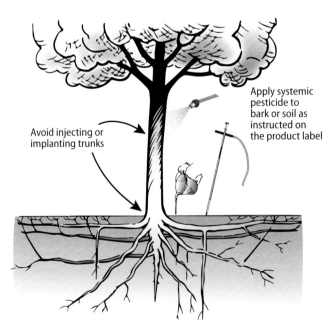

Avoid injecting or implanting trunks

Apply systemic pesticide to bark or soil as instructed on the product label

Figure 6-9.
When applying systemic pesticides to trees, spray bark or drench or inject soil whenever possible, as directed on product labels, instead of injecting or implanting trees with pesticide. Injecting or implanting trunks or roots injures trees and creates wounds that can provide entry sites for pests. Unless tools that contact internal parts of trees are cleaned and disinfected when treating multiple trees, contaminated tools can spread certain bacterial, fungal, and viral pathogens from one tree to another.

HOME INVASIONS BY NUISANCE PESTS

Ants, crickets, cockroaches, earwigs, lady beetles, millipedes, pillbugs, sowbugs, spiders, true bugs, and other invertebrates occasionally invade homes, sometimes in large numbers. For example, huge numbers of the introduced chamomile seed bug (*Metopoplax ditomoides*) sometimes invade buildings near riparian areas after migrating from brassbottoms (*Cotula coronopifolia*), which form carpets of yellow flowers in marsh tidal pools.

The best ways to stop invertebrates from coming inside are to

- Seal or screen gaps and openings in buildings.
- Install weather stripping and door sweeps (bottom barriers).
- Place tight-fitting screens over door and window openings.
- Use caulk, fine-mesh screen, steel wool, or expandable foam to seal up cracks around attic and basement vents, doors, windows, and other entry points.
- Modify or reduce habitat around buildings that favors invaders.
- Keep irrigation water and mulches, low-growing vegetation, debris, and other hiding places several feet back from foundations, especially near entranceways.
- Reduce outside lights' attraction of invertebrates: consider reduced brightness, switch to motion-activated security lights, and experiment with yellow or other-colored lighting.
- Keep plants pruned back from buildings and avoid specific plants that favor certain pests, such as female box elder trees that favor boxelder bugs.
- Manage honeydew-excreting insects, which favor abundant ants, and apply ant bait outdoors or otherwise manage ants near buildings.

Vacuum individual insects indoors. Acquire a hand-held, rechargeable battery-powered vacuum just for pest control. If you use your regular household vacuum you may stink up your home because certain pests (e.g., lady beetles, leaf beetles, and stink bugs) can leave a smelly residue inside. When possible, use a vacuum system that includes a HEPA filter, especially when vacuuming large numbers of pests and their leavings, to avoid releasing small particles of insect cuticle and feces into the air. See *Brown Marmorated Stink Bug Pest Notes* for information on trapping and vacuuming insects that come indoors.

Homeowners and pest control companies sometimes spray around foundations to kill or repel nuisance pests, but often the invasion period lasts for only a week or so and will have ended by the time the treatment is applied. Insecticides applied to hard surfaces (e.g., concrete and siding) partly wash off and move through gutters to pollute surface water. Instead of periodic spraying, seal buildings and modify or reduce habitat around foundations as described above to provide more long-lasting invasion prevention.

This Oriental cockroach (*Blatta orientalis*) lives and reproduces outdoors but enters poorly sealed structures, especially when dense ground covers, organic mulch, and debris occur near entranceways. To stop most invertebrates from coming indoors, modify and reduce habitat around foundations that favors invaders and seal or screen gaps and openings in buildings. *Photo:* J. K. Clark

Centipedes are found outdoors in damp, dark places and sometimes come indoors. Unlike most outdoor-dwelling invertebrates, this house centipede (*Scutigera coleoptrata*) can reproduce and persist inside. It feeds only on insects and other small invertebrates, so its presence indoors means that bugs, cockroaches, and other small pests that it preys on are entering homes through crevices and other openings that should be sealed or screened. *Photo:* J. K. Clark

Foliage, Fruit, and Root Chewers

Insects that commonly chew on foliage, fruit, or roots are discussed below. Snails are discussed at the end of this chapter.

CATERPILLARS

Caterpillars are the larvae of moths or butterflies (Lepidoptera). Hundreds of different caterpillar species feed on landscape plants, but most are so uncommon that they are not pests. Many species are important food for birds or mature into attractive butterflies. Management of pest caterpillars may differ from that of other caterpillarlike insects, including larvae of beetles, sawflies, and syrphid flies, so be sure to correctly identify larvae before taking action. New species are occasionally introduced, such as with the gypsy moth, which has repeatedly been found and eradicated in California, so take unfamiliar pests to your UC Cooperative Extension office or county agricultural commissioner for identification.

This parasitic wasp (*Hyposoter* sp.) is laying an egg in a caterpillar. Because the parasite larva feeds hidden within the host, this important biological control is easily overlooked. Most caterpillar species are uncommon and often well controlled by natural enemies. *Photo:* J. K. Cla

DAMAGE

Caterpillars chew irregular holes in flowers, fruit, or leaves; clip off entire leaves or flowers; or tunnel in buds or foliage. Some species fold or roll leaves together with silk and feed individually in these shelters. Others feed in a group on leaves beneath silk nests or tents.

The importance of the injury depends on the age, species, and health of the plant and the desired level of aesthetic quality. A relatively small number of caterpillars can retard the growth of plants that are young or already stressed from other causes. Severe defoliation, especially of conifers or over consecutive years, may cause branch dieback or lead to plant death. However, most otherwise healthy mature deciduous trees and shrubs tolerate extensive feeding by caterpillars, especially later in the growing season, with little or no loss in plant growth or vigor. This damage can be unsightly, but it often looks more serious than it is.

IDENTIFICATION AND BIOLOGY

The female moth or butterfly lays eggs singly or in a mass on the host plant. Eggs hatch after several days, except in species that overwinter as eggs. The emerging larvae feed singly or in groups on the plant. In addition to three pairs of legs on the thorax (immediately behind the head), caterpillars have pairs of prolegs (fleshy leglike tubercles) on at least some segments of the abdomen, but no prolegs on at least the first two abdominal segments. Legs and the presence and location of prolegs distinguish caterpillars from similar-appearing larvae of beetles, flies, and sawflies, as illustrated in Figure 6-10.

Most caterpillars eat voraciously and grow rapidly, shedding old skins three to five times before pupating (see Figure 6-1) in litter beneath a tree or on leaves or the trunk. Some species pupate within silken cocoons. The adult moth or butterfly emerges from the pupal case after several days to several months, depending on the species and season. Some species, such as the fruittree leafroller and most tussock moths, have one generation per year. Other species have several generations annually and can cause damage throughout the growing season.

MONITORING

Identify the species present and learn about their biology before taking action. For caterpillars that have only one genera-

Mature gypsy moth larvae are hairy, with a row of five pairs of blue spots and six pairs of red spots on their back. In the eastern United States, periodic outbreaks seriously defoliate many deciduous trees and some trees die. In California where it is not established, report suspected gypsy moths to the county agricultural commissioner. *Photo:* R. Zerillo, USDA-Forest Service

Prolegs are leglike tubercles on the abdomen, as with those on the rear end and abdominal segments 3 through 6 of this light brown apple moth larva and most other leafroller (tortricid) species. *Photo:* J. K. Clark

tion per year, it may be too late for control to be effective once mature caterpillars or the associated feeding damage is observed. On plants prone to caterpillar damage, monitor every week during the time of the year when caterpillars may be or have been present. Choose a method appropriate to your host plant and pest situation and be consistent in your method so that results are comparable among sample dates.

Visually inspect a set number of randomly selected leaves or terminals for damage or caterpillars and eggs or use the presence-absence sampling method described earlier in "Sampling." Sweep net shake or branch beat sample for caterpillars that are easily dislodged from foliage (neither technique is effective for species that form silken webs in leaves). Alternative Lepidoptera monitoring methods include timed counts, pheromone-baited traps, and frass collection.

Timed Counts. Timed counts can be used to monitor for any type of caterpillar that can easily be observed in foliage. Inspect foliage and record the number of caterpil-

lars, rolled leaves, or webbing "nests" seen in 1 or 2 minutes. Pull apart rolled leaves or webbing and count them only if they contain live caterpillars. Make several timed searches on different plants or on different parts of the same large plant. While counting insects, time your counts with an alarm watch or smartphone or work with a second person who can time and record. Timed counts are not useful if populations are so high that the number of insects recorded is limited by how quickly each can be seen and counted.

Pheromone-Baited Traps. Traps are commercially available for monitoring adults of various species of Lepidoptera, including American plum borer, certain armyworms, clearwing moths, cutworms, leafrollers, loopers, and Nantucket pine tip moth. Traps typically consist of a sticky surface and a dispenser containing a pheromone (sex attractant) to lure adults of a particular species.

To determine when specific moths are active in an area, hang one trap at chest height in each of two host trees, ideally

spaced several hundred feet apart. Deploy traps during the season when adults are expected and check them about once a week. Reapply sticky material or replace the traps when they are no longer sticky. Pheromone dispensers may need to be replaced about once monthly, especially if the weather has been hot. Check with trap distributors for specific recommendations.

Although traps generally catch only males, because both sexes are active around the same time, traps can be used to determine when females are also present and about when females will lay eggs. Traps do not reliably indicate numbers of an insect or whether control action is needed. Once moths are caught and when young larvae are anticipated to be present, inspect plants for caterpillars to determine the actual need for and timing of treatment. For some species, trapping moths in combination with monitoring temperatures using degree-days helps determine when to inspect plants or take control actions, as discussed later in "Pine Tip Moths." Be aware that fruittree leafroller and certain other species overwinter in egg masses, so young caterpillars may not be present until several months after the moths are active.

Frass Collection. Caterpillars excrete characteristic feces (frass), which drop from plants. Frass pellets increase in size as the larvae grow and are produced in greater amounts with increases in larval

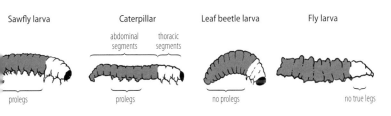

Figure 6-10.
Caterpillars (butterfly and moth larvae) can be distinguished from larvae of beetles, sawflies, and true flies by the number and arrangement of their appendages. Caterpillars and larvae of beetles and sawflies have three pairs of true legs, one pair on each thoracic segment. Larvae of most sawfly species also have leglike protuberances (called prolegs) on six or more of their abdominal segments. Caterpillars have prolegs on some (five or fewer) abdominal segments, and never on their first two abdominal segments. Beetle larvae have true legs but no prolegs. Fly larvae (such as predatory syrphids) have no true legs, but some species have prolegs or fleshy protuberances on their abdomen or on both their abdomen and thorax.

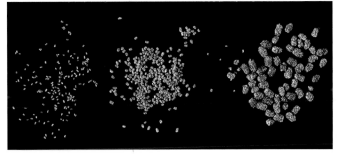

This frass (excrement) was collected to monitor the California oakworm. These three piles of caterpillar droppings (left to right) were produced by the smallest larvae (first-instar), third-instar, and fifth-instar larvae. Pellets in the right (largest pellet size) pile average $1/16$ inch long. *Bacillus thuringiensis* (Bt) will be more effective if foliage is sprayed when most droppings are of the smaller sizes, indicating the population is mostly early-instar oakworms that have not yet grown large enough to cause extensive defoliation. *Photo:* J. K. Clark

numbers or when temperatures are warmer, causing caterpillars to feed faster. Frass monitoring is not useful for tentmaking and leaf-rolling caterpillars; because they feed within silken webs, little of their frass falls to the ground.

Place several light-colored sticky cards, shallow trays, or cups beneath the canopy at regular intervals, such as for 24 hours once each week. Compare dropping density on cards or volume in cups among sample dates to help you decide whether control is warranted, as discussed in *California Oakworm Pest Notes*.

MANAGEMENT

Cultural Controls. Proper cultural care allows established plants to tolerate moderate defoliation. Provide irrigation if appropriate, depending on soil type, location, plant species, and weather. Protect roots and trunks from damage. Prune trees properly when needed.

Biological Controls. Predators, parasites, and natural outbreaks of disease sometimes kill enough caterpillars to control populations. Predators include assassin bugs, bigeyed bugs, birds, damsel bugs, ground beetles, lacewing larvae, mice and other small mammals, pirate bugs, and spiders. Certain wasps, such as paper wasps (*Polistes* spp., Vespidae) and ground-nesting Sphecidae, paralyze caterpillars and carry them to their nests, where the wasp larvae feed on them. Most caterpillar species are attacked by one or more species of parasitic wasp. Many caterpillar eggs are parasitized by tiny, naturally occurring *Trichogramma* wasps. *Trichogramma* are also commercially available for release in large numbers when moth eggs are present, but no research has demonstrated the effectiveness of *Trichogramma* releases in landscapes.

To benefit from natural enemies, avoid the use of broad-spectrum insecticides that destroy them. For instance, *Bacillus thuringiensis* (Bt) leaves most natural enemies unharmed because it kills only Lepidoptera larvae.

Caterpillars killed by naturally occurring bacteria, fungi, or viruses may turn dark, limp, and soft. These carcasses may degenerate into sacks of liquefied contents, which release more pathogen particles that may infect other caterpillars. Disease outbreaks can rapidly reduce populations under favorable conditions, although outbreaks are difficult to predict and may not occur until caterpillar populations have become high.

When monitoring caterpillars, also look closely for predators, parasites, and other evidence of biological control, including

- disease-killed caterpillars
- pupae or eggs with holes from which parasites emerged
- unhatched eggs that are darker than normal, indicating that they may contain parasites
- hatched caterpillar eggs with no evidence of caterpillars or damage

Adult tachinids are stout, black flies with many distinct hairs. This adult tachinid and its oblong pupal case are next to the larger amorbia pupal case within which the parasite developed as a larva. *Photo:* J. K. Clark

These white cocoons were made by larvae of an *Apanteles* wasp that emerged to pupate after feeding inside and killing these redhumped caterpillars. *Photo:* J. K. Clark

Caterpillars killed by fungi or viruses often hang limply from plants, as with the two silverspotted tiger moth caterpillars (*Lophocampa argenta*) underneath this Monterey pine twig. A healthy larva is on top. *Photo:* J. K. Clark

If you have an increasing number of pests but also many natural enemies, wait a few days before using insecticides. Monitor again to determine whether pest populations have declined or natural enemies are increasing to levels that may soon cause pest numbers to decline. Use Bt or the somewhat-selective insecticide spinosad. Where natural enemies are active, better long-term control is provided by using methods that conserve natural enemies.

Physical Controls. For the fall webworm, mimosa webworm, redhumped caterpillar, spiny elm caterpillar, tent caterpillars, and other species that feed in groups, clip and dispose of infested foliage. Use a pole saw (a blade on the end of a telescoping pole) or remove tents and the caterpillars inside by twirling or scrubbing the webbing into a ball using a toilet brush or similar tool attached to a telescoping pole, then squashing the larvae. Monitor regularly so you can locate and control infestations while the caterpillars are still young, because some group-feeding species disperse as the larvae mature. Clipping or removal are best done on cool, rainy, or overcast days when young caterpillars remain in tents or inactive groups. Heavily infested plants can be sprayed if necessary, preferably with a selective insecticide, and then monitored during subsequent seasons when populations are lower and physical control may be more practical.

Caterpillars such as fruittree leafroller and tussock moths overwinter in egg masses on bark or other objects. After leaves have dropped, inspect the bark and area around susceptible plants. Scrape any egg masses into a bucket of soapy water and dispose of them.

Insecticides. Entomopathogenic nematodes can control ground-dwelling larvae and pupae of armyworms and cutworms attacking certain low-growing ornamentals. Spinosad is highly effective against caterpillars. Viruses, each of which is toxic only to one pest species, are used mostly for a few crop pests (e.g., codling moth) or in certain government-managed programs

(e.g., gypsy moth control). Spraying trees during the dormant season with horticultural oil kills eggs overwintering on bark, including those of fruittree leafroller, tent caterpillars, and tussock moths.

Bacillus thuringiensis is effective against most exposed-feeding caterpillars, especially when larvae are young. Unlike broad-spectrum insecticides that kill on contact, caterpillars must eat Bt-sprayed foliage in order to be killed. Proper timing and thorough spray coverage is very important for effective application, so monitor caterpillar populations before treatment. Use a high-pressure sprayer (or hire a professional applicator) to provide adequate spray penetration when treating leafrolling and tentmaking species. Apply Bt during warm, dry weather when the younger stage caterpillars are feeding actively. Because Bt quickly decomposes, most caterpillars hatching after the application are not affected. A second application about 7 to 10 days after the first may be required.

Avoid carbamates, organophosphates, and pyrethroids. These broad-spectrum, residual insecticides also kill beneficial organisms, may cause outbreaks of other pests such as mites, and are contaminating surface waters. Do not apply broad-spectrum insecticides to flowering plants if honey bees are present.

CALIFORNIA OAKWORM

Phryganidia californica (Dioptidae) is the most important oak-feeding caterpillar along the coast through the coastal mountains of California. It most commonly defoliates coast live oak in the San Francisco Bay and Monterey Bay regions. Populations vary unpredictably year to year from very high to undetectably low. Healthy oaks generally tolerate extensive leaf loss, and insecticide application to control oakworms usually is not recommended.

The adult is a uniform gray, tan, or silvery moth with prominent wing veins. Females lay tiny, round eggs in groups of about two or three dozen, mostly on

the underside of leaves. Young larvae are yellowish green, with dark stripes on their sides and brown, overly large heads. Older larvae vary but commonly are dark with prominent lengthwise yellow or olive stripes. Pupae are pinkish, yellow, or white with black markings and hang from limbs, leaves, trunks, or objects near trees. Two generations per year typically occur in northern California; a third generation sometimes occurs at warmer and inland sites and during years of warmer, dry winters.

Fall-to-Spring Generation. Eggs laid during fall hatch within a few weeks, and young larvae overwinter on the underside of evergreen oak leaves. Eggs and larvae do not survive through winter on deciduous oaks because all the leaves drop. Larvae

The adult male (left) and female (right) of the California oakworm are tan to grayish moths with prominent wing veins, as seen here on coast live oak. *Photo:* J. K. Clark

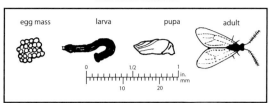

California oakworm eggs are whitish when laid but develop red centers that become pinkish, then brownish gray before hatching. See the chapter introduction for a photograph of an oakworm caterpillar. *Photo:* C. S. Koehler

California oakworm pupae usually occur on oak bark and are often killed by parasites. When parasitized by *Brachymeria ovata*, a stout black chalcid wasp emerges from this chrysalis instead of an adult moth. *Photo:* J. K. Clark

develop through five increasingly larger instars, and evergreen oaks may be extensively defoliated by the time larvae pupate, about May or early June in northern California.

Summer Generation. Adults emerge during June and July and may be seen fluttering around oaks in the late afternoon. Females lay eggs that hatch into larvae that feed (and may cause noticeable defoliation) from July through September, then pupate into oak moths. Development is more variable at warmer sites, where moths may be present almost any time from March through November.

Other Oak Leaf-Chewing Caterpillars. Other caterpillars on oaks include fruittree leafroller, which is the most common oak defoliator in the Central Valley of California. Unlike oakworms, larvae feed within foliage they web together. Larvae are green with brown or black heads, and when disturbed, they often wriggle vigorously and drop from leaves while suspended on silken threads.

Tussock moths and tent caterpillars chew oak leaves throughout the state. Unlike the greenish, relatively smooth-surfaced California oakworm and fruittree leafroller larvae, tent caterpillar and tussock moth larvae are hairy.

MANAGEMENT

Provide good cultural care to keep oaks healthy and conserve natural enemies that help to control oakworm. Insecticides usually are not warranted to protect oak health or survival. If defoliation is not aesthetically tolerable or trees need protection from leaf loss because they are stressed, regularly inspect foliage and apply least-toxic insecticides if young oakworms become abundant.

Cultural Controls. In comparison with pests, oaks are more often injured or killed by inappropriate irrigation, mechanical injury to trunks and roots, and soil compaction or changes in the soil grade. Many native oaks require no irrigation after they become established. Whether any irrigation is warranted depends on many factors, including the oak species, whether precipitation has been normal, soil conditions and type, and whether roots have been injured. Keep any irrigation away from the root crown.

Biological Control. Conserve and encourage predators, parasites, and natural outbreaks of disease that sometimes kill enough oakworms to control populations. Important oakworm predators include green lacewing larvae, pirate bugs, spined soldier bug (*Podisus maculiventris*), and yellowjackets. Birds, small mammals, and spiders also eat oakworms.

The most important parasites are two small wasps that leave a roundish hole in oakworm pupae they killed. The adult *Brachymeria ovata* (Chalcididae) is a stout, black and yellow parasite about ¼ inch long with enlarged rear legs. The *Itoplectis behrensii* (Ichneumonidae) adult has a black, slender body about ¼ to ⅔ inch long and orange legs with yellow and black bands. At least two tachinid flies (*Actia flavipes* and *Hyphantrophaga virilis*) feed as larvae inside oakworms and kill caterpillars as they pupate. The dry, deflated skins of tachinid-killed oakworms often remain attached to twigs or bark, and the oblong, reddish to dark brown tachinid pupal cases may be seen nearby.

After oakworms become abundant, they are often killed by a nuclear polyhedrosis virus, which causes larvae to become dark, limp, and soft. Oakworms killed by *Beauveria bassiana* may develop an unpleasant odor and be covered with whitish mycelia.

Chemical Control. If considering pesticide application, regularly inspect foliage to determine whether oakworm damage and populations are increasing and warrant control and to effectively time any application. Monitoring is important because oakworm outbreaks are unpredictable and do not occur most years.

At regular intervals inspect current-season leaf flush (lighter green, young shoot terminals) for young oakworms, diseased caterpillars, and predators, parasites, and other evidence of biological control (e.g., oakworm eggs, larvae, or pupae that are dead, discolored, or have holes). Record your findings for comparison among sample dates. If natural enemies are abundant or increasing, before deciding whether to apply insecticides wait a few days and monitor again to determine whether oakworm populations have declined or natural enemies are providing control.

Frass (excrement, pictured earlier) monitoring is another tool. Oakworms excrete dark brown pellets with a highly sculptured surface. Frass may be seen lodged in bark crevices, spider webs, and ground covers beneath infested oaks. Collect frass with cups or sticky cards under trees for several hours during warm weather when oakworms are present. Monitor every few days and compare the frass collections; greater amounts of frass indicate there is more caterpillar feeding.

If caterpillars become too abundant, apply *Bacillus thuringiensis* thoroughly to the underside of foliage during warm, dry weather when oakworms are feeding actively. A second Bt application about 7 to 10 days after the first may be warranted. Alternatively, certain insect growth regulators and spinosad (a microbial by-product) are highly effective insecticides with relatively low toxicity to many natural enemies.

Effectively treating large trees generally requires hiring a professional applicator. When hiring an applicator, discuss the specific pesticide to be applied and insist on an IPM-compatible one. For more information, see *California Oakworm Pest Notes*.

LEAFROLLERS

Leafrollers (Tortricidae) often feed and pupate within leaves they curl and web with silk. About 100 species occur in California, and many at least occasionally feed in landscapes. Many species are similar looking and for positive identification must be submitted to an expert.

Fruittree leafroller (*Archips argyrospila*) is especially damaging to oaks at warmer locations away from the coast. Its many additional hosts include ash, birch, box elder, California buckeye, elm, fruit and nut trees, locust, maple, poplar, rose, and willow. The obliquebanded leafroller (*Choristoneura rosaceana*), omnivorous leafroller (*Platynota stultana*), and light brown apple moth (LBAM, *Epiphyas postvittana*, pictured earlier) chew foliage on a wide variety of ornamental and fruit trees. See *Field Identification Guide for Light Brown Apple Moth in California Nurseries* to help you distinguish among leafroller pests.

Leafroller feeding gives foliage a ragged or curled appearance. Larvae prefer leaves but sometimes chew blossoms, fruits, and nuts, causing them to drop or become distorted or scarred. Unusually high populations can completely defoliate plants and cover them with silken threads. When disturbed, larvae wriggle vigorously and often drop to the ground on a silk thread. Older caterpillars construct a new leafroll nest frequently, often daily, eventually pupating inside a nest or on bark in a thin cocoon. Adults (moths) of most species are a mottled mix of colors and at rest appear bell shaped when viewed from above. Females lay eggs overlapping in flat masses (resembling fish scales) on leaves or small branches or twigs. The fruittree leafroller has only one generation per year and overwinters as eggs. Many other leafroller species have two or more generations and overwinter primarily as larvae in protected places.

For the major pests, species-specific pheromone-baited traps can be used to determine whether a particular leafroller is present. Adult captures coincide with egg laying and, in most species except for fruittree leafroller, the approximate time of egg hatching, so pheromone traps are useful to time insecticide applications targeting young larvae.

Managing Leafrollers. Conserve natural enemies, which often keep the populations

Fruittree leafroller larvae are green with a black, shiny head. As with many tortricids, caterpillars tie foliage together with silken threads and feed inside. When especially abundant, this species can entirely defoliate large trees. *Photo:* J. K. Clark

The fruittree leafroller lays its irregular, flat egg masses on twigs and small branches. Eggs are gray or brown but become paler or whitish after hatching, as with the lower egg mass with holes left by emerged larvae. A dormant-season oil application to egg-infested bark can control this pest. *Photo:* J. K. Clark

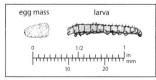

FRUITTREE LEAFROLLER

egg mass   larva

low. Parasitic wasps and tachinid flies are especially common in leafroller larvae and pupae. Assassin bugs, certain beetles, and lacewing larvae are common predators of leafrollers.

If fruittree leafroller has been a problem, apply horticultural oil in January or February to thoroughly cover limbs and small twigs infested with overwintering eggs. A properly timed foliar spray of Bt or spinosad controls most leafrollers. If larger larvae in rolled leaves are abundant, use a high-pressure sprayer so the insecticide penetrates into rolled foliage. Because fruittree leafroller has only one generation per year, by the time trees are severely defoliated, the caterpillar stage might be almost completed and spraying leaves may be of little benefit. For more information, consult *Leafrollers on Ornamental and Fruit Trees Pest Notes.*

LOOPERS

Over 300 species of loopers (Geometridae), or measuringworms, occur in California, but most are rarely or never pests. In addition to the three pairs of legs behind the head, true loopers have two pairs of prolegs near the rear on abdominal segments 6 and 10. This allows them to travel in the characteristic looping manner, pulling their rear forward as they arch up their back.

Bougainvillea looper (*Disclisioprocta stellata*) chews young foliage on *Bougainvillea* spp., four o'clock (*Mirabilis jalapa*), and some herbaceous species, causing severe defoliation and possibly killing young hosts. Young larvae are greenish and brown; older instars are mostly rusty brown and grow to about 1 inch long. Larvae may hide during the day. Pupae are dark brown, and the nocturnal adults are mainly brown with wavy bands of black and tan.

The omnivorous looper (*Sabulodes aegrotata =S. caberata*) feeds on several dozen species including acacia, box elder, California buckeye, chestnut, elm, eucalyptus, fruit trees, ginkgo, magnolia, maple, pepper tree, and willow. Young larvae are pale yellow

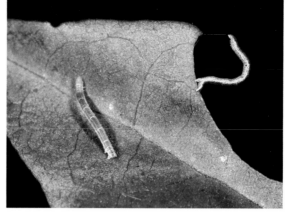

and chew the leaf surface, leaving a brown membrane. Older larvae grow to about 2 inches long and are yellow to pale green or pink, with dark brown, black, or green lines along the sides and a gold-colored head. They chew entirely through edges of leaves or shoots or between leaves tied with silk, often leaving only the midrib and larger veins. Mature larvae form pearly white to brown pupae, usually in webbing between leaves. Adults are tan moths with a narrow, black band across the middle of the wing and are active at night. The barrel-shaped, pale green to reddish brown eggs have a ring of tiny projections and are laid in clusters on the underside of leaves. Omnivorous looper has up to five generations each year; all stages may be found whenever foliage is present.

Conserve natural enemies that biologically control most looper species. Controls include thoroughly spraying infested foliage with Bt.

PALM LEAF SKELETONIZER

Larvae of this tiny moth (*Homaledra sabalella*, Coleophoridae) chew the surface of palm leaves, causing dark discolored blotches on leaves. Infested fronds become dry and brown and eventually die. Palms may look unsightly but are rarely killed.

The pale brown or cream-colored larva grows up to $3/5$ inch long and feeds hidden in a tube of dark brown excrement it gradually lengthens. Larvae feed in groups of up to 100, which gradually expand their area of chewed, tube-covered leaf surface. The tiny eggs and $1/4$-inch-long, gray to yellowish brown, night-flying moths are easily overlooked. The pest has several generations per year.

Palm leaf skeletonizer is reported in

only a limited area of southern California. In an effort to eradicate it, infested fronds are pruned off, bagged, and disposed of an... the palms are sprayed with insecticide. If you find suspected palm leaf skeletonizers, report this to the county agricultural commissioner. See *The Biology and Management Landscape Palms* for more information.

Palm leaf skeletonizer larvae feed within a dark tube of webbed excrement. The group-feeding larvae chew the surface of fronds, which turn brown, dry, and die. If you find caterpillars in California you suspect to be this species, report it to the county agricultural commissioner. *Photo: T. Broschat, University of Florida, Bugwood.org*

REDHUMPED CATERPILLAR

Schizura concinna (Notodontidae) most commonly feeds on liquidambar, plum, and walnut. Other hosts include cottonwood, redbud, willow, and most fruit and nut trees, especially where insecticides applied for other pests have killed the caterpillar's natural enemies. High populations are usually limited to the inland valleys.

Young larvae are yellow or reddish with a black head. Older larvae have a red head, and their yellowish body has lengthwise black, reddish, and white longitudinal lines and spinelike projections. The fourth segment behind the head is red and distinctly humped with two black projec-

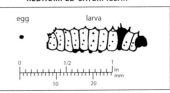

Older redhumped caterpillar larvae have two black projections on the reddish, enlarged fourth segment behind their head. This species usually does not seriously damage landscapes. Where it is undesirable, clip off infested shoots while larvae are feeding in groups. *Photo:* J. K. Clark

Spiny elm caterpillars are blackish with distinctive spines and a row of orangish spots down their back. Larval feeding does not harm the tree, and caterpillars mature into attractive morning cloak butterflies. Clip and dispose of the infested branch if caterpillars cannot be tolerated. *Photo:* P. Svihra

tions. Larvae feed in groups, particularly when young. They consume the entire leaf, except for the major vein, and often feed only on a single branch. Overwintering is as reddish brown pupae on the ground, often in a cell of soil or organic debris. Adults are reddish brown or gray moths, have a wingspan of up to 1⅜ inches, and begin to appear in April. Females lay pearly white, spherical eggs in masses of 25 to 100 on the underside of leaves. Redhumped caterpillar has three generations per year, sometimes four or five at warmer locations, and most any stage can be present from late spring through fall.

Redhumped caterpillar populations are often controlled by two parasitic wasps, *Hyposoter fugitivus* (Ichneumonidae) and *Apanteles schizurae* (Braconidae). Female wasps lay their eggs in caterpillars, and the wasp larvae feed inside. After killing their host, *Apanteles* larvae emerge and pupate in white, silken cocoons in groups on leaves near dead caterpillars, as pictured earlier. *Hyposoter* pupae are oblong and mottled black or purplish. Parasite populations can increase quickly, causing caterpillar outbreaks to crash.

Conserve parasites by avoiding the use of residual, broad-spectrum insecticide. Plant flowering species (see Figure 6-4) near commonly infested host trees because adult parasitoids live longer and can parasitize and kill more caterpillars when provided with nectar. The simplest control is to cut off and dispose of infested shoots, especially when caterpillars are young and grouped on a limited portion of the plant. If older caterpillars are abundant and spraying seems warranted, inspect plants regularly for the next generation and apply *Bacillus thuringiensis* or spinosad as eggs hatch and when caterpillars are young. For more information, consult *Redhumped Caterpillar Pest Notes.*

SPINY ELM CATERPILLAR

Nymphalis antiopa (Nymphalidae) feeds on elm, poplar, and willow. The group-feeding larvae cause ragged, chewed leaves, often on a single branch, which may be entirely defoliated. At maturity, larvae are mostly black with a row of orange to brown spots down the back and rows of tiny, white dots on each segment. The distinctive row of black spines around each larval segment can be irritating to human skin.

The spiny elm caterpillar pupae are black, brown, or gray. The adult, known as the mourning cloak butterfly, is mostly brownish black to purple, with a wingspan of about 2 inches. Adult wing margins are often ragged and have yellow bands bordered inwardly by blue spots. The tiny, almost cylindrical eggs are laid in masses of several dozen on leaves, limbs, or twigs. Spiny elm caterpillars have about two generations per year, but in southern California, both adults and larvae may be observed during almost any month.

Spiny elm caterpillar larvae do not harm the tree and no control is needed. Clip and dispose of the infested branch if caterpillars cannot be tolerated.

TENT CATERPILLARS

Malacosoma spp. (Lasiocampidae) feed on various deciduous trees and shrubs, including ash, birch, fruit and nut trees, madrone, oak, poplar, redbud, toyon, and willow. Adults are hairy, medium-sized, day-flying moths, usually dull brown, yellow, or gray. Tent caterpillars overwinter in pale gray to dark brown eggs encircling small twigs or in a flat mass on bark. The larvae hatch and begin feeding in the spring, and some species form silken webs on foliage. Mature tent caterpillars spin silken cocoons in folded leaves, on bark, or in litter. Adults emerge in midsummer. Tent caterpillars have one generation per year.

Western tent caterpillar (*M. californicum*) larvae are reddish brown with some blue spots and are covered with tufts of orange to white hairs. They spin large silken webs in which the larvae do most of their feeding.

The Pacific tent caterpillar (*M. constrictum*) looks very similar to *M. californicum*, except more blue is visible and the larvae usually feed only on oaks. Pacific tent caterpillars produce small tents a few inches wide. Larvae feed openly, in groups when they are young, and usually enter the tent only to molt.

Forest tent caterpillar (*M. disstria*) larvae are mostly dark blue, with wavy, reddish brown lines and distinct, white, keyhole-shaped markings down the back. Larvae feed in groups without webbing.

Inspect plants regularly, and when larvae are young, prune out tents or clip and

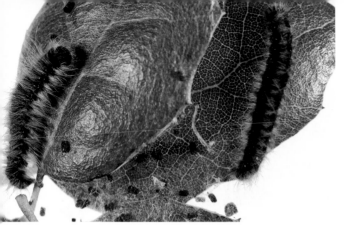

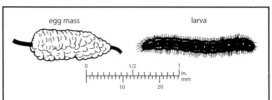

egg mass   larva

Western tent caterpillars are mostly reddish brown. Their dark frass, or droppings, are visible here, caught in their silken webbing on coast live oak leaves. *Photo:* J. K. Clark

dispose of infested branches if this can be done without cutting major limbs. *Bacillus thuringiensis* provides control if high-pressure spray equipment is used so that insecticide penetrates webbing and also thoroughly sprays foliage around webbing.

TUSSOCK MOTHS

Tussock moths (Lymantriidae) feed at least occasionally on most species of deciduous and evergreen trees. Adults are hairy and brown, gray, or white moths. Females of some species are flightless and appear to be wingless. Females produce pheromone to attract the night-flying males and lay tiny, whitish eggs in a mass of several hundred, covered with hairs. The tiny, dark caterpillars hatch in spring and can use a silk strand and their body hairs to float on the wind to other trees. Full-grown caterpillars have prominent hairs that protrude, sometimes in tufts, from colored tubercles along their body. These hairs readily detach from the larvae and are irritating to human skin. Pupation occurs on or near the host plant.

The western tussock moth larva has brightly colored spots and dense tufts of hairs. Tussock moths rarely become abundant enough in landscapes to warrant control. *Photo:* J. K. Clark

Western tussock moth (*Orgyia vetusta*) hosts include fruit and nut trees, hawthorn, manzanita, oak, pyracantha, toyon, walnut, and willow. Mature caterpillars are gray and have numerous bright blue, red, and yellow spots from which gray to white hairs radiate. They have four dense white tufts of hair on the back, two black tufts on the head, and a black and a white tuft at the rear. After emerging from the overwintering eggs and feeding during the spring, the larvae pupate on bark in hairy, brown or tan cocoons. Adults occur from late spring through early summer, and there usually is one generation per year. In southern California two generations may occur, and second-generation larvae are present from about late August to October.

Rusty tussock moth (*O. antiqua*) feeds on many different deciduous and ever-

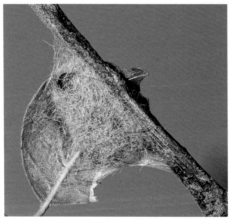

A brownish tussock moth cocoon on a coast live oak leaf. Tussock moth egg masses are covered with the females' brownish hairs and resemble this pupal cocoon. *Photo:* J. K. Clark

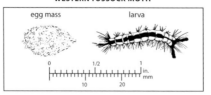

egg mass   larva

green plants. Its hairy, blackish larvae have three projecting tufts of black hair, two in front and one at the rear, and four orangish tufts along the back.

The Douglas-fir tussock moth (*O. pseudotsugata*) feeds primarily on Douglas-fir and true firs. Lighter-colored tufts of hair along the back, red spots on top, and an orange stripe along each side distinguish its mature larvae from those of the rusty tussock moth.

Naturally occurring diseases and parasites often keep tussock moth populations at low levels. For example, a tiny, purplish black wasp (*Telenomus californicus*) kills many western tussock moth eggs. *Bacillus thuringiensis* controls tussock moth larvae, especially if applied when most larvae are young.

WEBWORMS

The fall webworm (*Hyphantria cunea*, Arctiidae) makes prominent silk tents over outer terminal foliage, while tent caterpillars usually web more inner foliage, around the juncture of branches. Fall webworm hosts include aspen, birch, cottonwood, elm, fruit and nut trees, liquidambar, maple, mulberry, poplar, sycamore, and willow.

Adult moths emerge in the late spring or early summer and are white, sometimes with black wing spots. In June or July, the females lay globular white or yellow eggs in large masses. The larvae feed in groups in silken tents they gradually enlarge through late summer or early fall. Mature larvae are gray, orangish, or yellowish brown with long white or black hairs and orange projections along the body. Fall webworms overwinter in dark brown cocoons on the tree trunk or the ground and produce one or two generations per year.

Mimosa webworm (*Homadaula anisocentra*) is primarily a pest of *Albizia* spp.

(also called mimosa) in the Sacramento Valley. Young larvae feed in groups covered with silk, causing foliage to turn brown and die. Mature larvae feed singly, vary from gray to blackish brown, and have five longitudinal white stripes on the body. They pupate and overwinter in whitish cocoons on bark or in litter beneath trees. The small, silvery gray moths with black dots emerge and lay eggs in spring. The mimosa webworm usually has two generations per year.

Regularly inspect host plants for silken tents during late spring and summer. Prune out or "scrub" and dispose of caterpillar-infested tents, as discussed above in "Physical Controls." If nests are abundant and cannot be pruned or tolerated, apply Bt or spinosad with a high-pressure sprayer to penetrate webbed foliage. Inspect plants the next season, when any colonies should be fewer and easier to prune out.

Fall webworms typically feed on [sho]t terminals, as on this Lombardy poplar. This outer location of webbing helps distinguish them [fro]m tent caterpillars, which usually [web] foliage more inward in the tree canopy around branch junctures. *Photo: C. S. Koehler*

FALL WEBWORM

larva

SAWFLIES

Sawflies are not true flies; they are Hymenoptera, which includes ants, bees, and wasps. Named for the adult female's sawlike abdominal appendage for inserting eggs into plant tissue, sawflies are a diverse group with different species that feed openly on foliage, in leaf and stem mines, or in galls. Species with larvae that chew externally on foliage include conifer sawflies, cypress sawfly, pear sawfly, and roseslugs, discussed below. Hosts of other sawfly species include alder, birch, fruit trees, poplar, and oak. Willow gall sawflies are discussed in "Gall Makers," and the stem-boring raspberry horntail sawfly of caneberries and roses is discussed in "Twig, Branch, and Trunk Boring Insects."

Adults are usually $\frac{1}{2}$ inch long or shorter and mostly black with orange, red, or yellow and have two pairs of wings. The relatively wide abdomen is broadly

Fall webworm larvae are brown, gray, or orangish, with long hairs and orange tubercles. They chew terminal leaves within silk webbing, as on this willow. Inspect hosts regularly, and before colonies become abundant clip off webbed foliage where larvae feed together in a group. *Photo:* J. K. Clark

MIMOSA WEBWORM

larva

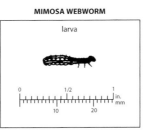

attached to the thorax, in contrast to most other adult wasps, which have a narrow "waist" between the thorax and abdomen.

Most exposed-feeding larvae (the pearslug is an exception) have six or more prolegs on the abdomen (see Figure 6-10) and one large "eye" on each side of the head. This distinguishes them from butterfly and moth caterpillars, which have five or fewer prolegs and small eyespots. Depending on the species, sawflies have from one to several generations per year.

Management. Healthy trees and shrubs tolerate moderate defoliation without significant loss in growth, flowering, or fruit yield. Natural enemies keep most sawfly populations low and can cause outbreak populations to soon decline. Fungal and viral diseases, insectivorous birds, parasitic wasps, predaceous beetles, and small mammals commonly kill sawflies.

Clip off infested foliage if larvae are on a small portion of the plant. Pearslugs and some other sawfly larvae that feed openly can be washed from plants with a forceful stream of water.

Most sawfly larvae that chew foliage are relatively easy to control if sprayed with almost any insecticide applied to achieve good spray coverage, including insecticidal soap, narrow-range oil, neem oil, or spinosad. Avoid broad-spectrum, residual insecticides because of their adverse affect on natural enemies. Bt is not effective against sawfly larvae because they are not Lepidoptera.

CONIFER SAWFLIES

Neodiprion spp. (Diprionidae) commonly feed on pines. Arborvitae, cypress, fir, hemlock, juniper, larch, and spruce are also hosts. Most conifer sawfly adults are yellowish brown to black with yellowish legs. Females lay eggs in niches carved in needles. Larvae are commonly yellowish or greenish and develop dark stripes or spots as they mature. Larvae often feed several to a needle with their heads pointed away from the twig or their bodies wrapped around the needle.

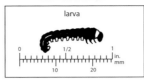

Young conifer sawflies, like these *Neodiprion fulviceps* larvae, often feed several to a needle with their heads pointed away from the twig. Unlike caterpillars and leaf beetle larvae, prolegs (appendages) can be seen on each abdominal segment on most species of sawflies. *Photo: J. K. Clark*

Pear sawfly larvae (pearslugs) skeletonize leaves. Their dark, slimy coating gives them a sluglike appearance, except the last instar (largest larva) lacks this coating. Pear sawfly larvae can be washed from plants with a forceful stream of water. *Photo: J. K. Clark*

PEAR SAWFLY
larva

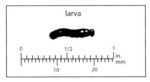

PEAR SAWFLY

Pear sawfly (*Caliroa cerasi*, Tenthredinidae) larvae, commonly called pearslugs or cherryslugs, skeletonize the leaf surface of cherry, pear, and occasionally hawthorn and quince. Larvae are dark olive-green and covered with slime, so they resemble slugs. Adults are shiny black with dark wings. There are generally two generations per year, with larvae most abundant in the mid to late spring and mid to late summer. Pear sawfly occurs mostly in coastal areas and is usually controlled by natural enemies.

A roseslug larva chewing a rose leaf. Although many sawfly larvae resemble moth larvae, *Bacillus thuringiensis* will not control roseslugs because they are not Lepidoptera. If intolerable, roseslugs can be controlled with almost any contact insecticide applied to achieve good spray coverage, including insecticidal soap, narrow-range oil, or neem oil. *Photo: J. K. Clark*

BRISTLY ROSESLUG
larva

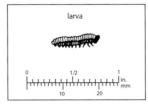

ROSESLUGS

Roseslug larvae skeletonize the leaf surface and may chew entirely through foliage, leaving just the large veins. Larvae grow to about ¾ inch long and are yellowish green with a brown head. Take care to distinguish roseslugs from beneficial syrphid larvae that feed on rose aphids. Roseslugs occur near chewed leaves, while syrphids occur among aphids, where they can be observed eating insects, not chewing leaves. Syrphid larvae have no true legs, while roseslugs have three pairs of legs. Depending on the species, roseslugs pupate in cocoons on leaves, in organic debris beneath plants, in a cell in twigs, or in the stub end of cut roses. Adults are stout, broad waisted, mostly black wasps with yellow or orange markings. Females lay small eggs in leaf tissue.

At least three species of sawflies (Tenthredinidae) are occasional pests of roses, including bristly roseslug (*Cladius difformis*), coiled or curled rose sawfly (*Allantus cinctus*), and the American, European, or common roseslug (*Endelomyia aethiops*), officially named "roseslug." Bristly roseslug has several generations per year and occurs mainly in coastal or cool areas. Its larvae are pale green, with many hairlike bristles. Roseslug and curled rose sawfly larvae do not have many distinct bristles or hairs. However, curled rose sawfly larvae have rows of pale dots and pale tubercles (short projections more stout than hairs), and larvae often coil their bodies while feeding.

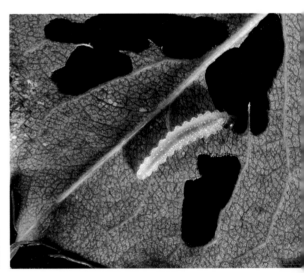

Several webspinning sawflies (*Acantholyda* spp., Pamphiliidae) occur on Monterey pine and other conifers. They spin nests or silken webs on foliage and feed inside in groups or singly. Unlike the numerous prolegs on most free-feeding sawflies (see Figure 6-10), these webspinning species have only one pair of mid-abdominal prolegs and another pair on their rear end. They pupate in an earthen cell in the ground.

CYPRESS SAWFLY

About a half dozen *Susana* spp. sawflies (Tenthredinidae) feed on broad-needled conifers in the western United States. The most important, primarily in southern California, is the cypress sawfly (*S. cupressi*), which primarily damages cypress but may also feed on arborvitae and juniper. Adult wasps are black and yellow. Larvae are grayish green with rows of whitish dots. The cypress sawfly spends the winter in a cocoon in the soil and has one generation per year.

LEAF BEETLES

About 500 species of leaf beetles and flea beetles (Chrysomelidae) occur in California. Only a few are woody landscape pests, including leaf beetles that feed on alder, aspen, cottonwood, coyote brush, elm, eucalyptus, hypericum, poplar, and willow. Adults and larvae of species that feed on woody plants typically scrape the surface or chew holes in leaves and foul leaves with their dark droppings. Damaged leaves may turn yellow or brown and drop prematurely. Repeated defoliation causes plants to decline, become susceptible to other problems, and in rare cases die. However, otherwise healthy deciduous plants tolerate extensive skeletonization or defoliation.

Most adult leaf beetles are less than 1/3 inch long, oval, and have threadlike antennae. The smallest species, called flea beetles, are metallic colored and often jump away when disturbed. Larger species may be colorful or blend with the colors of their host and usually drop when disturbed. In many species that feed on herbaceous plants and vegetables (e.g., cucumber beetles, *Diabrotica* spp.), larvae live in soil and chew roots, or they mine stems or leaves. Leaf beetle larvae are caterpillarlike but, unlike caterpillars and sawfly larvae, lack prolegs (see Figure 6-10).

COTTONWOOD LEAF BEETLE

The cottonwood leaf beetle (*Chrysomela scripta*) feeds on cottonwood and willow, as do other leaf beetles, including *Altica* and *Plagiodera* spp. Adult cottonwood leaf beetles are grayish, orangish, or yellowish, with variable black spots and stripes on the back. Females lay yellowish eggs in clusters of about 25 on the lower leaf surface. The young, black larvae feed in groups on the lower leaf surface. Mature larvae are yellowish, grayish, or reddish, often with rows of black tubercles. There are several generations per year.

Provide regular, deep irrigation for hosts planted in areas with hot, dry summers. Protect plants from injury, such as by avoiding compaction or other soil disturbances around roots. Vigorous host plants tolerate moderate leaf beetle feeding, and control is generally not warranted. If populations are not tolerable, foliar or systemic insecticides as discussed for elm leaf beetle may provide control.

ELM LEAF BEETLE

Xanthogaleruca luteola feeds on elms, especially European elm species. American and most Asian elm species are less severely fed upon. Chinese elm, zelkova, and many newer elm cultivars are infrequently fed upon.

Larvae skeletonize the leaf surface, while adults eat through the leaf, often in a shot hole pattern. High populations can entirely defoliate elms. Repeated extensive defoliation weakens elms, causing trees to decline.

Elm leaf beetle adults are olive-green with a black longitudinal stripe along each margin and down the center of their back. Females lay yellowish to gray eggs in double rows of about 5 to 25 on leaves. Newly hatched and recently molted larvae are black. Older larvae are dull yellow or greenish, with dense rows of tiny, dark tubercles that resemble

An elm leaf beetle adult, egg mass, and first-instar larva. This pest resembles the cedar leaf beetle (*Diorhabda elongata*), a beneficial species introduced to control invasive salt cedar, or tamarisk. *Photo:* J. K. Clark

Third-instar elm leaf beetle larvae on the lower surface of an English elm leaf they skeletonized. *Photo:* J. K. Clark

two black stripes, one along each side. After chewing leaves for several weeks, mature larvae crawl down the tree trunk and form yellowish pupae around the tree base. After about 10 days, adults emerge and fly to the canopy to feed and (except during fall) reproduce. Elm leaf beetle has at least one generation per year in northern California and one to three generations per year in central and southern California. Adults commonly overwinter in bark crevices, buildings, litter, and woodpiles.

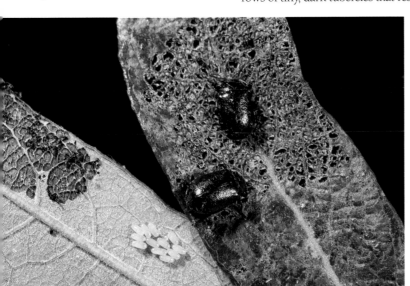

Adults, eggs, and chewing damage of the California willow leaf beetle (*Plagiodera californica*). Black larval frass covers part of the chewed underside of the left willow leaf. This species resembles another occasional pest, the imported willow leaf beetle (*P. versicolora*). *Photo:* J. K. Clark

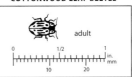

COTTONWOOD LEAF BEETLE

adult

0   1/2   1   in.
10   20   mm

MANAGEMENT

Provide elms good cultural care and conserve natural enemies. Correctly identify the cause of chewed leaves before taking management action. European flea weevil (*Orchestes alni*) adults also chew holes in elm leaves. This introduced beetle has become a serious elm pest in the eastern United States but has not been reported in California as of this writing.

Determine the need and effective timing for any pesticide application by visually inspecting leaves at about weekly intervals when the first generation of eggs or larvae is expected to be most abundant. Because beetle development and abundance vary dramatically from year to year, use the heat accumulation method in Table 6-4, as discussed above in "Degree-Day Monitoring," to determine the optimal time to inspect trees and (if beetles are too abundant) apply insecticide. For example, elm leaf beetles do not develop below about 52°F. First- and second-instar larvae of first-generation elm leaf beetle are most abundant at about 700 degree-days above 52°F accumulated from March 1. If populations are high and damage is anticipated, a foliar insecticide applied at about 700 degree-days will catch susceptible larvae

at their greatest abundance. Managers of large numbers of elms can improve their treatment decision making by using egg presence-absence sampling. See *Elm Leaf Beetle Pest Notes* for more information.

Cultural Control. Plant species or cultivars that resist both Dutch elm disease and elm leaf beetle, such as the elm hybrids Accolade, Emerald Sunshine, Frontier, and Prospector elms and most Chinese elms (except for Dynasty, which is highly susceptible to elm leaf beetle). Avoid planting European elms and other species that are especially susceptible to both Dutch elm disease and elm leaf beetle. Consult Table 5-11 in Chapter 5 for a more complete list of pest-resistant and -susceptible elms and their Latin names.

Provide elms with proper irrigation, especially American and European elms, which are adapted to summer rainfall. Protect trees from other pests and injury, such as by avoiding changes in grade and drainage around established trees. Check for dead or dying branches and promptly remove any limbs that pose a limb drop (failure) hazard. Otherwise, avoid unnecessary pruning and properly prune elms, especially when trees are young and preferably during late fall and winter.

Biological Control. Introduced parasites and native predators are least partly responsible for keeping elm leaf beetle populations generally low. The easiest-to-recognize parasite is a small black tachinid fly, *Erynniopsis antennata*. After an *Erynniopsis* larva feeds inside and kills beetle larvae, the parasite's $1/5$-inch-long, black to reddish pupae can be seen at the tree base among the bright yellow beetle pupae.

Adults of the tiny wasp *Oomyzus gallerucae* (Eulophidae) feed on elm leaf beetle eggs and the parasite's larvae feed inside eggs. The *Oomyzus* adult leaves a round hole when it emerges from beetle eggs, which remain golden. Unparasitized eggshells are whitish with a ragged hole after a beetle larva emerges. The tiny eulophid wasp *Baryscapus brevistigma* leaves one or more small, round holes in beetle pupae that it kills and emerges from around the tree base. Conserve these natural enemies by avoiding applications of broad-spectrum, residual, contact insecticides.

Insecticides. Monitor leaves regularly in early spring if you might inject trunks with systemic insecticide or spray foliage because you are targeting early-instar larvae. Foliar sprays for elm leaf beetle include azadirachtin, narrow-range oil, and spinosad. Avoid foliar application of

Table 6-4.

Timing of Elm Leaf Beetle Monitoring and Management.

ACTION	WHEN	WHEN, IF MONITORING DEGREE-DAYS (DD)	
		DD (F) FIRST GENERATION	DD (F) SECOND GENERATION
sample eggs once a week[1]	for several weeks after first-generation eggs appear in spring; repeat during second generation	329–689	1,535–1,895
single foliar spray	peak density of first and second instars combined	700	not recommended
bark banding or trunk spray	before earliest third instars crawl down trunk	700	2,000
systemic insecticide applied to soil	during late winter or early spring if pest damage or populations were high the previous season, especially if winter was colder or drier than average		
systemic insecticide applied to roots or trunks (e.g., implanted or injected)	as soon as possible during spring if egg sampling during 329–689 DD indicates thresholds are exceeded; because of the mechanical injuries to trees and potential spread of pathogens, use other methods when feasible		

Degree-days (DD) are accumulated above 52°F from March 1. See the website ipm.ucdavis.edu for more information.

1. Egg presence-absence sampling technique can be used to predict treatment need when managing large numbers of elms as explained in Dahlsten et al. 1993 in "Literature Cited" in the back of this book.

Tachinid flies during their larval stage are important parasites of many pests. This tachinid (*Erynniopsis antennata*) lays eggs into the second-instar larvae of elm leaf beetle, also shown here. *Photo: J. K. Clark*

ELM LEAF BEETLE PARASITE

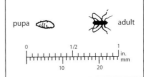

A close-up of elm leaf beetle prepupae (left) and pupae (center), and pupae of *Erynniopsis antennata* (right), a parasite of elm leaf beetle. *Photo: J. K. Clark*

carbamates, organophosphates, and pyrethroids because of the negative impact on natural enemies and potential for drift and other negative environmental impacts. Instead, use systemic insecticide or bark banding if management is warranted.

Bark Banding. A single application of a residual (persistent) contact insecticide (e.g., carbamate or pyrethroid) to encircle (band) an area several feet wide on the lower trunk each spring kills larvae that crawl over treated bark as they migrate down to pupate on soil. If you bark-band trees, do so before the earliest third instars crawl down the trunk. Spray bark at the rate labeled for elm bark beetles or other wood-boring insects. If rain occurs after application, trunks are repeatedly wet by sprinklers, or a less-persistent insecticide is used, a second application may be warranted. Effective products may not be available for home use and a licensed pest control operator may be needed for an effective treatment.

Bark banding alone does not provide satisfactory control in many situations. Treating one or a few elms is unlikely to be as effective as banding all nearby elms in a neighborhood because adult beetles can fly between treated and untreated trees. Expect little or no control during the first year when banding the more susceptible species (e.g., English and Scotch elms); banding all nearby elms over several consecutive years can provide control after the first year of treatment.

Systemic Insecticides. Translocated neonicotinoids can be sprayed on trunks or foliage, injected into trunks or soil, or drenched (watered) or injected into soil to provide season-long control. To avoid tree injury and potential spread of pathogens on contaminated tools, make a bark spray or soil application instead of injecting or implanting trees whenever possible.

A major disadvantage with systemic insecticide soil drenches is that there is a time delay between application and insecticide action. Therefore, the optimal treatment timing is generally late winter, which is before you know whether insects and damage will be abundant enough to warrant the application. Therefore, consider local weather and beetle density history when deciding whether to make systemic insecticide applications. If beetles and damage were low during the previous summer and fall, especially on untreated elms, it is unlikely that management will be needed the next spring. Relatively warm or wet winters also appear to reduce the likelihood that elm leaf beetle will be a problem in California the following spring.

EUCALYPTUS TORTOISE BEETLES

Two introduced leaf beetles (*Chrysophtharta m-fuscum* and *Trachymela sloanei*), also called tortoise beetles, chew semicircular holes and irregular notches along eucalyptus leaf edges. Adults and larvae can chew off most of a leaf's surface, leaving only the midvein. Adults also clip off young terminals, causing sparsely foliated trees. Well-established and properly maintained eucalyptus trees appear to better tolerate extensive leaf feeding.

These tortoise beetles have similar appearance and biology. Adults and mature larvae are about ¼ inch long. Larvae resemble caterpillars, and adults are hemispherical, superficially resembling a lady beetle: *Trachymela* adults are dark brown with blackish mottling, while *Chrysophtharta* adults are lighter colored and gray to reddish brown. *Trachymela* larvae are dark green to reddish brown with a black head and prothoracic shield (black area on the top and sides behind the head), while *Chrysophtharta* larvae are lighter greenish gray with a black head. Both adults and larvae feed mostly at night and are easily overlooked because their color blends with that of the foliage. During the day they may hide under bark or feed high in the canopy. Both species have several generations each year.

Distinguish tortoise beetles from eucalyptus snout beetle (*Gonipterus scutellatus*) adults, which are reddish brown weevils with an elongate head, as opposed to the hemispherical shape of the tortoise beetles. The legless weevil larvae are yellowish green with a slimy coating. Eucalyptus snout beetle is under good biological control from the introduced egg parasite *Anaphes nitens* and is rarely a pest; temporary outbreaks may occur where the weevil is newly introduced. Avoid applying pesticides and tolerate snout beetle leaf damage until natural enemies increase enough to provide biological control.

Tortoise beetle control usually is not needed, despite the tattered appearance of leaves. Tortoise beetles alone apparently do not kill trees, but their feeding in combination with other eucalyptus pests, especially if growing conditions or tree

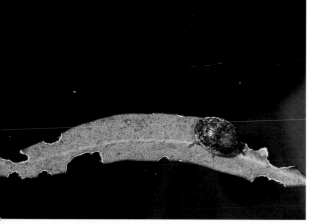

Two species of leaf beetles chew eucalyptus leaves. The adult *Trachymela sloanei* (shown here) is dark brown with blackish mottling. Adults of the other species, *Chrysophtharta m-fuscum* are lighter colored and gray to reddish brown. The hemispherical adults, also called tortoise beetles, are about ¼ inch long and superficially resemble a lady beetle. *Photo:* J. K. Clark

EUCALYPTUS TORTOISE BEETLE

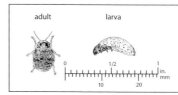

adult   larva

Trachymela sloanei larvae have a green to reddish abdomen with a black head and black prothoracic shield behind their head. If the cause of chewed eucalyptus leaves is not obvious, look under loose tree bark, where eucalyptus tortoise beetle adults, egg masses, larvae, and pupae may be found. *Photo:* J. K. Clark

A late-instar larva of *Chrysophtharta m-fuscum* on baby blue (silverleaved mountain) eucalyptus foliage it chewed. The light greenish to gray tortoise beetle larvae rest during the day and are easily overlooked as their color resembles that of host foliage. *Photo:* J. N. Kabashima

KLAMATHWEED BEETLE

Chrysolina quadrigemina was deliberately introduced into California during the 1940s to control *Hypericum perforatum*, a toxic rangeland weed. The Klamathweed beetle largely eliminated Klamath weed from several million acres, and each year saves ranchers millions of dollars in otherwise lost grazing land and poisoned livestock. However, certain other *Hypericum* spp. that subsequently became popular ground covers and shrubs in landscapes, especially *H. calycinum,* can be defoliated by this leaf beetle.

The metallic bluish green to brown adults feed on foliage the year around, except during the hot, dry summer. Eggs are laid from fall through spring singly or in clusters on leaves, where the grayish larvae feed. Larvae pupate just beneath the soil surface. Damage occurs during the spring, when plants produce most of their growth flush.

Insecticidal soap or another insecticide, applied when larvae or adults are feeding, can provide control. Removing litter accumulated beneath plants in hot areas may reduce the survival of adult beetles that rest there during the summer. Keeping soil beneath plants moist during the spring may increase disease and mortality

care practices are not optimal, could lead to tree death.

Providing trees a good growing environment and proper cultural care, especially appropriate irrigation, is the most effective strategy for keeping eucalyptus healthy. Choose well-adapted and pest-resistant eucalyptus or other species when planting. Certain *Eucalyptus* spp. are preferred or avoided by tortoise beetles and other pests, as summarized later in "Redgum Lerp Psyllid."

Foliar insecticide sprays are not recommended on large eucalyptus where concerns include pesticide drift, runoff into water, and toxicity to natural enemies. In unusual situations where leaf chewing

is intolerable, certain systemic insecticides (e.g., application to soil) may be effective. Before taking direct control actions, consult "Eucalyptus Longhorned Borers" and "Redgum Lerp Psyllid" later in this chapter. Consider the adverse effect of insecticide application on pollinators and natural enemies, such as the parasites of eucalyptus psyllids and eucalyptus snout beetle. For more information, consult *Eucalyptus Tortoise Beetles Pest Notes.*

KLAMATHWEED BEETLE

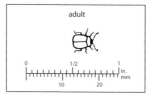

adult

Klamathweed beetle adults are metallic dark brown to bluish green. This important biological control agent of rangeland Klamath weed sometimes defoliates hypericum planted as ground cover. *Photo:* J. K. Clark

of immature beetles that pupate near the soil surface. Applying parasitic nematodes to soil beneath plants, as discussed below for weevils, may provide control if applications are made when most beetles are pupating, before adults emerge. Pupation often occurs during April and May, but populations vary, so monitor plants to determine when to treat; pupation occurs after mature larvae are observed on foliage.

WEEVILS

Over 1,000 species of weevils (Curculionidae), also called snout beetles, occur in California. Adults have an elongated head, and many species are flightless because their wing covers are fused. Females may feed for an extended period before laying eggs, and many species lack males and reproduce without mating. Larvae of most species are whitish grubs that feed hidden within plant parts or on roots in the soil.

Adults typically hide during the day and at night chew aboveground plant parts, causing leaves, flowers, or needles to appear notched or ragged or be clipped off from twigs. Although unsightly, this chewing by adults does not threaten the health of established woody plants and can be ignored. For example, the adults of Fuller rose beetle (*Naupactus godmani*), which are ¼ to ⅜ inch long and grayish to brown, commonly chew leaves, buds, and blossoms of many different landscape tree and shrub species, and their pale larvae chew roots. Plants tolerate this feeding without apparent permanent harm.

More serious pests are species with larvae that extensively chew on roots and root crowns or bore inside hosts. For example, larvae of the agave and yucca weevils tunnel in hosts, causing plants to decline and sometimes decay, collapse, and die. Larvae of introduced giant palm weevils bore in the apical meristem and kill palms. Soil-dwelling larvae of black vine weevil and diaprepes seriously damage roots, girdle root crowns, and can cause decline and sometimes the death of hosts.

Adult weevils characteristically chew irregular notches in leaf edges, as on this citrus. Collecting and identifying the species of weevils present is important because weevils that primarily chew foliage are relatively harmless in landscapes. However, feeding during the larval stage of some weevil species can seriously damage roots or other plant parts.
Photo: D. Rosen

MONITORING WEEVILS

Regularly inspect the foliage, buds, and flowers of host plants for evidence of weevil feeding. Damage may be due to weevils if foliage is clipped or notched or jagged edged, but no slime trails from snails or slugs and no leaf-feeding caterpillars, katydids, or other insects are found on foliage. To determine whether night-feeding adult weevils are present, sweep foliage with a net or lay a cloth or sheet beneath plants about 1 or 2 hours after dark and beat or shake several branches to dislodge any weevils.

Adults of many weevil species can be trapped when they seek shelter during the day. Band trunks with a strip of burlap (approximately 3 by 4 feet) folded lengthwise several times, then wrapped snugly around the basal trunk of each plant. Alternatively,

corrugated plastic tree wrap or corrugated cardboard with the smooth paper removed on one side, as pictured near the beginning of this chapter, can be wrapped around trunks with the corrugated side placed against the bark. Once or twice a week, gently remove the trunk wrap, carefully unfold it, count and record the number of weevils, then dispose of them.

Weevils of certain species may also be captured using a pitfall trap constructed from a several-inch-deep wide-mouthed plastic cup or dish and a funnel or smaller tapered cup. Cut off most of the funnel's spout or the bottom of the smaller cup and snugly insert it into the larger cup with the hole pointed down. Bury your trap so its top is flush with, or slightly below, the soil surface (Figure 6-11). Bury one or

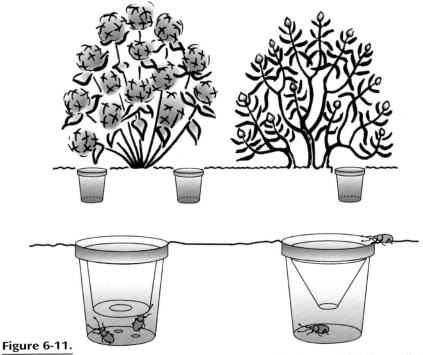

Figure 6-11.
Pitfall traps and trunk wraps (not shown) can be used to monitor adults of many soil-dwelling weevils. Construct traps from a funnel (at bottom right) or smaller cup with a hole in its bottom (left) fit inside another cup that is buried near plants so that weevils walking on the soil surface fall into it and are prevented from climbing out.

more traps as close to the trunk as possible beneath the canopy of each of several host plants. Check each trap about weekly and record the total number of weevils caught in all traps. Do not be surprised to find a variety of ground-dwelling creatures in your pitfall traps or trunk wraps, including beneficial predaceous carabid beetles. Release the predators, as they help to control pests.

MANAGEMENT OF WEEVILS

Plant resistant species, apply trunk barriers (e.g., sticky material), use traps, or apply systemic insecticide to soil or trunks to control weevils in many situations.

A residual, broad-spectrum, contact insecticide may be sprayed on foliage to control adults, but this can kill beneficial insects and wash off and pollute water. Monitor weevil populations beginning early in the spring, and if adult feeding cannot be tolerated, spray about a week after first detecting adults or damage. If weevil emergence is prolonged, a second foliar spray may be made about one month after the first application.

Soil-Dwelling Weevils. If plants have only one or a few trunks, prevent flightless weevils such as black vine weevil and diaprepes from feeding on foliage by trimming branches that provide a bridge to other plants or the ground, then apply a band of sticky material several inches wide to trunks, as pictured earlier in "Sticky

A healthy black vine weevil pupa and larva (left) and two darker pupae and a larva infected with beneficial *Steinernema feltiae* nematodes. *Photo:* J. K. Clark

Barriers." Persistent trapping year-round, as discussed under "Monitoring Weevils," may significantly reduce weevil populations in some situations.

Control soil-dwelling immature stages by applying parasitic nematodes (e.g., *Heterorhabditis* or *Steinernema* spp.) to soil. Apply nematodes when weevil larvae or pupae are expected to be present in midsummer to fall or in spring before adults emerge (about mid-March for most weevil species in northern California). In hot areas, apply nematodes in the early morning or evening. Soil must be warm (at least 60°F) and moist (well irrigated) but not soggy before application and for 2 weeks afterward.

AGAVE AND YUCCA WEEVILS

Scyphophorus spp. larvae bore in the base of hosts, causing plants to decline. Larval feeding in combination with decay microorganisms that colonize wounded tissue can cause infested plants to collapse and die. The dull, blackish adult snout beetles feed some on leaves and can cause punctured, discolored spots or small holes in foliage.

Adults and the pale larvae are about ½ inch long when mature. Agave weevil (*S. acupunctatus*) mostly bores in the plant base and upper roots of large species of agave that have wide leaf blades and bluish or gray leaf coloring, such as century plant (*Agave americana*). The more narrow-leaved, smaller agaves are less commonly infested. Yucca weevil (*S. yuccae*) mostly infests Joshua tree (*Hesperoyucca whipplei* =*Yucca whipplei*) and chaparral yucca, or Our Lord's candle (*H. whipplei*). Yucca weevil larvae tunnel in the base of

Two pale orangish agave weevil larvae exposed in the root crown of a foxtail agave. Agave weevils are serious pests because their tunnels become colonized by decay pathogens and cause infested hosts to decline and sometimes collapse and die. *Photo:* F. P. Wong

green flower stalks and in the apical meristem (central growing point) of hosts.

Control of agave and yucca weevils is difficult, and the insects are often not observed until the plant starts to collapse. To help protect nearby hosts, remove dying hosts and the immediately adjacent soil, where larvae may be pupating. Promptly dispose of infested material in ways that prevent weevils from emerging or escaping. Where agave weevil has been a problem, consider planting the smaller, narrower-leaved agave species. If applied before plants become infested or severely damaged, certain systemic insecticides may provide control.

BLACK VINE WEEVIL

Otiorhynchus sulcatus adults feed on many hosts, including yew (*Taxus* spp.) and broadleaf evergreens such as azalea, euonymus, and rhododendron. Adults are stocky, dark, about ⅜ inch long, and have sparse, fine, yellowish patches of hairs on their wing covers. They resemble cribrate weevil (*O. cribricollis*), which is somewhat smaller and also feeds on many plant species.

Larvae of these *Otiorhynchus* spp. feed on roots and bark near the soil surface and are the most damaging stage of these weevils, primarily in nurseries and on young landscape plants. Most feeding and damage occur in the fall and spring, and even a few weevils or slight foliage damage may warrant control actions to

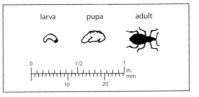

larva   pupa   adult

An adult black vine weevil chewing a euonymus leaf edge. Black vine weevil can be a severe pest of certain shrubs and young trees because its soil-dwelling larvae can extensively chew roots and girdle basal stems. *Photo: J. K. Clark*

Table 6-5.

Hybrid Rhododendrons Resistant to Feeding Injury by Adult Root Weevils.

RHODODENDRON HYBRID	RATING
P. J. Mezzitt	100
Jock	92
Sapphire	90
Rose Elf	89
Cilpimense	88
Lucky Strike	83
Exbury Naomi	81
Virginia Richards	81
Cowslip	80
Luscombei	80
Vanessa	80
Oceanlake	80
Dora Amateis	79
Crest	79
Rainbow	76
Point Defiance	76
Naomi	76
Pilgrim	76
Letty Edwards	76
Odee Wright	76
Moonstone	73
Lady Clementine Mitford	72
Candi	72
Graf Zeppelin	71
Snow Lady	71
Loderi Pink Diamond	71
Faggetter's Favourite	70

Ranked from highly resistant (100 rating) to moderately resistant (70) to *Sciopithes obscures, Otiorhynchus sulcatus, O. singularis, Nemocestes incomptus,* and *Dyslobus* spp. *Source*: Antonelli and Campbell 1984.

prevent larvae from developing and seriously damaging young plants. Black vine weevil has one generation per year. In California, adults emerge from pupae in the soil and feed during the night from March through September.

Plant less-susceptible species to avoid weevil damage. Many rhododendron hybrids resist damage from *Otiorhynchus* spp. weevils (Table 6-5). Provide proper cultural care and a good growing environment, especially appropriate irrigation and good soil conditions for roots, to keep plants vigorous and better able to tolerate damage.

CONIFER TWIG WEEVILS

Pissodes spp. feed on Douglas-fir, true fir, pine, and spruce. Most species are rarely serious pests. The adults chew foliage, causing minor damage, while the larvae of many species chew on roots and the trunk near the soil and can cause unsightly cankers on bark. Most species, such as Monterey pine weevil (*P. radiata*), are secondary pests that occur primarily on conifers that are already dying or injured.

The white pine weevil (*P. strobi*) is a serious pest of young pine and spruce. The pale, grublike larvae bore in inner bark, vascular tissue, and the centers of small branches, treetops, or stems, which can severely distort terminals and stunt plant growth. Other causes can produce similar damage, as summarized in Table 5-10 for insects and pathogens affecting pines. To confirm *Pissodes* spp. as the cause of damage, peel bark from affected shoots to expose the insects, their tunnels, or the pupal chambers covered in wood fiber that persist after the adults emerge.

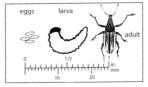

eggs   larva

adult

To help control white pine weevil, prune off and dispose of infested terminals before the weevils mature and emerge as adults.

DIAPREPES ROOT WEEVIL

The introduced diaprepes root weevil (*Diaprepes abbreviatus*), also called citrus root weevil or diaprepes, feeds on citrus and over 200 woody ornamental species. Larvae chew roots and the root crown, causing severe decline or death of hosts. The nocturnal adults chew mostly young, tender leaves from the leaf edges, causing irregular or semicircular notches and leaving dark frass (droppings) near chewing damage.

Adults are about ⅜ to ¾ inch long, with dark, lengthwise streaks on their back alternating with variable colors, commonly gray, orange, or yellow. Females lay a gelatinous mass of white to grayish eggs on leaves. The emerging grublike larvae drop to the soil and feed over about 4 to 16 months before maturing to a length of about 1 inch. Larvae pupate in a soil cell, and the emerging adult can feed and reproduce for several months.

To confirm weevil presence, shake chewed foliage over a light-colored collecting surface (such as a white sheet) to dislodge adults or deploy special diaprepes traps. You can also scrape away soil at the base of infested plants and look for chewed roots and root crowns and the soil-dwelling weevil larvae and pupae.

Adult diaprepes root weevils have dark streaks on their back but are otherwise variably colored, ranging from gray to yellow to orange. Larvae are the seriously damaging stage of diaprepes, as they chew the crown and roots of citrus and many other plants. *Photo: E. E. Grafton-Cardwell*

Diaprepes occurs in California only in Los Angeles, Orange, and San Diego Counties, as of this writing. If you find a weevil resembling *Diaprepes abbreviatus* in an area where it is not known to occur, place adults in a small jar filled with rubbing alcohol and take them to your county agricultural commissioner for positive identification. For more information, see the *Diaprepes Root Weevil* leaflet.

FILBERT WEEVILS AND ACORN WORMS

Oak acorns are commonly infested with larvae of filbert weevils (*Curculio* spp.) or the filbertworm (*Cydia latiferreana* =*Melissopus latiferreanus*, Tortricidae), and less commonly infested by the acorn moth (*Blastobasis* =*Valentinia glandulella*, Blastobasidae). Larvae of these moths also infest chestnut and other tree nuts. Although harmless to trees, larval feeding kills acorns and nuts that people collect for planting. Insect-damaged acorns may also become colonized by bacteria that cause liquid to drip from oaks, an annoyance discussed in Chapter 5 in "Drippy Nut Disease, or Drippy Acorn."

Larvae of the filbertworm moth are light brown to whitish with a dark head and three pairs of true legs. Filbert weevils are brown beetles with a long, thin snout. Their larvae are light brownish, yellow, or white without obvious legs, and they may curl into a C shape if disturbed.

Filbert weevils chew a small hole and insert their eggs into acorns on the tree. The female moths oviposit on nuts, and the emerging larvae bore inside. After acorns drop, larvae chew a hole and exit. They pupate and overwinter in the soil, or they pupate inside the nut, then emerge as adults in late spring to early summer. Filbertworms and filbert weevils have one generation per year.

No methods have been demonstrated as effective for controlling these insects, and no controls are recommended. If collected for planting, inspect acorns and dispose of those that have holes, feel unusually lightweight, or deform easily when squeezed, as these are often not viable. Another method is to place acorns in a tub of water and discard those that float; larval feeding creates air pockets, causing acorns to float.

GIANT PALM WEEVILS

Larvae of the red palm weevil (RPW, *Rhynchophorus vulneratus*) and South American palm weevil (SAPW, *R. palmarum*) feed in the apical growing point of Canary Island date palm and the root collar (basal trunk) of various date palms. Larvae damage or kill fronds by chewing in tissue where fronds attach to the trunk. This causes a sparse canopy of few fronds, seriously weakens the trunk structure, and can kill date palms. Heavily infested date palms may break at the trunk, or, with Canary Island date palm, the entire leaf crown may drop.

Unless the trunk and especially the growing point at the top of the trunk are regularly examined (e.g., from a lift bucket truck), infestations can be easily overlooked until palms are irrecoverably damaged. Indications of giant palm weevil infestation include

- a fermented odor from the upper trunk
- empty pupal cases or dead weevil adults around the base of palms
- gnawing sounds from trunks caused by the large larvae chewing inside
- truncated or cut-off leaf tips, which may be visible from the ground using binoculars
- tunnel openings on the trunk or base of fronds that exude chewed plant material (frass) or viscous liquid

An adult filbert weevil. The acorn has a hole chewed by a weevil larva, which emerged to pupate in soil after feeding inside. If collecting acorns for planting, the only known control is to inspect and discard damaged nuts. *Photo: J. K. Clark*

Palm weevil larvae chew cavities in the apical growing point of date palms, then pupate in a cocoon of palm fibers, as with the larva exposed here. Larval tunneling weakens date palms, sometimes killing them and causing the tops to fall to the ground. If dieback of the apical (newest, uppermost, or center) leaves is observed, carefully inspect it for larval mines and frass (excrement). When trimming palms, look for larval frass and feeding damage around the base of leaves and trunks. *Photo: M. Lewis, UC Riverside Center for Invasive Species Research*

The adult red palm weevil (*R. vulneratus*) is about 1½ inches long, with a distinct long snout. It varies greatly in color but commonly has a reddish brown to black body with (in some individuals) a red stripe on the pronotum behind the head. *Photo:* J. N. Kabashima

ROSE CURCULIO

		adult
0	1/2	1

These rose curculios chew blossoms and buds. Their larvae feed in flower buds, often killing buds before they open. Hand-pick adult weevils, especially if few in number. Clip and dispose of damaged buds and finished blossoms to remove larvae and reduce future populations of this pest *Photo:* J. K. Clark

- weevil adults, which can be caught in special traps

Adults are large snout beetles about 1½ inches long. Their color is variable, usually orange with black spots, black with a red stripe, black overall, or dark with contrasting colored blotches on the pronotum (topside behind the head). Larvae grow up to 2 inches long and are yellowish, legless grubs with a brown head. Mature larvae pupate in a chamber of coarse palm fibers in tissue where they fed. Palm weevils have several generations per year.

Whether these weevils or similar species (e.g., *R. ferrugineus*) will become permanently established in California is unknown. The RPW has been reported in a limited area of Orange County; the SAPW has been found in San Diego County because it is well established in Mexico and spreads northward.

Be sure to distinguish introduced *Rhynchophorus* weevils from the native giant palm borer (*Dinapate wrighti,* Bostrichidae), a relatively minor secondary pest of dying and severely stressed palms. The adults and trunk-boring larvae of all these beetles can be the same color and length. However, *Dinapate* adults are cylindrical

with a bulbous, snoutless head, and their larvae have prominent legs and closely resemble the larvae pictured in "White Grubs and Scarab Beetles."

Use good sanitation to prevent palm weevils from spreading from infested palms. Chip, burn, or deeply bury infested material. Plant resistant species, such as the native California fan palm, European fan palm (*Chamaerops humilis*), or non-palm species. See the discussions of palm diseases in Chapter 5 to help you decide which palm is most suitable for your situation. See *The Biology and Management of Landscape Palms* and cisr.ucr.edu for more information.

ROSE CURCULIOS

Merhynchites spp. adults are red to black snout beetles about ¼ inch long. Adults chew ragged holes in rose petals and make circular holes in buds and stems. The legless larvae are pale orange to whitish, up to ¼ inch long, and chew inside flower buds, often killing buds before they open. If rose curculios are numerous, they may kill terminal shoots.

Stem feeding below buds can cause terminals to bend, injury that may be confused with sawflies mining in canes, as discussed in "Raspberry Horntail." Rose curculio larvae resemble larvae of the rose midge (*Dasineura rhodophaga*). However, rose midges grow only to ¹⁄₁₆ inch long, are more slender, and are rarely pests in California.

Hand-pick adult weevils, especially if few in number. Adults drop from the plant when disturbed, so they can be collected by gently shaking canes over a tray or a bucket of soapy water. Prune out and dispose of damaged buds and finished flowers, which will remove larvae and help reduce future problems as well as improve plant appearance. Entomopathogenic nematodes applied to soil in late winter or early spring where mature larvae or pupae are overwintering may provide some control. A broad-spectrum, residual, contact insecticide can be applied to kill adults if the infestation is severe.

WHITE GRUBS AND SCARAB BEETLES

Certain scarab beetles (Scarabaeidae) and their larvae (white grubs) are occasional pests, feeding on flowers or fruit as adults or chewing roots during their larval stages. Adult beetles are medium to large, dull to brightly colored and metallic, and they have antennae that terminate in an oval club composed of several thin leaflike plates. The soil-dwelling larvae are commonly robust and yellowish to dirty white, with well-developed legs and an enlarged abdomen. When disturbed, larvae tend to curl into a C shape.

Pests include hoplia beetles, May and June beetles (e.g., *Cotinus*, *Phobetus*, and *Polyphylla* spp.), and rain beetles (*Pleocoma* spp.). Many species of scarabs are harmless or beneficial, including those that feed on and help to decompose animal dung or compost. Because larvae live underground or hidden in organic debris and since adults usually are nocturnal, many people observe scarabs only when adults are drawn to lights at night.

Japanese beetle (*Popillia japonica*) and some other scarab pests in the eastern United States and overseas do not occur in California. However, such exotic scarabs are occasionally introduced as larvae infesting roots or soil of imported plants. Larvae of many white grubs can be identified to genus or species by examining the arrangement of hairs and bare areas on their raster (the underside of their last abdominal segment) and comparing this pattern to illustrations in publications such as *Handbook of Turfgrass Insect Pests*. If you find unfamiliar scarabs, take them to the county agricultural commissioner or UC Cooperative Extension office for identification.

This *Pleocoma* species scarab is called a rain beetle because it emerges from the pupal stage in soil after precipitation in the fall. Adults are harmless, but larvae are root-feeders that sometimes damage young woody plants. *Photo: J. K. Clark*

Adult Japanese beetles are mostly shiny metallic green with coppery brown wing covers and tufts of short, whitish hairs along the side. One of these beetles has two white eggs of a tachinid parasite (*Hyperecteina aldrichi*) on its thorax. This serious pest is not established in California, so report suspected Japanese beetles to the county agricultural commissioner. *Photo: J. K. Clark*

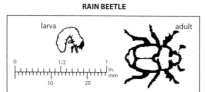

RAIN BEETLE

larva   adult

Most scarab larvae are dirty white to yellowish with well-developed legs and an enlarged abdomen. This tenlined June beetle (*Polyphylla decemlineata*) larva can girdle roots on small fruit and nut trees and ornamentals such as black locust, privet, and wisteria. Adults are mostly brown, with longitudinal white stripes on their wing covers. *Photo: J. K. Clark*

GREEN FRUIT BEETLE

Cotinis mutabilis is an occasional pest when adults chew ripening fruit. Green fruit beetle adults can fly a relatively long distance and are highly attracted to ripe fruit and the odors of fermentation and manure. The grublike larvae are dirty white or brownish and grow up to 2 inches long. They commonly burrow and feed in piles of compost, manure, organic fertilizers, and other decomposing organic matter. During spring, larvae mature and pupate in a cell of soil particles. Adults can be present from late spring through early fall. Green fruit beetle has one generation per year, spent mostly within the larval stage.

To prevent feeding by beetles, manage the grubs. Thinly spread or remove all piles of manure, lawn clippings, and leaves near fruit trees. To exclude beetles, compost within screened bins or turn compost

Green fruit beetle adults are ³/₄ to 1¹/₈ inches long and have metallic green wing covers with brownish margins. Adults sometimes chew ripe fruit and tomatoes, often in a large group. Most of their life cycle is as grublike larvae that feed harmlessly in compost and other decaying organic matter. *Photo:* J. K. Clark

piles frequently to speed decomposition and to expose grubs, which you can crush or leave to be eaten by birds.

Although managing compost and similar organic matter is more effective at reducing green fruit beetle populations, adults may be captured using homemade traps. Attract adults with a 1:1 mixture of grape or peach juice and water. Place several inches of this liquid bait in a 1 gallon container and in the opening insert a funnel of wire mesh with its widest opening facing up. Beetles attracted by the bait will land in the funnel and be guided to walk down into the container. Once inside, adults will be unable to escape. Additionally, reduce adult damage by harvesting fruit early and disposing of fallen fruit. Insecticides are of little value against this occasional pest. To help avoid this beetle's damage, consider planting cultivars that ripen early instead of those that ripen late.

Beetle Pets

Green fruit beetles are easily reared in a terrarium. Adults fed on banana, grapes, and other soft fruits will readily lay eggs in organic debris and soil, and their grubs will emerge in about a month. Entertainingly, the grubs curl into a C shape when disturbed, then crawl on their backs (move with their underside up) and rapidly burrow into litter. Children may enjoy helping to start a colony by collecting the grubs, which typically occur near fruit trees in uncovered piles of compost that have periodically been replenished over years. Place the grubs into a ventilated container with several inches of grass clippings, leaf litter, and ground-up, dry pet food on the bottom.

HOPLIA BEETLES

Adult *Hoplia* spp. chew light-colored blossoms of various herbaceous and woody plants, but they are a pest only in rose and sporadically in grape and strawberry. *Hoplia callipyge* especially chews holes in rose petals that are white or other light colors (e.g., pink, apricot, and yellow) but does not damage rose leaves. *Hoplia* larvae feed on decaying vegetation and roots of various herbaceous plants (e.g., strawberries and turfgrass) but apparently do not damage roots on rose or other woody plants.

Hoplia adults are about ¹/₄ inch long and reddish brown with silvery or coppery scales, making them appear iridescent in sunlight. *Hoplia callipyge* is primarily a problem in the Central Valley from Sacramento south to Bakersfield. *Hoplia* spp. have one generation per year, and their damage is confined to when the adults feed, which varies by location and spring weather and usually is about a 2- to 4-week period from mid-March to May.

Where hoplia beetles have been a problem, inspect light-colored rose blossoms regularly during spring. If flowers are infested, gently shake or knock the canes over a bucket of soapy water or hand-pick adults off flowers. Alternatively, clip off infested blooms into a bag or covered container and dispose of them. Typically, relatively few beetles are present, and regular hand-picking can control beetle populations in the immediate area.

Insecticide sprays should not be necessary in most landscape situations. If blossom chewing is intolerable, only residual, broad-spectrum insecticides (e.g., carbamates or pyrethroids) are likely to be effective. These insecticides are undesirable since they kill natural enemies and pollinators, can run off to contaminate water, and sometimes damage young blossoms. Systemic insecticides do not translocate into petals in sufficient concentrations to control beetles. Consult *Hoplia Beetle Pest Notes* for more information.

An adult hoplia beetle and its feeding damage on rose petals. Hoplia beetles cause only aesthetic damage in landscapes, and usually their feeding is obvious only on roses. Adults prefer light-colored blossoms, so concentrate monitoring and hand-picking of adults on apricot, pink, white, and yellow rose blossoms. Insecticide sprays are often not very effective. *Photo:* J. K. Clark

HOPLIA BEETLE

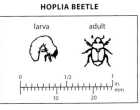

INDIAN WALKING STICK

Carausius morosus (Phasmatodea) sporadically defoliates azalea, camellia, geranium, hawthorn, hibiscus, ivy, jasmine, oak, privet, pyracantha, rose, and other species. The long, slender, slow-moving adults and nymphs resemble brownish twigs and are easily overlooked on plants, even when they have chewed off much of the foliage. To help detect them, spray chewed plants with water from a hose; this disturbance causes some of the insects to walk to the outer foliage, where they can more easily be seen.

Native to India, this insect is a sporadic pest in coastal locations because owners of pet walking sticks have released them into landscapes and carelessly discarded their eggs. Females can reproduce without mating, and they periodically drop their brownish, round eggs in their pet cages or containers. The eggs resemble the droppings (frass), so eggs and frass become mixed and the eggs become easily discarded outside when cages are cleaned.

Responsible owners of walking stick pets must understand the biology and not release these exotic creatures in California. This includes bagging and sealing cage debris and disposing of it in the trash. Unwanted walking sticks can be given to a responsible new owner or sealed in a plastic bag with the entire contents of the cage, including all debris, and placed in the freezer for at least 48 hours, then discarded in the trash. See *Indian Walking Stick Pest Notes* for more information.

CRICKETS, GRASSHOPPERS, AND KATYDIDS

Various species of crickets, grasshoppers, and katydids (order Orthoptera) chew holes or notch the edge of leaves. Katydids also chew circular, scabby scars on fruit, including apple, citrus, pear, and stone fruits. The devastating grasshopper (*Melanoplus devastator*), valley grasshopper (*Oedaleonotus enigma*), and less-common grasshoppers sometimes migrate in large numbers to feed in irrigated landscapes, such as after nearby unmanaged vegetation dries out when the rainy season ends. Although orthopterans' scattered chewing is often observed, these insects rarely cause serious damage to established woody plants. For more information, see *Grasshoppers Pest Notes*.

JERUSALEM CRICKETS

Because of their relatively large size and unusual appearance, Jerusalem crickets (*Stenopelmatus* spp., Stenopelmatidae) are commonly brought to entomologists for identification. They are given many common names, including "niñas de la tierra" (children of the earth) and sand crickets. They also are sometimes called potato bugs because they are occasional pests of potatoes, but the name "potato bug" is confusing because it is applied to several different invertebrates, including Colorado potato beetle, pillbugs, sowbugs, and Jerusalem crickets, none of which are true bugs (Heteroptera).

Jerusalem crickets are easily recognized by their bald round head, fat orangish abdomen with black rings, spiny hind legs, and a lack of wings. They molt up to about 10 times during their life span, which can be up to about 2 years. These insects are harmless to woody plants and only occasionally damage turf and vegetables. They live in soil, feeding mostly on nonwoody roots and succulent tubers, and may be important food for certain vertebrate predators. They are most often seen when exposed by gardeners turning the soil or

Katydids are easily overlooked on plants because their green color blends with the foliage. Adults are sometimes observed when they are drawn to lights at night. The oval flat eggs of this broadwinged katydid (*Microcentrum rhombifolium*) are laid overlapped in a group on twigs. Some other katydid species insert their eggs into leaf edges. *Photo: J. K. Clark*

Jerusalem crickets are harmless to people and woody plants. They live underground, feeding on succulent tubers and nonwoody roots. *Photo:* J. K. Clark

False, or small, honey ants (*Prenolepis imparis*) are tending woolly aphids on this shamel ash. This species resembles Argentine ant, but in comparison it is a relatively minor pest, in part because false honey ants are active mostly during cool weather (about 45° to 60°F) and are dormant during dry, warm weather. *Photo:* J. K. Clark

above ground after heavy irrigation or rainfall, at night, or at twilight during mild weather.

ANTS

Ants (Formicidae) are Hymenoptera closely related to bees and wasps. Of the 200 ant species in California, only a few are pests. Many ants are beneficial. For instance, some species improve soil and can be important predators of pest insects.

Pest ants can tunnel in wood (carpenter ants) or chew bark or plant parts (fire ants, pavement ants). Ants become landscape pests primarily when they feed on honeydew excreted by phloem-sucking insects, including aphids, mealybugs, soft scales, and whiteflies. Ants protect these honeydew producers from predators and parasites that might otherwise control the pests. Ants also sometimes disrupt the biological control of other pests, such as mites and armored scales. If ants are abundant on plants near structures, they are more likely to be a nuisance by coming indoors. When disturbed, certain ant species aggressively bite or secrete irritating formic acid. Red imported fire ants can severely sting people, pets, and wildlife.

IDENTIFICATION AND BIOLOGY

Ants are sometimes confused with termites. Ants have a narrow constriction between the thorax and abdomen, their antennae are distinctly elbowed, and winged ants have hind wings that are much shorter than the forewings. Termites

have a broad waist, antennae that are not elbowed, and wings of equal length (see Figure 6-12). It can be very helpful to identify the particular ant species present, as their biology and management often differs. For example, ant baits vary in composition and formulation and may be more effective on some species than others. To identify ants to species, view specimens under magnification and count the number and arrangement of antennal segments

and observe the number of nodes (projections) on the petiole, which is the first (narrow) segment of the abdomen. To identify common species, consult publications such as *A Key to the Most Common and/or Economically Important Ants of California, Integrated Pest Management for Citrus*, or the *Key to Identifying Common Household Ants* at ipm.ucanr.edu/ants.

Most ants are wingless workers (sterile females) that dig tunnels, defend the colony from natural enemies, forage for food outside the nest, and care for the tiny, pale, grublike immatures in the nest. Each colony also has one or more queens that spend most of their time in the nest laying eggs. Males occur only during the brief mating season, when winged males and queens mate, sometimes in flights or swarms outside of the nest. Most ants nest underground or beneath rocks, buildings, or other objects.

As each ant meets another from its nest, they exchange a tiny droplet containing food and colony communication chemicals (food sharing, or trophallaxis). Slow-acting baits take advantage of this food-sharing behavior so foraging ants spread insecticide throughout the colony and poison ants in their nests.

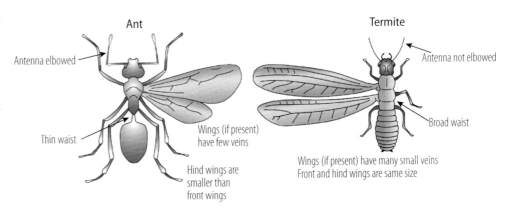

Figure 6-12.
Ants distinguished from termites. Ants have a narrow waist, elbowed antennae, and (if wings are present) hind wings that are much shorter than the forewings. Termites have beaded antennae, a broad waist, and wings (if present) that are of equal length.

ARGENTINE ANT

Linepithema humile (=*Iridomyrmex humilis*) is a common honeydew-feeding species in California and the southern states. The small workers are uniformly dark brown and travel in characteristic trails on bark or the ground. Argentine ants nest in moist soil and can quickly relocate nests in response to changes in food and weather. Colony size varies, often numbering in the thousands, and each Argentine ant colony may have many queens that contribute to this species' high reproductive capability. The winged reproductive males and females are about twice as long as workers and are rarely observed above ground, except sometimes in the spring. Argentine ant populations increase greatly in midsummer and early fall, so take control action (e.g., applying barriers or insecticide baits) beginning in late winter or early spring, before ants and the pests they tend become abundant. Slow-acting sweet liquid baits are the most attractive to Argentine ants.

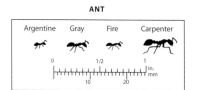

CARPENTER ANTS

Camponotus spp. usually nest underground or in wood but can sometimes be found in structures or the interior of living trees. Carpenter ants do not eat wood, but their tunneling weakens limbs, which may drop. Carpenter ants feed on insects, honeydew, and plant sap. Columns of these relatively large, black, or dark reddish ants may be observed foraging from nests, especially after sunset, when they become more active. Wood borings may also accumulate beneath nest entrances.

Help prevent carpenter ants and termites from attacking trees by providing plants with proper cultural care. Prune plants properly when needed, especially when trees are young, and prevent injuries to trunks and limbs, as detailed in Chapter 3. To reduce the likelihood of structural infestations, seal entry points, keep wood dry, and paint or otherwise treat wood. The effective baits and other pesticides can differ from those used for other ant species. Consult *Carpenter Ants Pest Notes* for more information.

PAVEMENT ANT

Tetramorium caespitum often nests in or under cracks in pavement and in lawns, near foundations, or under objects on the ground such as boards, bricks, stones, and wood. Workers are relatively slow moving, light to dark brown, and about ⅛ inch long. They resemble Argentine ants in size and color, but pavement ants have two distinct petiole nodes while Argentine ants have only one node.

Pavement ants feed on honeydew and prey on insects. They often nest near water, so populations can be reduced by improving soil drainage, repairing leaky irrigation systems, and reducing the frequency between irrigations to the extent this is compatible with healthy plant growth. Sweet baits placed near nests and trails are effective if using insecticides.

VELVETY TREE ANT

Liometopum occidentale is ¹/₁₀ to ¼ inch long. It has a glistening, blackish abdomen covered with fine hairs, a red thorax, and a brownish black head. Velvety tree ants are often observed in trails on limbs and trunks of trees with cavities because they nest inside decaying trees or in nearby soil. They feed primarily on honeydew and insects and usually do not damage landscapes. Because they are attracted to certain proteinaceous food and sweet liquids, their presence can be annoying at picnics and when occasionally drawn indoors. To reduce annoying populations, protect trees from injuries that can lead to wood decay, prune off dead limbs, and replace dead or hazardous hollow trees in which velvety tree ants nest.

These Argentine ants and other honeydew-feeding ant species increase populations of certain aphids and other phloem-sucking insects by protecting the pests from parasites and predators. *Photo:* J. K. Clark

The velvety tree ant has a reddish thorax and a black abdomen covered with fine hairs. This species is rarely a serious pest. Howeve because they often nest in decayed wood, trails of velvety tree ant on bark may indicate that trees should be inspected for potential hazards, such as dead or hollow limbs that should be pruned off. *Photo:* J. K. Clark

NATIVE FIRE ANTS

Several native *Solenopsis* spp. are important predators of insects. Most species rarely injure plants. However, the southern or California fire ant (*S. xyloni*) nests in mounds around trees and shrubs and can inflict painful bites or stings on people and pets. The southern fire ant also girdles and kills young trees by feeding on bark. If fire ants nest in landscapes, consider applying insecticide baits according to label directions. Be aware that the introduced red imported fire ant (discussed below) differs from native species and requires special attention. Fire ants are primarily attracted to insecticides mixed with protein baits.

RED IMPORTED FIRE ANT

Solenopsis invicta occurs in scattered locations in southern California and the San Joaquin Valley. This introduced pest is more prolific and aggressive than native ant species and will readily run up any object that touches their mound. Its painful sting can severely injure people and animals attacked by swarming ants. A single sting can be life-threatening to people who are allergic to the fire ant's venom. Although primarily predators of insects and other invertebrates, and therefore beneficial in some situations, fire ants also feed on seeds, tend honeydew-producing insects, and sometimes strip bark from young trees and shrubs.

The red imported fire ant workers are dark reddish and $\frac{1}{5}$ inch long, or much shorter, and workers of very differing size can be observed together in the same trail. Because they closely resemble the native southern fire ant, take suspected fire ants to the local UC Cooperative Extension office or county agricultural commissioner for identification.

Fire ants are primarily spread long distances by people moving ant-infested soil. Inspect incoming soil and container plants to avoid introducing fire ant colonies, especially if material is arriving from areas known to be infested. In areas where they are not established, do not attempt to con-

trol red imported fire ants yourself. In California, report suspected red imported fire ant infestations to the county agricultural commissioner or telephone 1-888-4FIRE-ANT. For more information, consult *Red Imported Fire Ant Pest Notes* or the website cdfa.ca.gov/plant/pdep/rifa.

MANAGEMENT OF ANTS

Where ants or honeydew producers have been a problem, inspect the trunks of plants and the surface of soil nearby for trails of ants beginning in late winter or early spring. During cool weather, look for ants during the warm times of day. During hot weather, monitor for ants during the cooler times of day. When numerous ants of a pest species are found, you may need to trace the trails to determine where the ants are actually feeding (e.g., on honeydew of sucking insects on leaves or twigs).

Use a combination of methods to control ants. It is not feasible to totally eliminate ants from an outdoor area. Focus your control efforts where ants are direct pests, such as when they are coming indoors or tending bothersome honeydew-producing insects, especially on plants near structures:

- Prune trees and shrubs back away from structures.
- Apply sticky material (e.g., Tanglefoot) or other barriers to trunks to exclude ants.
- Seal crevices and gaps in buildings through which insects enter.
- Keep organic debris (e.g., leaf litter and mulch), all vegetation, and irrigation water at least several inches back from foundations and walls.
- Apply effective, enclosed insecticide baits.

Insecticide Baits. An insecticide mixed with an attractant is the preferred chemical method for ant control. Effective insecticide baits are slow-acting, so that before

the workers die they will spread the toxicant among many other ants during food sharing and carry bait underground to kill ants in the nest. Abamectin and borates are examples of insecticides used in ant baits. Baits include solids or liquids that are prepackaged into ant stakes, small plastic bait station containers, or solid granules that are broadcast. Although baits require users to be patient, they can provide more effective long-term control than sprays. For the most effective and economical ant control, treat in late winter or early spring when ant populations are lowest. Place bait near nests, on ant trails beneath plants, and alongside building foundations.

Insecticide bait effectiveness varies with ant species, availability of alternative food, active ingredient, and the type of bait. For example, protein baits can attract Argentine ants in spring when colonies are producing young, while sweet liquid baits attract Argentine ants year-round.

Borate-sucrose water baits, such as 0.5 to 1% boric acid in 10 to 25% sucrose in water, controls Argentine ants and certain other ants. Properly mixed liquid boric acid bait

Red imported fire ant workers are mostly dark reddish brown. As with the native southern fire ant, red imported fire ants are variable in size; large and small workers occur together in the same trail. When disturbed, these highly aggressive ants often bite and sting in large numbers. If red imported fire ant is not established in your area, report suspected infestations to the county agricultural commissioner. *Photo:* J. K. Clark

This plastic bait station dispenses the right mixture of boric acid ($\frac{1}{2}$ to 1%) insecticide, sugar, and water to attract and kill Argentine ant. Unlike contact sprays that kill only ants foraging aboveground, effective insecticide baits are slow-acting, so the foraging ants will spread the toxicant among other ants during their food-sharing behavior and carry bait underground to kill ants in the nest. *Photo:* C. A. Reynolds

Adults of a psocid (top) and the acacia psyllid (bottom). In comparison with a psyllid, the adult psocid has longer antennae, a more narrow "neck," or separation between head and thorax, and chewing mouthparts (not visible here). Psocids do not damage plants but may be confused with psyllids and both insects sometimes occur together. *Photo: J. K. Clark*

can be applied in refillable, reusable bait stations. Slight changes in products, such as too high a concentration of active ingredient or preservatives in the product, can dramatically reduce their effectiveness. For example, if the boric acid concentration is higher than 1%, as in some commercial products, the ant bait is less effective. Consult *Ants Pest Notes* and *Urban Pest Management of Ants in California* for more information on managing ants.

Sucking Insects

Insects with piercing or sucking mouthparts include Hemiptera and thrips. The Hemiptera include true bugs (suborder Heteroptera) and the Homoptera, an informal group including aphids, leafhoppers, mealybugs, psyllids, scales, whiteflies, and others.

PSYLLIDS

Psyllids (Psyllidae) suck phloem sap and excrete honeydew, which promotes the growth of blackish sooty mold. Some species also excrete pale wax or crystallized honeydew. High psyllid populations can reduce plant growth and cause premature leaf drop. Certain species can cause terminals to distort, discolor, or die back. Many native shrubs such as manzanita and sugarbush (lemonade berry) can host native psyllid species, which are hardly ever serious problems. Several exotic introduced psyllids can be pests, including pittosporum psyllid (*Cacopsylla tobirae*), the rosewood tree or tipu psyllid (*Platycorypha nigrivirga*), and species discussed below.

Adult psyllids are commonly about ¹/₁₀ inch long and hold their wings rooflike over their bodies. When feeding, adult psyllids commonly tilt their abdomen at a 45° angle from the plant surface. In comparison with aphids, psyllids have short antennae, strong jumping legs, and

Sticky, waxy excretions of the introduced olive psyllid (*Euphyll olivina*). The small, flocculent bl(are individual psyllids covered w wax. *Photo: D. Rosen*

PSYLLID

adult

0	1/2	1
		in.
10	20	mm

Adults of the Asian citrus psyllid (*Diaphorina citri*) are brownish, with dark markings on their wings. Adults readily fly when disturbed and at rest tilt their rear end up at a 45° angle. This psyllid causes severe distortion of shoots and vectors the huanglongbing (citrus greening) pathogen that kills citrus and Rutaceae family ornamental plants. Repo any suspected Asian citrus psyllids or other new pests to the county agricultural commissioner if found where they not known to be established. For more information, see *Asian Citrus Psyllid. Photo:* M. E. Rogers, University of Florid

no cornicles. Do not confuse psyllids with similar-looking psocids (barklice), which feed on fungi, including sooty mold growing on psyllid honeydew. Mature psyllids commonly jump when disturbed, while psocids run or fly away. Unlike psyllids, psocids have a distinctly swollen, bulbous area in front of the head between the widely spaced antennae, which can be distinguished with a hand lens. Psocids also lack sucking mouthparts and do not damage plants.

Psyllids on deciduous trees overwinter as eggs or young nymphs in or around bud scales. In the spring, eggs hatch, and nymphs feed on developing buds and new leaves. On evergreen plants in mild-climate areas, all stages may be found year-round. Populations are usually highest when new plant growth is abundant, but high temperatures may reduce populations of some species. Consult *Psyllids Pest Notes* for more information.

ACACIA PSYLLID

Acizzia uncatoides is an introduced pest that feeds only on *Acacia* spp. plants. The adults are brownish, green, or orangish but often appear darker during cool weather. The tiny, orange eggs and orange to green, flattened nymphs are found primarily on growing tips, new leaves, and flower buds. The psyllids are most abundant in the spring when temperatures warm and host plants produce new growth flushes, but all life stages can occur throughout the year.

Avoid planting susceptible acacia species (see Table 6-6) and consider replacing severely infested acacias. Acacia psyllid is often kept under control by natural enemies, including a small black lady beetle (*Diomus pumilio*) introduced from Australia and a purplish pirate bug (*Anthocoris nemoralis*) pictured earlier in "Types of Natural Enemies."

During the months of March through

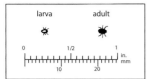

larva   adult

An adult lady beetle (*Diomus pumilio*) feeding on acacia psyllid eggs. Introduction of this ¹/₁₂-inch-long predator and a minute pirate bug (*Anthocoris nemoralis*) have greatly reduced acacia psyllid populations. See "Biological Control" early in this chapter for a photograph of the pirate bug eating an acacia psyllid nymph. *Photo:* J. K. Clark

Table 6-6.

Acacia spp. Susceptibility to the Acacia Psyllid in Northern California.

NOT INFESTED OR SLIGHTLY INFESTED (<0.5 NYMPH/TIP)
albida, armata, aspera, baileyana,[1] cardiophylla, collettioides, craspedocarpa, dealbata, gerardii, giraffae, karoo, podalyriifolia, pravissima
SLIGHTLY TO MODERATELY INFESTED (0.5–1.1 NYMPHS/TIP)
cultriformis, cunninghami, cyanophylla, decurrens, iteaphylla, mearnsii, robusta, triptera
MODERATELY TO SEVERELY INFESTED (>2 NYMPHS/TIP)
cyclops, implexa, longifolia, melanoxylon, obtusata, pendula, penninervis, retinodes, salingna, spectobilis

1. Infested by the baileyana psyllid (*Acizzia acaciaebaileyanae*).
Source: Koehler, Moore, and Coate 1983.

June, avoid practices that stimulate psyllid-preferred new growth, including fertilizing, irrigating, and pruning. Whenever possible, conduct any needed cultural practices from July through November. Natural enemies often do not become effective (abundant) until late spring, after psyllid populations have increased and weather warms. Tolerate psyllids until predators provide control, or temporarily reduce high populations with spot applications of insecticidal soap or oil.

EUCALYPTUS PSYLLIDS

At least eight species of introduced psyllids feed on eucalyptus in California, including the redgum lerp psyllid discussed separately below. Each psyllid species feeds on and damages only certain species of eucalyptus. The lemongum psyllid (*Cryptoneossa triangula*) and spottedgum lerp psyllid (*Eucalyptolyma maideni*) are abundant only on *Eucalyptus citriodora* and *E. maculata*. Both species can occur together on the same plants. Spottedgum lerp psyllid nymphs construct a covering (lerp) shaped like a cornucopia or funnel with openings that resemble skeletal ribs. Each lerp may harbor several nymphs that may be observed moving in and out from beneath their coverings. Lemongum psyllids do not construct any covering. They are free living but often occur beneath the coverings formed by spottedgum lerp psyllid. Lemongum psyllid and spottedgum lerp psyllid are substantially parasitized by *Psyllaephagus perplexans* and *P. parvus*, respectively, at least in coastal areas.

Bluegum psyllid (*Ctenarytaina eucalypti*) primarily infests *Eucalyptus pulverulenta* and juvenile foliage of *E. globulus*. This species does not produce hardened covers but can secrete extensive wax and honeydew. Pairs of adult bluegum psyllids mate tail-to-tail and may resemble a grayish moth unless examined more closely. Their pale eggs and orangish, gray, or green nymphs occur among flocculent wax on leaves and terminals. Bluegum psyllid is now under good biological control by the introduced parasitic wasp *Psyllaephagus pilosus*. Before taking any control actions, see "Management of Psyllids" below and consult the sections "Eucalyptus Longhorned Borers" and "Eucalyptus Tortoise Beetles" to learn how management efforts may affect other pest problems.

Certain psyllids form a hardened waxy cover called a lerp. The appearance of lerps varies according to the species of psyllid. Spottedgum lerp psyllid nymphs secrete this elongate, funnel-shaped, fluted cover, which can be abundant on *Eucalyptus citriodora* and *E. maculata*. *Photo:* J. K. Clark

An introduced wasp (*Psyllaephagus pilosus*) leaves a round hole when it emerges from psyllid nymphs that it killed. Bluegum psyllids are effectively controlled by this parasite, so eucalyptus should not be treated for this pest. *Photo:* J. K. Clark

OLIVE PSYLLID

Euphyllura olivina feeds on olive (*Olea europaea*), Russian olive (*Elaeagnus angustifolia*), and mock privet (*Phillyrea latifolia*). It produces abundant flocculent white wax on leaves and twigs and can cause premature leaf drop.

To improve tree health and resistance to pests, avoid frequent irrigation near tree trunks and provide good soil drainage. Prune off or thin interior limbs in hot locations to increase psyllid exposure to heat and increase air circulation, which suppresses olive psyllid populations. Where psyllids were intolerable the previous year and insecticide use is planned, target the first generation, typically present during March to April. Olive psyllids are more difficult to control during their second generation (May to June), when most of their waxy excrement and plant damage occur.

REDGUM LERP PSYLLID

Glycaspis brimblecombei secretes copious honeydew, and high populations cause extensive defoliation, increase tree susceptibility to other pests, and contribute to the premature death of highly susceptible hosts. At some locations, abundant yellowjackets feed on honeydew and can annoy or threaten people. Redgum lerp psyllid infests over two dozen *Eucalyptus* spp., especially river red gum (*E. camaldulensis*), flooded gum (*E. rudis*), and forest red gum (*E. tereticornis*). On certain other *Eucalyptus* spp., psyllid populations remain low and not bothersome.

Nymphs are yellow or brownish, resembling a wingless aphid, and form a roundish cover (lerp), which resembles an armored scale. Lerps are whitish, hemispherical caps on leaves that grow up to ⅛ inch in diameter. Nymphs enlarge their lerp as they grow or move and form a new covering. Adults are about ⅛ inch long, slender, and light green, with orangish and yellow blotches.

Females lay tiny, yellowish, ovoid eggs mostly on succulent leaves and shoots. Population increases often follow the production of new plant growth, but all psyllid life stages can occur on both new and mature foliage. Redgum lerp psyllid has several generations each year, and all stages can be present throughout the year in mild coastal areas. Because some nymphs form and abandon multiple lerps and the nymphs underneath are often parasitized, the number of lerps on leaves does not indicate the actual number of psyllids present.

Redgum lerp psyllid is attacked by many predators, including multicolored Asian lady beetle and other coccinellids, minute pirate bugs, larvae of lacewings and syrphid flies, spiders, and small birds. The introduced parasitic wasp *Psyllaephagus bliteus* is especially important. Adult *P. bliteus* are about 1/12 inch long, with metallic green bodies and yellowish legs. Females oviposit on young nymphs. After feeding as larvae, wasps pupate and emerge from beneath larger lerps, leaving a roundish emergence hole in parasitized nymphs and their lerps. Parasitism and various predators have greatly reduced redgum lerp psyllid populations, especially in coastal areas.

Management of Redgum Lerp Psyllid.

Plant resistant species to prevent redgum lerp psyllid from being a problem, as only a few species become highly infested (Table 6-7). Provide eucalyptus with proper cultural care and protect trees from injury. Drought stress and high nitrogen levels apparently increase redgum lerp psyllid populations. Consider providing trees with supplemental water during dry summer and fall weather. Avoid fertilizing eucalyptus. Use slow-release nitrogen products if other plants near eucalyptus

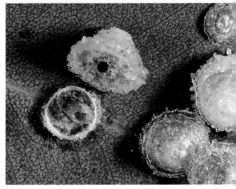

A redgum lerp psyllid cover with a parasite emergence hole is flipped over here to reveal a dead, mummified psyllid nymph, also with a parasite emergence hole. Inspecting beneath nearby lerps would reveal that some of the nymphs underneath them are parasitized. After feeding inside as a larva, the parasite (*Psyllaephagus bliteus*) pupates and emerges as a shiny, greenish wasp with yellow legs. *Photo:* J. K. Clark

require nitrogen fertilization.

Consider treating only those trees where the pest has been intolerable or tree health appears threatened by insects. Insecticide efficacy against redgum lerp psyllid has been variable, and the materials kill *P. bliteus* and psyllid predators. Before taking any action, consult the sections "Eucalyptus Longhorned Borers" and "Eucalyptus Tortoise Beetles" in this chapter. Consider the impact of controls on other eucalyptus pests, such as the potential of insecticides to disrupt biological control of the bluegum psyllid and eucalyptus snout beetle. For more information, consult *Eucalyptus Redgum Lerp Psyllid Pest Notes*.

EUGENIA PSYLLID

Trioza eugeniae damages only eugenia (*Syzygium paniculatum*). The adults are mostly dark brown, with a white band around the abdomen. The tiny, golden eggs are laid

The adult eugenia psyllid has a white band around its dark brown abdomen. *Photo:* J. K. Clark

Tiny golden eugenia psyllid eggs cause the edges of new eugenia leaves to glisten. This pest can be controlled by well-timed shearing and leaving clippings as mulch beneath plants; most of the psyllids on clipped foliage will die, but many of the parasites will be able to complete their development and emerge to attack psyllids on the plant. *Photo:* J. K. Clark

Eugenia psyllid nymphs feed in pits on the underside of eugenia leaves, causing the upper side of foliage to redden and distort. *Photo:* J. K. Clark

Table 6-7.

Eucalyptus spp. Relative Susceptibility to Introduced Pests in California.

COMMON NAME (GUM)	*EUCALYPTUS* SPP.[1]	LONGHORNED BORERS	REDGUM LERP PSYLLID	TORTOISE BEETLE
Australian beech	polyanthemos	–	L	L
baby blue, silverleafed mountain	pulverulenta	–	L[2]	–
blue	globulus	H	I–L[2]	H
dollar leaf, silver dollar	cinerea	–	L	–
flooded, or desert	rudis[3]	–	H	I
forest red	tereticornis	–	H	–
grand, or rose[3]	grandis	I	I	H
gray ironbark	paniculata	–	L	–
hybrid	trabutii	L	–	–
karri	diversicolor	H	I	–
lemon	citriodora	L	I[4]	L
long flowered	macrandra	–	I	–
manna	viminalis	H	I	H
mountain	dalrympleana	L	–	–
narrow leaved	spathulata	–	L	–
Nichol's willow leaved	nicholii	–	I	–
red flowering	ficifolia	–	L	L
red ironbark	sideroxylon	L	I–L	L
river red	camaldulensis	L	H	H
round leaved/red flowered	platypus/nutans	H	I–L	–
shining	nitens	H	H–I	–
silver	crenulata	–	–	L
spotted	maculata	–	–[4]	L
sugar	cladocalyx	L	I–L	–
swamp mahogany	robusta	L	L	–
Sydney blue	saligna	H	L	–
white ironbark	leucoxylon	–	I	–

KEY

– = information not available
H = highest susceptibility
I = intermediate susceptibility
L = less or least susceptible or reportedly not attacked
1. Redgum lerp psyllid = *Glycaspis brimblecombei*
Tortoise beetle = *Trachymela sloanei*
2. Susceptibility ratings based on *Phoracantha semipunctata* longhorned borer, which is believed to be similar to the susceptibility to *P. recurva*.
Susceptible to bluegum psyllid (*Ctenarytaina eucalypti*), but this psyllid is generally under good biological control.
3. *E. grandis* is also called flooded gum.
4. Susceptible to lemongum lerp psyllid (*Cryptoneossa triangula*) and spottedgum psyllid (*Eucalyptolyma maideni*).
Adapted partly from Brennan et al. 2001; Hanks et al. 1995.

primarily along the edges of young leaves, where the yellowish crawlers with orange-red eyes settle and feed and cause leaves to deform, discolor, and sometimes drop prematurely. As it forms a feeding pit, the nymph resembles a soft scale insect and appears flat when viewed from the lower leaf surface. Populations are highest when new foliage is produced in the winter and spring, but reproduction and all psyllid stages occur year-round.

The parasitic wasp *Tamarixia dahlsteni* introduced from Australia provides partial biological control of this pest. However, especially in cooler areas near the California coast, parasite populations often do not increase quickly enough in spring to provide satisfactory control.

If eugenia are regularly sheared (e.g., with topiary plants or neat hedges), well-timed pruning of new growth in combination with parasite conservation can be especially effective. Prune terminals after maximum spring growth appears or about 3 weeks after the first peak in adult psyllid density as determined by weekly branch beating, foliage inspection, or sticky traps, as discussed in "Monitoring Psyllids," below. Leave eugenia clippings as mulch near the shrubs for at least 3 weeks to allow parasites within psyllid nymphs to complete their development and emerge (see Figure 6-5). Eugenia psyllid eggs and nymphs on the cut foliage will die. Consider shearing eugenia tips at about 3-week intervals (and leaving clippings on-site) throughout the period of new plant growth or as long as adult psyllids are abundant. In addition to providing direct control, shearing terminals is the only way of eliminating damaged foliage (aside from waiting for old leaves to drop). Insecticide, especially systemics, can also provide control but will reduce natural enemy populations.

Feeding by psyllid nymphs pits and distorts California pepper tree foliage. This injury is relatively harmless to pepper trees and is not obvious unless plants are closely inspected. Excess irrigation and poor soil drainage are the primary problems that threaten pepper tree survival. *Photo: J. K. Clark*

PEPPERTREE PSYLLID

Calophya schini feeds only on the pepper tree *(Schinus molle)*, also called California or Peruvian pepper tree. Adult psyllids are greenish or tan and somewhat pear shaped. Females deposit their tiny eggs on growing tips throughout the year. The orangish nymphs feed on any succulent plant part, causing the plant to form a pit and deform and discolor around where each nymph settles. One psyllid generation requires only a few weeks during warm weather, and all life stages occur throughout the year.

The introduced wasp *Tamarixia schina* has reduced peppertree psyllid abundance and foliage distortion. Parasite larvae feed and pupate inside psyllid nymphs. Adult parasites leave a roundish emergence hole in dead nymphs that can be observed from the underside of leaves. Avoid using contact foliar insecticides on pepper trees, as they are not very effective at controlling psyllids and can substantially disrupt biological control.

Improve pepper tree health and tolerance to psyllids and other damage by providing proper cultural care. Pepper trees are adapted to well-drained, sandy soil and summer drought. Planting trees in heavy clay soils or in summer-watered landscapes, such as lawns, promotes root disease and causes trees to decline and die. Avoid psyl-

lid problems entirely by planting other species. For example, Australian willow myrtle or peppermint tree *(Agonis flexuosa)*, desert willow *(Pittosporum phillyraeoides)*, and Australian willow *(Geijera parviflora)* are relatively drought tolerant and have a weeping appearance resembling the pepper tree but are not affected by the psyllid.

Trees apparently tolerate psyllid damage so no treatment is needed to protect plant health. Conserving parasites generally provides adequate control in southern California and in warmer inland areas. In cooler areas, such as coastal northern California, parasites appear to be less effective. When adult psyllids are abundant and terminals are growing (usually winter), a systemic insecticide can provide long-term control where psyllids are intolerable.

MONITORING PSYLLIDS

If you plan to apply insecticide, monitor adults or eggs to help you time control actions to kill eggs or newly hatched nymphs before damage occurs. You may discover an annual cycle to psyllid abundance; population increases are typically associated with the availability of tender new growth.

Adult psyllids are easily monitored with yellow sticky traps or, on low branches, by beating foliage to dislodge insects. Consult the sections "Yellow Sticky Traps" and "Branch Beating" earlier in this chapter for more information on these techniques. Sticky traps are best for monitoring around tall trees or when the most important natural enemies are parasites (because parasitic wasps can also be monitored with yellow traps). Branch beating may be best for some species, such as acacia psyllid,

because it allows important predators to also be monitored. When sampling, count and record separately the number of adult (winged) psyllids and psyllid-feeding predators (dislodged by beating) or parasites (caught in yellow sticky traps).

Another approach is to visually inspect foliage near the ground and count the insects on four to eight or more new growth tips or branch terminals on each plant. Leaves should be of the same age at each sampling. Counting the small immature psyllids is tedious; counting only adults can be an efficient alternative.

Be sure that you distinguish psyllids from psocids, as discussed above. Sample every 1 or 2 weeks beginning before psyllids become a problem, typically in the spring. Depending on the situation, sample regularly until your control actions are effective, plants stop producing new growth, or psyllid populations decline.

MANAGEMENT OF PSYLLIDS

Plant species is a primary determinant of whether psyllids will be a problem. Choose species that are well adapted to the location. Consider replacing problem-prone plants with more pest-resistant species (Tables 6-6, 6-7). Discourage excessive flushes of succulent foliage that promote increased psyllid populations; for example, avoid fertilizing established woody plants and do not irrigate too frequently. Unless well timed and repeated, as discussed above for eugenia psyllid, minimize shearing or trimming terminals because this stimulates new growth preferred by psyllids.

Nonresidual contact insecticides (e.g., horticultural oil, insecticidal soap, and neem oil), and short-persistence products (spinosad) can provide temporary control if infested new growth is thoroughly covered with the insecticide spray. Because they do not leave long-lasting toxic residues that would kill natural enemies migrating in after their application, these are good choices where plants are small enough to be sprayed and have psyllid natural enemies that are only partially effective at that location or may not provide biological control until later in the season. Time sprays to kill eggs and young nymphs before damage or psyllids become abundant. Monitor susceptible new growth and spray soon after you observe a sharp increase in adult numbers on sticky traps or in beat samples or when you see significant numbers of nymphs on leaves and shoots.

Applying systemic insecticide is the most practical method for controlling psyllids in large trees. Wherever possible, make an application to bark or soil if the product is labeled for that use; this avoids the plant injury discussed in Chapter 2 under "Injecting or Implanting Pesticides." To minimize adverse effects on bees and natural enemies that feed on nectar and pollen, delay systemic insecticide application until after plants have finished flowering. Consult *Eucalyptus Redgum Lerp Psyllid Pest Notes* and *Psyllids Pest Notes* for more information.

APHIDS

Aphids (Aphididae) are small, soft-bodied insects that suck phloem juices. High aphid populations can slow plant growth and cause leaves to yellow, curl, or drop early. Some species distort stems or fruit or cause galls on roots, leaves, or stems. Aphids' whitish cast skins, honeydew and the resulting sooty mold growth, and (in some aphid species) waxy excrement can be unsightly and bothersome but are usually harmless to established woody landscape plants. Certain aphids transmit viruses that cause plant diseases, but this is usually not damaging to landscape trees and shrubs.

IDENTIFICATION AND BIOLOGY

Identification to species is often not necessary before determining how to manage

Shiny, sticky honeydew coats these flowering plum leaves infested with the waterlily aphid (*Rhopalosiphum nymphaeae*). *Photo:* J. K. Clark

aphids. Exceptions are that it can be very helpful to know whether your pest species is host-specific. Host-specific aphids attack only one species or a few closely related plant species and cannot spread to other plant species. Host-specific aphids commonly overwinter as eggs on their deciduous hosts. Controls such as dormant or delayed-dormant oil sprays or a soil-applied systemic insecticide in late winter need only be directed at that host. Conversely, aphid species that feed on many plant species migrate among hosts, may feed and reproduce throughout the year, and often lack a dormant-season egg stage in much of California where winters are mild. Most aphids on landscape trees and shrubs, including gall-making and woolly aphids, are host-specific.

Aphids often feed in dense groups on leaves or stems and do not rapidly disperse when disturbed. Individuals of most species are $^1/_{16}$ inch or less long and are pear shaped with long legs and antennae. They vary from green, yellow, white, brown, or red to black. Some species, such as the woolly aphids, are covered with a waxy, white to grayish coating. A pair of tubelike

Blackish sooty mold on California bay infested with California laurel aphids. Sooty mold can be rubbed or washed off foliage and does not damage plants, but if it is extensive it may slow plant growth by reducing photosynthesis. *Photo:* J. K. Clark

The manzanita leaf gall aphid (*Tamalia coweni*) caused these leaves to swell into harmless, pod-shaped galls. This aphid feeds only on manzanita and is prevalent on new growth, such as that stimulated by frequently irrigating and shearing plants. *Photo:* J. K. Clark

A dark, oblong aphid egg on a bud scale. Aphids typically produce eggs only during the fall, and aphid species that are host-specific on deciduous plants overwinter in the egg stage. A dormant-season oil spray to kill eggs can be a good management method for host-specific aphids on deciduous trees. *Photo:* J. K. Clark

projections (cornicles) near the hind end of the body distinguishes most aphids from other insects, except that gall-making, root-feeding, and woolly species of aphids typically have no distinct cornicles.

Most aphids have many generations per year, and populations can increase rapidly, especially under conditions of moderate temperatures. Extreme temperatures may retard aphid growth and reproduction. Throughout most of the year, adult aphids (which may be winged or wingless) give birth to live young without mating. Aphids may produce overwintering eggs, primarily in species that are host-specific on deciduous plants or that occur in locations with cold winters.

APHIDS WITH MANY HOSTS

BEAN APHID

Aphis fabae hosts include elderberry, euonymus, jacaranda, pyracantha, viburnum, and many herbaceous plants. Bean aphids are dark olive-green to black with black appendages. As with melon aphid, they can occur in dense colonies, sometimes with both green and black individuals.

GREEN PEACH APHID

Myzus persicae is one of the most common aphid species. It feeds on a large variety of woody and herbaceous landscape plants, including flowering plum, stone fruits, and other *Prunus* spp. Green peach aphid is green, yellow, or reddish, with black on the top of the head and thorax of winged adults. Because succulent foliage on some of its host plants is available year-round in mild-winter areas, this insect can have up to two dozen generations per year.

MELON APHID, OR COTTON APHID

Aphis gossypii hosts include apple, camellia, crape myrtle, euonymus, *Prunus* spp., willow, and many herbaceous plants. It feeds and reproduces all year long in areas of

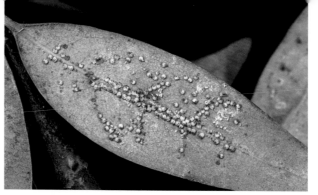

These round, gray California laurel aphids (*Euthoracaphis umbellulariae*) are sometimes mistaken for immature whiteflies or scale insects. They occur on the underside of California bay, often in rows along veins. Camphor tree and sassafras are occasionally infested. *Photo: J. K. Clark*

California with mild winters where hosts are available. Melon aphids are commonly blackish or dark green, but pale yellow to whitish forms also occur.

HOST-SPECIFIC APHIDS

Although many aphids look similar, most species on woody landscape plants feed on only one or several closely related plant species and cannot spread to plants of another species. Host-specific species

Melon, or cotton, aphids are commonly green to blackish, but pale yellow to whitish forms also occur. Unlike most aphid species, which feed only on a few closely related plants, the melon aphid infests many different hosts. *Photo: J. K. Clark*

include birch aphids (e.g., *Callipterinella calliptera*, *Betulaphis brevipilosa*, and *Euceraphis betulae*), linden aphid (*Eucallipterus tiliae*), Norway maple aphid (*Periphyllus lyropictus*), tuliptree aphid (*Illinoia =Macrosiphum liriodendri*), gall-making and woolly aphids, and the species discussed below. The rose aphid (*Macrosiphum rosae*) may spend part of the summer on nearby herbaceous plants, but it usually occurs on roses whenever succulent tissue is present.

CRAPEMYRTLE APHID

Sarucallis kahawaluokalani is occasionally so abundant on crape myrtle that plants may be defoliated. This aphid is mostly yellowish green, except for the winged adults, which have distinctive black marks on the abdomen, wings, and tips of the antennae.

GIANT CONIFER APHIDS

Cinara spp. aphids commonly occur on fir, pine, and spruce. They are among the largest aphids, up to 1/5 inch long. At first glance they are sometimes mistaken for ticks, but ticks have eight legs and no antennae, while aphids have two long, slender antennae and six legs. Giant conifer aphids have especially long legs and occur individually or in large colonies on foliage and bark. The purplish or black body may be covered with gray powder. Colonies on deodar cedar often

Palm aphid (*Cerataphis brasiliensis*) varies in color and resembles a whitefly pupa or scale insect. It commonly is greenish when young, darkens to brown, then blackens and develops a marginal ring of wax as it matures. This aphid feeds in dense colonies on unopened spears and young leaves of many palm species, causing yellow fronds and sticky honeydew that attracts ants. *Photo: D. R. Hodel*

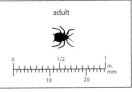

adult

0   1/2   1 in.
10   20   mm

Giant conifer aphids are among the largest aphids. At first glance they are sometimes mistaken for ticks, but ticks have eight legs and no antennae, while aphids have two long, slender antennae and six legs. This bow-legged fir aphid (*Cinara curvipes*) occurs on fir, spruce, and deodar cedar, often in large colonies. *Photo:* J. K. Clark

nfest only a single limb. These aphids may shift their bodies in unison when disturbed, apparently in response to an alarm pheromone they secrete. Each species of *Cinara* s specific to one or a few conifer species, but more than one aphid species may occur on the same plant. Giant conifer aphids give birth to live nymphs year-round in mild-winter climates but overwinter as eggs where winters are severe.

OLEANDER APHID

Aphis nerii is bright yellow or orangish with black appendages. It primarily infests oleander but occasionally occurs on milkweeds (*Asclepias* spp.) and vinca. Dense colonies can infest growing terminals, unopened flower buds, and foliage, especially when leaves are young and succulent or old and senescing.

Larvae of *Lysiphlebus testaceipes*, a braconid wasp, commonly feed inside oleander aphids, causing them to become papery, swollen, light brown to tan mummies. Lacewings, lady beetles, and syrphid lar-

vae are important predators on oleander aphid. Natural enemies sometimes control this pest, but their effectiveness is reduced by cultural practices that stimulate aphid-favoring succulent growth (e.g., frequent pruning and irrigation).

POPLAR GALL APHIDS

Several *Pemphigus* spp. cause galls on cottonwood and poplar leaves. The lettuce root aphid (*P. bursarius*) commonly galls Lombardy poplar leaves during the spring. Aphid feeding stimulates plant tissue to form a hollow gall around the aphid on leaves or leaf petioles. The enclosed aphid gives birth to about 100 to 250 waxy, grayish nymphs, many of which develop wings and emerge and migrate to feed on their alternate hosts, including lettuce and various Brassicaceae. The aphids feed on the

This hollow swelling on a poplar leaf petiole has been opened to reveal the waxy, grayish poplar gall aphids inside. These galls are harmless to poplar, but this aphid can damage its alternate hosts, primarily lettuce. *Photo:* J. K. Clark

Oleander aphids are yellowish with black appendages. Their most important biological control agents are tiny parasitic wasps, which cause parasitized aphids to become swollen mummies. Avoid frequent pruning and irrigation of oleander because these practices stimulate succulent plant growth, which increases aphid reproduction and survival and decreases the effectiveness of natural enemies. *Photo:* J. K. Clark

basal stem and roots of their alternate hosts, then in the fall fly back to poplars, where females lay overwintering eggs. These aphids can be significant pests on their vegetable crop alternate hosts but are harmless to poplar and no control is recommended on trees.

WOOLLY APHIDS

Woolly aphids cover themselves with white waxy material similar to that secreted by some adelgids and mealybugs. Some species feed in groups and cause gall-like swellings on bark or curled leaves. Many species alternate generations among plant species or different parts of the same plant, and the same species of aphid may look different depending on the host plant, season, and the part of the plant it is feeding on. Generally, no control is needed to protect the survival of otherwise healthy trees. Where intolerable, they are difficult to manage without using systemic insecticide.

The woolly apple aphid (*Eriosoma lanigerum*) feeds in wax-covered groups on leaves and bark and sometimes causes gall-like swellings on bark of apple, hawthorn, pyracantha, (infrequently) pear, and possibly other hosts. When its galls are abundant on roots it can slow or stunt plant growth.

Woolly apple aphids produced this pale flocculent wax on a pyracantha stem. Less obvious are the bean aphids, both black and green forms, infesting the flower bud stems. *Photo:* J. K. Clark

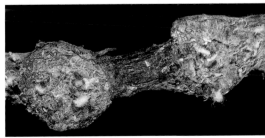

Woolly apple aphid feeding causes galls on branches and trunks of apple, hawthorn, pyracantha, and (less commonly) pear. Other woolly aphids (e.g., *Eriosoma* spp.) commonly gall bark and wood on elm trees. *Photo:* J. K. Clark

The parasitic wasp *Aphelinus mali* often kills virtually every aphid on aboveground parts, leaving a round emergence hole in each blackish, puffy mummy.

Woolly elm aphids (*Eriosoma* spp.) cause globular swellings on bark, stems, or roots and curling or galling of foliage on elms. Some elm-feeding aphids leave abundant wax on infested surfaces, while others, such as elm nipplegall aphid (*Tetraneura nigriabdominalis*), also called oriental grassroot aphid or grain aphid, excrete their wax inside pouchlike leaf galls.

Woolly oak aphids (*Stegophylla* spp.) curl leaves, produce copious flocculent wax on leaves and shoots, and occasionally cause scattered shoot tip dieback on at least coast live oak, interior live oak, and valley oak.

ASH LEAFCURL APHID

Prociphilus (=*Meliarhizophagus*) *fraxinifolii* causes ash leaves to curl, distort, form "pseudogalls," and drop prematurely. Officially named leafcurl ash aphid, it excretes copious honeydew and flocculent waxy material, making a mess beneath trees and promoting the growth of dark sooty mold. Aphids can be abundant on leaves from spring through fall and overwinter as eggs on bark and nymphs on roots. Abundant winged ash aphids can be a nuisance when they migrate between bark, foliage, or roots, especially in the fall. Certain ash species and cultivars appear to be less susceptible to leafcurl ash aphid, but there are

Ash leaves curled by woolly aphids (*Prociphilus* sp.). Many woolly aphids have a complex life history, migrating between leaves and roots, sometimes among different hosts. *Photo:* J. K. Clark

no research-based recommendations that identify resistant ash. The aphids apparently do not harm trees in landscapes, but their root feeding can damage small plants in nurseries.

HACKBERRY WOOLLY APHID

Shivaphis celti produces copious honeydew and small, roundish, waxy tufts on the underside of hackberry (*Celtis* spp.) leaves, especially on Chinese hackberry (*C. sinensis*). This Asian woolly hackberry aphid overwinters on terminals as eggs that hatch in spring when the new leaves flush. The aphid has multiple generations and the honeydew and wax excreted by adults and nymphs typically become most abundant and bothersome during late summer.

Insecticides apparently are not warranted to protect the health or survival of aphid-infested hackberries. If the excretions are intolerable and insecticide will be applied, avoid injecting or implanting pesticides or other materials into hackberry trunks or roots. Chinese hackberry are susceptible to an unexplained, tree-killing malady; the cause may be a vascular wilt pathogen mechanically spread by unsterilized tools that contact internal parts of multiple hackberry trees, such as tools used to prune trees or inject them with chemicals. See *Hackberry Woolly Aphid Pest Notes* for more information.

MONITORING APHIDS

Monitoring aphids or their honeydew helps you establish approximate thresholds so you can take action when aphid population levels are moderate and easier to control and before aphids or honeydew become too bothersome.

Monitor aphids and their natural enemies by visually inspecting 5 to 10 or more leaves, new growth tips, or 1-foot branch terminals on each of several plants. Aphids commonly occur on the lower leaf surface; clipping leaves may facilitate their inspection. Inspect foliage every 1 or 2 weeks during the period when aphids may be a problem. Record some measure of abun-

dance, such as the proportion of leaves or terminals with aphids (presence-absence sampling) and separately the proportion of samples with natural enemies (parasitized aphids or predators). Compare these numbers among sample weeks and years to evaluate control efficacy or establish thresholds as discussed under "Honeydew Monitoring," below.

To help locate infestations and where to monitor, inspect trunks for columns of ants, which often indicate that plant juice-sucking insects are abundant on plants. Most ants will feed on honeydew, and some species of ants aggressively tend phloem-sucking insects, chasing away natural enemies that would otherwise help to control these pests.

Honeydew Monitoring. Where substantial resources are spent on aphid control, monitoring honeydew provides a direct measure of this problem and helps managers establish thresholds, time control actions, and evaluate treatment efficacy.

Honeydew can be monitored with bright yellow, water-sensitive cards often used for monitoring pesticide sprayers. Where a water (or honeydew) droplet settles on a card, it produces a distinct blue dot. Attach the cards to a stiff background and place the cards beneath plants before honeydew typically becomes a problem, usually in the spring. Monitor for the same period about once each week, such as from 11 a.m. to 3 p.m. on a day of typical temperatures when no fog, rain, or overhead irrigation is anticipated. Handle the cards with forceps or gloves because they will change color from the moisture in your skin.

Label and save the cards from each monitoring date and when honeydew becomes intolerable (e.g., when complaints are received), note the approximate droplet density measured during that week and during previous weeks. Monitor honeydew the next season and take control action if honeydew density approaches the level that was previously found to be intolerable. Visually compare cards among sample dates before and after control actions to assess the effectiveness of management efforts. Note that if you apply systemic insecticide to soil, there will be a delay between the application and when it becomes effective.

To provide a quantitative threshold for treatment during subsequent years, count the drops from several subareas of each card (e.g., counting drops in three representative 1-sq-cm areas per card, then using the average of these numbers). Then calculate the overall average droplet density from all cards for each week (e.g., 4 drops/sq cm).

To save time, instead of counting drops, you can estimate honeydew density by visually comparing each card to reference cards with previously determined droplet densities (e.g., one card averaging 1 drop/sq cm, another with 2 drops/sq cm, and so forth in increasing increments to about 10 drops/sq cm if using a total of 10 reference cards). In some street tree situations, the complaint threshold for aphid honeydew collected during the 4-hour monitoring period has been found to range from 2 to 8 drops/sq cm. Thresholds acceptable for your situation may be different.

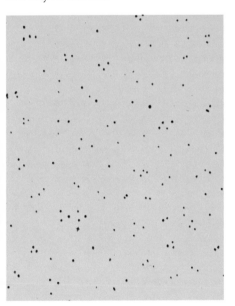

Special water-sensitive paper changes color at each spot where it is contacted by a droplet of honeydew or water. Placing several cards at intervals (for several hours about weekly) beneath trees is an efficient method of monitoring insect honeydew to establish thresholds, time control actions, and evaluate treatment efficacy. You can save sample cards exhibiting a range of known (counted) droplet density, then during subsequent monitoring, instead of counting drops, visually compare your samples to these reference cards to estimate honeydew density. *Photo:* J. K. Clark

A convergent lady beetle (*Hippodamia convergens*) eating a green peach aphid. This predator is named for the two converging white bars behind the adult's head; individuals vary in the number of black spots, and some have no spots. Although this lady beetle and certain other natural enemies can be purchased and released, conserving natural populations of parasites and predators is usually a more effective pest control method. *Photo:* J. K. Clark

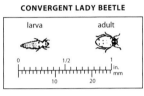

CONVERGENT LADY BEETLE

MANAGEMENT OF APHIDS

To manage aphids, use preventive cultural methods, conserve natural enemies, and apply least-toxic insecticides. If planting species especially susceptible to aphids (e.g., the plants named in "Host-Specific Aphids"), consider locating them where dripping honeydew will be less annoying (away from decks and pavement). Populations of most aphids are highest on plant parts with high nitrogen levels, such as new growth. Conversely, at least some predators (e.g., lacewings attacking oleander aphids) survive better when plants grow slowly or are not producing new growth. Minimize fertilization, pruning, and other cultural practices that stimulate succulent plant growth. For example, do not irrigate too frequently or with excess amounts. Avoid applying nitrogen fertilizers to established woody plants, except as discussed in Chapter 3. If fertilizing, apply no more than necessary and use slow-release materials to avoid stimulating aphid populations.

On small, sturdy plants, knock aphids off with a strong spray of water. Most dislodged aphids will not be able to return to the plant, and their honeydew will be washed off. Avoid wetting plants susceptible to foliar pathogens (e.g., black spot and rust on roses), or hose off plants early

Lady beetle eggs are typically oblong and (in aphid-feeding species) orangish and laid on their ends in a group. Many plant-feeding leaf beetles also lay groups of eggs resembling these. *Photo:* J. K. Clark

An orange convergent lady beetle pupa and several larvae among whitish skins of aphids on which the beetles fed. *Photo:* J. K. Clark

in the day so they dry more rapidly. Use baits or sticky barriers to control honeydew-seeking ants, since ants can disrupt biological controls by protecting aphids from their natural enemies.

If applying insecticide, choose products that are least toxic to natural enemies (e.g., insecticidal soap, horticultural oil, neem oil). Treat only in spots where aphids are most abundant to preserve natural enemies elsewhere. Do not spray foliage with oil on days when temperatures will exceed 90°F, and do not apply oil within 3 weeks of applying sulfur. Certain systemic insecticides can be highly effective, but where possible delay their application until after plants are done flowering to minimize killing of honey bees and natural enemies that feed on plant nectar and pollen.

If aphids were a problem the previous season, learn whether the pest species overwinters as eggs on bark; if so, apply narrow-range oil to shoot terminals after buds have swollen but before they burst into leaf. Oil applications may not be warranted for aphid infestations alone, but they may be a good choice if plants are also infested by mites, scale insects, or other pests overwintering on bark that will also be controlled. Consult the *Aphids Pest Notes* for more information.

Biological Controls. Low numbers of aphids are beneficial by producing honeydew that attracts and feeds adults of parasites and predators of aphids and other pests. Natural enemies may not appear in sufficient numbers until after aphids become abundant; however, their preservation is an essential part of a long-term IPM program.

Adult lady beetles and soldier beetles, and the larvae of lady beetles, lacewings, and certain flies (aphid flies, predaceous midges, and syrphids), are common aphid predators. Many small wasps, including *Aphelinus, Aphidius, Ephedrus, Praon,* and *Trioxys* spp., are important parasites that feed as immature wasps inside their host, causing the aphid to "mummify," or become slightly puffy and turn tan or black (see Figure 6-6). A round hole can be observed where the adult parasite has chewed its way out of the aphid it killed.

Aphids are very susceptible to fungal disease when the weather is humid. Look for dead aphids that have turned reddish or brown and have a fuzzy, shriveled texture, unlike the smooth, bloated mummies formed when aphids are parasitized. Fungus-killed aphids sometimes have fine whitish mycelium growing over their surfaces.

Some people release commercially available aphid predators. There is little information on their effectiveness. In most situations, conserving resident beneficials is more effective and economical than releasing purchased species.

The convergent lady beetle (*Hippodamia convergens*) is commonly sold for release to control aphids. Although resident lady beetles are important predators, purchased *Hippodamia* inherently disperse and most will soon fly away from the site where they are released, even if food is plentiful. Controlling aphids on roses can require 1,000 or more beetles per shrub released at 1- to 2-week intervals, so you may need to plan in advance and purchase the beetles through mail order to obtain large numbers of them. If beetles are stored in the refrigerator (do not freeze them) and some are released periodically, warm beetles weekly to room temperature and feed them very dilute sugar water by misting them using a trigger-pump spray bottle. Wetting plants first and releasing beetles on the ground near the trunk and under plants in the late evening when it is cooler may slow beetle dispersal.

In a futile effort to repel this syrphid larva (*Metasyrphus* sp.) grasping it, a rose aphid is secreting a droplet of noxious liquid from its cornicle. See "Predaceous Flies" earlier in this chapter for a photograph of an adult syrphid. *Photo:* J. K. Clark

Aphid mummies, the crusty skins of aphids killed by parasitic wasps, are commonly tan or black. After a parasite larva feeds inside and pupates, the adult wasp chews a round hole and emerges to seek other aphids. A parasitic wasp laying her egg in an aphid is pictured earlier in this chapter. *Photo:* J. K. Clark

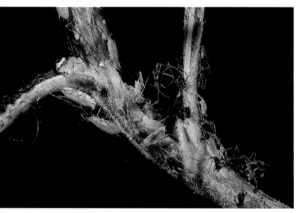

Orange aphid midge larvae (*Aphidoletes* sp.) are feeding on willow aphids (*Chaitophorus* sp.). The black, shriveled aphids have already been consumed. *Photo:* J. K. Clark

Rose aphids vary in color from green to red. However, the orangish rose aphids shown here are fuzzy because they were killed by a naturally occurring fungal disease. *Photo:* J. K. Clark

PREDACEOUS MIDGE

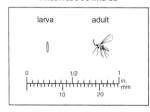

larva   adult

0   1/2   1 in.
          10   20   mm

ADELGIDS

Adelgids (Adelgidae) are small aphidlike insects that suck plant juices on conifers, including Douglas-fir, fir, hemlock, larch, pine, and spruce. *Adelges* and *Pineus* spp. adelgids cause white cottony tufts on the bark, branches, twigs, needles, or cones, and heavily infested trees may seem covered with snow. Cone-shaped galls or swollen twigs may also appear on infested spruce and fir. High adelgid populations cause foliage yellowing, early drop of needles, and drooping and dieback of terminals and can retard tree growth. Vigorous plants tolerate moderate populations, and most adelgid species are unlikely to seriously harm trees. An exception is the introduced balsam woolly adelgid. Be alert for this serious new pest and unusually damaging adelgid problems.

Adelgids commonly occur beneath cottony white or grayish material they secrete and sometimes are inside of galls. The adelgids themselves are small, dark, soft-bodied, oval insects. The insects can look different depending on the host, and several different-looking stages of the same species can occur together. Most adelgids have several generations per year and a complex life history that can include alternating generations between two different conifer species.

BALSAM WOOLLY ADELGID

Adelges piceae has seriously damaged or killed tens of thousands of true fir trees in the eastern United States and has become established in Oregon and northwestern California. This adelgid occurs only on *Abies* spp. and does not alternate hosts.

Introduced from Europe, the ¹/₂₅-inch-long, round, dark purple to black adelgids occur under a tiny blob of cottony wax on fir twigs and trunk bark. Infested shoots become stunted and develop globular galled wood (gouting) around buds and branch nodes. Shoots and limbs die back, and infested firs gradually decline and often die.

If you find an adelgid-infested dying fir or suspected balsam woolly adelgids, report them to the county agricultural commissioner.

COOLEY SPRUCE GALL ADELGID

Adelges cooleyi feeds on spruce and Douglas-fir, and some populations alternate generations between these hosts. The adelgids produce white, cottony tufts where they settle and feed on needles, shoots, and cones. High populations cause needles to become distorted, spotted, and drop prematurely.

Cooley spruce gall adelgids are causing new spruce terminals to become thick and short (right). The brown terminal is a previous year's shoot that was killed by adelgids. Clip and dispose of these galls to restore plant aesthetic quality. Pruning may provide some control if infested foliage is clipped when the galls are relatively young and still green. *Photo:* J. K. Clark

Cooley spruce gall adelgid feeding can cause spruce tissue to distort into a gall that eventually encloses the insects. The galls are from ¹/₂ to 3 inches long and light green to purple. About midsummer, the nymphs emerge from galls and molt into winged adults. The empty galls harden, turn brown, and may persist for years.

Galls in California are mostly seen along the North Coast and in mountainous areas. Cooley spruce gall adelgid produces galls on spruce only when Douglas-fir is nearby because this alternate host is the source of the adelgid's gall-forming stage.

The settled nymph (center with waxy margins) and several active first instars of Cooley spruce gall adelgid are less than ¹/₂₀ inch long. Adelgids' appearance varies greatly by life stage and host plant, and several different-looking stages of the same species can be present at the same time. In comparison with aphids, adelgids are usually smaller, lack obvious cornicles, and occur only on conifers. *Photo:* J. K. Clark

Pine bark adelgids produce this whitish gray wax where they feed on pine bark. Where intolerable, a forceful stream of water can dislodge and kill many of the adelgids; or, regularly inspect cottony masses in spring and apply oil when the tiny crawlers become abundant. *Photo: J. K. Clark*

PINE BARK ADELGIDS

The pine bark adelgid (*Pineus strobi*), also called the pine bark aphid, and the pine leaf adelgid (*P. pinifoliae*) produce abundant pale wax where they feed on pine bark and twigs. Some populations migrate between pine and spruce, and on spruce twigs form galls that are typically smaller and less prominent than those of the Cooley spruce gall adelgid. Adult females are purplish black, soft-bodied insects that lay yellowish pink eggs in cottony, wax-covered masses. The mobile nymphs (crawlers) that emerge resemble tiny pepper grains on the cottony egg masses.

MANAGEMENT OF ADELGIDS

Usually no control is needed to protect the health of established trees. An exception is the balsam woolly adelgid; seek a current source of information if you suspect you have this problem.

Predators on adelgids are especially important in natural forests, where most adelgids are uncommon. Adelgid predators include lacewings, the small maggotlike larvae of aphid flies and predaceous midges, and several species of small dark lady beetles.

Replace some spruce with other species to reduce adelgid populations that alternate hosts. A forceful stream of water directed at the cottony masses on conifers, especially on trunks, dislodges and kills many adelgids.

On spruce, to restore the plant's aesthetic quality and provide some control, break off or clip and dispose of infested foliage when the galls are green and before the insects have emerged. Avoid excess fertilization and quick-release fertilizers, which can promote adelgid populations.

High adelgid populations, especially on young trees, can be controlled by applying insecticidal soap, narrow-range oil, or certain other insecticides in the spring when crawlers are abundant. Be aware that oil spray may remove the desirable bluish color from foliage of certain conifers (e.g., blue spruce), but this phytotoxicity does not affect tree health. Residual, broad-spectrum insecticides (e.g., carbaryl) can cause an increase in mite populations unless a miticide is added. Insecticide sprays are more effective when a wetting agent is added and applications are made with high-pressure equipment so that the spray penetrates the insects' waxy secretions. Certain systemic insecticides can be effective.

Insecticide sprays are not effective in preventing galls unless application is correctly timed, based on careful and frequent monitoring. Examine the base of terminal buds on spruce and locate the overwintering females. Inspect females regularly during late winter and spring to determine when they begin to increase greatly in size and start to produce waxy strands; at this time thorough spraying with insecticide can provide control.

WHITEFLIES

Whiteflies (Aleyrodidae) are not true flies but are related to psyllids and aphids. They suck phloem sap, and high populations can cause leaves to yellow, shrivel, and

Pupae are used to distinguish among most whitefly spe These ash whitefly (*Siphoninus phillyreae*) pupae have a characteristic broad band of wax down the back and a of tiny tubes, each with a liquid droplet at the end. Ash whitefly is no longer a pest because it is biologically cor by two introduced natural enemies: a tiny black and ye parasitic wasp (*Encarsia inaron*) and a small, mostly bro lady beetle (*Clitostethus arcuatus*). *Photo: J. K. Clark*

drop prematurely. The honeydew excreted by nymphs makes a sticky mess, leads to blackish sooty mold growth, and attracts ants, which disrupt the biological control of whiteflies and other pests. Whitefly infestations usually do not threaten the health of established woody plants. They can be a serious problem in herbaceous plants, such as vegetables susceptible to whitefly-vectored viruses. Most species of whiteflies are uncommon because of parasites, predators, and other natural controls.

IDENTIFICATION AND BIOLOGY

Whiteflies usually occur in groups on the underside of leaves. Adults have a yellow body but are covered with mealy white wax, and their shape resembles a tiny moth or house fly. Adults of most species are similar in appearance, except some species have dark markings on their pale wings. The tiny oblong eggs hatch into barely visible, oblong, yellowish, mobile crawlers, which move a short distance then insert their mouthparts into plant tissue and remain settled until adulthood. Many species have several generations per year, with all stages present year-round on evergreen hosts in areas with mild winters.

Whitefly nymphs are flattened, oval, and in some species covered with waxy secretions. The mature nymph or last instar is commonly called a pupa, and the appearance of the pupa and the species of host plant help distinguish whitefly species. Depending on the species, pupae can be dark, pale colored, or translucent, and smooth surfaced or covered with filaments or curly whitish to transparent wax. Adult whiteflies usually leave a ragged or T-shaped slit in the pupal case from which they emerged; parasites generally leave a small, rounded emergence hole.

Most whitefly species prefer to feed and lay eggs on new plant growth, so regularly inspect succulent foliage and shoots on plants susceptible to these pests. Look for whiteflies whenever honeydew, sooty mold, or trails of ants are present. Inspect the underside of shriveled and yellow foliage and nearby healthy leaves for whiteflies. A cloud of tiny mothlike adults appears when you shake heavily infested foliage. Take the pupae of unfamiliar whitefly species to the county agricultural commissioner for identification as new species are periodically introduced. For more information, see the *Whiteflies Pest Notes.*

CROWN WHITEFLY

Aleuroplatus coronata occurs on oak and chestnut. Adults appear only during the spring and may become a nuisance on warm days when enormous numbers may

These crown whiteflies occur on oaks and chestnut. As nymphs mature into pupae, they change from pale to black and secrete broad, white, waxy plates. The similar Stanford whitefly (*Tetraleurodes stanfordi*) occurs on oaks and chinquapin (*Chrysolepis* spp.); its black pupae are surrounded by a narrow ring of white wax. Both whitefly species are apparently harmless to hosts. *Photo:* J. K. Clark

emerge, resembling swirling snowflakes. The adults soon disappear and can be ignored because they do no apparent damage to plants.

Tiny, white to pink eggs are found on the lower leaf surface. Nymphs are pale when young but blacken as they mature into pupae and become mostly covered with broad, white, waxy plates that spread out from the insects' sides, somewhat resembling a minute crown. Crown whitefly overwinters as a pupa and has one generation per year. Although high populations can be unsightly, generally no control is recommended.

GIANT WHITEFLY

Aleurodicus dugesii in California is a pest mostly in coastal areas. The species is larger than most whiteflies and attacks over 60 plant species, especially aralia, begonia, hibiscus, giant bird of paradise, mulberry, orchid tree, and xylosma. Introduced parasites have greatly reduced its populations.

Giant whitefly nymphs produce long, hairlike filaments of wax up to 2 inches long that give foliage a bearded appearance and may be mistaken for a leaf fungus. Most adults remain (and the females lay eggs) on the same leaf where they emerged from the pupal case. Many adults will remain on a dying or fallen leaf and perish as the leaf dries. This clustering tendency allows a relatively large number of whiteflies to be destroyed by removing relatively few leaves. Monitoring to detect infestations early allows new infestations to be largely eliminated by hand-picking and bagging foliage for disposal before whiteflies disperse. Where the whiteflies are a persistent problem in high-value areas, replacing plants with nonhost species is an option.

Directing a strong stream of water to the underside of infested leaves (syringing) is highly effective. Comparison studies with several pesticides found that syringing once every 1 to 3 weeks, depending on pest densities, performed as well or better than insec-

The dark giant whitefly pupae are parasitized. After feeding inside as a larva and pupating, a parasite (*Entedononecremnus krauteri*) left a round emergence hole in the upper right whitefly it killed. Because of introduced parasites, giant whitefly generally is a pest only on begonia, hibiscus, mulberry, and xylosma, especially near the coast, but outbreaks can occur on many other plant species if biological control is disrupted. *Photo:* J. K. Clark

Beardlike wax strands beneath leaves and blackish sooty mold caused by giant whitefly nymphs infesting hibiscus. Because this pest spreads slowly, infestations can be controlled by carefully monitoring hosts and promptly disposing of infested leaves where giant whiteflies are first found. *Photo:* J. K. Clark

ticide sprays. Unlike insecticides, syringing improves plant appearance by removing honeydew and will not have pesticides' negative effect on biological control.

Native predators, and especially three species of introduced parasitic wasps (*Encarsia noyesi* Aphelinidae, *Entedononecremnus krauteri* Eulophidae, and *Idioporus affinis* Pteromalidae) greatly reduce giant whitefly populations in most situations. After feeding inside as a larva, the adult parasite leaves a round exit hole in the whitefly pupa it killed. For more information, consult *Giant Whitefly Pest Notes*.

GREENHOUSE WHITEFLY

Trialeurodes vaporariorum is a pest primarily on vegetable crops and flowering annuals, but it can occur on woody species, including avocado, fuchsia, gardenia, lantana, redbud, and tree tomato (*Cyphomandra betacea*). Distinguish greenhouse whitefly by its pupae, which are oblong and have a transparent cover with vertical sides that have a fringe of filaments that protrude around the perimeter of the upper edge.

Encarsia formosa and other tiny wasps are important whitefly parasites. Female *Encarsia* also puncture nymphs and feed on the body fluids, killing the nymphs. In third- and fourth-instar whiteflies, *Encarsia* lay eggs that hatch into larvae that feed within, killing the whitefly. Parasitized nymphs resemble brownish or black scales, in contrast to the whitish or yellowish green color of healthy whitefly nymphs. *Encarsia* adults chew a round exit

hole in whiteflies and leave black deposits in the host, in contrast to the mostly clear skin with a ragged hole left by emerging whiteflies.

WOOLLY WHITEFLY

Aleurothrixus floccosus nymphs and pupae are covered with fluffy, waxy filaments. Eggs are laid in circles or partial circles on the lower surface of mostly full-sized leaves. Woolly whitefly occurs mostly on citrus and eugenia, usually at low densities. Parasitic wasps (especially *Amitus spiniferus* and *Cales noacki*) completely control woolly whitefly in most situations in California, except for occasional outbreaks when biological control is disrupted by ants or hot weather.

MANAGEMENT OF WHITEFLIES

Most whitefly species are satisfactorily controlled by natural enemies, unless these beneficials are disrupted by ants, dust, insecticides, or weather. Former pests now controlled by introduced natural enemies include ash whitefly (*Siphoninus phillyreae*), bayberry whitefly (*Parabemisia myricae*), giant whitefly, and woolly whitefly. Introduced parasitic wasps that attack only whiteflies provide the most important biological control. Bigeyed bugs, dustywings, lacewings, lady beetles, pirate bugs, and syrphid fly larvae are important native predators. For example, *Delphastus pusillus* and several other species of tiny lady beetles (1/16 inch long) prey entirely on whiteflies.

To conserve natural enemies, control ants with insecticide baits or sticky mate-

rial barriers, as discussed earlier under "Ants." Plant ground covers to reduce dust, which interferes with natural enemies. When managing whitefly-infested plants, trim only a portion of each plant at one time. Mature foliage with parasitized whiteflies and immature predators provides a refuge for natural enemies, which migrate to attack whiteflies that are more common on new growth.

To monitor populations, use sticky yellow cards to attract and trap adult whiteflies during the time of year when adults are active. Standard-sized or extra-large sticky traps may also help to reduce colonization of uninfested plants by adult whiteflies where many traps can be deployed around smaller, relatively isolated hosts. Traps will not help control immature whiteflies already on plants, and traps can also attract and kill important natural enemies.

A forceful stream of water applied to plant surfaces (syringing) washes away honeydew, which is often the most bothersome aspect of whiteflies. Regularly syringing small plants can control certain species, such as giant whitefly.

An insecticidal soap or narrow-range oil spray can provide temporary control; however, whiteflies are difficult to manage with foliar insecticides. Thorough coverage on the underside of leaves is essential. Applications made primarily to the upper leaf surface kill many beneficial insects while missing whiteflies and thus may do more harm than good. Broad-spectrum, residual, contact insecticides are generally

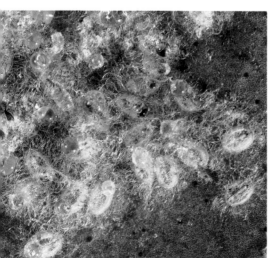

Woolly whitefly nymphs and pupae are covered with fluffy, waxy filaments and are found on eugenia and citrus. This species is under good biological control in most parts of California. It is an occasional pest in the San Joaquin Valley and southeastern desert, where hot weather disrupts its natural enemies. *Photo:* J. K. Clark

Greenhouse whitefly pupae have long, waxy filaments on top (submarginal filaments) and a short, marginal, waxy fringe. *Photo:* J. K. Clark

Small lady beetles are important whitefly predators, as with this *Delphastus pusillus* feeding on a whitefly nymph. *Photo:* J. K. Clark

Parasitized whitefly nymphs commonly darken or discolor and develop a round emergence hole. When unparasitized, the emerging adult whitefly typically leaves a ragged or T-shaped slit (right), as shown here for ash whitefly. *Photo:* J. K. Clark

Yellow sticky cards attract and capture a wide variety of flying insects, including adult aphids, psyllids, sharpshooters, thrips, and whiteflies. This sticky trap is being covered with clear plastic so its contents can be preserved and examined later. Trapping is a way to detect adults of certain insects and compare their numbers trapped over time to help determine whether pest populations are increasing, decreasing, or remaining about the same. *Photo:* J. K. Clark

not recommended. They are often more toxic to the actively searching beneficial insects than to whiteflies, which settle on the lower leaf surface where they are somewhat protected from insecticides. Whiteflies also have developed resistance to many broad-spectrum insecticides. Where populations are intolerable, certain systemic insecticides can be effective. Whenever possible, delay systemic application until after plants are done flowering to minimize poisoning of bees and natural enemies that feed on pollen and nectar.

MEALYBUGS

About two hundred species of mealybugs (Pseudococcidae) occur in California. Some, like the citrus mealybug and obscure mealybug, can occur on many different tree and shrub species in mild-winter areas.

Mealybugs tend to congregate in large numbers, forming white, cottony masses on plants. High populations slow plant growth and cause premature leaf or fruit drop and twig dieback. Honeydew production and blackish sooty mold are the primary damage caused by most mealybugs. A few species have saliva that is toxic to plants. Low populations of most species do not harm plants. High populations can cause plant decline, and young plants may be killed.

IDENTIFICATION AND BIOLOGY

Most mealybugs are wingless, soft-bodied insects covered with grayish or white

Certain sucking insects have saliva that deforms leaves and twigs, as with pink hibiscus mealybug (*Maconellicoccus hirsutus*) infesting this hibiscus in Imperial County, California. The pink hibiscus mealybug is now uncommon because it is under good biological control by an introduced parasitic wasp (*Anagyrus kamali*). *Photo:* J. K. Clark

mealy or cottony wax. Mature females are typically about 1/20 to 1/5 inch long, elongate, and segmented, and they may have wax filaments radiating from their body, especially at the tail. In most species, females lay tiny, yellow eggs in a mass intermixed with white wax. Male mealybugs are tiny, delicate, rarely seen insects with one pair of wings and two long tail filaments.

Most mealybug species feed on branches, twigs, or leaves, often in groups in protected places such as crevices or where foliage touches. Most mealybugs can move slowly and have several generations each

Many mealybug species form white cottony masses, and some aggregate in large numbers, as with these Gill's mealybugs (*Ferrisia gilli*) overwintering on bark. Because certain adelgids, aphids, scales, whiteflies, and other insects also produce wax, inspect masses to distinguish what insect is causing the waxiness. *Photo:* D. R. Haviland

year. Depending on the species, host, and climate, they overwinter only as eggs or females, or as all stages.

Many mealybugs superficially resemble each other, and records of which species attack what hosts are often unreliable. Adelgids, cottony cushion scales, and woolly aphids may sometimes be confused with mealybugs because they also cover their bodies with whitish wax. Where mealybugs are a problem, have them correctly identified to species by submitting specimens to an expert or consult publications such as *Mealybugs in California Vineyards*. Some species have very effective natural enemies, while others

Female mealybugs in most species lay orangish eggs in a wax-covered mass, as with this vine mealybug (*Planococcus ficus*). Vine mealybug has been reported only on grape, but it closely resembles the citrus and obscure mealybugs, which feed on many plant species. *Photo:* J. K. Clark

Adult females of Gill's mealybug have gray to pink bodies partly covered in wax, which gives the body surface a striped appearance. Sometimes they ar covered in long, thin filaments (as shown here). Gill's mealybug occurs mos fruit and nut trees, but it can potentially infest many woody broadleaves an been found in landscapes on escallonia and mulberry. *Photo:* D. R. Havilan

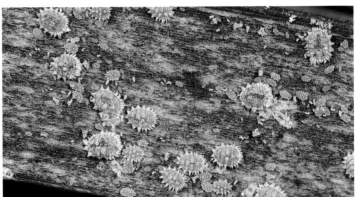

Coconut meal (*Nipaecoccus n* adult females nymphs secre wax and can fe on many diffe hosts, but the commonly oc palms. Adult f are oval shape and orangish marginal filam *Photo:* D. R. R

Bamboo mealybug (*Palmicultor lumpurensis*) feeds under the base of bamboo shoots, where they wrap around the main stem. When abundant, it kills young shoots and sometimes entire plants. Bamboo mealybug resembles pink hibiscus mealybug; both species lack marginal filaments and their pinkish body is covered with pale wax (removed here). However, *P. lumpurensis* infests only bamboo; pink hibiscus mealybug does not feed on bamboo. *Photo:* Florida Division of Plant Industry, Department of Agriculture and Consumer Services, Bugwood.org

Mealybug parasitism becomes apparent as wax weathers away, exposing the puffy, discolored bodies of dead mealybugs. These oblong, orangish citrus mealybug mummies formed as the parasite larva inside pupated into an adult. A hole or hinged cap is visible in some mummies where the adult wasp (*Leptomastix dactylopii*) emerged. *Photo:* J. K. Clark

apparently do not. Certain species move among many plant species, while management efforts for others can be narrowly focused on their relatively few hosts. New species become introduced, such as Gill's mealybug and vine mealybug, both of which feed on various woody deciduous and evergreen species.

CITRUS MEALYBUG

Planococcus citri is a pest in coastal California, mostly on citrus. It also occurs on ficus, fuchsia, gardenia, rose, and certain other ornamentals and is especially a pest in greenhouses and on ornamentals indoors. Citrus mealybug has short, waxy filaments of about equal length all around its margin. One grayish or dark

longitudinal stripe may be visible down its back, where less waxiness reveals the body color.

Citrus mealybug populations are often controlled by the mealybug destroyer lady beetle and parasitic wasps (e.g., *Leptomastix dactylopii* and *Leptomastidea abnormis*). The yellowish brown *L. dactylopii* female lays its eggs in late-instar nymphs and adult mealybugs, preferring hosts in warm, sunny, humid environments. At warm temperatures, *Leptomastix* can complete one generation in about 3 weeks. It is commercially available and has been released in combination with the mealybug destroyer to successfully control citrus mealybug in greenhouses and interior plantscapes. There is no information on the effectiveness of releases in outdoor landscapes.

CYPRESS BARK MEALYBUG

The cypress bark mealybug (*Ehrhornia cupressi*), sometimes called cypress bark scale, occurs beneath bark plates and in bark crevices on cedar, cypress, and juniper. Populations are usually innocuous in natural areas, but it can be a serious pest of Monterey cypress in urban areas. Foliage on infested plants becomes yellow and red and may die. Heavy populations cause dieback at the treetop, which gradually extends down the tree and may kill it.

Cypress bark mealybug nymphs and adults are round, bright red or orangish,

Cypress bark mealybugs revealed by removing a bark plate on Monterey cypress. Unlike most mealybugs, this species is round, bright red or orangish, and surrounded by white wax. The preadult female stage of incense-cedar scale closely resembles this mealybug, but incense-cedar scale is not common on urban conifers. *Photo:* J. K. Clark

The ground mealybug does not have obvious filaments along its sides or tail. It feeds on basal stems and roots. *Photo:* J. K. Clark

GROUND MEALYBUGS

Rhizoecus spp. live in the soil and feed on the roots of many different plants, including abutilon, acacia, boxwood, citrus, grape, palm, pine, *Prunus* spp., spruce, and syringa. Ground mealybugs may be covered with white wax and their short antennae and legs may be visible, but they do not have obvious filaments along their sides and tail. *Rhizoecus falcifer* is apparently the most common ground mealybug in California. High populations can cause a general decline of a plant and can kill young plants.

Minimize ground mealybug damage by providing adequate summer and fall irrigation to prevent drought stress. If high populations are damaging plants, a soil-applied systemic insecticide can be effective. Insecticidal soap in warm water poured on soil around small plants as labeled may reduce ground mealybug populations.

OBSCURE MEALYBUG

Pseudococcus viburni occurs on many hosts, including cactus, camellia, fruit trees, gardenia, magnolia, oak, oleander, palm, pine, walnut, willow, and yew. The light gray to white, powdery, wax-covered adult females have distinct filaments around the body. All stages occur year-round on bark, twigs, and leaves, and there are four to five annual generations. Obscure mealybug closely resembles several other species, including grape mealybug and vine mealybug.

MANAGEMENT OF MEALYBUGS

Provide proper cultural care so that plants are vigorous and can tolerate moderate mealybug feeding. Use good sanitation methods to avoid spreading mealybugs to uninfested plants. Low populations are easily overlooked in protected places on plants and spread when infested plants are moved. Mealybug crawlers are readily spread as contaminants on objects such as pruning tools that cut infested plants and then are used to cut other plants.

Naturally occurring predators and parasites provide good control of many mealybug species unless these beneficials are disrupted. To conserve natural enemies, avoid insecticides that disrupt biological control, reduce dustiness, and control ants that tend mealybugs. To control ants, trim

and surrounded by a ring of white wax. They do not move after settling as crawlers and have one generation per year. Incense-cedar scale, or Monterey cypress scale (discussed later), during its preadult female stage closely resembles the cypress bark mealybug. Both species can occur on the same hosts and are easily overlooked on the rough, reddish bark.

LONGTAILED MEALYBUG

Pseudococcus longispinus can occur in mild-winter locations on many hosts, including cactus, ficus, fuchsia, gardenia, hibiscus, jasmine, oleander, and palm. It is most commonly a pest on nursery stock and indoor ornamentals. It is distinguished by its two tail filaments, which are longer than its body. Unlike many mealybug species, no egg masses are found in longtailed mealybug colonies because females give live birth to nymphs. Several parasitic wasps, including *Acerophagus notativentris, Anarhopus sydneyensis,* and *Arhopoideus peregrinus,* often keep longtailed mealybug populations at low levels.

Most mealybugs are elongate, grayish or powdery white, and segmented. Many species have wax filaments radiating from their body, especially at the tail, as with these obscure mealybugs on grape. Similar-looking species include citrus mealybug, which occurs on many different plants, and aloe mealybug (*Vryburgia trionymoides*), which feeds on succulents (e.g., *Aloe, Crassula,* and *Echeveria* spp.). *Photo:* J. K. Clark

OBSCURE MEALYBUG

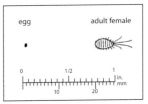

egg   adult female

0   1/2   1
| | | in.
                        mm
10   20

branches to eliminate ant bridges to structures or other plants, apply sticky barriers to trunks, or use enclosed pesticide baits, as discussed earlier under "Ants." Once ants and other disruptions of biological control are managed, several months (several insect generations) may be required before natural enemies become abundant enough to provide control, so be patient or make spot applications of insecticides that have minimal impacts on predators and parasites.

Avoid using residual, broad-spectrum insecticides, such as carbamates, organophosphates, and pyrethroids. Insecticidal soap, narrow-range oil, or a forceful stream of water can be applied to somewhat reduce populations with minimal harm to natural enemies that may migrate in later. Populations of cypress bark mealybug can be reduced by thoroughly spraying infested bark with narrow-range oil in August and again in September; however, furrowed bark and dense foliage make it difficult to thoroughly spray bark. Contact sprays are often not very effective because mealybugs are protected by a wax coating and feed within protected plant

parts. Certain systemic insecticides can be effective if conserving natural enemies and less-persistent insecticides do not provide sufficient control.

Mealybug Destroyer. The mealybug destroyer (*Cryptolaemus montrouzieri*) is an important predator of mealybugs. This mostly blackish lady beetle has a reddish brown head and rear. Mealybug destroyer larvae are covered with waxy white curls and resemble mealybugs, except that the lady beetle larvae grow larger and are more active. Both adult and larval lady beetles feed on all mealybug stages. The beetle has about four generations per year and lays its eggs into mealybug egg masses.

The mealybug destroyer survives poorly over the winter in cold areas of California and may need to be reintroduced locally in the spring to provide control. Some citrus growers purchase *Cryptolaemus* from commercial suppliers and release them in the spring. Adult mealybug destroyers may not reproduce or provide control unless mealybugs and their egg masses are relatively abundant, so make any releases on localized "hot spots."

SCALES

Scale insects feed by sucking plant juices, and some inject toxic saliva into plants. When numerous, some species can weaken a plant and cause it to grow slowly. Infested plants appear water stressed, leaves turn yellow, and may drop prematurely, and plant parts that remain heavily infested may die. The dead brownish leaves may remain on scale-killed branches, giving plants a scorched appearance. Plants infested by soft scales and certain other scales become sticky from honeydew and may blacken from the resulting sooty mold growth.

The importance of infestations depends on the scale species, the plant species and cultivar, environmental factors, and natural enemies. Populations of some scales can increase dramatically within a few months when weather is warm and honeydew-seeking ants protect scales from their natural enemies. Plants are not harmed by a few scales, and even high populations of certain species apparently do not damage plants.

This adult mealybug destroyer is an important predator of exposed mealybug species. *Photo:* J. K. Clark

MEALYBUG DESTROYER

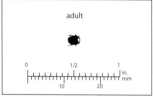

adult

0   1/2   1
|⊥⊥⊥⊥⊥|⊥⊥⊥⊥⊥|⊥⊥⊥⊥⊥|⊥⊥⊥⊥⊥| in.
         10   20   mm

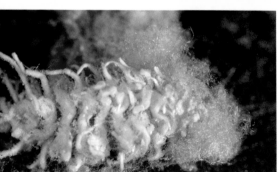

Mealybug destroyer larvae are covered with waxy white curls. They resemble mealybugs, except the lady beetle larvae move faster and can grow larger than a mealybug. This mealybug destroyer is feeding in a cottony sac on orangish eggs of the grape mealybug (*Pseudococcus maritimus*). *Photo:* J. K. Clark

Two mature female European fruit lecanium scales. The soft scale at left has been overturned to reveal hundreds of tiny eggs, which hatch into mobile first-instar nymphs (crawlers). *Photo:* J. K. Clark

This *Aphytis melinus* female parasite is laying her egg in a California red scale. Many scale species are biologically controlled if their natural enemies are conserved, such as by managing ants and avoiding the use of broad-spectrum residual insecticides. *Photo:* J. K. Clark

IDENTIFICATION AND BIOLOGY

Scales do not resemble most other insects. Adult females of most species are circular to oval, wingless, and lack a separate head or other easily recognizable body parts. Adult males are tiny, delicate, white to yellow insects with one pair of wings and a pair of long antennae. Adult males are rarely seen or are not known to occur in many species, and females of many scale species reproduce without mating.

Eggs are commonly produced beneath the body or cover of the female. The newly hatched nymphs (crawlers) walk along branches or are spread by the wind or inadvertently by people or animals. Crawlers are usually pale yellow to orange and about the size of a period. After crawlers settle and insert their strawlike mouthparts to feed on plant juices, armored scale nymphs and females remain immobile on the same plant part for the rest of their lives; nymphs of soft scales and certain other types can move a little, such as from foliage to bark before leaves drop from deciduous hosts in the fall.

It is often very helpful to have an expert positively identify the scale species present. Many obvious species are rarely, if ever, damaging pests on most hosts, while others are serious pests of certain plants. About one dozen scale families contain at least a few species important in landscapes. Scales in different families can vary greatly in appearance and biology. Most pest species are either armored scales or soft scales. For more information, see *Pests of the Garden and Small Farm, The Scale Insects of California,* and *Scales Pest Notes.*

ARMORED SCALES

Most armored scales (Diaspididae) are less than $1/10$ inch long and have a platelike shell or cover that can usually be removed to reveal the actual scale body underneath. The covers often develop concentric rings as nymphs grow, and many species have a slight protuberance (exuviae, or "nipple") of a different color that is formed from the covering of the first-instar nymph. Armored scales do not excrete honeydew, and once crawlers settle to feed, armored scales generally lose their legs and cannot move. Most species have several generations each year and develop through the life stages illustrated in Figure 6-13.

CYCAD SCALE

Furchadaspis zamiae infests cycads or sago palms (*Cycas* and *Zamia* spp.) and bird of paradise. High populations cause severe mottling or yellowing of foliage, stunt growth, and can eventually kill cycads, which are expensive and slow-growing plants. Scales occur on the underside of fronds and on the basal trunk and can resemble oleander scale. Both scale species occur on bird of paradise and cycads, but oleander scale does not harm plants. The yellow body of cycad scale females and nymphs is found beneath an oval to oblong cover that is moderately convex and translucent to white or cottony covered. Oleander scale females, pictured later, are mostly tan or yellowish with a roundish cover, and male oleander scales have white, translucent covers through which the male's body is partly visible. Cycad scale apparently lacks males. Submit samples to an expert for identification if it is uncertain what species is present. Cycad scale when abundant warrants control action.

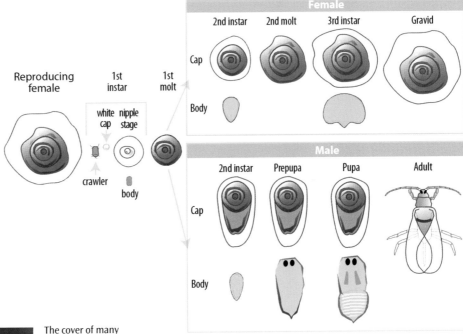

Figure 6-13.

Life cycle of a typical armored scale, California red scale. Eggs (not shown) hatch within the female's body and emerge as tiny first instars, which initially are mobile crawlers, then settle and secrete a cottony white cap (cover) and later a more solid cover (the nipple stage). Female and male scales develop differently, beginning with the second instar. The female cover and body underneath remain round and immobile. The male forms an elongated cover, and the body underneath develops eyespots, legs, and two wings. A tiny mobile adult male emerges and seeks a female to mate.

The cover of many armored scale species develops concentric rings as the scale grows. *Aphytis melinus* is laying an egg in this female California red scale. *Photo:* J. K. Clark

Cycad scales have convex, oblong, whitish to yellow covers and cast skins (exuviae) and are often partly covered with pale, woolly wax. On cycads, cycad scale and oleander scale are often confused because the normally flattened oleander scale females become convex on cycad fronds. Unlike these cycad scales, oleander scale females (pictured later) are more roundish, yellowish or tan, and have the exuviae (somewhat raised, differently colored area) near the center of the cover, not near one end as with these cycad scales. *Photo:* R. J. Gill, California Department of Food and Agriculture

Yellowing of fronds may indicate a cycad scale infestation. Distinguish cycad scale from oleander scale, which also occurs on sago palms. Oleander scale is found on many hosts and usually is innocuous, while cycad scale can seriously damage cycads. *Photo:* J. K. Clark

EUONYMUS SCALE

Unaspis euonymi when abundant is a serious pest of certain *Euonymus* spp. Scale feeding causes brownish to yellow leaf spots, severe yellowing and premature drop of foliage, and the gradual decline and sometimes the death of plants.

The immature male is felty or fuzzy, white, and elongated with three longitudinal ridges. The female is wider, oystershell shaped, slightly convex, and brown to black. Both sexes have brownish yellow areas on the narrow end of their covers. Scales overwinter as mature females, which in spring produce eggs that hatch into orangish crawlers that emerge over a several-week period. Euonymus scale has two or three generations per year.

Where this scale is a problem, replace and do not plant *Euonymus japonica*; it is extremely susceptible to euonymus scale. *Euonymus kiautschovica* (=*E. sieboldiana*) tolerates scales. *Euonymus alata* remains nearly scale-free even when heavy infestations occur on nearby susceptible hosts.

Various predators feed on euonymus scale, including an introduced lady beetle (*Chilocorus kuwanae*), which resembles the twicestabbed lady beetle (*Chilocorus orbus* =*C. stigma*) pictured later under "Management of Scales." Both predators are shiny black with two reddish spots. However, the spots of *C. kuwanae* tend to be deep red, somewhat rectangular, and located near the center of each wing cover. The spots of *C. stigma* tend to be more yellowish orange, round, and located more forward toward the head of the beetle. Larvae of both spe-

cies are brownish with black spines.

Conserve scale predators, such as by avoiding the use of broad-spectrum, residual insecticides. Narrow-range oil controls euonymus scale if foliage and shoots are thoroughly sprayed; time any application by monitoring crawlers. One annual application during several consecutive years may be necessary to reduce high populations.

GREEDY SCALE AND LATANIA SCALE

The many hosts of greedy scale (*Hemiberlesia rapax*) include acacia, bay, boxwood, ceanothus, fruit trees, holly, ivy, laurel, magnolia, manzanita, palm, pepper tree, pittosporum, pyracantha, redbud, strawberry tree, and willow. Latania scale (*H. lataniae*) has a similarly broad host list and in California may be more common than greedy scale. These scales occur on leaves, stems, and fruit. Though sometimes unsightly, these species rarely, if ever, cause serious damage to plants, except possibly in certain crops such as kiwifruit.

Females and nymphs are circular, convex, and gray, tan, or white. Distinct concentric rings often form on the cover as the nymphs grow. Greedy scale and latania scale cannot be distinguished in the field. These species also resemble oleander scale and olive scale, but the cover of greedy scale females is more globular or raised.

MINUTE CYPRESS SCALE

Minute cypress scale (*Carulaspis minima*), also called cypress scale, can occur on almost any conifer. Be aware that the name

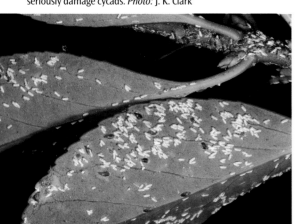

Euonymus scale male cocoons are oblong and white. Females are dark and wider on one end, and young nymphs are tiny and yellowish. This species is a serious pest of *Euonymus japonica*, but certain other *Euonymus* spp. are rarely attacked or are not damaged by this scale. *Photo:* J. K. Clark

Greedy scales are usually tan with an off-center yellow or brown nipple, like these infesting acacia. The covers become globular or distinctly raised as the scales mature. *Photo:* J. K. Clark

cypress scale is also applied to species discussed earlier under "Cypress Bark Mealybug" and later under "Incense-Cedar Scale, or Monterey Cypress Scale."

Minute cypress scale can be common on arborvitae, cypress, and juniper, but it usually is a serious pest only on Italian cypress (*Cupressus sempervirens*) and in nurseries. Infestations can cause foliage to turn yellow or brown and die, resulting in limb death. Other pests discussed later under "Juniper Twig Girdler" and "Cypress Tip Miners" can cause similar damage, so use a hand lens to inspect discolored foliage for the scales.

The minute cypress scale mature female is circular, convex, and whitish with a yellow center. The tiny immature male is elongate, oval, felty with longitudinal ridges, and yellow on the terminal of its mostly whitish cover. This scale has one to two generations per year. Apply oil if necessary or use another insecticide on oil-sensitive species when monitoring indicates that crawlers are active in the spring.

OLEANDER SCALE

Oleander scale (*Aspidiotus nerii*), also called ivy scale or white scale, is probably the most common armored scale species infesting landscapes. While often abundant and unsightly, it rarely, if ever, causes serious harm to plants. It prefers aucuba, cycad or sago palm, ivy, oleander, and olive. Its many other hosts include acacia, bay, boxwood, holly, laurel, magnolia, manzanita, maple, mulberry, pepper tree, redbud, yew, and yucca.

Oleander scale's cover is roundish and flattened and found on bark, foliage, and fruit. Female covers are mostly tan or yellowish. Immature males have white, translucent covers through which the body is partly visible. In comparison with the similar greedy scale, oleander scale is lighter colored, less globular, and the brown nipple (exuviae) is near the center on the cover.

No control is needed to protect plant health. If unsightly, prune out heavily infested branches. Oil can be applied during the dormant season or when monitoring indicates that crawlers are active in the spring.

OYSTERSHELL SCALE

Lepidosaphes ulmi usually damages only poplars and willows; high populations over several years can kill twigs and branches on young trees and cause stunted growth and decline of mature plants. Oystershell scale is usually not a serious pest in California on about 150 other plant species it feeds on, including alder, ash, aspen, box elder, boxwood, ceanothus, fruit trees, holly, maple, sycamore, and walnut. Oystershell scale does not attack citrus, but the similar-looking purple scale (*Lepidosaphes beckii*) can be found on citrus in coastal areas.

The gray to dark brown, elongated scales resemble miniature oysters and are found on bark, usually in clusters. Oystershell scale overwinters as whitish eggs under the cover of mature females, and the crawlers emerge in the spring after buds have burst. The scale usually has one generation per year in northern California and two generations in southern California.

SAN JOSE SCALE

Diaspidiotus perniciosus hosts include acacia, aspen, cottonwood, fruit trees, maple, mulberry, poplar, pyracantha, walnut, and willow. It can be a pest on roses, and more

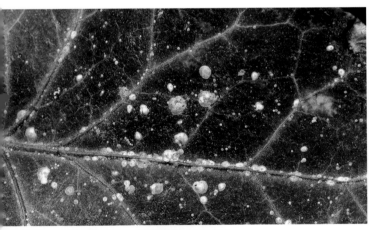

These oleander scales infesting ivy are tan to yellow and flattened to slightly convex. They are lighter colored and less raised or globular at maturity than the similar-looking greedy scale. Although obvious, both of these scale species are usually harmless to plants. *Photo:* J. K. Clark

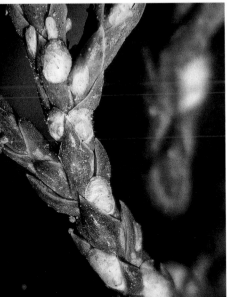

Minute cypress scale females are mostly whitish, roundish, and easily overlooked, as they are 1/25 inch long or less. Male cocoons are elongate and smaller at maturity. Minute cypress scale cannot be distinguished in the field from juniper scale (*Carulaspis juniperi*). Both scale species prefer cypress and juniper (shown here) but can occur on almost any conifer species. *Photo:* R. J. Gill, California Department of Food and Agriculture

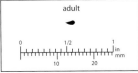

OYSTERSHELL SCALE

adult

0   1/2   1 in
|||||||||||||||||||||||||| mm
     10   20

Oystershell scale can be found on bark of dozens of plant species, including this Lombardy poplar. The similar-looking purple scale occurs only on citrus. *Photo:* J. K. Clark

commonly occurs on rose in landscapes when infested fruit or nut trees are nearby. San Jose scale is usually a serious pest only on nut trees and stone fruits, where high populations over several years can cause gradual decline and sometimes death of plants. After deciduous plants drop their leaves in the fall, brown dead foliage often remains on heavily infested branches that were killed by the scales.

Mature females have a smooth, yellow body beneath the round, grayish cover. San Jose scale looks like olive scale (*Parlatoria oleae*), except that the female olive scale's body beneath its cover is purple. Both species can occur on many of the same hosts, but olive scale is under good biological control and is rarely abundant unless its natural enemies are disrupted. Walnut scale (*Diaspidiotus juglansregiae*) also resembles San Jose scale, except that under its cover, the margin of the walnut scale female's yellow body has indentations. Walnut scale is usually abundant only on ash, birch, and walnut.

After settling on bark, San Jose scale crawlers secrete a white, waxy covering (called the white cap) and sometimes cause a red halo to form on young wood or fruit where they feed. A band of dark wax develops around the white cap, and most San Jose scales overwinter in this stage, called the black cap. In the spring, the roundish females become distinguishable from the elongate, immature males. Winged adult males begin emerging in March and April and females produce a sex pheromone that attracts the males. About a month after mating, the first generation of crawlers usually appears in May. The scale has two to five generations per year.

The twicestabbed lady beetle, a dark, pinhead-sized nitidulid beetle (*Cybocephalus californicus*), and tiny parasitic wasps, including *Aphytis* and *Encarsia* spp., are important natural enemies of San Jose scale. If spraying is warranted, use horticultural oil or other insecticides least disruptive to biological control. If spraying during the foliage season, monitor scale crawlers beginning in April with double-sided sticky tape and spray 2 to 3 weeks after the bright yellow, pinpoint-sized scale crawlers are first observed in the tape traps. Alternatively, on high-value trees, use pheromone traps and degree-day monitoring to time applications, as described in the online *Almonds: UC IPM Pest Management Guidelines.*

SOFT SCALES

Female soft scales (Coccidae) may be smooth or cottony and usually are hemispherical shaped and ¼ inch long or shorter. The scale's surface is the actual body wall of the insect and cannot be removed. Most immature soft scales retain their barely visible legs and antennae after settling and are able to move slowly. Soft

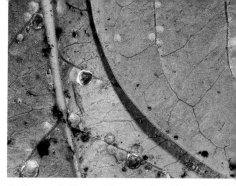

Pyriform scale (*Protopulvinaria pyriformis*) is a soft scale for its somewhat deltoid or pear (pyriform) shape. Matu... females are dark brown to orangish, with a white fringe include aralia, avocado, California bay, cypress, eugenia fatsia, gardenia, ivy, and, giant Burmese honeysuckle (*L... hildebrandiana*), shown here. *Photo: D. Rosen*

Lecanium scale nymphs tended by honey ants (*Prenolepi... imparis*). *Photo: J. K. Clark*

scales are prolific honeydew excreters, so they are often tended by ants that protect the scales from natural enemies. Most soft scale species have one generation per year.

BLACK SCALE

Saissetia oleae is especially abundant where ants are present. Its many hosts include aspen, bay, citrus, cottonwood, coyote brush, fruit trees, holly, maple, mayten, oleander, olive, palm, pepper tree, pistachio, poplar, privet, and strawberry tree.

Adult females are about ⅕ inch in diameter, dome shaped, and dark brown or black. Each female produces hundreds of white to orangish eggs beneath her hard shell body, mainly during May and June. The tiny, light brown crawlers settle and feed mostly on leaves. During the late second instar, an H-shaped ridge develops on the scale's back. After the second molt, scales migrate to twigs and become dark mottled gray and leathery. Once females start laying eggs, they darken and harden. Black scale has two generations per year along the coast and one generation in inland California.

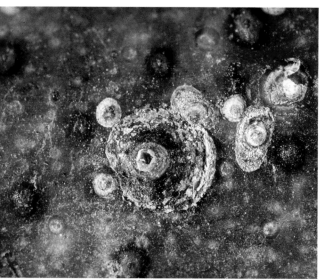

SAN JOSE SCALE

female	male
●	‑

0   1/2   1 in
|‖‖‖‖‖‖‖‖‖‖‖‖‖‖‖‖| mm
   10   20

The San Jose scale female cover is circular. The elongate covers shown here are immature male San Jose scales. Tiny, round first-instar nymphs, the white cap and black cap stages, are also present. *Photo: J. K. Clark*

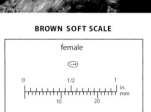

Black scales develop an H shape on the back that is most apparent in the second instars and preovipositional females. The H shape often disappears once females mature and become darker and more globular. *Photo:* J. K. Clark

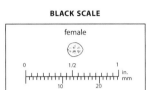

Brown soft scale has several generations per year, so different-sized life stages commonly occur at the same time, as with the yellowish first instars and orangish to dark brown older nymphs and females shown here. *Photo:* J. K. Clark

Various predators, and especially *Metaphycus* spp. parasitic wasps, can completely control black scale. However, black scale biological control is especially susceptible to disruption by ants. *Metaphycus helvolus,* the most important parasite, takes a relatively long time to deposit its eggs, and this parasite's long presence on scales greatly increases the likelihood that ants will locate and attack it. *Coccophagus* and *Metaphycus* spp. parasites kill young black scales by feeding on their body fluids as well as by laying eggs in them. Parasitized scales often darken, and their covers develop a round exit hole when the adult parasite emerges. *Scutellista* spp. wasps leave an exit hole in mature female scales after their larvae feed as egg predators underneath female scales.

Commercially available *Metaphycus* can be released when young scales with an H-shaped ridge are present, but there is little research on the effectiveness of parasite releases. Alternatively, foliage with parasitized scales can be collected in the field and relocated to infestations where they can emerge to supplement existing parasite populations.

Conserve natural enemies by controlling ants, reducing dust, and avoiding broad-spectrum, residual insecticides. Prune off heavily infested parts if they are limited to a small portion of the plant. In warm areas such as the Central Valley of California, prune to open up tree canopies and increase scale mortality from heat exposure. To directly control black scale, apply narrow-range oil during the dormant season or spray foliage in spring when crawlers and young nymphs are present.

BROWN SOFT SCALE

Coccus hesperidum seldom causes serious damage to landscapes, but its copious honeydew and the resulting sooty mold can be annoying. Brown soft scale prefers broadleaf evergreens, but it may be found on the leaves and twigs and occasionally fruit of almost any broadleaved plant. Hosts include aspen, cottonwood, fruit trees, holly, manzanita, palm, poplar, strawberry tree, and willow. Nymphs are mottled yellow-brown and rounded. Mature females are yellow to dark brown and can resemble citricola scale. Brown soft scale has three to five generations per year, and multiple life stages are usually present at once. Populations are usually highest from midsummer to early fall.

Metaphycus luteolus is an important parasite of brown soft scale, leaving one to several exit holes in larger nymphs or mature scales. Immature scales are also parasitized by *Coccophagus* spp. wasps, which cause parasitized nymphs to turn black. The twicestabbed lady beetle and various other predators commonly feed on brown soft scale.

Control honeydew-seeking ants and conserve beneficials. Prune out heavily infested branches. Apply oil if necessary during the dormant season or when monitoring indicates that crawlers are active in the spring.

Calico scale females are typically mottled white and brown or black, as on this box elder. On most hosts, calico scale populations in California remain low and do not damage the plants. *Photo:* J. K. Clark

CALICO SCALE

Eulecanium cerasorum occurs on many deciduous trees, usually on well-shaded plant parts. Hosts include box elder, liquidambar, maple, stone fruits, and walnut. Low populations are not damaging, but preferred hosts such as liquidambar may decline if highly infested over several years. Calico scale populations are rarely abundant enough in California to warrant control.

The mature female is relatively large and globular and has a mottled "calico" pattern of white or yellow with brown or black. The dark brown nymphs are flattened and covered with thick, elevated, transparent, waxy plates. The similar frosted scale secretes a powdery wax covering, as discussed later under "Lecanium Scales." Both species overwinter on twigs as nymphs that begin to mature in late winter. The crawlers emerge from females in early spring and move to feed on leaves. In the fall, the scales move back onto twigs.

CITRICOLA SCALE

Coccus pseudomagnoliarum is common in the Central Valley, where high populations over several years can cause decline of citrus and hackberry and reduced fruit on citrus. Citricola scales are mottled dark brown to gray and may be confused with brown soft scale, but the mature brown soft scale is smaller and yellow or dark brown, not gray. Citricola scale has only one generation per year, so most individuals are about the same size. Because brown soft scale has several generations per year, different-sized life stages commonly occur at the same time.

Citricola scale is well controlled by parasites in south coastal California, but not in California's Central Valley. Conserve natural enemies where they are present. Crawler density peaks in northern California at about 635 degree-days above 52°F accumulated from March 1. If applying oil, spray after this time, during about July through early September, after monitoring indicates that most crawlers have emerged and settled. Alternatively, hackberry can be sprayed with oil after leaves drop during the dormant season. On large trees, the most practical control is to apply systemic insecticide according to label directions, preferably as a soil application or bark spray.

Citricola scale females on a twig and their recently emerged, tiny, yellow first instars, which are mostly on the leaf. In comparison with the similar-looking brown soft scale, maturing citricola scales are more grayish and are all about the same size because citricola scale has only one generation per year. *Photo:* J. K. Clark

CITRICOLA SCALE

female

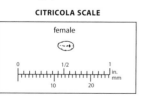

Mature female green shield scales are convex and oval and lay their eggs within flocculent wax. *Photo:* J. K. Clark

GREEN SHIELD SCALE

Pulvinaria psidii hosts include anthurium, aralia, begonia, camellia, croton, eugenia, ficus, hibiscus, gardenia, pittosporum, plumeria, and *Schefflera, Schinus,* and *Syzygium* spp. Heavy infestations produce copious honeydew and extensive sooty mold growth. Mature females cover plants with flocculent white egg sacs.

The brown, green, or yellowish female is about ⅛ to ⅙ inch long and produces a long, cottony egg sac. Green shield scale nymphs are relatively mobile, and after feeding in one spot for days can slowly move to infest adjoining plants.

The mealybug destroyer appears to be an important predator of green shield scale. Native parasitic wasps also help control this pest if you manage ants.

IRREGULAR PINE SCALE

Toumeyella pinicola is a serious pest of pines. High populations cause yellow foliage, seriously weaken trees, kill branches, and sometimes kill young pines, especially Monterey pine.

Irregular pine scale overwinters as females, typically in blackened and honeydew-encrusted colonies on 1- or 2-year-old branches. The yellowish, gray, and brown female is robust, dimpled, and irregularly circular. The male nymphs are elongate, flattened with a raised central ridge, and resemble grains of rice that sometimes occur in large numbers on needles. The crawlers are flattened, oval, and orange to yellow. They emerge and can be seen moving over shoots and needles during February through May in southern California and from late April through June

in the San Francisco Bay Area. One generation occurs each year.

At least one other *Toumeyella* sp. and the Monterey pine scale (*Physokermes insignicola*) also infest pines. In comparison with *Toumeyella* spp., the Monterey pine scale is more spherical, less irregular, darker, causes little or no damage, and is found mostly in the San Francisco Bay Area.

If scale populations are damaging, monitor crawlers with sticky tape traps or inspect foliage weekly with a hand lens beginning in February in southern California or in April in northern California. Apply oil when crawlers first become abundant and spray again about 3 weeks later if populations are especially high.

Irregular pine scale males look like grains of rice on these Monterey pine needles. Brownish, globular females covered with sooty mold encrust the twig. Monterey pine scale looks very similar, but its females are uniformly spherical and black. *Photo:* J. K. Clark

IRREGULAR PINE SCALE

female   male

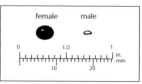

This oak lecanium scale resembles European fruit lecanium and frosted scale. A reddish brown, scale-feeding lady beetle (*Chilocorus bipustulatus*), which resembles its hosts, is visible atop the two left-most scales. Close examination would reveal that the lady beetle has three lighter spots on each wing cover. *Photo:* J. K. Clark

LECANIUM SCALES

At least five *Parthenolecanium* spp. occur in California. European fruit lecanium (*P. corni*) is the most common and may be a complex of more than one species. It feeds on a few evergreen species and many deciduous ornamentals, including alder, aspen, cottonwood, coyote brush, elm, fruit trees, pistachio, poplar, toyon, and walnut. European fruit lecanium is often innocuous, but very high populations sometimes develop, and the resulting honeydew and sooty mold become annoying.

Oak lecanium (*P. quercifex*) occurs on oak and other Fagaceae. If abundant it can weaken trees and kill twigs and branches, usually only on coast live oak.

Frosted scale (*P. pruinosum*), sometimes called globose scale, is usually a pest only on walnut. Insignificant populations occur on many other deciduous trees.

Female lecanium scales have a shiny, dark brown, oval-domed shell, often with several ridges along the back. Unlike the other lecanium scales, frosted scale females are covered by a white, waxy powder in spring that weathers away by early summer.

The yellow to brown lecanium scale nymphs emerge from May to July and feed mostly on leaves before moving in fall to overwinter on twigs. Lecanium scales have one generation per year.

Conserve natural enemies by controlling ants and dust and avoiding resid-

OAK LECANIUM SCALE

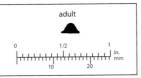

ual pesticides. Prune out heavily infested branches. If warranted, apply oil during the early dormant season (before mid-January), delayed dormant season (as buds begin to swell), or when monitoring beginning in the spring indicates that the peak density of crawlers has definitely passed and most scales are early-instar nymphs.

KUNO SCALE

Eulecanium kunoense sometimes damages plum and other *Prunus* spp. in the San Francisco Bay Area and Sacramento Valley. Rarely damaged hosts include cotoneaster, pyracantha, rose, walnut, and other woody Rosaceae. Kuno scale females are almost spherical, resembling beads on stems. Females are dark and shiny brown during most of the year, except they are yellow and orangish with black bands and blotches during a short period prior to egg production in late spring. The nymphs are yellow or brown and flattened. They feed on leaves during summer then migrate to twigs to overwinter. The elongate, translucent male cocoons can be prevalent on twigs.

Kuno scale has one generation per year. If applying oil to control it, spray twigs during the late dormant season or thoroughly drench the underside of leaves in early summer.

The Kuno scale female is shiny, brown, and almost spherical at maturity, resembling a bead on bark. *Photo:* L. L. Strand

TULIPTREE SCALE

Toumeyella liriodendri is a serious introduced pest in the San Francisco Bay Area. In addition to producing copious honeydew, it causes limb dieback and the decline and sometimes death of tulip tree (*Liriodendron tulipifera*) and deciduous magnolias. Its other hosts include linden.

Mature females are up to $\frac{1}{3}$ inch long, irregularly hemispherical, and variably colored, usually brown or gray with irregular black blotches and markings of green, orange, pink, red, or yellow. Females mature in late spring and are alive throughout the summer and fall (they exude liquid when their body is squashed), unlike the similar lecanium females, which mature earlier and by late June are dead and dried out.

Tuliptree scale females do not produce a large mass of eggs, and the crawlers emerge for a prolonged period from late summer to late fall. The scale overwinters on twigs as dark nymphs. In spring the males develop into elongate, gray to white cocoons.

If applying oil, spray overwintering nymphs during late winter or spray foliage during about September after most crawlers have emerged. Applying a systemic insecticide is the most practical method for managing this difficult-to-control pest; to minimize toxicity to pollinators and natural enemies, make the application after plants have completed their seasonal flowering.

Tuliptree scale females have irregular black blotches and varying markings of green, orange, pink, red, or yellow. Unlike the springtime crawlers of most soft scale species, tuliptree scale crawlers emerge for a prolonged period from summer through fall. *Photo:* A. S. Munson, USDA Forest Service, Bugwood.org

INSECTS, MITES, AND SNAILS • 195

Wax scales, which are hemispherical and mostly pale, are occasional pests of various ornamentals. These females were killed by parasites, as evidenced by the round emergence holes. *Photo:* J. K. Clark

WAX SCALES

Ceroplastes spp. wax scales can create a gooey mess on many different hosts by excreting copious honeydew. These infestations rarely, if ever, threaten plant health.

Chinese wax scale (*C. sinensis*) along the South Coast and in the San Francisco Bay Area is sometimes a pest on Australian willow, *Escallonia*, and mayten. Occasionally it is abundant on other hosts, including California bay, coyote brush, holly, *Mahonia*, and pepper tree. In southern California, barnacle scale (*C. cirripediformis*) is a pest primarily on gardenia.

Females are hemispherical and covered with thick, oily wax. Their predominant color is pale gray. Mature females have waxy ridges and white or dark-colored indentations or speckles. Crawlers are most abundant during fall but may be present anytime from fall through winter. There is apparently one generation per year.

OTHER COMMON SCALES

COCHINEAL SCALES

Dactylopius spp. (Dactylopiidae) feed only on cacti, usually on prickly pear (*Opuntia* and *Nopalea* spp.). At least three species are pests when they infest prickly pear planted as drought-tolerant ornamentals and thorny living fences and where the cacti are grown for their edible fruit and pads. Cochineal scales have been deliberately introduced in locations around the world to help biologically control introduced prickly pear where these cacti

are undesirable because they outcompete native vegetation and displace livestock forage plants.

All stages of *Dactylopius* spp. have a bright red body. In Mexico and South America, *D. coccus* has been cultured for centuries as the source of carmine dye. Except for crawlers, *Dactylopius* females are encased in white, sticky, filamentous wax, so their red color normally is not visible unless the body cover is pierced.

In addition to their fouling of plants with sticky wax and the red staining when their body is squished, cochineal scale feeding can cause severe decline and eventually the death of prickly pear. If a plant is heavily colonized, you can

- prune off and discard the most-damaged pads
- use a forceful spray of water to wash away scales' waxy covering
- spray insecticidal soap on the exposed scale bodies

COTTONY CUSHION SCALE

Icerya purchasi (Monophlebidae, formerly Margarodidae) can occur on many woody species, including citrus, cocculus, nandina, and pittosporum. The female body is orangish brown, and at maturity it is distinguished by its elongated, fluted, white cottony egg sac, up to ½ inch long. The crawlers are red with dark legs and antennae. All nymphal instars are reddish shortly after molting, before they produce cottony secretions. First- and second-instar nymphs settle on twigs and leaves, usually along veins. Third instars occur on branches, and mature scales occur mostly on large branches or the trunk. The loose,

white cocoon of immature males and the minute, red, winged male itself are rarely seen. Cottony cushion scales retain legs and can move throughout their life, with two to three generations per year.

Cottony cushion scale populations are usually well controlled by natural enemies unless biological control is disrupted, for example, by application of certain insecticides. An exception is on laurel-leaf snailseed (*Cocculus laurifolius*), which often is highly infested with cottony cushion scale, because vedalia lady beetles avoid this plant.

The red and black adults of vedalia beetle (*Rodolia cardinalis*) introduced from Australia feed voraciously on cottony cushion scales. The young, reddish vedalia larvae feed on scale eggs; more mature larvae feed on all scale stages. For photographs of each life stage, consult *Stages of the Cottony Cushion Scale* (Icerya purchasi) *and its Natural Enemy, the Vedalia Beetle* (Rodolia cardinalis).

The other important natural enemy is a parasitic fly (*Cryptochaetum iceryae*, Cryptochaetidae). The orangish *Cryptochaetum* larvae feed within the scale; later their dark, oblong pupal cases may be seen there. This fly produces up to eight generations per year.

Both *Cryptochaetum* and vedalia are important natural enemies of cottony cushion scale in coastal California, while vedalia predominates in inland areas. Conserve natural enemies by controlling ants and dust and by avoiding use of broad-spectrum, residual insecticides, which can have a severe adverse effect on vedalia. If scales cannot be tolerated until natural enemies become abundant, apply narrow-range oil to deciduous hosts during the dormant season or spray foliage

Black predominates on some adult vedalia, while others have more red. This lady beetle is the most important natural enemy of cottony cushion scale at interior California locations. A parasitic fly predominates near the coast. *Photo:* J. K. Clark

ny cushion scale females and nymphs encrust this grevillea This scale can occur on many hosts, but it is usually at very low ities because of effective natural enemies. *Photo:* J. K. Clark

VEDALIA

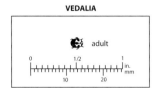

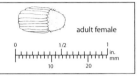
when the tiny, reddish scale crawlers and young nymphs are the predominate stages. See *Cottony Cushion Scale Pest Notes* for more information.

EHRHORN'S OAK SCALE

Mycetococcus ehrhorni (Asterolecaniidae) is bright red and $\frac{1}{25}$ inch long or less. On evergreen oaks in southern California, heavy infestations can be found on bark hidden under the white or grayish mycelia of the fungus *Septobasidium canescens*, which derives nourishment from honeydew produced by the scale. The scale and fungus give oak bark a whitewashed appearance and can severely stress oaks and slow their growth.

EUROPEAN ELM SCALE

Eriococcus (=*Gossyparia*) *spurius* (Eriococcidae) is a pest only on elm, especially Chinese elm (*Ulmus parvifolia*). It secretes copious honeydew and can cause extensive leaf yellowing, plant decline, and death of small branches.

The mature female is dark red, brown, or purple and surrounded by a white, feltlike wax fringe. Females encrust bark at the crotches of twigs and on the lower surface of limbs. Scales overwinter as yellow to brown second-instar nymphs on bark. In the spring, nymphs that are male develop into elongate, whitish cocoons. Females mature and produce crawlers by early summer. Nymphs feed on the underside of leaves, then migrate to bark in the fall. The insect has one generation per year.

Keep trees healthy so they tolerate insect feeding. American and European elms are adapted to summer rains; provide adequate irrigation in areas of prolonged summer drought. Several introduced parasites, primarily *Baryscapus coeruleus* and *Coccophagus insidiator,* attack European elm scale at some locations, as evidenced by their round emergence holes in mature female scales.

Increase the effectiveness of biological control by controlling honeydew-seeking ants. If scales are intolerable, apply narrow-range oil during the dormant season. In spring or early summer, apply oil when crawler abundance declines after it has peaked. Because female scale density peaks in northern California at about 540 degree-days above 52°F accumulated from March 1, begin once-a-week monitoring of crawlers by about 400 degree-days if oil spray is planned. Systemic insecticide also provides control.

European elm scale mature females are purplish, white fringed, and often tended by ants, as on this Chinese elm. At some locations, introduced parasites help to control this scale if ants are managed. *Photo:* J. K. Clark

Ehrhorn's oak scale on bark, revealed by removing the pale mycelia of an associated fungus that usually covers this scale on oaks. The mature female Ehrhorn's oak scale is bright red, and the nymphs are smaller and pale orangish to red. *Photo:* R. J. Gill, California Department of Food and Agriculture

EUROPEAN ELM SCALE

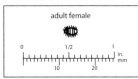

INCENSE-CEDAR SCALE, OR MONTEREY CYPRESS SCALE

Xylococculus macrocarpae (Xylococcidae) causes foliage discoloring and dieback on incense-cedar, Monterey cypress, and various *Cupressus* and *Juniperus* spp., especially on young plants growing at shaded locations. It usually is common only in stands of native conifers, not in urban areas.

Nymphs and females are oblong, dark reddish to brown, and wingless. The tiny, winged adult males are easily overlooked, but females grow to about ¼ inch long. The ⅛-inch-long preadult females closely resemble the cypress bark mealybug pictured earlier, and both species can occur on the same hosts.

Incense-cedar scale and cypress bark mealybug do not resemble the minute cypress scale also pictured earlier. All three of these species are sometimes called cypress scale, but they are in separate insect families and have different biology.

KUWANA OAK SCALE

Kuwania quercus (Kuwaniidae) feeding on blue oak causes extensive roughening and flaking off of the bark, exposing tissue underneath that is discolored and blackish. Adults and older nymphs are red, roundish, about 1/10 inch in diameter, and covered with grayish or white wax. They

occur mostly hidden under bark plates and in crevices on oak limbs and trunks. Kuwana oak scale closely resembles the oak xylococcus scale (*Xylococculus quercus*) and Ehrhorn's oak scale (*Mycetococcus ehrhorni*), which occur on the bark of various oaks.

OAK PIT SCALES

Asterodiaspis spp. (=*Asterolecanium* spp., Asterolecaniidae) cause ring-shaped depressions in bark, giving twigs a roughened, dimpled appearance and causing twig and branch dieback. Dead twigs and brown leaves that do not drop are apparent in winter on infested deciduous oaks. Infestations that cause dieback can lead to distorted terminal regrowth, severe tree decline, and eventual death of some oaks. High pit scale populations can also be associated with severe infection by fungi that cause twig dieback.

Least pit scale (*Asterodiaspis mina*) and drab pit scale (*A. quercicola*) prefer native

valley oak (*Quercus lobata*) and blue oak (*Q. douglasii*), and they also occur on coast live oak (*Q. agrifolia*) and California black oak (*Q. kelloggii*). Golden oak scale (*A. variolosa*), a less-common species, prefers English oak (*Q. robur*) and California black oak. Mature *Asterodiaspis* are brown, gold, or green, circular, immobile insects about the size of the head of a pin and resemble armored scales. Crawlers emerge from beneath the female from April through October, primarily in April through June.

Narrow-range oil provides control if applied to cover all bark and branch tips thoroughly in the delayed-dormant season, just before buds open. One annual application during several consecutive years may be necessary to reduce high populations to a low level. Certain insect growth regulators (IGRs) and systemic insecticides may also provide control. See *Oak Pit Scales Pest Notes* for more information.

Rough and exfoliating (flaking-off) bark on a blue oak trunk due to feeding by Kuwana oak scales. *Photo:* T. J. Swiecki, Phytosphere Research

Golden oak scale and pits made by scales that have dropped from this California black oak twig. Some of these pit scales have parasite emergence holes. *Photo:* J. K. Clark

OAK PIT SCALE

adult female

0   1/2   1   in.
       10   20   mm

Kuwana oak scales are bright red but easily overlooked, as they usually occur hidden under loose bark. Late-instar nymphs and adult females cover themselves with gray to whitish wax, as with these legless, immobile, nymphs called cysts. *Photo:* R. J. Gill, California Department of Food and Agriculture

Sycamore scale can verely spot and distort more leaves. Leaf bud nk is the most effective time to treat for this st. Once leaves mature and become damaged, as shown here, control ions are unlikely to be effective until the next son, when new leaves elop. *Photo: J. K. Clark*

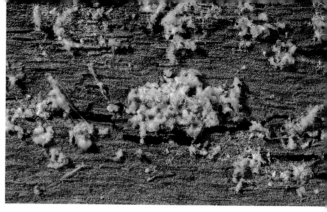

Orangish sycamore scale nymphs, eggs, and cottony wax on the underside of a bark plate. *Photo: J. K. Clark*

SYCAMORE SCALE

Stomacoccus platani (Steingeliidae) feeds only on *Platanus* spp., such as London plane (a commonly planted hybrid) and the native California sycamore (*P. race-mosa*). Sycamore scale reportedly is not a pest on American sycamore (*P. occidenta-lis*), which is native to the eastern United States.

Scale infestations cause numerous, small, yellow spots on infested leaves. Leaves infested during late winter and early spring as they are expanding become distorted and stunted and may turn brown and die. Infested leaves may drop prematurely, mostly during the late spring. Prolonged high populations of sycamore scale can stunt tree growth and can cause dieback of twigs and small branches. On young trees, infestations can cause abnormal bark roughening and sloughing off of bark.

Anthracnose and powdery mildew fungi also commonly cause premature drop of sycamore leaves. Distinguish sycamore scale by the tiny, discolored leaf spots it causes and by the scales' cottony wax, especially on bark.

The insects themselves are brown or yellow and less than $1/12$ inch long. The appearance of sycamore scale changes greatly during development, and except for the crawler stage it does not resemble most other scales. Sycamore scale adults resemble immature (wingless) thrips. The nymphs are elliptical and resemble insect eggs or tiny footballs. During certain stages, nymphs produce whitish wax and resemble whitefly pupae. Male cocoons, mature females, and egg masses can also be covered with cottony white tufts.

Sycamore scales can occur on foliage from leaf flush through fall and on tender bark anytime of year. They can occur on the undersurface of outer bark, the outer surface of inner bark, and underneath bark "scales" or plates (bark that cracks or sloughs off). The scales overwinter on and beneath bark, where females produce eggs that hatch during late winter coincident with sycamore bud break and early leaf flush. The crawlers (mobile first instars) move from bark to settle and feed on the underside of leaves and on leaf buds and petioles. Some scales develop on bark or move between leaves and bark during summer. Sycamore scale has several overlapping generations per year.

Lady beetles that feed on sycamore scale include the twicestabbed lady beetle (pictured below) and *Exochomus quadripustulatus* (which is black with four orange to yellow blotches).

Foliar sprays are not recommended because trees are often large and it is difficult to thoroughly wet the underside of leaves where scales feed. The lower leaf surface of California sycamore has dense mats of tiny hairs, which entwine scales and repel liquids. If scales were abundant the previous season and damage cannot be tolerated, apply narrow-range oil or another insecticide during the delayed dormant season. Inspect trees regularly beginning in late December and spray at bud break, when immature scales are highly susceptible to spray. Thoroughly spray branch tips and use a high-pressure sprayer to reach scales under the bark plates on trunks and large limbs. For additional pesticide recommendations (e.g., fungicide-insecticide combinations), consult *Sycamore Scale Pest Notes*.

MONITORING OF SCALES

Regularly inspect valued host plants for scales and their damage and associated ants. Because dead scales from previous generations may remain on the plant, distinguish the live scales by examining a sample of them. For example, squish some scales to see if liquid exudes or remove the cover or turn them over to look for a live body or eggs.

Control action may be warranted if
- honeydew is intolerably abundant
- there are high numbers of viable and nonparasitized scales
- scale-related plant dieback or decline is occurring

Consider using quantitative sampling if plants are especially valued or substantial resources will be devoted to managing scales. For example, count and record the number of mature female scales, and separately count the number of parasitized scales, on each of four to eight shoots on each of several plants. Compare the scale density on samples to that during previous years and other sample dates during the same year to determine whether scales are increasing and to evaluate whether any treatment was effective.

An alternative to counting each scale is determining the percentage of samples (e.g., branch terminals) with scales. This presence-absence sampling is quicker but less precise than counting each scale, and it is difficult to estimate parasitism.

For certain scale species, degree-day (temperature-monitoring) models are available to estimate when female scales

Wrap double-sided sticky tape tightly around twigs infested with female scales to monitor when scale crawlers are active. Double over the end so you can grasp and easily unwrap tapes to check them for captured scales. *Photo:* J. K. Clark

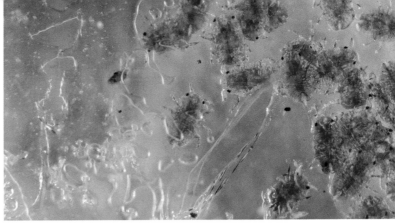

An enlargement of European fruit lecanium scale crawlers caught in a sticky tape trap. Many scales are most effectively controlled if plants are treated with oil or another insecticide after monitoring shows that crawler abundance has peaked and begun to decline, which is soon after most crawlers have settled. *Photo:* J. K. Clark

are likely to mature and crawlers are expected to be most abundant. Sticky tape traps and degree-day monitoring help determine the effective time to treat, but these methods do not tell you whether scales are abundant enough to warrant control.

Sticky Tape Traps. To effectively time a foliar insecticide application, choose several twigs or small branches for monitoring on each plant during the spring before crawlers begin to emerge. Tightly encircle each infested twig or branch with transparent tape that is sticky on both sides. Double over the loose end of the tape several times so you can pull the end to easily unwind it. Place a tag or flagging near each tape so you can readily find it. Change the tapes at regular intervals, about weekly. After removing the old tape, wrap the twig at the same location with fresh tape. Preserve the old sticky tapes by sandwiching them between a sheet of white paper and clear plastic. Label the tapes with the date, location, and host plant from which they were collected.

Scale crawlers get stuck on the tapes and appear as yellow or orange specks. Examine the tapes with a hand lens to distinguish the crawlers (which are round or oblong and have very short appendages) from pollen and dust. Use a hand lens to examine the crawlers beneath mature female scales on bark or foliage to be certain of crawler appearance. Other tiny creatures, including mites, may also be caught in the tapes.

Visually compare the tapes collected on each sample date. If a spring or summer foliar insecticide application is planned, unless another time is recommended for that species, spray after crawler production (abundance in traps) has peaked and definitely begun to decline, which is soon after most crawlers have settled.

MANAGEMENT OF SCALES

Conserving natural enemies and applying certain insecticides are primary tactics for managing scales. Appropriate plant selection can avoid many problems because a given scale species damages only certain plant species. Examples include scales discussed earlier that damage certain cypress, elm, euonymus, oak, pine, or sycamore but not other species or cultivars in the same plant genus. You can prune off infestations if they are limited to only a few parts of small plants. In areas with hot summers, pruning to open canopies can reduce populations of some species, such as black scale, citricola scale, and cottony cushion scale by increasing insect mortality from exposure to heat and parasites. For additional information, consult the *Scales Pest Notes.*

Biological Control. Scales are preyed upon by many predatory bugs, beetles, and lacewings. These include *Chilocorus, Hyperaspis,* and *Rhyzobius* spp. lady beetles that can easily be overlooked because many species are relatively small, the adults are colored and shaped like scales, and the larvae often feed beneath scales. Often, the most important natural enemies of scales are parasitic wasps, including species of *Aphytis, Coccophagus, Encarsia,* and *Metaphycus.*

To enhance the biological control of scales, prevent excessive dust and control honeydew-seeking ants, as discussed earlier

A parasitic wasp larva (*Coccophagus lecanii*) feeding inside has caused a lecanium scale nymph to darken (left) in comparison with the normal brownish color of an unparasitized nymph. *Photo:* J. K. Clark

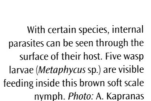

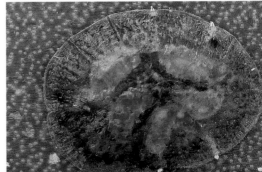

With certain species, internal parasites can be seen through the surface of their host. Five wasp larvae (*Metaphycus* sp.) are visible feeding inside this brown soft scale nymph. *Photo:* A. Kapranas

The adult *Rhyzobius lophanthae* has a reddish head and underside and black to grayish wing covers densely covered with tiny hairs. This 1/8-inch-long lady beetle is eating tiny, oblong, yellowish lecanium scale nymphs on this leaf underside. *Photo:* J. K. Clark

The adult twicestabbed lady beetle (*Chilocorus orbus* =*C. stigma*) has two reddish spots on its shiny, black body. The spots on *C. orbus* are roundish and occur somewhat forward of the wing cover's midpoint. Similar-looking species include *Chilocorus kuwanae*, which helps to control euonymus scale. The spots of *C. kuwanae* tend to be more rectangular and nearer the center of the wing covers. *Photo:* J. K. Clark

in "Ants." Avoid using broad-spectrum, residual insecticides for scales or other pests because these disrupt natural enemies. Apply narrow-range oil instead, preferably during the dormant season.

Insecticides. Dormant-season application of horticultural or narrow-range oil can control many scale species. When applied as a delayed-dormant spray, just before buds break, oil also kills overwintering mite, aphid, and caterpillar eggs on bark. Oil application in spring or summer controls exposed stages of many insects and is effective against scales if applied when crawlers and early-instar nymphs are the predominant stages. Adding a small amount of another contact insecticide to the oil can increase the application's effectiveness. Do not apply oil to the foliage of sensitive plant species, as identified on product labels or in *Managing Insects and Mites with Spray Oils,* and do not spray oils when plants are drought stressed or if temperatures are anticipated to soon be over 90° or under 32°F.

Avoid foliar sprays of residual, broad-spectrum carbamate, organophosphate, and pyrethroid insecticides because they disrupt the biological control of scales and other pests and can wash off and pollute surface water. Certain systemic insecticides (neonicotinoids) are effective against honeydew-excreting scales such as European elm scale and soft scales, but some (e.g., imidacloprid) do not control armored scales. Make a soil application or bark spray according to product labels whenever possible to avoid tree damage, as discussed under "Injecting or Implanting Pesticides" in Chapter 2. If applying a systemic insecticide, delay application until

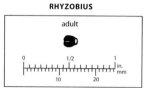

RHYZOBIUS

adult

0   1/2   1
||||||||||||||||||||||| in.
      10   20   mm

A twicestabbed lady beetle larva feeding on sycamore scale eggs and nymphs on the underside of a sycamore bark plate. *Photo:* J. K. Clark

after plants have completed their seasonal flowering to minimize toxicity to pollinators and adult natural enemies that feed on nectar and pollen. See *Scales Pest Notes* for more information.

CICADAS

Cicadas (Cicadidae) are known for the loud, shrill buzz or click sounds males make to attract females. In California, cicadas are relatively uncommon in landscapes and are rarely pests.

Females can cause terminal dieback where they insert eggs into the twigs of young woody plants. The emerging nymphs drop to the ground and use their enlarged clawlike forelegs to burrow into soil. Each nymph sucks on a root for several years, then emerges to climb an object, shed its skin, and become an adult. Except for netting young plants to exclude insects, cicada control efforts usually are not effective. Within a few weeks after they emerge, the relatively short-lived adults will die.

Adult cicadas are robust insects up to 1 1/2 inches long, with mostly clear wings. Cicadas are commonly blackish or dark brown, although some species have green, orange, or red markings. In California, these noisy insects are relatively uncommon and are not pests. *Photo:* J. K. Clark

LEAFHOPPERS AND SHARPSHOOTERS

Leafhoppers (Cicadellidae), including species that are called sharpshooters, suck the juices of landscape plants, but most never become significant pests. Leafhopper feeding causes leaves to appear stippled, pale, or brown, and shoots may curl and die. Leafhoppers secrete liquid that causes chalky residue or sticky honeydew on plants and some species vector plant pathogens.

Most adult leafhoppers are slender and $\frac{1}{4}$ inch long or less. Glassy-winged sharpshooter and a few other species can be up to about $\frac{1}{2}$ inch long. Some species are brightly colored, while others blend with their host plant. One or more long rows of spines on their hind legs and characters on their head distinguish leafhoppers from most other insects they resemble (Table 6-8). Leafhoppers are active insects that walk rapidly sideways or readily jump when disturbed. Adults and nymphs and their pale cast skins are usually found on the underside of leaves.

Females insert their tiny eggs in tender plant tissue. The wingless nymphs molt four or five times and mature in about 2 to 7 weeks. Leafhoppers overwinter as eggs on or in leaves or twigs or as adults on evergreen hosts or in protected places such as bark crevices. Most species have two or more generations each year.

With certain exceptions, you can ignore most leafhopper species because they do not seriously harm woody landscape plants. Exceptions include the glassy-winged sharpshooter and certain other non-native species. Be alert for the introduction of new species and report atypically severe damage or suspected invasive species to the county agricultural commissioner or UC Cooperative Extension. For example, the introduced Virginia creeper leafhopper (*Erythroneura ziczac*) is damaging Boston ivy, Virginia creeper (*Parthenocissus* spp.), and grape in at least northern California.

Where bothersome, insecticidal soap or narrow-range oil application can suppress high populations of leafhopper nymphs; thorough coverage of leaf undersides is important. It can be very difficult to effectively control adults, and generally no control is recommended.

Leafhopper feeding sometimes causes foliage to become bleached or stippled. This damage is usually harmless to plants. For example, rose leafhopper (*Edwardsiana rosae*) fed on these leaves, but it does not feed on rose blossoms. *Photo:* J. K. Clark

Leafhoppers have one or more long rows of spines along the entire length of their hind tibia (their longest leg segment), as apparent in this side view of an adult leafhopper on myoporum. *Photo:* J. K. Clark

Table 6-8.

Distinguishing Among Sucking Insects.

COMMON NAME (FAMILY)	ATTACHMENT OF SHORT ANTENNAE	BODY	PROTHORAX ENLARGED	OCELLI[1]	HIND TIBIAL SPINES[2]
cicadas (Cicadidae)	in front of or between the eyes[3]	≥1 inch long	no[4]	3	rows of spines along length
leafhoppers (Cicadellidae)	in front of or between the eyes[3]	usually ≤½ inch long, rounded, elongate	no	2 or none	1 or more long rows of small spines along entire length
planthoppers (many)	below eyes on side of head[5]	in U.S. usually ≤½ inch long, flattened broad	no[6]	2 or none	several large spines
spittlebugs (Cercopidae)	in front of or between the eyes[3]	usually ≤½ inch long, rounded, elongate	no	2 or none	1 or 2 stout spines along length and a whorl of many terminal spines
treehoppers (Membracidae)	in front of or between the eyes[3]	≤½ inch long, flattened, broad	yes[7]	2 or none	rows of spines along length

1. Small beadlike or colored spots, which are light receptors on the head between or above the larger true eyes.
2. The tibia is usually the longest leg segment, that between the femur and the terminal tarsi, or feet.
3. Antennae attached in front of the lowest point of the eye (as seen from the side) and between the eyes (as seen from the front).
4. Prothorax (area just behind the head) often prominent but does not extend over the abdomen or head.
5. Antennae attached below the eyes: behind the lowest point of the eye (as seen from the side) and definitely under the eyes (as seen from the front).
6. Head typically has enlargements, ridges, or projections (often snoutlike) in front of the eyes.
7. Greatly enlarged prothorax projects backward over the abdomen (and in some species also projects forward or above), so that treehoppers appear humpbacked or resemble a thorn or roughening on plants.

GLASSY-WINGED SHARPSHOOTER

Homalodisca vitripennis (formerly *H. coagulata*) feeds on over 300 plant species. When glassy-winged sharpshooter is abundant, it causes a whitish residue on leaves and fruit and surfaces beneath infested trees. Infested small plants may wilt.

Native to the southeastern United States, this introduced sharpshooter spreads a disease-causing bacterium (*Xylella fastidiosa*) from one plant to another. Less important vectors of the pathogen include blue-green sharpshooter (*Graphocephala atropunctata*), redheaded sharpshooter (*Carneocephala fulgida*), and smoke-tree sharpshooter (*Homalodisca lacerta*). Various strains of *X. fastidiosa* cause lethal plant diseases such as oleander leaf scorch, Pierce's disease of grape, and sweetgum dieback. Leafhopper-vectored *X. fastidiosa* also is the apparent cause of leaf yellowing, dieback, and death in southern California of crape myrtle, ginkgo, mulberry, *Nandina* spp., olive, ornamental plum, and western redbud. The relationships among leafhopper species, host plants, and pathogenic *Xylella* are not fully understood.

Glassy-winged sharpshooter adults are about ½ inch long and dark brownish with whitish and yellow spots. They closely resemble smoke-tree sharpshooters, but the head of the smoke-tree sharpshooter is covered with wavy, light-colored lines, not light spots. Nymphs of both species have prominent, bulging eyes and resemble small adults, except the immatures are wingless and uniformly olive-gray. Smoke-tree sharpshooter nymphs have blue eyes. Glassy-winged sharpshooter nymphs have red eyes.

Females insert a cluster (mass) of about 5 to 10 eggs into the lower surface of leaves. Eggs resemble a brown or greenish blister on the leaf, which females cover with a white, chalky secretion. Eggs leave a permanent brown to gray scar in leaf tissue after nymphs emerge. The insect overwinters as adults on evergreen plants and can infest them anytime of year but commonly moves to feed on deciduous hosts during spring and summer. It has two generations per year in southern California.

Glassy-winged sharpshooter eggs are more apparent when females leave white, chalky wax around eggs, as shown here. Eggs in leaves are easily overlooked and are a primary way glassy-winged sharpshooter is moved (e.g., in nursery stock) and introduced to new locations. *Photo:* J. K. Clark

Glassy-winged sharpshooter causes a greenish blister in leaves where it has inserted an egg mass. The left leaf contains two groups of unhatched eggs. After eggs hatch, a permanent brown to gray scar remains in leaf tissue. *Photo:* D. Rosen

GLASSY-WINGED SHARPSHOOTER

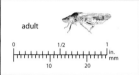

adult

0   1/2   1   in.
|++++++++|++++++++|++++++++| mm
        10   20

Glassy-winged sharpshooter (bottom) is much larger than most leafhoppers, such as the blue-green sharpshooter (top). Because it is the important vector of *Xylella* plant diseases, report suspected glassy-winged sharpshooters to the county agricultural commissioner if you find this insect outside of areas it is known to infest. *Photo:* J. K. Clark

There is no cure for the leafhopper-vectored *Xylella* diseases. Management relies on quarantines to stop the spread of glassy-winged sharpshooters, such as on nursery crops moved from infested areas to other locations where glassy-winged sharpshooter does not occur. Glassy-winged sharpshooter occurs in southern California at the time of this writing. Report suspected glassy-winged sharpshooters to the county agricultural commissioner if you find this insect outside of areas it is known to infest.

In infested areas, the only completely effective strategies are to plant nonhosts and gradually replace highly susceptible species (e.g., oleander and sweetgum). Where plants are performing acceptably

BLUE-GREEN SHARPSHOOTER

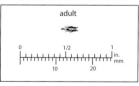

adult

0   1/2   1   in.
|++++++++|++++++++|++++++++| mm
        10   20

but sharpshooter excretions are intolerable, a systemic insecticide (e.g., imidacloprid) can be applied to soil beneath plants. Various foliar insecticides are available, but these require more reapplication. Most insecticides are toxic to the beneficial *Gonatocerus* spp. wasps that parasitize the eggs and help to reduce sharpshooter populations in landscape plantings. For more information, consult "Bacterial Leaf Scorch and Oleander Leaf Scorch" in Chapter 5, the *Pest Notes: Glassy-Winged Sharpshooter* and *Oleander Leaf Scorch,* and the website cisr.ucr.edu.

TWOSPOTTED LEAFHOPPER

Sophonia rufofascia can feed on dozens of plant species and has become especially common on guava in southern California. Depending on the host species, plants may be unaffected or their leaves may cup, become chlorotic, and develop dark dead patches. Damage may not become apparent until several weeks after initial feeding, so sometimes leafhoppers have moved and are not present when damage is first recognized.

Twospotted leafhopper adults and nymphs are light green to pale yellow and are found mostly on new shoots and the underside of terminal leaves. Adults are about ⅕ inch long and have a dark stripe down the center of their back that may be bordered by pink or red. Adults have two prominent, dark, eyelike spots near the rear tip of their wings, causing the leafhopper to appear to be moving backward when walking. Nymphs produce clear cast skins that can remain attached to the plant, resembling pale leafhoppers.

The twospotted leafhopper adult has two black, eyelike tail spots. It feeds on many different plants, causing severe leaf yellowing and distortion on some hosts, while causing no apparent injury to other plant species. *Photo:* D. Rosen

This Australian torpe bug (*Siphanta acuta,* Flatidae) occurs on various plants in coastal California. As with most planthoppers, it is a harmless curiosity in the landscape. *Phot* J. K. Clark

PLANTHOPPERS

Planthoppers (Fulgoroidea) occasionally are bothersome but usually do not damage landscapes. About a dozen planthopper families (e.g., Acanaloniidae, Delphacidae, Flatidae, Fulgoridae, Issidae) and hundreds of species occur in California. The location of the antennae (below each eye on the sides of the head) distinguishes planthoppers from most other sucking insects they resemble (see Table 6-8).

TREEHOPPERS

Treehopper (Membracidae) feeding damage is slight, although their plant sucking produces honeydew, and the resulting growth of sooty mold may blacken leaves and twigs. Treehoppers injure plants primarily by making numerous small slits or crescentlike punctures in bark where they lay their eggs. Egg punctures cause bark to appear roughened, and twigs may die back. Mature woody plants tolerate extensive egg-laying damage, but the growth of heavily infested younger plants may be retarded.

Treehopper adults are commonly greenish to brown and ½ inch long or shorter. They have an expanded hood covering the body that may be formed into hornlike projections. This enlarged prothorax distinguishes treehoppers from most other Hemiptera they resemble (Table 6-8). Nymphs have numerous spines on the back of the abdomen, and both immatures and adults readily jump.

The oak treehopper (*Platycotis vittata*) is common in the spring on the lower branches of deciduous and live oaks and occasionally on birch, chestnut, and certain other broadleaf trees. Adults are olive-green to bronze with reddish bands, and their surface is covered with tiny pits. They often scurry to the opposite side of the twig or leaf when approached. Females usually remain with their eggs and nymphs. The nymphs are black with yellow and red markings. The spring generation is colorful, and individuals typically aggregate in rows on twigs.

Hosts of the buffalo treehopper (*Stictocephala bisonia*) include ash, elm, fruit trees, hawthorn, locust, poplar, and many herbaceous plants (e.g., tomato). Adults are bright green or yellowish, with a yellowish underside. Nymphs are dark green with prominent spines on the back.

If treehopper populations were high on deciduous trees the previous season and damage cannot be tolerated, narrow-range oil can be applied to thoroughly cover terminals during the dormant season to kill overwintering eggs. For species known to feed on many different plants, removing some of the alternate hosts may reduce treehopper feeding on more-valued plants. High populations of nymphs and adults may be reduced by spraying exposed insects with insecticidal soap, narrow-range oil, or another insecticide.

Oak treehopper adults feeding on coast live oak. Tiny bark punctures are visible where eggs were laid. These bark punctures and some resulting twig dieback are this insect's only damage and can usually be tolerated. *Photo:* J. K. Clark

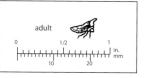

OAK TREEHOPPER

adult

SPITTLEBUGS

Spittlebugs (Cercopidae), or froghoppers, suck plant juices and can at least occasionally be found on almost any plant. Heavy infestations distort plant tissue and slow plant growth, but this is primarily a problem on herbaceous species. Spittlebugs' obvious and occasionally abundant masses of white foam on cones, foliage, and stems may be annoying, but spittlebugs do not seriously harm established woody plants.

Where spittlebug nymphs feed on plant tissue, they surround themselves with frothy white excrement, and more than one nymph may be found in a single spittle mass. Nymphs' body color may be orange, yellow, or green, and they grow through about five molts. Adult spittlebugs are inconspicuous, often brownish insects, about ⅓ inch long or less. They readily jump or fly when disturbed and resemble

leafhoppers, but spittlebugs lack the long rows of spines found along the hind tibia of leafhoppers (see Table 6-8).

The meadow spittlebug (*Philaenus spumarius*) feeds primarily on herbaceous plants but also on conifers and young woody deciduous species. Adults are robust and tan, black, or mottled brownish. Females lay white to brown eggs in rows at plant nodes. Nymphs are yellow to green and are hidden beneath a foaming mass of spittle.

The western pine spittlebug (*Aphrophora permutata*) is one of several *Aphrophora* spp. that feed on conifers and nearby herbaceous plants. It is especially common on Monterey pine and other pines in coastal California areas. Other hosts include Douglas-fir, fir, hemlock, spruce, and various broadleaves. Nymphs are mostly dark greenish, brown, or black. Adults are orangish to dark brown and may have an indistinct diagonal white line across the back. Overwintering occurs as tiny, pale yellow to purple eggs in a row on or in needles. Pine spittlebugs have one or two generations per year in California.

The white, frothy material conceals a spittlebug nymph sucking on this rose bud petiole. If bothersome, a forceful stream of water will wash away spittlebugs and their foam. *Photo:* J. K. Clark

Ignore spittlebugs on woody plants or wash nymphs off with a forceful stream of water. Spittlebugs are more likely to become abundant on woody plants when they migrate from nearby herbaceous species. Cut spittlebug-infested weeds in the spring before the insects mature and spread.

TRUE BUGS

Boxelder bugs, chinch bugs, lace bugs, plant bugs, and stink bugs are in the suborder Heteroptera, the only group of insects that entomologists call bugs. More than 600 species of true bugs occur in California. Many are aquatic or semiaquatic, and most of these are predaceous. Many terrestrial bugs feed on plants, although assassin bugs, bigeyed bugs, damsel bugs, pirate bugs, and some others pictured earlier in "Types of Natural Enemies" are important predators.

Plant-feeding bugs suck juices from leaves, fruit, or nuts. A pale white or yellow stippling forms around feeding sites, and plant tissue may become distorted. Lace bugs and plant bugs can leave dark specks of excrement on the underside of leaves. Large numbers of boxelder bugs and certain others can be a nuisance when they enter houses to overwinter. However, true bugs do not seriously harm established woody plants.

Leaffooted bugs, such as this *Leptoglossus clypealis*, are occasional or minor pests. Adults and nymphs suck plant tissues and sporadically cause foliage to yellow or become spotted. Young cones, fruit, and nuts can exude sap or shrivel and die when bugs feed on them. As with many true bugs, leaffooted bugs sometimes seek shelter inside buildings. See *Leaffooted Bug Pest Notes* for more information. *Photo:* J. K. Clark

IDENTIFICATION AND BIOLOGY

Bugs usually have thickened forewings with membranous tips. When they rest, the dissimilar parts of their folded wings overlap, forming the characteristic triangle or X shape on the back of true bugs. On plant-feeding species, the tubular mouthparts point downward, perpendicular to the plane of the insect's body. The mouthparts of predaceous bugs can be extended forward when attacking other insects. Depending on the species, eggs are laid exposed on foliage or bark or inserted in plant tissue. The flightless nymphs gradually change to winged adults without any pupal stage (see Figure 6-2).

MANAGEMENT

True bugs rarely cause serious harm to established woody plants. Provide proper cultural care so that plants are vigorous. Damaged foliage can be pruned out. Consider replacing especially susceptible plants with resistant species. Spraying for bugs is generally not recommended. Extreme populations may be reduced by applying an insecticide to foliage when nymphs are abundant in the spring; systemic insecticides may be the most effective. No treatment will restore distorted or stippled foliage, which remains until replaced by new growth. To help prevent bugs from coming inside, seal exterior cracks in buildings, screen windows and doors, and take other measures discussed earlier under "Home Invasions by Nuisance Pests."

ASH PLANT BUGS

Various plant bugs (Miridae) are occasional pests. *Tropidosteptes illitus* and *T. pacificus* feed mostly on ash in the interior valleys of California. The nymphs and adults suck juices from leaves, seeds, and twigs, but their damage usually is limited to leaf bleaching or stippling and dark spots of excrement on foliage. Extreme infestations can defoliate trees, although severe defoliation of ash is more often due to causes such as anthracnose fungus or drought stress.

The Pacific ash plant bug brown adults and green nymphs leave dark, liquid excrement on the underside of ash leaves. Trees usually tolerate plant bugs' sucking feeding, and damage is rarely severe enough to cause defoliation or warrant control. Defoliated ash is more commonly caused by anthracnose fungus or drought.
Photo: J. K. Clark

Tropidosteptes illitus nymphs are light brown and adults are yellow and brown or black. Pacific ash plant bug (*T. pacificus*) adults are brown, and the nymphs are green with black spots. Ash plant bugs overwinter as eggs in twig bark. In February or March the nymphs emerge and feed until they mature in April or May. The adults feed and females lay eggs until June or July. Ash plant bugs have one or two generations per year.

Trees apparently tolerate ash plant bug damage; it is rarely severe enough to cause defoliation or warrant control. If damage cannot be tolerated, an insecticide such as narrow-range oil or soap may be applied in the spring to thoroughly cover leaf undersides infested with nymphs.

BOXELDER BUG

The western boxelder bug (*Boisea rubrolineata,* Rhopalidae) usually becomes abundant only on female (seed-bearing) box elder trees (*Acer negundo*). They also feed on ash, maple, and fruit and nut trees. Boxelder bug feeding sometimes distorts and discolors or stipples leaves and fruit, but this does not seriously harm plants. Boxelder bugs are a pest primarily because high populations annoy people when

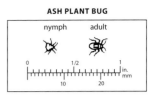

adults aggregate around buildings and seek shelter by entering homes.

Adults are mostly blackish, with narrow red lines marking the wing margins and veins. Three lengthwise red lines on the pronotum (one down the middle and on each margin) distinguish boxelder bugs from other bug species. Females lay oblong, reddish eggs in groups on bark or foliage. Nymphs are gray and bright red, and develop black marks as they mature. Boxelder bugs generally have two generations per year.

Adults and nymphs periodically aggregate in large groups, covering tree trunks, the ground, fences, or the sides of houses. During fall the adults seek overwintering shelter in dry, protected locations, including hollows of trees, under rocks or debris, and in buildings. During winter and early spring, boxelder bugs emerge from overwintering and can be found on light-colored, sunny surfaces, such as the south and west walls of houses.

Replace, and do not plant, the pod-bearing female box elder trees; they support high boxelder bug populations. Eliminate debris and litter around houses, especially in a 6- to 10-foot-wide strip along the south and west sides of foundations. Screen windows and doors, seal exterior cracks, and otherwise exclude insects, as discussed earlier in "Home Invasions by Nuisance Pests."

Vacuum up insects found indoors. Boxelder bugs trapped indoors do not reproduce and will eventually die. They do not injure people or pets but may spot curtains and furnishings with their excrement and when crushed they have an offensive odor.

Wash bugs off of walls or tree trunks with a forceful stream of water. Do not apply insecticides indoors for boxelder bugs. Insecticides also are not recommended outdoors for boxelder bugs

Boxelder bugs, like this adult, nymph, and eggs, are most common on female (pod-bearing) box elder trees. Three lengthwise red lines (down the middle and on each margin) on their pronotum distinguish boxelder bug adults from most other bugs they resemble. *Photo:* J. K. Clark

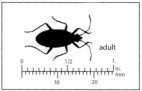
Look-Alike Bugs

Boxelder bugs can be confused with other species of true bugs that are blackish with red or orange. Correctly identify the species so you can learn how to effectively manage it. The adult appearance and host plants of some common look-alike bugs include the following.

Bagrada bug (discussed in "Stink Bugs") is five sided and mostly black with orange and whitish markings.

Bordered plant bug (*Largus cinctus,* Largidae) has a narrow red margin along the rear of its pronotum and entirely circling the margins of its blackish abdomen. It feeds on various herbaceous and shrub species.

Harlequin bug (*Murgantia histrionica,* Pentatomidae) has a calico or variegated pattern of bright red or orange on a black body that is five sided or shield shaped, as characteristic of stink bugs. Harlequin bug feeds on many plants but prefers Brassicaceae (crucifers, e.g., broccoli and cabbage).

Small milkweed bug (*Lygaeus kalmii,* Lygaeidae) has broad red areas along the rear of its pronotum and in a wide X pattern on its wing covers. It feeds on milkweeds and sometimes other plants.

Squash bug (*Anasa tristis,* Coreidae) has a narrow orangish margin along its abdomen and thorax, little or no orange on top of its pronotum, and is predominantly dark with some whitish markings. It feeds mostly on pumpkin, squash, and other Cucurbitaceae.

Bagrada bugs severely damage young crucifers in vegetable gardens and are a nuisance in landscapes, where they feed on sweet alyssum and other plants. Adults and nymphs of the Bagrada bug are mostly black and orange. Later-instar nymphs and adults also have white markings. *Photo:* S. K. Dara

These squash bugs are among the many similar-looking species of bugs you may need to distinguish so you can know how and where to control your specific pest. Adult squash bugs are mostly dark with some whitish markings, and they have little orange on top except for a narrow orange margin along the abdomen and thorax. *Photo:* J. K. Clark

because exclusion provides longer-lasting control and avoids pesticide wash-off that contaminates surface water. For more information, see *Boxelder Bug Pest Notes.*

LACE BUGS

Several *Corythucha* spp. (Tingidae) feed on one or a few closely related plant species, including alder, ash, birch, ceanothus, coyote brush, photinia, poplar, sycamore, toyon, and willow. Lace bug leaf stippling usually becomes obvious by late summer and can be distinguished from mite feeding by the lack of webbing and by the presence of dark specks of lace bug excrement on the underside of leaves. Certain true bugs, thrips, and other insects also produce leaf stippling and excrement, so examine the foliage for insects to distinguish these pests.

Adult lace bugs are about 1/8 inch long, and the expanded surfaces of their thorax and forewings have numerous semitransparent cells that give the body a reticulated or lacelike appearance, hence their name. Females insert tiny eggs partly in plant tissue, often hidden under excrement. The wingless nymphs commonly have body spines. All stages occur in groups on the underside of leaves and have several generations per year.

Tolerate lace bug damage, as even extensive leaf stippling does not threaten plants' health. Provide adequate irrigation and other care and grow species well adapted to that site's conditions. Certain plants (e.g., azalea and toyon) are more extensively damaged by lace bugs when drought stressed or grown in direct sun instead of the partial shade to which they are adapted.

Bleached photinia foliage caused by lace bug feeding. Other true bugs, leafhoppers, thrips, and mites cause similar leaf stippling. Look carefully for the insects or mites to diagnose the cause of this damage. *Photo:* J. K. Clark

Adults and a nymph of the California Christmas berry tingid (*Corythucha incurvata*), a common species on toyon. Leaf stippling and dark excrement from lace bugs can be prominent but are harmless to established woody plants. *Photo:* J. K. Clark

LACE BUG

adult

0   1/2   1   in.

10   20   mm

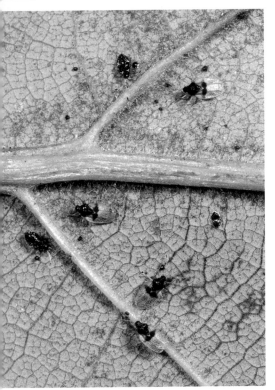

Adult avocado lace bugs (*Pseudacysta perseae*) are oval, with a dark head and thorax and lighter (brown, orangish, or white) appendages and wings. Their feeding causes avocado foliage to turn yellow or whitish, then brown; affected tissue then dies in blotchy patches between the leaf veins. *Photo:* D. Rosen

On toyon and possibly other shrubs, lace bug survival during winter and subsequent damage in spring may be reduced by keeping soil beneath host plants bare during December through February, shallowly cultivating the soil surface several times during this period, or using both practices. For example, during late fall, rake away and compost leaves beneath lace bug host plants. If organic mulch is reapplied in spring, avoid using leaves from the same plant genus as mulch near that plant because it may harbor adult lace bugs.

Many predators feed on lace bugs and in certain situations substantially reduce lace bug populations. If lace bugs cannot be tolerated, narrow-range oil or pyrethrins can be applied to infested lower leaf surfaces. Certain systemic insecticides applied early in the season can be effective. Avoid applying broad-spectrum, residual, contact insecticides that kill natural enemies and can wash off into surface water. Insecticides do not restore undamaged plant appearance. Consult *Avocado Lace Bug Pest Notes* or *Lace Bugs Pest Notes* for more information.

STINK BUGS

Stink bugs (Pentatomidae) are five sided or shield shaped with a large triangle shape (scutellum) on the adults' backs. They are wider than most other true bugs, and some species give off an offensive-smelling chemical when disturbed. Stink bug eggs are barrel shaped with circular caps and are usually laid in a group on leaves.

Most stink bugs are pests only on fruit trees, vegetables, or herbaceous plants or when they seek shelter indoors. Some are beneficial predators, including *Perillus* and *Podisus* spp., such as twospotted stink bug (*Perillus bioculatus*). The introduced brown marmorated stink bug (*Halyomorpha halys*) and the consperse stink bug (*Euschistus conspersus*) cause discolored depressions, blemishes, or dark pinpricks on fruit and fleshy vegetables. Damaged tissues become white and pithy and remain firm. See *Brown Marmorated Stink Bug Pest Notes* for more information.

The introduced Bagrada bug (*Bagrada hilaris*) severely damages young crucifers (Brassicaceae) and is a pest in vegetable gardens. It is a nuisance in landscapes containing bermudagrass turfgrass, sweet alyssum, mustards, or other hosts. *Bagrada* are mostly black and orange, and the adults and older nymphs have white markings, as pictured earlier. In comparison with adults of the similar-looking harlequin bug, Bagrada bug adults are much smaller, about 1/4 inch long. The nymphs superficially resemble those of several other stink bug species, including the consperse stink bug pictured here. See ipm.ucdavis.edu/PDF/pestalert/bagradabug.pdf and cisr.ucr.edu for more information.

Except for fruit and young plants (e.g., in nurseries), woody plants are not seriously harmed by stink bug feeding in most situations, and control is generally not recommended. Manage weeds and other

Consperse stink bug nymphs and the empty shells of the barrel-shaped eggs from which they hatched. Plant-feeding stink bugs are problems mostly when their puncture feeding damages fruit and vegetables. *Photo:* J. K. Clark

This introduced brown marmorated stink bug seeks shelter over winter by entering homes; it emits a foul odor when disturbed. Unlike native gray-brown, or rough, stink bugs (*Brochymena* spp.), the adult marmorated stink bug (top) is distinguished by alternating black and white on the margins of the abdomen. A fifth- (last-) instar nymph is also shown here. *Photo:* S. Aksum, USDA-ARS

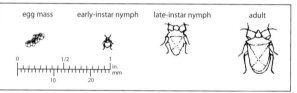

STINK BUGS

| egg mass | early-instar nymph | late-instar nymph | adult |

herbaceous vegetation to reduce stink bug migration to woody plants. See "Home Invasions by Nuisance Pests," above, for how to prevent bugs from coming indoors.

THRIPS

Thrips (Thysanoptera) are tiny, slender insects that puncture host tissue and suck the contents. Most species feed in flowers on pollen, young tissue in buds, or protected plant parts such as central terminals. Thrips feeding can stunt plant growth and cause leaves and terminals to become discolored (e.g., blotchy or stippled) and distorted (e.g., crinkled or tightly rolled). Some species cover the lower surface of leaves with black, varnishlike specks of excrement. Certain species vector viruses, but this primarily damages herbaceous plants in commercial production. Thrips infestations reduce the aesthetic quality of plants but rarely seriously damage or kill woody plants. Certain thrips are beneficial predators.

IDENTIFICATION AND BIOLOGY

Adults are narrow, less than $\frac{1}{20}$ inch long, and have long fringes on the margins of their wings. Females lay tiny eggs on or within leaf tissue. The nymphs (also called larvae) are commonly translucent white to yellowish. Pupae can occur on plants where active stages feed, or mature nymphs may drop and pupate near the soil surface. Thrips have several generations per year.

Banded thrips (*Aeolothrips* spp.), black hunter thrips (*Leptothrips mali*), sixspotted thrips (*Scolothrips sexmaculatus*), and vespiform thrips (*Franklinothrips* spp.) are beneficial predators that feed on pest insects and mites. Consult *Integrated Pest Management for Citrus* and *Thrips Pest Notes* for more information on distinguishing among thrips species.

Toyon thrips, *Liothrips* (=*Rhyncothrips*) *ilex*, feeding in young shoots caused this leaf distortion, which became apparent as the foliage matured. This damage does not seriously impair plant health. Because this thrips drops to pupate in leaf litter, keeping soil bare beneath plants during the winter may reduce thrips survival and damage. *Photo:* J. K. Clark

Greenhouse thrips bleaching and stippling on viburnum. Foliage discoloring begins in small patches, when thrips numbers are limited and relatively easy to control. As their numbers increase, they move throughout the plant and become increasingly difficult to control. No treatment will restore damaged foliage. *Photo:* J. K. Clark

Thrips are tiny, slender insects with fine fringes on their wings. The adult western flower thrips shown here varies greatly in color, from all dark to all pale to a mix of both. Although an occasional pest when it feeds on rose flower buds, it can be beneficial when it feeds on spider mites. *Photo:* J. K. Clark

THRIPS

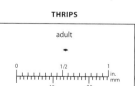

adult

CUBAN LAUREL THRIPS

Gynaikothrips ficorum, introduced from southern Asia, is found in southern California on *Ficus* spp., especially Indian laurel fig. All thrips stages occur year-round in leaf terminals, which become tightly rolled and podlike due to thrips' feeding. Adults are black, nymphs are yellow, and eggs are beige and laid on the leaf surface, commonly within the gall. Populations are highest from about October through December, and galled foliage is formed from midsummer through fall.

Cuban laurel thrips do not seriously harm ficus, so no control is needed if distorted foliage can be tolerated. Common predators on thrips in galls include green lacewing larvae and adults and nymphs of the pirate bug *Macrotracheliella nigra,* which is dark reddish brown to black and less than ⅛ inch long.

Rolled, podlike *Ficus microcarpa* terminals caused by Cuban laurel thrips feeding. To control this pest, conserve natural enemies and prune off galled terminals during winter. *Photo:* J. K. Clark

The Cuban laurel thrips adult is mostly black with some white on its wings, legs, and antennae. Larvae are yellow, and eggs are whitish. All thrips stages and their major predators (lacewing larvae and a dark minute pirate bug) occur within rolled leaves caused by thrips feeding. *Photo:* J. K. Clark

Pruning and disposing of infested terminals can provide effective control. Winter may be the best time to prune off tips because more galled tissue generally does not form until next summer and relatively few thrips can survive the winter outside of the protection provided by the rolled leaves. Where this problem occurs, consider planting *Ficus microcarpa* 'Green Gem,' which is resistant to Cuban laurel thrips.

FLOWER THRIPS

Frankliniella spp., most commonly western flower thrips (*F. occidentalis*), damage primarily herbaceous plants. Occasionally their feeding in landscapes damages flowers on roses and other woody plants. Flower thrips damage is of concern on exhibition blooms and in greenhouse production and certain field crops, but damage is usually not important in woody landscapes.

Flower thrips usually are found on or near blossoms. Adult *F. occidentalis* have pale wings but vary in body color from black, brownish, orange, yellow, or white to a mix, such as yellowish with a dark abdomen. The oblong immatures are usually orange or yellow.

Most rose cultivars are not noticeably blemished by one or two thrips per rose bud. Fragrant, light-colored, or white roses are most often damaged, while cultivars with sepals that remain tightly wrapped around the bud until blooms open have fewer problems. Because flowers can harbor thrips, removal and disposal of old,

Thrips feeding in buds occasionally kills portions of blossoms as with the dark, misshapen areas on this rose. Blossom damage in landscapes is sporadic and difficult to prevent. thrips often leave the flowers before they open fully and thrips feeding damage becomes apparent. *Photo:* J. K. Clark

spent flowers is sometimes recommended. However, the general benefit of this practice in landscapes is unknown and old blossoms also commonly shelter beneficial predators of thrips. Flower thrips are difficult to control with insecticides, and their use is generally not recommended for flower thrips in landscapes.

GREENHOUSE THRIPS

Heliothrips haemorrhoidalis was introduced from South America and occurs on many landscape plants. Usually it is a pest only on broadleaf evergreens, especially azalea and rhododendron. Other hosts include avocado, laurel (English and Grecian), photinia, and toyon.

Greenhouse thrips feed openly on foliage typically on the underside of leaves in dense colonies of adults and immatures, severely bleaching leaves and fouling foliage with blackish excrement. This species is relatively sluggish, so feeding damage often begins in one area and only gradually spreads throughout the plant. Adults are mostly black, except for their white wings. Nymphs are yellow, but black gut contents are often visible, and often they carry a black droplet at the tip of their abdomen. Greenhouse thrips has five to seven generations per year.

Predaceous mites, predatory thrips, and at least two parasitic wasps attack greenhouse thrips. The tiny female *Megaphragma mymaripenne* wasp oviposits inside greenhouse thrips eggs, which are hidden from view embedded in plant tissue. Eggs killed by *Megaphragma* develop a round hole where the *Megaphragma* adult emerged, usually in the raised middle of the egg blister. In contrast, when a greenhouse thrips emerges, part of the egg shell is often visible at the side of the egg blister.

When *Thripobius semiluteus* females lay an egg in a thrips nymph, the thrips becomes swollen around the head as the parasite matures inside. About 2 weeks before the wasp's emergence, parasitized nymphs turn black, in contrast to the yellow color of unparasitized nymphs. Unlike healthy black mature thrips, the black parasitized nymphs are smaller and do not move.

Where practical, frequent spraying of infested surfaces with a forceful stream of water controls greenhouse thrips. Insecticidal soap, oil, and pyrethrins provide control if thoroughly applied and are less disruptive of natural enemies than most insecticides.

Greenhouse thrips bleaches foliage and produces black excrement specks. Unlike most thrips species, greenhouse thrips feeds openly in groups on leaves. *Photo:* J. K. Clark

The yellow and black *Thripobius semiluteus* adult wasp (center left) lays its eggs in the yellowish nymphs of greenhouse thrips. The black, oblong bodies are parasite pupae, each of which developed inside of and killed a thrips nymph. *Photo:* J. K. Clark

MYOPORUM THRIPS

Klambothrips myopori, from New Zealand, severely distorts and galls foliage and terminals of *Myoporum laetum* (a shrub or small tree) and *Myoporum pacificum* (a ground cover) along coastal California. Thrips feeding causes foliage to curl and enclose the pests and can cause shoots to turn brown and die back. Foliage and surfaces under plants become fouled with black specks of thrips' excrement.

Adult myoporum thrips are shiny dark brown to black. The nymphs are white to yellowish and pupae are orangish.

Consider replacing myoporum that are unacceptably affected by this thrips. *Myoporum* cultivars vary in their susceptibility to thrips, and *Myoporum parvifolium* is apparently unaffected by *Klambothrips*. Because *Myoporum laetum* in coastal regions can spread to natural areas, replacing it also helps avoid this invasiveness problem. Minimize pruning of myoporum and avoid or minimize nitrogen fertilization around myoporum, as these practices stimulate susceptible plant growth and can increase thrips damage.

Avoid applying residual, broad-spectrum, contact insecticides as these adversely affect thrips predators, including green lacewings and *Orius* minute pirate bugs that commonly prey on myoporum thrips. Where damage is intolerable, certain systemic insecticides applied to soil or as a bark spray are the most practical pesticidal control; where possible, delay application until after plants have finished flowering to minimize toxicity to pollinators and natural enemies that feed on pollen and nectar. See *Myoporum Thrips Pest Notes* for more information.

MANAGEMENT OF THRIPS

Tolerate thrips damage on established woody plants where the injury is mostly aesthetic. Irrigate appropriately and provide proper cultural care to keep plants healthy and more tolerant of injuries. Avoid overirrigation and especially avoid growth-stimulating fertilization with

Thrips feeding when terminals were young and succulent caused these *Myoporum laetum* leaves and shoots to distort. *Photo:* D. Rosen

quick-release nitrogen products as these practices favor increased thrips population. Replace problem-prone plants with pest-resistant cultivars and native species or other plants well adapted to the local conditions. For example, plants adapted to grow in full sun can be stressed when planted in shady conditions and may be more susceptible to thrips damage. Conserve resident natural enemies by controlling dust and avoiding residual pesticides. When treatment is needed, use insecticides that are least toxic to natural enemies.

Prune and destroy injured and infested terminals when managing a few small specimen plants. Avoid shearing plants. Shearing—clipping the surface of dense foliage to maintain an even surface on formal hedges—stimulates susceptible new growth. New growth increases populations of Cuban laurel thrips, myoporum thrips, and other species, resulting in more damage. Prune by cutting plants just above growing points such as branch crotches and nodes instead of shearing off terminals.

Resident populations of predaceous arthropods, including dustywings, lacewings, predatory thrips, predaceous mites, pirate bugs, and spiders, may help to control plant-feeding thrips. The shiny, pear-shaped adults and nymphs of predatory mites range from translucent to tan to the color of whatever host on which they were feeding. Predatory mites often remain unnoticed because they are tiny and often are most abundant on the underside of leaves or in the interior area of trees.

Damage remains after insect populations have declined or disappeared. To be certain that thrips are present before any treatment, inspect susceptible plant parts or branch beat or shake foliage or flowers to dislodge thrips onto a sheet of paper or other light-colored surface.

Thrips can be difficult to control effectively with insecticides. Before using a pesticide, learn more about the biology of your pest species and the characteristics of available insecticides. Often you will learn chemical control of thrips cannot be effective until the next season when new plant growth develops. Insecticidal soap and oil can provide control if spray directly contacts the thrips. Most thrips are difficult to control effectively with insecticide sprays because they reproduce year-round and feed protected within plant parts that surround them. Certain systemics and the biological insecticide spinosad (which moves short distances into plant tissue) are most likely to be effective if applied before damage becomes extensive. Consult *Thrips Pest Notes* for more information.

Gall Makers

Galls are distorted, sometimes colorful swellings in branches, flowers, leaves, trunks, twigs, or roots. The many potential causes of galls include nematodes, parasitic mistletoe plants, pathogens, and certain plant-feeding insects and mites, as summarized in Table 6-9. Gall-causing invertebrates include certain species discussed in "Adelgids," "Aphids," and "Gall Mites, or Eriophyids." Ceanothus stem gall moth, ficus gall wasp, gall midges, oak gall wasps, and willow gall sawflies are discussed here.

Secretions of invertebrates apparently induce the growth of abnormal plant tissue inside which the invertebrates feed. Many galls harbor a single, legless larva. Other galls may harbor several larvae, some of which may be different species that are predators or parasites of the gall maker.

Most galls are not known to harm trees. Prune and dispose of galls if they are annoying. This may provide control of some species if pruning is done when the immatures are in plant tissue and before the adults begin to emerge.

CEANOTHUS STEM GALL MOTH

Periploca ceanothiella (Cosmopterigidae) larvae cause spindle-shaped swellings on stems, which may slow plant growth and reduce blooming. The small, dark moths emerge during spring and early summer, and females lay eggs on buds and flowers. The larvae tunnel and feed in shoot terminals through fall, then overwinter in galls. They have one generation per year.

Prune and dispose of galled shoots before the moths emerge in spring. Where galling cannot be tolerated, plant less-susceptible *Ceanothus* spp., as listed in Table 6-10, and consider replacing shrubs most susceptible to the ceanothus stem gall moth.

Anystis agilis is a relatively large, long-legged, reddish mite, as with this adult feeding on a thrips larva. It is called whirligig mite because of its fast, circular movements. *Anystis* also feeds on aphids, psocids, mites, and various other tiny invertebrates. *Photo:* J. K. Clark

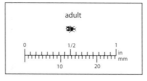

MINUTE PIRATE BUG

adult

0   1/2   1   in
                  mm
    10   20

This adult minute pirate bug (*Orius tricicolor*) and other predators help control many thrips species. Minute pirate bugs also feed on mites, insect eggs, aphids, mealybugs, psyllids, whiteflies, and small caterpillars. *Photo:* J. K. Clark

Eucalyptus are damaged by several introduced species of eulophid gall wasps, including *Selitrichodes globulus*, which severely distorted these blue gum shoots. Where larvae of this gall wasp fed inside, blue gum twigs hang downward and become dry and cracked. *Photo:* G. Arakelian, Los Angeles County Department of Agricultural Commissioner/Weights and Measures

Table 6-9.

Common Causes of Plant Galls and Characteristics for Distinguishing Them.

CAUSE	PLANT PARTS AFFECTED	DISTINGUISHING CHARACTERISTICS
broadleaf mistletoes	branches, trunks	Leafy, green, parasitic plant causes bark and wood to swell where mistletoe attaches to its host.
crown gall	basal stem, root crown; uncommon on more aerial parts	Surface of swelling same color and as firm as surrounding, ungalled tissue. On woody hosts, when gall is cut with a knife, internal tissue is softer than normal wood and lacks the typical pattern of annual growth rings.
ectomycorrhizae	roots	Roots may appear fuzzy or excessively forked or swollen, so they may appear to be diseased. However, there is no root decay associated with mycorrhizae. Their threadlike growth around feeder roots is usually too small to be visible to the naked eye; roots must be specially stained and examined under a microscope to confirm that these beneficial fungi are present.
eriophyid or tarsonemid mites	buds, stems, foliage	Tiny mites, cast skins, or excrement occur in and around distorted tissue. A microscope is needed to clearly discern these mites.
fasciation, bacterial or other causes	crown, stems, foliage, buds	Plant often becomes extensively distorted, and secondary decay may be present. Bacterial fasciation often develops under wet conditions.
fungal leaf galls	foliage, stems	Fungi, including *Exobasidium* and *Taphrina* spp., hosts include azalea, California buckeye, certain stone fruits, and oaks.
insects that bore, mine, or chew inside plant tissue	roots, stems, foliage	Cast skins, emergence holes, frass, tunnels, or the insects themselves (often maggotlike) may be observed in and around galls. Common groups are larvae of beetles, cynipid wasps, gall midges, moths, and sawflies. Certain insects are secondary, attracted to galls originally formed due to other causes.
nitrogen-fixing beneficial bacteria	roots	Galls easily rub off of roots; a thumbnail can easily be pressed into galls. They occur only on plants in certain groups, especially legumes.
root knot nematodes	roots	Surface of galls is as firm as surrounding ungalled tissue; swellings cannot be rubbed off; cutting into gall may reveal pinhead-sized, shiny, white female nematodes inside, which look like tiny, pear-shaped pearls that are visible if galls are inspected through a hand lens.
sucking insects	buds, stems, terminals	Aphids, thrips, true bugs, or other insects or their cast skins or excrement are present. Sucking insects commonly feed on new growth, which later distorts.
woolly aphids	roots, bark, foliage	Whitish flocculent material around galls during certain times of the year; aphids or cast skins commonly visible.

Terminals become swollen when ceanothus stem gall moth larvae feed inside. Infested shoots can become stunted (top left) and die back (bottom). If intolerable, prune off galled shoots during October to February or replace susceptible plants with ceanothus species resistant to stem gall moths. *Photo:* J. K. Clark

Table 6-10.

Relative Susceptibility of *Ceanothus* spp. and Cultivars to the Ceanothus Stem Gall Moth.

NOT INFESTED
americanus, Blue Cloud, *cuneatus, foliosus, gloriosus, gloriosus* var. *exaltatus, impressus, insularis, jepsonii*, Lester Rowntree, *masonii, megacarpus, papillosus, parryi, prostratus, purpureus, ramulosus* var. *fascicularis, C. rigidus* 'Albus,' *spinosus, verrucosus*
LIGHTLY INFESTED
aboreus, Concha, *diversifolius, integerrimus, lemmonii, leucodermis, lobbianus* var. *oliganthus*, Mary Lake, Mountain Haze, Royal Blue, Sierra Blue, Treasure Island
MODERATELY INFESTED
cyaneus, Marie Simon, Ray Hartman, *thyrsiflorus*
SEVERELY INFESTED
griseus, griseus var. *horizontalis*

Source: Munro 1963.

FICUS GALL WASP

Josephiella microcarpae (Agaonidae) infests only Indian laurel fig. The larvae of this leaf-galling wasp feed within brown, green, or reddish swellings or warty blisters in leaves. Infested leaves turn yellow, curl, and drop prematurely. Tree health and survival appears not to be threatened by leaf galls, but infestations are aesthetically undesirable. Because young leaves are especially susceptible, damage can be extensive on the new growth stimulated by recent pruning, especially on frequently pruned Indian laurel (e.g., topiary plants).

Adults are $1/10$-inch-long, dark brown wasps with pale yellow appendages. The pale larvae feed inside young terminals, causing expanding leaves to develop warty blisters about $1/20$ to $2/3$ inch in diameter on the upper and lower leaf surfaces. After pupating, the adult leaves a round exit hole in old galls.

Avoid this damage by planting other *Ficus* spp., including creeping fig (*F. pumila*), Moreton Bay fig (*F. macrophylla*), rubber plant (*F. elastica*), and weeping Chinese banyan (*F. benjamina*). Pruning, bagging, and disposing of young infested leaves helps reduce local gall wasp populations, but pruning can stimulate excess succulent new growth, which attracts egg-laying gall wasps. Promptly raking and disposing of fallen leaves may help reduce infestations by removing some wasp pupae before they emerge as adults. Where damage is intolerable, certain systemic insecticides may help reduce future damage.

GALL MIDGES

There are hundreds of species of gall midges (Cecidomyiidae), also called gall gnats or gall flies. Each species feeds inside only one or a few related hosts, including coyote brush, dogwood, Douglas-fir, ficus, honey locust, oak, pine, and willow. Certain midges (e.g., *Aphidoletes* and *Feltiella* spp.) are predators of small insects and mites.

Adult gall midges are tiny, delicate flies, often with long, slender antennae. Their tiny white, yellowish, reddish, or orange maggots bore into plant tissue and feed inside the distorted tissue that forms. Most species have several generations per year.

FICUS GALL MIDGE

Horidiplosis ficifolii infests Indian laurel fig in southern California. Larval feeding causes browning and curling of leaves resembling damage caused by Cuban laurel thrips, discussed earlier. Each brown lesion is about $1/5$ inch long and elliptical shaped, so this pest is also called the ficus eyespot midge. Manual removal of infested leaves can provide some control. To confirm the cause of damage, seal the removed leaves in a clear plastic bag. Within days, tiny orange larvae will crawl out of the lesions into the bag. After pupating, delicate, orange adults with one pair of wings will emerge.

HONEYLOCUST POD GALL MIDGE

Dasineura gleditchiae larvae cause honey locust leaflets to form brown, green, or reddish galls. Each distorted leaflet contains one to several pinkish white maggots. Heavily infested foliage turns brown and drops prematurely, leaving parts of branches leafless. Galls are most apparent during spring; by midsummer egg laying ceases, and plants often continue to produce new leaves that do not develop galls.

Adults emerge from overwintering pupae in soil and most egg laying occurs during spring. Larvae require succulent new terminals in order to feed and cause galls. Mature larvae drop from galls or crawl to pupate near the soil surface. There are about six generations per year.

Several species of parasitic and predaceous wasps feed on this pest. Because the gall midge has many generations and the larvae and pupae occur protected in galls or soil, this insect is not easily controlled with insecticides. On small plants, narrow-range oil sprayed to thoroughly cover terminals at intervals during about March and April kills gall midge eggs and can substantially reduce damage.

Established trees are rarely, if ever, killed by the galling, so damage can be tolerated. Avoid the Sunburst honey locust, which has bright yellow spring foliage. Sunburst readily defoliates in response to drought or temperature changes, as well as gall midge damage. Consider planting the Shademaster cultivar of honey locust, which appears to be less susceptible. Black locust and other *Robinia* spp. are not infested by this midge.

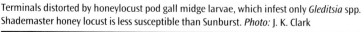

Terminals distorted by honeylocust pod gall midge larvae, which infest only *Gleditsia* spp. Shademaster honey locust is less susceptible than Sunburst. *Photo:* J. K. Clark

Warty blisters are conspicuous on both the upper and lower leaf surfaces of Indian laurel infested with ficus gall wasps. This pest infests only *Ficus microcarpa* and appears not to seriously threaten plant health. However, new growth stimulated by pruning apparently attracts the egg-laying gall wasps, so frequently or recently pruned plants can become extensively galled. *Photo:* J. K. Clark

OAK GALL WASPS

Gall wasps (Cynipidae) are especially abundant and varied on oaks. The size, shape, and color of the galls depends on the species of wasp and oak and the plant part infested. Several weeks or months after egg laying by the female wasp, a gall forms as one or more white larvae feed inside.

Adult cynipids are usually purple or black, small, stout insects that have clear wings with few veins. Oak gall wasps alternate between one sexual and one asexual generation each year. The appearance of galls and wasps and the part of the plant attacked often differ between generations. The abundance of cynipids and their new galls varies greatly from year to year, but galls on woody parts and evergreen foliage can persist long after the wasps have emerged.

Gall wasps are naturally killed by a complex of fungi, parasites, predators, and competing insects (primarily moth larvae and other wasps) that live within galls. Cynipids are also preyed upon by various small insectivorous birds, woodpeckers, and small mammals. Most gall wasps do not seriously damage oaks. In most situations, no controls have been shown to be effective and no management is necessary.

Certain gall wasps contribute to annoying problems, such as leaf or twig dieback or messy dripping. Certain *Andricus, Disholcaspis,* and *Dryocosmus* spp. gall wasps induce plants to secrete copious sticky nectar. If nectar-producing gall wasps are the problem, controlling ants can gradually (over several years) help to dramatically reduce nectar-inducing gall wasp populations by allowing gall wasp competitors and parasites to become more abundant.

Provide oaks with proper environmental conditions and good cultural care. No other management is recommended because gall wasps can be very difficult to control, and usually they do not seriously harm oaks. Certain leaf-galling species may be reduced by insecticides with systemic or translaminar (leaf-penetrating) activity applied in late winter to spring, or by a broad-spectrum, residual insecticide foli-

A stem gall made by *Callirhytis perdens* wasp larvae feeding in California black oak. Although sometimes unsightly, galls are generally harmless to oaks. *Photo:* J. K. Clark

Galls are diverse and sometimes quite attractive, like these of the pink spined turban gall wasp (*Antron douglasii*) and oak cone gall wasp (*Andricus kingi*) on the underside of valley oak leaves. *Photo:* J. K. Clark

SPINED TURBAN OAK GALL

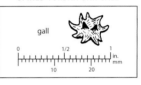

gall

0   1/2   1
|_____| in.
          mm
    10   20

The jumping oak gall wasp (*Neuroterus saltatorius*) causes discolored spots on the upper side of oak leaves, and these seedlike deformations on the underside, especially on valley oak. Galls drop from leaves in summer. Huge numbers of galls may be seen hopping an inch or more above the ground because of movement by a tiny wasp larva inside each gall. *Photo:* J. K. Clark

JUMPING OAK GALL

gall   o

0   1/2   1
|_____| in.
          mm
    10   20

Young oak apple galls are green. Galls turn reddish or brown, then black after larvae of the California gallfly (*Andricus californicus*), which are actually wasps, mature and leave the galls. *Photo:* J. K. Clark

Brown, round growths on valley oak twigs infested with larvae of a gall wasp (*Disholcaspis washingtonensis*). This oak cynipid and certain other gall wasps cause plants to secrete sticky nectar, which attracts ants. Because ants help protect nectar-producing gall wasps from parasites and predators, selectively controlling ants may gradually (over several years) help to reduce gall-related dripping. *Photo:* J. K. Clark

ar spray timed to coincide with leaf bud break or the early expansion of new leaves. However, gall wasps are unlikely to be well controlled by any single treatment; sometimes insecticide provides no control at all, and residual, broad-spectrum foliar sprays should be avoided because they can induce outbreaks of other oak pests and wash off and pollute surface water.

TWOHORNED OAK GALL WASP

Dryocosmus dubiosus feeds in leaves of coast live oak and interior live oak. Each larva feeds inside a brown, 1/8-inch-diameter, oblong gall that has two protuberances, hence the name "twohorned." Leaf margins beyond each gall often discolor and die. High populations of the gall wasp cause extensive leaf scorching, and some infested leaves drop prematurely. Although aesthetically undesirable, these oak gall wasps do not seriously threaten tree health.

Damage from twohorned oak galls is often confused with damage from fungi discussed in "Oak Branch Canker and Dieback" and "Oak Twig Blight" in Chapter 5 and beetles discussed below in "Oak Twig Girdlers." However, these other pests cause the entire leaf to die, and they kill many adjacent leaves in a group. Twohorned oak

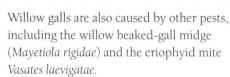

Willow leaf gall sawfly larvae feed within these red, berrylike galls. Several dozen sawfly species cause harmless, globular swellings on foliage, terminals, or twigs of various willows. *Photo:* J. K. Clark

gall wasps kill only portions of each leaf; each leaf remains partly green, and discolored leaves will be on the same twig as unaffected leaves.

The first-generation wasp larvae develop during late winter or early spring inside catkins. Affected flower parts turn into small reddish or brown galls, that when cut reveal pale larvae feeding inside. Second-generation larvae form galls in the veins of leaves. The wasps overwinter as pupae in galls, which often drop from the leaf. Gall wasp abundance varies naturally from year to year, although these population fluctuations may not be obvious because live oaks can retain galled and scorched foliage for several years.

WILLOW GALL SAWFLIES

Various *Euura* and *Pontania* spp. sawflies (Tenthredinidae) induce galls on the foliage, terminals, or twigs of willows. Adults are stout, broad-waisted wasps, and their larvae are legless and yellow to whitish.

Willow galls are also caused by other pests, including the willow beaked-gall midge (*Mayetiola rigidae*) and the eriophyid mite *Vasates laevigatae.*

The willow leaf gall sawfly (*Pontania pacifica*) commonly causes reddish berry-like galls on *Salix lasiolepis* foliage. Each gall is globular or elongate and about 1/3 inch long. Females insert their eggs in young willow leaves, and the leaf forms a gall at each location where a maggotlike larva feeds inside. When mature, larvae emerge from the galls and pupate on the ground. Willow leaf gall sawfly males are shiny black, and females are dull reddish. This insect apparently has several generations per year.

Willow leaf gall sawflies do not harm plants. No controls are recommended or known. The larvae of several wasps, and at least one weevil and moth, feed on the sawfly larvae or on the gall tissue, causing the sawflies to die. A *Eurytoma* sp. wasp appears especially important in reducing galling by *P. pacifica.*

Foliage Miners

Foliage-mining insects cause off-color patches, sinuous trails, or holes in leaves. Portions of a leaf or patches of foliage may turn yellow or brown and die. Tiny larvae may be seen dropping from foliage on silken threads. Severe infestations can slow plant growth, but established woody plants tolerate extensive foliage mining and are rarely, if ever, killed by these insects.

Larvae that feed inside of leaves, needles, shoots, or buds include certain beetles, flies, sawflies, and (most commonly) various small moths. In some species, only certain instars mine plants, and their

A twohorned oak gall on the underside of a coast live oak leaf vein. Where a gall wasp develops on a vein, portions of the leaf beyond the gall are killed. That cynipids caused this dieback may not be obvious because the scorched leaves remain on tree after the galls have dropped. *Photo:* J. K. Clark

Coast live oak leaf margins killed by twohorned oak galls. Other pests, such as twig blight fungi and twig girdler beetles, can be misdiagnosed as the cause of dead oak leaves. These other pests kill the entire leaf, and leaves are entirely killed in groups. Twohorned oak gall wasps kill only portions of each leaf, and partly brown leaves will be on the same twig as unaffected leaves. *Photo:* J. K. Clark

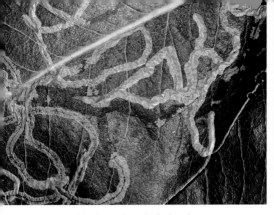

Citrus peelminer (*Marmara gulosa*) larvae caused these pale, winding mines in oleander leaves and rough, brown discoloring of stem bark (photo right). This damage is harmless to most hosts of this insect (e.g., fruit and nut trees, willows, and various garden vegetables). However, in citrus this species is a pest because it mines the rind, making fruit unmarketable. A different moth, the citrus leafminer (*Phyllocnistis citrella*), mines citrus foliage.
Photo: D. Rosen

...inuous trails caused by the madrone leaf miner (*Marmara arbutiella*). A yellowish moth larva is visible in its mine at the right edge of the leaf. This leaf tunneling is harmless to plants. *Photo:* J. K. Clark

...other larval stages externally chew the surface of buds, leaves, or other plant parts. Mature larvae may pupate inside leaves or in cocoons on bark, foliage, or soil.

The host species and characteristics of the larva's damage help to identify the insect species. The cypress tip miner, pine tip moth, oak ribbed casemaker, and shield bearers are discussed here.

Provide proper cultural care to keep plants vigorous. Prune out and dispose of foliage infested with immature leafminers to restore the plant's aesthetic appearance and provide some control. Plant resistant species or cultivars to avoid damage by some foliage miners. Other species can be effectively controlled by natural enemies; conserve these beneficials by avoiding broad-spectrum, residual, contact insecticides. Systemic insecticide applied during egg laying or when larvae are very young can reduce damage by certain species.

CYPRESS TIP MINERS

Cypress tip miner (*Argyresthia cupressella*, Yponomeutidae) is the most common of several *Argyresthia* spp. that tunnel in foliage of arborvitae, coast redwood, cypress, and juniper, mostly along the Pacific Coast. Infested foliage turns yellow in early winter and brown by late winter or early spring, then recovers its green color during the spring and summer. Although unsightly, even severe infestations apparently do not kill plants.

The adult tip miners are silvery tan moths. They can be found mostly from March through May in southern California and during April and May in northern Cali-

fornia. Females lay scalelike eggs on green tips. The yellow to green larvae feed inside branch tips until late winter or spring, then spin slender, white, silken cocoons externally between the twiglets. There is one generation per year.

Do not confuse damage with that discussed earlier under "Minute Cypress Scale" and later under "Juniper Twig Girdler." Where foliage mining cannot be tolerated, consider replacing plants especially susceptible to the cypress tip miner (Table 6-11). High populations of cypress tip miners and their damage can be reduced by applying a broad-spectrum, residual insecticide when adult moths are active. Beginning in early spring, examine foliage tips for the cocoons. When these appear, vigorously shake foliage and watch to see if

This silvery tan adult cypress tip miner moth emerged from the silken cocoon at the base of the dead leaflet. If damage is intolerable and control is planned, carefully monitor foliage and spray when new cocoons and adults are present. Spraying will not be effective unless it is very well timed, based on regular monitoring. *Photo:* J. K. Clark

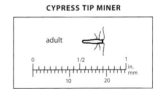

CYPRESS TIP MINER

adult

This juniper is brown because cypress tip miner larvae are feeding inside foliage, causing numerous, scattered leaflets to die. Planting less-susceptible juniper cultivars virtually eliminates tip miner problems. *Photo:* J. K. Clark

Table 6-11.

Susceptibility of Juniper (*Juniperus* spp.) Cultivars and Other Cupressaceae in California to the Cypress Tip Miner.

LEAST SUSCEPTIBLE, ½ TO 2½ TIP MINERS
Juniperus chinensis 'Kaizuka,' *J. scopulorum* 'Erecta Glauca,' *J. chinensis* var. *sargentii* 'Glauca,' *Thuja plicata*
MODERATELY SUSCEPTIBLE, 5 TO 8 TIP MINERS
J. chinensis 'Pfitzerana Aurea,' *J. sabina* 'Arcadia,' *J. sabina* 'Tamariscifolia,' *J. virginiana* 'Prostrata'
MORE SUSCEPTIBLE, 13 TO 19 TIP MINERS
Chamaecyparis lawsoniana 'Allumii,' *Juniperus chinensis* 'Pfitzerana,' *J. chinensis* 'Robust Green,' *J. virginiana* 'Cupressifolia'
MOST SUSCEPTIBLE, ABOUT 40 TIP MINERS
Thuja occidentalis

Numbers are tip miner larvae per 100 grams (3.5 oz) of foliage. *Source:* Koehler and Moore 1983.

silvery tan, tiny moths fly from the foliage. One application to foliage can be made when a large number of tip moths appear, between March and May. This reduces foliage browning the next year.

OAK RIBBED CASEMAKER

Bucculatrix albertiella (Lyonetiidae) is probably the most common species among many leafminers on oaks. Larval feeding between leaf veins causes damaged foliage to appear translucent. Its white, cigar-shaped cocoons with distinct longitudinal ribs are found on oak bark and leaves and on nearby plants and objects. Adults are mottled white, brown, and black moths. First-instar larvae mine inside the leaf. Later instars feed externally on the lower leaf surface and are olive-green with rows of pale spots. There are two generations per year. No control is generally warranted to protect the health of oaks.

PINE TIP MOTHS

Rhyacionia spp. (Tortricidae) larvae feed on pine shoot terminals and are pests primarily in southern California and on Monterey pine, especially when it is planted away from the coast. Nantucket pine tip moth (*R. frustrana*) and (near the coast) the Monterey pine tip moth (*R. pasadenana*) are the major pests, attacking most pine species with two or three needles per bundle. Ponderosa pine tip moth (*R. zozana*) mostly damages young, planted pines grown away from the coast.

Dead brown or reddish shoots very noticeable from a distance are the most obvious symptom of infestation. Silk webbing and boring frass are visible during close inspection of infested terminals. Tip moth damage to the central growing terminal can significantly alter tree shape, causing crooked or forked stems. Because other causes can produce similar damage, consult Table 5-10 in Chapter 5 and "Tree and Shrub Pest Tables" at the end of this book for help in distinguishing among insects and pathogens affecting pines.

The adults are reddish brown moths with silver-gray markings, and they begin emerging in January in southern California. The female lays tiny eggs singly on the new growth tips, where the young larvae feed on or inside the base of needles or buds. Older larvae are yellow to pale brown with a dark head and are found in terminals. They cover shoot tips with webbing and cause pitch to exude as they bore into the shoots and feed. Summer-

The adult Nantucket pine tip moth is reddish brown with s gray markings. If attempting control by pruning or spraying monitoring adults with pheromone traps to time control a will greatly improve management efficacy. *Photo:* J. K. Clar

A Nantucket pine tip moth pupa exposed in the terminal w it fed as a larva, killing needles and causing sap to exude. A ichneumonid wasp introduced to parasitize these pupae h improved the vigor and appearance of pines susceptible to moth. *Photo:* J. K. Clark

NANTUCKET PINE TIP MOTH

pupa   adult

0   1/2   1 in.
                          mm
      10   20

generation larvae pupate in the tips; over-wintering pupae commonly occur in the litter. The Nantucket pine tip moth has about four generations per year in southern California. Monterey pine tip moth and Ponderosa pine tip moth have one or two generations each year.

Management of Tip Moths. If little or no damage can be tolerated, do not plant species particularly susceptible to the tip moths (Table 6-12). Consider replacing susceptible species like Monterey pine if their performance is unacceptable. Alternatively, tolerate damage; pines appear to be less affected as they mature. Provide proper cultural care, especially appropriate irrigation, to increase tree tolerance to damage. The introduced ichneumonid wasp *Campoplex frustranae* parasitizes

Brown to translucent patches in coast live oak caused by oak ribbed casemaker larvae feeding inside the leaf and on its underside. No control is recommended, as this insect is apparently harmless to trees. *Photo:* J. K. Clark

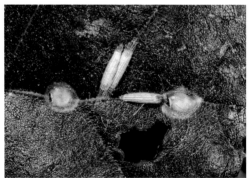

Oak ribbed casemaker spins a roundish web as it changes into a later instar, larger larva. The last instar forms a white, elongate cocoon with distinct longitudinal ribs from which the tiny adult moth emerges. The first instar feeds inside the leaf, and older larvae chew on the undersurface of leaves of various oak species. *Photo:* J. K. Clark

OAK RIBBED CASEMAKER

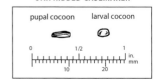

pupal cocoon   larval cocoon

0   1/2   1 in.
                          mm
      10   20

Table 6-12.

Relative Susceptibility of Pines (*Pinus* spp.) in California to the Nantucket Pine Tip Moth.

NOT INFESTED
amandii, attenuata, bungeana, canariensis, caribaea, coulteri, edulis, gerardiana, monophylla, montezumae, mugo, nigra, palustris, pinaster, pinea, thunbergiana, torreyana
UNDER 25% OF TIPS INFESTED
flexilis, halepensis, jeffreyi, oocarpa, ponderosa, rigida, taeda
ABOUT 30–40% OF TIPS INFESTED
brutia, cembroides, muricata, patula, pseudostrobus, roxburghii, sabiniana, sylvestris
ABOUT 50–85% OF TIPS INFESTED
contorta, densiflora, echinata, glabra, insularis, radiata, resinosa, virginiana

Source: Scriven and Luck 1980.

Nantucket pine tip moth pupae and has reduced moth populations in many locations, resulting in improved vigor and healthier appearance of infested pines.

Prune off and dispose of infested tips from October through January to prevent overwintering moths from emerging. If high-value pines must be pruned during other times, monitor adults with pheromone-baited traps and prune between the peaks representing each generation of moth flights to help eliminate larvae and pupae in tips.

To kill young larvae, apply a systemic insecticide or spray foliage with a broad-spectrum, residual insecticide soon after moths are observed flying during each generation. However, insecticide usually is not justified unless trees are of especially high aesthetic value. Insecticide spray can cause spider mite outbreaks and wash off and pollute surface waters. Spraying also kills natural enemies, which can provide substantial biological control.

Monitoring High-Value Pines. Insect activity varies with temperatures, and foliar spraying and pruning are more effective if they are performed between the peak flight periods. To monitor moth flights, hang one trap baited with tip moth pheromone at chest height in the outer canopy of each of two trees at least 50 feet apart from January through September. On properties with extensive pine plantings, deploy additional traps at approximately 500-foot intervals. Inspect each trap daily. Remove any debris and count, record, and remove any moths in the traps. Replace traps when the sticky surface becomes dirty and replace the pheromone lure about every 4 weeks or as recommended by suppliers.

If pruning or spraying a residual foliar insecticide, do so 10 to 14 days after the beginning of an overall decline in the number of first-generation moths are caught, or about 1 week after the peak number of each subsequent generation of moths. If spraying foliage, thorough coverage of all branch tips and the treetop is important for good control.

For the most effective control timing, use degree-day monitoring and moth traps. Spray foliage or prune at about 1,233 degree-days after the beginning of each moth generation, using a lower development threshold temperature of 42°F and an upper threshold of 99°F. The first generation begins in the spring when the first moth is caught. Subsequent generations start when moth catches first begin increasing after a dramatic decline in numbers from the previous peak.

SHIELD BEARERS

Coptodisca spp. (Heliozelidae) larvae feed within the leaves of apple, cottonwood, crape myrtle, oak, madrone, manzanita, poplar, and strawberry tree. At maturity, each larva cuts out a round or oval area of foliage approximately ¼ inch long. This portion of the leaf drops to the ground or is carried (the "shield") by the larva and fastened to bark, and the insect pupates inside this case. High populations of shield bearers cause leaves to develop numerous holes, and leaves may become partially necrotic and drop prematurely.

The madrone shield bearer (*C. arbutiella*) mines the foliage of madrone, manzanita, and strawberry tree. The ⅕-inch-long, silvery female moth emerges in the early spring and lays eggs, which apparently do not hatch until the fall, when the larvae begin mining. In the late winter, the mature black larva cuts an elliptical disk of foliage from the leaf and pupates inside it. The madrone shield bearer has one generation per year.

The cottonwood or poplar shield bearer (*Coptodisca* sp.) can be abundant on Fremont cottonwood and other *Populus* spp. It apparently can have three generations per year.

Plants tolerate even abundant leafminers, and no control is necessary to protect plant survival. Pick and dispose of infested leaves on small plants if damage cannot be tolerated.

Elliptical holes in manzanita foliage caused by the madrone shield bearer. This damage occurs when, after mining leaves, each mature larva cuts and carries away a portion of the leaf to pupate inside. *Photo:* J. K. Clark

Cottonwood shield bearer (*Coptodisca* sp.) larvae feeding in poplar leaf mines. No control is recommended or demonstrated to be effective for this pest on landscape trees. *Photo:* J. K. Clark

Twig, Branch, and Trunk Boring Insects

Insects that bore in heartwood, sapwood (xylem), cambium, and phloem tissues can create holes in bark and wood. Their chewing in tissue can cause sawdustlike powder or stains or oozing liquid on trunks and limbs, and discolored, wilted foliage that may drop prematurely. Many boring insects chew obvious holes in bark that help to identify the type of borer (see Figure 6-15). Be aware that horizontal rows of shallow holes in bark are typically caused by sapsucker birds and are commonly mistaken for holes of boring insects. For information on this problem, see the *Woodpeckers Pest Notes.*

Bark- and wood-boring insects can be serious pests because they damage plants' vascular system, weaken limbs and trunks, and can kill branches or entire plants. However, many borers are secondary pests that develop because trees lack proper care or are already stressed or dying from other causes.

Boring insects discussed here include bark beetles, carpenter ants, carpenter-worm, clearwing moths, flatheaded borers, juniper twig girdler, roundheaded borers, termites, and wood wasps. For information on borers that do not commonly infest live plants, see publications such as *Carpenter Bees Pest Notes, Drywood Termites Pest Notes, Wood-Boring Beetles in Homes Pest Notes,* and *Wood Preservation.*

MANAGEMENT OF WOOD BORERS

Prevention is the most effective method of managing bark- and wood-boring insects; in many instances it is the only available control. Plant only species well adapted for the conditions at that location. Prepare the site, plant properly, and protect tree roots and trunks from injuries, as discussed in Chapter 3. Protect trees from sunburn or sunscald and other abiotic disorders, as detailed in Chapter 4. Learn the cultural requirements of plants and provide proper care to keep them growing vigorously. Healthy plants are less likely to be attacked and are better able to survive the damage from a few boring insects. Appropriate irrigation is particularly important; plants are seriously damaged by irrigating too frequently or by providing too little water. Once plants are established, irrigate beneath their canopy but not near the trunk.

Remove hazardous limbs whenever they appear, but otherwise avoid pruning during the season when egg-laying adult borers are active and attracted to fresh tree wounds, as summarized in Table 6-13. Prune out and dispose of borers restricted to a few limbs and remove and dispose of dying trees to prevent boring insects from emerging and attacking other nearby trees. Replace old, declining trees so that future generations may enjoy mature trees.

Do not pile unseasoned, freshly cut wood near woody landscape plants. Freshly cut wood or trees that are dying or recently dead provide an abundant breeding source for many wood-boring pests. Solarize recently cut wood to kill beetles infesting it. Seal wood beneath heavy clear plastic tarps in a sunny location for several months through the warm season; after this time, any emerged borers will have died and the dry wood will no longer be suitable for most borers. Tightly seal the tarp edges (e.g., with soil) to prevent any insects from escaping. Instead of one large pile, use several smaller wood piles to promote quicker drying.

Except for pruning and general cultural practices that improve tree vigor, little can be done to control most boring insects beneath bark once trees have been attacked. Insecticides will not kill the boring larval stages of most insects.

With certain borers, especially valuable trees may be protected from further attack by a well-timed insecticide appli-

These horizontal rows of shallow holes in tree bark were caused by sapsucker birds, which are usually not managed. Do not confuse bird feeding with damage from boring insects. Unlike these holes from feeding on sap, when woodpeckers feed on boring insects they typically flake off patches of bark, leaving relatively large, rough-edged holes, as pictured later under "Bronze Birch Borer." *Photo:* J. K. Clark

Many bark- and wood-boring insects can continue to develop in cut wood and emerge to attack nearby trees. If logs may be infested with borers, tightly seal cut wood under UV-resistant clear plastic in a sunny location for several months to kill many borers infesti the wood and trap adults that emerge. Use the effective solarization techniques described in *Controlling Bark Beetles in Wood Residue and Firewood.* Be aware that certain species m not be well controlled. *Photo:* P. Svihra

Table 6-13.

Pruning To Control Wood-Boring Insects.

HOST PLANT	DAMAGE SYMPTOMS	PEST SPECIES	WHEN TO PRUNE
alder, ash, birch, cherry, oak, peach, plum, poplar, sycamore, willow[1,2]	dying limbs, rough bark, sawdustlike frass	clearwing moths (e.g., *Paranthrene*, *Podosesia*, and *Synanthedon* spp.)	fall through early winter (but avoid any pruning if possible)
alder, white[3]	D-shaped holes in bark, sudden midsummer branch death	flatheaded borer (*Agrilus burkei*)	fall through March (also remove bleeding, externally stained limbs)
birch[3]	D-shaped holes in bark, branches die back	bronze birch borer (*Agrilus anxius*)	fall through April
eucalyptus[3]	foliage off-color and light green to yellowish, branches die, broad galleries beneath bark	eucalyptus longhorned borers (*Phoracantha* spp.)	December or January (in southern California), November through March (in northern California)
juniper[3]	scattered dying or dead branches, entire plant is never dead	juniper twig girdler (*Periploca nigra*)	fall through April
pines[3]	pitchy masses 1 to several inches in diameter on limbs or trunk	Pitch moths (e.g., *Synanthedon sequoia*)	October to February (alternatively, excise pitch masses anytime and kill the larvae underneath)
many deciduous trees[2,3]	BB-shot-sized holes in bark, limbs decline or die	shothole borer (*Scolytus rugulosus*)	summer and fall
many[2,3]	branches die back, entire plant may be killed	Pacific flatheaded borer (*Chrysobothris mali*)	fall through February
many, including fruit trees, oak, and pine[1,2]	pitch tubes, sawdustlike frass, foliage yellows, limbs or top die back	Bark beetles (e.g., *Dendroctonus*, *Ips*, and *Scolytus* spp.)	November through February

Bury, chip, debark, solarize (by tightly sealing under clear plastic in a sunny location), or otherwise properly dispose of pruned wood. Diseases controlled by pruning are listed in Table 5-2.

1. Properly timed pruning reduces the likelihood that plants will become borer infested. Pruning is of uncertain benefit to already-infested plants but may reduce infestation of nearby hosts.

2. Stone fruits that are especially susceptible to Eutypa canker and dieback, such as apricot and cherry, should be pruned during July or August.

3. Properly timed pruning reduces the likelihood that plants will become infested. Timed pruning also removes and kills immature insects before pests emerge as adults and reinfest plants.

Adapted from Svihra 1994.

cation. However, unless trees are monitored regularly so that borer activity and attack can be detected early, and the type of boring insect is identified so an effective treatment method can be selected, pesticide use is likely to be too late and ineffective. Do not substitute insecticide applications for proper cultural care. Most borers are attracted to trees that are already unhealthy from other causes, and relying only on insecticide can allow trees to die from those other causes unless the growing environment and cultural practices are improved.

AMBROSIA BEETLES

Adult ambrosia beetles (Scolytinae) bore relatively deep into wood, where females lay eggs and the larvae feed on fungi the adults introduce. Most species are secondary pests that attack trees that are severely stressed or already dying. Several introduced ambrosia beetle species are serious pests. The beetles themselves resemble the true bark beetles discussed below. Ambrosia beetles and bark beetles sometimes occur together in the same tree, as with oaks.

Management of Ambrosia Beetles.
Provide trees with optimal cultural care (especially appropriate irrigation) to reduce their susceptibility to wood borers. Promptly cut down dead and dying trees and do not move potentially infested wood. Chip or grind the wood or cut it into logs and leave it on-site and solarize it in a sunny location and tightly covered with clear plastic. Because ambrosia beetles and plant-pathogenic fungi generally occur together in a tree, thoroughly clean and sterilize cutting tools with a registered disinfectant (e.g., bleach) before reusing them on other hosts. The cultural and chemical controls discussed later under "Management of Bark Beetles" can also be effective for ambrosia beetles.

These toothpicklike protrusions from bark are the frass (excrement) of the granulate ambrosia beetle, which aggressively attacks newly planted broadleaf trees. If you find suspected granulate ambrosia beetles or these pinnacles of excrement in California, report it to the county agricultural commissioner. *Photo:* L. Lazarus, North Carolina Division of Forest Resources, Bugwood.org

Ambrosia beetle frass usually is whitish, the color of sapwood and heartwood where ambrosia beetles tunnel and introduce fungi they eat. The oak ambrosia beetle adult shown here produces this frass and reproduces in large numbers in severely stressed and dying oaks. *Photo:* J. K. Clark

White, powdery, dry, or wet patches of discoloration on bar‍ are generally the first obvious symptoms of infestation by t‍ polyphagous shothole borer. This beetle introduces a *Fusar‍* fungus that is killing avocado, box elder, coast live oak, and‍ other trees in southern California. *Photo:* UC Riverside, Eska‍

GRANULATE AMBROSIA BEETLE

The introduced granulate ambrosia beetle (*Xylosandrus crassiusculus*), or Asian ambrosia beetle, aggressively attacks the trunks of young broadleaf trees in landscapes and nurseries. It is recognizable by its distinctive excrement—solidified pinnacles of frass that resemble short round toothpicks protruding from bark. These fragile structures are readily washed away by irrigation sprinklers or rain and may not be observed. It is thus possible to be unaware of an infestation of this beetle until the tree suddenly dies.

Granulate ambrosia beetle occurs in Oregon and the eastern United States. If you find a tree with suspected granulate ambrosia beetles in California, report it to the county agricultural commissioner.

OAK AMBROSIA AND BARK BEETLES

Oak ambrosia beetles (*Monarthrum dentiger, M. scutellare*) and the western oak bark beetle (*Pseudopityophthorus pubipennis*) infest injured and unhealthy tanoak and true oaks, such as those dying from *Phytophthora ramorum*, the cause of sudden oak death (SOD). Western oak bark beetle adults produce reddish frass on the surface of bark because they bore and lay eggs, mostly in oak's reddish inner bark (phloem). This distinguishes them from oak ambrosia beetles, which produce whitish frass around their tunnel openings because ambrosia beetles bore deeper into wood.

In northern California the beetles have about two generations per year. Most first-generation adults fly during about March, and the next generation emerges in about September. Because beetles develop at different rates and from eggs laid at various times, some adults may be flying and seeking stressed oaks to lay their eggs in anytime from about March through October.

POLYPHAGOUS SHOTHOLE BORER

An *Euwallacea* sp. ambrosia beetle in combination with a *Fusarium* sp. fungus kills avocado, big leaf maple, box elder, California sycamore, coast live oak, and certain other tree species. The complete host range is unknown. The adult beetle attacks a wide range of trees and spreads the *Fusarium* fungus on which its larvae feed, but only a subset of tree species are suitable to both infection by the fungus (discussed in "Fusarium Dieback" in Chapter 5) and feeding by the beetle larvae.

The first obvious symptoms of a polyphagous shothole borer infestation are patches of discoloration on bark, which can be dark, dry, water-soaked, or oily looking. These lesions are sometimes surrounded by wet discoloration (e.g., on oak and sycamore) or white crusty or powdery exudate (especially on avocado). Within bark lesions are tiny holes about $\frac{1}{30}$ inch in diameter made by the adult beetles. Scraping off the bark reveals dark, discolored wood. As the beetles and diseased wood become more abundant, and subsequent generations of beetles reinfest the host, foliage yellows and wilts,

branches die back, and the tree may die. Cutting the trunk in cross-section reveals black to brown discoloration up to about $1\frac{1}{2}$ inches deep into the wood around where the ambrosia beetles and fungus developed. Adults of this *Euwallacea* sp. beetle are about $\frac{1}{16}$ inch long and closely resemble the redbay ambrosia beetle (*Xyleborus glabratus*) discussed under "Laurel Wilt" in Chapter 5. See the website cisr.ucr.edu for more information.

BARK BEETLES

Most tree species are susceptible to bark beetles (Scolytinae), but whitewood conifers (e.g., firs, pines, and spruces) and certain broadleaves (black walnuts, elms, oaks, and stone fruits and other *Prunus* spp.) are especially susceptible to being infested and killed.

Adults chew into bark, where the females lay eggs and the larvae feed between inner bark and sapwood (the phloem-cambial region) of branches, trunks, twigs, or roots. Most bark beetles are secondary pests that attack trees weakened by factors such as

A red turpentine beetle pitch tube on Monterey pine. Pines tolerate a few red turpentine beetles, but attack often indicates that pines need improved growing conditions and better cultural care. *Photo:* J. K. Clark

Sawdustlike frass in lower bark crevices, spiderwebs, or on the ground like this indicate that bark beetles have been feeding in the tree. *Photo:* C. S. Koehler

drought stress, root diseases, wounds, or lack of proper cultural care. Some introduced species attack apparently healthy trees, and many species when very abundant can mass-attack a healthy tree. Many bark beetles vector fungi, and the beetle boring in combination with the pathogen infection and adverse abiotic conditions may kill the infested trees.

Bark beetle feeding damage to trees' vascular system causes foliage to fade to yellowish green, then to tan and red-brown. These symptoms of tree decline and death, which can also be due to other causes, are summarized in the "Tree and Shrub Pest Tables" and in Table 5-10 for pines. There may also be sawdustlike beetle frass in bark crevices or on the ground, reddish to whitish granular material (pitch tubes) or oozing pitch on the trunk, and holes in bark the size of BB shot ($\frac{1}{16}$ inch).

IDENTIFICATION AND BIOLOGY

Adults are small, cylindrical, hard-bodied beetles about the size of rice grains. Most species are dark red, brown, or black. Larvae and pupae of most species are cream or white, robust, and grublike and may have a dark head.

The species of tree attacked, the location of damage on bark, and the pattern of galleries (tunnels) revealed by peeling off a portion of bark help in identifying the bark beetle species present (Figure 6-14, Table 6-14). For example, red turpentine beetle adults mine out a wide cave-like gallery where their larvae feed as a group. In most other bark beetle species, the larvae feed

Table 6-14.

Bark Beetles Most Common in Landscapes.

SPECIES	TREES AFFECTED	COMMENTS
cedar, cypress, and redwood bark beetles (*Phloeosinus* spp.)	arborvitae, coast redwood, cypress, false cypress (*Chamaecyparis*), and juniper	tunnels form centipede pattern on inner bark and wood surface; adults feed on and kill twigs; egg-laying females attracted to trunk of dead or dying trees
elm bark beetles (*Scolytus multistriatus, S. schevyrewi*)	elms	tunnels form centipede pattern on inner bark and wood surface; cause BB-shot-size holes in bark; females lay eggs on injured or weakened trees and vector Dutch elm disease fungi
fir engraver (*Scolytus ventralis*)	red and white fir	adults excavate deep and long, two-armed galleries across the grain of the sapwood
mountain pine beetle (*Dendroctonus ponderosae*), Jeffrey pine beetle (*D. jeffreyi*)	pine, frequently Jeffrey, lodgepole, and sugar pine	adults make J-shaped gallery, from which larvae make many side tunnels; attack midtrunk of large trees, from 5 to about 30 ft aboveground
oak ambrosia beetles (*Monarthrum* spp.), oak bark beetles (*Pseudopityophthorus* spp.)	buckeye, oak, and tanoak	cause bleeding, frothy, bubbling holes in bark with boring dust that is whitish (*Monarthrum* spp.) or reddish (*Pseudopityophthorus* spp.); attack stressed trees
pine engravers (*Ips emarginatus, I. mexicanus, I. paraconfusus, I. pini, and I. plastographus*)	pine	attack pines near the top; adults often make wishbone-shaped (three-branched) or four-branched gallery, from which larvae make many side tunnels
pine twig beetles (*Pityophthorus carmeli, P. nitidulus, P. setosus*)	pine	chew on lateral shoots and twigs and mine the pith; spread pitch canker disease fungus
red turpentine beetle (*Dendroctonus valens*)	pine mostly, rarely in larch, spruce, and white fir	attacks lowest 10 ft of trunk and the large roots; pitch tubes appear on bark; by itself rarely kills tree
shothole borer (*Scolytus rugulosus*)	English laurel, fruit trees, hawthorn, and other broadleaf trees	infestation indicated by limb dieback and woody parts with gumming, boring dust, BB-shot-size holes
walnut beetle (*Pityophthorus juglandis*)	walnut, especially California black	adults make tiny holes in bark; spread lethal thousand cankers disease fungus
western pine beetle (*Dendroctonus brevicomis*)	Coulter and Ponderosa pines	attacks midtrunk, then spreads up and down; larvae chew winding galleries in inner bark and pupate in outer bark; attacks in conjunction with other pests

Cypress bark beetle feeding caused these dead terminals called flagging. This twig damage does not threaten tree survival and no control is recommended. *Photo:* J. K. Clark

Larvae of most bark beetles are small, pale grubs that mine inner bark and outer wood. Pines are often killed when infested by these California fivespined ips (*Ips paraconfusus*) or other *Ips* spp. larvae. Usually the trees were severely stressed and already unhealthy before beetles attacked them. *Photo:* J. K. Clark

individually in mines that lead away from the adult gallery.

New bark beetle species are periodically introduced, including the Mediterranean pine engraver (*Orthotomicus erosus*) and redhaired pine bark beetle (*Hylurgus ligniperda*) infesting Mediterranean pines (e.g., Aleppo and Italian stone pines). For more information, see *Bark Beetles Pest Notes*.

Centipedelike pattern: European elm bark beetle egg galleries

Winding mazelike egg galleries in pine: *Dendroctonus* spp. e.g., western pine beetle

Frass and pitch tubes on bark exterior at base of tree: many bark beetles, e.g., red turpentine beetle

Tuning-fork pattern of egg galleries: *Ips* spp., e.g., fivespined engraver beetle

Figure 6-14.
Bark beetle adults bore a tunnel, or gallery, in which they lay eggs, typically between inner bark (phloem) and the surface of sapwood (xylem). Larvae hatch and bore side tunnels. Tunnels packed with frass (excrement) are shown in black, while open portions of galleries are white. Adults of some species cause pitch tubes on bark. The location of tunnels and pitch tubes (e.g., height above ground), the pattern of adult and larval galleries, whether adult tunnels are open versus filled with frass, and the host plant help to identify the species of bark beetles. Illustrations by C. M. Dewees, A. Child, and D. Kidd.

CEDAR, CYPRESS, AND REDWOOD BARK BEETLES

Phloeosinus spp. are the only bark beetles commonly found in arborvitae, coast redwood, false cypress (*Chamaecyparis*), incense-cedar, and cypress. They primarily chew and kill twigs, causing dead tips, or "flags," of faded foliage hanging on the tree. Though sometimes unsightly, this twig damage does not threaten tree survival. Beetles also attack the limbs and trunk of stressed or injured trees, which may become girdled and die; on vigorous trees the limbs and trunk are rarely, if ever, attacked.

Phloeosinus spp. adults mine twigs mostly about 6 inches back from the tips. In unhealthy and declining limbs and trunks, females lay eggs along a relatively straight tunnel they chew between inner bark and the wood surface. The emerging larvae tunnel across (perpendicular to) the grain, and the pattern of adult and larval engravings resemble a centipede on the surfaces of inner bark and outer wood.

To help prevent infestations, provide trees proper cultural care, especially adequate water. Prune out and dispose of dead limbs where these beetles breed.

ELM BARK BEETLES

At least two introduced bark beetles (*Scolytus multistriatus* and *S. schevyrewi*) are serious pests of elms because they vector elm-killing *Ophiostoma* spp. fungi that cause Dutch elm disease (DED). Elm bark beetles occur throughout California, but DED in California has been limited to planted elms in the areas around the San Francisco Bay and Sacramento.

Elm bark beetle adults feed by chewing tender bark on elm twigs, occasionally causing twigs to die and drop. If the twig-feeding beetle emerged from elm wood infected with *Ophiostoma*, the beetle infects the elm with fungus. Infected limbs gradually decline, and the dying elm attracts egg-laying bark beetle females.

The female beetles bore through the bark and chew a straight egg-laying tunnel 1 to 2 inches long parallel to the grain of the wood, except that on Chinese elm the gallery is more meandering. Each larva bores a tunnel at right angles to the parent gallery. When inner bark and sapwood are separated, both surfaces appear engraved with a centipedelike pattern. After the larvae mature and pupate, they emerge as adults contaminated with fungi, which they spread to other elms.

The elm bark beetle adult feeds by chewing tender twigs. This feeding can introduce Dutch elm disease fungi that cause trees to decline and attract bark beetles to lay eggs. Fungi and beetle larvae together kill the trees. The larvae mature into adults contaminated with fungi, which they spread to other trees, continuing the cycle. *Photo:* P. Svihra

The female European elm bark beetle bores a linear tunnel where bark meets wood. The frass-filled side tunnels were made by larvae. *Photo:* J. K. Clark

To prevent Dutch elm disease, provide trees with optimal cultural care and good growing conditions. Prune off and dispose of dying and dead elm branches and properly dispose of elm wood. If limb death is due to DED, entirely removing the tree is the best way to minimize pathogen spread to other elms. Plant resistant elm cultivars as discussed in "Dutch Elm Disease" in Chapter 5.

PINE ENGRAVERS

Ips spp. are important pests of pines. *Ips pini* especially attacks Monterey pine planted outside its native coastal range. *Ips* populations develop in freshly cut or broken off pine limbs and in limbs dying or recently killed by other causes. In live pines, engraver beetles primarily bore in the treetops, although Monterey pine is often attacked along the entire trunk. *Ips* boring causes foliage color to fade and treetops to die within a few weeks or over several months or longer.

Engraver beetles make clean (frass-free) adult galleries that consist of three (wishbone-shaped) or four branches from a central cell. The larvae feed individually in mines at right angles from the adult (egg-laying) gallery. In southern California, engraver beetles can attack any time of year; in colder areas, the adults begin to emerge and attack pines in the late winter or spring.

RED TURPENTINE BEETLE

Dendroctonus valens infests mostly pines and, rarely, larch, spruce, and white fir. Adult boring produces pinkish brown to white pitch tubes, usually on the root crown and lower 10 feet of trunks. Brown, reddish, or white granular material collects around the tree base and in bark crevices. Pitch tubes higher on the trunk are usually caused by other species of bark beetles. Pitch blobs or ooze larger than about 1 inch are likely due to clearwing moths discussed in the section "Pitch Moths" or to injuries or pathogens.

Red turpentine beetle adults are reddish brown and larger than most other bark beetles, up to ¼ inch long. Larvae grow up to ⅜ inch long, feed in a group, and chew an excrement-filled cavity that grows to several square inches or more.

Red turpentine beetle alone is generally not a serious pest, and vigorous trees usually survive their boring. But attack by red turpentine beetle can indicate that trees are declining or stressed from old age, unfavorable growing conditions, injuries, or disease. Weakened trees, especially

RED TURPENTINE BEETLE

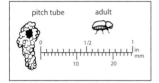

pitch tube   adult

This red turpentine beetle resembles adults of other bark beetle species, except that red turpentine beetle is somewhat larger than most other bark beetles. *Photo:* J. K. Clark

Monterey pines, are at high risk of being attacked and killed by a combination of other pests and abiotic factors in addition to bark beetles.

SHOTHOLE BORER

Scolytus rugulosus attacks the limbs and trunks of various woody broadleaves, including apple, English laurel, hawthorn, pear, and stone fruit trees. Clear to brownish sap often exudes from holes in bark bored by the egg-laying adult females. As larvae feed under bark, leaf color fades

Small holes in bark, some weeping sap, indicate that bark beetles have been boring into this tree. *Photo: J. K. Clark*

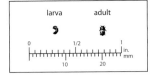
and foliage wilts. After pupation, the dark brown to black, $\frac{1}{10}$-inch-long adults emerge during spring and early summer and may be seen crawling on bark. Emerging adults leave numerous holes the size of buckshot or BB shot ($\frac{1}{16}$ inch) in bark without exuding sap.

Shothole borers attack trees weakened by root diseases, sunburn, insufficient irrigation, or stressful infestations of other pests. Populations that build up in weakened or injured trees may then move to attack nearby healthy limbs or trunks.

WALNUT TWIG BEETLE

Pityophthorus juglandis is a native species that bores into twigs on overshaded or stressed walnut trees and historically was not a pest. However, since 2008, *Juglans* spp., mostly California native black walnuts (*J. californica* and *J. hindsii*) have been dying from infection by a newly recognized fungus (*Geosmithia morbida*) spread by the $\frac{1}{15}$-inch-long adult beetles. In addition to small branch terminals, *P. juglandis* now bores into walnut limbs and trunks. At each site of boring, a small patch of phloem is killed and a small oozing canker develops, as pictured under "Walnut Thousand Cankers Disease" in Chapter 5. As beetle boring and cankered

tissue become abundant, the tree declines and dies.

If you find English walnut with suspected thousand canker disease, report it to the county agricultural commissioner or UC Cooperative Extension office.

WESTERN PINE BEETLE

Dendroctonus brevicomis attacks mostly pines stressed from old age, severe drought, root rot, or other injuries. More-vigorous trees may be attacked and killed due to the large numbers of beetles that emerge from nearby infested trees. Inconspicuous pitch tubes and boring dust appear on the main trunk of successfully attacked trees, often on the main trunk well above ground. Adult galleries are much-branched and run both laterally and longitudinally, crossing other galleries in a mazelike pattern. *Dendroctonus* spp. bark beetles pack at least part of the central egg-laying gallery with frass.

MANAGEMENT OF BARK BEETLES

To manage most species of bark beetles, reduce tree stress, keep plants healthy, and protect them from damage, as discussed in Chapters 3 and 4. Good cultural care, especially appropriate irrigation and keeping roots healthy, is particularly important to keep trees vigorous and resistant to most bark beetles.

Remove some trees when they become overly crowded, as bark beetles commonly kill trees in a group that outgrows the available resources (soil moisture and sunlight). Replace old, declining trees

with a mix of species well adapted to local conditions. Where *Ips* engravers or red turpentine beetle have been a problem, plant nonhost species such as deciduous trees, coast redwood, and atlas or deodar cedar.

Prune off infested limbs from November through February. Remove and dispose of dying trees so that boring insects do not emerge and attack other nearby trees. Obtain and use wood locally, and do not move firewood and logs between counties, as they can be infested with wood borers and spread them. Do not pile unseasoned, freshly cut wood near woody landscape plants, as beetles can breed or emerge from the wood. Solarize recently cut wood for at least several months by tightly sealing it beneath clear plastic in a sunny location.

Except for pruning off infested parts, nothing can be done to control bark beetle larvae beneath bark once trees have been attacked. Individual or small groups of uninfested trees may be protected from new infestations by properly spraying their bark with a residual, contact insecticide before trees become attacked. However, unless tree health is improved, there may be little or no long-term benefit to spraying trees that will continue to be at high risk of dying.

To protect uninfested pines from *Dendroctonus* spp. and certain broadleaf trees from fungus-vectoring bark beetles, have a professional applicator thoroughly drench the main trunk, exposed root crown (collar), and bark of large limbs with a residual insecticide (e.g., carbamate or pyrethroid) registered for that situation. In most cases, season-long control is provided by one thorough application per year in late winter to early spring in warm areas of the state and in late spring at cooler areas and higher elevations. If treating for red turpentine beetle, spray only the exposed root crown and lower 10 feet of the trunk, which is where this species attacks. Spraying to prevent attack by engraver beetles, shothole borer, and most other bark beetles is less common and generally not recommended. For more information, see the *Bark Beetles Pest Notes*.

FLATHEADED BORERS, OR METALLIC WOOD BORERS

Flatheaded borers (Buprestidae) can cause portions of bark to ooze, crack, and die. Limbs or entire trees, especially young trees, may be killed. Most flatheaded borer species do not attack vigorous plants.

Depending on the species, adults can be the color of bark (brown to grayish) or metallic and shiny (black, blue, coppery, or green). The streamlined bodies are flattened, elongate or oval, and typically have longitudinal grooves on the wing covers. The larvae bore beneath the bark and sometimes into the wood, and are found in winding tunnels filled with frass. Larvae typically are broad and flat in the front and narrow and tapered toward the rear. After pupation, the adult of many flatheaded borer species leaves a characteristic

Flatheaded borer larvae have a distinctly segmented body. Some species have several enlarged segments behind their head, as with this Pacific flatheaded borer. *Photo:* J. K. Clark

Rough, broken bark on a young apple tree trunk infested with Pacific flatheaded borer larvae. Protecting trees from injury and providing good cultural care are the primary controls for preventing most borer infestations. *Photo:* J. K. Clark

D-shaped emergence hole in bark and wood (see Figure 6-15).

MANAGEMENT OF FLATHEADED BORERS

Prevention, as detailed above in "Management of Wood Borers," is by far the most effective strategy for flatheaded borers. Correctly plant species that are well adapted to that site. Provide proper cultural care to keep trees vigorous. Protect trees from injuries, sunburn, and other abiotic causes of damage, as discussed in Chapters 3 and 4.

Prune out and dispose of dying limbs where borers breed, but prune when adult borers are not active, as listed in Table 6-13. Promptly remove dead trees. Do not pile freshly cut wood near trees. Solarize logs beneath clear plastic in the sun to prevent beetles from emerging from cut wood and attacking nearby hosts.

Larvae of the larger, shallow-boring species sometimes can be killed by probing tunnels with a sharp wire. This method is practical only in a small infestation, and it is often difficult to know whether the wire has reached and killed the larva.

Insecticides are generally not recommended for flatheaded borers. Although spraying the trunk and limbs with a residual, contact insecticide can be somewhat effective in protecting trees from egg laying by adult borers, spraying large trees in landscapes is often not practical, and there are concerns regarding drift and pesticide movement that contaminate surface waters. Preventive sprays are difficult to effectively time and may need to be repeated. The contact insecticide prod-

ucts available to home users are not effective against borers, and a licensed pest control operator must make the application. Systemic insecticides have not been shown to control most borer species in California. Insecticides cannot substitute for the lack of proper tree care. Borers are attracted to trees that are unhealthy from other causes, and trees may still die unless the growing environment and cultural practices are improved.

BRONZE BIRCH BORER

Agrilus anxius infests only birch. The many similar *Agrilus* spp. include bronze poplar borer (*A. liragus*), which infests *Populus* spp. in California, and emerald ash borer (*A. planipennis*), which occurs in the eastern and midwestern United States and has killed tens of thousands of ash trees.

Foliage of borer-infested trees turns pale green, yellow, and then brown. Leaves drop prematurely and scattered limbs die. Swollen ridges develop on branches and the trunk where larvae tunnel beneath bark. Liquid oozes from bark, creating stained blotches. Severely infested trees die prematurely.

Adult bronze birch borers are dark metallic, coppery beetles about ⅜ inch long. They cause inconspicuous chewing on leaf edges of alder, birch, and poplar, then females lay eggs on sunny portions of birch bark. Larvae chew into phloem and xylem tissue, packing frass (excrement) in the tunnels. Mature larvae tunnel in the outer sapwood, where they overwinter. In spring, larvae chew an oblong chamber beneath bark and pupate. Adults emerge in northern California beginning in April, leaving D-shaped, smooth-edged holes in bark.

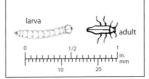

larva

adult

Adult flatheaded borers leave a D-shaped or elliptical emergence hole in bark, as with this bronze birch borer adult. The irregular holes (right and bottom left) are from woodpecker feeding on borers. *Photo:* J. K. Clark

Swollen ridges where borer larvae are feeding under the surface of wood, revealed by cutting and peeling back bark. Chiseling into ridges would reveal bronze birch borer larvae in frass-packed tunnels. White-barked birch are adapted to moist soils and are especially prone to borer attack when planted in hot, dry interior areas of California. *Photo:* J. K. Clark

Table 6-15.

Bronze Birch Borer Management Methods.

- Do not plant birch where they will be exposed to heat stress, such as the south and west sides of buildings.
- Avoid planting white-barked birch species (e.g., *Betula jacquemontii, B. pendula*), which are especially susceptible to bronze birch borer.
- If planting birch, consider using *B. alleghaniensis, B. lenta,* or *B. nigra*. Borers attack all birch, but these species appear to be less susceptible in California.
- Plant at the most suitable location, such as the east side of buildings where heat stress is typically less severe.
- Apply mulch to retain soil moisture and keep roots cooler.
- Provide sufficient irrigation, such as deep watering about every 1 to 2 weeks during prolonged dry weather.
- Do not fertilize birch unless nutrient deficiency has been confidently diagnosed as the cause of poor growth.
- Regularly inspect birch for bark swelling and wet stains from April through September; promptly prune off these borer-infested limbs and remove hazardous limbs whenever they appear. Otherwise, do not prune birch from April through August, the season when adult borers fly and are attracted to fresh pruning wounds.
- Promptly remove dead and dying birch. Chip the wood or cut it into logs and solarize it under clear plastic to prevent adults from emerging and infesting nearby birch.

Adapted from Svihra and Duckles 1999.

Management of Bronze Birch Borer.
To prevent infestations, provide trees with a good growing environment and proper cultural practices, as summarized in Table 6-15. Birch require more water than most other trees, so good soil conditions, healthy roots, and appropriate irrigation are especially critical. Avoid routine fertilization, which can increase birch's demand for water, thereby increasing tree stress. Birch typically have an especially short life span in California in hot, dry interior areas. Avoid planting white-barked birch species, which are especially susceptible to borers. Plant species that are adapted to the site's conditions.

Regularly monitor birches from mid-April through summer, looking for bark swelling and stains. Prune off and dispose of any infested branches before mid-September. Except for removing hazardous and borer-infested limbs whenever they appear, prune birches during fall through March when adult bronze birch borers are not flying.

If the critical cultural and environmental practices summarized in Table 6-15 have been followed and bronze birch borer is a local problem, spraying birch thoroughly about twice (once during April and again about late May) with a pyrethroid insecticide can kill adults and larvae hatching from eggs before larvae bore into wood. A professional applicator usually must be hired to apply an effective product using equipment that provides good spray coverage. Application of a systemic insecticide may provide some control if trees are relatively healthy and contain relatively few or no borers. Treatment can be warranted where nearby birch are infested and a source of borers. Severely infested trees will die regardless of treatment. Promptly remove birch that are highly infested, dying, or dead, and chip the wood or cut and solarize it under plastic to prevent borers from emerging to attack other birches.

FLATHEADED ALDER BORER

Agrilus burkei attacks only alders, primarily white alder (*Alnus rhombifolia*) in poorly irrigated landscapes. Wet spots, dark staining, and gnarled, ridged growth often appear on the bark of infested hosts. Adult emergence holes, often D shaped and about ⅛ inch in diameter, are left in bark. Infested limbs die and trees can be killed.

The metallic blue adult emerges from infested alders in April and May and feeds on foliage, although this slight damage is usually not apparent. In the late spring, females lay whitish egg masses on the trunk and main branches of alders, especially on stressed or declining trees. The larvae bore into the bark and chew winding tunnels through cambial tissue. There is one generation each year.

Management of Alder Borers. Provide proper cultural care and protect trees from injury, as discussed earlier in "Management of Wood Borers." Provide white alders with adequate irrigation throughout their lives if planted where drought prevails; these trees are native to sites near year-round surface water. Consider replacing problem trees and planting more borer-resistant species, such as black alder (*A. glutinosa*).

During late summer or fall, prune out and dispose of all branches showing bleeding, swelling, dieback, or other evidence of larval infestation. Avoid pruning white alder anytime between March and the end of May, as egg-laying adult beetles are attracted then to recent pruning wounds.

Beginning the end of March, inspect leaves for adult feeding holes and look for adult beetles during mid to late afternoon by examining foliage and branch beating. Foliage and wood can be thoroughly sprayed with a residual insecticide when adults are active, usually about mid-April. Do not substitute insecticide applications for proper cultural care or trees are still likely to die.

GOLDSPOTTED OAK BORER

The goldspotted oak borer, or GSOB (*Agrilus auroguttatus*), was introduced into

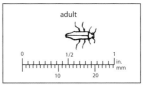

FLATHEADED ALDER BORER

Wet, dark staining on alder where flatheaded borer larvae are feeding under bark. Wood borer damage to vascular tissue also causes yellow foliage, fewer leaves than normal, and death of some branches or the entire tree. *Photo:* J. K. Clark

San Diego County in about 2002. It has killed thousands of oak trees in southern California and is expected to continue to spread. This flatheaded borer attacks only oaks, mostly those with a trunk diameter of about 1 foot or more. GSOB prefers mature oaks in the red oak group, especially California black oak and coast live oak. Canyon live oak can also be infested, while Engelmann oak is lightly attacked and not seriously damaged.

Signs and symptoms of GSOB infestation include

- D-shaped, adult emergence holes ⅙ inch wide in bark, like that in Figure 6-15 and the photograph in "Bronze Birch Borer"
- bark with black or red staining or cracking on the trunk or large branches
- bark chipped away and shallow holes where woodpeckers fed on GSOB larvae and pupae
- canopy thinning (fewer leaves than normal) and branch dieback

Larvae of GSOB feed in groups at the interface of the inner bark (phloem) and sapwood (outer xylem), causing dark discoloration and girdling of trunks and large limbs and tree death. The legless, whitish larvae grow up to ¾ inch long, have distinct segments, and two pincer-like spines at their rear. Larvae are present

mostly from late spring through fall, then overwinter as inactive mature larvae (pre-pupae). Pupae occur in outer bark mostly during spring and early summer and are oblong and pale to dark colored. The adult is about ½ inch long with six orange to yellowish spots on its dark green wing covers. Adults are present mostly from late spring to early fall. GSOB apparently has one generation per year.

Limiting GSOB's spread is the primary management method. Leave cut oaks on site unless wood has dried for 2 years, or grind wood into 3-inch particles. Unless handled in these or other ways that effectively kill borers, GSOB larvae and pupae can continue to develop and live in green wood under bark in dead oaks for months. When infested logs or firewood are moved, this serious pest is spread to new locations when the adult beetles emerge.

Remove dying oaks that are a hazard to people or property. Do not leave oak wood with intact bark in the open near uninfested oak trees. Debark logs, or screen or seal them tightly in a clear tarp for 2 years to help prevent emerging beetles from attacking nearby oaks (see "Management of Wood Borers" earlier for discussion of solarization). To avoid this borer problem, plant Engelmann oak or nonhost tree species. At the time of this writing, it is unknown whether residual bark sprays or

Bark bleeding and staining on an oak trunk infested with goldspotted oak borers. This introduced flatheaded borer has killed thousands of oak trees in southern California. It spreads to new locations when people move borer-infested oak logs and firewood. *Photo:* T. W. Coleman, USDA Forest Service

Older life stages of the goldspotted oak borer. From left: fourth-instar larva, two prepupae (hairpin configuration and constricted form), pupa, and adult. *Photo:* M. I. Jones

Figure 6-15.

Emergence holes of common wood borers illustrated here with those that bore into oaks.

	EMERGENCE HOLE			
Shape	Size (largest dimension)	Location	Borer family	Species example
round 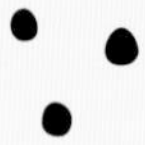	⅟₂₀ – ⅟₁₂ in	main trunk or branches, varying by beetle species and host tree	bark and ambrosia beetles (Scolytinae)	oak ambrosia beetles (*Monarthrum* spp.), western oak bark beetle (*Pseudopityophthoru* spp.)
round 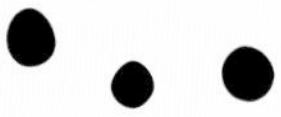	⅛ – ⅙ in	smaller branches	false powderpost beetles (Bostrichidae)	lead cable borer (*Scobicia declivis*)
D shape 	⅙ in	lower trunk	flatheaded borers (Buprestidae)	goldspotted oak bor (*Agrilus auroguttatus*)
round 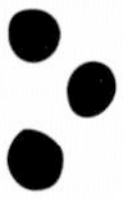	⅕ in	near cracked, gnarled, or otherwise injured bark	clearwing moths (Sesiidae)	western sycamore borer (*Synanthedon resplendens*)
oblong-crescent 	⅕ – ½ in	trunk and larger branches	flatheaded borers (Buprestidae)	appletree and Pacifi flatheaded borers (*Chrysobothris* spp.)
oval 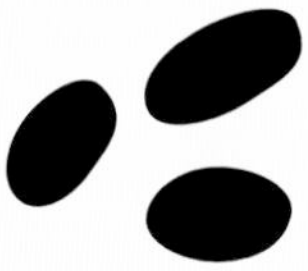	⅕ – ⅖ in	trunk, especially around wounds from sunburn, mechanical damage, or fire	roundheaded borers (Cerambycidae)	oak cordwood bore (*Xylotrechus nauticu*)
round 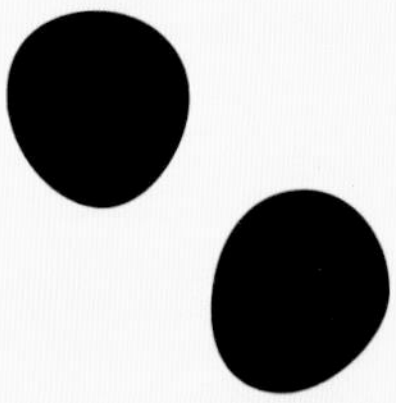	½ in	trunk and larger branches, especially in older trees	carpenterworm moth (Cossidae)	carpenterworm (*Prionoxystus robini*)

Adapted from Flint et al. 2013.

systemic insecticides can protect oaks from GSOB if applied early when trees are uninfested or only lightly infested. See *Goldspotted Oak Borer Field Identification Guide, Goldspotted Oak Borer Pest Notes,* and the website gsob.org for more information.

OAK TWIG GIRDLERS

Agrilus angelicus infests only small twigs of tanoak and true oaks, especially live oaks in southern California. Infestation by this flatheaded borer is indicated by scattered patches of whitish brown leaves throughout the canopy. Leaves are dead but have not been chewed and exhibit no surface scraping. Although its damage can be unsightly, oak twig girdler does not significantly harm trees.

The dark coppery brown adult emerges around June in coastal areas and in May farther inland. The eggs are laid singly on young twigs into which the whitish larva chews a linear mine several inches long for 3 to 6 months. It then mines spirally and girdles the twig, causing terminal foliage to die and turn brown. During the next season, it mines a foot or more down the branch toward the trunk, causing more foliage to die. It then bores back outward in the center of the branch, pupates just under the surface, and emerges as an adult. The development from egg to adult requires about 2 years.

Twig girdler damage may be confused with the disease symptoms discussed under "Oak Twig Blight" and "Oak Branch Canker and Dieback" in Chapter 5. To distinguish the twig girdler, peel back the bark of the larger twig at the junction of live and dead foliage and look for the girdler's flattened, spiral tunnel, possibly containing coarse, dark brown frass and a larva. Similar damage is also caused by less common species, including at least two roundheaded borers (*Aneflomorpha lineare* and *Styloxus fulleri*) and a false powder post beetle (*Scobicia suturalis*).

Provide trees with proper cultural care and good growing conditions. Drought-weakened oaks are especially prone to twig girdler attack. Because urbanization often reduces the natural availability of soil moisture, even drought-adapted trees may warrant infrequent, deep watering, as discussed in "Irrigation" in Chapter 3.

Prune infested branches to restore the oak's aesthetic quality. At least six species of parasitic wasps attack oak twig girdler; however, their importance in biological control has not been documented. Because damage by this pest does not affect tree survival and insecticide application kills natural enemies, no further management is recommended.

PACIFIC FLATHEADED BORER

Chrysobothris mali is attracted to diseased, stressed, or injured trunks or limbs and can infest dozens of woody landscape species. The adult has a dark bronze or gray body and mottled coppery wing covers. Adults chew the base of twigs and leaf petioles and when unusually abundant can partially defoliate young trees. Females lay eggs in bark wounds, such as those caused by sunburn, pruning cuts, staking trunks, or where the rootstock and scion are grafted together.

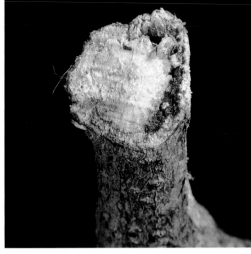

The frass-filled tunnel of an oak twig girdler larva. You may need to cut under bark at the junction between twigs with dead and green leaves to reveal this damage and distinguish twig girdler from problems such as oak twig blight fungi. Photo: J. K. Clark

Larvae excavate cambium beneath bark and may bore deeper into wood as they mature. Frothy, white sap may exude around boring sites and bark may crack. Mature larvae are pale yellow and enlarged just behind the head. Larvae form creamy white to dark pupae just under the bark surface in the spring. Adults emerge and fly from April through August. There is one generation per year.

To prevent damage from this pest, follow the recommendations earlier in "Management of Wood Borers."

PACIFIC FLATHEADED BORER

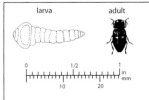

larva   adult

The adult Pacific flatheaded borer is dark bronze, gray, or a mottled coppery color. It is easily overlooked when resting or laying eggs on bark. Photo: J. K. Clark

LONGHORNED BEETLES, OR ROUNDHEADED WOOD BORERS

Most species of longhorned beetles (Cerambycidae), also called roundheaded wood borers, are relatively innocuous, attacking injured or stressed trees. The larvae chew inner bark and sometimes in the wood of limbs, trunks, and main roots. For example, the native poplar borer (*Saperda calcarata*) attacks unhealthy aspen, cottonwood, poplar, and willow. Important introduced species include eucalyptus longhorned beetles and, in the eastern United States, the Asian longhorned beetle.

Oval-shaped emergence holes in bark and stains or oozing liquid on limbs or

A last-instar eucalyptus longhorned borer exposed in wood. Each larva excavates a chamber and packs the tunnel behind it with frass. To manage eucalyptus longhorned borer damage, irrigate appropriately and properly handle infested firewood and logs. *Photo: P. Svihra*

trunks are common symptoms of roundheaded borer infestation. Exceptions include *Prionus californicus*, which chews the surface of roots, then bores inside roots and pupates in soil, so no bark damage or emergence holes may be evident aboveground. With any roundheaded borer species, foliage may discolor and wilt, limbs may die back, and branches or entire plants may die.

Adult roundheaded borers are medium to large, elongate, cylindrical beetles with long antennae. Their color varies by species from dull brownish or gray to brightly colored or a mix of pale and dark colors in a banded or spotted pattern. The larvae typically are creamy white, elongate, distinctly segmented, and cylindrical in cross-section.

MANAGEMENT OF LONGHORNED BEETLES

Prevention, as discussed earlier under "Management of Wood Borers," is the primary method of preventing longhorned beetle damage. Plant trees that are well adapted to the site conditions, provide proper cultural care, and protect trees from injuries, as discussed in Chapters 3 and 4. Prune off dead limbs during the season when adult borers are not active. Promptly

remove dead or dying trees. Do not store freshly cut wood near trees; debark or solarize logs beneath clear plastic in the sun to prevent beetles from emerging from and attacking nearby hosts.

Larvae of the larger, shallow-boring species can sometimes be killed by probing tunnels with a sharp wire. This method is practical only in a small infestation, and it is often difficult to know whether the wire has penetrated the tunnel far enough to kill the larva.

ASIAN LONGHORNED BEETLE

Anoplophora glabripennis aggressively attacks apparently healthy broadleaf trees. Tens of thousands of infested trees have been cut down in the eastern United States, but this pest is not established in California as of this writing.

The boring larvae cause liquid ooze and sawdustlike boring frass on bark and dying limbs, although the initially low populations of borers in a tree can easily be overlooked. After larvae pupate, the emerging adults leave a relatively large, smooth-edged hole about 1/2 inch in diameter in bark. Mature larvae, pupae, and adults are about 3/4 to 1 1/4 inches long. Adults have very long antennae with alternating

Table 6-16.

Characteristics for Distinguishing the Exotic Asian Longhorned Beetle from Some Native Longhorned Borers.

DISTINGUISHING CHARACTERS	COMMON NAME (SCIENTIFIC NAME)
Thorax entirely black. Antennae black-and-white banded. Wing covers have many irregular white blotches. Overall appearance smooth and shiny.	Asian longhorned beetle (*Anoplophora glabripennis*)
Thorax whitish, with one large black spot. Wing covers have several broad bands that are black and whitish.	banded alder borer (*Rosalia funebris*)
Thorax has white bands. Antennae all black.	cottonwood borer (*Plectrodera scalator*)
Wing covers have distinct white dot where they meet thorax. Overall appearance bumpy, rough, and dull colored.	Oregon fir sawyer (*Monochamus oregonensis*), or whitespotted pine sawyer (*M. scutellatus*)

Report suspected Asian longhorned beetles in California to the county agricultural commissioner.

This banded alder borer *(Rosalia funebris)*, or California laurel borer, feeds in declining broadleaf trees, including ash, California bay, and willow. Adults have wing covers with several black and pale bands (gray, light bluish, or white) and a pale thorax with one large black spot. This native species is not a significant pest, but it superficially resembles the invasive Asian longhorned beetle. *Photo: J. K. Clark*

The Asian longhorned beetle has killed thousands of apparently healthy trees since its introduction into the eastern United States. This exotic pest has about 20 variable-sized, whitish spots on its shiny, black body. Take any suspected Asian longhorned beetles to the county agricultural commissioner for identification. *Photo:* K. R. Law, USDA APHIS PPQ, Bugwood.org

IDENTIFICATION AND BIOLOGY

These *Phoracantha* spp. closely resemble each other. Adults have antennae as long as or longer than the body, and the body is a mix of dark (blackish and brown) and light (yellow to cream) shiny colors. By comparison, *P. recurva* have mostly cream to yellowish wing covers and long, dense, golden hairs along the underside of their antennae, while *P. semipunctata* have predominantly dark-colored wing covers and antennae with few to no distinct hairs on the underside. The adults and mature larvae are ¾ to 1¼ inches long.

Adult females lay groups of 3 to 30 eggs under loose bark. Hatching larvae bore into the cambium, and each pale larva chews a gradually widening gallery up to several feet long that can girdle a tree. Larvae can continue to develop after wood is cut, and after pupating emerge from logs or trees and fly to other eucalyptus, where females oviposit. Depending on temperatures and host quality, *Phoracantha* require 3 to 9 months to complete one generation.

MANAGEMENT

Many eucalyptus native to wetter areas of Australia have been planted in California

bands of black and white, and the body is smooth, shiny, and black with many irregular white blotches. The introduced Asian longhorned beetle can be distinguished from several native species by color differences, as summarized in Table 6-16.

Submit any suspected Asian longhorned beetles to the county agricultural commissioner. To avoid spreading boring pests, do not move firewood or logs to other counties.

EUCALYPTUS LONGHORNED BEETLES

Two introduced longhorned beetles have killed large numbers of eucalyptus trees in California. *Phoracantha semipunctata,* introduced in the 1980s, is now under good biological control. However, biological control has been less effective against a subsequently introduced species (*P. recurva*), which, in combination with several other introduced pests, can stress and kill eucalyptus.

Freshly cut wood, dying limbs, and eucalyptus suffering from stress, especially drought stress, attract these borers. Attacked trees may produce copious

amounts of resin, and the treetops, branches, or entire trees may be killed. Resprouting may occur from the tree base. Trees that receive proper cultural care, especially appropriate irrigation, are not readily attacked, including the more susceptible species.

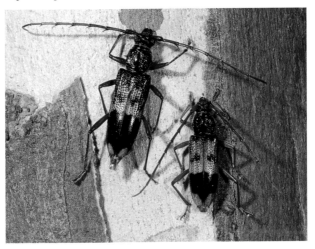

Two *Phoracantha* spp. longhorned beetles attack eucalyptus in California. In comparison with *P. semipunctata* (left), *P. recurva* (right) wing covers have more cream to yellowish color, with dark brown mostly limited to the rear third of the wing covers. *Photo:* J. K. Clark

EUCALYPTUS LONGHORNED BEETLE

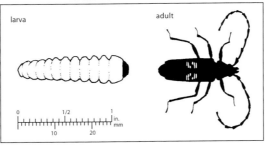

larva

adult

in minimally managed sites with no irrigation. They are especially susceptible to being attacked and killed by borers, as summarized and compared with other eucalyptus pests in Table 6-7 under "Redgum Lerp Psyllid." If planting eucalyptus, consider choosing species that resist longhorned borers, including *Eucalyptus camaldulensis, E. cladocalyx, E. robusta, E. sideroxylon,* and *E. trabutii.*

To manage both *Phoracantha* spp., reduce tree stress, properly handle eucalyptus wood, plant resistant species, avoid activities that disrupt biological control, and otherwise care for and protect trees, as discussed earlier under "Management of Wood Borers." Pesticide applications generally are not effective in managing these borers.

Inspect high-value trees regularly for damaging conditions and stress symptoms, including a sparse canopy, leaf yellowing, and shoots sprouting from the main trunk. As discussed in "Irrigation" in Chapter 3, consider supplemental watering during prolonged dry periods, particularly if seasonal rainfall has been below normal. Avoid frequent watering, as this promotes Armillaria and Phytophthora root diseases.

Prune dead branches and remove dead trees immediately; conduct other pruning during December and January (in southern California) and November through March (in northern California) when adult beetles are inactive. Chip, bury, or burn dead wood (where permitted) to prevent beetles from breeding in it. Alternatively, remove the bark or solarize logs under clear plastic in a sunny location for at least 6 months.

A tiny introduced wasp (*Avetianella longoi,* Encyrtidae), commonly parasitizes *Phoracantha* eggs. Parasitized eggs develop brown shells through which the body and dark eyespots of developing wasps can be observed. *Avetianella* typically kills over 90% of *P. semipunctata* eggs in the field, but is less effective against *P. recurva.* Conserve parasites by avoiding application of broad-spectrum, residual insecticides to eucalyptus.

Before applying any insecticide, consult the sections "Redgum Lerp Psyllid" and "Eucalyptus Tortoise Beetles." Consider the potential of insecticides to disrupt biological control of these and other pests, including the bluegum psyllid and eucalyptus snout beetle. For more information, consult *Eucalyptus Longhorned Borers Pest Notes.*

CLEARWING MOTHS

Larvae of clearwing moths (Sesiidae) bore in the vascular system of numerous hosts, including alder, ash, aspen, birch, cottonwood, Douglas-fir, oak, olive, pine, poplar, stone fruits, sycamore, and willow. With the species attacking pine, sycamore, and willow, feeding is tolerated by trees and apparently causes no serious harm. Most other mature trees can tolerate feeding by a few clearwing moth larvae, but high larval populations of certain species can girdle and kill plant parts so that limbs may drop and entire plants may die. The presence of clearwing moths often indicates that trees lack appropriate cultural care or have otherwise been injured. Sunburn and drought stress commonly lead to attack by borers, although in hot interior locations birch, poplar, and willow often become infested despite sufficient irrigation.

IDENTIFICATION AND BIOLOGY

Copious sawdustlike frass on bark or around trunks, trunk and branch swellings, dying limbs, and rough or gnarled bark are key signs of infestation. Clearwing moths often leave an empty brown pupal case protruding from bark or lying on soil around the tree base, except for peachtree borer, which pupates in soil.

Clearwing moth adults are recognized by their narrow, mostly clear wings. They are day-flying moths that resemble paper wasps but have a broad waist. The two sexes usually have different amounts of clear wing area and are differently colored, often with yellow, orange, or red on black. Adult female moths emit a pheromone that attracts the males.

The female deposits eggs in cracks, crevices, or rough areas on bark, and the larvae bore into the bark, cambium, or heartwood. Larvae have a dark brown head and whitish to pink bodies up to 1½ inches long. The similar larvae of American

Scattered dead branches may indicate a clearwing moth infestation, as with the ash borer larvae feeding in these Raywood ash. Infestation by clearwing moths often indicates that trees lack appropriate cultural care or have otherwise been injured.
Photo: J. K. Clark

Gnarled and wounded bark and frass on a white poplar trunk infested with larvae of two species of borers, the western poplar clearwing and carpenterworm. *Photo:* J. F. Karlik

Clearwing moth larvae are usually whitish to pink with a reddish brown head, as with this sycamore borer. Examining the pattern of hooks beneath their prolegs can distinguish clearwings from some other larvae, as illustrated in Figure 6-16. *Photo:* J. K. Clark

plum borer (discussed below) attack some of the same hosts (e.g., fruit trees, olive, and sycamore); be sure to distinguish these pests because some monitoring and management methods for them differ.

AMERICAN HORNET MOTH

Sesia tibialis closely resembles western poplar clearwing discussed below, and it infests many of the same plants. Hosts include aspen, cottonwood, poplar, and willow. The adult American hornet moth is mostly blackish blue with some brown, orange, or yellow.

ASH BORER

Larvae of the ash borer (*Podosesia syringae*), also known as the lilac or lilac-ash borer, infest ash, olive, privet, and syringa (also called lilac). In California populations, the male ash borer resembles a paper wasp with long brownish legs and a black body with narrow yellow bands. It occurs primarily in the Central Valley, mostly in host tree trunks and small limbs within 5 to 10 feet of the ground. Infestations often occur where bark has been injured, such as by improper staking, lawn mowers, string trimmers, or previous generations of *Podosesia*.

Adults emerge and females lay eggs on bark from April to early June. Larvae are creamy white with a brown head. They periodically expel sawdustlike frass, which accumulates on bark around their tunnel opening. There is one generation per year.

PEACHTREE BORER

Synanthedon exitiosa attacks all stone fruit trees, including apricot, cherry, peach, and plum. The peachtree borer, also called greater peachtree borer, occurs in California mainly in coastal areas and in the northern San Joaquin Valley and is not the same species as the lesser peachtree borer (*S. pictipes*) in the eastern United States.

Peachtree borer adults are mostly bluish black. Males have narrow yellow bands on the abdomen; females have a single orange band. Virtually all larval tunneling occurs within a few inches of the ground near the base of the main trunk, after which larvae emerge and pupate in soil. Therefore, to monitor for peachtree borers, inspect the basal trunk for small masses of reddish brown frass.

Keep soil bare around the tree base, especially in California's Central Valley, to increase peachtree borer egg and larval mortality due to dryness and heat. Also use the methods described below in "Management of Clearwing Moths."

REDBELTED CLEARWING

Synanthedon culiciformis is common around Sacramento, California. It infests red and white ash, birch, and alder. The adult is mostly brownish black with an orangish red band on the anterior of the abdomen. Its biology and management are similar to those of the ash borer.

Each peachtree borer larva causes a small mass of reddish brown frass on lower trunk bark. Keeping soil bare around the tree base in California's Central Valley can increase peachtree borer egg and larval mortality from dryness and heat. *Photo:* J. K. Clark

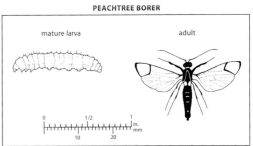

PEACHTREE BORER

mature larva   adult

Rough bark and copious, sawdustlike frass in crevices and around the tree base caused by sycamore borers. Sycamore trees tolerate even very high populations of clearwing moths, shown here, without apparent threat to tree health. This differs from peachtree borer, where even a few larvae and small amounts of frass on the basal trunk can indicate severe damage to stone fruit trees. *Photo: J. K. Clark*

SYCAMORE BORER

Synanthedon resplendens is prevalent in sycamore and also infests ceanothus and oak. Adults are present from May through July and may be seen intermittently running while rapidly fluttering their wings. The male is mostly yellow with a brownish black head, black bands on the body, and mostly yellow legs with black. The mostly clear wings have orangish to yellow margins. Sycamores tolerate extensive boring by this insect, and generally no control is recommended.

WESTERN POPLAR CLEARWING

The western poplar clearwing (*Paranthrene robiniae*), also called the locust clearwing, is found throughout warm, low-elevation sites in southern California and the Central Valley. It is a pest of birch, poplar, and willow in landscapes and nurseries. Willows apparently tolerate infestations without serious harm, and there is large variability in susceptibility among poplar varieties.

The adult moth greatly resembles a hornet or vespid wasp, except unlike the narrow, threadlike waist and filamentous antennae of a wasp, the adult clearwing moth has a thick waist and feathery antennae. The forewings are an opaque pale orange to brownish; the hind wings are clear. The thorax is black with a yellow hind border, and the abdomen is yellow with three broad black bands. The body of the desert form of this insect is entirely pale yellow. To complete one generation requires 1 or 2 years.

MANAGEMENT OF CLEARWING MOTHS

Mature woody plants usually tolerate the feeding of a few clearwing moth larvae. However, the presence of this pest often indicates that plants are injured, neglected, or stressed. Protect roots and limbs from injuries and provide proper cultural care and good growing conditions. In hot inte-rior areas, key preventive measures are to provide adequate irrigation and whitewash the trunks of young trees—paint trunks with white interior latex paint diluted with an equal amount of water.

Prune and dispose of infested and dying limbs. Damaging infestations may warrant application of beneficial nematodes to kill larvae or bark spraying to kill emerging and egg-laying adults, thereby reducing reinfestations. Clearwing moths infesting conifers have somewhat different management, as discussed below in "Pitch Moths."

Monitoring. Other wood-boring insects produce similar damage, so identify the cause correctly before taking control action. Excavate larvae and compare them with the photographs in this book; use a hand lens to examine the arrangement of small hooks on the bottom of larval prolegs

A western poplar clearwing pupal cast skin (lower left) and a frass-covered larval tunnel entrance on a poplar tree. Regular trunk inspection for new pupal skins and fresh frass are key methods for timing control actions and evaluating treatment effectiveness. *Photo: J. K. Clark*

Bark swelling and frass (borer excrement) around the tunnel opening of a western poplar clearwing borer larva feeding in a poplar tree branch. *Photo: J. F. Karlik*

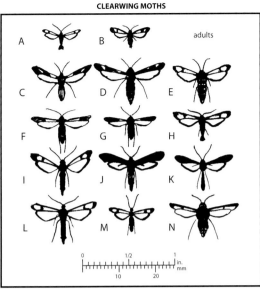

A B adults

C D E

F G H

I J K

L M N

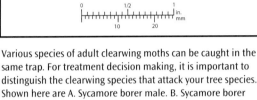

Various species of adult clearwing moths can be caught in the same trap. For treatment decision making, it is important to distinguish the clearwing species that attack your tree species. Shown here are A. Sycamore borer male. B. Sycamore borer female. C. Western poplar clearwing male. D. Western poplar clearwing female. E. Sequoia pitch moth female. F. Ash borer male from the western U.S. G. Ash borer male from the eastern U.S. H. Douglas-fir pitch moth male. I. Peachtree borer male from the eastern U.S. J. Peachtree borer female from the eastern U.S. K. Pitch mass borer (*Synanthedon pini*) from the eastern U.S. L. Peachtree borer male from the western U.S. M. Redbelted clearwing male. N. Oak borer (*Paranthrene simulans*) from the eastern U.S. *Photo:* J. K. Clark

to distinguish clearwing moth larvae from American plum borer (Figure 6-16). Time any direct control actions and assess treatment effectiveness by using pheromone-baited traps, or use a combination of the methods described below.

Regularly inspect hosts for fresh clearwing moth pupal cases protruding from bark and around the base of trees. Old pupal cases can persist for months, so remove these when found and monitor frequently with care during spring and summer to ensure that any pupal cases observed are new. Trunk inspection does not work for peachtree borer because larvae drop from tunnels and pupate in soil.

For clearwing moth species with larvae that expel sawdustlike frass from bark openings, brush away the frass from spring through fall and plug tunnel entrances with rope putty or grafting wax. Spray the plug with brightly colored paint to make it easier to relocate that spot for monitoring. Check the plug 1 week later. If the plug is gone, a larva is still feeding beneath the bark. This method is especially useful to confirm whether larvae were killed after applying nematodes or probing tunnels with wire.

Pheromone Traps. Sticky traps baited with a sex attractant capture male moths

of at least the ash and peachtree borers and the redbelted and western poplar clearwings. Place traps to capture moths for species identification, determine when egg-laying adults are active, and especially to effectively time insecticide application to kill adults.

In spring, hang one or two pheromone

traps of the appropriate type in separate locations within about 25 yards of clearwing host trees. Once each week, check traps for moths and identify whether any are a species that might attack your plants. Maintain and replace traps or the pheromone dispenser as recommended by the manufacturer. When male moths are captured, females of that

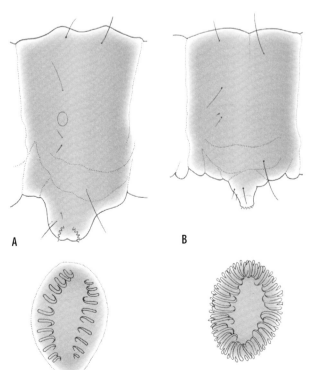

A B

Figure 6-16.

Distinguish American plum borer larvae from clearwing moth larvae by using a hand lens to examine the crochet pattern (arrangement of small hooks) on the bottom of the prolegs on their fourth abdominal segment: A. Clearwing larva have crochets in two transverse rows (rows are oriented perpendicular to the insect's body length). B. American plum borer crochets are more dense and form a complete circle or oval. The plum borer's prolegs are also more narrow than clearwing prolegs when compared with the width of their abdominal segment. Distinguishing these larvae can be important because they infest some of the same plants (e.g., fruit trees, olive, sycamore), but some of the cultural and physical controls for them are different. Adapted from Peterson 1956.

species are also present and laying eggs.

Clearwing moths may be captured almost any time during the growing season, but each species typically flies in numbers during only a few months each year:

- Ash borer and redbelted clearwing adults fly from April through July.
- Peachtree borer, sycamore borer, and western poplar clearwing adults are active primarily from May through July.
- Western poplar clearwing adults in southern California occur mostly from February through July but can also be present during fall.

When using pheromone traps to time spraying, it is important to correctly distinguish the adults of the species that are a pest of your plants from the adults of other clearwing moth species that may be caught. To help you identify whether the moths are a species that attacks your plants, compare them to the adult descriptions and photographs in this publication, consult photographs in the *Clearwing Moth Pest Notes*, or take the trap containing your moths to your county agricultural commissioner or UC Cooperative Extension office. Save the identified moths for comparison when additional moths are captured.

Cultural Control. Make sure trees receive appropriate irrigation and that roots have an adequate volume of uncompacted soil that provides sufficient aeration (see Chapter 3). Provide good soil conditions. Protect roots, trunks, and limbs from injury. Keep weed trimmers and lawn mowers away from trunks, for example, by maintaining mulch or a several-foot-wide area around trunks free of turf and other vegetation. Stake young trees only if needed to protect or support the trunk or anchor the root ball during the first year or so after planting.

Egg-laying clearwing moth females are attracted to tree wounds. Avoid pruning live branches unless necessary to develop tree structure or remove severely infested, dying, or hazardous limbs. Except for hazardous limbs, which should be removed whenever they appear, decide when to prune based on that tree species' susceptibility to pruning-related problems. To avoid borer infestations, prune only during fall through early winter when egg-laying adult borers are not active. However, stone fruits that are especially susceptible to Eutypa canker and dieback, such as apricot and cherry, should be pruned during July or August. Consult Tables 5-2 and 6-13 and "Pruning" in Chapter 3 to learn the best pruning time for your situation.

Physical Control. Kill peachtree borers and possibly larvae of other species by carefully using a stiff wire to probe the trunk during spring or fall where gummy frass exudes from bark. It is difficult to know whether the larva has actually been killed; reinspect trunks in a week and probe tunnels again if fresh gum exudate is observed, indicating that a live larva is present. Do not create large wounds when probing tunnels.

Where peachtree borer is a problem, remove suckers and keep vegetation and mulch away from the base of the tree. Bare soil around trunks increases the likelihood that any tunneling will be observed, and in the Central Valley of California it increases heat exposure and dryness and reduces survival of borer eggs and larvae.

Biological Control. Clearwing moths are killed by various predators (e.g., woodpeckers) and parasites. For example, larvae parasitized by *Apanteles* spp. wasps have many small, oblong maggots (parasite larvae) or white cocoons (pupae) adhering to their bodies. A minute, blackish brown braconid wasp emerges from each cocoon after killing the clearwing larva. Avoid disrupting natural enemies whenever possible, for example, by not spraying trees that tolerate borers (e.g., sycamores) and by using physical and preventive cultural controls.

Nematodes. Larvae of species that maintain a tunnel open to the outside can be controlled with *Steinernema carpocapsae* and *S. feltiae* insect-parasitic nematodes. Nematodes have been shown to control the peachtree borer, redbelted clearwing, sycamore borer, and western poplar clearwing. See the section "Entomopathogenic Nematodes" earlier in this chapter for nematode biology and proper handling.

Placing a clearwing moth pheromone lure in a wing-type trap. Pheromone traps are a critical tool for timing insecticide spraying of trunks to kill egg-laying adults. *Photo:* J. K. Clark

A cluster of white cocoons left by *Apanteles* sp. wasps that killed a sycamore borer larva. *Photo:* J. K. Clark

Apply nematodes with a hand pump or squeeze bottle applicator at a concentration of 1 million or more invasive-stage nematodes per ounce of distilled water. First clear the tunnel entrance, then insert the applicator nozzle as far as possible into each gallery. Inject the suspension until the gallery is filled or liquid runs out another hole, then plug the tunnel entrances with rope putty or grafting wax. Agitate the applicator frequently to keep nematodes suspended in the liquid. Adding 2% red or orange latex pigment marks treated tunnels.

Apply nematodes during warm weather (at least 60°F) from spring through fall when clearwing larvae are actively feeding. Nematode-treated larvae continue to feed and push frass from their tunnels for several days before dying. A second application 1 or 2 weeks after the first increases the likelihood that borers will become infect-

ae of clearwing moth species that maintain a tunnel n to the outside can be controlled by squirting in mercially available insect-killing nematodes. To age borers, any short-term control such as this ld be combined with improving the tree's growing ronment and cultural practices. *Photo:* J. K. Clark

ed. After application, plug the opening of each gallery and inspect openings a week later, as discussed above. If the gallery opening is no longer plugged, the larva has not died. Re-treat the gallery.

Bark Spraying. If trees are of high value, making two or more applications of a broad-spectrum, residual, contact insecticide to the trunk and base of limbs when egg-laying adults are active may limit attack by the ash borer, peachtree borer, sycamore borer, western poplar clearwing, and perhaps other species. Spraying may not be practical on large trees in landscapes in part because of drift, pesticide movement off-site that can contaminate surface waters, and the difficulty in timing applications effectively. A licensed pest control operator is needed because effective products are not available to home users. Soil or trunk application of systemic insecticides have not been shown to be effective in California against clearwing borers.

Bark spraying must be properly timed to be effective. To determine when moths are emerging, frequently examine trunks and limbs or inspect pheromone-baited traps for that species, or both, as discussed above. About 10 to 14 days after you first detect adults (observe fresh pupal cases or catch pest moths in traps), spray an effective insecticide so that it thoroughly wets bark on the main trunk and the base of limbs where they join the trunk. For peachtree borer, allow spray to run down the trunk base and wet soil within several inches of the root crown. There is no need to spray foliage or the upper canopy. If moths continue to be caught or fresh pupal cases appear for longer than a month after the application, a second spray may be warranted.

Trees may continue to decline unless insecticides are used in combination with improved tree care practices. For more information, see *Clearwing Moths Pest Notes.*

PITCH MOTHS

Sequoia pitch moth (*Synanthedon sequoiae*, Sesiidae) larvae bore in Douglas-fir and

pines throughout California, especially in Monterey pine in northern California. The Douglas-fir pitch moth (*S. novaroensis*) infests pines, spruce, and Douglas-fir. Larval feeding causes copious amounts of resin to exude from conifer bark and sometimes causes limbs to die or break off, especially if infested trees are young. However, larvae cause little injury to cambium and wood and their damage is primarily aesthetic.

Infestations are recognized by the gummy gray, pink, whitish, or yellow masses that protrude from trunks and limbs. As the dirty white or creamy larvae excavate shallow cavities just below the bark surface, the pitch mass grows up to several inches in diameter and hardens and darkens. A brownish, papery pupal case may protrude from the mass after the larva matures and the adult emerges. Old pitch masses can remain on bark for several years and are often reinfested because female pitch moths are attracted to lay eggs where bark is injured.

In comparison with a bark beetle pitch tube (pictured earlier), pitch moth masses are more variable in shape, from roundish to elongate oval, and can be several inches long. Bark beetle pitch tubes are usually ½ inch or smaller and typically have a distinct round hole in the center. Canker fungi, western gall rust, and mechanical wounds typically cause a relatively thin layer of resin on bark instead of protruding pitch masses. Consult Table 5-10 in Chapter 5 for help in distinguishing the insects and pathogens affecting pines.

The sequoia pitch moth's mostly clear wings have bluish black margins with yellow at the base. The head and thorax are mostly brownish black, the abdomen is broadly banded yellow, and the legs are mostly bright yellow. Instead of yellow, the markings on the Douglas-fir pitch moth are orangish. The day-flying adults can be present anytime from May through September but are most abundant in June and July. Females lay their eggs on injured bark, commonly around pruning wounds. Most individuals require 2 years to develop from egg to adult and most of their life span is as larvae.

Sequoia pitch moth causes gummy masses, which are usually elongate to roundish and grow to several inches in diameter. In contrast, bark beetle pitch tubes are usually less than ¹/₂ inch in diameter and often have a distinct round hole near their center. Certain fungi and mechanical wounds also cause pines to ooze, but usually only as thin layers of pitch, not the distinct protruding masses from boring insects. *Photo:* J. K. Clark

Management of Sequoia Pitch Moth.

Pines' susceptibility to sequoia pitch moth varies based on tree species and cultural practices. Where sequoia pitch moth has been a problem, consider planting only less-susceptible pines (Table 6-17). Stake newly planted pines for no longer than 2 years, and only if needed and properly done, as discussed in "Staking" in Chapter 3. Remove any stakes that came from the nursery and do not tie or fasten trunks firmly; allow the main stem to flex without rubbing on stakes.

Provide trees with proper cultural care, especially appropriate irrigation. Protect trees from injury. Infested small branches can be cut off, but this may lead to future infestations around the pruning wound. Make any cuts properly (just outside the branch bark ridge; see Figure 3-8) from October through January, before the egg-laying females appear in the spring.

Insecticide applications have not been effective for pitch moths. The only direct control is to scrape away or pry off the

pitch masses. If you carefully excise the mass, you can find and kill the larva or pupa underneath. Larvae are easily overlooked because their color closely resembles the color of pitch. Young larvae (those found in smaller, paler masses) typically occur below the bark surface within a small cavity they chew. Pupae and older, larger larvae usually are found nearer the surface of the mass, somewhat outward from the bark cavity they created when young. If you simply scrape the pitch away without actually locating and killing the larva or pupa, the insect can survive and cause a new pitch mass to develop.

Once the borer is removed, sap flow will decrease, and the wound can gradually close. Properly removing pitch masses from all nearby trees, along with appropriate cultural practices, can reduce reinfestations and control local pitch moth populations. No other control except minimizing injuries is recommended, as pines apparently are not seriously harmed by this insect. For more information, see the *Pitch Moths Pest Notes*.

OTHER BORERS

AMERICAN PLUM BORER

Euzophera semifuneralis (Pyralidae) is occasionally found boring in wood of fruit and nut trees, mountain ash, olive, and sycamore, mostly in young trees. The female moths are attracted to lay eggs in bark wounds around main branch crotches, the lower trunk, and root crown. Larvae bore

Sequoia pitch moth larvae are easily overlooked because their pink to gray color closely resembles the color of pitch. If pitch masses are simply scraped away without definitely locating and killing the larva or pupa, the insect can survive and cause a new pitch mass to develop at that site. *Photo:* J. K. Clark

Scraping away or prying off masses is the only way to eliminate unsightly pitch moth masses. This helps control the pest if done to all nearby pines and the lar[va] underneath is killed. *Photo:* J. K. Clark

Table 6-17.

Relative Susceptibility of Pines (*Pinus* spp.) to Sequoia Pitch Moth.[1]

MOST SUSCEPTIBLE
Afghan, Aleppo, Brutia, Calabrian, and Mondel (*Pinus brutia*, *P. eldarica*, and *P. halepensis*);[2] Bishop (*P. muricata*), Japanese black (*P. thunbergiana*), Mexican (*P. patula*), Monterey (*P. radiata*), ponderosa (*P. ponderosa*), shore or beach (*P. contorta*)
LEAST SUSCEPTIBLE
Canary Island (*P. canariensis*), Italian stone (*P. pinea*)

1. Pines are more susceptible to pitch moths if pruned or otherwise injured.

2. Various common and scientific names are used for these closely related Asian and European natives. Many species are quite susceptible, but their susceptibility varies, and confusion among names makes them difficult to distinguish.

Adapted from Frankie, Fraser, and Barthell 1986.

in cambium, causing extensive gumming, reddish orange frass, and webbing. This weakens limbs, which often break during windy conditions, and sometimes kills young trees.

Adults are gray moths with brown and black wing markings. Larvae are dull green, pinkish, or white. Both are up to about 1 inch long. Overwintering is as pupae in bark crevices or damaged wood on trees. Adults emerge and the females lay eggs beginning in about April. American plum borer has three to four generations per year.

American plum borer attacks some of the same hosts as clearwing moths, but their cultural and physical controls differ. Before taking control action, distinguish among these pests, such as by using a hand lens to examine the arrangement of small hooks on the bottom of larval prolegs (Figure 6-16).

Avoiding bark wounds is the most critical management strategy for American plum borer. Larvae can enter cambium only through relatively fresh bark wounds. Pruning cuts, injuries from sunburn or weed trimmers, and bacterial galls and canker fungi make trees highly susceptible to infestation. Therefore, protective measures such as painting lower trunks with interior white latex diluted with an equal amount of water as discussed under "Sunburn" in Chapter 4 may reduce borer-susceptible bark injuries.

Where damage is unavoidable or already present, a residual, contact insecticide can be applied from about 1 foot above the main lower branch crotch to the soil line. Use pheromone-baited traps to determine the peak in adult flights, which coincides with the peak egg laying. Peak egg laying during the first generation is an optimal treatment time. Reapplication during peak flight in later generations may also be warranted.

CARPENTERWORM

Prionoxystus robiniae (Cossidae) larvae bore in sapwood and deeply into heartwood of broadleaf trees, including ash, birch, cottonwood, elm, locust, oak, willow, and fruit trees. Older declining trees and certain species (e.g., Modesto ash) are more often infested.

Extensive gumming from American plum borer larvae in the cambium of tree branch crotches. Avoiding bark wounds is the most critical management strategy for this pest, as larvae can enter cambium only through existing injuries. *Photo:* J. K. Clark

CARPENTERWORM

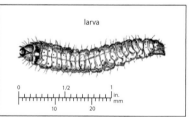

larva

Sawdustlike frass around the tree base and dark liquidy discoloration or bleeding sap on limbs or trunks can indicate carpenterworm feeding. The bark often becomes corky, gnarled, or rough where several carpenterworms are feeding underneath. Vigorous trees can apparently tolerate some carpenterworms. Continued attacks over several years can cause branch dieback and make trees hazardous and prone to fail.

The adult is a mottled grayish moth up to 1¾ inches long that blends with the color of bark. Adults emerge in the evening during the spring and summer, and the female lays eggs on creviced, rough, or injured bark, such as where carpenterworms are already feeding. Larvae feed under bark for 2 to 4 years and periodically come to the bark surface to expel frass. Larvae are light greenish to white with a dark brown head and grow 2 to 3 inches

Despite their large size (up to 1¾ inches), carpenterworm adults on bark are easily overlooked because of their mottled gray coloring. *Photo:* J. K. Clark

AMERICAN PLUM BORER

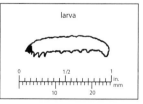

larva

An adult carpenterworm emerged from this ½-inch-diameter hole in bark (center) after feeding inside during its larval stage and pupating. To clearly observe this evidence of infestation, you may need to chip away the surface of large limbs or the trunk where bark appears distorted or gnarled or contains frass. *Photo:* H. Kimber, City of Fresno

long. A mature larva's tunnel is about ½ inch in diameter and 6 to 10 inches long. After pupation, the emerging adult often leaves its brown, papery pupal case up to 2 inches long protruding from the bark.

MANAGEMENT

Provide trees with proper cultural care and protect them from injuries, as detailed in Chapters 3 and 4. Appropriate irrigation is especially important; for example, California native oaks are adapted to summer drought and can be stressed or killed by frequent nearby irrigation. If trees are heavily infested, they may be a failure hazard and warrant removal; have infested trees inspected by a professional arborist or other tree care expert.

No direct control actions are known to be effective on large trees. Squirting entomopathogenic nematodes into tunnel openings or inserting a sharp wire and probing tunnels to puncture larvae as discussed earlier under "Management of Clearwing Moths" can be effective where tunnels can be reached.

Carefully monitoring bark at least once a week beginning in late winter and promptly spraying bark when the first new pupal case appears with a residual, broad-spectrum, contact insecticide labeled for that use can kill emerging and egg-laying adults. This requires hiring a professional with the equipment, experience, and access to pesticide for treating that situation. Because of carpenterworm's prolonged life cycle and varying development rates among individuals, it will be necessary to repeatedly inspect bark and re-spray at intervals over a period of about 4 years or longer. Do not spray trees unless substantial resources will also be devoted to improving the tree's cultural care and growing environment to reduce the likelihood of carpenterworm reinfestation. For more information, consult *Carpenterworm Pest Notes.*

Juniper twig girdler caused these scattered dead canopy patches. Foliage miners, scale insects, various pathogens, and dog urine can cause similar symptoms, so be sure to correctly diagnose the cause of damage before taking any action. *Photo:* J. K. Clark

JUNIPER TWIG GIRDLER

Periploca nigra (Cosmopterigidae) is a pest of juniper primarily in southern California and the warm interior valleys. It causes foliage on small limbs to become yellow, then turn brown and die, resulting in a checkerboard of green and brown limbs by late summer. Twig girdler feeding does not kill entire juniper plants.

Do not confuse twig girdler damage with the causes discussed earlier in "Cypress Tip Miners" and "Minute Cypress Scale" and in Chapter 5 in "Cypress Canker" and "Root and Crown Diseases." Dieback caused by root disease fungi typically occurs where soil moisture has been high for prolonged periods. Dying juniper branches caused by mice have bark chewed away in bands from lower parts of branches. Twig dieback at the edges of plantings may be the result of dog urine, which can be diagnosed by the characteristic odor.

To confirm twig girdler as the cause of damage, peel bark from the branch where dead and living tissue meet. Inspect the wood for girdling tunnels, the cream-colored larvae with a brown head, and the shiny, black to brown pupae. The small, shiny, brownish black twig girdler moth is not often observed; it flies primarily from May through June in the San Francisco Bay Area and March through May in southern California. The females lay tiny eggs on woody stems into which larvae tunnel and feed for about 8 to 9 months. There is a single generation per year.

MANAGEMENT

To improve plant appearance, prune out and dispose of affected branches. Avoid planting Tam juniper (*Juniperus sabina* 'Tamariscifolia'), which is very susceptible to twig girdler. Instead of Tam juniper or Hollywood juniper (*J. torulosa* or *J. chinensis* 'Kaizuka' or 'Torulosa'), which is rarely planted now, consider planting one of the less-girdler-susceptible junipers.

If damage cannot be tolerated, it can possibly be reduced by spraying foliage thoroughly about twice with a broad-spectrum, residual, contact insecticide to kill adult moths and prevent them from laying eggs. Effective control with a contact insecticide is difficult. Boring larvae are not affected, and spraying does not restore the appearance of damaged foliage, which remains brown until new growth occurs. In southern California, spray in late March and early May; in northern California, spray in early June and mid-July.

RASPBERRY HORNTAIL

Hartigia cressoni (Cephidae) larvae boring in rose canes cause drooping or "flagging" young rose shoots in spring. Raspberry horntail boring damage is usually severe only in rose, but caneberries (including blackberry and raspberry) are also hosts.

Inspect canes for drooping or dead tips during April through June. A small dark or reddish spot or oviposition scar where an adult laid an egg can often be observed in canes several inches below the flower bud.

This rose shoot is wilting and dying because a raspberry horntail larva is feeding inside. To control this boring sawfly, inspect roses during spring and prune off infested shoots below any noticeable damage. *Photo:* J. K. Clark

The larva tunneling inside is white or yellowish, segmented, and up to 1 inch long. The black with yellow adult wasps are about ½ inch long with thick waists.

Prune off and dispose of infested terminals. Make cuts in healthy pith below any noticeable oviposition scar or hole or any stem swelling (which indicates a larva inside). No insecticide spraying is recommended.

WOOD WASPS, OR HORNTAILS

Wood-boring wasps (Siricidae) are large, nonstinging Hymenoptera that are attracted to dying or recently killed conifers. The large wasps look threatening and make a noisy buzz when flying but are harmless to people. Adults have a cylindrical body about 1 inch long that is black or metallic dark blue, often with red or yellow markings. An ovipositor that is ¾ inch long projects from the rear of females, which lay eggs through bark into wood where the pale, legless larvae grow up to 1 inch long. Larvae are cylindrical, with a rear-end spine (horntail); they chew in sapwood and heartwood for one to several years, while making a gallery up to 1 foot long.

Wood wasps are a problem because larvae can complete development in cut wood and emerge as wasps from firewood and lumber. Emerging wood wasps leave a ¼- to ½-inch-diameter hole in wood (e.g., cabinets or flooring) and any covering materials, then noisily fly around inside the structure. Wood wasps attack only trees and will not bore into wood in buildings or furniture. See the *Wood Wasps and Horntails Pest Notes* for more information.

Adult wood wasps are typically cylindrical, about 1 inch long, and mostly dark with orange or yellow, as with this *Sirex californicus* female. Despite their noisy buzz when flying, adults are harmless to people. *Photo:* N. M. Schiff, USDA Forest Service

TERMITES

Termites (Isoptera) sometimes chew tunnels and form large colonies in heartwood in trees. Termite biology and life history resemble that of ants, but unlike ants' appearance, termites have a broad waist, equal-length wings, and antennae that are not elbowed (see Figure 6-12). Each type of termite differs in biology, such as the preferred wood and other habitat conditions. Drywood termites (*Incisitermes* spp.) are the type most commonly found in trees.

Termites enter trees through existing wounds. They are not the primary cause of tree damage and they do not attack living cambium or sapwood. To avoid termite infestations, prevent wounds to trees, as discussed in Chapters 3 and 4. Prune properly when needed during a time of year when pruning will not attract other boring pests (see Table 6-13). Provide proper cultural care to keep trees vigorous. Keep mulch (both organic and inorganic materials) at least several inches away from trunks. Because termite boring can make limbs or trunks hazardous (likely to fail), have termite-infested trees inspected by a certified arborist or other tree care professional.

Spraying surfaces with insecticide is not effective against termites in trees. There is no research demonstrating that drilling holes and injecting insecticide controls termites infesting trees in California. Trees may be a hazard from dead wood in trunks or limbs (i.e., structural weakness) whether insecticide is applied for termites or not. For more information, consult publications such as *Drywood Termites Pest Notes* and *Subterranean and Other Termites Pest Notes.*

BEE AND YELLOWJACKET NUISANCE

Most Hymenoptera (ants, bees, and wasps) are innocuous or are beneficial pollinators, predators, or parasites. Certain Hymenoptera are both beneficial and a nuisance, including carpenter bees that chew in wood and behave aggressively, leafcutter bees that chew leaves, and honey bees that migrate

Males of some carpenter bee species are golden or light brown, as with this valley carpenter bee *(Xylocopa varipuncta)* male. Carpenter bee females are usually blackish or dark metallic blue or purple. Their large size, loud buzzing, and bumble bee appearance can be frightening, but males cannot sting and females rarely sting people. *Photo: J. K. Clark*

and swarm in a mass. Certain ant species (discussed above), bees, and social wasps (primarily yellowjackets) can be especially hazardous to persons allergic to their venom.

Management options vary by situation and nuisance species. For example, controlling honeydew-producing insects reduces yellowjacket attraction to infested plants. Managing turf to reduce flowering weeds (e.g., clovers) reduces lawns' attractiveness to bees. For more information, see *Bee Alert, Africanized Honey Bee Facts* and the *Pests Notes: Bee and Wasp Stings, Carpenter Bees, Removing Bee Swarms and Established Hives,* and *Yellowjackets and other Social Wasps.*

Honey bees (*Apis mellifera*) sometimes migrate in a noisy and temporarily aggregate on tree limbs or other surfaces. When swarming, honey bees are usually unlikely to sting alone. However, swarms could potentially be Africanized bees, which are unpredictable and may attack and sting in numbers. For assistance, contact beekeepers or pest control professionals who advertise bee removal services or the health department or vector control agency. *Photo: D. Ro*

Various traps are available to help manage aggressive yellowjackets (*Dolichovespula* and *Vespula* spp.). This lure trap and water traps (not shown) are best used during late winter and spring to capture newly emerged female wasps before they reproduce. Poison bait traps (not shown) work best later in the season when wasps' insect prey is less available. *Photo: J. K. Clark*

Adult leafcutter bees (*Megachile* spp.) leave semicircular holes in leaf margins where they cut out pieces as nesting material for their larvae. This foliage removal is harmless to plants, and these bees are important pollinators that should not be killed. *Photo: J. K. Clark*

Unlike ants, termites have a broad waist and antennae that are not elbowed, as with the western subterranean termite (*Reticulitermes hesperus*) soldier (lower left) and workers. Termites enter trees through existing wounds, so protect trees from injury and provide proper cultural care to prevent termite infestation. *Photo: J. K. Clark*

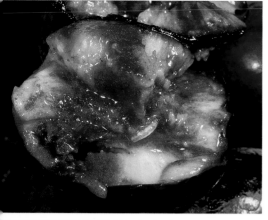

A pale maggot of the olive fruit fly infests this olive. Most fruits are susceptible to one or more species of exotic tephritid fruit flies, which cause fruits to soften, decay, and drop prematurely. *Photo: J. K. Clark*

An adult tephritid fruit fly is about ¹/₅ inch long, considerably larger than the vinegar fly (Drosophilidae), or common fruit fly, also shown here. The Mediterranean fruit fly (photo left) has a mostly tan abdomen with darker stripes and light brown bands across each wing and another along the outer front edge. Unless you are confident the species found is already known to be established, such as olive fruit fly or walnut husk fly, report any maggots in fruit or tephritid flies you find to the county agricultural commissioner. *Photo: J. K. Clark*

Stippled, bleached foliage caused by twospotted spider mites feeding on the underside of rose leaflets. Drought stress, dustiness, and applying certain broad-spectrum, residual insecticides can cause a spider mite outbreak. Because pests such as leafhoppers, thrips, and various true bugs produce similar damage, identify the pest to determine the effective control actions. *Photo: J. K. Clark*

FRUIT FLIES

Larvae of various Tephritidae feed in the pulp of fruits and certain other crops. Introduced pest species in California include apple maggot (*Rhagoletis pomonella*), olive fruit fly (*Bactrocera oleae*), and walnut husk fly (*Rhagoletis completa*). Adult tephritids and mature larvae (pale maggots) are about ¼ inch long, considerably larger than the vinegar flies (Drosophilidae), or common fruit flies, that mostly feed in fermenting or injured fruit but include a few pests such as the spotted wing drosophila (*Drosophila suzukii*).

Exotic tephritids such as the Mediterranean, Mexican, and Oriental fruit flies are periodically introduced and have apparently been eradicated. Take any tephritid flies or maggots you find in fruit to the county agricultural commissioner for identification unless you are confident the species found is already known to be established. See the *Pest Notes: Olive Fruit Fly, Spotted Wing Drosophila,* and *Walnut Husk Fly* for more information.

Mites

Mites are common, but in most landscapes in urban coastal areas they are not serious

pests. In interior valleys of California, spider mites are significant pests during the hot dry season.

Certain species are plant feeders, but many mites are beneficial predators. Mites often go unnoticed because they are tiny and natural controls such as weather and predators frequently keep their populations low. Their damage to plants can usually be observed before you notice the mites themselves.

Spider mites and false spider mites are the most common pest mites. They cause leaves to appear stippled or flecked with pale dots where their feeding kills tiny areas of leaf tissue. Damage may merge into large patches of discolored foliage. Mite feeding on fruit appears as a silvery or brownish sheen called russeting. Certain spider mite species cover leaves, shoots, or flowers with fine webbing. Other types of mites mostly cause plant tissues to become distorted, thickened, or galled.

Prolonged heavy infestations slow plant growth, cause leaves or fruit to drop prematurely, and may kill young plants. Severe spider mite infestations often occur on plants that are drought stressed and where mite predators are disrupted by pesticide applications or excessive dust.

Vigorous woody plants tolerate extensive stippling or tissue distortion with little or no loss in plant growth or fruit yield.

IDENTIFICATION AND BIOLOGY

Mites are arachnids (spider relatives) and have an oval, unsegmented body. Most mites pass through an egg stage, a six-legged larval stage, and two eight-legged nymphal stages before becoming an eight-legged adult (Figure 6-17). Mites overwinter as eggs or adult females on bark or in litter. Where winters are mild, all stages of mites can be present year-round on evergreen plants. At moderate temperatures, some species can complete a generation in 1 or 2 weeks.

The most common pest groups are the spider mites and red mites (Tetranychidae), false spider mites (Tenuipalpidae), gall mites (Eriophyidae), and cyclamen and broad mites (Tarsonemidae). Common predatory mites (Phytoseiidae) are often slightly larger than spider mites and false spider mites, but individuals of all these groups are only about ¹/₅₀ inch long. Eriophyids and tarsonemids are about one-fourth the size of spider mites. The species of most mites can be positively identified only by an expert. Determining what taxonomic family mites are in is usually sufficient because mites within these groups commonly have similar biology and management.

Where spider mites or false spider mites are suspected causes of damage, inspect the underside of stippled or distorted plant foliage and nearby healthy plant parts with a hand lens to determine whether mites are present. Or hold a sheet of paper beneath the plant, tap the foliage sharply, and inspect the paper for any dislodged mites; to the naked eye they resemble moving specks. Some of these mites may be beneficial predators helping to control pest mites. Stippled and distorted foliage can also be caused by lace bugs, plant bugs, and thrips. However, these other pests often leave specks of dark excrement on leaf surfaces and they do not make webbing.

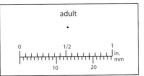

SPIDER MITE

adult

·

0   1/2   1 in.
└┴┴┴┴┴┴┴┴┴┴┴┴┴┴┴┴┴┘ mm
        10   20

SPIDER MITES

Many different spider mites (e.g., *Oligonychus* and *Tetranychus* spp.) occur in landscapes and typically are most abundant in groups on the underside of leaves during summer. For example, twospotted spider mite (*Tetranychus urticae*) and Pacific spider mite (*T. pacificus*) can be common on azalea, fuchsia, maple, rose, fruit and nut trees, and other broadleaves. Sycamore spider mite (*O. platani*) is common in hot, dry areas of California on the upper surface of sycamore leaves and other hosts, including loquat, oak, and pyracantha.

Spider mite adults have eight legs and tiny globular or spherical bodies that are translucent or colored. Immatures and adults are commonly yellowish or greenish, with irregular dark blotches on the sides of the body. Overwintering females and colonies of stressed mites may lack blotches and turn red or orange overall. Most tetranychid mites have long bristles on their body, and spider mites produce silken webbing. Eggs are translucent and spherical. For more information, see *Spider Mites Pest Notes*.

RED MITES

The European red mite (*Panonychus ulmi*) is especially common on apple and *Prunus* species such as stone fruits. It is less common on ash, elm, locust, rose, and other plants. Southern red mite (*Oligonychus ilicis*) is most common on broadleaf evergreens such as azalea, camellia, holly, and rhododendron. Citrus red mite (*Panonychus citri*) occurs on many plants but is a pest only on citrus.

Most species of red mites produce little webbing and closely resemble each other. Adults and nymphs are mostly red with irregular, dark blotches on their oval, globular, tiny body. Their tiny, red eggs have a whitish, vertical stalk that is visible with a hand lens. The citrus red mite egg also has 10 to 12 threads radiating from the tip of this stalk to the leaf surface. Hot weather reduces citrus and southern red mite populations. Southern red mite populations are promoted by cooler weather.

Inspect foliage with a hand lens when mites or tiny insects are suspected to be present. Hold the lens close to your eye and move the object being viewed closer or further away until it is in focus. *Photo:* J. K. Clark

Twospotted spider mites have two irregular dark blotches on the sides. The body can be pink or orange during winter and when populations are overcrowded or stressed. Most commonly, the twospotted mite body is yellowish or greenish. *Photo:* J. K. Clark

Figure 6-17.
Spider mites, false spider mites, and predatory phytoseiid species develop through five life stages. Eggs hatch, producing six-legged larvae. The two nymphal stages and adults have eight legs. At moderate temperatures, some species can complete one generation in about 10 days.

Egg

Larva

Protonymph

Deutonymph

Adult

An adult male and egg of southern red mite on an azalea leaf. This species thrives under cool conditions and is most common on broadleaf evergreens. *Photo:* J. K. Clark

European red mite eggs shown close up on a leaf. Citrus red mite eggs look similar, except that they have 10 to 12 threads radiating from the tip of the vertical stalk protruding from each egg. The tiny fibrils resemble guy wires, the tension cables attached to the top of utility towers. *Photo:* J. K. Clark

PERSEA MITE

Oligonychus perseae is a common pest of avocado and to a lesser extent camphor tree in coastal areas. Mite feeding causes small, circular dead spots that are yellowish or purplish to brown on the underside of infested leaves; each spot is covered with fine silvery webbing. Spots initially develop along leaf veins and are distinctly visible on the upper leaf surface. The spots can coalesce, and highly infested leaves drop prematurely.

Adults and immatures are oval shaped, slightly flattened, and elongated. Their bodies are yellow to green, often with several tiny, dark spots on the abdomen. Populations increase during midsummer and decline in the fall.

PINE AND SPRUCE SPIDER MITES

Several mite species cause conifer foliage to become discolored bronze, reddish, or yellow. *Oligonychus subnudus* and *O. milleri* occur on pines, especially Monterey pine. The spruce spider mite (*O. ununguis*) can occur on virtually any conifer, including arborvitae, *Chamaecyparis*, coast redwood, Douglas-fir, fir, giant sequoia, juniper, spruce, and occasionally pine.

Depending in part on the host plant and mite species, these mites can be green, pink, or brown. *Oligonychus ununguis* produces webbing, especially around the base of needles; neither *O. subnudus* nor *O. milleri* produces obvious webbing. *Oligonychus subnudus* is most common in spring, *O. ununguis* is usually most abundant in the spring and fall, and populations stop reproducing under prolonged high temperatures. *Oligonychus milleri* can be abundant anytime from spring through fall.

Feeding on young foliage may not produce obvious damage until months later, when foliage matures. Therefore, the effectiveness of any current treatment (e.g., thoroughly covering foliage with narrow-range oil) may not become apparent until the following season.

These circular spots visible on the upper leaf surface were caused by persea mites, which feed in silk-covered groups on the underside of leaves on avocado and camphor tree (shown here). Similar blotchy-spot damage on various bamboos is caused by bamboo spider mites, *Stigmaeopsis* (=*Schizotetranychus*) spp. *Photo:* D. Rosen

Spruce spider mite webbing on male (pollen-producing) cones and the base of ponderosa pine needles. In California, spruce spider mite is usually most abundant in the spring and fall; reproduction stops under prolonged high temperatures. *Photo:* J. K. Clark

Pale, bleached needles caused by pine spider mites (*O. subnudus*) feeding when needles were young and developing. Once damage becomes obvious, it is too late to take effective control action. If discolored foliage cannot be tolerated, monitor new growth the next season and if mites are abundant, apply oil or other effective pesticide. *Photo:* W. Cranshaw, Colorado State University, Bugwood.org

FALSE SPIDER MITES

False spider mites (Tenuipalpidae), also called flat mites, include privet mite (*Brevipalpus obovatus*) and various other *Brevipalpus* and *Tenuipalpus* spp. The biology and damage of false spider mites resemble those of spider mites, except that tenuipalpids produce no silk webbing. Most false spider mites are orange to red with dark spots. Their bodies are somewhat smaller and more flattened than spider mites or red mites, and their eggs are oval. False spider mite damage includes faint brown flecks or large chlorotic areas on the upper leaf surface, brown areas on the lower leaf surface, and stunted plant growth.

MANAGEMENT OF SPIDER MITES AND FALSE SPIDER MITES

Mite damage is usually not as serious as it looks. Most plants tolerate extensive leaf stippling or distorted tissue without being seriously harmed. However, high spider mite numbers on fruit trees and shrubs such as roses can cause premature leaf drop; this can result in sunburn and, if defoliation occurs during spring or early summer, reduced flowering or stunted growth the following season.

Generally, the most important mite management actions are to conserve natural enemies and provide proper cultural care, especially appropriate irrigation to keep plants healthy. Drought-stressed plants experience more mite outbreaks and are less able to tolerate pest feeding.

Regularly sprinkling foliage and washing plants during hot weather helps control some mites, in part by removing dust and improving the environment for predaceous mites. On shrubs and small trees where it is practical, forceful spraying of plants with water can reduce spider mite numbers adequately. Make sure to get good coverage, especially on the underside of leaves. Regularly repeat this foliage wetting at least once a day during hot weather, preferably in the morning.

High foliar nitrogen levels can favor outbreaks of some mites by increasing their reproduction. Do not apply nitrogen unless fertilization is truly necessary, avoid excess rates, and use less-soluble forms, such as urea and most organic fertilizers.

Natural enemies, especially predaceous mites, frequently control plant-feeding mites. Other important predators include the spider mite destroyer lady beetle (*Stethorus picipes*), sixspotted thrips (*Scolothrips sexmaculatus*), and brown and green lacewings.

A spider mite destroyer lady beetle eating a red mite. This pinhead-sized predator is highly efficient at locating spider mites. In coastal areas, a tiny, blackish predatory rove beetle (*Oligota oviformis*) also preys on spider mites. In comparison with *Stethorus* (shown here), *Oligota* is more elongated and has a pointed abdomen that curves upward at the rear. *Photo:* J. K. Clark

Sixspotted thrips is an important mite predator. This natural enemy is named for the three dark spots on each forewing. *Photo:* J. K. Clark

Predaceous mites are the most important natural enemies of plant-feeding mites. Shown here are two elongate, translucent western predatory mites, a dark-blotched spider mite (center), and spider mite eggs. *Photo:* J. K. Clark

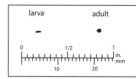

SPIDER MITE DESTROYER

larva   adult

0   1/2   1
|+++++++++++++++++| in.
   10   20   mm

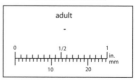

SIXSPOTTED THRIPS

adult

-

0   1/2   1
|+++++++++++++++++| in.
   10   20   mm

The *Stethorus* spider mite destroyer larva is dark gray to brownish and covered with numerous fine hairs. Here it is feeding on twospotted mites. *Photo:* J. K. Clark

Larvae of predaceous midges, such as this *Feltiella occidentalis*, feed on spider mites mostly in coastal areas. Larval color varies from whitish or pale yellow to brown, orange, or pink. *Photo:* J. K. Clark

Phytoseiulus persimilis eating a twospotted spider mite egg. A spider mite nymph and more eggs are to its left. Photo: J. K. Clark

Predaceous mite eggs, such as those of Galendromus annectens (bottom) are typically colorless and oblong. Plant-feeding spider mite eggs (top) are generally smaller, spherical, and colored to opaque. Photo: J. K. Clark

Avoid using broad-spectrum, residual pesticides (e.g., organophosphates, pyrethroids, and carbaryl) for mites or other pests. Even though their labels may say they control mites, these pesticides kill mite predators and can induce physiological changes in mites or host plants and, especially during hot weather, sometimes cause dramatic outbreaks of mites. If spraying is necessary, use selective miticides (also called acaricides) or horticultural (narrow-range) oil. Oil thoroughly applied to the underside of foliage at ½ to 1% active ingredient is very effective against spider mites that feed exposed on plants and has relatively low toxicity to predatory mites. Liquid sulfur and sulfur and soap combination products are effective in reducing populations of some mites on some plants, but sulfur can be phytotoxic to plants during warm temperatures. Avoid sulfur dust, which poses an inhalation hazard. Do not apply sulfur within 3 weeks of an oil spray.

Control dust and ants; they also disrupt natural enemies. For example, plant ground covers or other herbaceous plants to reduce dustiness. Place ant pesticide baits at the tree base or trim branches to eliminate ant bridges and apply sticky material barriers to tree trunks. The release of commercially available predaceous mites may be helpful on small plants in some situations, such as after reducing the number of pest mites with an oil application.

PREDACEOUS MITES

Predaceous mites include Euseius, Neoseiulus, and Phytoseiulus spp. and the western predatory mite (Galendromus =Metaseiulus occidentalis). Most species are long-legged, pear shaped, shiny, and fast moving. Many are translucent, although after feeding they often take on the color of their host and may be bright red, yellow, or green. Predaceous mite eggs are colorless and oblong in comparison with the eggs of most plant-feeding mites, which are commonly spherical and colored to opaque.

Predaceous mites commonly can be seen on the underside of leaves in the interior of trees. One way to distinguish plant-feeding mites from predaceous species is to observe mites closely with a good hand lens. Predaceous species appear more active and move faster than plant-feeding species; the predators stop only to feed. Many predaceous mites feed not only on all stages of plant-feeding mites, but also on insects such as thrips and scale crawlers or on pollen and fungi.

OTHER TYPES OF MITES

BROAD MITES AND CYCLAMEN MITE

Tarsonemids are primarily pests of herbaceous ornamentals in greenhouses or outdoors in humid growing areas. They are much smaller than spider mites and usually feed hidden within protected plant parts, such as flowers or terminal buds. Tarsonemids cause stunting and distortion of terminal shoots and leaves. Leaf margins often become thickened, leathery, brownish, and downward cupped.

Broad mite (Polyphagotarsonemus latus) has a tapered body that is widest between the second pair of legs and narrower toward the rear. Cyclamen mite (Phytonemus pallidus) has sides that are more nearly parallel, not sharply tapered.

GALL MITES, OR ERIOPHYIDS

Aceria, Eriophyes, Phytoptus, Trisetacus, and Vasates spp. are among the many eriophyids that can blister, discolor, or distort flowers, leaves, or shoots. Common hosts include alder, elm, fuchsia, grape, linden, maple, walnut, and willow. Eriophyids are named for a primary type of damage they cause, such as blister mites, bud mites, gall mites, and rust mites. For example, fuchsia gall mite distorts fuchsia blossoms. The cottonwood gall mite (Eriophyes parapopuli) causes dark, warty, woody swellings on twigs near the buds of cottonwoods and poplars. Feeding by erineum mites causes pale, felty or hairy patches of distorted leaf growth called erineum.

Adult and immature eriophyids have four legs, which appear to be coming out of the head. Their body is carrot shaped or wormlike and commonly yellow, pinkish, or white. Eriophyids are minute, and a microscope is required to clearly distinguish them. One detection method is to place infested plant tissue in a container with 90% ethyl alcohol. Shake this for about 10 seconds so tissue is thoroughly coated and eriophyids are killed and dislodged. At a magnification of about 25 to 30×, examine the fluid for pale, elongate eriophyids.

Eriophyids cause aesthetic damage and may reduce fruit yield, but most do not seriously harm woody landscape plants and can be tolerated. No controls are recommended for most eriophyid mites. If gall mites have been intolerable, narrow-range oil or wettable sulfur thoroughly sprayed on terminals just before bud break can reduce populations of some species. Do not apply oil within about 3 weeks of spraying sulfur.

Eriophyids (e.g., erineum mites and gall mites) are tiny and wormlike or wedge shaped. Eriophyids have four legs, which appear to be coming out of their head. *Photo:* J. K. Clark

FUCHSIA GALL MITE

Aculops fuchsiae causes fuchsia leaves and shoots to thicken, distort, and form irregular galls. The mites occur on growing tips year-round and in flowers. Because fuchsias grow best where summers are cool,

Gall mites caused these *Fuchsia magellanica* leaves to thicken and distort. Growing resistant fuchsia cultivars may be the only way to avoid the frequent tip pruning and pesticide application often needed to control fuchsia gall mites. *Photo:* J. K. Clark

Live oak erineum mites feed in yellow to orange felty masses in depressions on the underside of coast live oak leaves, causing the top of infested leaves to look blistered. A fungus also galls oak leaves and produces spore masses that resemble erineum mite damage. However, the spore masses can occur on either leaf surface, while the felty colonies of erineum mites usually occur only on the underside of oak leaves. *Photo:* J. K. Clark

Walnut purse gall mites, or pouch gall mites (*Eriophyes brachytarsus*), caused these harmless growths on California black walnut leaves. *Photo:* J. K. Clark

this mite is a particular problem in coastal California.

To reduce damage, plant only resistant fuchsias (Table 6-18) and consider replacing susceptible species. Prune or pinch off and destroy infested terminals. Pruning may be followed with two applications of a miticide, applied 2 to 3 weeks apart. Soap or oil sprays provide some control, but for eriophyids are less effective than synthetic miticides, which may only be available to professional applicators.

Table 6-18.

Susceptibility of *Fuchsia* spp. and Cultivars to Fuchsia Gall Mite Damage in California.

LOW SUSCEPTIBILITY OR RESISTANT[1]
Baby Chang, *boliviana*, Chance Encounter, Cinnabarina, Isis, Mendocino Mini, *microphylla* ssp. *Hidalgensis*, Miniature Jewels, *minutiflora*, Ocean Mist, *radicans*, Space Shuttle, *thymifolia, tincta, venusta*
MODERATE SUSCEPTIBILITY[2]
aborescens, denticulata, Dollar Princess, Englander, *gehrigeri*, Golden West, Lena, Machu Picchu, *macrophylla*, Pink Marshmallow, Postijon, *procumbens*, Psychedelic, *triphylla*
HIGH SUSCEPTIBILITY[3]
Angel's Flight, Bicentennial, Capri, China Doll, Christy, Dark Eyes, Display, Firebird, First Golden Anne, Jingle Bells, Kaleidoscope, Kathy Louise, Lisa, Louise Emershaw, Love, *magellanica*, Manrinka, Novella, Papoose Raspberry, South Gate, Stardust, Swingtime, Tinker Bell Troubadour, Vienna Waltz, Voodoo, Westergeist

1. No control needed.
2. Merely pruning off galled tissue whenever it occurs provides adequate control.
3. Pruning galled tissue followed by spraying may be necessary every several weeks to provide high aesthetic quality.
Sources: Koehler, Allen, and Costello 1985; Costello, Koehler, and Allen 1987.

LIVE OAK ERINEUM MITE

Eriophyes mackiei causes green to brown raised blisters on the leaves of coast live oak and other evergreen oaks. The mites occur in yellow to orange felty masses (erineum) in depressions on the underside of blistered leaves. Infested leaves may become curled or distorted but this damage is harmless to oaks. No control is known or needed.

Oak leaf blister fungus (*Taphrina coerulescens*) causes similar swellings on oak leaves and produces masses of pale spores on leaf surfaces that resemble the felty mass caused by erineum mites. However, *Taphrina* spores (asci) can occur on either leaf surface while the felty colonies of erineum mites usually occur only on the underside of oak leaves. Microscopic examination may be required to confidently distinguish between erineum mites and *Taphrina* spores. A dissecting binocular microscope is required to clearly distinguish whether eriophyids are present.

Pillbugs and Sowbugs

Sowbugs and pillbugs are not insects; they are soil-dwelling crustaceans related to crayfish. They feed primarily on decaying plant material, but also chew seedlings and succulent plant parts where they touch damp soil. If pillbugs or sowbugs are a problem, reduce the amount of decaying organic matter and minimize soil surface wetness. For example, keep compost and mulch back from plants, use drip irrigation instead of sprinklers, and water early in the day so surfaces are drier by evening. If these invertebrates are coming indoors, see "Home Invasions by Nuisance Pests."

Snails

Snails are mollusks that glide along on a muscular "foot," which secretes mucus that later dries to a silvery trail. They need moist conditions and have similar management and biology as slugs, except that snails have a conspicuous spiral shell. Because snails and slugs prefer succulent foliage near the ground, they are primarily pests of seedlings, herbaceous plants, and other low-growing vegetation. Much of the discussion below on snails also applies to slugs, but slugs especially are rarely pests of landscape trees and shrubs.

IDENTIFICATION AND BIOLOGY

Snails chew irregular holes in leaves, clip off succulent plant parts, and can also chew fruit and tender or young bark. Dried silvery trails on and around foliage, as well as chewed plants, indicate snail (and slug) activity.

Search protected places near the ground as described below to find snails during the day or inspect chewed plants at night using a flashlight. Snails are most active during mild, damp periods during the night and early morning. In mild-winter areas such as southern California and coastal locations, snails are active throughout the year. During cold weather, snails hibernate in the topsoil. During hot, dry periods, snails seal themselves off with a parchmentlike membrane and often attach themselves to shaded tree trunks, fences, or walls.

The brown garden snail (*Cornu aspersum*, formerly *Helix aspersa*) is the most

The brown garden snail and its feeding damage on citrus. Snails and slugs are primarily pests of seedlings and low-growing herbaceous plants, but they also chew fruit and the tender bark of young woody plants. *Photo:* J. K. Clark

Brown garden snail eggs are about 1/5 inch long, pale to dark orange or brown, and round or teardrop shaped with a protuberance at one end. Eggs occur in damp locations, such as on soil near this drip irrigation hose. If applying baits, make only spot applications in moist locations, preferably in areas shaded from the sun. *Photo:* J. K. Clark

common pest snail. Its mostly brown shell can include black, tan, and yellow in bands, flecks, and swirls and can grow up to 1 1/4 inches in diameter. Adult brown garden snails lay spherical to teardrop-shaped eggs that range from brown, yellow, to white. Mature snails deposit eggs in a loosely clumped group in sheltered locations on the soil surface or in a slight depression.

Other introduced pest species include the white garden snail (*Theba pisana*), which is established in parts of San Diego County, and milk snail (*Otala lactea*) in San Diego and Orange Counties. If you find these snails in other counties, report them to the county agriculture commissioner. Do not bring snails or slugs into California or move them to new locations. For more information, consult *Slugs* and the *Snails and Slugs Pest Notes*.

MANAGEMENT

Use a combination of methods that are effective against most species of snails and slugs. Avoid watering too frequently and irrigate early in the day so surfaces dry by evening. Reduce the places around susceptible plants where mollusks can hide during the day. The survivors congregate in the remaining shelters, where they are more easily located and controlled. Snail harborage includes boards, debris, dense ground covers such as ivy, leafy branches growing near the ground, stones, and weedy areas around tree trunks.

During the rainy season, or year-round in well-irrigated locations, regularly inspect for snails hiding in shelter that cannot be eliminated, such as low ledges under fences or decks, in sprinkler valve boxes, and near the ground on walls adjacent to vegetation. Frequent hand-picking can be effective; wearing rubber gloves may be desirable. Wooden squares about 12 inches on a side, raised off the ground by 1-inch runners, can be used to monitor and trap snails. Place one or two trap boards beneath each tree or group of shrubs. Check the boards and other hiding places every day the first week, every other day the second week, every 3 to 4 days the third week, and weekly thereafter. Crush or dispose of these pests.

Snails and slugs are repelled by copper. Copper foil or screen wrapped around planting boxes, headers, or trunks can prevent mollusks from crossing for up to several years. *Photo: J. K. Clark*

This predatory decollate snail can effectively control brown garden snails in southern California. However, it is illegal to import it into other areas of the state because it attacks ecologically important native snails and slugs. *Photo: J. K. Clark*

Containers of beer also attract slugs, and to a lesser extent, snails. Fill relatively flat containers, about the dimensions of an 8-ounce can of tuna, with beer and bury them with their tops about level with the soil surface. Replace the beer at least twice a week because fresh (not flat) beer is more attractive to snails and slugs, as are certain brands of beer. Keep irrigation sprinkler water out by covering each trap with an inverted pot with legs cut from the pot's rim so the snails have access to the trap. Because the beer must be replaced every few days and each trap attracts only those slugs and snails that are within a few feet, beer traps are not very effective relative to the amount of labor involved. Avoid their use where decollate snails are common predators or are being introduced, because these predators are killed in beer traps.

Barriers. Copper flashing, screen, or foil can be wrapped around planting boxes, headers, or trunks to repel snails for several years. Prune lower branches that touch the ground or other objects to eliminate "bridges" and keep snails out of trees.

Alternatively, brush Bordeaux mixture (a copper sulfate and hydrated lime mixture) on trunks to repel snails. One treatment should last about a year and can be prepared as discussed in *Bordeaux Mixture Pest Notes*. Sticky material applied to trunks to exclude ants and flightless species of weevils can also help to exclude snails, as pictured earlier under "Sticky Barriers." Materials including diatomaceous earth, wood ashes,

and sand have limited effectiveness as barriers if they remain dry, but do not provide long-term control.

Biological Controls. Predators of snails include birds, decollate snail, ground beetles, reptiles, and small mammals. Certain flies parasitize snails and slugs. To help improve natural enemies' contribution to pest control, avoid the use of broad-spectrum insecticides with long residuals.

The predatory decollate snail (*Rumina decollata*) consumes young to half-grown brown garden snails and has been effective in controlling snails in citrus orchards. However, because they also feed on succulent young plants to a limited degree, releasing decollate snails may not be desirable in herbaceous ornamentals or newly planted landscapes.

Decollate snail introductions in California are permitted only in certain counties in central and southern California. Releases in other areas are illegal because of the potential impact on native snail and slug populations of ecological importance in natural areas. Do not release decollate snail in the San Francisco Bay Area or northern California. Check with your local wildlife protection agency to determine whether decollate snail introductions are permitted in your area. For more information on releasing decollate snails, see *Citrus: UC IPM Pest Management Guidelines* or *Integrated Pest Management for Citrus*.

Pesticide Baits. Molluscicide baits can temporarily control snails and slugs but are

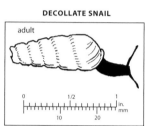

DECOLLATE SNAIL

most effective when used in an integrated program that includes reducing hiding places, using barriers, and conserving natural enemies. Baits are toxic to the decollate snail, and products containing metaldehyde can poison and even kill dogs and other mammals that might feed on it. Iron-based baits (e.g., iron phosphate) are effective and relatively safe for pets and wildlife.

The time of baiting is critical; it should be during a cool, damp period—when snails are most active—before dry, warm weather begins. If the soil surface is dry, irrigate before applying bait to promote snail activity. Make spot applications of bait instead of widespread applications. Apply bait lightly in a narrow strip around sprinklers or in other moist and protected locations. Plant-eating snails are drawn to these locations, but predaceous snails are less affected than when bait is widely dispersed. Do not make molluscicide applications if heavy rain is expected soon, and avoid irrigating overhead for several days after application, as baits decompose upon exposure to excessive moisture.

Chapter 7   Weeds

We think of weeds as unattractive or noxious species,
but some plants can be desirable in one setting and a weed in
another. For example, turf and vigorously spreading ground covers need
to be controlled when they invade areas under trees and shrubs. Certain
trees and shrubs produce abundant seed that disperse widely and often
grow where they were not planted and are not desired. The potential
problems with weeds around trees and shrubs include the following.

- Plants near trunks are commonly trimmed; the mowers or string trimmers often
  wound trunk bases, which can girdle and kill plants or make them susceptible to
  certain diseases and boring insects.
- Weeds compete with desirable landscape species for light, moisture, and nutrients;
  newly planted trees and shrubs especially can be stunted because of nearby competing vegetation.
- Weedy herbaceous species provide habitat for snails, rodents, and other pests.
- Certain weeds injure people, such as causing a rash from exposure to poison oak
  and allergic reactions to weed pollen.
- Weeds can be a visual blight in the landscape.
- Certain introduced ornamentals, such as arundo, blue gum eucalyptus, brooms,
  Himalayan blackberry, iceplant, pampas grass, and tamarisk, have spread from
  landscapes and become invasive pests in parks and natural wildlands.

Prevent and manage weeds using a combination of effective methods, as discussed in
this chapter and in the *Weed Management in Landscapes Pest Notes* and the *Pest Notes* on
specific weed species online at ipm.ucanr.edu. Weed management in annual bedding
plants, ground covers, nurseries, and turfgrass is discussed in publications listed in "Suggested Reading" at the back of this book.

Turf has stunted this
tree's growth and (in the
background) invaded the
shrub beds. Keeping other
plants away from the
trunks of shrubs and trees
improves their growth and
appearance and reduces
the risk that trunks will be
wounded by string trimmers
or mowers. *Photo:* J. K. Clark

Weed Management before Planting

For the most effective management, control weeds before planting, before weeds emerge, and before weeds produce perennial structures, seed, or vegetative propagules (Figure 7-1). Established perennial weeds can be especially difficult to manage after landscapes are planted. If you wait for weeds to mature and become a problem in landscapes, your management options will be more limited and results will be less satisfactory than if you take preventive measures.

Before planting landscapes
- Assess and monitor the site to identify problems.
- Prepare the site. Remove weeds (especially perennials) and amend, loosen, and grade the soil before planting to provide for good landscape plant growth.
- Design or redesign landscapes to avoid weeds. For example, separate plants with growth characteristics or cultural care needs that are incompatible with nearby plants.
- Avoid introducing weeds and manage those that develop during site preparation, such as when cultivation brings buried seeds closer to the surface, where they are more likely to germinate.
- Encourage the rapid establishment of desired plants. Plant correctly, take steps to prevent weed growth, and adequately care for trees and shrubs so they fill in the site and displace and exclude weeds.

ASSESS AND MONITOR THE SITE

Identify and remedy any adverse site conditions so desirable species will grow well and outcompete weeds and problems will be easier to manage. Before planting and when the weeds are visible, evaluate soil conditions, mulch, and the exposure, shading, and slope of the site so problems can be corrected or anticipated before

planting. Investigate and correct drainage, soil compaction, and water infiltration rate problems as discussed earlier in Chapter 3, "Growing Healthy Trees and Shrubs."

Identify the species and abundance of weeds present, especially perennial weeds, and control them before grading and developing the site. Perennials and summer annuals are easiest to identify in mid to late summer; the best time to look for winter annual weeds is mid to late winter. Maintain written records of monitoring information, especially when managing commercial or large-scale landscapes. Weed species and their abundance change seasonally and in response to your management actions. Good records help you evaluate and improve your management program. Good records include
- the species, age (seedling or established), location, and density of weeds
- time of year when weeds emerge
- what action you took and the date, location, and approximate size of any area treated

- comparison and evaluation of monitoring results from different dates to determine
  * whether problems are increasing, decreasing, or remaining about the same
  * how effective were previous management activities
  * whether control action is needed or practices should be modified to increase the effectiveness of weed management

Landscape Maps. One monitoring method for weeds and other pests is to use landscape maps. Draw a sketch of the landscape area and mark the location of landscape plants, mulch, bare soil, and ground coverings. Identify individual large plants or groups of smaller plants. Survey the landscape periodically (such as every fall and spring and before and after control actions) and use a copy of the map to record the location, date, and severity of weed (or other pest) problems. For example, rate each different landscape area from 0 to 4, where 0

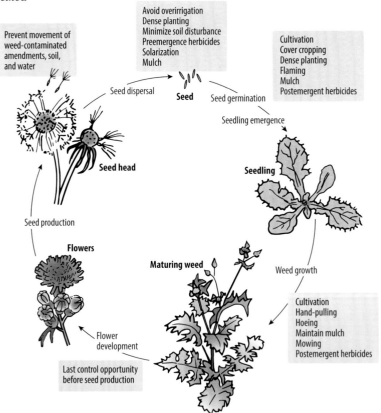

Figure 7-1.
Methods and timing for annual weed control, illustrated here with annual sowthistle. The most effective times for weed management are before planting, before weeds emerge, and before weeds produce seed or vegetative reproductive parts.

ct landscapes regularly to detect problems early and mine whether actions such as reapplying mulch are nted. Learn to identify young plants of common weeds u can control them before they flower. *Photo:* J. K. Clark

means very few weeds and 4 means a very heavy infestation (Figure 7-2). For greater detail, you can use a transect count method and assign quantitative weed ratings. Be sure to date all maps and notes.

For large or commercial sites, it may be helpful to record the location and abundance of weeds or other pests using GPS (global positioning system) or a computerized database or GIS (graphical information system). Smartphones, tablet computers, or other mobile devices with appropriate software may facilitate collection of this information.

PREPARE THE PLANTING SITE

Eliminate established weeds and reduce future weed growth before starting grading, development (e.g., installing an irrigation system), or planting (Table 7-1). Especially

reduce or eliminate perennials, such as bermudagrass, field bindweed, and nutsedges before planting or moving soil.

To control emerged annuals and certain perennial weeds, you can apply a broad-spectrum (nonselective) postemergence herbicide (e.g., glyphosate). In some situations, you can use cultivation, sheet mulching, or soil solarization. To reduce annual weeds and the soil seed bank, irrigate soil and apply a preemergence herbicide or shallowly cultivate (or spray with herbicide) the weeds when they are young. For established perennials, often the most practical method is to apply a translocated (systemic) herbicide to foliage.

SHEET MULCHING

After controlling weeds (e.g., by applying herbicide or mowing or knocking down

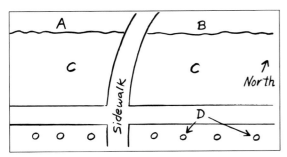

Date 10 April, 2015 _____ Landscape monitored by Jane Smith _____

Location 400 Main Street, City Hall _____

Weed infestation level ratings:

0 no weeds   **2** moderate, 3-10% weeds   **4** very heavy, >25% weeds

1 light, ≤1-2% weeds   **3** heavy, 11-25% weeds

Landscape Area	Weed infestation level	Most common or problematic weed species present	Management action planned / comments
A. Roses	1	Annual bluegrass, Annual sowthistle	Hoe weeds then apply more mulch
B. Roses	0		Apply more organic mulch
C. Turf	1	California burclover	Dethatch, aerate, fertilize, overseed
D. Young trees	4	Kikuyugrass	Remove back from trunks

Notes: Monitor again in May after completing control actions.

Figure 7-2.

A landscape map for recording weed infestations. Mark the location of large plants, groups of plants, and other major features. Monitor the landscape regularly and use a copy of the original map to record information such as weed species, infestation levels, and locations. You can also use this method to record when and where other problems occur, such as diseased plants and insect infestations. For large landscapes you can use Google maps (e.g., satellite view) and overlay a grid so you accurately identify and locate problem areas. Paper records from the field can be entered into a premade spreadsheet or database for electronic record keeping.

Solarization before planting controls insects, nematodes, many soilborne diseases, and certain weeds within several inches of the surface. To solarize, cover bare, moist, smooth soil with clear plastic for at least 4 to 6 weeks during warm, sunny weather. *Photo:* J. K. Clark

weeds so all vegetation lies flat), heavy paper commercially available in rolls (sheet mulch) can be applied to the soil surface to exclude light, causing emerged annuals to die and at least suppressing the growth of perennials. After application, you can make holes in it and plant through the sheets, similar to the methods used for geotextiles discussed later under "Inorganic and Synthetic Mulches." This sheet mulching can be useful to decrease weed pressure until newly planted ground cover fills in open spaces and grows enough to shade out weeds. However, paper hinders air and water movement, so rain and overhead irrigation tend to run off, which can favor weed growth at the paper's edge. Before applying sheet mulch you may need to install a drip irrigation system beneath the paper; and it can be difficult to tell whether soil moisture beneath sheet mulch is being maintained at levels appropriate for healthy plant growth.

CULTIVATION

Soil cultivation cuts, uproots, or buries weeds. Rototill with a shallow setting or manually cultivate before planting to kill young annual weeds. Reduce the number of weed seeds by irrigating the soil to promote germination, then very shallowly cultivate again in about 1 to 3 weeks after seedlings emerge. Repeat these steps (water, wait, then cultivate) two or more times before planting. Avoid using heavy equipment on moist soil, which can compact the soil. Do not till more than 2 inches deep, as this brings buried weed seeds to the surface, where they are more likely to germinate.

Make each subsequent cultivation shallower than the last to minimize movement of buried seeds to the surface. To minimize subsequent soil disturbance, the area must already be at final grade. If possible, install the irrigation system before cultivation.

Although most effective against annuals, cultivation also controls young perennial weeds if it is conducted before they have stored much carbohydrate or produced rhizomes, stolons, or tubers. In general, plants should be cut before new top growth exceeds 6 inches, and cultivation must be repeated when plants regrow. This repeated damage and disruption requires perennials to continually draw on carbohydrate reserves, which will eventually be depleted, killing the plant.

Some established perennials in dry inland valleys can be controlled by cultivating every 1 to 3 weeks during the summer without any irrigation and allowing the soil and weeds to dry. This controls bermudagrass, johnsongrass, kikuyugrass, and nutsedges. Repeated cultivation is generally required for more than 1 year to control field bindweed. Cultivation to control perennials is effective only when soil is warm and dry, so use other methods during cool, rainy weather and in coastal areas. Cultivating perennial weeds when soil is moist or followed by irrigation or rain can increase their spread. Do not dry-cultivate perennials unless the process can be repeated as necessary; otherwise, tillage can increase weed problems by spreading propagules that can become established in the landscape.

SOLARIZATION

Solarization (tarping soil to retain the sun's heat and raising the soil temperature to a level that is lethal to seeds) before planting can effectively control most annual and certain perennial weeds for 6 months to 1 year (see Table 7-2). Solarization also reduces nematodes and many soilborne

Table 7-1.

Weed Management Methods Summary.

PLANTING TYPE AND COMMENTS	MANAGMENT RECOMMENDATIONS
Trees and Shrub Beds. Dense shade under established plantings reduces weeds. Preplant weed control is less critical than in other types of plantings. A combination of methods is generally needed to completely control weeds.	• Control perennial weeds before planting, although control may be possible after planting. • Apply a thick layer of organic mulch or use geo textile fabrics under a shallow mulch layer. • Apply preemergence herbicide if needed. • Manually remove (e.g., hand-pull or hoe) weeds, make spot applications of postemergence herbicides (e.g., translocated glyphosate for perennials), or both.
Mixed Plantings of Woody and Herbaceous Species. Site preparation and weed control before planting are especially critical. Weed management after planting is complex because different types of desirable species are present. Different areas of the planting often require different control actions. Because of their phytotoxicity to certain desirable plants, relatively few postplant herbicides are available.	• Control perennial weeds and reduce the soil seed bank before planting and during at least the first two growing seasons. • Plant the woody species first, then plant the herbaceous species after two growing seasons. • Plant close together to shade the entire soil area. • Group together plants that tolerate the same management practices (e.g., the amount and frequency of irrigation they need).

Adapted from Wilen and Elmore 2007.

plant pathogens. For effective solarization, users must recognize and avoid its shortcomings.

Cover the soil with clear plastic for about 4 to 6 weeks during the warmest times of the year. Solarization is most rapid and effective if conducted during a period of intense sunlight with little wind. Solarization requires more time and is less effective in cloudy, windy coastal areas. In California's Central Valley, solarize soil from late May through September. In south coastal California, mid-July through August is usually the sunniest, calmest time of year. Along coastal areas of central and northern California, August to October and May through June are usually best; fog or wind can be at a minimum then.

To solarize effectively (Figure 7-3):

- Clear any vegetation from the area to be solarized. Scrape vegetation off, mow weeds closely to ½ inch tall and rake the soil free of cuttings, or rototill no deeper than 4 inches.
- Grade soil and otherwise prepare it so it is ready for planting and any subsequent disturbance will be minimized.
- Irrigate soil thoroughly just before covering it. Lightly work the soil surface to even it, then irrigate again if the soil surface has dried.
- Be sure the ground is smooth and free of clods and trash so that the plastic lies very close to the soil surface. Air gaps created by clods or air pores in dry soil are poor conductors of heat.
- Cover moist, smooth, debris-free soil with intact, clean, clear plastic tarps, preferably UV-inhibiting treated plastics 1.5 to 2 mil thick.
- Seal seams with clear plastic tape and seal the edges with soil to retain heat and to keep the tarp from blowing off.
- Protect the tarp surface from punctures during solarization to achieve the highest soil temperatures. If holes form in the plastic while on the soil, seal them with clear patching tape.
- Keep sprinkler irrigation of nearby plants away from the plastic to prevent evaporative cooling from reduc-

ing the effectiveness of solarization.
- Plant soon after removing the plastic and minimize soil disturbance as much as possible.
- After solarization, avoid deep cultivation. Working the soil deeper than about 3 inches may bring weed seeds to the surface that were buried too deeply to have been exposed to temperatures high enough to kill them.

Using a double layer of plastic with an air space of at least ½ inch between layers can raise soil temperatures from 2° to 10°F higher than when using a single layer. Using a double layer requires more preparation and expense but can make solarization more feasible in areas with a cooler climate. Consult *Soil Solarization for Gardens & Landscapes Pest Notes* and *Soil Solarization* for more information.

Table 7-2.

Weed Susceptibility to Soil Solarization.

WEEDS	CONTROL EFFECTIVENESS
annual broadleaves and grasses	most species well controlled (see exceptions below)
bermudagrass and johnsongrass	controlled if rhizomes are not buried deeply
common purslane, large crabgrass, lovegrass, wild oat	partially controlled
field bindweed, purple and yellow nutsedges, some clovers	inconsistent or little or no control, even under favorable conditions

See the *Soil Solarization for Gardens & Landscape Pest Notes* online at ipm.ucanr.edu for a more complete list of the species controlled by solarization.

REDESIGN LANDSCAPES TO AVOID WEEDS

Consider redesigning and replanting landscapes that are difficult to manage or frequently reinfested with undesirable plants. Grow landscape species that are well adapted to local conditions, including light, moisture, soil, and temperature and choose species or cultivars that resist common insect and disease problems (see Table 3-2). Plants that are appropriate for their location grow more vigorously and are better able to compete with and exclude weeds.

Group together (cluster) plants with compatible cultural requirements and avoid planting species together that have incompatible growth characteristics. For example, bermudagrass and kikuyugrass

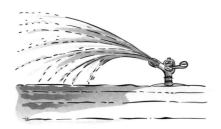

Figure 7-3.

Follow these four steps to effectively solarize soil: cultivate and remove plant material; level and smooth the soil surface; thoroughly wet the soil; and cover soil with clear plastic for 4 to 6 weeks during warm, sunny weather.

This concrete barrier allows regular mowing to help keep rhizomatous (running) turfgrass from invading the shrubs. Use driveways, sidewalks, headers, or edging to separate plants with different cultural requirements and to contain invasive species.
Photo: C. A. Reynolds

Avoid planting species that will create ongoing management problems. This vining ivy grows prolifically in irrigated landscapes. Unless the ivy is cut back regularly, its growth is likely to cause the tree to gradually decline and die prematurely.
Photo: J. K. Clark

are hardy, drought-tolerant turfgrasses, but they spread by rhizomes and stolons and invade nearby areas planted with shrubs. Minimize invasiveness by planting ground covers and shrubs in larger clusters rather than several smaller groups, each bordered by turf. Avoid planting in ways that do not allow weeds to be managed (e.g., leaving insufficient space to use a trimmer). Install landscape edging to reduce spread into a landscape bed. For example, a concrete

strip or header between a planting bed and turfgrass facilitates closer mowing.

Install enough plants so that desirable species shade most of the soil; denser plantings will have fewer weed problems, but be sure each plant receives enough sunlight and soil moisture to grow well. Interplant annuals or short-lived or low-growing perennials between longer-lived or taller-growing species with the expectation that some plants will die out as the others grow and fill the site. Using mulch to cover bare ground between shrubs and trees is a very effective method of managing weeds.

Choose drip irrigation or microsprinklers instead of high-volume or flood sprinklers so water is placed at or near your plants. Watering large areas where there are no desirable plants will encourage weeds to grow there.

Edging and Headers. Install deep barriers, headers (e.g., mowing strips),

planter beds, borders (e.g., sidewalks or other hardscapes), or other types of edging (e.g., flexible plastic) to separate plant types and reduce weed encroachment (Figure 7-4). Within each area bordered by edging, headers, or pavement, grow a single plant type (e.g., all shrubs or all turf) so the entire section can be managed the same way and plantings are less likely to become weeds by overgrowing adjacent landscapes.

PREVENT WEED INTRODUCTIONS

Weeds are often inadvertently introduced in landscape material and with new plants. Container plants and manure, mulch, organic fertilizers, topsoil, and certain other soil amendments are often highly contaminated with weed seeds or vegetative propagules (parts that can grow into new plants), such as rhizomes, stolons, or tubers. Inspect the potting media surface of container plants for weeds. Check to see whether commercially obtained manure, mulch, or soil containing plant-derived material has been pasteurized, such as by proper composting or steam treatment, to reduce the likelihood of being contaminated with weeds. Do not bring in or move around the site anything unless it is known to be weed-free.

Dispose of weeds carefully so they do not become problems. For example, where

Figure 7-4.
Separate shrub areas from invasive turfgrasses and ground covers by installing edging, headers, or mow strips: concrete, metal, plastic, or wood barriers extending at least several inches belowground and about an inch aboveground.

Header

This French broom produces large numbers of seed and vigorously resprouts from roots after shoots are cut. Brooms were introduced as ornamentals, but they spread and crowd out more desirable native vegetation, create a wildfire hazard, and are poisonous to cattle and horses. When selecting landscape species, consult resources such as the California Invasive Plant Council (cal-ipc.org) and Plant Right (plantright.org). *Photo:* J. M. DiTomaso

Weeds will grow where soil is bare, moist, and exposed to sunlight. Good weed control before planting, covering soil with mulch, and using drip irrigation instead of sprinklers can prevent weeds from competing with young trees and shrubs and result in the desired plants quickly shading out weeds. *Photo:* D. Rosen

greenwaste is reused, cut off any reproductive parts, seal them in a plastic bag, and dispose of the weed propagules in the trash before composting the nonreproductive parts or allowing the greenwaste to be collected and moved off-site.

Weeds can spread as windblown seed or in soil and water runoff. Manage weeds in adjacent sites where feasible, plant windbreaks as barriers to airborne seeds, and prevent water drainage into landscapes from infested sites. Surface water (from ponds or rivers) is often highly contaminated with weed seeds; filter surface water to exclude seeds before using it for irrigation.

Mowers and other landscape implements and even footwear may be contaminated with weed seeds or vegetative propagules. Clean equipment, tools, and shoes well before moving them from weedy to weed-free areas.

EXCLUDE INVASIVE PLANTS

Prevent new pest introductions during planting and travel:

- Don't plant a pest, such as invasive brooms, pampas grass, and certain eucalyptus and ivy.
- Use alternatives to invasive garden plants, such as those recommended by the California Invasive Plant Council (cal-ipc.org) and Plant Right (plantright.org).
- Obtain plants and landscaping materials from reputable, local nurseries.

- Do not bring fruit, plants, seeds, or soil into California unless you know they are certified to be pest-free or were inspected by agricultural officials.

Take any unfamiliar pests to your county agricultural commissioner or University of California (UC) Cooperative Extension office for identification or telephone the California Department of Food and Agriculture's Pest Hotline, 1-800-491-1899. See the *Pest Notes: Invasive Plants* and *Woody Weed Invaders* for more information.

ENCOURAGE ESTABLISHMENT OF DESIRED PLANTS

Provide trees and shrubs with optimal cultural care and growing conditions so plants establish as quickly as possible and develop a healthy canopy that outgrows and shades out nearby weeds. Use mulch, preemergence herbicide, and other methods to prevent weeds from sprouting while new plantings are getting established.

Weed Management in Plantings

Weed management depends on the types of landscape plants and the species, growth stages, and life cycles (annual, biennial, perennial) of weeds. Control options include hand-pulling, heating, hoeing, mowing, mulching, and herbicide applications. A combination of appropriate methods usually provides the best control. For instance, after hand-pulling, hoeing, or postemergence herbicide application, discourage subsequent growth of

annual weeds by minimizing soil disturbance, using low-volume irrigation, and carefully choosing, applying, and maintaining mulch in the landscape. See Table 7-1 for a summary of weed IPM methods. If you take these preventive actions and control perennial weeds before planting, regular hand-pulling or hoeing of occasional weeds may be all that is needed once landscape plants become established. If landscapes have chronic problems or require excessive management, consider redesigning and replanting them as discussed above.

HAND-PULLING

Pull weeds by hand to selectively remove undesirable plants. Although it can be time consuming, hand-pulling is a preferred method when weeds are scattered throughout landscapes, relatively few in numbers, or will produce seed unless promptly removed. Remove the entire

Organic mulches can provide a pleasant, natural appearance and improve plant growth as well as control weeds. *Photo: J. K. Clark*

Good long-handled tools for removing individual weeds have a small fork or other soil-penetrating device, a fulcrum portion for leverage (e.g., a curve between the fork and the handle), and a comfortable handle grip. Shown here are example weed pullers and twisters. *Photo:* K. Windbiel-Rojas

crown and roots by pulling weeds when they are young and soil is loose or wet. Removing just the aboveground part temporarily improves landscape aesthetics, but perennials and other established weeds often regrow from the parts remaining underground. To make it easier to remove roots

- Irrigate dry soil 1 or 2 days before hand-pulling.
- Grasp weeds firmly near their base and rock them back and forth several times to loosen roots.
- Steadily pull (do not jerk) weeds upward.
- Place weeds into covered containers or bags and dispose of them away from landscapes to remove any seeds and prevent discarded weeds from rerooting.

Various weed poppers and long-handled tools are available (Figure 7-5) to remove individual weeds and their roots by hand while minimizing soil disturbance. The best ones have a small fork (or even no fork), a fulcrum for leverage (the curve between the fork and the handle), and a handle that is comfortable to grip. Long-handled tools with nail-like tines that go into the soil to grab the weed and a retractable plunger to shoot the weed off the end allow you to remove and dispose of individual weeds without soiling your hands. A drawback of these tools is that they take a large soil plug and leave a hole where the weed was removed.

Various manually operated, mechanical pullers are available to uproot undesirable shrubs and sapling trees with minimal soil

disturbance. A popular type was originally designed to remove invasive brooms from uneven terrain without vehicle access. It has steel jaws to grip the trunk, a basal support stand, and a lever bar used to pry the basal roots from the soil. To effectively control established perennials that reproduce vegetatively, it may be necessary to dig up and destroy all underground stems (rhizomes) and tubers that can grow into new plants or to hand-pull regrowth repeatedly until the carbohydrate reserves underground are depleted.

HOEING

Use a hoe to cut weeds at or slightly below the soil surface. Use scuffle hoes, also called shuffle or hula hoes, to cut weeds

off at their basal stem by scraping back and forth with a push-pull motion on the surface of dry soil. Keep blades sharp so that weeds are easily cut. Unlike soil cultivation used before planting, do not loosen or dig in the soil. Hoe at the soil surface or no deeper

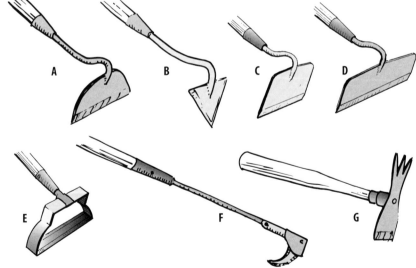

Figure 7-5.

Hand-weeding tools. Hand-weeding is more effective and less work when you use tools designed for that purpose and you keep the tools sharp. A–D. Hoes for shallowly chopping or cutting off weeds' basal stems. E. Hula, scuffle, or shuffle hoe for scraping back and forth using a push-pull motion on the surface of dry soil to cut weeds. F. Briar hook for pulling weeds. G. Mattock, which has heads like an ax and pick for digging out roots and small stumps.

than about ½ inch to reduce the likelihood of injuring roots of desirable plants, minimize damage to any layer of preemergence herbicide, and limit exposing new weed seeds to the surface for germination.

The best time to hoe is when the soil surface is dry and no rain is expected for at least several days after hoeing. Allow 3 to 5 days for weeds to fully dry out before irrigating.

Hoeing is most effective against annual species, since many perennial weeds can regrow from severed roots or stems. Young broadleaf weeds are easily controlled by cutting them at the soil surface. Before hoeing grasses, hand-pull several of them and determine the depth of their crown (the point where roots and stems meet, as in the photograph). Cut weeds at or just below their crown, which for grasses is often about ¼ inch below the surface.

MOWERS AND STRING TRIMMERS

If you choose to tolerate herbaceous species around trees and shrubs, mow or cut them (e.g., use a string or blade trimmer) before they bloom and form seed. Most annual broadleaf weeds are effectively controlled, but annual grasses may regrow because their growing points are usually just under the soil surface. Most perennials are not well controlled, but cutting will make them less noticeable. Be aware that repeated mowing without other control methods can induce weed shifts to low-growing perennial broadleaves and grasses. Some weeds (e.g., smooth crabgrass and spotted spurge) produce seed on plant parts below mowing levels.

Prevent mower and string-trimmer wounds to trunks and root crowns as this can girdle and kill young trees or promote attack by pathogens and wood-boring insects. Make a trunk guard by splitting a segment of flexible plastic pipe lengthwise and place it around small trunks. Instead of mowing or cutting, hand-pull, hoe, or apply herbicides and mulch to keep other plants at least 6 inches away from the base of trunks.

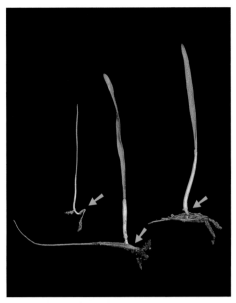

Grass seedlings carefully removed from soil to reveal their crowns (where stems meet roots, marked by arrows). To control grasses by hoeing, you must cut them at or below crown level, which is often about ¼ inch below the soil surface. These are seedlings of littleseed canarygrass (left), ripgut brome (center), and wild oat (right). *Photo:* J. K. Clark

FLAMERS AND OTHER WEED HEATERS

Herbaceous weeds can be controlled using special hand-held flamers (most fueled by propane) or infrared heaters without an open flame. Heating controls annual weeds in bare earth, along fencerows, in pavement cracks, and in certain other areas.

With flamers, only briefly touch the basal stem area of herbaceous weeds with the tip of a flame; heating the stems damages their cells and causes the plants to die. Proper flaming technique should not visibly char or burn plants or create smoldering vegetation or air pollution. It is not necessary to flame the foliage because all aboveground parts die if the basal stem is killed. Flaming is best done in early morning or late evening when winds are low and any open flame is more visible.

Plants may wilt, change color, or appear unaffected soon after heating them. Even if no change in the weeds is evident immediately, heating causes plants to yellow and

Cut weeds near the soil surface when soil is dry by using a sharp scuffle hoe. Avoid digging under the soil surface, as this exposes seeds, which germinate. *Photo:* J. K. Clark

String trimmers cut or break herbaceous species and small-diameter woody weeds. *Photo:* J. K. Clark

Use trunk guards to protect young trees and thin-barked species when using weed trimmers or mowers nearby. Plastic pipe that is several inches in diameter and split lengthwise can be spread and placed around small trunks. *Photo:* J. K. Clark

Weeds being killed by heating them with a propane-fueled flamer. Only brief contact with heat on the basal stem is needed to damage herbaceous plants, as with these that wilted after being flamed and will die within a few days. Do not flame dry wood chips, dry vegetation, and other flammable materials. Wet conditions (e.g., after rain or irrigation) can be good times to flame. Keep fire suppression materials handy, and use flamers properly to avoid causing a fire. Photo: J. K. Clark

die within several days. Broadleaf annuals and seedlings are most susceptible to flaming; grasses and established perennials are only partially controlled and often regrow. In general, flaming does not kill weed seeds in the soil seed bank. Flame or heat weeds when they are less than a few inches tall, especially less-susceptible species such as grasses.

Fire is a serious hazard when flaming weeds. Wet conditions can be good times to flame: during the rainy season or after a thorough irrigation, when surfaces are moist and humidity is high. Use good judgment to identify hazardous situations where flaming should be avoided. Do not flame dry vegetation, dry wood chips, other flammable materials, or immediately adjacent to structures. Keep the flame away from desired plants. Keep fire suppression equipment handy, such as a fire extinguisher, shovel, and water, in case of accidents.

HOT WATER OR STEAM

Weeds can be killed by applying steam or superheated hot water or using proprietary equipment that applies a hot surfactant foam to foliage and stems. Treatment is most effective against annuals and young weeds but will not control weed seeds. Grasses and established perennials can regrow, requiring retreatment. Spot application is most feasible, such as weeds growing along fence lines, edges of rights-of-way and paths, or in pavement cracks. Treatment of large open areas is more time consuming than with herbicides. In comparison with flamers, work proceeds at a slower pace because the temperatures are lower and weeds must be exposed for a longer time to kill them. Most equipment is for commercial situations or professional users, not for homeowner use, and commonly requires large amounts of water and good site access. Boiling water from a stove is not very effective because water rapidly loses heat and may not be hot enough to kill weeds by the time it is applied to the weed.

MULCH

Mulch is a layer of material placed on soil to exclude sunlight and suppress weed germination and growth. In addition to good weed control, mulch conserves soil moisture by reducing evaporation and reducing water use by weeds. Mulch moderates the soil wetting and drying cycle between irrigations and moderates soil temperatures around roots, improving plant growth. Mulch also reduces compaction and erosion from irrigation, rainfall, and foot traffic.

Effective mulching is initially time consuming, but it greatly reduces ongoing weed management costs. When mulch is promptly applied to an adequately prepared site, especially after removing perennial weeds and their propagules, occasional hand-pulling or shallow hoeing of weed seedlings and periodic reapplication of organic mulch may be the only weed management activities necessary. If perennials

become established, other practices such as herbicide use may be necessary.

Organic (plant-derived) mulches include bark, lawn clippings, leaves, greenwaste, and wood chips (Table 7-3). Nonorganic mulches include crushed rock, polyethylene plastic, and geotextiles (landscape fabrics). In comparison with organic mulches, nonorganic mulches are generally longer lasting but more expensive. In plantings of trees and shrubs, a combination of geotextile covered with bark or wood chips provides relatively long-lasting weed control. Where landscapes are periodically replanted (e.g., annual beds), organic mulch alone is often the best choice because problems occur when geotextile or plastic partially breaks down or needs to be removed and when rock mulch mixes with soil. Keep flammable mulches away from structures in fire-prone areas.

Organic mulches have the major advantage over nonorganic mulches of gradually improving soil quality as they decompose and they typically improve water penetration by reducing water runoff and increasing infiltration. Many organic mulches are attractive, contrasting nicely with foliage and flower colors and providing a pleasant "natural" appearance and aroma. Organic mulches are often readily available on-site or are available for free or minimal cost from landfills, arborists, or tree maintenance companies.

Mulch can also be combined with a preemergence herbicide. The mulch type and thickness, herbicide placement on top versus beneath mulch, and preemergence herbicide efficacy are among the important considerations when combining mulch and preemergence herbicide. See *Weed Management in Landscapes Pest Notes* for specific herbicide recommendations for use with mulch.

Organic Mulches. Apply organic mulch no deeper than about 3 to 4 inches to minimize air and water penetration problems. The thickness or depth of mulch needed to adequately suppress weed growth depends on the mulch type and weed pressure.

The larger the particle size of the mulch, the greater the depth of mulch required to effectively exclude light from the soil surface. Particles larger than about 1½ inches generally do not provide good control because weeds commonly grow through the spaces between pieces.

Wood chips or bark nuggets with particle sizes ½ to 1½ inches resist decomposition and provide long-term weed control when applied up to 4 inches deep. Medium-sized bark chips and ground wood (particles of about ¼ to ½ inch) are moderately resistant to decomposition and provide good weed control when applied 2 to 3 inches deep.

Fine particles (e.g., sawdust and well-composted greenwaste) are less suitable for mulching and should be applied no deeper than about 2 inches because fine particles pack tightly. When watered overhead, fine-textured mulches can remain undesirably soggy and waste water because they require frequent irrigation to adequately wet the root zone. Fine particles decompose relatively quickly and, if too decomposed, favor weed seed germination and growth. For example, composts used for mulch may initially suppress weeds, but as they break down they can soon become a nutrient-rich bed for seed germination and weed growth.

Organic mulches gradually move, settle, and decompose into the soil, which improves soil quality (e.g., tilth) and improves plant growth. Periodically reapply organic mulch to a depth needed to maintain good weed suppression, unless plants grow together and shade the soil. For example, if 3 inches of mulch is used initially, reapply an additional 1 inch if the depth of the mulch has been reduced by that much. Do not wait until you see bare soil to reapply. If weeds grow through or germinate in mulch, hand-pull them or lightly rake or shallowly hoe the mulch to remove the weeds when they are young.

Inorganic and Synthetic Mulches.
Among these types (Table 7-4), geotextiles, also called landscape fabrics, are the most recommended for long-term weed control in woody plantings. Although relatively expensive, geotextiles are durable, and their spun or woven polyesters and polypropylenes allow air and water to pass through them. Do not use them in periodically replanted areas (e.g., annual flower beds) or where fabric would inhibit the rooting and spread of desirable plants (ground covers). Tree and shrub roots penetrate landscape fabrics; if fabric that has been in place longer than about 5 years is removed, tree and shrub root systems can be damaged.

Before installing fabrics, smooth the soil and remove any sharp objects. To plant first and install the fabric afterward, cut an X through the fabric, place this opening over the plant, and pull the fabric down over the plant. After laying the cloth close to the ground, use U-shaped nails to peg it down. If laying down fabric before planting, cut an X through the fabric and dig the planting hole. Avoid leaving soil on top of the fabric, then plant through the hole.

Fold the cut fabric back down as continuously as possible and secure flaps

Table 7-3.

Organic Mulches.

MATERIAL	COMMENTS
bark chips and ground bark	Attractive, slowly improves soil as it gradually decomposes. Relatively expensive. Medium- to coarse-sized particles are long-lasting and resistant to wind movement; fine-textured products favor weed seed germination. Can be placed over plastics or landscape fabric as a decorative material, but these synthetic materials can be difficult to remove once they begin to deteriorate.
compost[1]	Excellent source of organic matter; readily available or can be made. Because of its smaller particle size, applying and maintaining a 2-inch layer can be effective. May harbor weed seeds, especially if not properly composted. May promote crown disease if applied to contact trunk. When irrigated overhead, holds a significant amount of water that is not available to plants. Can break down quickly and readily become infested by weeds germinating from windblown seeds.
grass clippings and leaves	Readily available, generally must be reapplied often. May contain weed propagules, e.g., seeds or bermudagrass stems. Mats and reduces water penetration, especially if not dried before application. Better if composted before use.
greenwaste	Uncomposted yard and tree trimmings. A variable mixture including bark, grass, ground wood, and leaves. Relatively inexpensive. May contain weed or pathogen propagules if not well composted. Can increase problems from small vertebrate pests (e.g., voles).
sheet mulching, heavy paper	Easy application. Must be purchased, typically in rolls. May be useful to decrease weed pressure until newly planted ground cover fills in open spaces. Tends to break or tear after transplanting or if walked on. May create a hydrophobic (water-repellent) barrier.
wood chips	Medium- to coarse-textured products are long-lasting and resistant to wind movement. Sometimes inexpensive or free. May not stay in place on slopes.

1. Commonly composed of greenwaste properly composted for 30 days or more.

Using a combination of methods provides more effective weed control, as illustrated in these 15 × 15 foot plots of *Pittosporum tobira* photographed 4 years after planting in Davis, California. All plants were drip-irrigated, with no weed control except at the time of planting. All plots (area inside yellow box) are surrounded by unmanaged weeds. The top left plot received a single postplant preemergence herbicide (oryzalin) application to bare soil. In the top right plot, redwood chip mulch was applied 2 to 3 inches deep once after planting. In the left plot, landscape fabric was applied and covered with 2 to 3 inches of redwood chips. *Pittosporum* within the left plot are larger and weed-free in comparison with the other plots because of the superior weed control and greater moisture availability provided by landscape fabric covered with wood chip mulch. *Photos: J. K. Clark*

water penetration and moisture retention characteristics of your mulched landscape. Keep organic mulch and waterproof synthetics 6 inches or more back from trunks to avoid promoting root and crown diseases. Regularly inspect mulch and remove any weeds soon after they appear.

Avoid black plastic (solid polyethylene) because it is not permeable to air or water and tends to tear and break down relatively quickly. Apply geotextile underneath natural inorganic mulches (e.g., rock) to prevent soil and mulch from mixing, which favors weed growth and contaminates soil with rocks.

Be sure organic mulch is not contaminated with seeds or perennial weeds such as bermudagrass stolons and nutsedge tubers. Apply organic mulch no deeper than about 4 inches (2 inches for fine-particle materials) to help maintain proper soil aeration and moisture levels. Avoid

tightly to avoid gaps, or cover gaps with another piece of fabric. Apply organic mulch at least 1 inch deep on top of fabric to improve landscape appearance and retard the fabric's photodegradation by sunlight.

Geotextile fabrics vary significantly in cost, performance, and special features. For example, yellow nutsedge can grow through all geotextiles, but some fabrics are better at suppressing it than others. Some products are impregnated with herbicide, and many incorporate compounds that resist photodegradation by UV light. Geotextiles generally are not recyclable and are typically disposed of in landfills.

Mulching Problems. If not properly selected or used, mulches have disadvantages, such as favoring root diseases and certain weeds and interfering with irrigation. Before applying mulch, properly grade the soil, control perennial weeds, and install any needed irrigation equipment. Be sure your type of irrigation (e.g.,

drip versus overhead sprinklers) and how you operate it provide appropriate root zone moisture and do not favor water-related root diseases. How much water to apply, how often to irrigate, and the best method vary by growing situation and the

Table 7-4.

Inorganic and Synthetic Mulches.

MATERIAL	COMMENTS
black plastic (solid polyethylene)	Very effective, widely available, relatively inexpensive but has many drawbacks: Restricts air and water movement. Drip irrigation under the plastic usually must be used since overhead irrigation cannot penetrate it. Breaks down in a few months and is unattractive unless a top mulch is applied. Tends to tear and break readily, allowing weeds to grow through holes. Not the best choice for long-term weed control.
geotextiles, or landscape fabrics (spun or woven polypropylene and polyester)	Very effective, relatively long-lasting if covered with bark or other suitable mulch. Allow air and water penetration. Expensive, may be unattractive without a top mulch. Brands differ in effectiveness and resistance to UV light.
gravel and crushed stone	Avoid using as mulch because over time they tend to become mixed in soil and are time consuming and difficult to remove.
plastic, clear	Not recommended for weed control because it encourages weed growth unless the soil is solarized and the clear plastic is left in place.

A covering layer of crushed rock can improve the appearance of the site and protect landscape fabrics applied to control weeds. Rock is difficult to separate from soil, so always use landscape fabric underneath rock mulches. *Photo:* J. K. Clark

The particle size of organic mulch determines how thick a layer is needed to provide good weed control. Bark and wood chips $1/2$ to $11/2$ inches in size shown here provide relatively long-lasting weed control when applied 3 to 4 inches thick. The tan cups holding seedlike structures are the fruiting bodies of bird's nest fungus (*Cyathus olla*), a beneficial antagonist to *Phytophthora* spp. *Photo:* J. K. Clark

mulches with a pH less than 4 or that have an "off odor" such as ammonia, vinegar, or rotten egg; these may injure plants, especially herbaceous plants. Before applying fresh bark, leaves, or wood chips around herbaceous species or young plants, it may be a good idea to thoroughly soak the material with several inches of water to leach out potentially phytotoxic plant chemicals. Periodically check the thickness and reapply organic mulch as needed to adequately cover soil and maintain a sufficiently deep layer to provide good weed control.

Mushrooms sometimes grow in decaying mulch, and their appearance can alarm people or pose a hazard if children are tempted to eat them. Unless consumed, mushrooms are generally harmless to people, and many are beneficial, as discussed in Chapter 5 and *Mushrooms and Other Nuisance Fungi in Lawns Pest Notes*. Slime mold, sometimes called dog vomit fungus, commonly grows in mulch, as pictured in Chapter 5. It is not harmful and will dry out within 1 or 2 days.

IRRIGATION

Avoid overwatering, which primarily results from irrigating too frequently. Run irrigation systems long enough to wet desirable plants' entire root system and then allow the upper topsoil to dry somewhat before irrigating again. Weeds will establish in areas where water drains or puddles and where soil near the surface

remains wet for a long time. Install drip irrigation or another efficient system that allows appropriate watering, as discussed in Chapter 3. Unless the water source is high in minerals, irrigating beneath mulch with porous hoses or emitter heads can be especially good because this keeps water away from the surface so weed seeds near the surface are less likely to be wetted and germinate. Alternatively, use low-volume aboveground emitters to deliver water slowly near landscape plant roots. High-volume (high-pressure) sprinklers widely disperse water, much of which becomes available to weeds. Seasonally adjust irrigation schedules according to the weather and plants' changing need for water.

BIOLOGICAL CONTROL

Biological control is mostly used to control certain invertebrates, as discussed in Chapter 6. Although many naturally occurring microorganisms, seed-eating birds, insects, and other small animals feed on and destroy weed seeds near the soil surface, few biological control agents are available to control most weeds in landscapes.

For example, two introduced *Microlarinus* spp. weevils help control puncturevine (*Tribulus terrestris*), which has spiny seed capsules that injure people and puncture tires. The weevils provide much better control of puncturevine if the weeds are water stressed, so avoid irrigation or water drainage into areas with puncturevine.

Domesticated goats and sheep are used in some situations to control weeds and create firebreaks, for example, on steep hillsides that cannot be mowed or easily sprayed. Goats or sheep can also be used to clear weeds before planting new landscapes. They eat many different plants, but goats prefer woodier species, including bamboo, blackberry, and poison oak. Goats and sheep may be rented and confined to a weedy location using portable fences or herd dogs. They are most effective when temporarily crowded into a comparatively small area, where they consume nearly all available vegetation before being moved to a new location. Provide livestock with water and protection from dogs and vandals. Because of the difficulty and expense in maintaining, protecting, and confining these animals, they have limited applications in most landscapes.

For more discussion of biological control, consult the *Natural Enemies Handbook*.

HERBICIDES

Herbicide applications are a convenient and generally inexpensive and effective way to control weeds and enhance desirable plant growth in certain landscape situations (Table 7-5). Properly applied, they are especially useful during the establishment of new plantings, particularly at sites infested with perennial weeds that should be controlled before installing the landscape plants.

A well-designed, mature landscape should require little or no ongoing use of herbicides. To avoid common problems

Two introduced weevils (*Microlarinus* spp.) provide partial biological control of puncturevine in some areas of California. Weevil presence can be recognized by feeding scars, the lighter patches on the stem, and brownish areas on the green seed capsule. Cutting into these plant crowns, stems, or seed may reveal weevil larvae, pupae, or frass inside. A weevil egg-laying hole (at tip of arrow) is visible in this stem.
Photo: J. K. Clark

from herbicides, rely on nonchemical methods where possible and minimize herbicide use in established landscapes. Herbicides can damage desirable plants and leach or run off during irrigation or rain and pollute water. Certain soil-applied herbicides can damage desirable plants when their roots share the same soil as weed roots, or they can persist in the soil and may injure species planted later or nearby plants whose roots grow into treated soils. Repeated use of the same herbicide will result in a shift of species to those that tolerate that herbicide, or the current weed population may develop resistance to the frequently used herbicide.

Understanding the biology of the target weed and how different herbicides affect plants is important to help you choose the best herbicide and effective application timing for your particular situation. Always follow the product's label directions when using herbicides.

When herbicides are used, it is best to combine them with nonchemical methods. For example, after spraying to control existing weeds, apply mulch to reduce new weed growth from seeds. Properly plant and care for desirable species to increase competition and shading that reduces reinvading weeds. See *Pest Notes* on specific weed species and *Weed Management in Landscapes Pest Notes* for herbicide information.

HERBICIDE TYPES

Herbicides differ in their active ingredient, formulation (manufacturer-added ingredients to improve product mixing and application), application timing relative to plant growth (preemergence versus postemergence), and how they control weeds. Knowing the herbicide's mode of action (how it kills weeds) is important

for choosing a product that will effectively control weeds without damaging desirable landscape plants, and for preserving herbicide effectiveness to minimize the development of herbicide resistance. Different herbicides may be appropriate for different situations or require different application equipment, such as granules (typically applied dry, then watered in) versus emulsifiable concentrates, flowables, and wettable powders (mixed with water and sprayed).

Certain products contain multiple herbicides (e.g., two or more postemergence herbicides) to control more species and life stages of weeds with a single application. Products containing multiple herbicides can have more limitations on their use to prevent damage to desirable plants or avoid environmental problems.

Preemergence Herbicides. Preemergence herbicides must be applied before

Table 7-5.

Herbicide Selection Checklist.

❑ **Weed Species.** What weed species are causing problems, and what life stages are present?

❑ **Alternatives.** Can some practical method other than herbicides be used?

❑ **Timing and Registration.** Are the site or situation and the weed types and life stages you plan to treat listed on the herbicide label? Read and follow the label thoroughly before purchasing, mixing, or applying any herbicide so the application is effective and you know how to avoid injuring desirable plants, people, or the environment.

❑ **Regulation.** Will the planned application comply with all applicable pesticide regulations, as discussed in Chapter 2? For example, individuals applying herbicides or other pesticides for hire must be trained and may need to be certified or licensed by the California Department of Pesticide Regulation or supervised by a certified or licensed person.

❑ **Residual Control.** What is the desired length of herbicide activity (effectiveness)? Preemergence herbicides with a longer residual activity prevent weed germination for a longer period. They may also preclude desirable seedlings or young plants from being planted until the herbicide is no longer active. Avoid using herbicides that may affect future plantings.

❑ **Surroundings.** Could desirable plants be inadvertently damaged by herbicide application at that site? Could roots, green bark, or foliage of nearby trees and shrubs be affected if a broadleaf or nonselective herbicide is applied? Use only products that can be applied safely at that site. Avoid herbicide drift or runoff.

❑ **Application Equipment.** Do you have the necessary equipment to apply the herbicide correctly, and is it in good working order?

❑ **Calibration.** Can you accurately measure and apply the correct amount of that herbicide? It is easier to measure and apply certain formulations. Proper calibration and application are essential for safe and effective herbicide use.

❑ **Incorporation Requirements.** Must the material be watered or mechanically incorporated into the soil?

are young. Because some materials are persistent, they may affect plants that are placed into a treated site for up to a year or more after application. This is especially true for plants grown from seed directly planted in a treated bed; their growth will likely be inhibited if there is any residual preemergence herbicide in the bed.

Postemergence Herbicides. Postemergence herbicides are applied to emerged weeds. Contact postemergence herbicides usually kill only those green plant parts on which spray is deposited, usually leaves and stems, so thorough coverage is important for good control (Figure 7-7). Contact herbicides are most effective against annual weeds, especially young broadleaves; perennial weeds and some grasses often regrow (e.g., from roots or tubers) after their aboveground parts are killed. For example, an iron compound (FeHEDTA) can be applied to lawns as a contact herbicide to kill broadleaf weeds without injuring most species of lawn grasses.

Translocated (systemic) herbicides are taken up by the plant and are transported to the growing tips of roots and shoots, so it may not be necessary to spray the entire plant to kill it. The principal advantage of translocated herbicides is that they can control or suppress perennial weeds that otherwise regrow from roots or underground vegetative propagules. For systemic herbicide to be transported to other portions of the plant, it must be

weed seeds germinate. They kill germinating seeds for several weeks or months after application. Some preemergence herbicides must be cultivated into the soil to control weeds. Others may be placed on the surface and followed with appropriate sprinkler irrigation or rainfall. Because herbicides differ in the extent to which they stick to soil particles and dissolve in water, some soil-applied herbicides require you to know your soil type and adjust the application rate accordingly. Soil-specific rates ensure the herbicide is effective and reduce the likelihood that it will damage the roots of desirable plants or move and pollute water (Figure 7-6).

Preemergence herbicides generally do not kill established weeds and are relatively safe for application around existing landscapes. However, if applied before transplanting or soon after planting and before soil is settled around roots, some preemergence herbicides may retard the root growth of desirable plants, especially when plants

SANDY SOIL

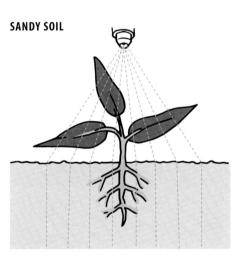

Pesticides applied to a sandy soil move quickly through the soil as more water is applied.

CLAY SOIL

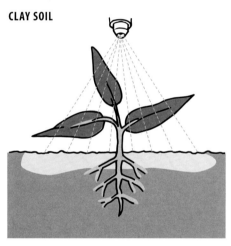

Water moves more slowly through clay soils. Pesticides are more likely to cling to soil particles near the surface and be washed into runoff by rain or irrigation water.

Figure 7-6.
Soil type and herbicide characteristics affect pesticide movement, effectiveness, and the risk of causing problems, so the label for some preemergence herbicides specifies different application rates according to soil type. In comparison with clay soils, less herbicide sticks to sand particles, so more of the herbicide can contact weed seedling roots and be effective at lower rates. Sandy soils also pose a higher risk that excess chemical will leach downward, so a lower rate reduces the risk of polluting groundwater. Herbicide dissolved in water or bound to soil silt can move in runoff water and damage desirable plants along the drainage path or pollute waterways.

To increase the growth and survival of this young tree, you can apply selective herbicides that kill only grasses and preemergence herbicides that prevent weed seeds from germinating. To improve the effectiveness of control, use a combination of methods, such as applying mulch after using the herbicide. *Photo:* J. K. Clark

Selectivity. Some herbicides are nonselective; they can kill weeds as well as desirable plants or broadleaves as well as grasses. Nonselective herbicides such as glyphosate are used where there are no desirable plants nearby or where they can be applied so that they contact only weeds and not the green bark or foliage of desirable plants. Do not apply nonselective or broadleaf herbicides unless you can prevent them from contacting desirable plants, such as by using a sponge, wick, wiper, or low-pressure controlled droplet applicator, as illustrated in Chapter 2, Tables 2-2 and 2-3.

Selective herbicides kill only certain types of plants and can be used around some desirable plants, as directed on the label. For example, fluazifop and sethoxydim used as directed can selectively control most annual grasses and bermudagrass without injuring trees or shrubs. Phenoxy herbicides, such as dicamba and 2,4-D in some weed-and-feed products for lawns, kill most broadleaf weeds but not grasses when used as directed. However, phenoxy herbicides commonly injure trees and shrubs with roots growing under the grass or when the material drifts onto leaves or volatilizes after application and is moved by air.

Fumigants are nonselective pesticides applied to soil before planting to kill weeds such as yellow nutsedge, seeds, nematodes, and soilborne pathogens. Fumigants are available only to professional applicators. They can cause air and water pollution problems and are generally not available for use in urban landscapes because of health hazards and regulatory restrictions (e.g., buffer zones). Solarization is a broad-spectrum pest control alternative to soil fumigation in certain circumstances.

Natural and Organic Herbicides. Some people prefer organically acceptable herbicides, including those containing acetic acid, citric acid, d-limonene, or botanical oils (e.g., cinnamon, clove, and lemongrass oil). These are contact, non-

moved with the flow of vascular fluids, so the time of application is critical. For best control, perennial weeds should be growing vigorously and have an abundance of mature leaves when they are sprayed. Poor control can result when weeds are stressed or stunted, such as from lack of water. Depending on the season and stage of plant growth, the effect of some translocated herbicides may not be apparent until well after they are applied. For example, the effect of glyphosate may not be apparent until a week or more after application.

Many postemergence herbicides can damage desirable plants they contact. Landscape plants may be damaged by direct spraying or from herbicides that reach plants through drift, volatilization after application, contact with roots in treated soil, and by translocation through natural root grafts. Natural root grafts are common between adjacent trees of the same species, and herbicides can translocate from treated stumps through roots to nearby trees.

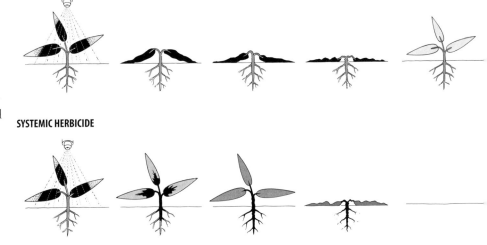

CONTACT HERBICIDE

SYSTEMIC HERBICIDE

Figure 7-7.
When treating weeds, make sure they are at the growth stage recommended on the pesticide label. Contact herbicides usually kill only those green plant parts on which spray is deposited. Thorough coverage is vital for good control, and certain weeds (e.g., established perennials) may regrow from underground plant parts. Systemic herbicides are taken up by sprayed leaves and translocated within the plant to growing tips of roots and shoots to kill weeds, but the effective time to apply them varies among weed species and growing situation. For example, in the spring the growth of many perennials is primarily in aboveground parts and a systemic herbicide will not be moved to underground perennial structures, making a systemic less effective.

selective herbicides that damage any green vegetation they touch; if used as directed they generally will not damage woody stems and trunks. They can control small, young weeds, especially broadleaves. Weeds that are small but older, such as regrowth after mowing, may not be well controlled because they generally have a thicker cuticle and a more extensive root system. These contact herbicides provide no residual control; perennials and many grasses will regrow from underground parts. When using these products, wear proper protective equipment, including a face shield or goggles, as some products can severely damage eye tissue.

Corn gluten meal, sold as an organic preemergence herbicide, is not an effective herbicide according to UC research. Because it contains nitrogen, corn gluten meal can improve turfgrass growth and competitiveness to weeds. However, corn gluten meal is much more expensive than other products, may attract pests, and can result in an unpleasant mess and odor when it gets wet.

Safe Use and Handling of Herbicides. If you decide to use herbicides or other pesticides, strictly follow products' label instructions. The label is a legal document you are required to follow. Minimize exposing yourself to pesticides and do not expose other people to them. Wear proper protective equipment as directed on the label; chemical-resistant gloves and protective eye wear are required when handling or applying most pesticides in California, even if the label does not say so. Avoid using pesticides or application methods that might injure nontarget organisms, property, or the environment. Persons who handle herbicides or other pesticides as part of their job must be formally trained and may need to be certified or licensed by the California Department of Pesticide Regulation or supervised by a certified or licensed person.

Minimize the potential for herbicides to move off-site and contaminate water. For example, avoid excessive irrigation and runoff into storm drains or surface water as discussed in "Pesticides Can Contaminate Water" in Chapter 2. For preemergence herbicides that must be incorporated by irrigation, apply only enough water, about $\frac{1}{2}$ inch, to move herbicide into soil during the first irrigation after application.

Carefully read the product label each time before you buy or use an herbicide. Know how much to use, how and when to use it, how long it lasts, which ornamental plants or locations it is registered for, and what weeds it kills. Be aware that some commonly used herbicides, such as 2,4-D, glyphosate, and triclopyr, can severely damage or kill desirable plants, as illustrated in "Pesticides and Phytotoxicity" in Chapter 4. Some products are highly irritating to eyes and skin. Some products will injure aquatic species, so avoid runoff, drift, and mixing excess solution. Even the water used to rinse empty containers and the mixing equipment must be disposed of properly, such as by adding the rinsate to the sprayer before filling the tank.

Use accurate rates; overdosing even with "selective" herbicides may injure desirable plants or the environment. Keep children and pets out of treated areas until the herbicide is mixed or irrigated into the soil, or has dried. **Do not use herbicide spray equipment to apply any other pesticides.** Some herbicides leave difficult-to-remove residues in spray tanks or hoses that can damage desirable plants. Check herbicide labels for specific information about how to clean sprayers after using herbicide.

Purchase and mix only the amount of herbicide needed. Store herbicides and other pesticides in their original labeled container in a locked cabinet out of the reach of children and pets. For more information see *Landscape Maintenance Pest Control*, *Lawn and Residential Pest Control*, and *Weed Management in Landscapes Pest Notes.*

Herbicide Resistance. Tolerance and resistance prevent some herbicides from controlling certain weeds. Tolerant plant species have a natural lack of susceptibility to certain herbicides. Tolerance can be desirable because it allows the use of selective herbicides, such as those that control either grasses or broadleaf weeds but not both. However, because of a lack of competition from susceptible species, repeated application of selective herbicides can allow tolerant weeds to increase unless other methods are used to control them.

Resistance occurs when weeds are no longer controlled by herbicides that previously provided control. Individuals in a species vary in their susceptibility to pesticides. Repeated applications of the same herbicide or other herbicides with the same mode (or site) of action over several generations favors the survival and reproduction of the nonsusceptible weeds (see Figure 2-5). Eventually the weed population consists mostly of nonsusceptible weeds and is no longer controlled by (is resistant to) the herbicide. An herbicide with a different mode of action or other measures must be substituted to control that weed species.

When an herbicide fails to control weeds, the cause is usually that the herbicide was not used according to its label directions or was the wrong herbicide for that situation. However, resistant weeds are an increasing problem in California and include some populations of rigid ryegrass (*Lolium rigidum*) and horseweed (*Conyza canadensis*), or mare's tail, that are resistant to glyphosate, and perennial ryegrass (*Lolium perenne*) and Russian thistles (*Salsola* spp.) that are resistant to ALS (acetolactate synthase) inhibitors or sulfonylurea herbicides (e.g., chlorsulfuron and halosulfuron). Where herbicides are frequently applied, take steps to avoid development of resistance by rotating applications among herbicides with different modes of action, maximizing reliance on nonchemical methods, and other methods listed in Table 7-6. See *Herbicide Resistance: Definition and Management Strategies* for more information.

Table 7-6.

Herbicide Resistance Avoidance Steps.

- Scout landscapes before applying any herbicide to determine what weed species and life stages are present and what methods are likely to control them.
- Combine nonherbicidal controls (e.g., hoeing and mulching) with herbicide applications.
- Use nonchemical methods as the first choice to manage weeds instead of herbicides.
- Learn herbicide modes of action and rotate (alternate) applications using herbicides with different modes of action.[1]
- Tank-mix herbicides with different modes of action that each provide good control and apply them at the same time.
- Avoid spreading weeds and their propagules (e.g., seed) from infested areas, for example, by controlling runoff water and cleaning equipment before moving to another site.
- Monitor landscapes regularly and identify where weeds are not being controlled as expected. If most weed species listed on the label are controlled but one species is not, suspect that this species may be resistant or the herbicide or other method may not have been properly used.
- Contact your local UC Cooperative Extension office if you suspect herbicide resistance, since further testing may be warranted. Do not try to control resistant weeds with higher rates or other herbicides. Use nonherbicidal methods if these weeds must be controlled.

1. To learn the mode (site, or mechanism) of action of specific herbicides, see the website of the Weed Science Society of America (wssa.net/weed/resistance).
Adapted from Retzinger and Mallory-Smith 1997; Wilen 2012.

Types of Weeds and Their Management

Plants typically develop through four life stages: seed, seedling, vegetative growth (leaves, stems, and roots), and flowering (reproduction). These stages occur during different times of the year and are completed within a period that ranges from a few months to many years, depending on the growing location and species, as discussed in the species-specific sections with scientific names below. Plants can be grouped as annuals, biennials, or perennials; knowing the life stages and development cycle of target weeds is critical so you can select the effective methods and timing to control them.

Annuals. Annual plants begin each growing season as seeds and complete their life cycle and die within one year. Annuals are classified as summer annuals or winter annuals, depending on when they most commonly grow. Winter annuals normally germinate in the fall or early winter and flower and produce seed and die by early summer. Annual bluegrass, annual ryegrass, annual sowthistle, common chickweed, common groundsel, and certain mustards (e.g., London rocket and black mustard) are winter annual weeds. In coastal areas with a moderate climate, winter annuals may germinate at any time

of the year if the site is moist. Annuals such as annual bluegrass and little mallow can behave as biennials or short-lived perennials in some areas of California.

Summer annuals germinate in the spring or early summer and flower and produce seed in summer or fall before dying by winter. Major species of summer annuals include crabgrass, pigweed, purslane, and spotted spurge. In most areas of California with a moderate climate, many summer annuals will germinate in the fall if irrigation or other surface moisture is present. These weeds can grow vigorously until cold temperatures arrive, at which time they stop growing but do not die. When the weather warms, these "off-season" summer annuals resume growth and are not affected by the preemergence herbicides that are typically applied during late winter to control summer annuals.

Most annuals can be controlled in mature landscape plantings by an integrated program of mulching, hoeing, hand-weeding, or spot application of an herbicide. If herbicides are used, winter annuals may be controlled by a preemergence herbicide applied during fall and a contact or translocated herbicide applied during winter. Most summer annuals can be controlled by a preemergence herbicide applied during winter and application of a foliar contact or translocated herbicide in late winter through spring.

Biennials. Biennial weeds complete their life cycle in two growing seasons. They pro-

duce vegetative parts in the first growing season, and flowers and seed develop during the following year. In some mild-weather parts of California, the plants can grow as annuals and develop flowers and seed the same year that they germinate. Bristly oxtongue and milk thistle are biennials that often grow as annuals.

Control biennials during the seedling stage, when most management methods are effective. During the early vegetative rosette-formation stage, first-year biennials are susceptible to most mechanical methods and herbicides that control that weed type. Definitely control biennials before they grow a flower stalk. In the second year, once the seed head is produced, herbicides are generally of little benefit. At this point, cutting or killing the flower stalk may be somewhat effective.

Perennials. Perennial plants can live for more than two years and include trees and woody shrubs. Most perennials initially develop from seed, but many herbaceous perennials reproduce and spread mostly by vegetative parts, such as stolons (stems that creep along the ground), and some do not even produce seed, for example Bermuda buttercup oxalis. The aboveground portion of herbaceous perennials may die back during the winter, then regrow during the spring or early summer from bulbs, rhizomes (underground stems), roots, or tubers, as illustrated in Figure

7-8. For example, a single tuber of yellow nutsedge can grow and reproduce so that tens of plants and hundreds of tubers can be produced within a year.

Common perennial weeds include bamboo, bermudagrass, blackberry, creeping woodsorrel oxalis, field bindweed, kikuyugrass, nutsedges, and poison oak. Because of their underground food reserves, established perennials are more difficult to control than annual weeds.

Preventing establishment of perennial weeds should be an important focus of your weed management program. Most chemical, cultural, and mechanical methods control seedlings of perennial plants. Preemergence herbicides are generally ineffective against perennial regrowth from vegetative propagules even though the emerging shoots may resemble seedlings. Once herbaceous perennials have developed more than about five leaves (as early as 4 weeks after seed germination for some perennial species), vegetative propagules will have begun to form, and they often will not be killed by a one-time destruction of aboveground parts.

You can control broadleaved perennials by digging them out, but you must remove all the bulbs, rhizomes, stolons, tubers, and roots to prevent weed regrowth. Persistent repeated cultivation can control most perennials if you avoid spreading their vegetative propagules. Translocated herbicides are generally the easiest-to-use, most effective control method. The best application time for many species is when the perennials are moving food reserves to the roots, but the timing for that differs among species. For example, for bermudagrass and nutsedges the most effective spray timing is soon after full leaf expansion (generally in late spring), whereas field bindweed and perennial pepperweed should be sprayed during early flowering.

For well-established and difficult-to-control perennials, multiple management techniques may be necessary. For field bindweed, you generally have to apply a translocated herbicide, then cultivate, treat the regrowth during early flowering, and repeat these steps over several years.

WEED IDENTIFICATION

Learn the identity and life cycle of weeds in the landscape so you can choose effective and appropriate management methods if control action is warranted. For example, most herbicides do not control all species, so if weeds are not identified accurately, herbicide applications may be ineffective and wasteful.

Flowers are the primary structures used to reliably identify the species of plants. However, because weeds should be controlled before they flower, the shape and arrangement of vegetative plant parts such as leaves, stems, and veins are key characteristics used to identify weeds. Grasses (Poaceae) and sedges (Cyperaceae) resemble each other, but grass leaves alternate on each side of the stem, and sedge leaves are joined to the stem in groups of three. Grass stems are hollow, rounded, and have nodes (joints) that are hard and closed; sedges have a solid stem that is triangular in cross-section.

Vegetative plant parts used in identifying weeds are illustrated in Figure 7-8. Many different weeds can infest landscapes; some common troublesome species are described below. For help identifying weeds, see Figure 7-9, the *Weed Photo Galley* online at ipm.ucanr. edu, *Weed ID Tool* at wric.ucdavis.edu, *Weed Pest Identification and Monitoring Cards, Weeds of California and Other Western States,* and *Weeds of the West.* For more assistance contact UC Cooperative Extension Advisors, UC Master Gardeners, county agricultural commissioners, knowledgeable pest control advisers, certified nursery professionals, and botanical garden personnel.

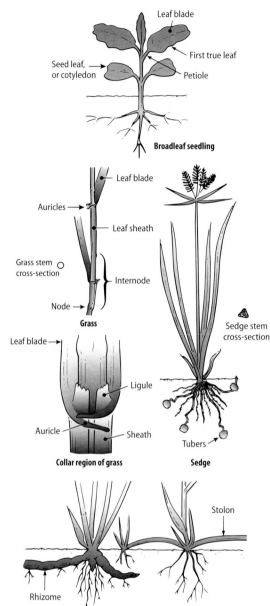

Figure 7-8.
Vegetative parts of weeds and terms used in identification. Grasses and sedges superficially resemble each other, but their biology and management often differ. Correctly identify the species or plant type so you can choose the effective methods and timing of control action.

Bermudagrass
Scientific name: *Cynodon dactylon* (Grass Family: Poaceae)

Figure 7-9.
The *Weed Photo Gallery* for pest identification is online at ipm.ucanr.edu.
Photos: J. K. Clark

Bermudagrass is a perennial grass that is frequently used for lawns but also is a troublesome weed in many gardens. The plant grows rapidly when temperatures are warm and moisture is abundant. It grows throughout California, except the Great Basin area, to an elevation of about 2,900 ft (900 m). Bermudagrass invades agricultural crops and other disturbed locations.

Habitat
Gardens, agronomic crops, orchards, turf, landscaped and forestry areas, and other disturbed sites.

Seedling
Seedling stems and leaves are similar to those of the mature plants.

Collar
The ligule consists of a fringe of short, white hairs. There are no auricles.

ANNUAL BROADLEAVES

ANNUAL SOWTHISTLE

Sonchus oleraceus (Asteraceae) is a common summer or winter annual that can grow year-round in California's Central Valley and coastal areas. Annual sowthistle seed leaves are stalked and covered with a powdery, gray bloom. They have smooth edges and are spoon shaped. True leaves have wavy edges and are pointed at the tips of the lobes. Upper leaf bases clasp the stems with clawlike lobes. Mature plants may reach a height of 3 to 6 feet. Its hollow stems secrete milky juice when cut or crushed. The yellow flowers of sowthistle mature into white, fluffy seed heads.

Annual sowthistle superficially resembles certain other weeds, including common groundsel and dandelion. However, dandelion is a perennial that regrows from a taproot, and it has only basal leaves that all grow from very near the ground. Older annual sowthistle and common groundsel plants have leaves at their base and on stems well aboveground. Common groundsel has conspicuous black tips on the green bracts surrounding each yellow flower; the green bracts of sowthistle are not black tipped. Closely related species include perennial sowthistle (*S. arvensis*) and spiny sowthistle (*S. asper*), also called prickly sowthistle.

Annual sowthistle seed germinate only in the top ½ inch of soil, so cultivation or solarization before planting can provide control. Mulching is effective, but periodic reapplication of organic mulch and spot treatments (e.g., hoeing or hand-pulling) are usually required. Because seed of annual sowthistle and related species (e.g., common groundsel and prickly lettuce) are easily dispersed by wind, these weeds frequently germinate in decaying organic mulch or soil that accumulates on top of synthetic mulch.

COMMON GROUNDSEL

Senecio vulgaris (Asteraceae) is a winter annual, but it can grow anytime of year in coastal California. Infestations are most

Annual sowthistle develops yellow flowers that mature into fluffy, white seed heads. Because these windblown seed germinate in decaying organic matter, organic mulch requires regular maintenance (such as periodic reapplication) and spot treatments (such as hoeing or hand-pulling) to control sowthistle. *Photo:* J. K. Clark

Annual sowthistle seed leaves have smooth edges and are spoon shaped, while true leaves have wavy edges and prickles that become more prominent on older leaves. *Photo:* J. K. Clark

This common groundsel resembles several other weedy Asteraceae in landscapes, such as annual sowthistle and prickly lettuce. Common groundsel can be distinguished by the conspicuous black tips on the green bracts at the base of the unopened flower head, as at the tip of the arrows (right). *Photo:* J. K. Clark

problematic during cool, moist periods. Plants die during extended hot, dry periods.

Cotyledons (seed leaves) are narrowly football shaped to oblong, ⅖ inch long or less, and often have a purplish underside. The first true leaves are egg shaped with shallow teeth, while later leaves are deeply lobed with toothed edges.

At the base of flower buds and blossoms, the bracts (green leaflike structures) have conspicuous black tips that hug the base of the flower head, distinguishing groundsel from others in the thistle family. The yellow flower head matures into a white puffball—a seed head similar to that of dandelion. Control groundsel as described for annual sowthistle.

CUDWEEDS

Most cudweeds (*Gnaphalium* spp., Asteraceae) are annual species, but purple cudweed (*G. purpureum*) is either a summer or winter annual or a short-lived biennial. Creeping cudweed (*G. collinum*) is a perennial. Seed leaves, first true leaves, and the flower clusters and fruit are covered with whitish to pale gray woolly hairs. Their seed leaves are oblong or spoon shaped, and true leaves are narrow, lance shaped, and several times longer than wide. Mature cudweeds are sparsely branched, mostly erect, 8 to 20 inches tall, and develop numerous clusters of small, woolly flower heads.

Control cudweeds by cultivating, flaming, hand-pulling, hoeing, mulching, and applying certain herbicides.

Cudweed flower heads are bristly, crowded, and densely arranged on the stem or at the base of leaf stalks. *Photo:* J. K. Clark

HAIRY FLEABANE

Conyza bonariensis (Asteraceae) is a common summer annual or biennial. The gray-green cotyledons are nearly hairless, about twice as long as wide, and difficult to distinguish from horseweed cotyledons.

Hairy fleabane leaves are alternate to one another along the stem. The first true leaf is oval to oblong and sparsely or densely covered with hairs. Older leaves are covered with short, stiff hairs and long, soft hairs. Leaf edges range from smooth to weakly toothed or lobed. Upper leaves are linear to lance shaped. Lower leaves are generally egg shaped and taper to a short stalk.

Mature hairy fleabane is less than 4 feet tall and develops white flowers. Its leaves are gray-green and covered with both short, stiff hairs and long, soft hairs. This differs from horseweed, which has darker green foliage and only short, stiff hairs on the leaves. *Photo:* J. K. Clark

Young horseweeds grow as a rosette (circular cluster) of abundant dark green lea near the ground. *Photo:* J. M. DiTomaso

Although similar to horseweed, hairy fleabane has more branches and more basal leaves and its stems usually branch near the base. Horseweed stems usually branch only in the upper half of the plant. Horseweed leaves have only short, stiff hairs; hairy fleabane leaves have a combination of short, stiff hairs and long, soft hairs. Horseweed leaves are dark green, while hairy fleabane leaves are gray-green. Horseweed can reach over 9 feet tall, while fleabane grows 4 feet tall or less.

Fleabane blooms from June through September, and its upper branches contain many flower heads. Each flower head has a cup-shaped base formed of green leaflike bracts that overlap. The top encloses tiny, yellow flowers (disk flowers) surrounded by small, cream-colored bristles. As the flower head opens, it matures into a fuzzy, spherical, brownish white seed head that often reddens with age and resembles the puffball of a dandelion, but much smaller. Control hairy fleabane as described for cudweeds. Some populations of fleabane are resistant to glyphosate.

HORSEWEED, OR MARE'S TAIL

Conyza canadensis (Asteraceae) is a summer annual or biennial broadleaf. Cotyledons are oval to egg shaped, hairless to densely covered with short stiff hairs, and up to $\frac{1}{10}$ of an inch long. The first leaf is shaped like an egg or football and has a hairy stalk.

Later leaves are more-or-less football shaped with slightly toothed edges.

Young plants grow as a rosette (circular cluster) of dark green leaves near the ground. Older plants grow a single main stem up to 10 feet tall that develops many branches, mostly from the upper half. In similar-looking hairy fleabane, the main stem usually branches near the base. However, cutting the main stem of horseweed can cause several branches to regrow from the base, and short, mowed horseweed can resemble hairy fleabane. Distinguish these species as described in "Hairy Fleabane."

Horseweed's dark green leaves alternate along the stem and branches. Many flower heads grow and bloom at the ends of branches from June through September. The small, daisy-like flower heads have a green, urn-shaped base and a top of white to pinkish petals around a yellow center. Each flower head matures into a dirty, whitish seed head that resembles the puffball of dandelion but is much smaller.

Control horseweed as described for cudweeds. Be aware that in some locations horseweed is resistant to glyphosate.

PRICKLY LETTUCE

Lactuca serriola (Asteraceae) is a common winter annual or biennial broadleaf that sprouts from seed, beginning with winter rains. Its egg- to football-shaped seed leaves often have slightly indented tips and a few fine hairs along the

Older horseweeds grow a tall stem that branches near the top, where shoot terminals develop yellow flowers Undisturbed horseweed can grow up to 10 feet tall, about twice the mature height of similar-looking hairy fleabane. *Photo:* J. K. Clark

edges. True leaves have smooth, prickly, or weakly toothed edges, rounded tips, and a row of prickly bristles on the lower midvein. Leaves alternate along the stem and are at least two times longer than wide. Lower (older) leaves are 2 to 10

Two forms of prickly lettuce—one has deeply lobed leaves and the other unlobed, oblong leaves. It and related *Lactuca* spp. are distinguished by the row of bristles along the lower midvein of leaves. This broadleaf winter annual or biennial is a prolific colonizer of landscapes and grows up to 6 feet tall. *Photo: J. K. Clark*

inches long, either deeply lobed or oblong and unlobed. Plants are a basal rosette of leaves until they grow a flowering stem, which develops multiple side branches and can grow to 6½ feet tall. From April through October, prickly lettuce develops numerous small, pale yellow, dandelionlike flowers on branches that extend out from the main stalk. In comparison with the similar-looking bitter lettuce and willow-leaf lettuce (other *Lactuca* spp.) and spiny sowthistle, prickly lettuce has much stiffer bristles on the lower midvein. Control it as described for cudweeds.

BLACK MEDIC

Medicago lupulina (Fabaceae) is a low-growing summer or winter annual, or sometimes a short-lived perennial, that invades planting beds and turf. Its seed leaves are oblong and about ⅙ to ⅓ inch long, with smooth edges. True leaves are three egg- or heart-shaped leaflets that alternate along the stem. Leaflets are ⅖ to ⁷⁄₁₀ inch long and have hairy and toothed edges. Stems are hairy and grow prostrate up to 1⅓ feet long. From April through July, black medic produces 10 to 20 yellow, slender, pealike flowers in a dense cluster. Each flower matures into a green or black, smooth, kidney-shaped pod ¹⁄₁₂ to ⅛ inch long that contains one yellowish to green seed. Similar-looking California burclover has several seeds in each pod and nearly hairless leaves. Control black medic by hand-pulling, hoeing, flaming, mulching, and applying certain herbicides.

CALIFORNIA BURCLOVER

Medicago polymorpha (Fabaceae) is an annual broadleaf common in ornamental plantings and turfgrass. It has oblong seed leaves and rounded cloverlike true leaves composed of three leaflets, which usually have serrated edges and reddish midveins. Stems may be upright but usually grow along the ground up to 2 feet long. At stem tips, California burclover produces a cluster of small, bright yellow flowers from March to June. Each flower becomes a green or brown pod that is usually coiled two to six times and is either hairless and smooth or has two to three rows of distinct prickles. Each pod contains several yellowish or tan, kidney-shaped seeds. Similar-looking black medic has one seed per pod and its leaves, and especially stems, are hairy. Control burclover by hand-pulling, hoeing, flaming, mulching, and applying certain herbicides.

COMMON PURSLANE

Portulaca oleracea (Portulacaceae) is a summer annual that produces copious seed and rapidly colonizes warm, moist sites. Its prostrate, succulent, reddish stems radiate out from a central root, resembling the spokes of a wheel. Common purslane leaves are spoon shaped, smooth, succulent, and ½ to 2 inches long. The yellow, five-petalled flowers are ⅜ inch long, occur singly, open only in sunshine, and mature into small pods. Lifting mature purslane in late summer can reveal thousands of the tiny, reddish brown to black, oval seed that

As with other *Medicago* spp., the true leaves of black medic are three egg-shaped leaflets that are alternate to one another along the stem. Black medic has hairy leaves and stems, and it blooms during spring and early summer. *Photo: J. M. DiTomaso*

California burclover produces bright yellow flowers at the stem tips during March to June. These mature into green pods, which turn brown when dry. The pods have two or three rows of prickles, which often have tiny hooks at the tips. *Photo: J. K. Clark*

Common purslane has smooth, succulent leaves on shoots that grow from a central root outward up to 1 foot long. After it flowers, lifting a mature purslane can reveal thousands of seed on the soil surface, so promptly control plants (e.g., hand-pull or hoe them) before they flower. *Photo:* S. A. Parker

Little mallow has heart-shaped seed leaves. True leaves are more roundish with in edges, and they often have a red spot at the base of the blade where the leaf joins petiole. Take control action when mallow is young. Mature plants have a deep tap from which they can regrow, and they produce persistent seed. *Photo:* J. K. Clark

germinate mostly during February and March in the southern desert and in late spring in cooler areas of California.

Control purslane primarily by preventive and cultural methods, especially hand-weeding and mulching. Make sure new plants and soil are weed-free. Clean mowers, cultivation equipment, and other tools before moving them from infested sites. Monitor landscapes and shallowly cultivate topsoil, hand-pull, or hoe weeds before they produce seed. Dispose of purslane away from the site after removal to prevent rerooting. You can also solarize the soil before planting or apply preemergence herbicide before seedlings emerge, as discussed in *Common Purslane Pest Notes*.

LITTLE MALLOW

Malva parviflora (Malvaceae) is an annual (primarily a winter annual) or biennial, depending on location and weather. Little mallow and other *Malva* spp. that infest landscapes are also called cheeseweeds because their fruit are disk shaped, with flattened lobes resembling an uncut block of cheese.

True leaves of little mallow are roundish and crinkled with wavy, shallow-toothed margins, and they often have a red spot at the leaf base. The mature plant forms dense bushes that trail along the ground or grow upright, 1 to 4 feet tall. It has a tough, woody stem and quickly develops a deep taproot, making it hard to remove

by hand even when young. Its flowers are white with a bluish or pinkish tinge and are held in clusters in the leaf axils.

Cultivating, flaming, hand-pulling, and hoeing are effective against mallows during the seedling stage. Mallows are susceptible to solarization and mulching before plants emerge. Certain preemergence herbicides can be effective but preemergence herbicides should be incorporated or irrigated in 1 to 2 inches deep because seed can germinate even if buried to that depth. Postemergence herbicides provide best control if applied while the weeds are young, before plants are about 4 to 6 inches tall. Later treatments are less effective due to mallow's strong taproot. Glyphosate provides only partial control of mallow. Certain other postemergence herbicides will provide better control.

SPURGES

About two dozen species of spurge (*Euphorbia =Chamaesyce* spp., Euphorbiaceae) occur in California. Spotted spurge (*E. maculata*) is the major weedy spurge, but about five other spurges also occur in landscapes, including creeping spurge (*E. serpens*), ground spurge (*C. prostrata*), and petty spurge (*E. peplus*).

Spotted spurge is a many-branched, low-growing, mat-forming summer annual. Seed leaves are oval and tinged reddish underneath. True leaves are egg shaped, often have a reddish spot in the middle, and grow opposite each other on stalks up to 20 inches long.

Creeping spurge and ground spurge grow prostrate, like spotted spurge. Petty spurge grows upright and can reach 20

Spotted spurge is named for the dark reddish area in the center of true leaves. Milky, sticky sap exuding from the broken stem (top center) is characteristic of plants in the family Euphorbiaceae. *Photo:* J. K. Clark

inches tall. These spurges have no markings on their leaves, but foliage otherwise resembles that of spotted spurge. All spurges produce milky, sticky sap and small, inconspicuous flowers. Characteristics of the seed capsules (fruit) and seed are used to distinguish the species.

Seeds germinate around February or March. Each plant can produce several hundred seed, which can survive for up to 12 years in the soil. Spotted spurge thrives in areas of mostly bare soil, in cracks in pavement, and along edges and survives even very close mowing. Because it has a low-growing habit, some landscape managers tolerate spurge growing under woody plants, unlike more visible weed species that can grow up through or over shrubs.

Apply mulch before seed germinate. Hoe or otherwise cultivate seedlings when small. Larger plants are easy to hand-pull if soil is moist and spurge is not too abundant; grasp a mat of stems and pull slowly to remove the whole plant, including its central taproot. Because spurge is capable of setting seed within a few weeks of emergence, regularly monitor landscapes and promptly remove weeds before they produce seed. Preemergence herbicides applied in early spring are effective and especially useful where infestations and seed production have been heavy. For more information, including herbicide recommendations, see *Spotted Spurge and Other Spurges Pest Notes.*

Mature spotted spurge is low growing. Leaves grow opposite on short stalks, are finely toothed on the margin, and are covered with soft hairs. *Photo:* J. K. Clark

WILLOWHERBS

Fringed willowherb, or northern willowherb (*Epilobium ciliatum,* Onagraceae), and tall annual willowherb (*E. brachycarpum* =*E. paniculatum*), also called panicle-leaf willowherb, are similar-looking weeds that infest landscapes. Also called fireweeds, these *Epilobium* spp. vary at maturity from about ½ to 6 feet tall. Their cotyledons (seed leaves) are hairless and almost round. Their true leaves are narrow, oblong, and about ⅖ to 6 inches long on fringed willowherb (a perennial) and ⅖ to 2 inches long on tall annual willowherb (a summer annual). Their small pink, purplish, or white flowers have four petals and occur from about June through September.

Control willowherbs before they flower or produce seed. Each seed is attached to a tuft of long, white hairs and is easily dispersed by wind. Provide good drainage and maintain irrigation systems to prevent water puddles where windblown seed often collect and germinate. To control willowherbs, cultivate, hand-weed, solarize soil, and apply mulch or certain preemergence herbicides. Flaming also controls the seedlings.

Be aware that willowherbs have a high degree of tolerance to glyphosate. Additionally, summer annual willowherbs may germinate in the fall in mild-climate areas when moisture is present and can survive over winter. In those situations they are not controlled by the winter application of a preemergence herbicide.

Fringed willowherb develops tiny, pinkish flowers on the end of stalks. Each of these stalked flowers will mature into a ¹/₂-inch-long brown seed pod. *Photo:* J. K. Clark

Fringed willowherb seeds at the stalk terminal in a peeled-back capsule or pod. Each seed is attached to a tuft of long, white hairs and is easily dispersed by wind. *Photo:* J. M. DiTomaso

ANNUAL GRASSES

ANNUAL BLUEGRASS

Poa annua is a winter annual adapted to cool, moist sites. It can grow any time of year in coastal areas especially in shady areas if the soil is moist. It grows 3 to 12 inches tall at maturity. Its flower heads are branched with clusters of three to six flowers at the tip of each branch.

Annual bluegrass has light green, short, smooth leaf blades. Leaves are often folded at the tip, resembling the prow of a boat, especially on young plants. Seedlings such as this germinate in the fall or anytime where warm soil is frequently irrigated. *Photo:* J. K. Clark

In ornamental plantings bluegrass can form a dense patch that is aesthetically distracting, but probably has little detrimental effect on established shrubs and trees. For information on identification and management, consult *Annual Bluegrass Pest Notes.*

CRABGRASSES

Large, or hairy, crabgrass (*Digitaria sanguinalis*), commonly called crabgrass, is a pale green summer annual that typically is found in gardens and planting beds. It usually has many branches at the base and spreads from roots growing at swollen joints in the stem. Leaves are 2 to 5 inches long. Smooth crabgrass (*D. ischaemum*) is similar to large crabgrass but is smaller, not hairy, and usually found in turfgrass. The bract covering the smooth crabgrass grain is brownish black, compared with pale yellow in large crabgrass.

Seed begin germinating about February or March, when soil at a depth of 1 to 2 inches is 50° to 55°F for 3 to 7 days. Crabgrass flower stalks resemble the long

claws of a bird. Flower stalks arise in a group near the stem tip, but unlike bermudagrass, often there are additional flower spikes branching from beneath the stem tip. Crabgrass thrives under hot conditions, often where there is frequent irrigation or poor drainage.

Manage crabgrass with a regular program of cultivation, hand-weeding, mulching, and avoiding excess irrigation. Vigilance is required as plants are difficult to remove once they grow and develop an extensive root system. It is especially important to remove or otherwise control plants before they flower. Late-season control is not recommended because plants will die naturally as the temperature drops. Crabgrass seed is only partially controlled with solarization, but it is easily controlled by many preemergence herbicides. For more information and specific herbicide recommendations, see *Crabgrass Pest Notes.*

Most large crabgrass flower stalks arise in a group near the stem tip, but often there are additional flower spikes branching from beneath the stem tip, as shown here. This summer annual reproduces from seed. Its flowers and seed heads resemble bermudagrass (pictured later). However, bermudagrass spreads primarily by rhizomes and stolons, which are lacking in crabgrass. *Photo:* J. K. Clark

SEDGES

Sedges (Cyperaceae) resemble grasses but have solid stems that are triangular in cross-section (Figure 7-8) and leaves that usually radiate out in three directions from the stem. True grasses have hollow, round stems and leaves that alternate along the stem. The sedge family includes both annual and perennial species.

GREEN KYLLINGA

Kyllinga brevifolia is a weedy perennial sedge that invades ornamental beds and turf. It is sometimes confused with young purple or yellow nutsedge because it is similar in size and growth habits. Green kyllinga grows best during warm weather and in moist or wet areas that receive full sun. It is dormant in winter but remains green or becomes yellowish, turning brown after freezing weather.

Green kyllinga can reach a height of about 15 inches if unmanaged. Its leaves are long, narrow, and from 1 to 5 inches long or longer. It is easily identified by its flower stalk, which usually occurs from May to October, sometimes earlier, in warm locations. Flower stalks are triangular in cross-section, 2 to 8 inches long, and terminate in a round, green flower head about ³/₈ inch in diameter. Directly below the flower are three or four leaflike bracts that radiate out. Green kyllinga seed are oval, flat in cross-section, and about ¹/₁₆ to ¹/₈ inch wide. Seed are highly viable, readily spread, and germinate at or very near the soil surface when soil is moist and reaches about 65°F.

Seedlings may require several weeks to become established, after which they develop a vigorous system of underground stems, or rhizomes. If mowed, green kyllinga becomes a low-growing, thick mat that can survive and produce seed even if mowed to ¾ inch tall. If chopped into pieces, new plants can be produced from each node or stem section. Seed and rhizomes are spread by cultivation, foot traffic, mowing, and in infested soil or container plants that are moved.

Green kyllinga is a relatively low-growing sedge that spreads by rhizomes (underground stems) and seed. It is distinguished from purple and yellow nutsedges by the absence of underground tubers, as seen in this plant removed from landscape. *Photo:* J. K. Clark

Several characteristics differentiate green kyllinga from other plants. In turf, green kyllinga stands out due to its different foliage texture and faster growth rate once established. Its flowers, foliage texture, growth habit, and absence of underground tubers distinguish it from nutsedges. Yellow and purple nutsedge commonly occur as individual plants and have much wider leaves than the finer-bladed green kyllinga. Green kyllinga's round seed head is less than $1/2$ inch in diameter, whereas nutsedges have an inflorescence that is several inches wide in an open spikelet with multiple flowers. Green kyllinga usually grows in continuously enlarging patches, similar to rhizomatous turfgrasses.

Prevention is very important because there are few control options for green kyllinga in ornamental plantings. Maintain landscapes to promote vigorous, dense turfgrass and ornamentals; this will exclude weeds and shade the soil surface, making it difficult for green kyllinga seedlings to establish. Avoid overirrigation and excessively wet soil. Thoroughly clean mowers and cultivation tools before moving from infested to weed-free areas. Inspect new plants and soil to be sure they are weed-free. If solitary green kyllinga plants appear, promptly hand-remove the entire plant, including roots and rhizomes, or spot-spray an herbicide. It may be easiest to remove the plant and soil and refill the hole with clean soil. Mark, isolate, and monitor the area for several months to make sure that removal was complete. Newly emerged solitary green kyllinga may be hand-pulled or hoed carefully to avoid breaking rhizomes into smaller pieces and "transplanting" them to new areas, especially in irrigated areas.

Combined with hand-removal, geotextiles can be effective if they are overlapped and no light is allowed to penetrate to the soil. Wood chips or bark may be placed on top. Organic mulches alone will not be effective because green kyllinga will grow through the mulch.

Certain preemergence herbicides can be effective if green kyllinga is not already established. Spot treatment with certain postemergence herbicides can reduce growth, but it can be difficult to avoid phytotoxicity to desirable plants. For more information, see *Green Kyllinga Pest Notes.*

NUTSEDGES

Nutsedges are common weeds in the coastal valleys, Central Valley, and southern areas of California. They thrive in waterlogged soil, where drainage is poor, irrigation is too frequent, or sprinklers are leaky. Once established, however, they will tolerate normal irrigation conditions or drought.

Yellow nutsedge (*Cyperus esculentus*) grows up to 3 feet tall and has tan or yellowish flowers and light brown seeds. Purple nutsedge (*C. rotundus*) grows to $1\frac{1}{3}$ feet tall and has purplish flowers and blackish brown seeds. These nutsedges spread primarily from tubers, or "nutlets," that form on rhizomes growing as deep as 8 to 12 inches below the surface. Yellow nutsedge tubers are formed at the ends of rhizomes and have an almond taste when eaten. Purple nutsedge tubers occur like beads on a chain and have a bitter taste. Tubers can remain viable for several years, even in dry soil.

Do not confuse purple or yellow nutsedge with tall flatsedge (*C. eragrostis*).

Green kyllinga produces a globular, green flower about $3/8$ inch in diameter immediately above three or four leaflike bracts that radiate outward. Kyllinga flowers are distinctly smaller than the fully open spikelets (stalked flower heads) of tall flatsedge, purple nutsedge, and yellow nutsedge. *Photo:* J. K. Clark

Tall flatsedge does not produce tubers; it spreads by seed or short, thick rhizomes. It can be distinguished from purple and yellow nutsedges by its wider leaves and stems, its shorter and thicker rhizomes, its lack of tubers, and especially by its much wider spikelets. In mowed turf, it can be distinguished by its tendency to grow in tight clumps that are less than 1 foot in diameter. If uncut, tall flatsedge grows up to 4 feet, taller than purple or yellow nutsedges.

Irrigate properly, maintain irrigation equipment, and provide good drainage. You will eventually control purple and yellow nutsedge if you limit their tuber production, such as by frequent, repeated cultivation or hand-weeding when plants are young. Control individual plants before they produce five or six leaves; older purple and yellow nutsedges produce tubers from which they can resprout.

Planting competing species that grow tall quickly can suppress nutsedges because they do not grow well in shade. Prior to planting the landscape, purple nutsedge can be controlled with repeated summer tillage of dry soil because (unlike yellow nutsedge) its tubers are readily killed by drying. Tubers are not killed if soil is moist or has large clods. Commonly

Yellow nutsedge has tan or yellowish flowers and brown seed. It can grow up to 3 feet tall and spreads to form a dense, weedy clump that becomes increasingly difficult to control as the plant develops more underground propagules. *Photo:* J. K. Clark

used black polyethylene plastic mulches do not control nutsedges because they grow through the plastic, but mulching with thick polypropylene polymer geotextiles controls nutsedges. Certain herbicides can reduce nutsedge populations, but herbicides alone generally do not provide good control. It is extremely difficult to control established nutsedges without an integrated program involving several tactics. Consult *Nutsedge Pest Notes* for more information.

PERENNIAL GRASSES

BAMBOO

Bamboos (many genera) are evergreen grasses that grow as woody perennials. Their hollow, woody stems are divided into sections by obvious joints, and plants reproduce via rhizomes. Bamboo leaves are narrowly attached to stems, i.e., the base of leaves tapers down to a stalk-like petiole. This distinguishes bamboo from giant reed (*Arundo donax*); *Arundo* leaves have a heart-shaped base that broadly wraps around the stem, so the foliage and stem resemble a corn stalk. *Arundo* is a bamboo-like perennial with a tough (but not woody) main stem that can grow up to 30 feet tall and has extensively invaded low-elevation streamsides and coastal riparian areas.

Bamboos are often planted, but frequently they become weeds. Some "running" species (e.g., *Pleioblastus* and *Phyllostachys* spp.) spread beyond where they are desired unless they are regularly trimmed and their root zone is confined.

In most situations, plant only clumping bamboos (e.g., *Bambusa* and *Fargesia* spp.). If running species are used, provide a barrier to confine spreading rhizomes by planting bamboo only in areas bordered by pavement or fences that extend beneath the soil. Rhizomes can grow under fences that are not deep enough. Install edging or headers (narrow concrete, wood, or metal barriers extending 18 inches or more below the soil surface) to help confine the rhizomes.

To eliminate established clumps of bamboo or reduce their spread, regularly cut

The *Phyllostachys* sp. of running bamboo in the background is sprouting new plants from rhizomes that can spread a long distance from the main clump into turf and shrubs. *Photo:* J. K. Clark

This young nutsedge resembles grass, but its leaves are thicker and stiffer than most grasses. Nutsedge leaves are V shaped in cross-section and grow from the base in sets of three; grass leaves grow alternate to one another along the stem. *Photo:* J. K. Clark

back or mow aboveground parts to near the ground, dig up and remove all underground parts, or apply an appropriate herbicide to foliage or freshly cut stumps. A combination of management methods and repeated action against regrowth is often needed to provide good control.

BERMUDAGRASS

Common bermudagrass (*Cynodon dactylon*), usually called bermudagrass, spreads readily and is often a weed in planting beds, shrubs, and lawns where other grass species are preferred. It is found along roadsides, sidewalks, and in vacant urban lots, but it is generally absent from California's inland valley and foothill slopes that are unirrigated and dry during the summer. Bermudagrass becomes dormant and turns brown during the winter in cold areas and is not as aggressive in dense shade. Certain cultivars are used as a hardy turfgrass that is well adapted to drought and alkaline soils.

Mature common bermudagrass plants form a dense mat with spreading, branching stolons and rhizomes. Unmowed stems grow 4 to 18 inches tall. Bermudagrass has a conspicuous ring or fringe of short, whitish hairs at the base of each blade. Plants produce inflorescences consisting of three to seven slender spikes radiating from one point. Common bermudagrass reproduces from seed as well as rhizomes and stolons.

With persistent effort, bermudagrass can be controlled by consistently removing plants as they emerge and by preventing additional seed and stems from being introduced. Rhizomes are readily killed when exposed to the sun and allowed to dry, so repeated cultivation can provide control if soil is dry and remains dry for a week or more after each cultivation. Bermudagrass can grow through most organic mulches. Thick geotextile can be effective if the fabric is handled and maintained to prevent holes, but will not stop bermudagrass stolons from growing over it.

Properly performed solarization controls bermudagrass in warmer areas of California, but it is most effective when conducted for at least 6 weeks during the sunniest time of the year. Before solarizing, closely mow bermudagrass to about ½ inch and rake the soil free of cuttings and trash before irrigating and tarping. Alternatively, rototill then irrigate soil before tarping it. Do not rototill deeper than 4 inches, as deeper cultivation may bury rhizomes too deeply to be effectively controlled. For more information, see "Solarization" earlier in this chapter.

Well-timed systemic herbicide applications can control bermudagrass, but retreatment will be necessary to kill plants that sprout later. Follow all label instructions and absolutely ensure that nonselective herbicides do not contact desirable plants. Glyphosate is most effective when bermudagrass is flowering or soon after seed heads form, when it is actively growing. There should be sufficient leaves present to allow good spray coverage. Herbicides selective for grasses (e.g., fluazifop or sethoxydim) are most effective if applied when new growth is about 4 to 6 inches tall. Once mature plants are controlled, monitor regularly and cultivate, apply mulch, or use other methods to control seedlings that will emerge when irrigation resumes. Once you have removed bermudagrass, avoid reintroduction with infested compost, mulch, sod, or soil or contaminated mowers and tools. For more information, consult *Bermudagrass Pest Notes*.

KIKUYUGRASS

Pennisetum clandestinum is sometimes grown as a hardy turfgrass. It is also an invasive perennial in ornamental plantings, rights-of-way, and lawns of other turfgrass species. It interferes with irrigation equipment, overgrows fences and shrubs, and invades ground covers and flower beds.

Kikuyugrass has light green leaves about 1 to 10 inches long and ⅛ to ¼ inch wide with pointed tips. St. Augustine grass has a similar appearance and growth habit, but St. Augustine grass has rounded leaf tips. An identifying characteristic of

Bermudagrass is an invasive, creeping perennial that spreads as branching stolons with rootlets and green shoots (shown here) and subsurface rhizomes. *Photo: J. K. Clark*

kikuyugrass is the long fringe of hairs that parallels the stem in the leaf collar region. Kikuyugrass flowers are low to the ground and the white stalks holding the anthers are often mistaken for a fungus.

Kikuyugrass spreads primarily by thick, fleshy stolons that can form a mat on the surface and by rhizomes that grow horizontally under the soil surface. It can also spread by seed, which most varieties of kikuyugrass abundantly produce when mowed.

The best way to control kikuyugrass is to prevent its introduction into new areas. Small stem pieces can produce new shoots and roots, so clean equipment and tools such as mowers to remove any kikuyugrass seed and stems before moving from infested areas. Make sure that incoming soil, sod, and planting stock are weed-free. Regularly inspect landscapes and adjoining areas for invading weeds and take prompt action to control them.

Mulching with a thick geotextile fabric can be effective if the edges are overlapped and the seams tightly sealed and no light is allowed to penetrate to the soil. Organic mulches alone may not be effective because

These thick, fleshy kikuyugrass stolons are overgrowing a Natal plum hedge. Prevent kikuyugrass spread beyond turfgrass because infestations established in other ornamentals are difficult to manage. *Photo:* J. K. Clark

Kikuyugrass flowers grow low to the ground and have long stamens, the white stalks with brown terminals (pollen-producing anthers) shown here. *Photo:* J. K. Clark

plants sprouting from rhizomes can grow through the mulch. Dense turfgrass and ornamental plantings that shade the soil surface make the establishment of kikuyugrass, especially by seed, more difficult. Avoid irrigation runoff and overwatering, as kikuyugrass thrives with excess moisture and nitrogen.

Hand-pulling is the primary control method when kikuyugrass first appears in residential landscapes. Cultivation or hoeing are sometimes effective, but often

are detrimental because they break stolons and rhizomes and transplant them to new areas, especially if followed by irrigation.

Certain preemergence herbicides applied in about March can limit germination of kikuyugrass seed. Preemergence herbicides are of little benefit if kikuyugrass is already established. Multiple applications of post-emergence grass-selective herbicides are required to control established infestations. Careful spot-spraying with nonselective herbicides (e.g., using a sponge or wick applicator) may be feasible for solitary plants where contact with desirable plants can absolutely be avoided. For more information, see *Kikuyugrass Pest Notes.*

PERENNIAL BROADLEAVES

DODDER

Cuscuta spp. (Convolvulaceae) are parasitic plants that penetrate host plant tissues to obtain water and nutrients. After seedlings attach to a suitable host, their "roots" die and the dodder is no longer connected to the ground. Spaghettilike, leafless dodder stems twine around their host, covering it with a tangled yellow or orangish net or mat. Flowers and seed capsules are about ⅛ inch long and develop in clusters. Each dodder plant can produce thousands of hard seed that can remain dormant in the soil for years.

Established dodder is difficult to manage, so promptly eliminate seedlings and isolated infestations before dodder reproduces and spreads. Cultivation, flaming, hand-pulling, mulching, or applying non-

selective herbicides kills seedlings before they attach to hosts. Certain preemergence herbicides control dodder if applied before seedlings emerge (beginning in mid-February in most of central and southern California). More than one application a season may be necessary, as dodder continues to germinate through most of the growing season.

On woody plants, prune off infested host tissue below the point of dodder attachment. On heavily infested hosts, it may be necessary to cut down the entire host and take ongoing action over several years until the soil seed bank is depleted. Rotating the site for several years into nonhost plants such as turfgrass may be an efficient strategy. Planting turfgrass around a shrub or tree may also reduce dodder's impact. Dodder seedlings try to attach to the first plant they encounter, and seedlings die if that plant is not a suitable host.

Be alert for exotic Japanese dodder (*C. japonica*) and report any suspected infes-

Spaghettilike dodder stems are growing as a tangled orangish mat in this California pepper tree. If uncontrolled, dodder can spread to cover the entire canopy of a tree, severely stressing and gradually killing its host. Be alert for invasive Japanese dodder and report any suspected infestations to the county agricultural commissioner. Unlike the distinctly orange stems pictured here, Japanese dodder stems are vibrant yellow-green. *Photo:* J. K. Clark

Dodder seed capsules and stems shown close up. Each dodder plant can produce thousands of seed that can remain dormant in the soil for years, so eliminate isolated patches as soon as they appear. *Photo:* J. K. Clark

tations or unusual dodder problems to the county agricultural commissioner. Unlike the distinctly orange stems of other dodder, Japanese dodder stems are vibrant yellow-green and somewhat thicker. Consult *Dodder Pest Notes* for more information.

FIELD BINDWEED

Convolvulus arvensis (Convolvulaceae) is a very common herbaceous perennial that dies back aboveground during fall and regrows from rhizomes and roots in spring. Field bindweed, also called perennial morningglory, has slender, twining stems up to 5 feet long when mature. Leaves vary greatly, but are often arrowhead shaped or rounded with a blunt tip. The white, pink, or reddish funnel-shaped flowers open on sunny mornings. Seed pods are roundish and light brown. Plants produce abundant seed, which can remain dormant in the soil for many years. Many new plants can develop from field bindweed's extensive rhizomes and root system, which grow to a depth of 10 feet or more and spread several feet wide. Plants often occur in heavy soils, in hardpan or crusty soils, and less often in sandy soils.

Established field bindweed is not controlled by a single treatment or in a single season. Seedlings are easily controlled with cultivation before young plants develop beyond the five-leaf stage. Three to four weeks after germination, they are difficult to control. At unirrigated sites, repeated cultivation of dry soil at about 2- to 3-week intervals for as long as plants regrow can control established plants, but lack of

persistence in cultivation only spreads the weed as it sprouts from severed pieces of root. Geotextile fabrics and sometimes other mulches (e.g., black plastic) are effective if seams are tightly sealed and no light reaches the soil and plants; it might take more than 3 years of light exclusion before the bindweed dies. Shade from shrubs and trees reduce bindweed growth if bindweed is prevented from climbing above the foliage of overstory plants.

Various preemergence herbicides are effective before seedlings emerge but do not control bindweed that regrows from rhizomes and roots. Monitor emerged bindweed regularly to effectively time an herbicide application: the best control is achieved if a translocated herbicide (e.g., glyphosate) is applied just after the first flowers open when bindweed is growing vigorously but before full bloom. To avoid damage to desirable plants, use only spot applications (e.g., a sponge or wick applicator or directed spray). Multiple treatments are often necessary to control a well-established infestation. For more information, see *Field Bindweed Pest Notes.*

IVY

Some ornamental *Hedera* spp. (Araliaceae) become weeds unless they are well maintained. These vigorous vines overgrow nearby structures and plants and can gradually kill trees by shading them and physically injuring hosts with their twining stems and aerial roots. Ivy often harbors cockroaches, rats, and snails.

Algerian ivy (*H. canariensis*) and English

ivy (*H. helix*) are the common pest species. Do not confuse these with less-invasive plants that are also called ivy, including Boston ivy (*Parthenocissus tricuspidata*), grape ivy (*Cissus rhombifolia*), and Swedish ivy (*Plectranthus australis*). Algerian and English ivy are evergreen woody vines with self-clinging branches. These vines climb vertical objects and can reach heights of 25 to 50 feet.

Cultivated varieties vary in leaf color, size, and shape, and juvenile and adult ivy foliage differ markedly in appearance. Juvenile foliage occurs on flexible stems and develops into palmate leaves with three to five shallow lobes. Adult foliage occurs on stiff stems with leaves that generally are not lobed; when present, adult foliage is often at the terminals of tall vines

Field bindweed seed leaves (bottom) are nearly square, with an indented tip. New shoots from rhizomes resemble this seedling but lack cotyledons. Unlike with seedlings, preemergence herbicides or a single cultivation will not control bindweed regrowing from rhizomes or roots. *Photo:* J. K. Clark

Field bindweed has blunt-tip, arrowhead-shaped leaves and white- to reddish-tinged, funnel-shaped blossoms. The flowers can be present any time from spring through fall, opening on sunny mornings and often closing overnight. *Photo:* J. K. Clark

Algerian ivy (shown here) and certain cultivars of English ivy are vigorously spreading, evergreen vines with woody branches that cling to and climb objects and other plants. Without frequent management, they often overgrow and gradually kill trees and shrubs. Many *Hedera* spp. are not invasive, but their variable foliage makes them difficult to distinguish, and ivy is commonly mislabeled in nurseries. *Photo:* J. M. DiTomaso

where it is not readily observed. Mature foliage produces clusters of small greenish flowers that develop into bluish or black berries about ¼ inch in diameter.

Consider planting less-vigorous species as ground covers instead of Algerian or English ivy. If there is any chance that plants will not be pruned regularly, remove the ivy within a couple years of planting, after which it becomes increasingly invasive. Regularly shear terminals to contain their spread, clip off flowers or fruit if any are seen, and remove seedlings that appear where ivy is not desired. To maintain ivy's appearance, do most pruning in spring when new growth will more quickly cover bare stems.

Retard growth of unwanted ivy by increasing its exposure to light and decreasing its access to water, to the extent compatible with the needs of surrounding landscapes. Remove ivy by pulling up vines and digging out the roots. Repeat this until ivy no longer resprouts. Where ivy climbs trees, cut stems as high above ground as you can reach and remove them. Ivy readily develops from cuttings, which may already have aerial roots. Dispose of ivy in a manner that does not allow prunings to develop into new infestations.

Some broadleaf and nonselective herbicides control ivy, but it can be difficult to use them in landscapes without damaging desirable plants. One technique is to cut stems and promptly (within minutes) apply an appropriate systemic herbicide (e.g., triclopyr) to fresh wounds with a brush, sponge, or wick applicator.

MISTLETOES

Mistletoes are parasitic perennials that grow on woody plants, extracting moisture and nutrients from their host. The most common species in California landscapes are true mistletoes (*Phoradendron* spp.), also called American, broadleaf, large leaf, or leafy mistletoes. An otherwise healthy tree can tolerate a few mistletoes, but individual branches may be killed. Bark often swells or forms galls around where mistletoe attaches to its host. Host plants can suffer reduced vigor or become stunted, especially if they are stressed by other problems such as drought or disease. Dwarf mistletoes (*Arceuthobium* spp.) can infect only conifers and are found mostly in forested areas. In comparison with true mistletoes, dwarf mistletoes have more severe impact, often slowly killing their conifer hosts.

Several *Phoradendron* spp. occur in California. Oak mistletoe (*Phoradendron leucarpum* ssp. *tomentosum*) occurs primarily on *Quercus* spp. and less often on other hosts including California bay laurel and manzanita. Bigleaf mistletoe (*P. leucarpum* ssp. *macrophyllum*) does not attack oaks, but it infects many landscape plants, including alder, ash, birch, black walnut, box elder, California buckeye, cottonwood, fruit and nut trees, locust, maple, mesquite, and willow. Other *Phoradendron* spp. occur only on conifers, including fir, incense-cedar, and juniper. European mistletoe (*Viscum album*) resembles *P. leucarpum*. European mistletoe in the United States occurs only in Sonoma County, California, primarily on alder, apple, black locust, cottonwood, and maple.

American and European true mistletoes have succulent green stems that become woody at the base. The green leaves are thick and nearly oval. Plants often develop a roundish form up to 2 feet or more in diameter. These true mistletoes produce white to pink, pea-sized fruit that grow in easily seen clusters. The seed are dispersed in bird excrement, which is why mistletoes often are most abundant around the treetops.

Dwarf mistletoes' mature stems are 6 to 8 inches long or less, nonwoody, segmented, and have small, scalelike leaves or appear leafless. Dwarf mistletoe seed are spread mostly by their forcible discharge from fruit. All mistletoe seeds are sticky and can be spread attached to animals or tree-trimming equipment.

Consider replacing severely infested trees with different species. For example, Chinese pistache, crape myrtle, eucalyptus, ginkgo, golden rain tree, liquidambar, and sycamore are rarely infested. Conifers are also less often attacked by leafy mistletoes, but incense-cedar and white fir are significantly infested in their native mountainous habitats.

The most effective control is to prune out infected branches as soon as mistletoe appears. Cut limbs infected with true mistletoe at least 1 foot below the point of mistletoe attachment, but make pruning cuts properly at crotches but leaving the branch bark ridge intact (see Figure 3-8). In conifers with dwarf mistletoe, entirely prune off infected limbs as well as the branches within a few feet of the infected branch. Do not remove more than two-thirds of a tree's canopy (its foliage and leafy branches). Severe heading (topping) is often used to remove heavy tree infestations; however, such pruning weakens

True (leafy) mistletoes are evergreen plants that are most apparent on deciduous trees during the host plant's dormant season, as with these green clumps of bigleaf mistletoe. The most effective control is to prune off infested branches as soon as mistletoe appears. *Photo:* J. K. Clark

Bigleaf mistletoe has thick and nearly oval leaves. The small, sticky, orangish to white berries are spread by birds and tree-trimming equipment. *Photo:* J. K. Clark

Stems and seed of a female dwarf mistletoe (*Arceuthobium occidentale*) growing from a pine branch. The branch swelling visible here is caused by the mistletoe. *Photo:* J. K. Clark

a tree's structure and destroys its natural form. In some cases it is best to remove severely infested trees entirely because they are a source of mistletoe seed.

True mistletoe infesting a main branch or trunk where it cannot be pruned may be controlled by cutting off the mistletoe flush with the limb or trunk of its host, then covering the bark. Wrap the attachment point with several layers of wide, black polyethylene or landscape fabric. Tie it with twine or flexible tape to exclude light. True mistletoes require light and die within a year or more after they are cut and wrapped. It may be necessary to repeat this treatment, especially if the plastic becomes detached. If mistletoe is cut but not covered, it will grow back. However, cutting can reduce the spread of mistletoe by tem-

porarily preventing seed production and may reduce damage to its host.

The plant growth regulator ethephon may be applied during the late dormant season as directed on the label to control true mistletoe on some species of severely infested deciduous trees. To be effective, spray must thoroughly wet the mistletoe foliage when daytime temperatures are above 65°F in the spring before the tree begins to grow new leaves. Spray only the individual mistletoe plants, not the entire

tree. Spraying provides only temporary control by causing some mistletoe plants to fall off, and mistletoe may soon regrow at the same point. For more information, consult *Mistletoe Pest Notes*.

Bermuda buttercup is attractive when flowering, but this *Oxalis* sp. can be a nuisance when it spreads into shrubs like this juniper. *Photo:* J. K. Clark

Creeping woodsorrel infests these container plants. Its pods forcefully eject tiny seed up to severa[l] so woodsorrel oxalis is readily spread in new plants and soil that have been near mature weeds s[uch] as these. Yellow woodsorrel also has seed-ejecting capsules and similar foliage. Creeping woodso[rrel] foliage color ranges from green to dark purple, including plants with these purple-tinged green l[eaves.] The other *Oxalis* spp. common in landscapes have only green foliage. *Photo:* J. K. Clark

OXALIS

Bermuda buttercup, or buttercup oxalis (*Oxalis pes-caprae,* Oxalidaceae), creeping woodsorrel (*O. corniculata*), and yellow woodsorrel (*O. stricta*) are pests in lawns, planting beds, and nurseries. They have yellow, five-petalled flowers and compound leaves. Each leaf consists of three heart-shaped leaflets resembling clover, and the foliage is green, except that creeping wood-sorrel foliage color ranges from green to purple-tinged green, to deep purple.

Oxalis spp. prefer moist, shady situations, but grow in full sun in cool, coastal locations. The woodsorrels are more tolerant of sun even in hot areas if soil remains moist.

Their low-growing or underground vegetative reproductive parts distinguish the species:

- Bermuda buttercup has bulblets– small, whitish, roundish propagules found up to 1 foot deep in soil.
- creeping woodsorrel has stolons– above-ground, low-growing, horizontal stems that root at stem joints.
- yellow woodsorrel has rhizomes–long, slender, tough, horizontal underground stems.

Bermuda buttercup leaflets are each spotted with two purple dots, and the leaves are larger and thicker than those of the woodsorrels. Bermuda buttercup's bright yellow flowers are ¾ to 1½

inches in diameter and bloom from late fall through spring. These flowers are wider and have taller stalks than those of creeping woodsorrel. Bermuda buttercup is sometimes grown as an ornamental because this hardy plant flowers profusely. It does not produce viable seed and spreads when bulb-contaminated soil is moved to uninfested areas.

Creeping woodsorrel flowers are about ¼ inch in diameter and can be present throughout much of the year. Flowers mature into hairy, cylindrical, pointed pods about ⅓ to 1 inch long. Each pod contains 10 to 50 reddish seed, each about ¹⁄₂₅ inch long. Creeping woodsorrel reproduces primarily by seed, which are forcefully ejected up to several feet by the pods. Seed germinate when conditions are warm and moist. Seedlings rapidly form a fleshy taproot, a prostrate mass of foliage, and shallow roots that spread outward. If plants are pulled, roots often break and remain in the soil, allowing plants to regrow. Creeping woodsorrel is the only oxalis with a stipule, a scalelike projection at the base of each leaf petiole; however, you need high magnification to clearly see these tiny stipules. In comparison with yellow woodsorrel, creeping woodsorrel also typically has a more prostrate growth habit and some plants have purplish leaves.

Yellow woodsorrel has somewhat larger flowers (about ¾ inch in diameter)

and taller growth (up to 1 foot high) than creeping woodsorrel. Both woodsorrels have seed-ejecting capsules. Look for creeping woodsorrel's stipules and stolons as above to distinguish the species.

Hand-pull young oxalis plants before vegetative propagules and flowers form. Use shallow cultivation to kill young seedlings. Apply landscape fabric (geotextiles) or thick organic mulch to prevent most oxalis growth.

Solarization before planting provides partial control of creeping woodsorrel. Multiple applications of a postemergence herbicide may be needed to eliminate well-established infestations and can be used in combination with preemergence herbicides and mulches to reduce reinfestation of the area from seed. Because Bermuda buttercup reproduces only vegetatively, the most effective methods are hand-removal, digging up the area to try to remove the bulbs, and repeated applications of translocated herbicides. Absolutely avoid broad-spectrum and nonselective herbicide contact with desirable landscape plants. For more information, consult *Creeping Woodsorrel and Bermuda Buttercup Pest Notes.*

POISON OAK

Contact with poison oak (*Toxicodendron toxicarium,* Anacaridaceae) or its oil, which rubs off onto clothing or pets, causes many

Poison oak has small, whitish flowers in spring that become white or pale greenish berries by late summer. Leaves are clusters of three leaflets, each 1 to 4 inches long and resembling an individual leaf. The most terminal leaflet usually has a short stem; the side leaflets have no distinct stem. *Photo: J. K. Clark*

people to develop a very bothersome skin rash. Poison oak (also called western poison oak) in California is common from sea level to elevations of about 5,000 feet. It occurs in open chaparral and woodlands, coniferous forests, and grassy hillsides.

Poison oak is an erect, deciduous woody shrub or vine that often climbs trees and shrubs and loses its leaves over winter. Its leaves are green or light red in the spring, glossy green in late spring and summer, and yellow or red in the fall. Leaves are clusters of three leaflets (or occasionally five, seven, or nine leaflets), each leaflet being 1 to 4 inches long and resembling an individual leaf. The most terminal or central leaflet has a petiole or stem; the side leaflets have no distinct stem. The leaves of true oaks grow singly, and each leaf has a distinct petiole. In spring, poison oak produces clusters of small, white flowers that by late summer develop into white or pale greenish berries.

Poison oak in high-use areas warrants aggressive control action because of its severe skin hazard. Manual removal and herbicide application are the primary controls for poison oak. Goat or sheep grazing can be effective if the livestock are confined in infested, small areas and prevented from feeding on desirable plants. A combination of chemical and physical control may be needed more than once on well-established clumps of poison oak. Once an area has been cleared of poison oak, plant desirable species and provide proper care to desirable species to exclude reinfestation.

Mechanical control (e.g., rototilling) can spread root pieces that resprout and is generally not effective unless done repeatedly and without spreading plant parts. Mowing at least four times a season may provide control. Do not burn poison oak; burning causes hazardous oils to be transported with the smoke. Breathing smoke from poison oak causes severe respiratory irritation.

Physical removal can control a few plants, but individuals who are sensitive to poison oak should not do this work. Carefully cut and remove all top growth and, when soil is moist, grub or dig out roots to a depth of 8 to 10 inches and remove horizontal runners. Wear tightly woven protective clothing, including washable cotton gloves worn over plastic gloves. Wash tools, rinse them in alcohol, and then dry and oil tools to prevent rust. Separately launder all clothing thoroughly and shower immediately after working around poison oak; use isopropyl alcohol, a product especially for removing poison oak oils, or a large amount of cold water and soap to promptly wash poison oak oils from the skin.

Poison oak can be controlled with postemergence systemic herbicides that are nonselective (e.g., glyphosate) or are selective for broadleaves (triclopyr), but do not allow spray to contact desirable plants. Glyphosate is most effective after fruit form but before leaves lose their green color. Triclopyr is effective from full foliage to flowering (spring to midsummer) when plants are rapidly growing. Certain herbicides can be applied to basal bark or freshly cut stumps. Where poison oak climbs tree trunks, you can cut woody stems near the base and apply systemic herbicide to the fresh wounds; the vines in the tree will die and gradually decay. Follow all herbicide label directions carefully. See *Poison Oak Pest Notes* for more information.

Blackberry leaflets grow in clusters of three to five on long, thorny, somewhat woody canes. Wild blackberries can spread aggressively over shrubs, like this flowering *Rhaphiolepis*. *Photo:* J. K. Clark

WILD BLACKBERRIES

Rubus spp. (Rosaceae) have long, trailing, somewhat woody canes. Leaves are up to several inches wide and compound, with three to five leaflets. The blackberry species in California produce white to reddish flowers and have thorny stems. The red to shiny black fruit are enjoyed by people and wildlife. Blackberry hedges can provide an effective barrier to restrict access by people and pets, but they can overgrow surrounding vegetation and sometimes harbor rats.

Non-native weedy blackberries include cutleaf blackberry (*R. laciniatus*) and Himalaya blackberry (*R. discolor* =*R. procerus*). Two native species, thimbleberry (*R. parviflorus*), which is nonvining, and California blackberry (*R. ursinus* =*R. vitifolius*) can also be weeds.

Blackberries regrow from the crown or rhizomes each year and following burning, mowing, or herbicide application; controls must be repeated or combined to eliminate blackberries. Keep desired hedges well pruned. Kill established clumps by repeatedly pruning stems or rototilling canes until root reserves are exhausted. Although difficult work, established clumps can be killed by digging crowns from soil and allowing them to dry in the sun. Alternatively, apply a broadleaf or nonselective herbicide to foliage during early summer when leaves and canes are rapidly expanding. Professional applicators can use certain bark- or soil-applied herbicides. For more information and specific herbicide recommendations, consult *Wild Blackberries Pest Notes*.

ALGAE, LICHENS, AND MOSSES

Woody plants sometimes have green, gray, or orangish tissue growing on or hanging from bark. Algae, lichens, and mosses are common causes of these growths on bark and other moist, shady locations such as on decks, pavement, and roofs. Algae and mosses are relatively simple (primitive) green plants. Lichens are an association between certain algae and fungi.

These organisms are generally harmless to trees. Many are epiphytes, green plants that derive their nutrients and moisture from the air and grow on the host surface

Lichens are growing harmlessly on this hackberry bark. Algae, lichens, and mosses often develop as green, gray, or orangish tissue growing on or hanging from older bark. *Photo:* J. K. Clark

only for support. Some are saprophytes, absorbing nutrients from soil and dead organic debris that lodge in bark crevices, and minimally consuming the outer, dead bark surface. Heavy growth is common on older trees in part because the more rapidly expanding bark on younger trees spreads and "dilutes" the appearance of these surface growths. However, profuse epiphytes or saprophytes on bark may indicate that a host plant is growing more slowly than desired and would benefit from improved environmental conditions and appropriate cultural care. Abundant algae and moss sometimes indicate that soil drains poorly or landscapes are being overwatered, and these poor growing conditions can seriously damage plants.

Moss and algae thrive under damp conditions; their growth may be retarded by reducing humidity (such as preventing sprinklers from hitting bark), improving air circulation around host plants (such as by trimming branches), improving soil drainage, and increasing the frequency between irrigations to the extent compat-

ible with healthy plant growth. Moss prefers partly shaded locations, so it may be retarded by increasing light around plants, such as by thinning branches and nearby plants. Conversely, growth of lichens and sometimes algae is stimulated by increased light.

Some people consider epiphytes and saprophytes in landscapes to be interesting and desirable. However, mosses and algae on hard surfaces make them slippery and unsafe for walking. Where growths are not tolerable, herbicidal soaps, certain copper sprays, iron HEDTA, or certain other pesticides may be applied to provide control. Keep these sprays out of waterways and away from desirable, succulent plant parts, which may be damaged. Overall, the best management for mosses and algae is to reduce the frequency of watering, reduce the use of quick-release nitrogen fertilizer, and minimize the amount of eroding soil flowing over concrete or other impermeable surfaces.

Chapter 8 Nematodes

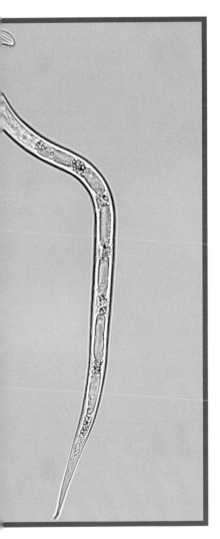

Nematodes are tiny (usually microscopic) unsegment-ed roundworms. Depending on the species, nematodes feed on bacteria, fungi, plants, vertebrates, or insects or other invertebrates. Some nematodes are beneficial because they kill pest insects or plant-parasitic nematodes. For example, *Mononchus* spp. nematodes feed on other nem-atodes. *Heterorhabditis* and *Steinernema* spp. nematodes can be purchased and applied to control certain insects that feed on roots or bore in trees, as discussed in "Entomopathogenic Nematodes" in Chapter 6.

The most important plant-parasitic (plant-feeding) species are root knot nematodes (*Meloidogyne* spp.), which attack many woody and herbaceous ornamentals, fruit and nut trees, and vegetables. Other root-feeding species, in approximate order of their importance in woody landscape plants, include root lesion, dagger, ring, stunt, and citrus nematodes. Foliar nematodes (*Aphelenchoides* spp.) and pine wilt nematode (*Bursaphelenchus xylophilus*) feed in aboveground plant parts but usually are not pests in California landscapes. Consult *Nematodes Pest Notes* online at ipm.ucanr.edu and other publications listed in "Suggested Reading" at the back of this book for more information.

DAMAGE

Nematodes can damage boxwood, fruit and nut trees, roses, and various herbaceous orna-mentals and garden vegetables. They also infest the roots of many other tree and shrub species, but the extent to which nematodes damage these other woody plants in California landscapes is not well known. Their microscopic size, hidden feeding habits, and some-times subtle damage make it difficult to diagnose whether nematodes are damaging plants.

This adult root lesion nematode (*Pratylenchus penetrans*) punctures the cell wall and feeds outside the root, attached by its mouthparts. The nematode's syringelike stylet (mouthpart) for puncturing host cells and a muscular bulb at the base for pumping out plant cell contents are visible here under high magnification. *Photo:* A. T. Ploeg

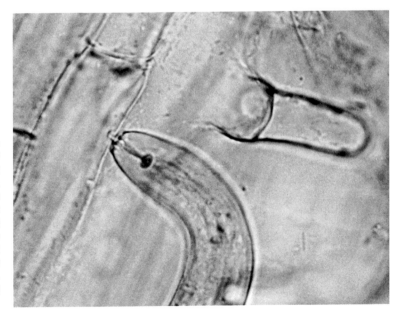

Nematodes affect tree crops primarily by reducing fruit or nut yields and slowing plant growth. Slowed growth may not be a serious concern in ornamental trees and shrubs unless damage occurs during plant establishment or to plants being stressed by other maladies.

Most plant-feeding nematodes live in soil and feed on or in roots. These plant-parasitic nematodes insert their entire body or only their syringelike stylet (mouthpart) into plant tissue and pump out cell contents. This feeding directly injures or kills plant cells and inhibits the plant's ability to obtain nutrients and water. Nematode damage symptoms that can be visible aboveground include slow growth or yellow foliage that may drop a few weeks prematurely. Plant tip dieback may be visible on infected trees and shrubs that are not regularly pruned. Similar decline symptoms can be caused by root decay or vascular wilt diseases, certain insects and mites, and a lack of proper cultural care, such as inappropriate irrigation or excess fertilizer. Nematodes commonly act in combination with other pests, especially microorganisms that cause root disease (e.g., *Phytophthora* spp.). Certain nematodes vector plant-pathogenic viruses.

The severity of nematode damage depends on the host age, vigor, scion and rootstock cultivar, cultural practices, soil conditions (e.g., soil type and temperature), and the nematode species, race (strain), and abundance. Damage usually results only from heavy populations on the nematode-parasitized roots. Nematode feeding causes plant stress, but established woody plants are rarely killed by nematodes. However, if woody plants become heavily infected with nematodes, it may be difficult to grow susceptible annual and herbaceous plants nearby because nematodes have become so prevalent in the soil.

IDENTIFICATION AND BIOLOGY

Plant-feeding nematodes develop through six stages: egg, four juvenile stages, and adult. Many species can develop from egg to egg-laying adult in as little as 3 to 4 weeks when the soil is warm and moist. Juveniles, such as the second stage pictured at the beginning of this chapter, and adult males are long, slender worms. The mature adult females of some species, such as root knot nematode, change to a swollen, pearlike shape, whereas females of other species, such as lesion nematode, remain slender worms. In some, such as root knot nematodes, only adult males and second-stage juveniles are mobile in soil or roots; the other juvenile stages and adult females are immobile. Most plant-feeding nematodes are too small to be seen without the aid of a microscope.

Nematodes require moist environments to feed and reproduce. To move they need a layer of water (e.g., wet soil particles). During adverse conditions, such as dry soil, cold temperatures, or the lack of host plants, some species develop stages resistant to drying out, which become inactive and can survive for a year or more.

Many landscape tree and shrub species are susceptible to nematodes (Table 8-1), and nematode infestations should be suspected whenever a general decline of a particular plant species is observed, including stunting or yellow leaves. If no other causes for the unhealthy plant are obvious, remove soil from around some roots and examine them for signs of nematodes. Ease the roots out of the soil gently so that the smaller feeder roots are not broken off and can be examined. Nematodes can cause root galls or stubby, stunted, or proliferating roots. Roots that are darkened or have lesions and plants with fewer roots than normal can also indicate a nematode infestation. Similar symptoms can be produced by various other causes, including *Phytophthora* infection of roots (pictured at the end of Chapter 5), prolonged excess soil moisture, and phytotoxicity from soil-applied herbicide. Disorders or root disease pathogens often occur in combination with nematodes.

Not all nematodes produce obvious symptoms on roots. To confirm a nematode infestation, collect several samples of small roots and soil from plants showing poor growth and send them to a laboratory that can identify nematode presence. Be aware that the mere presence of root-feeding nematodes does not mean that they are the cause of plant damage. If possible, separately collect one or two small, apparently healthy plants or root pieces from one large healthy plant of the same species growing nearby and send them along with surrounding soil for comparison testing. Contact the local county agricultural commissioner or University of California (UC) Cooperative Extension office or a UC nematology specialist (listed online at the ucanr.edu

A soil-dwelling predatory mononch (*Mononchus* sp.) nematode swallowing a small nematode, the two ends of which are protruding from the predator's mouth. *Photo:* J. O. Becker

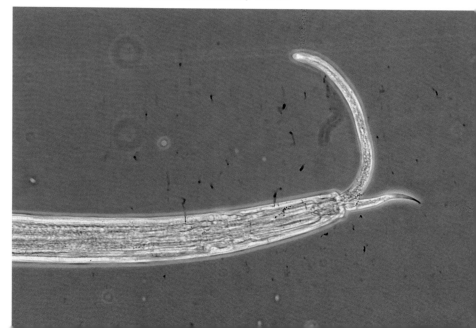

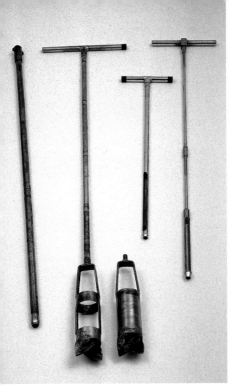

...gers (center) or a shovel are good tools for collecting ...d feeder roots for laboratory analysis of nematodes. ...ld soil tubes (right) and the two-piece Veihmeyer tube ...re useful for monitoring soil type and moisture. The ...eyer tube has a slotted hammer for driving the tube ...il and removing it. The other types are pushed into ...ile turning their handles back and forth. These tools ...llect soil cores down to about 2 feet. *Photo: J. K. Clark*

website) for help in determining where to submit plant samples for analysis.

You also can use a simple bioassay to detect certain species of root knot nematodes. Plant melon seed in pots containing moist field soil. If *Meloidogyne arenaria, M. incognita,* or *M. javanica* are present, visible galls will develop on the melon roots in about 3 weeks when pots are kept at about 80°F. However, this test is not useful for the northern root knot nematode (*M. hapla*) because it can be present without inducing galls on melon seedlings. *Meloidogyne hapla* is common on roses and certain fruit and nut trees and herbaceous plants.

MANAGEMENT OF NEMATODES

Nematodes are difficult to manage because in most landscape situations you cannot directly reduce their numbers. The following practical nematode management methods are preventive:

- Use good sanitation to avoid introducing or spreading nematodes.
- Grow plant cultivars or species that

Table 8-1.

Trees and Shrubs Known to Be Hosts of Plant-Parasitic Nematodes in California.

HOST PLANT		PLANT-PARASITIC NEMATODES		
COMMON NAME	SCIENTIFIC NAME	ROOT KNOT	ROOT LESION	OTHER
abelia, glossy	*Abelia grandiflora*	*		
albizia	*Albizia*	*		
alder	*Alnus*	*		
aralia	*Fatsia (=Aralia)*	*		
azalea, rhododendron	*Rhododendron*	*		foliar, spiral, stubby root, stunt
bird of paradise	*Strelitzia*			spiral
boxwood	*Buxus*	*		
cactus	several genera	*		cyst
camellia	*Camellia japonica*			spiral, stunt
catalpa	*Catalpa*	*		
cedar	*Cedrus*	*		
cryptomeria	*Cryptomeria japonica*	*	*	
dracaena	*Cordyline*	*		
echium	*Echium*	*		
euonymus	*Euonymus*	*		
ferns	many genera			foliar
fig	*Ficus*			foliar
fir	*Abies*			dagger
gardenia	*Gardenia jasminoides*	*		
hibiscus	*Hibiscus*	*		foliar
hydrangea	*Hydrangea*	*	*	foliar
juniper	*Juniperus*	*		
lilac	*Syringa*			citrus
maidenhair	*Ginkgo*	*		
melaleuca	*Melaleuca*	*		
mulberry	*Morus*	*		
oak	*Quercus*	*		
olive	*Olea*	*	*	citrus
orchid tree	*Bauhinia*	*		
palm	several genera	*		
pear	*Pyrus*		*	
pittosporum	*Pittosporum*	*		
plum	*Prunus*		*	ring
poinsettia	*Euphorbia*	*		
rose	*Rosa*	*	*	dagger
saltbush	*Atriplex*	*		
sweet shade	*Hymenosporum flavum*	*		
tea tree	*Leptospermum*	*		
walnut	*Juglans*	*	*	ring

KEY

spiral = *Heliocotylenchus* spp.

stubby root = *Paratrichodorus, Trichodorus* spp.

See the text for the scientific names of the other nematodes.

Contact your local UC Cooperative Extension office and suppliers for current recommendations on nematode-resistant rootstocks. Consult *Nematodes Pest Notes* for the nematode susceptibility of fruit and nut trees and *Integrated Pest Management for Floriculture and Nurseries* and the online *Floriculture and Ornamental Nurseries: UC IPM Pest Management Guidelines* for the nematode susceptibility of herbaceous ornamentals.

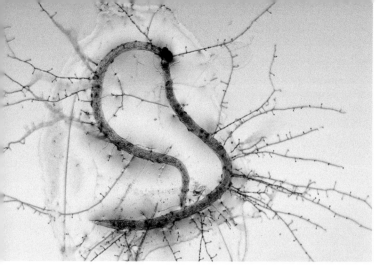

Many plant-parasitic nematodes are killed by natural enemies. This cyst nematode (*Heterodera* sp.) was killed by a beneficial fungus (*Hirsutella rhossiliensis*) that is producing numerous tiny spores on hyphal strands growing from the nematode's body. Other passing nematodes will become infected and killed and produce more infective spores. *Photo:* J. K. Clark

are nonhosts or are resistant or tolerant to nematode damage.

- Provide good cultural care and appropriate growing conditions to reduce stress on plants and increase their tolerance of infestations.

To avoid introducing nematodes into landscapes, use only amendments, soil, and plants obtained from a reliable supplier that presumably are free of pathogens. Plant-parasitic nematodes can be introduced with plants or soil from infested sites. Do not obtain rooted plants from neighbors, friends, or others; start your own root cuttings or grow plants from seeds. Nematode-free plants for certain species can be obtained from participants in the California Nursery Stock Nematode Certification program. Proper pasteurization (e.g., heating in a kiln or applying steam) eliminates nematodes from soil amendments that tolerate being heated.

Do not allow irrigation water from around infested plants to run off onto soil around healthy plants as this spreads nematodes. Do not transfer soil from around infested plants to healthy plants. Thoroughly wash soil and plant parts from all equipment and tools used around infested plants before leaving that site. Work first in uninfested areas before working around plants or soil suspected of being infested with nematodes.

Consider replacing severely damaged plants and replant with species or cultivars more tolerant of the specific nematodes present; for example, do not replant the same plant genera into the old site. If you intend to grow susceptible woody plants

in a nematode-infested area, consider fallowing the soil (keeping it free of plants and weeds) for 4 years before planting to reduce nematode populations; this can help improve the establishment of new plants.

Modest amounts of composted greenwaste, peat, or other decomposed organic amendments can be incorporated into soil before planting, as discussed in "Prepare the Soil" in Chapter 3, to improve plant growth, especially in sandy soils. Adding organic matter to increase the soil's water-holding capacity and irrigating more frequently can lessen the effects of nematode injury because nematodes cause more damage in plants that are water stressed. Provide proper cultural care so that plants are vigorous and better able to tolerate feeding by nematodes and other pests.

No chemical nematicides or soil fumigants are available to home gardeners or professional landscapers because of safety restrictions such as buffer zone requirements. Pesticides may be available to certified applicators for use in certain commercial situations, but these are costly, rarely used, and generally are not warranted or recommended for nematodes in landscapes. Use alternatives, such as planting species or cultivars tolerant to the nematodes present at that location. Plant susceptible species only in locations where nematode populations are low and where soil or conditions are not conducive to nematode buildup. Nematodes generally cause more damage in sandy soils and are less of a problem in silty and clay loams.

Alternative Methods Not Demonstrated to Be Effective in Woody Landscapes.
Solarization before planting (covering moist, bare soil with clear plastic for 4 to 6 weeks during warm weather, as discussed in Chapter 7) temporarily reduces nematodes in about the upper 6 inches of soil. This is where most root growth occurs when plants are young and most susceptible to nematode damage. When used in combination with planting only species that are tolerant of or resistant to nematodes, preplant solarization may be of some value during the establishment of woody plants. However, soil solarization for nematode control has been shown to be effective only for annual or short-lived plants.

Brassicaceae plant residue (e.g., mustard meal, *Brassica juncea*) chopped and incorporated at high rates into soil before planting can also control nematodes affecting annual plants. Incorporating *Brassica* residue and solarizing soil are sometimes used in combination, but there is little research on their usefulness before planting trees or shrubs.

You can suppress root knot and lesion nematodes by growing nematode-suppressive plants as a solid planting free of weeds for an entire season, then mowing off the tops and cultivating them into the soil before planting. Certain marigolds (*Tagetes* spp.), especially French marigolds (including Nemagold, Petite Blanc, Queen Sophia, and Tangerine), are the most effective. Growing nematode-suppressive marigolds beneath trees or shrubs (intercropping) is not usually very effective.

Natural enemies of nematodes are common in many soils. Beneficial species include predaceous nematodes, predatory mites, and nematode-trapping and parasitic fungi. However, biological control is not sufficient to prevent damage from nematodes on susceptible cultivars. Efforts to enhance biological control of nematodes, such as by manipulating cultural practices and soil conditions, have not been successful.

A heavy root knot nematode infestation has caused roots to form these numerous galls. Beneficial nitrogen-fixing bacteria nodules also grow on some roots, but bacterial nodules rub off and a thumbnail can be pressed into them easily, while root knot nematodes cause roots to develop firm swellings. *Photo:* J. K. Clark

ROOT KNOT NEMATODES

Root knot nematodes (*Meloidogyne* spp.) are the most common nematodes attacking annual and perennial landscape plants, especially in warm, irrigated, coarse-textured soils (sand, sandy loam, and loamy sand). Boxwood, rose, and most fruit and nut trees can be damaged in landscapes by *Meloidogyne* spp., and many other landscape tree and shrub species are hosts of root knot nematodes (Table 8-1), as are many herbaceous ornamentals and common weeds. A plant species resistant to one species of root knot nematode may still be susceptible to other *Meloidogyne* spp., and more than one root knot nematode species can occur at the same site.

Root knot nematodes (Figure 8-1) cause galls or swellings on roots of many broadleaf plants. Some infected plants, especially annual grasses and certain legumes, may exhibit no galls even though nematodes still reproduce on them. When roots infected with root knot nematodes are washed they may appear gnarled and restricted. Infected roots may become attacked by other pathogens, including those causing crown gall, root and crown decays, and certain vascular wilt diseases.

Look for root knot nematode galls on plant roots. Beneficial nitrogen-fixing bacteria often form nodules on the roots of beans and other legumes, but these rub off roots easily, while galls caused by root knot nematodes are truly swellings of the roots. Also, a thumbnail can be pressed into a bacterial gall easily but not into a root knot gall. Other causes of plant galls are summarized in Table 6-9.

To confirm a root knot nematode infestation, you must collect root and soil samples and send them to a laboratory that can provide positive identification of the infesting species.

ROOT LESION NEMATODES

Root lesion nematodes (*Pratylenchus* spp.) can occur on many tree and shrub species (Table 8-1). On roses and some other hosts, infected plants are stunted, leaves may turn yellow, and root systems are smaller and darker than on healthy plants.

In the early stages of infestation, a lack of feeder roots and well-developed major roots may be observable if you carefully remove soil around roots of unhealthy plants and compare them with roots from around healthy plants. Root lesion nematodes occasionally cause brown or black lesions to appear on roots. Lesions are usually apparent only on the larger roots of older trees, especially on walnut. Lesions, when present, become apparent when roots are scraped with a knife. However,

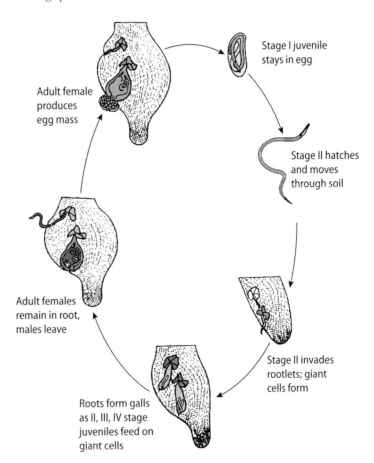

Stage I juvenile stays in egg

Adult female produces egg mass

Stage II hatches and moves through soil

Adult females remain in root, males leave

Stage II invades rootlets; giant cells form

Roots form galls as II, III, IV stage juveniles feed on giant cells

Figure 8-1.

Root knot nematode life cycle and stages. Root knot nematodes spend most of their active life cycle in galls on roots. Second-stage juveniles invade new sites, usually near root tips, causing root cells to grow into giant cells where the nematodes feed. As feeding continues, the plant produces a gall around the infected area. Mature females produce eggs in a small gelatinous mass on the root surface (as shown) or inside the root.

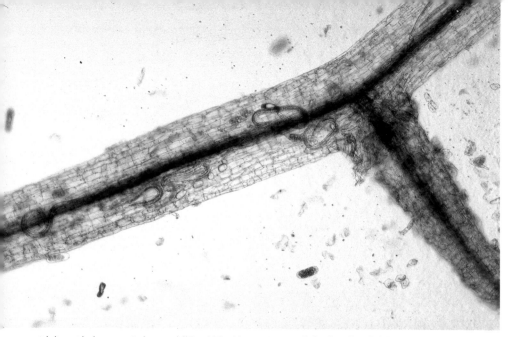

Adult root lesion nematodes are visible within this root. Nematode feeding directly injures or kills plant cells and inhibits the plant's ability to obtain nutrients and water. *Photo:* J. K. Clark

lesions can develop on roots for other reasons unrelated to nematodes, for example, on walnuts if they are given too much water.

Many plant species may be infected with high populations of root lesion nematodes without any evidence of necrosis (dead tissue) or lesions. A laboratory analysis of samples is the only sure method of diagnosing root lesion nematodes.

Pin nematodes (*Paratylenchus* spp.) are related species that have been reported to cause lesions and death of some plum and prune roots. Although pin nematodes are frequently found in these fruit trees, they currently are not thought to significantly stunt plant growth or reduce fruit yield.

DAGGER NEMATODES

Dagger nematodes (*Xiphinema* spp.) can vector viruses and cause a sparseness of feeder roots, but this is difficult to recognize. Terminal galls can form on the roots of grapes, each with hundreds of dagger nematodes of the same species feeding on it. A laboratory analysis of soil surrounding affected roots is usually needed to confirm the presence of this pest, which mostly lives in soil and feeds from outside, externally on roots.

RING NEMATODES

Ring nematodes are named for their annulated (ringed) body, which is visible under a microscope; most nematode species have a relatively smooth body. Ring nematodes kill tips on the smallest roots and, if extensive, this damage weakens trees. For example, *Mesocriconema xenoplax* can pre-

dispose *Prunus* spp. and occasionally apple to branch dieback and springtime death by bacterial canker disease.

STUNT NEMATODES

Stunt nematodes (*Tylenchorhynchus* spp.) cause roots and aboveground plant parts to grow slowly and be undersized. Leaves may turn yellow and drop prematurely, and shoots can die back.

CITRUS NEMATODE

Citrus nematode (*Tylenchulus semipenetrans*) can reduce fruit size and number on citrus, grape, olive, and persimmon, and it also infests lilac. Serious infestations cause undersized leaves and twig dieback. Provide trees with proper cultural care so they are better able to tolerate nematode feeding. Purchase trees from a nursery that sells nematode-free plants and choose nematode-resistant or -tolerant rootstock if available. For more on citrus nematode biology and management, see *Integrated Pest Management for Citrus*.

In comparison with healthy roots (left), citrus nematode and certain other nematodes cause infested roots to appear dark or dirty (as at right) in part because soil clings to the damaged roots. Roots affected by *Phytophthora* (pictured near the end of Chapter 5), root-infecting fungi, and abiotic disorders also exhibit similar symptoms; these maladies often occur in combination with nematodes. Laboratory testing of properly collected samples is needed to confirm whether nematodes are the cause of damage. *Photo:* J. O. Becker

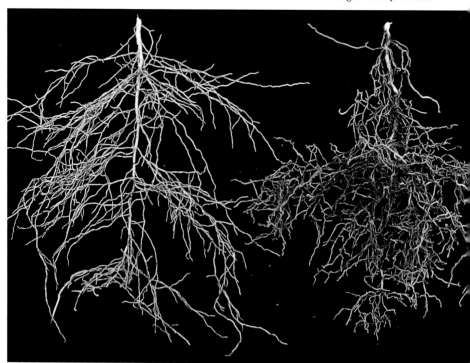

Chapter 9 Problem-Solving Tables

This chapter contains two complementary tables that must be used together when diagnosing problems. If you are not certain of the cause of a problem and do not know which chapter to consult for solutions, go to the "Problem-Solving Guide." This briefly summarizes common damage symptoms that can occur on many woody landscape plants and directs you to the section(s) of the book that discuss these problems. The "Tree and Shrub Pest Tables" are organized according to host plants. These list problems of about 200 genera of landscape plants, excluding many common problems (primarily abiotic or noninfectious disorders) that affect many different species. Problems such as too much or too little water that affect many different plants are usually included only in the "Problem-Solving Guide." Use both tables when diagnosing problems.

The "Tree and Shrub Pest Tables" are organized alphabetically by plant genus or scientific name. Related species and some genera are grouped together. For plants with several common names, only some of the names may be listed. Check the index at the back of the book to find the page numbers in the "Tree and Shrub Pest Tables" where the plant species of interest to you is located.

The "Problem-Solving Guide" and "Tree and Shrub Pest Tables" refer you to other parts of the book where you can find more information on identification, biology, and management. For pests not prominently named or specifically discussed elsewhere in the book, their scientific name and sometimes a brief description of management practices are provided in these tables; for more information, the reader may be referred to earlier sections on similar pests or maladies. Scientific names of pests are not listed in the tables if those names are prominently presented in major text sections earlier in the book.

Because of the broad scope of this publication, its California emphasis, and because new plant and pest species are periodically introduced from elsewhere, some of the pests you encounter may not be pictured or described here. For identification and biology of pests affecting herbaceous and nonwoody flowering ornamentals, consult the online guides at ipm.ucanr.edu and publications such as *Integrated Pest Management for Floriculture and Nurseries*. For vertebrate pests, see publications such as *Wildlife Pest Control Around Gardens and Homes*. Other sources of information are listed in the "Suggested Reading." Some pest problems can be diagnosed reliably only with the use of special tools such as laboratory tests or by experienced professionals. Your UC Cooperative Extension Advisor, UC Master Gardener, certified arborist, certified nurseryperson, or qualified horticultural consultant may be able to diagnose the cause of problems or direct you to other professional diagnostic services.

Problem-Solving Guide

Common disorder symptoms and pest damage on many woody plant species, by plant part on which they frequently appear

WHAT THE PROBLEM LOOKS LIKE	POTENTIAL CAUSE	COMMENTS
Entire Plant May Exhibit Symptoms		
Foliage fades, discolors, then wilts, sometimes initially in scattered portions of the canopy. Foliage drops prematurely. Branches, treetop, or entire plant may die.	Aeration deficit.	*See* 44.
	Armillaria root disease.	Possible mushrooms near trunk in fall. *See* 120.
	Boring insects.	*See* 220.
	Dematophora (Rosellinia) root rot.	May be white growths on soil. *See* 121.
	Freeze damage.	*See* 56.
	Fusarium wilt.	*See* 115.
	Gall makers.	*See* 212.
	Heterobasidion root disease.	Affects mostly conifers. *See* 121.
	Nematodes.	*See* 289.
	Phytophthora root and crown rot.	May have bark cankers or ooze. *See* 122.
	Root weevils, e.g., black vine.	*See* 159.
	Scale insects.	*See* 188.
	Sunburn or sunscald.	*See* 57, 58.
	Twig blight fungi.	*See* 83.
	Verticillium wilt.	Vascular tissue may turn brown. *See* 117.
	Water deficit or excess.	*See* 44.
Older needles or leaves drop. Lower or inner branches may die back.	Normal maturation, not a disease; may be aggravated by poor pruning or improper irrigation.	Evergreen plants periodically drop old foliage. Lower and inner, shaded branches naturally die and drop. Certain plants become dormant during summer or naturally drop foliage sooner than others.
Yellow, dead, or prematurely dropping leaves or needles. Foliage discolored, flecked, sparse, stunted, distorted, or may have irregular yellow patterns. Plant grows slowly.	Air pollution.	*See* 63.
	Herbicide phytotoxicity.	*See* 54.
	Nutrient deficiency.	*See* 45.
	Salt damage.	*See* 50, 52.
	Viruses.	*See* 92.
	Yellows or phytoplasmas.	*See* 96.
Leaves, blossoms, or fruit have black or brown lesions or spots. Leaf veins may darken. Terminals may die back. Bleeding cankers or lesions may occur on stems.	Bacterial blast, blight, and canker.	*See* 83.
Grayish, yellowish, or brownish encrustations form on plant. Leaves may yellow, branches may die back.	Scale insects.	*See* 188.
Leafless, orange stems entangle host plant. Clusters of small, white flowers in summer yield seed that germinate from soil next season.	Dodder, *Cuscuta* spp., parasitic annual plants sprouting from seeds.	*See* 282.
Bark, Limb, and Trunk Symptoms, Primarily		
Fleshy or woody growths occur on bark, may be bracketlike or seashell shaped.	Trunk and limb rots, fungi decaying internal tissue.	Decay often undetected except where bark cut or injured. Limbs or tree may fail (fall). *See* 111.
Bulging bark outgrowths or galls form.	Crown gall, a bacterial infection.	Damage often around root crown. *See* 108.
	Gall makers, e.g., cynipids.	*See* 212.
	Mistletoe, a parasitic plant.	*See* 284.
	Western gall rust, a fungal disease.	Affects only pines. *See* 96.
	Woolly aphids, e.g., woolly apple aphid.	*See* 177.

WHAT THE PROBLEM LOOKS LIKE	POTENTIAL CAUSE	COMMENTS
Bark splits or cracks.	Cambium killed by *Armillaria* or *Phytophthora*. Freeze damage. Inappropriate water. Lightning damage. Rapid growth. Sunburn or sunscald. Wind damage.	*See* 120, 122. *See* 56. *See* 44. *See* 62. Provide proper cultural care. *See* Chapter 3. *See* 57, 58. *See* 62.
Areas of dead bark (canker) may be surrounded by callus tissue layers. Material may ooze from bark. Limbs or entire plant may die back.	Canker diseases, fungal infections. Inappropriate water. Mechanical injuries e.g., pruning wounds. Sunburn or sunscald.	*See* Table 5-8, page 97. *See* 44. *See* 60. *See* 57, 58.
Stained bark exudes dark liquid, often around crotches, or whitish frothy material, often from cracks.	Foamy canker or alcoholic flux. Wetwood or slime flux.	Liquid has pleasant alcoholic or fermentative odor. *See* 101. Liquid odor, typically rancid. *See* 110.
Gummy ooze or pitch masses on bark. Foliage may discolor, and branches or entire plant may die.	Bark- or wood-boring insects. Canker diseases, fungal infections.	*See* 220. *See* 97.
Horizontal rows of shallow holes in bark.	Sapsucker birds.	Trees generally tolerate this. Usually not managed. *See* 220, *Woodpeckers Pest Notes*.
Green plants with smooth stems and thick, roundish leaves infest branches.	Broadleaf mistletoe, a parasitic, evergreen plant on host.	*See* 284.
Small, leafless, orangish, upright plants grow on conifer stems. Distorted and slow plant growth. Branches die back.	Dwarf mistletoe, *Arceuthobium* spp., host-specific parasitic plants that infect only conifers and extract nutrients from host plant.	Prune out infected branches. Replace heavily infested plants with species from other genera; dwarf mistletoes will not spread to unrelated species. *See* 284.
Grayish, greenish, or orangish tissue grows on or hangs from bark.	Algae, lichens, and mosses.	Often grow on older trees. Generally harmless to host plant, but may indicate overwatering of the landscape. *See* 288.
Foliage and Terminal Symptoms, Primarily		
Sudden wilting, blackening, or browning of shoots, blossoms, or fruit. Plant appears scorched.	Bacterial blast, blight, and canker. Fire blight.	Severely affects lilacs (*Syringa*) and *Prunus* spp. *See* 83. Affects only rose family plants. *See* 84.
Conspicuous spots or irregular dead areas form on leaves. Foliage or flowers curl, turn brown or black, may drop prematurely.	Anthracnose. Frost damage. Leaf spots. Root diseases. Root injury. Rust fungi. Sunburn. Water deficit or excess. Wind or hot weather damage.	*See* 81. *See* 56. *See* Table 5-1, page 68. *See* 119. *See* 60. *See* 94. *See* 57, 58. *See* 44. *See* 62.

Foliage and Terminal Symptoms, Primarily (continued)

WHAT THE PROBLEM LOOKS LIKE	POTENTIAL CAUSE	COMMENTS
Foliage is flecked, yellowed, bleached, or bronzed by pests that suck plant juices.	Leafhoppers, or sharpshooters. Mites. Sycamore scale. Thrips. True bugs.	*See* 201. *See* 245. *See* 199. *See* 209. *See* 205.
Dark, varnishlike specks form on leaves.	Greenhouse thrips. House flies and certain other flies. True bugs.	*See* 210. Locate and manage flies' larval stages. *See* 205.
Light-colored powdery growth on plant. Leaves distort, discolor, or drop prematurely.	Powdery mildew, a fungal disease.	*See* 90.
Dry orangish or yellowish pustules occur, usually on leaf undersides.	Rust fungi.	*See* 94.
Dark, sooty growth on leaves and stems, washes from plant.	Sooty mold, a fungus growing on honeydew excreted by insects on that plant or nearby plants.	*See* 92.
Clear, sticky substance appears on plant. May be white to clear cast skins on foliage or twigs. On oaks, liquid may be brownish.	Honeydew, excreted by plant-sucking insects, including aphids, scales, leafhoppers, mealybugs, psyllids, and whiteflies.	Identify insect, then see that section. If on oaks, also see Drippy Nut Disease, or Drippy Acorn, 80.
Copious, misty, nonsticky liquid raining from plant. Surfaces covered with whitish residue.	Glassy-winged sharpshooter.	*See* 203.
Whitish, frothy material on foliage.	Spittlebugs.	*See* 205.
Whitish, cottony, waxy material on plant.	Adelgids. Cottony cushion scales, some others. Mealybugs. Whiteflies. Woolly aphids.	*See* 181. *See* 196. *See* 185. *See* 182. *See* 177
Winding or blotched tunnels or mines in foliage.	Foliage miners or shield bearers.	*See* 216, 219.
Chewed, tattered, or scraped foliage, shoots, or blossoms.	Grasshoppers or katydids. Leaf beetles or flea beetles. Moth or butterfly larvae (caterpillars). Sawfly larvae. Snails or slugs. Weevils.	*See* 166. *See* 155. *See* 144. *See* 153 Slimy or clear trails present. *See* 251. Adult weevils feed at night. *See* 159.
Silken tents, mats, or webbing occur on chewed foliage or terminals.	Tent caterpillars, webworms, or leafrollers.	*See* 149–153.
Discolored foliage. Excessive or spindly growth.	High or low light. Herbicide phytotoxicity. Nutrient deficiencies.	*See* 58. *See* 54. *See* 45.
Distorted, curled, swollen, or galled leaves, flowers, stems, or branches.	Aphids. Gall makers. Gall mites. Herbicide phytotoxicity. Moth larvae (caterpillars). Nutrient deficiencies. Psyllids. *Taphrina* spp. fungi. Thrips.	*See* 175. *See* 212. *See* 249. *See* 54. *See* 144. *See* 45. *See* 170. *See* 213. *See* 209.

Tree and Shrub Pest Tables

Common Problems and Their Causes, by Plant

WHAT THE PROBLEM LOOKS LIKE	PROBABLE CAUSE	COMMENTS
Abelia spp., Abelia		
Galls or swellings on roots. Plant may grow slowly.	**Root knot nematodes.** Tiny, root-feeding roundworms.	*See* 293.
Abies spp., Fir, True fir[1]		
Foliage discolors, wilts, stunts, or may drop prematurely. Discolored bark or cankers may ooze sap. Branches or plant may die.	**Phytophthora root and crown rot.** Disease favored by excess soil moisture and poor drainage.	*See* 122.
Foliage browns, yellows, wilts, and needles drop prematurely. Branches die back. Entire plant may die.	**Heterobasidion root disease.** Fungal disease spreads through natural root grafts and airborne spores.	*See* 121.
Foliage may yellow and wilt. Bark or wood may have bracketlike or fan-shaped fungal fruiting bodies. Limbs or entire plant may die.	**Wood decay,** or **Heart rot,** including *Ganoderma applanatum, Laetiporus gilbertsonii (=L. sulphureus), Pleurotus ostreatus, Schizophyllum commune, Stereum* sp., *Trametes* spp.	Fungi that attack injured, old, or stressed trees. *See* 66, 111.
Foliage turns red, brown, then fades. May be small, pimplelike growths or brownish cankers on bark. Limbs die back.	**Cytospora canker.** Fungal disease primarily affecting injured or stressed trees.	*See* 100.
Foliage browns, and needles drop prematurely. Slow plant growth.	**Needle casts,** *Virgella robusta, Lirula abietis-concoloris.* Fungal diseases favored by cool, wet conditions in spring.	*See* Needle Blight and Cast, 85.
Foliage browns or yellows, and needles drop prematurely. Treetop or entire tree dies.	**Engraver beetles.** Adults are small, brown bark beetles. White larvae bore under bark.	*See* Pine Engravers, 225.
Brown to purplish insects clustered on foliage. Sticky honeydew and blackish sooty mold may be on foliage.	**Giant conifer aphids.** Dark, long-legged, ≤$^1/_5$ inch long.	Apparently harmless to trees. *See* 176.
Chewed foliage. Tree may be defoliated.	**Rusty tussock moth**, *Orgyia antiqua;* **Douglas-fir tussock moth.** Hairy caterpillars, ≤1 inch long.	*See* 152.
Chewed needles.	**Conifer sawflies.** Green larvae, ≤1 inch long, on needles.	*See* 153.
Bleached or stippled foliage. Foliage color abnormally light green or yellowish. May be fine webbing at foliage base.	**Spider mites**, including **Spruce spider mite.** Tiny arthropods, often green, suck sap.	*See* Pine and Spruce Spider Mites, 247.
Interior needles turn brown or yellow and drop prematurely, leaving only young terminal needles. Tree may die.	**Spruce aphid**, *Elatobium abietinum.* Small, pear-shaped insects, dark to light green, in groups on older foliage.	Unlike most aphids, this can be a serious pest. *See* Aphids, 175.
Needles yellow, curl, swell at their base, and drop prematurely.	**Balsam gall midge**, *Paradiplosis tumifex.* Larvae feed on young needles. Pupae overwinter in soil. Tiny flies lay eggs in spring.	*See* Gall Midges, 214.
Needles are chewed. Roots or the basal trunk may be cankered or injured.	**Conifer twig weevils.** Small black to brown weevils chew needles and shoots. Pale larvae bore in the root crown of dying or injured conifers.	Except for white pine weevil, most are secondary pests of minor importance. *See* 161.
Powdery, white or grayish material on cones, limbs, needles, or the trunk.	**Adelgids.** Aphidlike insects suck sap. Certain species alternate hosts with other conifers.	Balsam woolly adelgid seriously damages or kills firs. *See* 181.
Wet, white, frothy masses on needles or twigs.	**Spittlebugs,** including **Western pine spittlebug.** Green to black sucking insects secrete spittle.	Tolerate; spittlebugs cause no apparent harm to trees. *See* 205.
Pitchy masses 1 to 4 inches in diameter on trunks and limbs.	**Douglas-fir pitch moth.** Dirty whitish larvae, ≤1 inch long, in pitch.	*See* 239.
Distorted, stunted twigs or needles. Needles may drop prematurely.	**Balsam twig aphid**, *Mindarus abietinus.* Tiny, greenish yellow, powdery insects.	Vigorous plants tolerate. *See* Aphids, 175.

[1] For more information, see *Pests of the Native California Conifers.*

Abies spp. (continued)

WHAT THE PROBLEM LOOKS LIKE	PROBABLE CAUSE	COMMENTS
Bark stained brownish, exudes rancid fluid, often around crotches, wounds.	**Wetwood**, or **Slime flux**. Bacterial infections.	Usually do not cause serious harm to trees. *See* 110.
Small leafless, orangish, upright plants on host stems. Distorted and slow plant growth. Branches die back.	**Dwarf mistletoe**, *Arceuthobium* spp. Host-specific parasitic plants that extract nutrients from host plant.	Prune out infected branches. Replace heavily infested plants with broadleaf species or conifers from other genera; dwarf mistletoes don't spread to unrelated host species. *See* 284.

Abutilon spp., Abutilon, Chinese bellflower, Chinese lantern, Flowering maple

WHAT THE PROBLEM LOOKS LIKE	PROBABLE CAUSE	COMMENTS
Leaves with yellowish blotches.	**Abutilon mosaic virus**. Virus that is mechanically spread during propagation.	Considered attractive. Infected plants cannot be cured. *See* 7, 93.
Leaves or blossoms chewed.	**Fuller rose beetle**, *Naupactus godmani*. Pale brown adult snout beetle, about $^3/_8$ inch long.	Adults hide during day and feed at night. Larvae feed on roots. *See* Weevils, 159.
Sticky honeydew and blackish sooty mold on foliage.	**Black scale**, **Brown soft scale**. Black, brown, or yellowish bulbous or flattened insects.	*See* 192, 193.
Sticky honeydew and blackish sooty mold on foliage. Leaves may yellow and wither.	**Whiteflies**, including **Bandedwinged whitefly**, *Tetraleurodes abutilonea*. Tiny, whitish, mothlike adult insects.	*See* 182.
Brownish, grayish, tan, or white encrustations on twigs.	**Oleander scale**. Tiny, flattened, circular insects, $\leq^1/_{16}$ inch long.	Rarely if ever causes serious damage to plants. *See* 191.

Acacia spp., Acacia[2]

WHAT THE PROBLEM LOOKS LIKE	PROBABLE CAUSE	COMMENTS
Leaves discolored, wilted, stunted, may drop prematurely. Discolored bark may ooze sap. Branches or plant may die.	**Phytophthora root and crown rot**, **Pythium root rot.** Diseases favored by excess soil moisture and poor drainage.	*See* 122, 123.
Leaves discolor, wilt, and drop prematurely. Fewer leaves than normal. Limbs may die back.	**Wood decay**, *Ganoderma applanatum*. Fungus produces whitish to brown, globular or bracketlike basidiocarps.	Especially common in *Acacia baileyana* and *A. melanoxylon. See* 66, 111.
Leaves turn brown or yellow, especially along margins and at tip. Leaves may drop prematurely.	**Leaf burn**, or **Scorch**. Abiotic disorders commonly caused by frost, inappropriate irrigation, or poor drainage; but many other potential causes.	Provide plants with a good growing environment and proper cultural care. *See* Abiotic Disorders, 43.
Sticky or waxy honeydew and possibly blackish sooty mold on foliage. Terminals may brown or die.	**Acacia psyllid**. Tiny brown, green, or orange, flattened or winged insects on new growth.	*See* 138, 170.
Sticky honeydew and blackish sooty mold on plant. May be cottony bodies (egg sacs) on bark.	**Cottony cushion scale**. Orangish, flat immatures or cottony females on bark.	Normally controlled by natural enemies. *See* 196.
Copious, misty, nonsticky liquid raining from plant. Surfaces covered with whitish residue.	**Glassy-winged sharpshooter**. Active, dark brown or gray leafhoppers, $\leq^1/_2$ inch long, suck xylem fluid.	Vectors *Xylella* pathogens. Report suspected glassy-winged sharpshooters to agricultural officials if found in areas where this pest is not known to occur. *See* 203.
Wet, white, frothy masses on foliage.	**Spittlebugs**, including *Clastoptera arizonica*. Greenish bugs in spittle, suck sap.	Tolerate, does not damage plants. Hose plants with water. *See* 205.
Stippled, flecked, or bleached foliage.	**Leafhopper**, *Kunzeana kunzii*. Green insects, $\leq^1/_{16}$ inch long, suck sap.	Damage commonly minor, tolerate. *See* 201.
Chewed leaves. Foliage may be rolled and tied together with silk.	**Omnivorous looper**. Yellow, green, or pink larvae, $\leq 1^1/_2$ inches long, with green, yellow, or black stripes.	*See* 150.
Chewed leaves webbed with silk.	**Orange tortrix**, *Argyrotaenia citrana*. Larvae whitish with brown head and dark "shield" on back. Adults orangish to gray moths, $\leq^3/_4$ inch long.	Larvae wriggle vigorously when touched. Vigorous plants tolerate moderate defoliation. *See* Caterpillars, 144.
Chewed leaves or blossoms.	**Fuller rose beetle**, *Naupactus godmani*. Pale brown adult snout beetle, about $^3/_8$ inch long.	Adults hide during day and feed at night. Larvae feed on roots. *See* Weevils, 159.

[2] Some species are invasive weeds. Other species may be better choices when planting.

WHAT THE PROBLEM LOOKS LIKE	PROBABLE CAUSE	COMMENTS
Brownish, gray, tan, orangish, or white immobile encrustations (insects) on bark. Rarely, declining or dead twigs or terminals.	**California red scale, Greedy scale, Latania scale, Oleander scale, San Jose scale.** Tiny, circular to oval, flattened insects.	*See* 189–192.
Dieback of occasional twigs. Tunnels in twigs or branches, often at crotch.	**Leadcable borer**, *Scobicia declivis*. Black or brown beetles, $1/4$ inch long.	Prune out affected parts. Eliminate nearby dead wood in which beetles breed.

Acer negundo, Box elder

Foliage fades, yellows, browns, or wilts, often scattered throughout canopy. Branches die. Entire plant may die.	**Verticillium wilt.** A soil-dwelling fungus that infects through roots.	*See* 117.
Foliage yellows and wilts. Bark may have lesions that are dark, dry, crusty whitish, oily looking, or water-soaked. Branches or entire plant dies.	**Fusarium dieback.** Fungal disease spread by the polyphagous shothole borer, an ambrosia beetle, or bark beetle.	Occurs at least in southern California. *See* 115.
Discolored leaf spots, often between veins. Large spots chlorotic, then tan or brown.	**Sunburn.** Noninfectious disorder, appears during or after drought stress and high temperatures.	*See* 57.
Small, discrete spots to irregular, large blotches on leaves. Leaves may drop prematurely if severe.	**Leaf spots**, including *Cylindrosporium, Phyllosticta, Septoria* spp. Fungi spread by air or splashing water.	Favored by wet conditions. *See* 86.
Oval or irregular, glossy black, thick, tarlike raised spots on upper leaf surface.	**Tar spot**, *Rhytisma punctatum*. Fungus most prevalent in moist environments.	Rake and dispose of leaves in the fall. *See* Leaf Spots, 86.
Powdery, white growth on leaves. Tiny, black overwintering bodies may develop later.	**Powdery mildew**, *Phyllactinia guttata =P. corylea*. Fungal disease favored by moderate temperatures, shade, and poor air circulation.	Generally not severe enough to warrant control. *See* 90.
Chewed leaves. Foliage may be rolled and tied together with silk.	**Caterpillars**, including **Fruittree leafroller, Omnivorous looper**. Moth larvae, $\leq 1\frac{1}{2}$ inches long.	*See* 144.
Chewed leaves or blossoms.	**Fuller rose beetle**, *Naupactus godmani*. Pale brown adult snout beetle, about $3/8$ inch long.	Adults hide during day and feed at night. Larvae feed on roots. *See* Weevils, 159.
Foliage discolored, stippled, or bleached and may drop prematurely. Terminals may distort. Plant may have fine webbing.	**Mites**, including *Oligonychus* sp., **Twospotted spider mite**. Tiny, greenish or yellowish arthropods, may have two dark spots.	*See* 245.
Spotted or yellow foliage, usually severe only on female trees.	**Boxelder bugs.** Gray and red adults, about $1/2$ inch long. Nymphs red.	Trees tolerate damage. Adults invading houses are the primary problem. *See* 206.
Sticky honeydew, blackish sooty mold, and whitish cast skins on leaves.	**Aphids.** Tiny, pear-shaped insects, often brown, green, or yellowish, clustered on leaves.	*See* 175.
Sticky honeydew and blackish sooty mold on foliage. White popcornlike material on twigs (female scale egg sacs).	**Cottony maple scale**, *Pulvinaria innumerabilis*. Immatures flattened, yellow to tan, $\leq 1/16$ inch long, on leaves, suck sap.	Vigorous plants tolerate moderate populations. *See* similar Green Shield Scale, 194.
Sticky honeydew and blackish sooty mold on foliage.	**Calico scale.** Adults globular, black with white or yellow spots.	*See* 193.
Brown to gray encrustations on bark. Twigs or limbs may die back.	**Oystershell scale.** Individuals about $1/16$ inch long, oyster shaped, suck sap.	*See* 191.
Woody parts die back. Wet spots or sawdustlike frass on bark.	**Flatheaded appletree borer.** Whitish larvae, $\leq 3/4$ inch long, under bark.	*See* 227.
Bark stained brownish, exudes rancid fluid, often around crotches, wounds.	**Wetwood**, or **Slime flux**. Bacterial infections.	Usually do not cause serious harm to trees. *See* 110.

Acer spp., Maple

Foliage fades, yellows, browns, or wilts, often scattered throughout canopy. Branches die. Entire plant may die.	**Verticillium wilt.** A soil-dwelling fungus that infects through roots.	*See* 117.

WHAT THE PROBLEM LOOKS LIKE	PROBABLE CAUSE	COMMENTS
Foliage turns red, brown, then fades. Leaves drop prematurely. Cankers develop on bark.	**Annulohypoxylon canker, Botryosphaeria canker and dieback, Cytospora canker, Nectria canker.**	Fungal diseases that commonly infect injured or stressed trees. *See* 98–101.
Foliage yellows and wilts. Branches or entire plant dies.	**Phytophthora root and crown rot**. Disease favored by too much water or poor drainage.	*See* 122.
Foliage yellows and wilts. Bark may have lesions that are dark, dry, crusty whitish, oily looking, or water-soaked. Branches or entire plant dies.	**Fusarium dieback**. Fungal disease spread by the polyphagous shothole borer, an ambrosia beetle, or bark beetle.	Occurs on bigleaf maple, at least in southern California. *See* 115.
Powdery, white growth on leaves. Tiny, black overwintering bodies may develop later.	**Powdery mildews**, *Phyllactinia guttata*, *Sphaerotheca fuliginea*.	Fungal diseases favored by moderate temperatures, shade, and poor air circulation. *See* 90.
Discolored leaf spots, often between veins. Large spots chlorotic, then tan or brown.	**Sunburn**. Abiotic disorder, commonly appears during or after drought stress and high temperatures.	*See* 57.
Oval or irregular, glossy, black, thick, tarlike raised spots on upper leaf surface.	**Tar spot**, *Rhytisma punctatum*. Fungus most prevalent in moist environments.	Rake and dispose of leaves in the fall. *See* Leaf Spots, 86.
Small, discrete spots to irregular, large blotches or holes over most of leaf surface. Leaves may drop prematurely if severe.	**Leaf spots**, including *Cylindrosporium*, *Phyllosticta*, *Septoria* spp. Fungi spread by air or splashing water. Diseases favored by wet conditions.	*See* 86.
Leaves with brown, light green, or yellow lesions or dead spots, especially along leaf margins.	**Ramorum blight**, *Phytophthora ramorum*. Pathogen spreads via airborne spores and contaminated plants and soil. Infects bigleaf maple, *Acer macrophyllum*.	Primarily a problem in wildlands, killing many oaks there. *See* Sudden Oak Death and Ramorum Blight, 105.
Leaves with dark blotches or black veins. Leaves may drop prematurely. Terminals may die back.	**Bacterial blast, blight, and canker**. Bacteria persist in bark and infect wet leaves and small twigs.	*See* 83.
Leaves with black to brown blotches. Bulges or blisters on upper leaf surface. Leaves curled, may drop prematurely.	**Leaf blight**, or **Leaf curl**, *Taphrina* spp. Fungal diseases promoted by wet foliage during leaf flush.	Provide proper cultural care. Usually does not seriously harm trees. No control generally recommended.
Leaves stippled, bleached, or reddened. May be fine webbing on leaves or terminals. Leaves may drop prematurely.	**Spider mites**, including **Pacific spider mite, Twospotted spider mite**. Tiny arthropods, commonly greenish or yellowish, may have two dark spots.	*See* 246.
Leaves crumpled or with swollen galls, often along midvein. Leaves may brown and drop prematurely.	**Pod gall midge**, *Dasineura communis*. Adults are tiny flies. Pale larvae occur in galls, then drop and pupate beneath plant.	Plants tolerate extensive galling. *See* Gall Midges, 214.
Leaves with reddish, felty patches on underside. May be yellow blotches on upper surface of leaves.	**Erineum mites**, *Aceria* and *Vasates* spp. Tiny, elongate eriophyid mites that feed on underside of leaves.	These mites do not seriously harm trees. No control is recommended. *See* Gall Mites, or Eriophyids, 249.
Spotted or yellow foliage, usually severe only on female trees.	**Boxelder bugs**. Gray and red adults, $1/2$ inch long. Nymphs red.	Trees tolerate damage. Adults invading houses is the primary problem. *See* 206.
Leaves bleached or stippled with spots larger than mite stippling. Cast skins on underside of leaves. May be sticky or whitish honeydew on foliage.	**Leafhoppers**, including **Rose leafhopper**, *Edwardsiana rosae*; **Potato leafhopper**, *Empoasca fabae*. Greenish, yellow, or whitish wedge-shaped insects, $\leq 1/8$ inch long.	Plants tolerate moderate stippling. Apply insecticidal soap or another insecticide if intolerable. *See* 201.
Sticky honeydew, blackish sooty mold, and whitish cast skins on leaves.	**Aphids**, including *Periphyllus* spp.; **Painted maple aphid**, *Drepanaphis acerifolii*. Tiny, green insects clustered on leaves.	*See* 175.
Sticky honeydew and blackish sooty mold on foliage. May be dead or dying twigs and branches. Cottony white material (*Pulvinaria* egg sacs) on plant.	**Black scale; Calico scale; Cottony maple scale**, *Pulvinaria innumerabilis*. Yellow, brown, black, or white with spots, flattened to bulbous.	*See* 192, 193.

WHAT THE PROBLEM LOOKS LIKE	PROBABLE CAUSE	COMMENTS
Leaves chewed. Foliage may be webbed, defoliated, or contain silk tents.	**Fall webworm, Fruittree leafroller, Omnivorous looper**. Caterpillars, ≤1½ inches long.	*See* 149–153.
Brownish, grayish, tan, or white encrustations on twigs and branches. Rarely, dead or dying twigs or branches.	**Oleander scale, Oystershell scale, San Jose scale**. Tiny, circular to elongate individuals, often in groups.	*See* 191.
Dieback of woody plant parts. Tunnels and larvae in wood.	**Flatheaded appletree borer, Pacific flatheaded borer**.	Whitish larvae, ≤1 inch long, chew under bark. *See* 227.
Bark gnarled, exuding frass or liquid, and may have holes ≤½ inch diameter. Slow tree growth. Limbs may die and fall.	**Carpenterworm**. Whitish larvae, ≤2½ inches long, with brown head, tunnel in wood.	*See* 241.
Bark stained brownish, exudes rancid fluid, often around crotches, wounds.	**Wetwood**, or **Slime flux**. Bacterial infections.	Usually do not cause serious harm to trees. *See* 110.
Foliage may yellow and wilt. Bark may have bracketlike or fan-shaped fungal fruiting bodies. Branches or entire plant may die.	**Wood decay**, or **Heart rot**, including *Ganoderma* spp., *Laetiporus gilbertsonii* (=L. sulphureus), *Pleurotus ostreatus*, *Stereum* spp., *Trametes* spp.	Fungi that attack injured, old, or stressed trees. *See* 66, 111.

Aesculus californica, California buckeye; *Aesculus* spp., Buckeye, Horse chestnut

Bleeding or frothy material bubbling from tiny holes in trunk or limbs. Fine boring dust may surround holes.	**Oak ambrosia** and **Oak bark beetles**. Adults brown bark beetles, ⅛ inch long. White larvae tunnel beneath bark.	*See* 222.
Leaves yellow and brown. Dead leaves may hang on tree.	**Blight**. Normal dormancy, leaves die during summer or sooner under drought conditions.	Early leaf drop is normal drought adaptation. No control.
Leaves with brown spots. Leaves drop prematurely.	**Anthracnose**, or **Leaf blotch**, *Guignardia aesculi*. A fungal disease favored by wet conditions.	*See* 81.
Leaves and leaf petioles with brown lesions or dead spots.	**Ramorum blight**, *Phytophthora ramorum*. Pathogen spreads via airborne spores and contaminated plants and soil.	Primarily a problem in wildlands, killing many oaks there. *See* Sudden Oak Death and Ramorum Blight, 105.
Leaves with pale blisters, spots, or felty masses of pale spores.	**Yellow leaf blister**, *Taphrina aesculi*. Fungus favored by wet conditions during spring.	Rarely serious enough to threaten tree health or warrant control effort.
Powdery, white growth on leaves, tiny, black, overwintering bodies may develop later.	**Powdery mildew**, *Phyllactinia guttata*. Fungal disease favored by moderate temperatures, shade, and poor air circulation.	Generally not severe enough to warrant control. *See* 90.
Chewed leaves. Foliage may be rolled and tied together with silk.	**Fruittree leafroller, Omnivorous looper**. Yellow, green, or pink larvae, ≤1½ inches long.	*See* 149, 150.

Agave spp., Agave; *Hesperoyucca* spp., Yucca

Leaves discolor, stunt, wilt, or drop prematurely. Plants grow slowly and may die. Roots or plant base dark, decayed.	**Pythium root rot**. Pathogens promoted by excess soil moisture and poor drainage.	*See* 123.
Leaves with discolored or dead blotches or spots. Leaves may wilt and die prematurely if severe.	**Leaf spots** and **Blights**, including *Kellermania* spp., *Leptosphaeria* sp., *Microsphaeropsis concentrica, Stagonospora gigantea*.	Fungi, some of which are secondary invaders of dead or injured tissue. *See* 86.
Leaves turn brown or yellow, especially along margins and at tip. Leaves may wilt and die prematurely.	**Leaf burn**, or **Scorch**. Abiotic disorders with many potential causes, including frost, overirrigation, or poor drainage.	Provide plants with a good growing environment and proper cultural care. *See* Abiotic Disorders, 43.
Sticky honeydew, blackish sooty mold, and cottony waxy material on plant.	**Large yucca mealybug**, *Puto yuccae*. Powdery, white oval insects, ≤⅛ inch long.	Conserve natural enemies that provide control. Apply soap, oil, or another insecticide if not tolerable. *See* Mealybugs, 185.

Agave spp. (continued)

WHAT THE PROBLEM LOOKS LIKE	PROBABLE CAUSE	COMMENTS
Sticky honeydew and blackish sooty mold on foliage. Possible yellowing and dieback of foliage.	**Hemispherical scale**, *Saissetia coffeae*. Yellowish or brown, oval, flattened, or bulbous insects.	*See* Soft Scales, 192.
Brownish, grayish, tan, or white encrustations on bark. May be declining or dead plant parts.	**Latania scale, Oleander scale, Oystershell scale**. Tiny, circular to elongate insects.	*See* 189–192.
Decline of plant, which may decay at base, collapse, and die. Small, discolored spots or holes in leaves.	**Agave weevil, Yucca weevil**. Adults are black snout beetles about ¹/₂ inch long. Larvae are pale grubs that tunnel in basal plant parts.	*See* 160.
Agonis spp., Australian willow myrtle, Juniper willow myrtle, Peppermint tree		
Leaves discolor, stunt, or drop prematurely. Plants grow slowly and may die. Roots or basal stem may be dark or decayed.	**Phytophthora root and crown rot**, *Phytophthora cinnamomi*. Pathogen promoted by excess soil moisture and poor drainage.	*See* 122.
Albizia =*Albizzia* spp., Albizia, Mimosa, Silk tree		
Foliage yellows and wilts. Bark may have lesions that are dark, dry, crusty whitish, oily looking, or water-soaked. Branches or entire plant dies.	**Fusarium dieback**. Fungal disease spread by the polyphagous shothole borer, an ambrosia beetle, or bark beetle.	Occurs at least in southern California. *See* 115.
Leaves yellowing and may die, with older foliage affected first. Browning of vascular tissue. Limbs may die back.	**Mimosa wilt**, *Fusarium oxysporum* f. sp. *perniciosum*. Fungus infects plants through spores in soil.	Avoid injuring living tissue. Provide good drainage and appropriate irrigation. *See* Fusarium Wilt, 115.
Leaves turn yellow and drop prematurely.	**Natural senescence**. Leaves often naturally drop sooner than on other deciduous species.	Providing trees with a good growing environment and appropriate cultural care may delay leaf drop.
Sticky or waxy honeydew and possibly blackish sooty mold on foliage. Terminals may brown or die.	**Acacia psyllid**. Tiny orange, green, or brown, flattened or winged insects on new growth.	*See* 138, 170.
Chewed leaves. Silk tents in trees.	**Mimosa webworm**. Gray to brown larvae, ≤¹/₂ inch long, with white stripes.	*See* Webworms, 152.
Holes in wood. Boring dust on bark. Branches may break and fall. Tree may decline.	**Carpenterworm**. Dark whitish larvae, ≤2¹/₂ inches long, boring in wood.	*See* 241.
Alnus spp., Alder		
Foliage yellows and wilts. Branches or entire plant may die. May be white fungus beneath basal trunk bark.	**Armillaria root disease**. Fungus present in many soils. Favored by warm, wet soil. Persists for years in infected roots.	*See* 120.
Foliage yellows and wilts. Scattered branches die back. Bark may ooze.	**Botryosphaeria canker and dieback**, *Botryosphaeria dothidea*. A fungal disease.	Typically affects drought-stressed hosts, especially at warmer locations. *See* 98.
Leaves with whitish, powdery growth. Shoots or leaves may be stunted and distorted.	**Powdery mildews**, including *Erysiphe aggregata, Microsphaera penicillata, Phyllactinia guttata*.	Fungi favored by moderate temperatures, shade, and poor air circulation. *See* 90.
Leaves with brown to yellow blotches or spots and orangish pustules. Leaves may die and drop prematurely.	**Rusts**, including *Melampsora betulinum* =*M. alni*. Fungi infect and develop when leaves are wet.	*See* 94.
Leaves smaller and less abundant than normal, canopy declines. Basal trunk with small patches or extensive areas of wet, dark cankers and decay, can girdle trunk.	**Alder decline, Collar rot**, *Phytophthora siskiyouensis*.	In southern California has killed many red and white alder species, including *Alnus cordata, A. rhombifolia*. *See* 122.
Leaves may discolor and wilt, usually in spring. Branches may die back. Stems may have dark cankers and fungal fruiting bodies.	**Annulohypoxylon canker, Nectria canker**. Fungi primarily affect declining, injured, or stressed trees.	*See* 98, 101.

WHAT THE PROBLEM LOOKS LIKE	PROBABLE CAUSE	COMMENTS
Foliage may yellow and wilt. Bark or wood may have bracketlike or fan-shaped fungal fruiting bodies. Limbs or entire plant may die.	**Wood decay**, or **Heart rot**, including *Ganoderma applanatum*, *Pleurotus ostreatus*, *Stereum* sp., *Trametes* spp. Fungi that attack injured, old, or stressed trees.	*See* 66, 111.
Leaves may discolor and wilt. Dieback of branches. Gnarled, ridged bark with wet spots and D-shaped, $1/8$-inch-diameter emergence holes.	**Flatheaded borers**, including **Flatheaded alder borer**. Whitish larvae, $\leq 1/2$ inch long, beneath bark.	*See* 227.
Leaves may discolor and wilt. Branches wilted or dying. Boring dust, swellings, or holes on trunk or branches.	**Redbelted clearwing**. White larvae, ≤ 1 inch long, with brown head. Adults are wasplike moths.	*See* 235.
Leaves may discolor and wilt. Bark may have boring frass, elliptical holes, or oozing liquid. Branches and entire tree may die.	**Longhorned beetles**, including **Banded alder borer**, *Rosalia funebris*. Larvae are white, ≤ 1 inch long. Adult *Rosalia* are whitish with black.	Usually secondary pests that attack injured or severely stressed trees. *See* 232.
Leaves cupped, curled, folded, or thickened (galled) along edges or midvein.	**Alder gall midge**, *Dasineura* sp. Tiny, whitish to pink larvae in distorted tissue.	Apparently harmless to plant. Tolerate or prune out affected tissue. *See* Gall Midges, 214.
Leaves or catkin scales curl, enlarge, or twist and turn reddish or purplish. Shiny or pale fungal spores may cover infected tissue.	**Catkin hypertrophy**, or **Leaf curl**, *Taphrina japonica*, *T. occidentalis*. Fungi favored by wet conditions during spring.	Apparently harmless to plants. No control recommended.
Sticky honeydew, blackish sooty mold, and whitish cast skins on leaves.	**Aphids**, including *Euceraphis gillettei*, *Pterocallis alni*. Yellowish green, tiny insects on leaves, suck sap.	*See* 175.
Sticky honeydew and blackish sooty mold on leaves. Plant decline or dieback may occur.	**Cottony maple scale**, *Pulvinaria innumerabilis*. Females cottony. **European fruit lecanium scale**.	Flat to bulbous, brown, $\leq 1/4$ inch long. Plants tolerate moderate populations. *See* similar Green Shield Scale, 188; Lecanium Scales, 195.
Cottony white, waxy tufts on leaves.	**Cottony alder psyllid**, *Psylla alni*. Yellow to green insects, $1/16$ inch long, beneath wax, suck sap.	Vigorous plants tolerate moderate populations. *See* Psyllids, 170.
Stippled or bleached leaves. Dark, varnishlike specks and cast skins on undersides of leaves.	**Lace bugs**, *Corythucha* spp. Adults $\leq 1/8$ inch long, wings lacy. Nymphs spiny.	*See* 207.
Leaves skeletonized.	**Alder flea beetle**, *Altica ambiens*. Metallic blue adults, $1/4$ inch long. Larvae are brown to black.	Vigorous trees tolerate moderate defoliation. Provide plants proper cultural care. *See* Leaf Beetles, 155.
Webbing or tents on branch terminals. Chewed foliage.	**Fall webworm**. Larvae white to yellow, hairy, ≤ 1 inch long.	*See* 152.
Brown to gray encrustations on bark. May be stunting or dieback of woody parts.	**Oystershell scale**. Immobile, tiny, oyster-shaped insects, often in colonies.	*See* 191.
Galls or swellings on roots.	**Root knot nematodes**. Tiny, root-feeding roundworms.	*See* 293.
Amelanchier spp., Serviceberry, Shadbush		
Leaves, blossoms, or shoots suddenly blacken, brown, shrivel, or wilt. Plant appears scorched.	**Fire blight**. Bacterium infects plants through blossoms.	*See* 84.
Leaves with tiny brown, reddish, or purple blotches or spots, may have yellow halos. Leaves may drop prematurely.	**Entomosporium leaf spot**. A fungal disease promoted by wet foliage.	*See* 87.
Leaves with brown to yellow blotches and orangish pustules. Leaves may drop prematurely.	**Rusts**, *Gymnosporangium* spp. Fungi favored by wet foliage.	Avoid overhead watering. Vigorous plants tolerate moderate infection. *See* Cedar, Cypress, and Juniper Rusts, 95.
Leaves discolor and wilt. Woody parts die back. Wet spots or sawdustlike frass on bark.	**Borers**, including **Flatheaded appletree borer**. Whitish larvae, $\leq 3/4$ inch long, chew under bark.	*See* 227.

Amelanchier spp. (continued)

WHAT THE PROBLEM LOOKS LIKE	PROBABLE CAUSE	COMMENTS
Leaves discolor and wilt. Dead branches, limbs, or twigs. Plant declines, may die. Tiny, BB shot–sized holes in bark, which may ooze.	**Shothole borer**. Small, brown adult bark beetles, whitish larvae, tunnel beneath bark.	*See* 225.
Leaves stippled, bleached, or reddened. May be fine webbing on leaves or terminals. Leaves may drop prematurely.	**Spider mites**, including **McDaniel spider mite**, *Tetranychus mcdanieli*. Brown, green, reddish, or yellow specks.	*See* 246.
Leaves scraped, skeletonized, may be slime covered.	**Pear sawfly**, or **Pearslug**. Green, slimy, sluglike larvae, ≤$^1/_2$ inch long.	*See* 154.
Sticky honeydew, blackish sooty mold, and whitish cast skins on leaves.	**Aphids**, including **Bean aphid**. Small pear-shaped insects, often green, yellowish, or blackish.	*See* 175.

Araucaria spp., Araucaria, Bunya-Bunya tree, Monkey puzzle tree, Norfolk Island pine		
Gray to reddish encrustations on shoots or needles (scale insect bodies). Plant may decline and die back.	**Dictyospermum scale**, *Chrysomphalus dictyospermi*. Tiny, immobile, flattened insects suck plant juices, have several generations per year.	Occasional problem in California, mostly in the south. Conserve introduced natural enemies, which can provide excellent biological control. *See* Armored Scales, 189.
Black, dark brown, or gray encrustations on shoots or needles. Needles may yellow. More a problem in California in south.	**Black araucaria scale**, *Lindingaspis rossi*. Tiny, circular to oval armored scales.	Conserve natural enemies that provide control. Apply narrow-range oil or another insecticide when crawlers are numerous in the spring. *See* Armored Scales, 189.
Needles discolored. Sticky honeydew, blackish sooty mold, and grayish, flocculent material on plant. Plant growth may slow.	**Golden mealybug**, *Nipaecoccus aurilanatus*. Females globular, $^1/_{10}$ inch long, purplish with felty or off-white, golden band marginally and on back.	Distinguish the waxy, whitish, active larvae of the beneficial mealybug destroyer that commonly feed on this pest. Natural enemies, soap or oil sprays, or another insecticide can help to control. *See* Mealybugs, 185.
Sticky honeydew and blackish sooty mold on plant. Elongate, whitish material (mature female egg sacs) at leaf axils.	**Araucaria scale**, *Eriococcus araucariae*. Females elongate and yellowish brown with a dark stripe.	Vigorous plants tolerate moderate populations. *See* Scales, 188.

Arbutus menziesii, Madrone; *Arbutus unedo*, Strawberry tree		
Leaves discolor, stunt, or drop prematurely. Cankers may develop on basal stems. Plants grow slowly and may die. Roots or basal stem may be dark or decayed.	**Phytophthora root and crown rot**, *Phytophthora cactorum*. Pathogen promoted by excess soil moisture and poor drainage.	*See* 122.
Leaves brown, yellow, and wilt. Scorched leaves may remain on plant. May be dying and dead branches or cankers on large branches and the trunk. Dead branches may turn black.	**Cankers, Twig dieback**, including *Botryosphaeria dothidea, Fusicoccum aesculi, Nattrassia mangiferae, Phomopsis* spp. Fungi that primarily attack injured or weakened trees.	Prune out dead and diseased branches in summer. Provide appropriate water and other cultural care to improve tree vigor. Manage as with Botryosphaeria Canker and Dieback, 98.
Leaves yellow, wilt, and drop prematurely. Branches die back, and entire plant may die.	**Heterobasidion root disease**. Decay fungus that spreads through natural root grafts and airborne spores.	*See* 121.
Leaves with brown, light green, yellow, or dead spots or blotches. Twigs and small branches may canker, ooze, or die back.	**Ramorum blight**, *Phytophthora ramorum*. Pathogen spreads via airborne spores and contaminated plants and soil. Infected *Arbutus menziesii* saplings can be killed.	Primarily a problem in wildlands, killing many oaks there. *See* Sudden Oak Death and Ramorum Blight, 105.
Leaves with small discrete spots to irregular, large blotches or holes over most of surface. Tree may defoliate if severe.	**Leaf spots**, including *Cryptostictis arbuti, Mycosphaerella arbuticola, Phyllosticta fimbriata, Sphaceloma* sp. Fungi spread in water and favored by cool, wet, conditions.	Diseases apparently promoted by wet winters. Plants tolerate extensive leaf spotting. Little or no effective controls in landscapes or wildlands. *See* 86.
Leaves with oval or irregular, glossy black, thick, tarlike, raised spots on upper surface.	**Tar spot**, *Rhytisma arbuti*. Fungus most prevalent in moist environments.	Rake and dispose of leaves in the fall. *See* Leaf Spots, 86.
Leaves with mined blotches or dead patches. Lower leaf surface scraped.	**Leafblotch miner**, *Gelechia panella*; **Leafminer**, *Marmara arbutiella*. Moth larva, ≤$^2/_3$ inch long, feed within leaves.	Do not harm plant. Tolerate or clip and dispose of infested leaves. *See* Foliage Miners, 216.

WHAT THE PROBLEM LOOKS LIKE	PROBABLE CAUSE	COMMENTS
Leaves with elliptical blotches or holes, $1/8$ to $1/4$ inch long, or winding tunnels.	**Madrone shield bearer**, *Coptodisca arbutiella*. Larvae, $\leq 1/4$ inch long, mine leaves. Adult is tiny moth.	Vigorous plants tolerate moderate leaf damage. No management known. *See* Shield Bearers, 219.
Leaves have pinkish to pale blisters. Leaves are partly or all crisp, distorted, thickened, and brown or white. White or pinkish spores may cover infected tissue.	**Blister blight**, or **Leaf gall**, *Exobasidium vaccinii*. Fungus spreads by air only during wet weather.	Avoid overhead watering. Prune only when dry. Improve air circulation. Vigorous plants tolerate extensive leaf galling. Hand-pick or prune out galls. Rake and dispose of fallen leaves.
Sticky honeydew and blackish sooty mold on foliage. Dark, immobile bodies (pupae), about $1/16$ inch long, on underside of leaves.	**Madrone whitefly**, *Trialeurodes madroni*. Fringe of white filaments on sides of dark pupal cases.	Tolerate. Apparently does not damage plants. *See* Whiteflies, 182.
Sticky honeydew and blackish sooty mold on foliage. White felty sacs mostly on twigs.	**Madrone psyllid**, *Euphyllura arbuti*. Tiny, reddish, winged adults or flattened, gray to whitish, waxy nymphs.	After initial abundance, populations decline due to predators and parasites, such as *Psyllaephagus arbuticola*. Conserve natural enemies. *See* Psyllids, 170.
Sticky honeydew and blackish sooty mold on foliage. Twigs or branches may decline or die back.	**Black scale, Brown soft scale**. Black, brown, orangish, or yellow, flattened or bulbous insects.	*See* 192, 193.
Sticky honeydew, blackish sooty mold, and whitish cast skins on plant.	**Aphids**. Small green, black, brown, or yellowish pear-shaped insects, often in groups.	*See* 175.
Leaves stippled, bleached. Varnishlike specks on underside.	**Greenhouse thrips**. Tiny, slender, black adults or yellowish nymphs.	*See* 210.
Chewed leaves. Tents or mats of silk on leaves.	**Western tent caterpillar**. Hairy larva, mostly brown, ≤ 2 inches long.	*See* Tent Caterpillars, 151.
Trunk or limbs with roughened, wet, or oozing area. Cracked bark and dieback.	**Flatheaded borers**. Larvae whitish, with enlarged head, under bark.	Adults bullet shaped, metallic, coppery, gray, greenish, or bluish. *See* 227.
Brownish, grayish, tan, or white encrustations on twigs or bark. Rarely, declining or dead twigs or branches.	**Greedy scale, San Jose scale**. Circular, flattened, $\leq 1/16$ inch long, often in colonies.	Rarely if ever causes serious damage to trees. *See* 190, 191.

Archontophoenix, Brahea, Butia, Caryota, Chamaedorea, Dactylifera, Dypsis, Hedyscepe, Howea, Jubaea, Kentia, Livistona, Phoenix, Rhapis, Sabal, Syagrus =Arecastrum, Trachycarpus, Washingtonia spp., Palm[3]

Fronds bend downward. New leaves deformed, twisted. Entire leaf crown curves downward.	**Leaning crown syndrome**, or **Palm bending**. Boron deficiency (especially in *Dactylifera, Kentia* spp.), eriophyid mites, or an unknown cause in some cases.	Provide palms with good growing conditions and appropriate cultural care. *See* Boron and Other Specific Ions, 52; Gall Mites, or Eriophyids, 249.
Upper trunk and all fronds break off and drop from the top of the remaining trunk.	**Crown drop**. Internal trunk decay, apparently due to fungal pathogens spread by contaminated pruning tools and climbing spikes.	Lethal disease of Canary Island date palm and, less often, other *Phoenix* spp. *See* Sudden Crown Drop of Palms, 110.
Trunk breaks off at base. Trunk toppling may be preceded by foliage discoloring.	**Banana moth**, *Opogona sacchari*. Brownish, dirty white, or semitransparent larvae, $\leq 1 1/8$ inches long, cause brown excrement, chewed roots, and decay around basal trunk.	Provide palms with good cultural care and growing conditions. Avoid wounding basal trunks. Inspect the lower trunk of new palms and reject those with moth larvae or damage.
Fronds wilt and die. Basal trunk discolored dark and may crack.	**Ganoderma butt rot**, *Ganoderma* spp. Trunk rot fungi produce large brown to white, globular or bracketlike fruiting bodies.	No specific management known. Palms with conks may be hazardous and warrant removal. *See* 66; Wood Decay, 111.
Yellowing and death of fronds, often older fronds or leaflets on one side die first. Vascular tissue brown.	**Fusarium wilt**. Fungus infects through roots and contaminated pruning tools.	Lethal disease of Canary Island date palm. *See* 115.

[3] For more information, see *The Biology and Management of Landscape Palms*.

Archontophoenix spp. (continued)

WHAT THE PROBLEM LOOKS LIKE	PROBABLE CAUSE	COMMENTS
Yellowing and death of one or two older fronds or leaflets on one side of rachis.	**Dothiorella blight**, *Dothiorella* sp. May be dark vascular streaking, but only on older fronds, never on new or young fronds.	Attacks *Phoenix canariensis*. Uncommon and does not severely damage plants.
Yellowing of main leaves. Plant decline. Water-soaked decay around basal trunk.	**Phytophthora trunk or collar rot**. Disease pathogens most active during warm, moist conditions.	Provide good soil drainage and proper cultural care. Fungicides have generally been ineffective in large palms. *See* Phytophthora Root and Crown Rot, 122.
Premature yellowing of lower fronds.	**Abiotic disease**. Many potential causes, including excess or deficient water; magnesium, nitrogen, or potassium deficiency; and pesticide or other chemical injury.	Provide a good growing environment and proper cultural care. Apply special palm fertilizer to prevent or remedy nutrient deficiency. *See* Abiotic Disorders, 43.
Yellow and dark mottling and necrosis of fronds. Black or greasy diamond-shaped or elongate fungal bodies on fronds. Fronds may die back.	**Diamond scale**. A fungal disease. Symptoms superficially resemble nutrient disorders such as magnesium or potassium deficiency.	Primarily affects California fan palm, *Washingtonia filifera*. Where a problem, plant *W. robusta* or other species. *See* 87.
Yellow and brown leaves, which may die. Dark, elongate lesions on main stem (rachis, or petiole) of fronds.	**Rachis blight**, *Cocoicola californica* (on *Phoenix*), *Serenomyces virginiae* (on *Washingtonia*), and other fungi.	Mostly a nuisance and typically important only on stressed palms or when other diseases are present.
Leaf stalk bases rot and die. Terminal bud dies. Infected tissue may be covered with pink spores. Trunk cankers on Queen palm (*Syagrus romanzoffianum*).	**Pink Rot**. Fungal disease most serious on plants of low vigor, near the coast, and when fronds are wet.	Select species appropriate for that location. Provide good growing conditions and appropriate cultural care. *See* 109.
Yellowing fronds and plant death. Tissue may contain tunnels or boring insects.	**Giant palm borer**, *Dinapate wrighti*. Brown to black adult beetles and stout, yellowish larvae, both ≤1¹/₂ inches long, tunnel in wood.	Primarily a secondary pest attacking dead and dying palms. Provide good growing conditions and proper cultural care. Dispose of dead palms, in which beetles breed. *See* similar Giant Palm Weevils, 162.
Sparse canopy, fewer fronds than normal. Yellowing fronds. Canopy may die or drop.	**Giant palm weevils**. Yellowish larvae, ≤2 inches long, tunnel in apical growing point and base of fronds.	Especially infest Canary Island date palm. Reported in limited area of southern California. *See* 162.
Brownish, grayish, tan, orange, or white encrustations on fronds. May be yellowing or dieback of fronds.	**Boisduval scale**, *Diaspis boisduvalii*; **California red scale; Greedy scale; Oleander scale**.	Tiny, circular to oval, flattened insects. *See* Armored Scales, 189.
Frond tissue discolored, yellowish around each spot where scale insects are feeding.	**Fern scale**, *Pinnaspis aspidistrae*. Adult female oystershell shaped with brown cover; immature male cover white, elongated.	*See* Armored Scales, 189.
Browning and scraping of upper surface of fronds. Fronds webbed and covered with reddish frass.	**Palm leaf skeletonizer**. Creamy, orangish to whitish larvae, ≤¹/₂ inch long, in groups or within silken tube, chew leaf surface. Adults are small tan moths.	Attacks many palms, including *Phoenix* and *Washingtonia* spp. Palms are seldom killed, but become unsightly. *See* 150.
Sticky honeydew and blackish sooty mold on fronds.	**Black scale; Brown soft scale; Hemispherical scale**, *Saissetia coffeae*. Flattened to bulbous, black, brown, or yellowish insects.	*See* 192, 193.
Sticky honeydew, blackish sooty mold, and cottony waxy material on plants.	**Mealybugs**, including **Longtailed mealybug, Obscure mealybug, Vine mealybug.**	Powdery, gray insects. *See* 185.
Sticky honeydew and blackish sooty mold on foliage. Popcornlike bodies (egg sacs) on bark.	**Cottony cushion scale**. Orangish, flat immatures or cottony females on bark.	Normally controlled by natural enemies. *See* 130, 196.
Foliage has sticky honeydew and blackish sooty mold. Leaves yellow and wither. Tiny, whitish, mothlike adult insects present.	**Whiteflies**. Oval, flattened, yellow to greenish nymphs.	*See* 182.

WHAT THE PROBLEM LOOKS LIKE	PROBABLE CAUSE	COMMENTS
Sticky honeydew and blackish sooty mold may occur on fronds. Small, flattened, dark disks with pale, waxy fringes (insects) on fronds.	**Palm aphid**, *Cerataphis brasiliensis*. Atypical aphid, is flattened, dark, and surrounded by white wax, resembling certain whitefly nymphs.	*See* 176.
Sticky honeydew, blackish sooty, and pale wax on fronds.	**Coconut mealybug**, *Nipaecoccus nipae*. Oval-shaped and orangish insects. Adult females have marginal wax filaments.	*See* 186.
Foliage discolored, stippled, bleached, or reddened. Fronds may decline prematurely. Plant may have fine webbing.	**Mites**, including *Oligonychus* spp. Tiny greenish, red, or yellowish mites, suck sap.	*See* 245.
Stippled or bleached fronds with varnishlike specks on undersides.	**Greenhouse thrips**. Tiny, slender, black adults or yellow nymphs.	On palm reported only on *Chamaedorea* spp. *See* 210.
Plants stunted and may decline or slowly die. Powdery, waxy material may be visible on roots and around crown.	**Ground mealybugs**, *Rhizoecus* spp. Small, slender, pale insects, may be lightly covered with powdery wax, but lack marginal filaments.	*See* 187.
Galls or swellings on roots.	**Root knot nematodes**. Tiny, root-feeding roundworms.	*See* 293.

Arctostaphylos spp., Manzanita, Bearberry, Kinnikinnick

Leaves discolor, stunt, wilt, or drop prematurely. Plants grow slowly and may die. Roots or basal stem dark, decayed.	**Phytophthora root and crown rot**. Disease promoted by excess soil moisture and poor drainage.	*See* 122.
Leaves discolor and wilt. Branches or treetop dying. Branches may canker or exude reddish pitch. Branches may blacken.	**Cankers** or **Dieback**, including **Botryosphaeria canker and dieback**, *Botryosphaeria arctostaphyli*; *Fusicoccum* spp. Fungi primarily affect injured or stressed trees.	Prune out dead and diseased branches in summer. Provide appropriate water and other cultural care to improve tree vigor. *See* 98.
Leaves with brown, light green, yellow, or dead spots or blotches. Twigs and stems may canker, ooze, or die back.	**Ramorum blight**, *Phytophthora ramorum*. Pathogen spreads via airborne spores and contaminated plants and soil.	Primarily a problem in wildlands, killing many oaks there. *See* Sudden Oak Death and Ramorum Blight, 105.
Leaves with dark or discolored blotches. Leaves may yellow and drop prematurely.	**Leaf spots**, including *Phyllosticta amicta, Cryptosporium candidum*.	Fungal diseases that infect and develop when foliage is wet. *See* 86.
Leaves have black or yellow blotches. Leaves may yellow overall and drop prematurely. Terminals may die back.	**Bacterial leaf spot and blight**, *Xanthomonas* sp. Bacteria spread by water. Favored by cool, wet conditions. Persist in plant debris.	Avoid overhead irrigation. Improve air circulation. Rake and dispose of fallen leaves. *See* Leaf Spots, 86.
Leaves with dark or yellow spots. Leaf undersides with orangish pustules. Leaves may drop prematurely.	**Rusts**, *Chrysomyxa arctostaphyli, Pucciniastrum sparsum, P. evadens*. Fungi infect and develop when foliage is wet.	Avoid overhead watering. Plants tolerate moderate populations. *See* 94.
Leaves, buds, or shoots bushy, distorted, galled, reddish, or thickened. Leaves with red blisters.	**Leaf gall**, or **Kinnikinnick**, *Exobasidium vaccinii, E. vaccinii-uliginosi*. Fungi infect and develop in wet new growth.	Avoid overhead watering. Prune only when dry. Improve air circulation. Vigorous plants tolerate extensive leaf galling. Hand-pick or prune out galls. Rake and dispose of fallen leaves.
Fleshy red galls on leaves. Plants may grow slowly.	**Manzanita leaf gall aphid**. Tiny, gray or greenish insects in leaf galls.	*See* 175.
Leaves with elliptical blotches or holes, $1/8$ to $1/4$ inch long, or winding tunnels.	**Madrone shield bearer**, *Coptodisca arbutiella*. Larvae, $\leq 1/4$ inch long, mine leaves. Adult is tiny moth.	Vigorous plants tolerate extensive leaf damage. *See* Shield Bearers, 219.
Leaves chewed, drop prematurely. May be silken webbing on plant.	**Tent caterpillar, Western tussock moth**. Hairy brown or colorful caterpillars, ≤ 2 inches long.	*See* 151, 152.
Sticky honeydew, blackish sooty mold, whitish cast skins on leaves. Shoots may die back.	**Aphids**, including *Wahlgreniella nervata*. Pink to green insects in colonies on leaves and terminals.	Plants tolerate moderate populations. *See* 175.

Arctostaphylos spp. (continued)

WHAT THE PROBLEM LOOKS LIKE	PROBABLE CAUSE	COMMENTS
Sticky honeydew and blackish sooty mold on foliage. Possible twig and branch dieback.	**Brown soft scale**. Yellow to brown, flattened insects in groups.	*See* 193.
Sticky honeydew and blackish sooty mold on foliage. May be whitish wax.	**Psyllids**, including **Manzanita psyllid**, *Neophyllura* (=*Euphyllura*) *arctostaphyli*. Adult *Neophyllura* are brown, reddish, and yellow insects resembling leafhoppers.	Usually not abundant and rarely if ever harmful to manzanita. *See* 170.
Sticky honeydew and blackish sooty mold on foliage. Immobile, dark or pale, oval bodies (immatures), ≤1/16 inch long, on leaf undersides.	**Whiteflies**, including **Crown whitefly; Greenhouse whitefly; Iridescent whitefly**, *Aleuroparadoxus iridescens*. Tiny, mothlike adults.	Conserve natural enemies that help control. *See* 182.
Sticky honeydew and blackish sooty mold on foliage. Cottony waxy material on plant.	**Mealybugs**, including **Manzanita mealybug**, *Puto arctostaphyli;* **White mealybug**, *Puto albicans*. Powdery, white insects, ≤1/4 inch long, with waxy fringe.	*See* 185.
Brownish, grayish, tan, or white encrustations on twigs and branches. Possible dead or dying twigs or branches.	**Greedy scale; Manzanita scale**, *Diaspis manzanitae;* **Oleander scale**. Tiny, oval to circular insects.	Vigorous plants tolerate moderate populations. *See* 189–192.
Trunk or limbs with roughened, wet, or oozing area. Cracked bark and dieback.	**Flatheaded borers** including *Chrysobothris* spp. Larvae whitish with enlarged head, under bark. Adults are bullet shaped.	*See* 227.
Artemisia spp., Artemisia, Mugwort, Sagebrush, Tarragon, Wormwood		
Sticky honeydew and blackish sooty mold on leaves or twigs. Plants may decline.	**Black scale**. Brown to black, bulbous to flattened insects, ≤3/16 inch long. Raised H shape often on back.	*See* 192.
Sticky honeydew, blackish sooty mold, and whitish cast skins on plant.	**Aphids**. Small green, orange, or black insects, often in groups on leaves or stems.	*See* 175.
Chewed foliage. Plant may be defoliated by larvae that crawl up trunk in spring.	**Leaf beetles**, *Trirhabda flavolimbata, T. pilosa*. Black to brown larvae, ≤1/2 inch long, overwinter as eggs in soil. Adults are metallic, blue to green beetles, head yellowish.	One generation each year. Vigorous plants tolerate damage. Sticky material around trunk may exclude larvae if foliage is trimmed back from ground. *See* 155.
Leaves or stems thickened, distorted, galled, or have felty patches.	**Eriophyid mites**, *Aceria* spp. Microscopic mites living in groups.	*See* Gall Mites, or Eriophyids, 249.
Atriplex spp., Saltbush[4]		
Discolored spots and orangish pustules on leaves. Leaves may drop prematurely.	**Rusts**, *Puccinia* sp., *Uromyces shearianus*. Pathogens favored by wet conditions. Fungal spores persist in plant debris.	Avoid overhead watering. Plants tolerate moderate populations. *See* 94.
Stems have brownish cankers, sometimes with grayish centers. Dead tissue has small, black, spore-forming fungal pycnidia.	**Coniothyrium stem blight and canker**, *Microsphaeropsis olivaceae*. Fungus infects through wounds. Spread and favored by wet conditions.	Provide plants with proper cultural care and good growing conditions to keep them vigorous. Prune off and dispose of diseased tissue; otherwise avoid wounding plants, especially when wet.
Sticky honeydew and blackish sooty mold on plant. Bulbous, irregular, brown, gray, or white bodies (scales) on twigs.	**Wax scales**. Bulbous to hemispherical waxy insects, suck sap.	*See* 196.
Aucuba japonica, Aucuba, Gold-dust plant, Gold spot aucuba, Japanese aucuba		
Leaves with black spots, blotches, or terminals. Black cannot be scraped off as with sooty mold.	**Leaf spot**. Abiotic disorder due to excess light and high temperature, may be aggravated by drought stress.	Provide appropriate growing environment and cultural care. Provide some shade. Plant tall species and darken light-colored surfaces nearby.
Yellowish or brownish spots on leaves, leaves may drop prematurely if severe.	**Phyllosticta leaf spot**, *Phyllosticta aucubae*. A fungal disease promoted by wet foliage.	Avoid overhead watering. *See* Leaf Spots, 86.

[4] Some species are invasive weeds. Other species may be better choices when planting.

WHAT THE PROBLEM LOOKS LIKE	PROBABLE CAUSE	COMMENTS
Brownish, grayish, tan, reddish, yellowish, or whitish encrustations (insect bodies) on twigs or leaves. Plant may decline or die back.	**Dictyospermum scale**, *Chrysomphalus dictyospermi*; **False oleander scale**, *Pseudaulacaspis cockerelli*; **Greedy scale**; **Oleander scale**. Small, flattened, oval to round insects.	*See* Armored Scales, 189.
Sticky honeydew and blackish sooty mold on plant. Waxy, cottony material on leaves or twigs.	**Obscure mealybug**. Powdery, whitish insects, ≤1/8 inch long, with filaments, longest at tail. Suck sap.	*See* 187.
Sticky honeydew and blackish sooty mold on plant. Leaves distorted, may drop prematurely. Plant growth slow.	**Foxglove aphid**, *Aulacorthum solani*. Small, dull greenish or brownish to shiny yellowish insects, usually in groups.	*See* Aphids, 175.
Baccharis pilularis, Coyote brush, Chaparral broom		
Leaves with dark or yellow spots. Leaf undersides with orangish pustules. Leaves may drop prematurely.	**Rusts**, *Pucciniastrum baccharidis*, *P. evadens*. Fungi infect and develop when foliage is wet.	Avoid overhead watering. Plants tolerate moderate populations. *See* 94.
Leaves or shoots with powdery growth. Shoots or leaves may be stunted and distorted.	**Powdery mildews**, including *Erysiphe cichoracearum*, *Phyllactinia guttata*.	Fungi favored by moderate temperatures, shade, and poor air circulation. *See* 90.
Leaves and stems have dark to yellow spots. Foliage may shrivel and drop prematurely.	**Leaf and stem spots**, including *Cercospora baccharidis*. Fungi spread by air or splashing water. Disease favored by prolonged wet conditions.	Avoid wetting foliage. Use drip irrigation where feasible. Promptly remove and dispose of plant debris and infected leaves. *See* Leaf Spots, 86.
Brownish, grayish, tan, or white bark encrustations. May be decline or dieback of branches, twigs.	**Greedy scale**. Tiny, circular to elongate individuals on twigs, branches.	*See* 190.
Leaves with brown to pale blotches and irregular mines.	**Baccharis leaf miners**, *Bucculatrix* spp. Larvae of tiny moths chew foliage.	No control is generally needed to protect plant health. *See* similar Oak Ribbed Casemaker, 218.
Chewed foliage. Plant may be defoliated.	**Baccharis leaf beetle**, *Trirhabda flavolimbata*. Larvae are brown to black, ≤1/2 inch long. Adults are metallic, blue to green beetles, with yellowish head.	Insect has one generation each year. Vigorous plants tolerate moderate damage. Insecticide can be applied in the spring. *See* Leaf Beetles, 155.
Chewed foliage. Plant may be defoliated.	**Looper**, *Prochoerodes truxaliata*. Brown to purplish caterpillars, ≤1/2 inch long, hide on ground during day.	Vigorous plants tolerate moderate damage. If intolerable, apply *Bacillus thuringiensis* or other insecticide when young larvae are abundant. *See* 150.
Chewed leaves. Foliage may be webbed. Plant may be defoliated.	**Omnivorous looper**. Yellow, green, or pink larvae, ≤1 1/2 inches long, with green, yellow, or black stripes.	Larvae crawl in "looping" manner. *See* 150.
Fleshy, knoblike swellings (galls) on shoot tips. Galled shoots stop growing.	**Gall fly**, *Rhopalomyia californica*. Orange maggots, ≤1/16 inch long, in galls. Adults tiny, delicate flies, lay tiny, reddish eggs on terminals.	Plants are not killed. Tolerate galling. No known artificial controls. Many species of beneficial parasites attack gall fly larvae. Conserve natural enemies. *See* Gall Midges, 214.
Bead galls on leaves, open on leaf underside. Leaves may be deformed.	**Baccharis gall mite**, *Aceria baccharices*. A tiny, elongate eriophyid mite.	Plants apparently tolerate extensive galling. Control difficult. *See* Gall Mites, or Eriophyids, 249.
Twigs distorted, swollen, and pitted. Leaves dwarfed. Shoots may die back.	**Pit-making pittosporum scale**, *Planchonia* (=*Asterolecanium*) *arabidis*. Brown to white insects, ≤1/8 inch long, on twigs, often in pits.	In California, an occasional problem in the north. Management not investigated. *See* similar Oak Pit Scales, 198.
Sticky honeydew and blackish sooty mold on leaves or twigs. Plants may decline.	**Black scale**. Brown to black, bulbous to flattened insects, ≤3/16 inch long. Raised H shape often on back.	*See* 192.

Baccharis pilularis (continued)

WHAT THE PROBLEM LOOKS LIKE	PROBABLE CAUSE	COMMENTS
Sticky honeydew and blackish sooty mold on plant. Bulbous, irregular, brown, gray, or white bodies (scales) on twigs.	**Wax scales**. Bulbous to hemispherical waxy insects, suck sap.	*See 196.*
Copious, misty, nonsticky liquid raining from plant. Surfaces covered with whitish residue.	**Glassy-winged sharpshooter**. Active, dark brown or gray leafhoppers, ≤$1/2$ inch long, suck xylem fluid.	Vectors *Xylella* pathogens. Report suspected glassy-winged sharpshooters to agricultural officials if found in areas where this pest is not known to occur. *See 203.*
Leaves stippled or bleached, with cast skins and varnishlike specks.	**Lace bug**, *Corythucha* sp. Brown adults, ≤$1/8$ inch long, wings lacelike.	*See 207.*
Dead or declining branches or plant. Tunneling in wood.	**Flatheaded borer**, *Chrysobothris* sp. Whitish larvae, ≤$1^1/4$ inches long, with enlarged head, tunnel in wood.	*See 227.*

Bambusa, Fargesia, Pleioblastus, Phyllostachys spp., Bamboo[5]

Stems die back. Entire plant may die. May be white fungus beneath basal trunk bark.	**Armillaria root disease**. Fungus present in many soils. Favored by warm, wet soil. Persists for years in infected roots.	*See 120.*
New foliage yellows beginning along margins, but veins green. Leaves may develop brown blotches.	**Iron deficiency**. Abiotic disorder usually caused by poor soil conditions or unhealthy roots.	*See 47.*
Pale, brown, or purplish blotches or streaks on foliage. Foliage may drop prematurely.	**Spider mites**, including **Bamboo spider mites**, *Stigmaeopsis =Schizotetranychus* spp. Tiny, green to yellow mites, often with dark spots.	*See 246.*
Yellowing leaves. Sticky honeydew, blackish sooty mold, and tiny, whitish cast skins on leaves.	**Bamboo aphids**, *Takecallis* spp. Pale yellow insects, about $1/16$ inch long, with black marks.	Vigorous plants tolerate many aphids. *See* Aphids, 175.
Sticky honeydew and blackish sooty mold on plant. Young shoots or entire plant may die.	**Bamboo mealybug**, *Palmicultor lumpurensis*. Pinkish, elongate, waxy insects feed under the base of bamboo shoots, where they wrap around the main stem.	*See* Mealybugs, 185.
Sticky honeydew and blackish sooty mold on plant. Cottony material in leaf axils. Dead foliage.	**Noxious bamboo mealybug**, *Antonina pretiosa*. Black, brown, or reddish, elongate to round insects that secrete grayish wax.	Plants tolerate moderate populations. *See* Mealybugs, 185.
Groups of dark to pale, flattened insects encrusting plant. Possible dieback of plant parts.	**Bamboo scale**, *Asterolecanium bambusae*. Black, orange, or yellow, circular to oval insects, ≤$1/16$ inch long, on leaves, stems, or under sheaths.	Usually unimportant in landscapes. *See* Scales, 188.

Bauhinia spp., Orchid tree, Brazilian butterfly tree

Leaves with black, brown, tan, or yellow blotches or spots. Spots may have dark or yellowish margins. Foliage may die and drop prematurely.	**Anthracnose, Blights**, and **Leaf spots**. Fungi spread in water and favored by cool, wet, conditions.	*See 81, 86.*
Foliage has sticky honeydew and blackish sooty mold. May be copious white, waxy material. Leaves may yellow, wither, and drop prematurely.	**Whiteflies**, including **Giant whitefly**. Adults are tiny, whitish, mothlike insects. Nymphs and pupae are immobile and oval.	*See 182.*
Sticky honeydew, blackish sooty mold, and cottony or waxy material on plant.	**Mealybugs**. Powdery, grayish insects, ≤$1/4$ inch long.	*See 185.*
Galls or swellings on roots. Plants may grow slowly.	**Root knot nematodes**. Tiny, root-feeding roundworms.	*See 293.*

Betula spp., Birch

Leaves discolor, wilt, stunt, or drop prematurely. Plant may die. May be white fungus beneath basal trunk bark.	**Armillaria root disease**. Fungus present in many soils and favored by warm, wet soil. Persists for years in infected roots.	*See 120.*

[5] Some species are invasive weeds. Other species may be better choices when planting.

WHAT THE PROBLEM LOOKS LIKE	PROBABLE CAUSE	COMMENTS
Leaves discolored, wilted, stunted, may drop prematurely. Discolored bark may ooze sap. Branches or plant may die.	**Phytophthora root and crown rot.** Decay pathogen favored by excess irrigation and poor drainage.	*See* 122.
Leaves discolor and wilt. Branches or treetop dying. Branches may have cankers or brownish ooze.	**Annulohypoxylon canker, Botryosphaeria canker and dieback, Cytospora canker, Nectria canker.**	Fungi primarily affect injured or drought-stressed trees. *See* 98–101.
Foliage may yellow and wilt. Bark or wood may have bracketlike or fan-shaped fungal fruiting bodies. Limbs or entire plant may die.	**Wood decay**, or **Heart rot**, including *Ganoderma* spp., *Laetiporus gilbertsonii* (=*L. sulphureus*), *Pleurotus ostreatus*, *Schizophyllum commune*, *Stereum* sp., *Trametes* spp.	Fungi that attack injured, old, or stressed trees. *See* 66, 111.
Leaves with irregular, brown to yellow blotches. Leaves may die and drop prematurely.	**Anthracnose.** Favored by cool, wet, conditions. Fungal pathogen persists in plant debris.	*See* 81.
Leaves spotted, may yellow and drop prematurely. Reddish yellow pustules on lower leaf surface (spores that re-infect birch).	**Rust**, *Melampsoridium betulinum*. Fungal spores overwinter around bud scales.	Rake and dispose of all birch leaves in the fall. *See* 94.
Leaves discolor and wilt. Branches wilted or dying. Boring dust, ooze, and holes on trunk or branches.	**Borers,** including **Bronze birch borer, Carpenterworm, Redbelted clearwing, Western poplar clearwing.**	Whitish larvae with brown head, bore under bark. *See* 227–241.
Brown, gray, or tan encrustations on bark. May be dead or declining twigs and branches.	**Oystershell scale, Walnut scale**. Individuals elongate to round, ≤$^3/_{16}$ inch diameter, often in colonies.	*See* 191.
Sticky honeydew, blackish sooty mold, and tiny, whitish cast skins on leaves.	**Aphids.** Yellow to green insects, ≤$^1/_{16}$ inch long, on leaf undersides.	*See* 175.
Sticky honeydew and blackish sooty mold on plant.	**Frosted scale.** Hemispherical to flattened, brown or whitish insects, may be waxy, ≤$^1/_4$ inch long, suck sap.	*See* Lecanium Scales, 195.
Chewed leaves. Foliage may be webbed or contain silken tents.	**Fall webworm, Fruittree leafroller, Tent caterpillars.** Larvae ≤$1^1/_2$ inch long, may be hairy.	*See* 149–153.
Stippled, flecked, or bleached leaves with whitish cast skins on undersides.	**Leafhoppers,** including *Empoasca* sp., *Alebra albostriella*. Green, wedgelike insects, ≤$^1/_8$ inch long.	Plants generally tolerate, control rarely warranted. *See* 201.
Leaves stippled or bleached, with cast skins and varnishlike specks.	**Lace bug,** *Corythucha* sp. Brown adults, ≤$^1/_8$ inch long, wings lacelike.	*See* 207.
Leaves with large blotchy or irregular mines. Larvae may be visible through leaf surface. Foliage turns brown, wilts, and may drop prematurely.	**Leafminers,** including **Birch leafminer sawfly**, *Fenusa pusilla*. Larvae feed in leaf tissue. Adults are black, broad-waisted, stout wasps.	Consider planting less-susceptible black birch (*Betula lenta*), monarch birch (*B. maximowicziana*), river birch (*B. nigra*), or yellow birch (*B. alleghaniensis*). If severe, systemic insecticide can be applied during spring. *See* Foliage Miners, 216.
Leaves chewed along edges. Entire branches may be defoliated.	**Sawflies,** including **Birch sawfly**, *Arge pectoralis*; **Dusky birch sawfly**, *Croesus latitarsus*.	Green, orangish, or yellow larvae with black dots, often feed in groups on leaves. Adults are black, stout wasps. *See* 153.
Bougainvillea spp., Bougainvillea		
Foliage or flowers curl, turn brown or black, and die. Branches may die back.	**Leaf burn**, or **Scorch**. Abiotic disorders with many potential causes, including freeze or cold damage.	Provide plants with a good growing environment and proper cultural care. *See* Abiotic Disorders, 43.
Sticky honeydew and blackish sooty mold on foliage. Cottony or waxy material on plant.	**Citrus mealybug, Longtailed mealybug**. Powdery, grayish, ≤$^1/_4$ inch long, with waxy filaments.	Vigorous plants tolerate moderate populations. Conserve natural enemies that help in control. *See* 186, 187.
Sticky honeydew and blackish sooty mold on foliage. Leaves cupped, curled, or twisted.	**Cowpea aphid**, *Aphis craccivora*; **Melon aphid**. Brown, black, or green pear-shaped insects, usually in groups.	Conserve natural enemies that can provide control. Hose forcefully with water. *See* 175.

WHAT THE PROBLEM LOOKS LIKE	PROBABLE CAUSE	COMMENTS
Sticky honeydew and blackish sooty mold on foliage. Possible twig or branch decline or dieback.	**Brown soft scale**. Small brown, orangish, or yellowish insects, flattened or bulbous.	*See* 193.
Leaves chewed. Plants may be severely defoliated.	**Loopers**, *Asciodes gordialis, Disclisioprocta stellata*. Caterpillars ≤1 inch long.	At least two species of small moths. *See* Loopers, 150.
Shoots clipped, bark gnawed, branches girdled, limbs die back. May be nest of leaves and twigs.	**Roof rat**, *Rattus rattus*. Typically feeds at night and nests in dense trees and shrubs.	*See Rats Pest Notes* online at ipm.ucanr.edu.
Brachychiton spp., Bottle tree		
No serious invertebrate or pathogen pests have been reported in California. If plants are unhealthy, investigate whether they have been injured or lack good growing conditions or appropriate cultural care.		
Buddleia spp., Buddleia		
Leaves webbed, skeletonized. Leaves and terminal buds chewed.	**Buddleia budworm**, *Pyramidobela angelarum*. Whitish to pale yellow to green larvae with black head, ≤1/3 inch long. Adult is tiny, grayish moth.	Prune out infested foliage. Vigorous plants tolerate moderate defoliation. *See* Caterpillars, 144.
Buxus spp., Boxwood, Box		
Foliage discolors, wilts, stunts, and may drop prematurely. Branches or plant may die. Plant base or roots dark or decayed.	**Root and crown rots, Foliage blights, Stem necrosis**, *Phytophthora* spp., *Pythium* spp. Decay pathogens favored by excess soil moisture and poor drainage.	*See* 122, 123.
Leaves on terminals turn upward and become red, bronze, or yellow. Branches may have cankered bark and terminal shoot dieback.	**Pseudonectria canker**, or **Volutella canker and blight**, *Pseudonectria rousseliana =Volutella buxi*. Fungal pathogen persists in infected wood, is favored by wet conditions.	Provide adequate summer irrigation. Prune out infected branches during dry weather. *See* similar Nectria Canker, 101.
Foliage may yellow and wilt. Bark or wood may have bracketlike or fan-shaped fungal fruiting bodies. Limbs or entire plant may die.	**Wood decay**, or **Heart rot**, including *Ganoderma lucidum*. Fungi that attack injured, old, or stressed plants.	*See* 66, 111.
Leaves with discolored blotches or spots, which may be dotted with dark, fungal fruiting bodies.	**Leaf spots, Blights**, including *Cercospora* sp., *Macrophoma candollei, Phyllosticta* sp.	Fungal diseases that infect and develop when foliage is wet. *See* 86.
Foliage bleached, discolored, stippled, or streaked. Plant may have fine webbing.	**Spider mites**, including **Boxwood mite**, *Eurytetranychus buxi*. Tiny greenish, reddish, or yellowish arthropods.	*See* 246.
Brownish, grayish, orange, tan, or white encrustations on bark. Rarely, twigs or branches die back or growth is stunted.	**California red scale, Greedy scale, Oleander scale, Oystershell scale**. Tiny, circular to elongate insects.	*See* 189–192.
Leaves have blisters, blotchy mines, or yellow spots. Foliage sparse. Leaves drop prematurely. Shoots may die back. Plant growth may be stunted.	**Boxwood leafminer**, *Monarthropalpus flavus =M. buxi*. Greenish or orange fly larvae mine leaves. Adults are tiny, delicate, and resemble mosquitoes. *Buxus harlandii, B. microphylla*, and some *B. sempervirens* cultivars can be heavily damaged.	Consider replacing susceptible species. Resistant plants include *Buxus sempervirens* 'Argenteo-variegata', 'Pendula', and 'Suffruticosa'. Rake and dispose of infested fallen leaves, which may contain fly pupae. *See* Foliage Miners, 216.
Sticky honeydew and blackish sooty mold on foliage. Popcornlike bodies (egg sacs) on bark.	**Cottony cushion scale**. Orangish, flat immatures or cottony females on bark.	Normally controlled by natural enemies. *See* 130, 196.
Cupping of leaves.	**Boxwood psyllid**, *Psylla buxi*. Greenish adults, 1/8 inch long, nymphs flattened.	American boxwood more susceptible than English boxwood. Tolerate, psyllids apparently do not harm shrubs. *See* Psyllids, 170.

WHAT THE PROBLEM LOOKS LIKE	PROBABLE CAUSE	COMMENTS
Plants stunted, decline, may die. Powdery, waxy material may be visible on roots and around crown.	**Ground mealybug**, *Rhizoecus falcifer*. Small, slender, pale insects, that may have powdery wax covering but no marginal filaments.	*See* 187.

Caesalpinia, Poinciana spp., Caesalpinia, Bird of paradise bush, Barbados pride, Dwarf poinciana

Foliage discolors and wilts. Branches develop cankers and die back. Seashell-shaped fruiting bodies may develop on injured tissue.	**Sapwood rot**, *Schizophyllum commune*. A decay fungus that aggressively invades injured tissue and unhealthy plants.	Protect plants from injuries and stress. Provide good growing conditions and appropriate cultural care. *See* Wood Decay, 111.

Callistemon spp., Bottlebrush

Foliage or flowers curl, turn brown or black, and die. Terminals die back.	**Leaf burn**, or **Scorch**. Abiotic disorders with many potential causes, including freeze or cold damage.	Provide plants with a good growing environment and proper cultural care. *See* Abiotic Disorders, 43.
Foliage yellows or exhibits nutrient disorder symptoms. Plant may grow slowly.	**High pH**. Soil is calcareous (high in calcium carbonate), soil pH >7.5.	A problem where soil or irrigation water is alkaline. *See* pH Problems, 53.

Calodendrum capense, Cape chestnut

No serious invertebrate or pathogen pests have been reported in California. If plants are unhealthy, investigate whether they have been injured or lack good growing conditions or appropriate cultural care.

Camellia spp., Camellia

Leaves discolor, wilt, stunt, and may drop prematurely. Discolored bark or cankers may ooze sap. Branches or plants may die.	**Root and crown rots**, *Phytophthora* spp., *Pythium* sp. Pathogens favored by excess soil moisture and poor drainage.	*See* 122, 123.
Leaves turn brown or yellow, especially along margins and at tip. Leaves may drop prematurely.	**Leaf burn**, or **Scorch**. Abiotic disorders. Causes include boron toxicity, excess light, frost, overfertilization, poor soil conditions, saline irrigation water or soil, or too much or too little water.	Provide plants with a good growing environment and proper cultural care. *See* Abiotic Disorders, 43.
Leaves with brown, dead areas, mostly at the tips.	**Ramorum blight**, *Phytophthora ramorum*. Pathogen spreads via airborne spores and contaminated plants and soil.	Primarily a problem in nurseries and on other hosts in wildlands, killing many oaks there. *See* Sudden Oak Death and Ramorum Blight, 105.
New leaves and shoots chlorotic, except for green veins. Plants may be stunted.	**Iron deficiency**. Abiotic disorder usually caused by poor soil conditions or unhealthy roots.	*See* 47.
New leaves light green to yellow, especially at sunny sites.	**Light damage**. Noninfectious disorder caused by high-intensity light.	Provide partial shade e.g., plant overstory species. *See* High and Low Light, 58.
Leaves with irregular, yellow mottling. Blossoms mottled whitish.	**Viruses**, or **Variegation**, including **Camellia yellow mottle virus**. Introduced during plant propagation, not spread by insects.	Generally harmless to plants and often considered attractive. Infected plants cannot be "cured." *See* 93.
Leaf underside has pimplelike blisters or water-soaked spots. Leaves brown, harden, appear corky, especially on underside. Foliage yellows and drops prematurely.	**Edema**. Abiotic disorder. Leaves accumulate excess water when soil is warm and moist and air is cool and moist. Exact cause unknown.	Irrigate only in morning. Avoid irrigation if cool and cloudy. Provide good air circulation. Keep humidity low. Provide good drainage. *See* 60.
Leaves or shoots cupped, distorted, or swollen, and whitish or reddish. Galled tissue may develop discolored blotches.	**Leaf gall**, *Exobasidium camelliae*. Fungus infects and spreads when plant is wet.	Avoid overhead watering. Prune only when dry. Improve air circulation. Vigorous plants tolerate extensive leaf galling. Hand-pick or prune out galls. Rake and dispose of fallen leaves.
Flower bud edges turn brown, then entire bud browns and drops. Petals turn brown or drop prematurely.	**Camellia bud mite**, *Cosetacus* (=*Aceria*) *camelliae*. Translucent to white eriophyids, $\leq^{1}/_{100}$ inch long, that infest inner surface of bud scales or petals.	*See* Gall Mites, or Eriophyids, 249.

Camellia spp. (continued)

WHAT THE PROBLEM LOOKS LIKE	PROBABLE CAUSE	COMMENTS
Buds drop prematurely.	**Premature bud drop**. Abiotic disorder caused by poor cultural practices.	Provide appropriate irrigation and good drainage. Spring bud drop is caused by inadequate cultural care the previous summer and fall when buds were developing.
Blossoms rot. Brown lesions develop on petals, centers discolor first. Blossoms drop prematurely.	**Camellia petal blight**. Fungal disease promoted by rainy weather.	*See* 79.
Sticky honeydew and blackish sooty mold on foliage. Leaves may be cupped, curled, or twisted.	**Aphids**, including **Black citrus aphid**, *Toxoptera aurantii*; **Green peach aphid; Melon aphid**. Brown, black, or green insects in groups on growing points.	Conserve natural enemies that can provide control. *See* 175.
Sticky honeydew and blackish sooty mold on foliage. Twigs or branches may decline or die back. May be cottony material (egg sacs) on plant.	**Black scale, Brown soft scale, Green shield scale**. Small brown, green, yellowish, or blackish insects, flattened or bulbous.	*See* 192, 193, 194.
Sticky honeydew, blackish sooty mold, and cottony waxy material on plant.	**Mealybugs**, including **Longtailed mealybug, Obscure mealybug**. Grayish, oval, waxy, and slow moving, with marginal filaments.	*See* 185.
Sticky honeydew and blackish sooty mold on plant. Plant has elongate, slender, cottony material (egg sacs).	**Cottony camellia scale**, *Pulvinaria floccifera*. Oval, flattened, yellow or brown insects and cottony eggs.	*See* similar Cottony Cushion Scale, 196.
Sticky honeydew and blackish sooty mold on plant. Leaves yellow and wither. Tiny, whitish, mothlike adult insects.	**Whiteflies**, including **Greenhouse whitefly**. Oval, flattened, yellow to greenish nymphs and pupae with waxy filaments.	*See* 182.
Copious, misty, nonsticky liquid raining from plant. Surfaces covered with whitish residue.	**Glassy-winged sharpshooter**. Active dark brown or gray leafhoppers, ≤$1/_2$ inch long, suck xylem fluid.	Vectors *Xylella* pathogens. Report suspected glassy-winged sharpshooters to agricultural officials if found in areas where this pest is not known to occur. *See* 203.
Brownish, grayish, tan, or white encrustations on bark. Plants may grow slowly or are stunted. May be dieback of twigs or branches.	**Greedy scale, Oleander scale, Oystershell scale**. Individuals oval to elongate, ≤$1/_{16}$ inch long.	*See* 189–192.
Chewed leaves and blossoms. Plants may decline or grow slowly.	**Black vine weevil; Fuller rose beetle**, *Naupactus godmani*. Adults black, pale brown, or gray beetles, $3/_8$ inch long, with a snout.	*See* 160.
Leaves chewed. Plant can be extensively defoliated.	**Indian walking stick**. Brown, twiglike insects, up to 4 inches long, easily overlooked.	In California in at least the Central and South Coast. *See* 166.
Foliage may yellow, wilt, die back, or drop prematurely. Plant grows slowly.	**Spiral nematode, Stunt nematode**. Tiny, root-feeding roundworms.	*See* 294.

Carissa grandiflora, *C. macrocarpa*, Natal plum

WHAT THE PROBLEM LOOKS LIKE	PROBABLE CAUSE	COMMENTS
Foliage turns red or yellow. Leaves wilt and may drop prematurely. Branches or entire plant dies. Roots or basal stem dark, decayed.	**Phytophthora root and crown rot, Pythium root rot**. Diseases favored by excess soil moisture and poor drainage.	*See* 122, 123.
Brown, black, tan, or yellow spots or blotches on leaves. Spots may have dark or yellowish margins. Foliage may die and drop prematurely.	**Leaf spots**, including *Alternaria* sp. Spread by air or splashing water. Favored by prolonged cool, wet, conditions. Persist in plant debris.	*See* 86.
Sticky honeydew, blackish sooty mold, and cottony or waxy material on plant.	**Mealybugs**. Powdery, grayish insects, ≤$1/_4$ inch long.	*See* 185.
Sticky honeydew and blackish sooty mold on foliage. May be declining or dead twigs or branches.	**Black scale; Hemispherical scale**, *Saissetia coffeae*. Small black, brown, or yellowish, flattened or bulbous insects.	*See* 192, 194.

WHAT THE PROBLEM LOOKS LIKE	PROBABLE CAUSE	COMMENTS
Carnegiea gigantea, Saguaro; *Cereus, Echinocactus, Echinopsis, Epiphyllum, Lobivia, Mammillaria* spp., Cactus; *Opuntia* spp., Prickly pear		
Leaves or stems discolor, wilt, and may die back. Plant base or roots dark, decayed, or discolored.	**Root and crown rots** and **Stem rots**, *Phytophthora* spp., *Pythium* spp. Decay pathogens favored by excess soil moisture and poor drainage.	*See* 122, 123.
Leaves or stems discolor, wilt, or die. May be white fungus beneath basal trunk bark.	**Armillaria root disease**. Fungus present in many soils. Favored by warm, wet soil. Persists for years in infected roots.	*See* 120.
Leaves or stems discolor, wilt, and may die back. Tissue browns and may decay.	**Leaf burn**, or **Scorch**. Abiotic disorders with many potential causes, including frost, too much or too little water, and sunburn, which kill tissue.	Provide proper growing conditions and good cultural practices. *See* Abiotic Disorders, 43.
Leaves or stems with brown to yellow spots or blotches. Foliage or stems may shrivel and die.	**Anthracnose, Blights**, and **Leaf spots**, including *Alternaria, Cercospora, Colletotrichum*, and *Phyllosticta* spp.	Fungal pathogens favored by cool, wet, conditions. *See* 81, 83, 86.
Leaves or stems with brown or yellow spots. Basal stem or roots may brown and rot. Plant may yellow, wilt, and die.	**Fusarium wilt, Basal stem rot**, and **Leaf spot**, *Fusarium oxysporum, Fusarium* sp. Fungi persist in soil and infect through roots or spores on wounded leaves or stems.	Avoid injuring live tissue. Provide good drainage. *See* 115.
Soft, brown decay and/or gray, woolly growth on buds, flowers, leaves, and dead tissue.	**Botrytis blight**, or **Gray mold**. Fungus develops in injured or inactive tissue and is favored by wet conditions and moderate temperatures.	*See* 78.
Foliage or stems yellow and wilt. Branches or entire plant may die. Fungal fruiting bodies may be on wood or around roots.	**Wood decay**, or **Heart rot**, including *Phanerochaete* spp. Fungi that attack injured, old, or stressed plants.	*See* 111.
Whitish to brownish encrustations on plant. May be dieback of plant parts.	**Armored scales**, including **Cactus scale**, *Diaspis echinocacti*. Tiny, oval to circular, flattened insects.	*See* 189–192.
White, sticky, filamentous wax on pads, which may discolor and die back.	**Cochineal scales**. Insects exude reddish liquid when their body is punctured.	Common on prickly pear. *See* 196.
Sticky honeydew and blackish sooty mold on plant. Waxy, cottony material on plant.	**Mealybugs**, including **Longtailed mealybug**. Powdery, whitish insects, ≤$^1/_8$ inch long, with filaments longest at tail, suck sap.	*See* 185.
Stunting of plant. Pinhead-sized white, yellow, or brown projections on roots.	**Cyst nematode**, *Heterodera cacti*. Microscopic root-feeding roundworms.	*See* Nematodes, 289.
Stunting of plant. Galls or swellings on roots.	**Root knot nematodes**. Tiny, root-feeding roundworms.	*See* 293.
Carpenteria californica, Bush anemone		
Sticky honeydew, blackish sooty mold, or whitish cast skins. Leaves and shoots may curl or distort.	**Aphids**. Small green, brown, black, or yellowish insects, often in groups.	*See* 175.
Thickening of sections of twigs. Shoots may be killed or distorted.	**Pit-making pittosporum scale**, *Planchonia* (=*Asterolecanium*) *arabidis*. Brown to white insects, ≤$^1/_8$ inch long, on twigs.	Occasional problem only. No known management. *See* similar Oak Pit Scales, 198.
Carpinus caroliniana, Hornbeam		
Powdery, white growth on leaves. Leaves or shoots may distort.	**Powdery mildews**, including *Phyllactinia guttata, Microsphaera penicillata*. Fungal diseases favored by moderate temperatures, shade, and poor air circulation.	*See* 90.

Carpinus caroliniana (continued)

WHAT THE PROBLEM LOOKS LIKE	PROBABLE CAUSE	COMMENTS
Foliage yellows and wilts. Bark or wood may have bracketlike or fan-shaped fungal fruiting bodies. Branches or entire plant may die. Trees may fall over.	**Wood decay**, or **Heart rot**, including *Ganoderma applanatum*, which produces large, globular or bracketlike basidiocarps on bark. Fungi that infect injured or stressed trees.	*See 66, 111.*
Foliage discolors and wilts. Branches develop cankers and die back. Dark, hemispherical fruiting bodies develop on bark or wood.	**Annulohypoxylon canker**, *Annulohypoxylon mammatum*. A decay fungus that attacks injured or stressed trees.	*See 98.*
Sticky honeydew and blackish sooty mold on foliage. Possible plant decline or dieback.	**European fruit lecanium**. Brown or yellow, flat or bulbous, immobile scale insects.	See Lecanium Scales, 195.
Cassia spp., Cassia, Gold medallion tree, Senna		
Stippled or bleached leaves, varnishlike excrement specks on undersides.	**Thrips**. Tiny slender, blackish or yellowish insects.	*See 209.*
Castanea spp., Chestnut		
Leaves discolored, wilted, stunted, may drop prematurely. Discolored bark may ooze sap. Branches or plant may die.	**Phytophthora root and crown rot**. Decay pathogens favored by excess irrigation and poor drainage.	*See 122.*
Leaves turn yellow or brown on scattered branches, which die. Foliage drops prematurely. Orange cankers develop on limbs and the trunk. Bark splits. Trunk becomes girdled and the tree dies.	**Chestnut blight**, *Cryphonectria* (=*Endothia*) *parasitica*. An introduced fungus that infects through bark wounds and has killed many thousands of chestnuts.	Primarily affects American chestnut (*C. dentat*a) and European chestnut (*C. sativa*) in the eastern U.S. Apparently eradicated from the western U.S. In the West, promptly report suspected infections to agricultural officials.
Leaves turn yellow or brown on scattered branches. Foliage dies and may drop prematurely. Cankers or lesions may develop on limbs and possibly on the trunk.	**Cankers, Stem blight**, including *Amphiporthe castanea*; **Botryosphaeria canker and dieback**; *Coryneum* sp. Fungal diseases most often affecting injured or stressed trees.	Protect trees from injury and provide good cultural care, especially appropriate irrigation. *See 98.*
Foliage may yellow and wilt. Bark or wood may have bracketlike or fan-shaped fungal fruiting bodies. Limbs or entire plant may die.	**Wood decay**, or **Heart rot**, including *Laetiporus gilbertsonii* (=*L. sulphureus*), *Pleurotus ostreatus, Stereum* sp., *Trametes* spp.	Fungi that attack injured, old, or stressed trees. *See 111.*
Chewed leaves. Foliage may be rolled and tied together with silk.	**Omnivorous looper**. Yellow, green, or pink larvae, ≤1$^{1}/_{2}$ inches long, with green, yellow, or black stripes.	*See 150.*
Sticky honeydew and blackish sooty mold on foliage. Tiny, mothlike adults.	**Crown whitefly**. Black, oval nymphs with spreading, whitish, waxy plates.	*See 183.*
Holes in nuts. Tunnels inside nuts may contain whitish larvae.	**Acorn moth, Filbertworm**. Adults are tiny bronze, coppery, gray, or reddish brown moths.	May reduce natural tree regeneration, but established trees are not damaged. *See* Filbert Weevils and Acorn Worms, 162.
Casuarina spp., Beefwood, Coast beefwood, She-oak		
Foliage may discolor and wilt. Branches or entire tree may die. May be white fungus beneath basal trunk bark.	**Armillaria root disease**. Fungus present in many soils. Favored by warm, wet soil. Persists for years in infected roots.	*See 120.*
Leaves may discolor, wilt, and drop prematurely. Branches or plant may die. Trees may fall over.	**Wood decay**, or **Heart rot**, including *Ganoderma* sp. Fungi that attack injured, old, or stressed trees.	*Ganoderma* produces large, globular or bracketlike basidiocarps on lower trunks. *See 66, 111.*
Grayish encrustations on bark. Rarely, dead or declining branches.	**Latania scale**. Individuals circular, ≤$^{1}/_{16}$ inch long, usually in groups.	*See 190.*
Sticky honeydew and blackish sooty mold on plant. Popcornlike bodies (egg sacs) on bark.	**Cottony cushion scale**. Orangish, flat immatures or cottony females on bark.	Normally controlled by natural enemies. *See 130, 196.*
Catalpa spp., Catalpa		
Foliage fades, yellows, browns, or wilts, often scattered throughout canopy. Branches die. Entire plant may die.	**Verticillium wilt**. A soil-dwelling fungus that infects through roots.	*See 117.*

WHAT THE PROBLEM LOOKS LIKE	PROBABLE CAUSE	COMMENTS
Foliage may yellow and wilt. Bark or wood may have bracketlike or fan-shaped fungal fruiting bodies. Limbs or entire plant may die.	**Wood decay**, or **Heart rot**, including *Stereum* sp., *Trametes hirsuta, Trametes versicolor*.	Fungi that attack injured, old, or stressed trees. *See* 111.
Sticky honeydew, blackish sooty mold, and whitish cast skins on foliage.	**Melon aphid**. Small, greenish, blackish, or yellowish insects in groups.	*See* 176.
Sticky honeydew and blackish sooty mold on foliage. Cottony or waxy material on plant.	**Grape mealybug**, *Pseudococcus maritimus*. Powdery, grayish insects, ≤1/4 inch long, with waxy filaments.	Vigorous plants tolerate moderate populations. Conserve natural enemies that help in control. *See* Mealybugs, 185.
Galls or swellings on roots.	**Root knot nematodes**. Tiny, root-feeding roundworms.	*See* 293.
Ceanothus spp., Ceanothus, Wild lilac		
Leaves discolor, stunt, wilt, or drop prematurely. Stems discolor, canker, and die. May be white fungus beneath basal trunk bark.	**Armillaria root disease**. Fungus present in many soils. Favored by warm, wet soil. Persists for years in infected roots.	*See* 120.
Leaves discolor, stunt, wilt, or drop prematurely. Plants grow slowly and may die. Roots dark, decayed.	**Phytophthora root and crown rots**. Disease favored by excess soil moisture and poor drainage.	*See* 122.
Leaves wilted, discolored, may drop prematurely. Branches or entire plant may die. May be white fungal growth or dark crust on basal trunk, roots, or soil.	**Dematophora (Rosellinia) root rot**. Fungus favored by mild, wet conditions. Infects primarily through roots growing near infested plants.	Less common than *Armillaria* or *Phytophthora*. *See* 121.
Leaves turn red, brown, then fade. May be cankers or oozing sap on bark.	**Cankers** including **Botryosphaeria canker and dieback, Cytospora canker**.	Fungal diseases that primarily attack injured or stressed plants. *See* 97–100.
Leaves or shoots turn reddish, then yellow, brown, and drop prematurely. Powdery growth on tissue. Shoots or leaves may be stunted and distorted.	**Powdery mildews**, including *Erysiphe polygoni, Microsphaera penicillata*.	Fungi develop on living tissue, are favored by moderate temperatures, shade, and poor air circulation. *See* 90.
Leaves with brown to yellow spots or blotches. Spots may have dark or yellowish margins. Foliage may shrivel and drop prematurely.	**Leaf spots**, including *Cercospora ceanothi, C. macclatchieana, Phloeosporella ceanothi, Phyllosticta* sp. Spread by air or splashing water.	Favored by prolonged cool, wet conditions. *See* 86.
Leaves turn brown or yellow, especially along margins and at tip. Leaves may drop prematurely.	**Leaf burn**, or **Scorch**. Abiotic disorders with many causes. Too much water after establishment or poor drainage are most common causes.	Provide plants with a good growing environment and proper cultural care. *See* Abiotic Disorders, 43.
Leaves may discolor and drop prematurely. Trunk or limbs with roughened, wet, or oozing area. Cracked bark and dieback.	**Flatheaded borers**. Larvae under bark, whitish with enlarged head. Adults are bullet shaped, metallic, coppery, gray, greenish, or bluish.	*See* 227.
Bark on trunk roughened, may exude liquid or boring dust. Slow plant growth.	**Sycamore borer**. Pink larvae, ≤3/4 inch long, bore under bark.	*See* 236.
Spindle-shaped swellings (galls) on green stems. Reduced flowering.	**Ceanothus stem gall moth**. Gray larvae, ≤1/4 inch long, inside gall.	*See* 212.
Stippled or bleached leaves with dark specks of excrement on undersides.	**Ceanothus tingid**, *Corythucha obliqua*. Adults brown, 3/16 inch long, wings lacelike. Nymphs smaller, flattened.	*See* Lace Bugs, 207.
Brownish, gray, tan, or white encrustations on bark. Rarely, dieback of twigs or branches.	**Greedy scale, Oystershell scale**. Tiny, flattened, round or elongate insects.	*See* 189–192.
Sticky honeydew, blackish sooty mold, and whitish cast skins on foliage. Reduced shoot growth.	**Ceanothus aphid**, *Aphis ceanothi*. Small black to reddish brown insects.	Vigorous plants tolerate. Conserve beneficials. *See* Aphids, 175.
Sticky honeydew, blackish sooty mold, and waxy, cottony material on plant.	**Mealybugs**. Powdery, elongate to oval insects, ≤1/8 inch long.	*See* 185.

Ceanothus spp. (continued)

WHAT THE PROBLEM LOOKS LIKE	PROBABLE CAUSE	COMMENTS
Cottony spots, ≤¹⁄₄ inch long, on underside of leaves.	**Psyllid**, *Euphalerus vermiculosus*. Small greenish nymphs beneath cottony material. Suck sap.	Apparently do not damage plants. *See* 170.
Leaves chewed, drop prematurely. May be silken webbing on plants.	**Tent caterpillars**. Hairy, colorful larvae, ≤2 inches long.	*See* 151.
Cedrus spp., Cedar, Atlas cedar, Deodar cedar; *Calocedrus decurrens*, Incense-cedar[6]		
Stunted, bushy branches. Orangish, gelatinous masses or galls on bark. Stems may die back.	**Gymnosporangium rusts**. Fungi that infect and thrive under wet conditions.	*See* Cedar, Cypress, and Juniper Rusts, 95.
Foliage pinkish tan, then browns, dies, and drops prematurely. Shoots die back.	**Needle blight and cast**, including *Kabatina* sp., *Sirococcus conigenus*. Fungi spread by splashing water, favored by cool, wet springs.	Prune out and destroy infected tissue during dry weather. Rake and dispose of plant debris in and beneath limbs. *See* 85.
Foliage discolors, wilts, stunts, may drop prematurely. Discolored bark or cankers may ooze sap. Branches or plant may die.	**Phytophthora root and crown rot**. Decay pathogens favored by excess soil moisture and poor drainage.	*See* 122.
Leaves yellow and wilt. Entire plant may die. May be white fungus beneath basal trunk bark.	**Armillaria root disease**. Fungus present in many soils. Favored by warm, wet soil. Persists for years in infected roots.	Incense-cedar is susceptible. *See* 120.
Branches, treetop dying. Some branches reddish, pitchy or grayish, bare. Wood cankered.	**Botryosphaeria canker and dieback, Phomopsis canker**. Fungi primarily affect injured and drought-stressed trees.	Provide proper irrigation. *See* 98.
Twigs killed back about 6 inches from tips. Dead foliage hanging on tree.	**Cedar bark beetles**. Small, dark adults bore twigs. Pale larvae bore in broken or recently killed limbs.	Terminal feeding does not seriously harm trees. *See* Cedar, Cypress, and Redwood Bark Beetles, 224.
White, waxy threads on bark. Foliage may redden, yellow, and eventually die. Treetop or entire tree may die.	**Cypress bark mealybug**, or **Cypress bark scale; Incense-cedar scale**, or **Monterey cypress scale**. Tiny, immobile, reddish, waxy insects on bark beneath wax.	Primarily pests on Monterey cypress and incense-cedar. Only heavily infested young trees may be harmed. *See* 186, 198.
Sticky honeydew and blackish sooty mold on plant.	**Giant conifer aphids**, *Cinara* spp. Small, black, long-legged insects.	Often infest only a single branch. More common in California in south than in north. *See* 176.
Foliage yellow or brown. Brownish, gray, tan, or whitish encrustations on bark or foliage.	**Armored scales**, including **Latania scale; Juniper scale**, *Carulaspis juniperi*. Tiny, circular to elongate, insects, ≤¹⁄₁₆ inch long, on stems or leaves.	Juniper scale in the field cannot be distinguished from similar minute cypress scale. *See* 189.
Galls or swellings on roots.	**Root knot nematodes**. Tiny, root-feeding roundworms.	*See* 293.
Celtis spp., Hackberry		
Foliage yellows beginning along margins, but veins green. Leaves may be undersized or necrotic.	**Iron deficiency**. Abiotic disorder usually caused by poor soil conditions or unhealthy roots, often from excess irrigation and poor drainage.	*See* 47.
Foliage brown, yellow, undersized, or sparse. Limbs die back. Vascular tissue may be dark, stained. Tree may die.	**Hackberry dieback**. Unexplained malady. Symptoms resemble those of a vascular wilt disease. Reported only on *Celtis sinensis*, occurs at least around Davis, California.	Provide appropriate cultural care and a good growing environment, especially adequate drainage. Avoid excess irrigation. Sterilize tools after working on each hackberry to avoid mechanically spreading pathogens.
Foliage fades, yellows, browns, or wilts, often scattered throughout canopy. Branches die. Entire plant may die.	**Verticillium wilt**. A soil-dwelling fungus that infects through roots.	*See* 117.
Foliage wilts, discolors, and may drop prematurely. Branches die back. Entire tree may die. May be white fungus beneath basal trunk bark.	**Armillaria root disease**. Fungus present in many soils. Favored by warm, wet soil and frequent irrigation. Persists for years in infected roots.	Uncommon in hackberry if conditions highly conducive to *Armillaria* development are prevented. *See* 120.
Foliage may yellow and wilt. Bark or wood may have bracketlike or fan-shaped fungal fruiting bodies. Limbs or entire plant may die.	**Wood decay**, or **Heart rot**, including *Ganoderma lucidum, Laetiporus gilbertsonii* (=*L. sulphureus*), *Pleurotus ostreatus, Trametes versicolor*.	Fungi that attack injured, old, or stressed trees. *See* 66, 111.

WHAT THE PROBLEM LOOKS LIKE	PROBABLE CAUSE	COMMENTS
Sticky honeydew and blackish sooty mold on foliage. Plant growth may slow. Foliage may yellow.	**Citricola scale, European fruit lecanium**. Brownish or grayish, flattened or bulbous insects.	Avoid injecting or implanting roots or trunks with systemic insecticide for scale control, as this may mechanically spread plant pathogens. *See* 194, 195.
Sticky honeydew and blackish sooty mold on foliage. Cottony bluish white wax or small, fuzzy balls on underside of leaves.	**Hackberry woolly aphid**. Tiny insects suck phloem sap from the underside of leaves. Overwinter as eggs on hackberry twigs.	Does not threaten survival of otherwise healthy hackberry. Avoid injecting or implanting roots or trunks with systemic insecticide for control, as this may mechanically spread plant pathogens. *See* 178.
Ceratonia siliqua, Carob		
Brownish, grayish, tan, or white encrustations on twigs or foliage.	**Oleander scale**. Tiny, oval, immobile, tan to yellow insects.	Rarely if ever causes serious damage to plants. *See* 191.
Foliage may yellow and wilt. Bark or wood may have bracketlike or fan-shaped fungal fruiting bodies. Limbs or entire plant may die.	**Wood decay**, or **Heart rot**, including *Ganoderma* sp., *Laetiporus gilbertsonii* (=*L. sulphureus*).	Fungi that attack injured, old, or stressed trees. *See* 66, 111.
Cercidiphyllum japonicum, Katsura tree		
Brown to gray encrustations on bark. Possibly declining or dead twigs or branches.	**Purple scale**, *Lepidosaphes beckii*. Small, elongate insects, often in colonies.	Plants tolerate moderate populations. Conserve beneficials. *See* Armored Scales, 189.
Cercis spp., Redbud		
Leaves fade, yellow, brown, or wilt, often scattered throughout canopy. Branches die. Entire plant may die.	**Verticillium wilt**. A soil-dwelling fungus that infects through roots.	*See* 117.
Leaves discolor, stunt, wilt, or drop prematurely. Plants grow slowly and may die. Roots dark, decayed.	**Phytophthora root and crown rot**. Disease favored by excess soil moisture and poor drainage.	*See* 122.
Leaves turn red, brown, then fade. May be small, pimplelike growths or brownish cankers on bark.	**Cytospora canker**. Fungal disease that primarily affects injured or stressed trees.	*See* 100.
Foliage yellows and wilts. Bark or wood may have bracketlike or fan-shaped fungal fruiting bodies. Branches or entire plant may die.	**Wood decay**, or **Heart rot**, including *Trametes hirsuta, Trametes versicolor*. Fungi that attack injured, old, or stressed trees.	*See* 111.
Leaves with discolored spots or irregular blotches. Leaves may drop prematurely.	**Anthracnose, Leaf spots**, including *Mycosphaerella cercidicola*. Fungi infect and develop on wet tissue.	*See* 81, 86.
Brownish, grayish, tan, or white encrustations on bark. Rarely, declining or dead twigs and branches.	**Greedy scale, Oleander scale**. Tiny, oval insects, often in colonies.	Rarely if ever cause serious damage to redbud. *See* 189–192.
Sticky honeydew and blackish sooty mold on foliage. Tiny, powdery, white mothlike insects.	**Whiteflies**, including **Greenhouse whitefly**. Flat, oval, yellow to green nymphs on leaf undersides.	*See* 182.
Chewed leaves. May be silken tents or mats on plants.	**Tussock moths, Tent caterpillars, Redhumped caterpillar**.	Larvae, ≤2 inches long, hairy or colorful. *See* 150–152.
Leaves chewed, tied together with silk. Tree may be defoliated.	**Fruittree leafroller**. Green larvae, ≤³/₄ inch long, with black head.	*See* Leafrollers, 149.
Cercocarpus spp., Mountain mahogany		
Powdery, white growth and yellow blotches on leaves or shoots. Leaves or shoots may be undersized or distorted.	**Powdery mildew**, *Sphaerotheca macularis*. Fungus favored by moderate temperatures, shade, and poor air circulation.	*See* 90.
Leaves with small discrete spots to irregular, large blotches. Leaves may drop prematurely.	**Anthracnose, Leaf spots**, including *Phloeospora cercocarpi, Sphaceloma cercocarpi*. Fungi spread in water and are favored by cool, wet, conditions.	Diseases apparently promoted by wet winters. Plants tolerate extensive leaf spotting. Little or no effective controls in landscapes or wildlands. *See* 81, 86.

[6] For more information, see *Pests of the Native California Conifers*.

Cercocarpus **spp. (continued)**

WHAT THE PROBLEM LOOKS LIKE	PROBABLE CAUSE	COMMENTS
Leaves discolor and wilt. Branches or treetop may die. Bark may canker or ooze.	**Cankers**, including *Annulohypoxylon mori*, *A. rubiginosum*. Fungi primarily affect injured or drought-stressed trees.	*See* Annulohypoxylon Canker, 98.
Foliage yellows and wilts. Branches or entire plant may die. May be bracketlike or fan-shaped fungal fruiting bodies on bark or wood.	**Wood decay**, or **Heart rot**, including *Phellinus* spp., *Trametes* spp. Fungi that infect injured or severely stressed trees.	*See* 111.
Leaves yellow, wilt, and drop prematurely. Branches die back and entire plant may die.	**Heterobasidion root disease**. Decay fungus that spreads through natural root grafts and airborne spores.	*See* 121.
Leaves discolor and wilt. Trunk or limbs with cankered, roughened, wet, or oozing area. May be cracked bark and dieback.	**Flatheaded borers**, including **Flatheaded appletree borer**. Pale larvae, ≤$1/_2$ inch long, with enlarged head, bore under bark.	*See* 227.
Blackish sooty mold, sticky honeydew, and whitish cast skins on plant.	**Aphids**. Small greenish to black insects, often in groups on leaves and stems.	*See* 175.

Chilopsis linearis, Chilopsis, Desert catalpa, Desert willow

Foliage fades, yellows, browns, or wilts, often scattered throughout canopy or first on one side of the plant. Branches die. Entire plant may die.	**Verticillium wilt**. A soil-dwelling fungus that infects through roots.	*See* 117.
Sticky honeydew, blackish sooty mold, and whitish cast skins on foliage.	**Aphids**. Small pear-shaped insects, usually in groups.	*See* 175.

Chionanthus retusus, Chinese fringe tree; *Chionanthus virginicus*, Fringe tree

No serious invertebrate or pathogen pests have been reported in California. If plants are unhealthy, investigate whether they have been injured or lack good growing conditions or appropriate cultural care.

Choisya ternata, Choisya, Mexican orange, Mock orange

Foliage yellows and wilts. Branches or entire plant dies.	**Phytophthora root and crown rot**. Disease favored by too much water or poor drainage.	*See* 122.
Sticky honeydew and blackish sooty mold on leaves and twigs. Plant may grow slowly.	**Black scale**. Brownish, orangish, or yellow, flattened (immature) or blackish, bulbous (adult) insects.	*See* 192.
Stippled, flecked, or bleached leaves. Leaves may drop prematurely.	**Mites**, including **Citrus red mite**. Tiny, sand-sized specks on leaves, often reddish. Suck sap.	*See* Red Mites, 245.

Cinnamomum camphora, Camphor tree

Foliage fades, yellows, browns, or wilts, often scattered throughout canopy. Branches die. Entire plant may die.	**Verticillium wilt**. A soil-dwelling fungus that infects through roots.	*See* 117.
Leaves discolor, wilt, and die. Plants grow slowly and may die. Basal stem or roots dark, decayed.	**Phytophthora root and crown rot**. Disease favored by excess soil moisture and poor drainage.	*See* 122.
Foliage may yellow and wilt. Bark or wood may have bracketlike or fan-shaped fungal fruiting bodies. Limbs or entire plant may die.	**Wood decay**, or **Heart rot**, including *Schizophyllum commune*. Fungi that attack injured, old, or stressed trees.	*See* 111.
Sticky honeydew and blackish sooty mold on plant.	**California laurel aphid**, *Euthoracaphis umbellulariae*. Grayish insects, $1/_{16}$ inch long, resembling immature whiteflies or scales on undersides of leaves.	Ignore; even heavy populations do not harm tree. *See* 176.
Small brown, purplish, or yellow, circular dead spots on leaves, initially near veins.	**Persea mite**. Tiny, yellow to green mites with dark spots on underside of leaves beneath silvery webbing.	*See* 247.

WHAT THE PROBLEM LOOKS LIKE	PROBABLE CAUSE	COMMENTS
Cistus spp., Cistus, Rock rose[7]		
Leaves yellow and wilt. Entire plant may die. May be white fungus beneath basal trunk bark.	**Armillaria root disease**. Fungus present in many soils. Favored by warm, wet soil. Persists for years in infected roots.	*See* 120.
Leaves brown, fade, yellow, or wilt, often scattered throughout canopy. Foliage may appear sparse, undersized. Plants may grow slowly. Branches die. Entire plant may die.	**Verticillium wilt**, *Verticillium albo-atrum*. Soil-dwelling fungus that infects through roots.	*See* 117.
Sticky honeydew, blackish sooty mold, and whitish cast skins on foliage.	**Aphids**. Small insects, usually in groups, often black.	*See* 175.
Clematis spp., Clematis		
Leaves chlorotic and necrotic and may be undersized.	**Alkaline soil**. Symptoms of various mineral deficiencies and toxicities develop if irrigation water or soil has high pH.	*See* pH Problems, 53.
Conspicuous spots or irregular dead areas form on leaves. Foliage or flowers curl and turn brown or black.	**Sunburn**. Abiotic disorder most prevalent under bright, hot, water-stressed conditions.	*See* 57.
Leaves and stems have brown to yellow spots or blotches. Spots may have dark or yellowish margins. Foliage may shrivel and drop prematurely.	**Leaf and stem spots**, *Cercospora squalidula, Ramularia clematidis*. Fungi spread by air or splashing water. Disease favored by prolonged wet conditions. Pathogens persist in plant debris.	Avoid wetting foliage. Use drip irrigation where feasible. Promptly remove and dispose of plant debris and infected leaves. *See* Leaf Spots, 86.
Leaves discolor, wilt, and die. Plants grow slowly and may die. Basal stem or roots dark, decayed.	**Phytophthora root and crown rot**. Disease favored by excess soil moisture and poor drainage.	*See* 122.
Reddish yellow pustules on lower leaf surface. Leaves discolor, spot, and may drop prematurely.	**Rusts**, *Puccinia pulsatillae, P. recondita*. Pathogens favored by wet conditions. Fungal spores persist in plant debris.	*See* 94.
Foliage or shoots turn reddish, then yellow, brown, and drop prematurely. Powdery growth on tissue. Shoots or leaves may be stunted and distorted.	**Powdery mildew**, *Erysiphe polygoni*. Fungus develops on living tissue, favored by moderate temperatures, shade, and poor air circulation.	*See* 90.
Foliage bleached, stippled, or with silvery or whitish blotches and dark excrement. Leaves or flower buds distorted.	**Thrips**. Tiny, slender insects.	*See* 209.
Sticky honeydew, blackish sooty mold, and whitish cast skins on foliage.	**Aphids**, including **Green peach aphid**. Groups of small insects, often green or yellowish.	*See* 175.
Sticky honeydew and blackish sooty mold on foliage. Possibly twig or branch decline or dieback.	**Brown soft scale**. Small brown, orangish, or yellowish insects, flattened or bulbous.	*See* 193.
Sticky honeydew and blackish sooty mold on foliage. Tiny, whitish, mothlike insects (adult whiteflies) on foliage.	**Whiteflies**, including **Greenhouse whitefly**. Nymphs oval, flattened, yellowish to greenish, suck sap.	*See* 182.
Chewed leaves. Foliage may be webbed. Plant may be defoliated.	**Omnivorous looper**. Yellow, green, or pink larvae, ≤1$^1/_2$ inches long, with green, yellow, or black stripes.	Larvae crawl in "looping" manner. *See* 150.
Brown to gray encrustations on bark. Twigs or limbs may die back.	**Oystershell scale**. Individuals about $^1/_{16}$ inch long, oyster shaped, suck sap.	*See* 191.
Cocculus laurifolius, Cocculus		
Sticky honeydew and blackish sooty mold on plant. Popcornlike bodies (egg sacs) on bark.	**Cottony cushion scale**. Orangish, flat immatures or cottony females on bark.	Can be a severe pest on *Cocculus*, especially when planted away from the coast. Vedalia lady beetle important in biological control avoids this plant. *See* 130, 196.

[7] Some species (e.g. *Cistus salviifolius*) are invasive weeds. Other species may be better choices when planting.

WHAT THE PROBLEM LOOKS LIKE	PROBABLE CAUSE	COMMENTS
Convolvulus cneorum, Bush morning glory, Silver bush		
Leaf undersides with orangish pustules. Leaves may be spotted, drop prematurely.	**Rust**, *Puccinia convolvuli*. Pathogen favored by wet conditions. Fungal spores persist in plant debris.	Avoid overhead watering. Plants tolerate moderate populations. *See* 94.
Sticky honeydew or blackish sooty mold on leaves or twigs. Plants may decline.	**Black scale**. Brown to black, bulbous to flattened insects, ≤$^3/_{16}$ inch long. Raised H shape often on back.	*See* 192.
Cordyline, *Dracaena* spp., Dracaena, Corn plant, Dragon tree, Ti tree[8]		
Leaves yellow and wilt. Entire plant may die. May be white fungus beneath basal trunk bark.	**Armillaria root disease**. Fungus present in many soils. Favored by warm, wet soil. Persists for years in infected roots.	*See* 120.
Leaves fade, yellow, brown, or wilt, often scattered throughout canopy or first on one side of plant. Branches die. Entire plant may die.	**Verticillium wilt**. A soil-dwelling fungus that infects through roots.	*See* 117.
Leaves have tan to brown circular spots or blotches. Leaf tips die. Plants turn slightly yellow, wilt.	**Anthracnose**, *Glomerella cingulata*. Fungus persists in infected tissue. Spores spread by splashing water. Favored by wet conditions.	*See* 81.
Plants may stunt, decline, or (rarely) die. Powdery, waxy material may be visible on roots and around crown.	**Ground mealybug**, *Rhizoecus falcifer*. Small, slender, pale insects; may have powdery wax covering but no marginal filaments.	*See* 187.
Sticky honeydew and blackish sooty mold on plant. Foliage may yellow. Plants may grow slowly.	**Mealybugs**, including **Longtailed mealybug**. Powdery, gray insects with waxy marginal filaments.	*See* 185.
Sticky honeydew and blackish sooty mold on plant. Elongated, whitish material (egg sacs) on stems or leaves.	**Cottony cushion scale**. Females brown, orange, red, or yellow with elongated, white, fluted egg sacs when mature.	Natural enemies usually provide good control. *See* 130, 196.
Leaves chewed.	**Fuller rose beetle**, *Naupactus godmani*. Pale brown adult snout beetle, about $^3/_8$ inch long.	Adults hide during day and feed at night. Larvae feed on roots. *See* Weevils, 159.
Foliage discolored, stippled, or bleached. Terminals may distort. Plant may have fine webbing.	**Spider mites**. Tiny greenish, reddish, or yellowish mites; may have two dark spots.	*See* 246.
Foliage discolored, bleached, or stippled. Leaf underside has dark, varnishlike excrement.	**Dracaena thrips**, *Parthenothrips dracaenae*; **Greenhouse thrips**.	Tiny, slender, black, brown, or yellowish insects. *See* 209.
Cornus spp., Dogwood		
Leaves with large, irregular, brown or purplish blotches. Infected leaves may drop prematurely or dead gray leaves may remain on twigs overwinter. Infected twigs develop discolored, sunken spots, cankers, then die back.	**Dogwood anthracnose**, *Discula destructiva*. A fungal disease favored by wet conditions.	A serious problem in susceptible hosts when conditions favor disease development. Plant resistant cultivars and species. *See* 81.
Leaves, flowers, or young shoots with small, circular, dirty yellow spots with purple margins.	**Spot anthracnose**, *Elsinoe corni*. A fungal disease favored by wet conditions.	Generally not a serious problem and is tolerated by plants. *See* Anthracnose, 81.
Leaves with brown, gray, reddish, yellow, or whitish, circular to angular spots.	**Leaf spots**, including *Phyllosticta*, *Ramularia*, and *Septoria* spp. Fungal diseases that infect and develop when foliage is wet.	*See* 86.
Leaves discolored, wilted, stunted, may drop prematurely. Discolored bark may ooze sap. Branches or plant may die.	**Phytophthora root and crown rot**. Disease favored by excess soil moisture and poor drainage.	*See* 122.
Leaves discolor and wilt. Branches die back. Entire tree may die. May be white fungus beneath basal trunk bark.	**Armillaria root disease**. Fungus present in many soils. Favored by warm, wet soil. Persists for years in infected roots.	*See* 120.

[8] Some species are invasive weeds. Other species may be better choices when planting.

WHAT THE PROBLEM LOOKS LIKE	PROBABLE CAUSE	COMMENTS
Leaves or shoots turn yellow or brown and wilt. Leaves drop prematurely. Limbs may canker or die back.	**Cankers**, including *Cytospora, Nectria, Phoma,* and *Phomopsis* spp. Fungal diseases that primarily affect injured or stressed trees.	Protect trees from injury. Provide good cultural care, especially appropriate irrigation. *See* 97–101.
Leaves turn brown at the margins and tip. Leaves may be curled or puckered. Leaves may drop prematurely.	**Leaf burn**, or **Scorch**. Abiotic disorders. Causes include drought, excess heat, injury, overirrigation, and poor drainage.	Protect trees from injury. Provide good cultural care, especially appropriate irrigation. *See* Abiotic Disorders, 43.
Leaves or shoots turn yellow, then brown. Powdery, whitish growth on buds, leaves, or shoots. Shoots or leaves may be stunted and distorted.	**Powdery mildews**, *Microsphaera penicillata, Phyllactinia guttata*. Fungi favored by moderate temperatures, shade, and poor air circulation.	*See* 90.
Soft, brown decay and/or gray, woolly growth on buds, flowers, leaves, and dead tissue.	**Botrytis blight**, or **Gray mold**. Fungus develops in plant debris or inactive tissue. Favored by wet conditions and moderate temperatures.	*See* 78.
Stippled, flecked, or bleached leaves with whitish cast skins on undersides. Leaves may drop prematurely.	**Leafhoppers**. Pale green, yellow, or white, wedgelike insects, ≤$^1/_8$ inch long.	Plants generally tolerate, control rarely warranted. *See* 201.
Sticky honeydew and blackish sooty mold on foliage. Possible decline or dieback of twigs or branches.	**Brown soft scale, European fruit lecanium scale.** Flattened to hemispherical, brown to yellow insects, suck sap.	*See* 193, 195.
Sticky honeydew and blackish sooty mold on foliage.	**Greenhouse whitefly.** Tiny, powdery, white, mothlike adult insects.	Immatures are green to yellow, flattened and oval. *See* 184.
Sticky honeydew, blackish sooty mold, and whitish cast skins on foliage.	**Aphids**. Small green, yellowish, brown or blackish insects, often in groups.	*See* 175.
Trunk or limbs with roughened, wet, or oozing area. Cracked bark and dieback.	**Flatheaded borers**. Whitish larvae with enlarged head, under bark.	Adults are bullet shaped, metallic, coppery, gray, greenish, or bluish. *See* 227.
Brown to gray encrustations on bark. May be declining or dead twigs or branches.	**Oystershell scale**. Tiny, elongate to oval insects on bark.	*See* 191.
Corylus spp., Hazelnut, California hazel, Western hazel		
Foliage yellows and wilts. Entire plant may die. May be white fungus beneath basal trunk bark.	**Armillaria root rot**. Fungus present in many soils. Favored by warm, wet soil. Persists for years in infected roots.	*See* 120.
Treetop or branches dying. Some branches reddish and pitchy or grayish and bare.	**Botryosphaeria canker and dieback**, *Botryosphaeria obtusa*.	Fungus primarily affects injured or stressed trees. *See* 98.
Blossoms, leaves, and terminals turn brown, wilt, and die. Cankers or lesions may occur on twigs.	**Bacterial blast, blight, and canker**. Disease favored by wet conditions.	*See* 83.
Blossoms, leaves, and terminals turn brown, wilt, and die. Leaves discolor and drop prematurely. Cankers or lesions may occur on limbs or twigs. Cambial tissue darkens.	**Bacterial blight**, *Xanthomonas campestris* pv. *corylina*. Disease favored by wet conditions.	*See* similar Bacterial Blast, Blight, and Canker, 83.
Blossoms, leaves, and terminals turn brown, wilt, and die. Leaves discolor and drop prematurely. Cankers or lesions may occur on limbs or twigs, especially in upper canopy.	**Eastern filbert blight**, *Anisogramma anomala*. A fungus that infects buds and young leaves and shoots via airborne spores produced during wet conditions. Occurs at least in Oregon and the eastern U.S.	Prune out infected limbs during dry weather. Remove escaped seedlings. Plant resistant cultivars and species, such as *Corylus cornuta* var. *californica*.
Discolored blotches on foliage, often brown, reddish, or yellow. Leaves may drop prematurely.	**Leaf spots**, *Anguillospora coryli, Gnomonia gnomon, Mamianiella coryli, Septoria ostryae*.	Fungal pathogens favored by wet conditions. *See* 86.
Leaves with brown, light green, or yellow spots or dead blotches.	**Ramorum blight**. Pathogen spreads via airborne spores and contaminated plants and soil. Infects *Corylus cornuta* var. *californica*.	Primarily a problem in wildlands, killing many oaks there. *See* Sudden Oak Death and Ramorum Blight, 105.

Corylus spp. (continued)

WHAT THE PROBLEM LOOKS LIKE	PROBABLE CAUSE	COMMENTS
Powdery, white growth on leaves, tiny, black overwintering bodies later.	**Powdery mildew**, *Phyllactinia guttata*. Fungal disease favored by moderate temperatures, shade, and poor air circulation.	Generally not severe enough to warrant control. *See 90.*
Sticky honeydew and blackish sooty mold on foliage.	**Aphids**, including *Macrosiphum* spp.; **Filbert aphid**, *Myzocallis coryli*. Pear-shaped, greenish to yellow insects, usually in groups on leaves and terminals.	Conserve effective natural enemies, including the filbert aphid parasite, *Trioxys pallidus. See 175.*
Leaves chewed. Foliage webbed with silk. Naked or hairy larvae, ≤1½ inches long.	**Caterpillars**, including **Filbert leafroller**, *Archips rosanus*; **Obliquebanded leafroller**, *Choristoneura rosaceana*; **Tent caterpillars**.	Vigorous plants tolerate moderate defoliation. *See 149–151.*
Holes in nuts. Tunnels inside nuts may contain whitish larvae.	**Filbertworm**. Adults are tiny bronze, coppery, or reddish brown moth. Larvae bore in nut crops and oak acorns.	Established trees are not damaged. *See* Filbert Weevils and Acorn Worms, 162.

Cotoneaster spp., Cotoneaster[9]

WHAT THE PROBLEM LOOKS LIKE	PROBABLE CAUSE	COMMENTS
Leaves with tiny, reddish to brown spots, may have yellow halos. Larger, dark areas on leaves. Leaves may drop prematurely.	**Entomosporium leaf spot**. A fungal disease promoted by wet foliage.	*See 87.*
Leaves or fruit with dark scabby or velvety spots.	**Scab**, *Venturia* sp. A fungal disease promoted by wet foliage.	*See 88.*
Leaves or shoots with powdery, white growth and yellow blotches. Shoots and leaves may be undersized or distorted.	**Powdery mildew**, *Podosphaera* sp. Fungal disease favored by moderate temperatures, shade, and poor air circulation.	*See 90.*
Leaves discolor, stunt, wilt, or drop prematurely. Stems discolored, cankered, and die. May be white fungus beneath basal trunk bark.	**Armillaria root disease**. Fungus present in many soils. Favored by warm, wet soil. Persists for years in infected roots.	*See 120.*
Leaves wilted, discolored, may drop prematurely. Branches or entire plant may die. May be white fungal growth or dark crust on basal trunk, roots, or soil.	**Dematophora (Rosellinia) root rot**. Fungus favored by mild, wet conditions. Infects primarily through roots growing near infested plants.	Less common than *Armillaria. See 121.*
Sudden wilting, then shriveling and blackening of shoots, blossoms, and fruit. Plant appears scorched.	**Fire blight**. Bacterium enters plants through blossoms.	*See 84.*
Flower buds, petals, or stems turn brown and die. Shoots and stems may darken, canker, exude resin, and die.	**Bacterial blast, blight, and canker**. Disease favored by wet conditions.	*See 83.*
Leaves wilted, discolored, may drop prematurely. Trunk or limbs with roughened, wet, or oozing area. Cracked bark and dieback.	**Flatheaded borers**. Whitish larvae with enlarged head, under bark. Adults are bullet shaped, metallic, coppery, gray, greenish, or bluish beetles.	*See 227.*
Bronzing, darkening, or spotting of fruit or leaves. Buds or fruit may blacken or fail to develop and drop prematurely.	**Pearleaf blister mite**, *Phytoptus pyri*. A microscopic elongate eriophyid mite.	*See* Gall Mites, or Eriophyids, 249.
Leaves bleached, discolored, or stippled. Terminals may distort. Plant may have fine webbing.	**Spider mites**. Tiny greenish, reddish, or yellowish mites; may have two dark spots.	*See 246.*
Leaves bleached, discolored, or stippled with dark varnishlike specks on undersides.	**Lace bug**, *Corythucha* sp. Pale brown adults, ≤⅛ inch long, wings lacelike. Nymphs flattened.	*See 207.*
Brownish, grayish, tan, or white encrustations on bark. May be declining and dead twigs.	**Greedy scale, Oystershell scale, San Jose scale**. Tiny, circular to elongate individuals, often in colonies.	*See 189–192.*
Sticky honeydew, blackish sooty mold, and whitish cast skins on plant.	**Aphids**, including **Apple aphid**, *Aphis pomi*. Tiny insects, often green, clustered on new growth.	*See 175.*

[9] Some species are invasive weeds. Other species may be better choices when planting.

WHAT THE PROBLEM LOOKS LIKE	PROBABLE CAUSE	COMMENTS
Sticky honeydew and blackish sooty mold on leaves and twigs. Possible dieback.	**Kuno scale.** Adult scales beadlike and dark shiny brown.	*See* 195.
Foliage covered with silken webs. Leaves skeletonized.	**Cotoneaster webworm,** *Athrips rancidella.* Larvae brownish black. Tiny, grayish moths, active at night.	Vigorous plants tolerate moderate defoliation. Prune out infested foliage. Tolerate or spray if larvae abundant. *See* Webworms, 152.

Crataegus spp., Hawthorn[10]

WHAT THE PROBLEM LOOKS LIKE	PROBABLE CAUSE	COMMENTS
Sudden wilting, then shriveling and blackening of shoots, blossoms, and fruit. Plant appears scorched.	**Fire blight.** Bacterium enters plants through blossoms.	*See* 84.
Tiny, reddish to brown leaf spots, may have yellow halos. Larger, dark areas on leaves. Leaves may drop prematurely.	**Entomosporium leaf spot.** A fungal disease promoted by wet foliage.	*See* 87.
Leaves yellow, spotted, and may have orangish pustules. Leaves may drop prematurely. Swellings possible on leaves, twigs.	**Rusts,** *Gymnosporangium* spp. Fungi alternate hosts, often on juniper or cedar, and spread by windblown or water-splashed spores.	Avoid overhead watering. Vigorous plants tolerate moderate populations. *See* Cedar, Cypress, and Juniper Rusts, 95.
Powdery, white growth on leaves. Tiny, black overwintering bodies may develop later.	**Powdery mildews,** including *Phyllactinia guttata.* Fungal diseases favored by moderate temperatures, shade, and poor air circulation.	*See* 90.
Black to dark olive spots on fruit, leaves, or stems. Fruit may have scabby blotches or become misshapen.	**Scab,** *Venturia inaequalis.* A fungal disease spread by water from infected leaves and twigs.	*See* 88.
Leaves chewed, may be tied with silk. Plant may be defoliated.	**Caterpillars,** including **Tent caterpillars, Tussock moths.**	Green or hairy larvae, ≤1¹/₂ inches long. *See* 144.
Leaves chewed. Plant can be extensively defoliated.	**Indian walking stick.** Brown, twiglike insects, up to 4 inches long, easily overlooked.	In California in at least the Central and South Coast. *See* 166.
Leaves scraped, skeletonized, and webbed together with silk.	**Leafminers,** including **Apple-and-thorn skeletonizer,** *Choreutis pariana.* Adult *Choreutis* is a dark brown moth, <¹/₂ inch long. Larvae are ≤1 inch long, and greenish or yellow.	Occurs throughout the northern U.S., including northern California. Plants usually tolerate damage. *See* Foliage Miners, 216.
Leaves scraped, skeletonized, may be slime covered.	**Pear sawfly,** or **Pearslug.** Green, slimy, sluglike insects, ≤¹/₂ inch long.	*See* 154.
Roughened twig bark. Possible twig dieback.	**Treehoppers,** including **Buffalo treehopper,** *Stictocephala bisonia.*	Bright yellow to green insects, horny or spiny. *See* 204.
Sticky honeydew, blackish sooty mold, and whitish cast skins on foliage. Leaves may curl or distort.	**Aphids,** including **Apple aphid,** *Aphis pomi.* Groups of small green or green with black insects on leaves.	*See* 175.
Pale wax on bark of limbs, trunk, twigs, or roots. Bark may be gnarled, swollen.	**Woolly aphids,** *Eriosoma lanigerum, Eriosoma crataegi.* Small gray or purplish, waxy insects, mostly on bark.	*See* 177.
Sticky honeydew and blackish sooty mold on foliage. May be whitish cottony bodies on bark.	**Cottony maple scale,** *Pulvinaria innumerabilis;* **Frosted scale.** Brown, yellow, white, or waxy flattened to bulbous insects.	*See* Lecanium Scales, 195.
Sticky honeydew and blackish sooty mold on plant. Whitish cottony bodies on bark. Bark rough. Branches may deform. Possible dieback.	**Azalea bark scale,** or **Woolly azalea scale,** *Eriococcus azaleae.* Dark, reddish, oval insects underneath white wax, on bark crevices and in crotches.	Uncommon in California. If found in California, report to county agricultural commissioner. *See* similar European Elm Scale, 197.
Leaves bleached. May be sticky honeydew or dark excrement specks on foliage.	**Leafhoppers,** including **White apple leafhopper,** *Typhlocyba pomaria;* **Rose leafhopper,** *Edwardsiana rosae.*	Greenish to whitish, wedge-shaped insects, ≤¹/₈ inch long. Leafhoppers do not seriously threaten plant health. *See* 201.

[10] Some species are invasive weeds. Other species may be better choices when planting.

Crataegus spp. (continued)

WHAT THE PROBLEM LOOKS LIKE	PROBABLE CAUSE	COMMENTS
Stippled, bleached, or reddened foliage.	**Mites**, including **European red mite**; *Oligonychus* spp. Tiny green, reddish, or yellow arthropods.	*See* 245.
Grayish encrustations on bark or twigs. Rarely, declining or dead twigs or branches.	**San Jose scale**. Circular, flattened, ≤$^1/_{16}$ inch long, often in colonies.	Usually harmless to hawthorn. *See* 191.
Trunk or limbs with roughened, wet, or oozing area. Bark cracks. Limbs die back.	**Flatheaded borers, Longhorned beetles.** Whitish larvae tunneling under bark.	*See* 227, 232.
Dead branches, limbs, or twigs. Plant declines, may die. Tiny, BB shot–sized holes in bark, which may ooze.	**Shothole borer**. Small brown, adult bark beetles, whitish larvae, tunnel beneath bark.	*See* 225.
Cryptomeria japonica, Cryptomeria		
Leaves turn yellow, then brown, beginning at the tip. Lower, older foliage is most affected.	**Blight**, *Pestalotiopsis funerea*. Fungal disease that affects stressed and weakened plants.	Provide good growing conditions and appropriate cultural care.
Foliage may yellow or wilt. Roots may be darker or fewer than normal. Galls or swellings may be on roots. Plants may grow slowly.	**Root lesion nematode, Root knot nematodes**. Tiny, root-feeding roundworms.	*See* 293.
Cupaniopsis anacardioides, Cupaniopsis, Carrot wood[11]		
No serious invertebrate or pathogen pests have been reported in California. If plants are unhealthy, investigate whether they have been injured or lack good growing conditions or appropriate cultural care.		
Cupressus spp., Cypress; *Cupressus macrocarpa*, Monterey cypress; *Chamaecyparis* spp., False cypress; *Chamaecyparis lawsoniana*, Port Orford cedar[12]		
Foliage brown or yellow and dying. Cankers and resinous exudate on limbs or trunk.	**Cankers**, including **Cypress canker**, *Cytospora cardinale*; **Cytospora canker**, *Cytospora cenisia*; *Phomopsis* sp. Fungi that infect cypress bark.	Primarily infect *Cupressus* spp., especially Leyland cypress and Monterey cypress planted away from the coast. *See* 97–100.
Foliage discolors, wilts, stunts, may drop prematurely. Discolored bark or cankers may ooze sap. Branches or plant may die.	**Phytophthora root and crown rot; Port Orford cedar root disease**, *Phytophthora lateralis*. Decay pathogens infect through wounds or roots in moist soils.	*P. lateralis* infects primarily Port Orford cedar. Do not move soil in areas where *Phytophthora* occurs. *See* 122.
Stunted, bushy branches or witches' brooms. Orange masses or gall on bark. Stems die back.	**Gymnosporangium rusts**. Fungi infect and develop during wet conditions.	Occur on *Cupressus* spp. *See* Cedar, Cypress, and Juniper Rusts, 95.
Foliage browning at tips beginning in fall, worst late winter to spring.	**Cypress tip miners**. Adults are small, silvery tan moths. Green larvae, ≤$^1/_8$ inch long, tunnel in foliage.	*See* 217.
Foliage yellow or brown. Whitish to brownish encrustations on foliage.	**Juniper scale**, *Carulaspis juniperi*; **Minute cypress scale**. Circular to elongate insects, ≤$^1/_{16}$ inch long.	Populations in California rarely warrant control, except on Italian cypress. *See* 190.
Stippled, flecked, or yellow foliage.	**Spruce spider mite**. Tiny green specks, may be fine webbing on foliage.	Occurs on *Chamaecyparis* spp. *See* Pine and Spruce Spider Mites, 247.
Stickiness and blackening of foliage from honeydew and sooty mold.	**Arborvitae aphid**, *Dilachnus tujafilinus*. Brown to gray insects, about $^1/_8$ inch long. Suck sap.	Plants tolerate moderate populations. *See* Aphids, 175.
Chewed needles.	**Conifer sawflies, Cypress sawfly**. Green larvae, ≤1 inch long, on needles.	*See* 153, 154.
Dead and living foliage tied together with silk. Foliage may turn brown.	**Cypress leaf tier**, *Epinotia subviridis*; **Cypress webber**, *Herculia phoezalis*. Pink to dark larvae, ≤$^3/_4$ inch long, feed singly (*Epinotia*) or grouped in "nests" (*Herculia*).	In California more common in the south. Vigorous plants tolerate moderate defoliation. Prune out infested foliage or tolerate. Apply insecticide in March or April if intolerable. *See* Caterpillars, 144.
Twigs killed back about 6 inches from tips. Dead foliage hanging on tree.	**Cypress bark beetles**. Small, dark adults bore twigs. Pale larvae bore in broken or recently killed limbs.	Terminal feeding does not seriously harm trees. *See* Cedar, Cypress, and Redwood Bark Beetles, 224.

[11] Some species are invasive weeds. Other species may be better choices when planting.

[12] For more information, see *Pests of the Native California Conifers*.

WHAT THE PROBLEM LOOKS LIKE	PROBABLE CAUSE	COMMENTS
Foliage discolors. Branches may be killed, sometimes to trunk. Coarse boring dust at trunk wounds and branch crotches.	**Cypress bark moths**, *Laspeyresia cupressana; Epinotia hopkinsana*. Larvae, ≤$^1/_2$ inch long, feed under bark, in cones, or on foliage. Often colonize cypress cankers.	Provide plants proper cultural care. Avoid excess water and fertilizer, which promote rapid growth and susceptible thin bark. Avoid wounding bark. Insects are secondary, not cause of dieback.
Branches dead or dying. Holes about $^1/_4$ inch diameter in wood. May be large larvae boring beneath bark.	**Western horntail**, *Sirex areolatus*. Dark, metallic blue, broad-waist adult wasps and yellowish or white larvae, both ≤1$^1/_2$ inches long.	Attacks only dead, dying, or injured trees. Protect plants from injury. Provide plants with a good growing environment and appropriate cultural care. *See 243.*
White, waxy threads or tufts on bark. Foliage may redden, yellow, and eventually die. Treetop or entire tree may die.	**Cypress bark mealybug,** or **Cypress bark scale; Incense-cedar scale,** or **Monterey cypress scale**, *Xylococcus macrocarpae*. Tiny, immobile, reddish, waxy insects on bark beneath wax.	On Monterey cypress and incense-cedar. Only heavily infested young trees may be damaged. *See 186, 198.*
Plant has sticky honeydew and blackish sooty mold. Plant grows slowly. Foliage may yellow.	**Pyriform scale**, *Protopulvinaria pyriformis*. Insects triangular, $^1/_8$ inch long, brown, yellow, or mottled red.	*See 192.*
Bright orange bark, mostly along the coast on the bark side facing the ocean.	**Lace lichen**, *Trentepohlia aurea*. A nonparasitic green algae containing orange carotenoid pigments. Organism thrives in salty California coastal winds.	Mostly on Monterey cypress, but also on downed wood, rocks, and other surfaces. Tolerate as it is harmless to trees.
Cycas spp., Sago palm, Cycad		
Brown, yellow, or wilted leaves. Roots or basal stem may be dark, decayed. Fronds or entire plant may die.	**Phytophthora root and crown rot, Pythium root rot**. Diseases favored by excess soil moisture and poor drainage.	*See 122, 123.*
Brown, yellow, or wilted leaves. Fronds or entire plant may die.	**Leaf burn**, or **Scorch**. Abiotic disorders with many potential causes, including excess light, overirrigation, poor drainage, and sunburn.	Provide plants with a good growing environment and proper cultural care. *See Abiotic Disorders, 43.*
Sticky honeydew, blackish sooty mold, and pale, waxy material on leaves.	**Mealybugs**, including **Longtailed mealybug**. Powdery, whitish insects, ≤$^1/_8$ inch long, with filaments.	*See 185.*
Mottling or yellowing of leaves. White to tan encrustations on plant. Growth stunted. Plants may die back.	**Cycad scale, Oleander scale**. Scales feed on the underside and base of fronds and on the basal trunk.	Distinguish between these species. Oleander scale is innocuous, but cycad scale is a serious pest. *See 189, 191.*
Daphne spp., Daphne		
Leaves discolored, wilted, stunted, may drop prematurely. Discolored bark may ooze sap. Branches or plant may die.	**Phytophthora root and crown rot**. Disease favored by excess soil moisture and poor drainage.	*See 122.*
Yellowing of foliage. Decline and death of branches or entire plant.	**Euonymus scale**. Tiny, elongate, white male and purplish, oyster-shaped female insects encrusting leaves and stems.	Primarily a pest of euonymus. *See 190.*
Dodonaea viscosa, Dodonaea, Hopbush, Hopseed tree		
Foliage fades, yellows, browns, or wilts, often scattered throughout canopy. Branches die. Entire plant may die.	**Verticillium wilt**. A soil-dwelling fungus that infects through roots.	*See 117.*
Sticky honeydew and blackish sooty mold on leaves or twigs. Twigs and branches may die back.	**Black scale**. Brown to black, bulbous to flattened insects, ≤$^3/_{16}$ inch long. Raised H shape often on back.	*See 192.*
Echium spp., Echium, Pride of Madeira		

No serious invertebrate or pathogen pests have been reported in California. If plants are unhealthy, investigate whether they have been injured or lack good growing conditions or appropriate cultural care.

WHAT THE PROBLEM LOOKS LIKE	PROBABLE CAUSE	COMMENTS
Erica spp., Heath;[13] Calluna vulgaris, Heather		
Leaves discolored, wilted, stunted, or may drop prematurely. Discolored bark may ooze sap. Branches or plant may die.	**Phytophthora root and crown rot.** Disease favored by excess soil moisture and poor drainage.	See 122.
Leaves discolor, wilt, stunt, or drop prematurely. Stem bases discolor and die. May be white fungus beneath basal trunk bark.	**Armillaria root disease.** Fungus present in many soils. Favored by warm, wet soil. Persists for years in infected roots.	See 120.
Leaves or shoot tips turn brown, reddish, yellow, and may drop prematurely. Shoots may be bushy, distorted, or stunted. Leaves or shoots may be covered with whitish growth.	**Powdery mildew**, Erysiphe polygoni. Fungus develops on living tissue, favored by moderate temperatures, shade, and poor air circulation.	Infects Erica persoluta. See 90.
Leaves have brown or yellow blotches and may have powdery, orange pustules. Leaves may drop prematurely.	**Rust**, Uredo ericae. Spores from foliage may be carried for miles by wind. Favored by low temperatures, dew, and rain.	Infects Erica hirtiflora and E. persoluta var. alba. See 94.
Foliage discolored, stippled, brownish, or bleached, and may drop prematurely. Terminals may distort. Plant may have fine webbing.	**Spider mites**, including **Twospotted mite.** Tiny greenish, reddish, or yellowish mites, may have two dark spots.	See 246.
Bark or leaves have gray, brown, tan, or yellow encrustations (colonies of scales). Foliage may yellow. Rarely, plant dies back.	**Greedy scale, Oleander scale, Oystershell scale**. Circular flat or elongate to oval insects, ≤$1/16$ inch long.	See 190, 191.
Sticky honeydew and blackish sooty mold on foliage. Possible plant decline or dieback.	**European fruit lecanium.** Brown or yellow, flat or bulbous, immobile scale insects.	See Lecanium Scales, 195.
Foliage notched around margins. Wilting or dead plants. Some roots stripped of bark or girdled near soil.	**Black vine weevil.** Adults are black or grayish snout beetles, about $3/8$ inch long. Larvae are white grubs with brown head.	Larvae feed on roots. Adults hide during day and feed at night. See 160.
Eriogonum spp., Buckwheat		
Chewed flowers. Leaves chewed or surfaces scraped.	**Common hairstreak butterfly**, Strymon melinus. Larvae greenish with short, brown hairs. Adult 1 inch long, gray above with marginal red wing spot.	Plants tolerate extensive feeding by larvae, which mature into attractive butterflies. See Caterpillars, 144.
Flowers chewed.	**Tumbling flower beetle**, Mordella sp. Dark, $1/6$-inch-long, narrow adults with tapered abdomen.	Feed mostly on pollen, damage usually minor. Adults often become very active when disturbed. No known controls.
Sticky honeydew, blackish sooty mold, or whitish cast skins. Flowers may drop prematurely or distort.	**Aphids**, Braggia spp. Small, grayish green to black insects in groups.	Plants tolerate many aphids. Flowers and seed can be reduced. See 175.
Erythrina spp., Coral tree		
Foliage discolors and wilts. Branches die back. Entire tree may die. May be white fungus beneath basal trunk bark.	**Armillaria root disease.** Fungus present in many soils. Favored by warm, wet soil. Persists for years in infected roots.	See 120.
Foliage fades, yellows, browns, or wilts, often scattered throughout canopy or first on one side of plant. Branches die. Entire plant may die.	**Verticillium wilt.** A soil-dwelling fungus that infects through roots.	See 117.
Foliage yellows and wilts. Bark may have lesions that are dark, dry, crusty whitish, oily looking, or water-soaked. Branches or entire plant dies.	**Fusarium dieback.** Fungal disease spread by the polyphagous shothole borer, an ambrosia beetle, or bark beetle.	Occurs on E. corallodendron at least in southern California. See 115.
Sticky honeydew and blackish sooty mold on plant. Elongated, whitish material (egg sacs) on twigs or leaves.	**Cottony cushion scale.** Females brown, orange, red, or yellow, with elongated, white, fluted egg sacs when mature.	Natural enemies usually provide good control. See 130, 196.
Sticky honeydew and blackish sooty mold on foliage. Cottony or waxy material on plant.	**Obscure mealybug.** Powdery, grayish, insects, ≤$1/4$ inch long, with waxy filaments.	Vigorous plants tolerate moderate populations. Conserve natural enemies that help in control. See 187.

[13] Some species are invasive weeds. Other species may be better choices when planting.

WHAT THE PROBLEM LOOKS LIKE	PROBABLE CAUSE	COMMENTS
Escallonia spp., Escallonia		
Sticky honeydew and blackish sooty mold on plant. Bulbous, irregular, brown, gray, or white bodies (scales) on twigs.	**Chinese wax scale**. Bulbous to hemispherical, waxy insects, suck sap.	*See* 196.
White cottony masses, sticky honeydew, and blackish sooty mold on bark or leaves.	**Gill's mealybug**, *Ferrisia gilli*. Gray to pinkish, oblong insects.	*See* 186.
Foliage yellows, browns, and wilts. Branches die. Entire plant may die.	**Escallonia dieback**. Unexplained malady affecting *Escallonia* 'Fradesii.'	Provide a good growing environment and appropriate cultural care. No known specific remedy.
Foliage yellows and drops prematurely. Plant declines.	**High pH, Improper pruning**. Alkaline soil, pH >7.5, and shearing terminals can cause severe decline.	*See* Pruning, 35; pH Problems, 53.
Eucalyptus spp., Eucalyptus, Gum[14]		
Leaves yellow and wilt. Branches or entire plant dies. Roots may be dark or decayed.	**Phytophthora root and crown rot.** Disease favored by excess soil moisture and poor drainage.	*See* 122.
Leaves brown or yellow. Branches die back. Entire plant may die. May be white fungus beneath basal trunk bark.	**Armillaria root disease**. Fungus present in many soils. Favored by warm, wet soil. Persists for years in infected roots.	*See* 120.
Leaves yellow and wilt. Stems may have dark, sunken cankers. Limbs or entire tree may die.	**Cankers,** including **Botryosphaeria canker and dieback; Diaporthe stem canker and dieback,** *Diaporthe eucalypti; Harknessia* spp.; **Nectria canker**, *Nectria eucalypti*. Fungi primarily affect injured or stressed trees.	Avoid wounding plants, except to prune off infected limbs. Provide a good growing environment and appropriate cultural care. *See* 98, 101.
Foliage may yellow and wilt. Bark or wood may have bracketlike or fan-shaped fungal fruiting bodies. Limbs or entire plant may die.	**Wood decay**, or **Heart rot**, including *Ganoderma applanatum, Laetiporus gilbertsonii (=L. sulphureus), Pleurotus ostreatus, Schizophyllum commune, Stereum* sp., *Trametes* spp.	Fungi that attack injured, old, or stressed trees. *See* 66, 111.
Leaves and twigs with small to large discolored blotches, spots, or streaks. Leaves may drop prematurely. Cankers may develop on twigs.	**Anthracnose, Leaf spots**, and **Tar spots**, including *Colletotrichum gloeosporioides, Heterosporium eucalypti, Mycosphaerella molleriana, Phyllachora eucalypti, Septoria* spp.	Fungi infect and develop on wet tissue. *See* 81, 86.
Leaves turn brown or yellow, especially along margins and at tip. Leaves may drop prematurely.	**Leaf burn**, or **Scorch**. Abiotic disorders with many potential causes, including direct injury to trunks or roots, frost, poor soil conditions, and too much or too little water.	Provide plants with a good growing environment and proper cultural care. *See* Abiotic Disorders, 43.
Leaves or shoots with powdery growth. Shoots or leaves may be stunted and distorted.	**Powdery mildews**, including *Erysiphe cichoracearum, Phyllactinia guttata*.	Fungi favored by moderate temperatures, shade, and poor air circulation. *See* 90.
Leaves may discolor, wilt, or drop prematurely. Dead tree or dying limbs have broad galleries beneath bark.	**Eucalyptus longhorned beetles**. Adults about 1 inch long, reddish brown with yellow on the back. Larvae whitish.	*See* 233.
Leaf underside has pimplelike blisters or water-soaked spots. Leaves brown, harden, appear corky, especially on underside. Foliage may yellow, drop prematurely.	**Edema**. Noninfectious disorder. Leaves accumulate excess water when soil is warm and moist and air is cool and moist. Exact cause unknown.	May not be manageable in landscapes. Provide good air circulation. Provide good drainage. *See* 60.
Leaves, shoot terminals, or stems with black, purple, red, or tan bumps or wartlike swellings. Infested parts develop a roughened appearance.	**Eucalyptus gall wasps**, *Epichrysocharis (=Aprostocetus) burwelli, Epichrysocharis* sp., *Ophelimus* sp., *Selitrichodes globulus*, and others. Tiny, mostly dark wasps with pale larvae that feed within leaves or terminals.	Occur on various *Eucalyptus* spp., and new wasp species are periodically introduced. Cause only aesthetic damage and do not threaten plant health. No control known or recommended. *See* 212

[14] Some species are invasive weeds. Other species may be better choices when planting.

WHAT THE PROBLEM LOOKS LIKE	PROBABLE CAUSE	COMMENTS
Sticky honeydew and blackish sooty mold on foliage. May be tiny, whitish caps or funnel-shaped waxiness on leaves. New shoots may be distorted, covered with whitish, waxy strands.	**Psyllids,** including **Bluegum psyllid, Lemongum lerp psyllid, Redgum lerp psyllid; Spottedgum psyllid.**	Tiny gray, green, or orange sucking insects. Some species feed beneath waxy covers on leaves. *See 171–172.*
Leaves chewed. Trees may drop leaves prematurely.	**Eucalyptus tortoise beetles.** Adults hemispherical and dark brown with black or gray to reddish brown. Larvae dark greenish or light greenish gray, all ≤³/₈ inch long.	At least two species. Insects may be overlooked, as they often feed at night and hide beneath bark during the day. *See 157.*
Leaves chewed. Leaves with scraped surface, winding discolored trails, or elongate holes.	**Eucalyptus snout beetle,** *Gonipterus scutellatus.* Reddish brown adult weevils and legless, yellowish green larvae with a slimy coating.	Uncommon, as under good biological control from *Anaphes nitens* egg parasite. Holes in leaves distinguish this species from tortoise beetles (above), which chew only along leaf edges.
Distorted bark, galls or swellings around trunk base.	**Ligno-tubers.** Abiotic disorder. Galls are latent buds from which shoots sprout in response to stress.	Protect trees from injury. Provide good growing conditions and proper cultural care. Ligno-tubers are not harmful, but may indicate tree stress or injury.
Euonymus spp., Euonymus		
Discolored leaf blotches, often between veins. Large spots chlorotic, then tan or brown.	**Sunburn,** or **Summer leaf scorch.** Noninfectious disorder, appears during or after drought stress and high temperatures.	*See 57.*
Yellowing of foliage. Decline and death of branches or entire plant.	**Euonymus scale.** Tiny, elongate, white male and purplish, oyster-shaped female insects encrusting leaves and stems.	*See 190.*
Brownish, gray, tan, or orange encrustations on bark or foliage. Rarely, stunted, declining, or dead branches.	**California red scale, Greedy scale, Latania scale.** Tiny, circular, flattened insects on stems or leaves.	Rarely if ever cause serious damage to euonymus. *See 188–190.*
Whitish patches of growth on foliage. Leaves may yellow, distort, or drop prematurely.	**Powdery mildew,** *Microsphaera euonymi-japonici.* A fungal disease favored by moderate temperatures, shade, and poor air circulation.	*See 90.*
Copious, misty, nonsticky liquid raining from plant. Surfaces covered with whitish residue.	**Glassy-winged sharpshooter.** Active, dark brown or gray leafhoppers, ≤¹/₂ inch long, suck xylem fluid.	Vectors *Xylella* pathogens. Report suspected glassy-winged sharpshooters to agricultural officials if found in areas where this pest is not known to occur. *See 203.*
Sticky honeydew, blackish sooty mold, or whitish cast skins on plant.	**Aphids,** including **Melon aphid, Bean aphid.** Groups of small green, black, or yellow insects.	*See 175.*
Wilting or dead plants. Some roots stripped of bark or girdled near soil. Foliage may be chewed, ragged.	**Black vine weevil.** Adults are black or grayish snout beetles, about ³/₈ inch long. Larvae are white grubs with brown head.	*See 160.*
Galls or swellings on trunk and roots, usually near soil, may be on branches. Branches or entire plant may die.	**Crown gall.** Bacteria that infect plant via wounds.	*See 108.*
Galls or swellings on roots.	**Root knot nematodes.** Tiny, root-feeding roundworms.	*See 293.*
Fagus spp., Beech		
Foliage may yellow and wilt. Bark or wood may have bracketlike or fan-shaped fungal fruiting bodies. Limbs or entire plant may die.	**Wood decay,** or **Heart rot,** including *Laetiporus gilbertsonii* (=*L. sulphureus*), *Pleurotus ostreatus, Trametes versicolor.*	Fungi that attack injured, old, or stressed trees. *See 111.*
Sticky honeydew, blackish sooty mold, and whitish cast skins on foliage. Cottony, waxy material on leaves.	**Woolly beech leaf aphid,** *Phyllaphis fagi.* Small, greenish insects in groups on underside of leaves.	Plants tolerate abundant aphids. *See* Woolly Aphids, 177.

WHAT THE PROBLEM LOOKS LIKE	PROBABLE CAUSE	COMMENTS
Brown to gray encrustations on bark. Twigs or limbs may die back.	**Oystershell scale.** Individuals about $1/16$ inch long, oyster shaped, suck sap.	*See* 191.
Fatsia japonica =Aralia sieboldii, Aralia, Angelica		
Conspicuous spots or irregular dead areas on leaves. Foliage or flowers curl and turn brown or black.	**Sunburn.** Abiotic disorder most prevalent under bright, hot, water-stressed conditions.	*See* 57.
Brown to yellow blotches or spots on upper surface and orange or yellowish pustules on underside of leaves.	**Rust,** *Nyssopsora clavellosa.* Fungus favored by wet conditions.	*See* 94.
Sticky honeydew and blackish sooty mold on foliage. Twig or branch decline or dieback.	**Black scale; Brown soft scale; Hemispherical scale,** *Saissetia coffeae.* Small black, brown, or yellowish insects, flattened or bulbous.	*See* 192, 193.
Plant has sticky honeydew and blackish sooty mold. Plant grows slowly. Foliage may yellow.	**Pyriform scale,** *Protopulvinaria pyriformis.* Triangular insects, $1/8$ inch long, and brown, yellow, or mottled red.	*See* 192.
Blackish sooty mold, sticky honeydew, and whitish cast skins on plant.	**Aphids.** Small, pear-shaped insects on leaves, often in groups.	*See* 175.
Brownish, grayish, tan, or white encrustations on bark. Rarely, plant parts may die back.	**Greedy scale, Latania scale, Oleander scale.** Tiny, circular, flattened insects, usually in groups.	*See* 189–192.
Leaves chewed. Leaves may be tied together with silk.	**Orange tortrix,** *Argyrotaenia citrana;* **Omnivorous looper.** Caterpillars, $\leq 1^1/2$ inches long.	Vigorous plants tolerate moderate leaf damage. Prune out infested foliage. If intolerable, apply *Bacillus thuringiensis* or another insecticide when young larvae are feeding. *See* Caterpillars, 144.
Galls or swellings on roots. Plants may grow slowly.	**Root knot nematodes.** Tiny, root-feeding roundworms.	*See* 293.
Ficus spp., Ficus, Fig;[15] *Ficus microcarpa,* Indian laurel fig, Laurel fig		
Foliage yellows and wilts. Branches or entire plant dies. Roots or plant base may be dark or decayed.	**Phytophthora root and crown rot.** Disease favored by excess soil moisture and poor drainage.	*See* 122.
Foliage browns or yellows. Branches die back. Entire plant may die. May be white fungus beneath basal trunk bark.	**Armillaria root disease.** Fungus present in many soils. Favored by warm, wet soil. Persists for years in infected roots.	*See* 120.
Foliage yellows and drops prematurely. Branches die back. Tree declines and dies.	**Ficus canker, Sooty canker,** or **Ficus branch dieback.** Fungal disease of stressed and mechanically injured Indian laurel fig.	Commonly affects pruned, drought-stressed hosts. *See* 100.
Limbs die back, bark gnawed, or branches girdled. May be nest of leaves and twigs.	**Roof rat,** *Rattus rattus.* Typically feeds at night and nests in dense trees and shrubs.	*See Rats Pest Notes* online at ipm.ucanr.edu.
Foliage bleaches out to yellow or almost white, especially in bright, hot locations.	**Excess light.** Abiotic disorder induced by excess light or changes in lighting.	Avoid changing light conditions around established plants. Provide partial shade. *See* High and Low Light, 58.
Foliage drops prematurely. Dropped leaves may be yellow or a healthy-looking green. Entire plant may defoliate.	**Abiotic disorders.** Caused by any severe environmental change, including extreme light or temperature, inappropriate irrigation, and root injury.	Protect plants from injury. Avoid rapid environmental changes. Provide a good growing environment and appropriate cultural practices. *See* 43.
Foliage underside has pimplelike blisters or water-soaked spots. Leaves brown, harden, appear corky, especially on underside. Foliage may yellow and drop prematurely.	**Edema.** Abiotic disorder. Leaves accumulate excess water when soil is warm and moist and air is cool and moist. Exact cause unknown.	Irrigate only in morning. Avoid irrigation if cool and cloudy. Provide good air circulation. Keep humidity low. Provide good drainage. *See* 60.
Leaves or stems with warty blisters or swellings. Leaves brown or yellow, curl, and drop prematurely.	**Ficus gall midge, Ficus gall wasp.** Females oviposit in leaves or stems, where maggots cause swellings or brown lesions.	Infest new growth of *Ficus microcarpa. See* 214.

[15] Some species are invasive weeds. Other species may be better choices when planting.

Ficus spp. (continued)

WHAT THE PROBLEM LOOKS LIKE	PROBABLE CAUSE	COMMENTS
Curling and purple pitting of terminal leaves. In California, problem mostly in south.	**Cuban laurel thrips.** Slender, black adults, or yellow nymphs, ≤$^1/_8$ inch long, in curled leaves.	*Ficus microcarpa*, also sometimes called *F. nitida* or *F. retusa*, is preferred host; cv. 'Green Gem' is mostly resistant. *See* 210.
Sticky honeydew, blackish sooty mold, and cottony waxy material on plant.	**Mealybugs,** including **Citrus mealybug, Longtailed mealybug, Vine mealybug.** Grayish, oval, waxy, slow-moving insects, may have waxy filaments.	*See* 185.
Sticky honeydew and blackish sooty mold on foliage. Tiny, whitish, mothlike adults.	**Whiteflies,** including **Fig whitefly,** *Singhiella simplex*; **Greenhouse whitefly.** Nymphs oval, flat, and green, tan, or yellow.	*See* 182.
Sticky honeydew and blackish sooty mold on foliage. May be cottony material (egg sacs) on plant.	**Green shield scale.** Brownish, green, orange, red, or yellowish convex or flattened insects on bark or leaves.	*See* 194.
Sticky honeydew and blackish sooty mold on leaves and twigs. May be dieback of twigs or branches.	**Black scale, Brown soft scale.** Black, brown, or yellow, flattened or bulbous insects, ≤$^1/_5$ inch long.	*See* 192, 193.
Gray to reddish encrustations on shoots or needles (scale insect bodies). Plant may decline and die back.	**Dictyospermum scale,** *Chrysomphalus dictyospermi*. Tiny, immobile, flattened insects that suck plant juices. Several generations per year.	Occasional problem in California, mostly in south. Conserve introduced natural enemies, which can provide excellent biological control. *See* Armored Scales, 189.
Forsythia spp., Forsythia		
Leaves yellow and wilt. Plants wilt and may die, often suddenly. Roots and stem near soil dark, decayed. Leaves may have dark blotches.	**Phytophthora root and crown rot.** Disease favored by wet, poorly drained soil.	*See* 122.
Leaves with black, brown, tan, or yellow spots or blotches. Foliage may die and drop prematurely.	**Leaf spots,** including *Alternaria* spp., *Phyllosticta* spp. Fungi spread by air or splashing water. Favored by prolonged cool, wet, conditions.	*See* 86.
Buds, flowers, leaves, or shoots darken or wilt and die.	**Bacterial blast, blight, and canker.** Disease favored by wet conditions.	*See* 83.
Galls or swellings on roots or stems, usually near the soil.	**Crown gall.** Bacteria infect plant via wounds.	*See* 108.
Galls or swellings on twigs or stems, often high on the plant in comparison with crown galls. Twigs or stems may die back.	**Stem gall,** *Phomopsis* sp. A fungal disease most damaging to injured or stressed plants.	Protect plants from injury. Provide a good growing environment and appropriate cultural care.
Fraxinus spp., Ash		
Foliage fades, yellows, browns, or wilts, often scattered throughout canopy. Branches die. Entire plant may die.	**Verticillium wilt.** A soil-dwelling fungus that infects through roots.	*See* 117.
Foliage fades, yellows, browns, or wilts, often scattered throughout canopy. Branches die. Entire plant may die.	**Raywood ash canker and decline.** Affects drought-stressed, mechanically injured raywood ash.	Provide a good growing environment and appropriate cultural care. *See* 104.
Foliage may yellow and wilt. Bark or wood may have bracketlike or fan-shaped fungal fruiting bodies. Limbs or entire plant may die.	**Wood decay,** or **Heart rot,** including *Ganoderma* spp., *Laetiporus gilbertsonii* (=*L. sulphureus*), *Pleurotus ostreatus*, *Schizophyllum commune, Trametes* spp.	Fungi that attack injured, old, or stressed trees. *See* 66, 111.
Leaves and branches wilt in spring. Stems may have dark cankers, callus tissue, or coral-colored pustules.	**Nectria canker,** *Nectria* sp. Fungus primarily affects injured or stressed trees.	*See* 101.
Leaves with irregular, brown, tan, or white areas. Premature leaf drop. Twigs die back. Large branches may die if repeatedly defoliated.	**Anthracnose,** *Discula fraxinea*. Fungus infects leaves and twigs. Splashing rain spreads spores.	Affects only flowering ash (*Fraxinus ornus*) and Modesto ash (*F. velutina* var. *glabra*). *See* 81.

WHAT THE PROBLEM LOOKS LIKE	PROBABLE CAUSE	COMMENTS
Leaves with brown spots with pale or yellow borders. Leaves may drop prematurely.	**Leaf spots**, including *Mycosphaerella effigurata, M. fraxinicola*. Fungi promoted and spread by wet conditions.	*See* 86.
Leaves turn brown or yellow, especially along margins and at tip. Leaves may drop prematurely.	**Leaf burn**, or **Scorch**. Abiotic disorders. Causes include drought, dry wind, excess fertilization, excess light, and saline irrigation water or soil.	Provide plants with a good growing environment and proper cultural care. *See* Abiotic Disorders, 43.
Leaves or shoots with powdery growth. Shoots or leaves may be stunted and distorted.	**Powdery mildews**, including *Erysiphe cichoracearum, Phyllactinia guttata*.	Fungi favored by moderate temperatures, shade, and poor air circulation. *See* 90.
Leaves may discolor and wilt. Branches wilted or dying. Boring dust, ooze, or holes on trunk or branches.	**Ash borer, Redbelted clearwing**. White larvae, ≤1 inch long, head brown.	Adult moths wasplike. *See* 235.
Leaves may discolor and wilt. Holes, ≤¹/₂ inch in diameter, in wood. Boring dust. Limbs may die and drop.	**Carpenterworm**. Dark whitish larvae, ≤2¹/₂ inches long, bore in wood.	*See* 241.
Sticky honeydew, blackish sooty mold, and whitish wax on foliage. Leaves curl, gall, and distort. Leaves may drop prematurely.	**Ash leafcurl aphid**. Gray to green, waxy insects in distorted leaves.	*See* 178.
Sticky honeydew and blackish sooty mold on leaves and twigs. Tiny, white, mothlike adult insects.	**Ash whitefly**. Tiny, oval nymphs are flattened and clear. Older nymphs have a band of white wax on the back.	Under good biological control and uncommon unless natural enemies are disrupted. *See* 182.
Sticky honeydew, blackish sooty mold, or whitish wax on foliage. Leaves or terminals curled or distorted.	**Ash psyllid**, *Psyllopsis fraxinicola*. Aphidlike insects, suck sap.	*See* Psyllids, 170.
Sticky honeydew and blackish sooty mold on plant.	**Frosted scale**. Bulbous or flattened, brown or whitish insects, may be waxy, ≤¹/₄ inch long, suck sap.	*See* Lecanium Scales, 195.
Copious, misty, nonsticky liquid raining from plant. Surfaces covered with whitish residue.	**Glassy-winged sharpshooter**. Active, dark brown or gray leafhoppers, ≤¹/₂ inch long, suck xylem fluid.	Vectors *Xylella* pathogens. Report suspected glassy-winged sharpshooters to agricultural officials if found in areas where this pest is not known to occur. *See* 203.
Spotted, stippled, or yellow foliage. Shoot terminals and leaves may distort.	**Boxelder bugs**. Gray and red adults, about ¹/₂ inch long. Nymphs red.	Trees tolerate damage. Adults invading houses are the primary problem. *See* 206.
Coarse, yellow stippling (flecking) of leaves. Dark, varnishlike excrement on undersides of leaves. Leaves may drop prematurely.	**Ash plant bugs**. Brown, green, or yellowish true bugs, ≤³/₁₆ inch long.	*See* 206.
Stippled, bleached leaves with dark specks of excrement on undersides.	**Arizona ash tingid**, *Leptoypha minor*. Pale brown insect, ≤¹/₈ inch long, wings lacelike.	*See* Lace Bugs, 207.
Leaves chewed in spring, usually only on inner canopy near bark where the night-feeding larvae hide during the day.	**Ash moth**, or **Bentley ash moth**, *Oncocnemis punctilinea*. Dark gray to brown larvae, ≤¹/₂ inch long, with lighter longitudinal stripe. Head dark with light patterns.	Pest in Central Valley and southern deserts of California. One generation per year. Ash generally tolerates damage. Apply *Bacillus thuringiensis* or another insecticide if intolerable. *See* Caterpillars, 144.
Leaves chewed, scraped, covered with silken strands, or webbed together.	**Moths**, including **Fall webworm; Omnivorous leafroller**, *Platynota stultana; Zelleria* sp. Larvae are brownish, greenish, orangish, tan, or dirty whitish with a dark head.	Vigorous plants tolerate moderate defoliation. *See* Caterpillars, 144.
Tan or gray encrustations on bark. May be dead or declining twigs and branches.	**Oystershell scale, Walnut scale**. Tiny, round to elongate insects, often in colonies.	*See* 191.
Roughened twig bark. Possible twig dieback.	**Treehoppers**, including **Buffalo treehopper**, *Stictocephala bisonia*.	Bright yellow to green insects, horny or spiny. *See* 204.
Green plants with smooth stems, thick roundish leaves infesting branches.	**Broadleaf mistletoe**. Parasitic evergreen plant on host.	*See* 284.

WHAT THE PROBLEM LOOKS LIKE	PROBABLE CAUSE	COMMENTS
Fremontodendron =Fremontia spp., Flannel bush		
Brownish, grayish, tan, or white bark encrustations. May be decline or dieback of branches, twigs.	**Greedy scale.** Tiny, circular to elongate individuals on twigs, branches.	*See* 190.
Foliage yellows and wilts. Branches or entire plant dies.	**Phytophthora root and crown rot.** Disease favored by wet, poorly drained soil.	*See* 122.
Fuchsia spp., Fuchsia		
Leaves brown, fade, yellow, or wilt, often scattered throughout canopy. Branches die. Entire plant may die.	**Verticillium wilt.** A soil-dwelling fungus that infects through roots.	*See* 117.
Leaves yellow. Plants wilt and die, often suddenly. Roots and stem near soil dark, decayed, girdled by lesions. Leaves may have dark blotches.	**Phytophthora root and crown rot, Pythium root rot.** Pathogens survive in soil. Diseases favored by wet, poorly drained soil.	*See* 122, 123.
Leaves discolor, wilt, stunt, or drop prematurely. Stems discolored, cankered, and die. May be white fungus beneath basal trunk bark.	**Armillaria root disease.** Fungus present in many soils. Favored by warm, wet soil. Persists for years in infected roots.	*See* 120.
Soft, brown decay and/or gray, woolly growth on buds, flowers, leaves, and dead tissue.	**Botrytis blight,** or **Gray mold.** Fungus develops in plant debris or inactive tissue.	Favored by high humidity and moderate temperatures. Spores airborne. *See* 78.
Leaves yellow or with yellow blotches, lines, or intricate patterns. Leaves may distort. Plants may be stunted.	**Bean yellow mosaic virus, Cucumber mosaic virus.** Spread by aphids or in infected plants. **Impatiens necrotic spot virus** and **Tomato spotted wilt virus.** Spread by thrips or in infected plants.	Usually not damaging or important in landscapes. Once infected, there is no control except to replace plants. *See* Mosaic and Mottle Viruses, 92.
Leaf undersides with orangish pustules. Leaves may be spotted, drop prematurely.	**Rust,** *Pucciniastrum pustulatum.* Fungus requires moisture to develop.	Avoid overhead watering. Plants tolerate moderate populations. *See* 94.
Leaves and shoots distorted, galled, or thickened.	**Fuchsia gall mite.** Microscopic, wormlike eriophyid mite.	*See* 250.
Leaves blotched or dark streaked. Leaves or terminals distorted. Flowers small. Plants stunted. Buds may darken or drop.	**Cyclamen mite,** *Phytonemus pallidus.* A pinkish orange mite, ≤$1/100$ inch long, which feeds protected in buds and distorted tissue.	Primarily a problem during propagation and in greenhouses. Relatively uncommon in landscapes. *See* Broad Mites and Cyclamen Mite, 249.
Stippled or bleached leaves with varnishlike specks on undersides.	**Greenhouse thrips.** Tiny, slender, black adults or yellow nymphs.	*See* 210.
Leaves discolored, stippled, or bleached, and may drop prematurely. Terminals may distort. Plant may have fine webbing.	**Mites,** including **Privet mite,** *Brevipalpus obovatus;* **Twospotted spider mite.** Greenish or yellowish, tiny arthropods, which may have two dark spots.	*See* 245.
Sticky honeydew and blackish sooty mold on foliage. Tiny, whitish, mothlike adult insects.	**Bandedwinged whitefly,** *Trialeurodes abutilonea;* **Greenhouse whitefly; Iris whitefly,** *Aleyrodes spiraeoides.* Nymphs oval, flattened, yellow to greenish.	*See* 182.
Sticky honeydew, blackish sooty mold, and whitish cast skins on foliage.	**Aphids,** including **Crescent-marked lily aphid,** *Aulacorthum circumflexum;* **Green peach aphid; Potato aphid,** *Macrosiphum euphorbiae.*	Groups of small black, green, or yellowish insects on succulent foliage and shoots. *See* 175.
Sticky honeydew and blackish sooty mold on foliage. Waxy, cottony material on plant.	**Citrus mealybug; Longtailed mealybug; Mexican mealybug,** *Phenacoccus gossypii.*	Oblong, waxy, slow-moving insects, ≤$1/8$ inch long. *See* 186, 187.
Sticky honeydew and blackish sooty mold on foliage. Possible decline or dieback of twigs and branches.	**Black scale.** Black, brown, orange, or yellow, flattened or bulbous insects on leaves or twigs.	*See* 192.
Chewed or notched leaves or blossoms.	**Fuller rose beetle,** *Naupactus godmani.* Pale brown adult snout beetle, about $3/8$ inch long. Larvae feed on roots.	Adults hide during day and feed at night. *See* Weevils, 159.
Brownish, grayish, tan, or white encrustations on bark. Rarely, dead or declining branches.	**Greedy scale, Latania scale.** Individuals circular, flattened, ≤$1/16$ inch long, usually in groups.	*See* 190.

WHAT THE PROBLEM LOOKS LIKE	PROBABLE CAUSE	COMMENTS
Gardenia augusta, Gardenia, Cape jasmine		
Foliage yellows. Plants wilt and die, often suddenly. Roots and stem near soil dark, decayed, girdled by lesions. Leaves may have large dark spots.	**Phytophthora root and crown rot**. Disease favored by wet, poorly drained soil.	*See 122.*
Leaves discolor, wilt, stunt, or drop prematurely. Stems discolored, cankered, and die. May be white fungus beneath basal trunk bark.	**Armillaria root disease**. Fungus present in many soils. Favored by warm, wet soil. Persists for years in infected roots.	*See 120.*
New growth chlorotic, except for green veins. Plants may be stunted.	**Iron deficiency**. Abiotic disorder usually caused by poor soil conditions or unhealthy roots.	*See 47.*
Buds or leaves drop prematurely. Fewer flowers than normal.	**Leaf drop** and **Bud blast**. Noninfectious disorder caused by poor cultural practices and poor growing conditions.	Provide appropriate irrigation and good drainage. Bud drop in spring is caused by poor conditions and practices the previous summer and fall, when buds developed.
Leaves bleached, stippled, may turn brown and drop prematurely. Terminals may be distorted. Plant may have fine webbing.	**Spider mites**, *Tetranychus* spp. Tiny, often green, pink, or red pests; may have two dark spots.	*See 246.*
Sticky honeydew, blackish sooty mold, and whitish cast skins on foliage.	**Aphids,** including **Green peach aphid, Melon aphid**. Groups of small, pear-shaped insects, commonly green.	*See 175.*
Sticky honeydew and blackish sooty mold on foliage. Cottony or waxy material on plant.	**Mealybugs,** including **Citrus mealybug, Longtailed mealybug, Obscure mealybug**. Powdery, grayish insects, $\leq 1/4$ inch long, with waxy filaments.	Vigorous plants tolerate moderate populations. Conserve natural enemies that help in control. *See 185.*
Sticky honeydew and blackish sooty mold on foliage. Leaves may yellow and wither. Tiny, whitish, mothlike adult insects.	**Whiteflies,** including **Bayberry whitefly**, *Parabemisia myricae;* **Citrus whitefly; Greenhouse whitefly.**	Oval, flattened, translucent, yellow to greenish nymphs and pupae. *See 182.*
Sticky honeydew and blackish sooty mold on foliage. Possible decline or dieback of twigs and branches. May be cottony whitish material (egg sacs) on bark.	**Black scale; Brown soft scale; Cottony cushion scale; Green shield scale; Hemispherical scale**, *Saissetia coffeae;* **Wax scales.**	Black, brown, gray, green, orange, yellow, waxy, or whitish, flattened or bulbous insects, on leaves or twigs. *See* Soft Scales, 192.
Plant has sticky honeydew and blackish sooty mold. Plant grows slowly. Foliage may yellow.	**Pyriform scale**, *Protopulvinaria pyriformis.* Triangular insects, $1/8$ inch long, brown, yellow, or mottled red.	Occurs in southern California. *See* Soft 192.
Foliage chewed or notched around margins. May be wilting or dieback of young plants. Some roots stripped of bark or girdled near soil.	**Black vine weevil; Cribrate weevil,** *Otiorhynchus cribricollis;* **Fuller rose beetle,** *Naupactus godmani.*	Adults are black, brown, or gray snout beetles, $\leq 3/8$ inch long. Larvae are white grubs with brown head. *See* 160.
Galls or swellings on roots. Plants may grow slowly.	**Root knot nematodes**. Tiny, root-feeding roundworms.	*See 293.*
Geijera parviflora, Australian willow		
Leaves stunt, yellow, wilt, or drop prematurely. Plants grow slowly and may die. Roots dark, decayed.	**Phytophthora root and crown rot**. Disease promoted by excess soil moisture and poor drainage.	*See 122.*
Sticky honeydew and blackish sooty mold on plant. Bulbous, irregular, brown, gray, or white bodies (scales) on twigs.	**Chinese wax scale**. Bulbous to hemispherical, waxy insects, suck sap.	*See 196.*
Ginkgo biloba, Ginkgo, Maidenhair tree		
Chewed leaves.	**Omnivorous looper**. Yellow, green, or pink larvae, $\leq 1 1/2$ inches long, with green, yellow, or black stripes.	Larvae crawl in "looping" manner. *See 150.*
Galls or swellings on roots.	**Root knot nematodes**. Tiny, root-feeding roundworms.	*See 293.*

Ginkgo biloba (continued)

WHAT THE PROBLEM LOOKS LIKE	PROBABLE CAUSE	COMMENTS
Foliage may yellow and wilt. Bark may have bracketlike or fan-shaped fungal fruiting bodies. Branches or entire plant may die.	**Wood decay**, or **Heart rot**, including *Trametes hirsuta, Trametes versicolor.* Fungi that attack injured, old, or stressed trees.	*See* 111.
Gleditsia triacanthos, Honey locust		
Leaflets turn yellow and drop prematurely.	**Leaf drop**, normal senescence. Leaves often drop in fall sooner than on other species.	Providing trees with a good growing environment and appropriate cultural care may delay leaf drop.
Leaflets turn brown or yellow, especially along margins and at tip. Leaflets may drop prematurely.	**Leaf burn**, or **Scorch**. Abiotic disorders, commonly caused by extreme temperature, inappropriate irrigation, poor drainage, and sunburn.	Provide plants with a good growing environment and proper cultural care. *See* Abiotic Disorders, 43.
Leaflets yellow, wilt, and drop prematurely. Stems may have dark, sunken cankers. May be orangish, red, or yellow fungal fruiting bodies on wood.	**Cankers**, including **Nectria canker**, *Nectria cinnabarina; Phomopsis* sp. Fungi primarily affect injured or stressed trees.	*See* 101.
Leaflets or shoots with white, powdery growth. Leaflets may drop prematurely. Terminals may be distorted or stunted.	**Powdery mildew**, *Microsphaera ravenelii.* A fungal disease favored by moderate temperatures, shade, and poor air circulation.	*See* 90.
Leaflets brown, yellow, stippled, or distorted and may drop prematurely. Branch tips may die back.	**Honeylocust plant bug**, *Diaphnocoris chlorionis.* Greenish or yellow bugs, $\leq 3/16$ inch long.	*See* similar Ash Plant Bugs, 206.
Leaflets bleached or stippled with spots larger than mite stippling. Cast skins on underside of leaflets.	**Leafhoppers**, including **Potato leafhopper**, *Empoasca fabae.* Greenish to whitish, wedge-shaped insects, $\leq 1/8$ inch long.	Plants tolerate moderate stippling. *See* 201.
Leaflets terminate in podlike galls. Foliage browns and drops prematurely.	**Honeylocust pod gall midge**. Adult is tiny, delicate fly. White larvae occur in galls.	*See* 214.
Leaflets discolored, stippled, or bleached. Terminals may distort. Plant may have fine webbing.	**Spider mites**. Tiny greenish, reddish, or yellowish mites; may have two dark spots.	*See* 246.
Chewed leaflets. Silken tents in tree.	**Mimosa webworm**. Larvae gray to brown with white stripes, $\leq 1/2$ inch long.	*See* Webworms, 152.
Sticky honeydew, blackish sooty mold, and whitish cast skins on plant. Foliage may yellow.	**Aphids,** including **Green peach aphid; Leaf curl plum aphid**, *Brachycaudus helichrysi.*	Pear-shaped, often brown, green, reddish, or yellowish insects. *See* 175.
Sticky honeydew and blackish sooty mold on foliage. Tiny, whitish, mothlike adult insects.	**Greenhouse whitefly**. Nymphs are oval, flattened, yellow to greenish insects.	*See* 184.
Sticky honeydew and blackish sooty mold on foliage. Twigs or branches may decline or die back.	**Black scale, European fruit lecanium**. Black, brown, or yellow, flattened or bulbous insects.	*See* 192, 195.
Roughened twig bark. Possible twig dieback.	**Treehoppers**, including **Buffalo treehopper**, *Stictocephala bisonia.*	Bright yellow to green insects, horny or spiny. *See* 204.
Dieback of branches. Gnarled, ridged bark with wet spots. Oval to D-shaped emergence holes in wood.	**Flatheaded borers**, *Agrilus* spp.; **Longhorned beetles**, including **Locust borer**, *Megacyllene robiniae.*	Whitish larvae, ≤ 1 inch long, beneath bark. *See* 227, 232.
Foliage may yellow and wilt. Bark or wood may have bracketlike or fan-shaped fungal fruiting bodies. Limbs or entire plant may die.	**Wood decay**, or **Heart rot**, including *Ganoderma lucidum, Laetiporus gilbertsonii* (=*L. sulphureus*).	Fungi that attack injured, old, or stressed trees. *See* 66, 111.
Grevillea spp., Grevillea, Silk oak		
Sticky honeydew and blackish sooty mold on plant. Elongated, whitish material (egg sacs) on bark.	**Cottony cushion scale**. Females brown, orange, red, or yellow, with elongated, white, fluted egg sacs when mature.	Natural enemies usually provide good control. *See* 130, 196.
Sticky honeydew, blackish sooty mold, and cottony or waxy material on plant.	**Mealybugs**. Powdery, grayish insects, $\leq 1/4$ inch long, may have waxy filaments.	*See* 185.

WHAT THE PROBLEM LOOKS LIKE	PROBABLE CAUSE	COMMENTS
Grayish encrustations on bark. Rarely, dead or declining branches.	**Scales**, including **Latania scale**. Individuals circular, $\leq^1/_{16}$ inch long, usually in groups	*See* 190.
Gymnocladus dioica, Coffee tree, Kentucky coffee tree		
Tan or gray encrustations on bark. May be dead or declining twigs and branches.	**Walnut scale**. Individuals $\leq^3/_{16}$ inch diameter, often in colonies.	*See* similar San Jose Scale, 191.
Hebe, Veronica spp., Hebe		
Foliage discolors, wilts, stunts, or may drop prematurely. Discolored bark or cankers may ooze sap. Branches or plant may die.	**Phytophthora root and crown rot**. Disease favored by excess soil moisture and poor drainage.	*See* 122.
Foliage fades, yellows, browns, or wilts, often scattered throughout canopy. Branches die. Entire plant may die.	**Verticillium wilt**. A soil-dwelling fungus that infects through roots.	*See* 117.
Yellowing and death of foliage, older foliage affected first. Browning of vascular tissue.	**Fusarium wilt**, *Fusarium oxysporum*. Fungus persists in soil and infects plant through roots.	Avoid injuring live tissue. Provide proper drainage, irrigation, and fertilization. Avoid planting hebe where *Fusarium* was previously a problem. *See* 115.
Leaves with reddish, brownish, or yellowish spots or blotches.	**Septoria leaf spot**. A fungal disease favored by wet foliage.	Avoid overhead irrigation. *See* 88.
Twigs distorted, swollen, and pitted. Leaves dwarfed. Shoots may die back.	**Pit-making pittosporum scale**, *Planchonia* (=*Asterolecanium*) *arabidis*. Brown to white insects, $\leq^1/_8$ inch long, on twigs, often in pits.	In California, an occasional problem in the north. Management not investigated. *See* similar Oak Pit Scales, 198.
Hedera spp., Ivy, Algerian ivy, English ivy[16]		
Leaves have black, brown, tan, or yellow blotches. Spots usually angular and vein-limited where veins are large, round where veins are absent. Spots may have yellowish halo. Leaves may drop prematurely.	**Bacterial leaf spots and blight**, *Pseudomonas cichorii, Xanthomonas campestris* pv. *hederae*. Bacteria spread by water. Favored by cool, wet conditions. Persist in plant debris.	Avoid overhead irrigation. Improve air circulation. Plant resistant ivy cultivars. *See* Leaf Spots, 86.
Leaves have brown to yellow spots or blotches. Spots may have dark or yellowish margins. Foliage may shrivel and drop prematurely.	**Anthracnose, Blights**, and **Leaf spots**, *Alternaria, Colletotrichum, Phyllosticta* spp. Fungi spread by air or splashing water. Favored by wet conditions.	Spots often less angular and have less of a chlorotic halo than with bacterial-caused spots. *See* 81, 83, 86.
Discolored spots between veins. Large spots chlorotic, then tan or brown.	**Sunburn**. Abiotic disorder commonly caused by high temperatures, inadequate irrigation, poor soil conditions, or unhealthy roots.	*See* 57.
Foliage yellows. Plants wilt and die, often suddenly. Roots and stem near soil dark, decayed, girdled by lesions. Leaves have large dark spots.	**Root and crown rots**, *Pythium, Phytophthora* spp. Pathogens survive in soil, favored by wet, poorly drained soil.	*See* 122, 123.
Foliage discolored, stippled, or bleached and may drop prematurely. Terminals may distort. Plant may have fine webbing.	**Spider mites,** including **Privet mite, Twospotted spider mite**. Tiny greenish, yellowish, or red arthropods; may have two dark spots.	*See* 246.
Plant has sticky honeydew and blackish sooty mold. Plant grows slowly. Foliage may yellow.	**Pyriform scale**, *Protopulvinaria pyriformis*. Triangular, $^1/_8$ inch long, brown, yellow, or mottled red.	Occurs in southern California. *See* 192.
Foliage yellows. Rarely, plant may decline or die back. Bark or leaves have gray, brown, tan, white, or yellow encrustations.	**Dictyospermum scale**, *Chrysomphalus dictyospermi;* **Greedy scale; Latania scale; Oleander scale; Yellow scale**. Circular, flattened insects, $<^1/_{16}$ inch long.	*See* Armored Scales, 189.
Plant has sticky honeydew and blackish sooty mold. Foliage may yellow.	**Nigra scale**, *Parasaissetia nigra;* **Brown soft scale**. Oval, flat or convex, black, brown, orange, or yellow insects, $\leq^3/_8$ inch long.	*See* 193.

[16] Some species are invasive weeds. Other species may be better choices when planting.

Hedera spp. (continued)

WHAT THE PROBLEM LOOKS LIKE	PROBABLE CAUSE	COMMENTS
Plant has sticky honeydew, blackish sooty mold, and whitish cast skins. Foliage may yellow.	**Aphids**, including **Bean aphid; Green peach aphid; Ivy aphid**, *Aphis hederae*. Small pear-shaped insects, often green, yellowish, or blackish.	*See* 175.
Plant has sticky honeydew, blackish sooty mold, and cottony material (egg sacs). Foliage may yellow.	**Grape mealybug**, *Pseudococcus maritimus*. Oval, soft, powdery, waxy insects, ≤$1/_8$ inch long.	*See* Mealybugs, 185.
Foliage has sticky honeydew and blackish sooty mold. Leaves may yellow and wither. Tiny, whitish, mothlike adult insects.	**Whiteflies**, including **Citrus whitefly**. Oval, flattened, translucent, yellow to greenish nymphs and pupae.	*See* 182.
Leaves chewed. Foliage may be rolled or tied together with silk.	**Omnivorous looper**. Tan to brownish adult moths and yellow, green, or pink larvae, ≤$1^1/_2$ inches long, with green, yellow, or black stripes.	*See* 150.
Foliage or shoots chewed, ragged, or clipped. May be slimy or silvery trails on or around plants.	**Snails**. Mollusks move slowly on slimy, muscular foot and have a spiraled shell.	*See* 251.
Leaves chewed. Plant can be extensively defoliated.	**Indian walking stick**. Brown, twiglike insects up to 4 inches long, easily overlooked.	In California in at least the Central and South Coast. *See* 166.

Heteromeles arbutifolia, Toyon, Christmas berry

Sudden wilting, then shriveling and blackening of shoots and blossoms. Plants appear scorched.	**Fire blight**. Bacteria enter plants through blossoms.	*See* 84.
Leaves with brown, light green, yellow, or dead spots or blotches. Twigs and stems may canker, ooze, or die back.	**Ramorum blight**, *Phytophthora ramorum*. Pathogen spreads via airborne spores and contaminated plants and soil.	Primarily a problem in wildlands, killing many oaks there. *See* Sudden Oak Death and Ramorum Blight, 105.
Foliage discolored, wilted, stunted, drops prematurely. Discolored bark may ooze sap. Branches or plant may die.	**Phytophthora root and crown rot**, *P. cactorum*. Pathogen favored by excess irrigation and poor drainage.	*See* 122.
Tiny, reddish to brown leaf spots, may have yellow halos. Larger, dark areas on leaves. Leaves may drop prematurely.	**Entomosporium leaf spot**. A fungal disease promoted by wet foliage.	*See* 87.
Dark scabby or velvety spots on leaves or fruit.	**Scab**, *Spilocaea photinicola*. Fungal disease promoted by moist spring.	*See* 88.
Foliage may discolor and wilt. Dieback of branches or entire plant.	**Pacific flatheaded borer**. Whitish larvae with enlarged head in tunnels.	*See* 227.
Terminal leaves severely curled and twisted. Damage occurs early in season.	**Toyon thrips**, *Rhyncothrips ilex*. Tiny, slender, black (adult) and orangish (immature) insects in new terminals.	Insect has one annual generation. Tolerate damage. Keeping soil bare beneath plants may reduce damage. *See* 209.
Stippled, bleached leaves with varnishlike specks on undersides.	**Greenhouse thrips**. Tiny, slender, black adults or yellowish nymphs.	*See* 210.
Stippled, bleached leaves with varnishlike specks on undersides.	**Lace bugs**, *Corythucha* spp. Adults ≤$1/_8$ inch long, wings lacy. Nymphs spiny.	*See* 207.
Sticky honeydew and blackish sooty mold on leaves and twigs. Tiny, white, mothlike insects (adults) present.	**Ash whitefly; Crown whitefly; Iridescent whitefly**, *Aleuroparadoxus iridescens*. Tiny, oval, flattened nymphs, often white wax on fringe or back.	Conserve natural enemies, no other control generally recommended, plants tolerate. *See* Whiteflies, 182.
Sticky honeydew and blackish sooty mold on foliage. Possible plant decline or dieback.	**European fruit lecanium**. Brown to yellow, flat or bulbous, immobile scale insects on leaves and twigs.	*See* Lecanium Scales, 195.
Chewed leaves. May be silken tents or mats of silk in plant.	**Western tent caterpillar, Western tussock moth**. Hairy caterpillars, ≤2 inches long, dark or colorful.	*See* 151–152.

WHAT THE PROBLEM LOOKS LIKE	PROBABLE CAUSE	COMMENTS
Chewed leaves.	**Fuller rose beetle**, *Naupactus godmani*. Pale brown adult snout beetle, about $^3/_8$ inch long.	Adults hide during day and feed at night. Larvae feed on roots. *See* Weevils, 159.

Hibiscus spp., Hibiscus

WHAT THE PROBLEM LOOKS LIKE	PROBABLE CAUSE	COMMENTS
Leaves yellow. Plant grows slowly. Plants wilt and die. Roots and stem near soil dark and decayed.	**Root and crown rots**, *Pythium* spp., *Phytophthora* spp. Pathogens persist in soil. Favored by wet, poorly drained soil.	*See* 122, 123.
Leaves discolor, wilt, stunt, or drop prematurely. Stems discolored, cankered, and die. May be white fungus beneath basal trunk bark.	**Armillaria root disease**. Fungus present in many soils. Favored by warm, wet soil. Persists for years in infected roots.	*See* 120.
Leaves have brown to yellow spots or blotches, mostly on older foliage. Spots may have dark or yellowish margins. Foliage may shrivel, drop.	**Leaf spots**, including *Cercospora* spp. Spread by air or splashing water. Favored by prolonged wet conditions. Uncommon on hibiscus in western U.S.	Avoid overhead irrigation. Do not overwater or crowd plants. Use good sanitation. *See* 86.
Leaves have dark, angular spots or blotches, may have reddish margin or yellow border. Foliage may yellow and drop.	**Bacteria leaf spots**, *Pseudomonas* spp., *Xanthomonas* spp. Spread by air and splashing water.	Avoid overhead irrigation. Do not overwater or crowd plants. Use good sanitation. *See* Leaf Spots, 86.
Leaves with yellow to brownish blotches, rings, spots, or vein-banding.	**Hibiscus chlorotic ringspot virus**. Mechanically spread pathogen, not insect vectored. Often introduced with new plant.	Keep plants vigorous by providing proper cultural care. No other treatment. *See* Mosaic and Mottle Viruses, 92.
Soft, brown decay and/or gray-woolly growth on buds, flowers, leaves, and dead tissue. Lower leaves, growing points may decay.	**Botrytis blight**, or **Gray mold**. Fungus favored by mild, wet conditions. Spreads by airborne spores.	*See* 78.
Chewed leaves or blossoms.	**Fuller rose beetle**, *Naupactus godmani*. Pale brown adult snout beetle, about $^3/_8$ inch long.	Adults hide during day and feed at night. Larvae feed on roots. *See* Weevils, 159.
Leaves chewed. Plant can be extensively defoliated.	**Indian walking stick**. Brown, twiglike insects up to 4 inches long, easily overlooked.	In California in at least the Central and South Coast. *See* 166.
Foliage has sticky honeydew and blackish sooty mold. May be copious, white, waxy material on plant.	**Whiteflies**, including **Bandedwinged whitefly**, *Trialeurodes abutilonea;* **Giant whitefly; Greenhouse whitefly; Nesting whitefly**, *Paraleyrodes minei*.	Nymphs are oval and flattened. Adults are tiny, whitish, mothlike insects. *See* 182.
Sticky honeydew and blackish sooty mold on foliage. May be cottony material (egg sacs) on plant.	**Black scale, Brown soft scale, Green shield scale**. Brownish, black, green, orange, red, or yellowish, convex or flattened insects on bark or leaves.	*See* 192, 193, 194.
Sticky honeydew and blackish sooty mold on foliage. Waxy, cottony material on plant.	**Mealybugs** including **Longtailed mealybug**. Powdery, grayish insects, $\leq^1/_8$ inch long, may have waxy filaments.	*See* 185.
Sticky honeydew, blackish sooty mold, and whitish cast skins on foliage.	**Aphids**, including **Melon aphid**. Small greenish, blackish, or yellowish insects in groups.	*See* 175.
Copious, misty, nonsticky liquid raining from plant. Surfaces covered with whitish residue.	**Glassy-winged sharpshooter**. Active, dark brown or gray leafhoppers, $\leq^1/_2$ inch long, suck xylem fluid.	Vectors *Xylella* pathogens. Report suspected glassy-winged sharpshooters to agricultural officials if found in areas where this pest is not known to occur. *See* 203.
Galls or swellings on roots.	**Root knot nematodes**. Tiny root-feeding roundworms.	*See* 293.

Hydrangea spp., Hydrangea

WHAT THE PROBLEM LOOKS LIKE	PROBABLE CAUSE	COMMENTS
Plants stunted. Root system reduced, small roots rotted.	**Pythium root rot**, *Pythium* spp. Pathogens spread by spores in soil and water. Favored by excess soil moisture and poor drainage.	*See* 123.

Hydrangea spp. (continued)

WHAT THE PROBLEM LOOKS LIKE	PROBABLE CAUSE	COMMENTS
Leaves discolor, wilt, stunt, or drop prematurely. Stems discolored, cankered, and die. May be white fungus beneath basal trunk bark.	**Armillaria root disease.** Fungus present in many soils. Favored by warm, wet soil. Persists for years in infected roots.	*See* 120.
Brown to orangish, powdery pustules and yellowish spots on leaves. Leaves may yellow and drop prematurely.	**Rust,** *Pucciniastrum hydrangeae.* Fungus survives on living tissue. Spores are airborne. Favored by high humidity and splashing water.	*See* 94.
Soft, brown decay and/or gray, woolly growth on buds, flowers, leaves, and dead tissue.	**Botrytis blight,** or **Gray mold.** Fungus develops in plant debris or inactive tissue. Favored by wet conditions and moderate temperatures.	*See* 78.
White, powdery patches on leaves and stems. Brown patches may be on upper surface of leaves. Basal leaves may yellow, then brown, and die. Flowers may be deformed or spotted.	**Powdery mildew,** *Erysiphe polygoni.* Fungus survives on living plants. Spores spread by splashing water. Favored by moderate temperatures, shade, and crowding.	*See* 90.
Foliage may yellow or wilt. Roots may be darker or fewer than normal. Galls or swellings may be on roots. Plants may grow slowly.	**Root lesion nematode, Root knot nematodes.** Tiny root-feeding roundworms.	*See* 293.
New growth chlorotic, except for green veins. Plants may be stunted.	**Iron deficiency.** Abiotic disorder usually caused by poor soil conditions or unhealthy roots.	*See* 47.
Sticky honeydew, blackish sooty mold, and whitish cast skins on plant. Foliage may yellow.	**Aphids,** including *Aulacorthum circumflexum;* **Green peach aphid, Melon aphid.** Pear-shaped, often green or yellowish insects.	*See* 175.
Sticky honeydew and blackish sooty mold on plant. Copious, white, waxy material may be present. Leaves may yellow and wither. Tiny, whitish, mothlike adult insects.	**Whiteflies,** including **Giant whitefly, Greenhouse whitefly.** Oval, flattened, translucent, yellow to greenish nymphs and pupae.	*See* 182.
Sticky honeydew and blackish sooty mold on plant. Elongate, slender, cottony material (egg sacs) on plant.	**Cottony hydrangea scale,** *Pulvinaria hydrangeae.* Oval, flattened, yellow or brown insects and cottony eggs.	*See* similar Green Shield Scale, 194.
Dark brown encrustations on bark. Foliage may yellow. May be decline or dieback of plant.	**Oystershell scale,** *Lepidosaphes ulmi.* Elongate to oval insects, ≤$1/16$ inch long.	*See* 191.
Unusual branching. Premature flower bud formation. Distorted flowers. Delayed or no flowers. Dead leaf patches. Leaf tips may wilt.	**Plant bugs,** *Lygus* spp. Brown, green, or yellowish true bugs, ≤$1/4$ inch long, suck plant juices.	*See* Ash Plant Bugs, 206.
Leaves bleached, stippled, may turn brown and drop prematurely. Terminals may be distorted. Plant may have fine webbing.	**Spider mites,** *Tetranychus* spp. Tiny, often green, pink, or red pests; may have two dark spots.	*See* 246.
Foliage chewed and webbed together with silk. Stems may be mined. Larvae wriggle vigorously when touched.	**Caterpillars,** including **Greenhouse leaftier,** *Udea rubigalis. Udea* larvae are ≤$3/4$ inch long, yellowish green with three longitudinal green to white stripes.	*Udea* has about six generations per year. *See* Caterpillars, 144.

Hymenosporum flavum, Sweet shade

No serious invertebrate or pathogen pests have been reported in California. If plants are unhealthy, investigate whether they have been injured or lack good growing conditions or appropriate cultural care.

Hypericum spp., Hypericum, Gold flower, St. Johnswort[17]

Leaves wilted, discolored, may drop prematurely. Branches or entire plant may die. May be white fungal growth or dark crust on basal trunk, roots, or soil.	**Dematophora (Rosellinia) root rot.** Fungus favored by mild, wet conditions. Infects primarily through roots growing near infested plants.	*See* 121.

[17] Some species are invasive weeds. Other species may be better choices when planting.

WHAT THE PROBLEM LOOKS LIKE	PROBABLE CAUSE	COMMENTS
Leaves with orangish powder or pustules. Leaves spotted, discolored, and may drop prematurely.	**Rusts**, including *Melampsora hypericorum, Uromyces triquetrus*. Fungal diseases favored by wet foliage.	*Hypericum calycinum* is especially susceptible. Avoid overhead watering. Vigorous plants tolerate moderate rust infections. *See 94.*
Chewed leaves. Entire plant may be defoliated.	**Klamathweed beetle**. Adults metallic, oval, bluish, about $1/4$ inch long.	*See 158.*
Leaves stippled, bleached, with varnishlike specks on undersides.	**Greenhouse thrips**. Tiny, slender, black adults or yellowish nymphs.	*See 210.*
Ilex spp., Holly[18]		
Leaves discolored, wilted, stunted, may drop prematurely. Discolored bark may ooze sap. Branches or plant may die.	**Phytophthora root and crown rot**. Disease favored by excess soil moisture and poor drainage.	*See 122.*
Leaves wilted, discolored, may drop prematurely. Branches or entire plant may die. May be white fungal growth or dark crust on basal trunk, roots, or soil.	**Dematophora (Rosellinia) root rot**. Fungus favored by mild, wet conditions. Infects primarily through roots growing near infested plants.	Less common than *Phytophthora. See* 121.
Foliage yellows and wilts. Bark or wood may have bracketlike or fan-shaped fungal fruiting bodies. Branches or entire plant may die.	**Wood decay**, or **Heart rot**, including *Pleurotus ostreatus, Trametes hirsuta, Trametes versicolor*.	Fungi that attack injured, old, or stressed trees. *See 111.*
Sticky honeydew and blackish sooty mold on foliage. Possible decline or dieback of twigs and branches.	**Black scale, Brown soft scale**. Yellow, orange, brown, or black, flattened or bulbous insects, often in groups.	*See 192, 193.*
Sticky honeydew and blackish sooty mold on plant. Bulbous, irregular, brown, gray, or white bodies (scales) on twigs.	**Wax scales**. Bulbous to hemispherical, waxy insects, suck sap.	*See 196.*
Leaves bleached, stippled, may turn brown and drop prematurely. Terminals may be distorted. Plant may have fine webbing.	**Mites**, including **Southern red mite**. Tiny arthropods, often green, pink, or red.	*See Red Mites, 246.*
Brownish, grayish, or tan encrustations on twigs, branches, or leaves. May be dead or declining twigs or branches.	**Greedy scale, Oleander scale, Oystershell scale.** Tiny, circular to oval insects, often in colonies.	*See 189–192.*
Slender winding or blotched mines. Pinpricklike scars in leaves. Mines occur in American holly, puncture scars on American and Japanese holly.	**Native holly leafminer**, *Phytomyza ilicicola*. Adults are tiny, black flies, active about April to June. Flattened, pale larvae are in mines in leaves.	Ignore or prune out damage. Plants tolerate abundant mines. Can apply certain insecticides if not tolerable. *See* Foliage Miners, 216.
Galls or swellings on trunk and roots, usually near soil.	**Crown gall**. Bacterium that persists in soil and infects plant via wounds.	*See 108.*
Jacaranda spp., Jacaranda		
Foliage browns or yellows. Branches die back. Entire plant may die. May be white fungus beneath basal trunk bark.	**Armillaria root disease**. Fungus present in many soils. Favored by warm, wet soil. Persists for years in infected roots.	*See 120.*
Leaves with blackish sooty mold, sticky honeydew, and whitish cast skins.	**Aphids**, including **Bean aphid**. Small, green or black insects in groups.	*See 175.*
Juglans spp., Walnut, Black walnut, California black walnut[19]		
Leaves discolor, wilt, stunt, may drop prematurely. Discolored bark may ooze sap. Branches or plant may die.	**Phytophthora root and crown rot**. Disease favored by excess soil moisture and poor drainage.	*See 122.*
Leaves yellow and drop prematurely. Limbs die back and entire tree dies. Bark with dark staining, ooze, and BB shot–sized holes, but these symptoms are often difficult to observe.	**Walnut thousand cankers disease**. Fungal pathogen spread by walnut twig beetle, a tiny bark beetle.	Commonly kills black walnuts. *See* 108.
Leaves discolor, wilt, stunt, may drop prematurely. Branches die back. Entire tree may die. May be white fungus beneath basal trunk bark.	**Armillaria root disease**. Fungus present in many soils. Favored by warm, wet soil. Persists for years in infected roots.	*See 120.*

[18] Some species are invasive weeds. Other species may be better choices when planting.

[19] For pests of walnut nuts, see *Pests of the Garden and Small Farm* and ipm.ucanr.edu.

Juglans spp. (continued)

WHAT THE PROBLEM LOOKS LIKE	PROBABLE CAUSE	COMMENTS
Foliage yellows and wilts. Bark or wood may have bracketlike or fan-shaped fungal fruiting bodies. Branches or entire plant may die.	**Wood decay**, or **Heart rot**, including *Laetiporus gilbertsonii* (=*L. sulphureus*), *Pleurotus ostreatus*, *Schizophyllum commune*, *Trametes* spp.	Fungi that attack injured, old, or stressed trees. *See* 111.
Leaves discolor and wilt. Branches die. Bark may have cankers or brownish ooze.	**Cankers, Twig dieback**, including *Botryosphaeria dothidea*, *Diplodia juglandis*, *Nectria cinnabarina*. Fungi primarily affect injured or stressed trees.	*See* Botryosphaeria Canker and Dieback, 98; Nectria Canker, 101.
Leaves with irregular, brown, tan, yellow, or white blotches or spots. Leaves may die and drop prematurely.	**Anthracnose, Blight**, and **Leaf spots**, including *Cylindrosporium juglandis*, *Marssonia californica*, *M. juglandis*.	Fungi favored by wet conditions and persist in infected leaves and twigs. *See* 81, 83, 86.
Leaves may yellow or wilt. Roots may be darker or fewer than normal. Galls or swellings may be on roots. Plants may grow slowly.	**Ring nematode, Root lesion nematode, Root knot nematodes**. Tiny root-feeding roundworms.	*See* 293, 294.
Leaves discolor or wilt. Dieback of woody plant parts. Tunnels and larvae in wood.	**Pacific flatheaded borer**. Whitish larvae, ≤1 inch long, with enlarged head.	*See* 227.
Leaves discolor or wilt. Brown to gray encrustations on bark. May be declining or dead twigs or branches.	**Italian pear scale**, *Epidiaspis leperii*; **Walnut scale**, *Diaspidiotus* (=*Quadraspidiotus*) *juglansregiae*; **Oystershell scale; San Jose scale**. Tiny, circular to elongate insects.	Italian pear scale survives only under shelter such as lichens and moss on bark. The female's body is reddish purple beneath its cover. *See* Armored Scales, 189.
Sticky honeydew, blackish sooty mold, and whitish cast skins on foliage.	**Walnut aphid**, *Chromaphis juglandicola*; **Duskyveined aphid**, *Callaphis juglandis*. Tiny, yellowish to brown insects.	Conserve natural enemies that provide control. Plants tolerate aphids, control is rarely needed. *See* Aphids, 175.
Sticky honeydew and blackish sooty mold on foliage. Woody parts may decline and die back.	**Black scale, Calico scale, Citricola scale, European fruit lecanium, Frosted scale, Kuno scale.**	Oval, flat or bulbous, brown, yellow, spotted, or whitish, waxy insects. *See* 192–195.
Sticky honeydew and blackish sooty mold on foliage. Cottony or waxy material on plant.	**Obscure mealybug**. Powdery, grayish insects, ≤1/4 inch long, with waxy filaments.	Conserve natural enemies that help in control. *See* 187.
Leaves stippled, bleached, or reddened.	**European red mite, Pacific spider mite, Twospotted spider mite.**	Tiny greenish, reddish, or yellowish specks. *See* 246.
Leaves with swollen, globular growths on upper surface and small pits on underside. Leaves may curl or distort.	**Walnut purse gall mite**, or **Walnut pouch gall mite**, *Eriophyes brachytarsus* =*Aceria brachytarsus*. Tiny, elongate eriophyid mites.	Occurs on *J. californica* and *J. hindsii*. Harmless to walnut. No control recommended. *See* Gall Mites, or Eriophyids, 249.
Chewed leaves. Silken tents may occur on terminals. Single branch or entire tree may be defoliated.	**Fall webworm, Redhumped caterpillar, Tussock moths**. Smooth to hairy caterpillars, ≤1 1/2 inches long.	*See* 150–153.
Galls or swellings on trunk and roots, usually near soil, may be on branches.	**Crown gall**. Bacteria infect plant via wounds.	*See* 108.
Juniperus spp., Juniper[20]		
Foliage yellows and wilts. Branches or entire plant dies. Roots may be dark or decayed.	**Phytophthora root and crown rot.** Disease promoted by excess soil moisture and poor drainage.	*See* 122.
Foliage brown, reddish, yellow, or wilted. Shoots die back. Shoots may be bushy, stunted, witches' brooms.	**Twig blights**, including *Kabatina juniperi*, *Phomopsis juniperovora*. Fungal pathogens that infect wet foliage through wounds.	Avoid wounding twigs, except to prune out and dispose of infected shoots. Keep foliage dry. Use drip irrigation and provide good air circulation. Avoid fertilization.
Foliage browns or yellows. Branches die back. Entire plant may die. May be white fungus beneath basal trunk bark.	**Armillaria root disease**. Fungus present in many soils. Favored by warm, wet soil. Persists for years in infected roots.	*See* 120.
Brown or yellow dying foliage on branches. Cankers and resinous exudate on limbs or trunk.	**Cypress canker**. A fungal disease.	Not common on juniper. Primarily infects cypress. *See* 99.

[20] For more information, see *Pests of the Native California Conifers*.

WHAT THE PROBLEM LOOKS LIKE	PROBABLE CAUSE	COMMENTS
Orangish, gelatinous masses or galls on bark. Stem dieback. Shoots may be discolored, bushy, or stunted.	**Juniper rusts**, *Gymnosporangium* spp. Fungi favored by wet conditions, alternate on deciduous hosts.	*See* Cedar, Cypress, and Juniper Rusts, 95.
Leaves yellow and wilt. Branches or entire plant may die. Trees may fall over.	**Wood decay**, or **Heart rot**, including *Schizophyllum commune, Stereum* spp., *Trametes* spp.	Fungi that infect injured, old, or stressed trees. *See* 111.
Browning of shoot tips beginning in fall, worst late winter to spring.	**Cypress tip miners**. Adults are small, silvery tan moths. Green larvae, ≤$^1/_8$ inch long, tunnel in foliage.	*See* 217.
Browning of shoot tips. Plant appears brown most of the year.	**Juniper needle miner**, *Stenolechia bathrodyas*. Green larvae, ≤$^1/_8$ inch long, mine leaflets. Adults are silvery moths, about $^1/_4$ inch long. In California, a problem along the South Coast.	Plants tolerate extensive needle mining. If not tolerable, adults may be sprayed during three annual flights of moths. Shake foliage regularly March–October, spray only if moths abundant. *See* similar Cypress Tip Miners, 217.
Browning and webbing of needles, initially in patches and then overall if severe.	**Juniper webworm**, *Dichomeris marginella*. Brown, reddish, purplish, or yellowish larvae mine needles when young and feed externally in webbing when mature. Adults are small, coppery, brown and white moths.	Plants tolerate extensive needle mining. If not tolerable, apply an effective insecticide during spring or fall when larvae or adults are observed. *See* similar Cypress Tip Miners, 217.
Foliage turns brown or yellow. Needles may drop prematurely.	**Leaf burn**, or **Scorch**. Abiotic disorders with many potential causes, including inappropriate irrigation, poor drainage, saline water, unfavorable soil pH.	Older, inner needles are unavoidably shed naturally. Provide plants with a good growing environment and proper cultural care. *See* Abiotic Disorders, 43.
Pale greenish or yellow needles. Foliage stippled.	**Spider mites**, including **Spruce spider mite**. Tiny arthropods, often greenish, may be in fine webbing at foliage base.	*See* Pine and Spruce Spider Mites, 247.
Yellow or brown foliage. Whitish to brownish encrustations on foliage.	**Juniper scale**, *Carulaspis juniperi;* **Minute cypress scale**. Circular to elongate insects, ≤$^1/_{16}$ inch long.	Populations in California rarely warrant control, except on Italian cypress or in nurseries. *See* 190.
Yellow or brown foliage on scattered dying or dead branches. Entire plant never dead.	**Juniper twig girdler**. Off-white larva with brown head, ≤$^3/_8$ inch long, occurs in tunnel beneath twig bark.	*See* 242.
Yellow or brown foliage. Dieback of branches.	**Flatheaded borer**, *Chrysobothris* sp. Whitish larva, ≤1$^1/_4$ inch long, with enlarged head, tunnels in wood.	In California affects mostly *Juniperus chinensis* 'Kaizuka' or 'Torulosa' in San Joaquin Valley. Keep plants vigorous. *See* 227.
Tufts of cottony material protruding from bark. Foliage may redden, yellow, and eventually die. Treetop or entire tree may die.	**Cypress bark mealybug; Incense-cedar scale**, or **Monterey cypress scale**, *Xylococcus macrocarpae*. Reddish insects, ≤$^1/_{16}$ inch long, under cottony wax.	*See* 186, 198.
Brown to purplish insects clustered on foliage. May be sticky honeydew or blackish sooty mold on foliage.	**Giant conifer aphids**. Dark, long-legged, ≤$^1/_5$ inch long.	Harmless to trees. *See* 176.
Chewed needles.	**Conifer sawflies, Cypress sawfly**. Green larvae, ≤1 inch long, on needles.	*See* 153, 154.
Galls or swellings on roots.	**Root knot nematodes**. Tiny root-feeding roundworms.	*See* 293.

Kalmia spp., Kalmia, Mountain laurel, Western laurel

Leaves with black to dark spots. Leaves may drop prematurely.	**Leaf spot, Scab**, including *Gibbera kalmiae, Phyllosticta* sp. Fungal diseases spread by water from infected leaves and twigs.	*See* 86, 88.
Leaves with brown, dead blotches, commonly around midvein and leaf tip.	**Ramorum blight**, *Phytophthora ramorum*. Pathogen spreads via airborne spores and contaminated plants and soil.	Primarily a problem in wildlands, killing many oaks there. *See* Sudden Oak Death and Ramorum Blight, 105.

WHAT THE PROBLEM LOOKS LIKE	PROBABLE CAUSE	COMMENTS
Koelreuteria spp., Golden rain tree, Chinese flame tree		
Trunk or limbs with roughened, wet, or oozing area. Cracked bark and dieback.	**Flatheaded borers**. Whitish larvae with enlarged head, tunneling under bark.	*See 227.*
Bark may ooze and have tiny holes. Foliage may yellow and wilts. Branches may die.	**Polyphagous shothole borer**. An ambrosia beetle, or bark beetle.	Occurs at least in southern California. *See 222.*
Laburnum spp., Golden-chain tree		
Foliage yellows and wilts. Limbs may have lesions and dieback.	**Diaporthe stem canker and dieback**, *Diaporthe rudis*. Fungus infects through wounds, persists in infected tissue.	Avoid wounding plants, except to prune off infected limbs. Provide a good growing environment and appropriate cultural care.
Sticky honeydew and blackish sooty mold on foliage. Cottony or waxy material on plant.	**Grape mealybug**, *Pseudococcus maritimus*. Powdery, grayish insects, ≤$1/4$ inch long, with waxy filaments.	Vigorous plants tolerate moderate populations. Conserve natural enemies that help in control. *See Mealybugs, 185.*
Chewed leaves. Plants may be defoliated.	**Genista caterpillar**, *Uresiphita reversalis*. Caterpillars, ≤$1^{1}/4$ inches long, green to orange with black and white hairs.	Introduced to control brooms, which are often considered weeds. Apply *Bacillus thuringiensis* or another insecticide. *See Caterpillars, 144.*
Lagerstroemia hirsuta, L. indica, Crape myrtle		
Foliage may yellow and wilt. Bark or wood may have bracketlike or fan-shaped fungal fruiting bodies. Limbs or entire plant may die.	**Wood decay**, or **Heart rot**, including *Trametes versicolor*.	Fungi that attack injured, old, or stressed trees. *See 111.*
Foliage covered with whitish growth. Shoots may be stunted, distorted.	**Powdery mildews**, including *Erysiphe lagerstroemiae*. Fungal diseases favored by moderate temperatures, shade, and poor air circulation.	Plant resistant cultivars. *See 90.*
Sticky honeydew, blackish sooty mold, and whitish cast skins on plant.	**Crapemyrtle aphid**. Yellowish green, pear-shaped insects with black wing markings.	*See 175.*
Copious, misty, nonsticky liquid raining from plant. Surfaces covered with whitish residue.	**Glassy-winged sharpshooter**. Active, dark brown or gray leafhoppers, ≤$1/2$ inch long, suck xylem fluid.	Vectors *Xylella* pathogens. Report suspected glassy-winged sharpshooters to agricultural officials if found in areas where this pest is not known to occur. *See 203.*
Lantana montevidensis, Lantana		
Soft, brown decay and/or gray, woolly growth on buds, flowers, leaves, and dead tissue.	**Botrytis blight**, or **Gray mold**. A fungal disease favored by moist conditions and moderate temperatures.	Avoid overhead watering. Thin canopy to improve air circulation. Provide proper care. Prune out dying tissue, make cuts in healthy stems. *See 78.*
Sticky honeydew and blackish sooty mold on foliage. Tiny, whitish, mothlike adult insects.	**Greenhouse whitefly**. Nymphs oval, flattened, yellowish to translucent.	*See 184.*
Leaves with brown to gray, irregular, roundish botches. Heavily infested leaves may drop prematurely.	**Lantana leafblotch miner**, *Liriomyza* sp. Yellow to orange larvae feed within leaves, then drop and pupate in soil. Adults are small, black, bright yellow flies.	Occurs in south coastal California. No documented control. Insecticides that have translaminar movement may control larvae. *See Foliage Miners, 216.*
Larix spp., Larch		
Foliage discolors and drops.	**Normal dormancy.**	Larch are deciduous conifers that normally drop all needles in the fall.
Leaves yellow and wilt. Bark or wood may have bracketlike or fan-shaped fungal fruiting bodies. Branches or entire plant may die. Trees may fall over.	**Wood decay**, or **Heart rot**, including *Fomitopsis officinalis, Laetiporus gilbertsonii* (=*L. sulphureus*), *Phaeolus schweinitzii*, and *Phellinus weirii*.	Fungi that infect injured, old, or stressed trees. *See 111.*
Powdery white or grayish material on cones, limbs, needles, or the trunk.	**Adelgids**. Aphidlike insects suck sap and alternate hosts with other conifers.	*See 181.*
Brown, reddish, or white gummy or granular material exudes from lower trunk.	**Bark beetles**, including **Red turpentine beetle**. Adults ≤$1/4$ inch long, dark, cylindrical. Whitish maggotlike larvae under bark.	Attack only stressed trees. Provide trees proper care. *See 225.*

WHAT THE PROBLEM LOOKS LIKE	PROBABLE CAUSE	COMMENTS
Chewed needles.	**Conifer sawflies**. Green larvae, ≤1 inch long, on needles.	See 153.
Laurus nobilis, Grecian laurel, Sweet bay		
Stippled or bleached leaves with varnishlike specks on undersides.	**Greenhouse thrips**. Tiny, slender, black adults or yellow nymphs.	See 210.
Leaf margins cupped or rolled inward, forming galls, which turn red, then brown.	**Laurel psyllid**, *Trioza alacris*. Nymphs about ¹/₁₆-inch-long, powdery insects, within galls.	Conserve natural enemies that help control. Vigorous plants tolerate. If not tolerable, well-timed, repeated shearing of terminals may provide some control. *See* Psyllids, 170.
Grayish encrustations on leaves or bark. Rarely, declining or dead twigs.	**Oleander scale**. Tiny, circular to oval insects on twigs, branches, and leaves.	Rarely if ever cause serious damage to plants. *See* 191.
Leptospermum spp., Tea tree		
Foliage yellows and wilts. Branches or entire plant dies. Rootlets decayed or sparse.	**Phytophthora root and crown rot**. Disease favored by wet soil and poor drainage.	See 122.
Branches die back. Entire tree may die. May be white fungus beneath basal trunk bark.	**Armillaria root disease**. Fungus present in many soils. Favored by warm, wet soil. Persists for years in infected roots.	See 120.
Galls or swellings on roots. Plants may grow slowly.	**Root knot nematodes**. Tiny root-feeding roundworms.	See 293.
Ligustrum spp., Privet[21]		
Leaves yellow, wilt, and may drop prematurely. Branches die back. Entire tree may die. May be white fungus beneath basal trunk bark.	**Armillaria root disease**. Fungus present in many soils. Favored by warm, wet soil. Persists for years in infected roots.	See 120.
Leaves wilt, discolor, may drop prematurely. Branches or entire plant may die. May be white or dark fungal growth on basal trunk, roots, or soil.	**Dematophora (Rosellinia) root rot**. Fungus favored by mild, wet conditions. Infects primarily through roots growing near infested plants.	Less common than *Armillaria*. *See* 121.
Leaves discolor, stunt, wilt, and remain dead on plant or drop prematurely. Plants grow slowly and may die. Root crown dark, decayed.	**Phytophthora root and crown rot**. Disease favored by excess soil moisture and poor drainage.	See 122.
Leaves may yellow and wilt. Plants grow slowly or decline. Limbs or entire tree may die.	**Tenlined June beetle**, *Polyphylla decemlineata*. Cream-colored grubs, ≤2 inches long, chew roots.	More of a problem in sandy soils. *See* White Grubs and Scarab Beetles, 164.
Leaves fade and wilt, often scattered throughout canopy. Branches die. Boring dust, ooze, or holes on trunk or branches.	**Ash borer**, or **Lilac borer**. White larvae, ≤1 inch long, bore in wood. Adults are wasplike moths.	See 235.
Leaves or shoots stunted, brown. Buds distorted, galled, dead.	**Privet rust mite**, *Aceria ligustri*; **Privet bud mite**, *Vasates ligustri*. Microscopic eriophyids, suck plant juice.	Vigorous plants tolerate mite feeding. Eriophyid mites are difficult to control. *See* Gall Mites, or Eriophyids, 249.
Leaves discolored, stippled, or bleached. Foliage may drop prematurely. Terminals may distort. Plant may have fine webbing.	**Spider mites**, including **Privet mite**, *Brevipalpus obovatus*. Greenish or yellowish, tiny arthropods, which may have two dark spots.	See 246.
Sticky honeydew and blackish sooty mold on foliage.	**Black scale**. Orangish, flat immatures or brown to black, bulbous adults, may have raised H shape on back.	See 192.
Brown, gray, or orangish encrustations on bark. Rarely, declining or dead twigs or branches.	**Armored scales**, including **California red scale, San Jose scal**e. Tiny, circular to oval insects on bark.	Rarely if ever cause serious damage to privet. *See* 189–192.
Foliage chewed or notched around margins. May be wilting or dieback of young plants. Some roots stripped of bark or girdled near soil.	**Black vine weevil**. Adults are black or grayish snout beetles, about ³/₈ inch long. Larvae are white grubs with brown head.	See 160.

[21] Some species are invasive weeds. Other species may be better choices when planting.

Ligustrum spp. (continued)

WHAT THE PROBLEM LOOKS LIKE	PROBABLE CAUSE	COMMENTS
Leaves chewed. Plant can be extensively defoliated.	**Indian walking stick.** Brown, twiglike insects up to 4 inches long, easily overlooked.	In California in at least the Central and South Coast. *See 166.*
Twigs distorted, swollen, and pitted. Leaves dwarfed. Shoots may die back.	**Pit-making pittosporum scale,** *Planchonia* (=*Asterolecanium*) *arabidis.* Brown to white insects, ≤$^1/_8$ inch long, on twigs, often in pits.	In California, an occasional problem in the north. Management not investigated. *See* similar Oak Pit Scales, 198.
Liquidambar spp., Liquidambar, Sweet gum		
Limbs drop, usually during or after hot weather.	**Summer limb drop.** Abiotic disorders caused by tree injury or stress, such as drought.	Protect trees from injury. Provide good growing conditions and appropriate cultural care. Have tree inspected by arborist.
Leaves discolor and wilt. Branches or treetop dying. Branches may canker or exude reddish pitch.	**Botryosphaeria canker and dieback.** Fungus primarily affects injured or drought-stressed trees.	Provide proper irrigation. *See 98.*
Leaves turn yellow then brown. Upper canopy dies back. Entire plant dies rapidly or slowly over several years.	**Bacterial leaf scorch,** or **Oleander leaf scorch.** Bacteria spread by certain leafhoppers, especially glassy-winged sharpshooter.	*See 112.*
Dieback of leaves, twigs, and limbs, progressively.	**Dieback.** Unexplained malady resembling dieback from Botryosphaeria canker.	Occurs at least in Marin County. Frequency and severity of symptoms vary greatly year-to-year.
Bark exudes white, frothy material, often around wounds, has pleasant odor.	**Foamy canker.** Unidentified cause, possibly a bacterium.	Foamy material appears for only short time during warm weather. *See 101.*
Bark may have lesions that are dark, dry, crusty whitish, oily looking, or water-soaked. Foliage yellows and wilts. Branches or entire plant dies.	**Fusarium dieback.** Fungal disease spread by the polyphagous shothole borer, an ambrosia beetle, or bark beetle.	Occurs at least in southern California. *See* 115.
Leaves may discolor, wilt, and drop prematurely. Bark may have bracketlike or fan-shaped fungal fruiting bodies. Branches or plant may die. Trees may fall over.	**Wood decay,** or **Heart rot,** including *Ganoderma* spp., *Schizophyllum commune,* *Stereum* spp., *Trametes* spp. Fungi that attack injured, old, or stressed trees.	*See 66, 111.*
Leaves with dark or discolored blotches. Leaves may drop prematurely.	**Leaf spots,** including *Cercospora* spp. Fungi favored by wet conditions.	*See 86.*
Sticky honeydew and blackish sooty mold on foliage. Trees may decline and eventually die.	**Calico scale.** Adults globular, black with white or yellow spots.	Can be a serious pest on liquidambar. *See* 193.
Chewed foliage. Webbing or tents on branch terminals.	**Fall webworm.** Larvae white to yellow, hairy, ≤1 inch long.	*See 152.*
Leaves chewed, may be tied with silk. Plant may be defoliated.	**Fruittree leafroller, Tussock moths.** Larvae green or hairy, ≤1$^1/_2$ inches long.	*See 149–152.*
Chewed foliage. Typically only single branches are defoliated.	**Redhumped caterpillar.** Larvae ≤1 inch long, with red head, body yellowish with reddish and black stripes.	*See 150.*
Foliage may be notched around margins. Some roots stripped of bark or girdled near soil. Young trees may wilt or die.	**Black vine weevil.** Adults are black or grayish snout beetles, about $^3/_8$ inch long. Larvae are white grubs with brown head.	Larvae feed on roots. Adults hide during day and feed at night. *See 160.*
Liriodendron tulipifera, Tulip tree, Tulip poplar, Yellow poplar		
Leaves wilt, discolor, and may drop prematurely. Branches die back. Entire tree may die. May be white fungus beneath basal trunk bark.	**Armillaria root disease.** Fungus present in many soils. Favored by warm, wet soil. Persists for years in infected roots.	*See 120.*
Leaves discolor, stunt, wilt, or drop prematurely, often on one side of plant. Stem xylem discolored. Stems or entire plant may die.	**Verticillium wilt.** Fungus persists in soil, infects through roots.	*See 117.*

WHAT THE PROBLEM LOOKS LIKE	PROBABLE CAUSE	COMMENTS
Leaves yellow and wilt. Branches die back.	**Cylindrocladium root rot**, *Cylindrocladium* spp. Fungi persist in soil. Disease is favored by excess soil moisture and poor drainage.	Usually affect only young trees. Provide good drainage. Reduce irrigation if disease develops. *See* similar Root and Crown Diseases, 119.
Black, raised, tarlike blotches on leaves.	**Tarspot**, *Rhytisma liriodendron*. Fungus most prevalent in moist environments.	Occasional problem that does not threaten tree health. Rake and dispose of fallen leaves. *See* Leaf Spots, 86.
Dark spots on leaves. Leaves may yellow and drop prematurely.	**Physiological leaf spotting, Leaf burn**. Abiotic disorders, commonly caused by drought stress, dry wind, excess heat, low humidity, or poor cultural care.	Provide a good growing environment and appropriate cultural care, especially good drainage and adequate summer irrigation. *See* Abiotic Disorders, 43.
Foliage turns reddish, then yellow, brown, and may drop prematurely. Powdery growth on tissue. Shoots or leaves may be stunted and distorted.	**Powdery mildew**, *Erysiphe polygoni*. Fungal disease favored by moderate temperatures, shade, and poor air circulation.	*See* 90.
Sticky honeydew, blackish sooty mold, and whitish cast skins on leaves. Possible premature leaf yellowing.	**Tuliptree aphid**, *Macrosiphum liriodendri*. Tiny, green insects in colonies on underside of leaves.	Plants tolerate extensive aphid populations. *See* Aphids, 175.
Sticky honeydew and blackish sooty mold on leaves. Twigs and limbs may die back.	**Tuliptree scale**. Females ≤$1/_3$ inch, irregularly hemispherical, and variably colored brown to gray with other-colored blotches.	In California present at least in San Francisco Bay Area. *See* 195.
Galls or swellings on trunk and roots, usually near soil, may be on branches.	**Crown gall**. Bacteria infect plant via wounds.	*See* 108.
Lonicera spp., Honeysuckle, California honeysuckle		
Foliage yellows and wilts. Stems may have decay or lesions and die back.	**Diaporthe stem canker and dieback**, *Diaporthe eres*; **Eutypa dieback**, *Eutypa lata*; **Phoma stem and leaf blight**, *Phoma xylostei*. Fungi infect through wounds, persist in infected tissue. Spread during wet conditions.	Avoid wounding plants, except to prune off infected tissue only during prolonged dry weather, making cuts in healthy tissue. Provide a good growing environment and appropriate cultural care. *See* Canker Diseases, 97.
Foliage discolors and wilts. Branches develop cankers and die back. Dark, hemispherical fruiting bodies develop on bark or wood.	**Annulohypoxylon canker**, *Annulohypoxylon rubiginosum*. A decay fungus that attacks injured or stressed trees.	*See* 98.
Leaves with brown, light green, yellow, or dead spots or blotches.	**Ramorum blight**, *Phytophthora ramorum*. Pathogen spreads via airborne spores and contaminated plants and soil. On honeysuckle, hosts include at least *L. hispidula*.	Primarily a problem in wildlands, killing many oaks there. *See* Sudden Oak Death and Ramorum Blight, 105.
Foliage or shoots turn reddish, then yellow, brown, and drop prematurely. Powdery growth on tissue. Shoots or leaves may be stunted and distorted.	**Powdery mildew**, *Erysiphe polygoni*. Fungus develops on living tissue, favored by moderate temperatures, shade, and poor air circulation.	*See* 90.
Brownish, grayish, tan, or white encrustations on bark or twigs. Rarely, declining or dead twigs or branches.	**Greedy scale**. Circular, flattened, ≤$1/_{16}$ inch long, often in colonies.	Rarely if ever causes serious damage to plants. *See* 190.
Sticky honeydew, blackish sooty mold, and whitish cast skins on plant. Blossoms may be distorted.	**Aphids**, including **Potato aphid**, *Macrosiphum euphorbiae*. Small, pear-shaped insects in groups, often greenish, but may be pink to yellow.	*See* 175.
Plant has sticky honeydew and blackish sooty mold. Plant grows slowly. Foliage may yellow.	**Pyriform scale**, *Protopulvinaria pyriformis*. Triangular, $1/_8$-inch-long, brown, yellow, or mottled red insects.	*See* 192.

Lonicera spp. (continued)

WHAT THE PROBLEM LOOKS LIKE	PROBABLE CAUSE	COMMENTS
Chewed leaves. Plants may be defoliated.	**Genista caterpillar**, *Uresiphita reversalis*. Caterpillars ≤1¼ inches long, green to orange with black and white hairs.	Introduced to control brooms, which are often considered weeds. Apply *Bacillus thuringiensis* or another insecticide. *See* Caterpillars, 144.
Lophostemon (=*Tristania*) *conferta*, Tristania, Brisbane box; *Tristaniopsis* (=*Tristania*) *laurina*, Water gum		
No serious invertebrate or pathogen pests have been reported in California. If plants are unhealthy, investigate whether they have been injured or lack good growing conditions or appropriate cultural care.		
Lyonothamnus floribundus, Ironwood, Catalina ironwood		
Leaves chlorotic and necrotic and may be undersized.	**Alkaline soil**. Symptoms of various mineral deficiencies and toxicities develop when grown in soil with high pH.	*See* pH Problems, 53.
Sticky honeydew and blackish sooty mold on leaves. White, popcornlike bodies on bark.	**Cottony cushion scale**. Orangish, flat immatures or cottony females on bark.	Usually under good biological control. *See* 130, 196.
Magnolia spp., Magnolia, Tulip tree		
Leaves and blossoms with dark blotches. Blossoms, leaves, and terminals turn brown, wilt, and die. Cankers or lesions may occur on twigs.	**Bacterial blast, blight, and canker**. Disease favored by wet conditions.	*See* 83.
Leaves and branches wilt in spring. Stems may have dark cankers, callus tissue, or coral-colored pustules.	**Nectria canker**, *Nectria* spp. Fungi primarily affect injured or stressed trees.	*See* 101.
Brown, black, tan, or yellow spots or blotches on leaves. Spots may have dark or yellowish margins. Foliage may die and drop prematurely.	**Leaf spots**, including *Cladosporium* sp., *Phyllosticta* sp. Fungi spread in water and are favored by prolonged cool, wet, conditions.	*See* 86.
Leaves yellowish, except along veins. Leaves may be small or drop prematurely.	**Iron deficiency**. Abiotic disorder usually caused by poor soil conditions or unhealthy roots.	*See* 47.
Leaves turn brown or yellow, especially along margins and at tip. Leaves may drop prematurely.	**Leaf burn**, or **Scorch**. Abiotic disorders with many potential causes, commonly caused by high temperatures, inappropriate irrigation, or poor drainage.	Provide plants with a good growing environment and proper cultural care. *See* Abiotic Disorders, 43.
White, powdery growth on leaves or shoots. Terminals may be distorted or stunted.	**Powdery mildew**. Fungal disease favored by moderate temperatures, shade, and poor air circulation.	*See* 90.
Sticky honeydew and blackish sooty mold on leaves. White, popcornlike bodies on bark.	**Cottony cushion scale**. Orangish, flat immatures or cottony females on bark.	Usually under good biological control. *See* 130, 196.
Sticky honeydew and blackish sooty mold on leaves.	**Obscure mealybug**. Powdery, gray insects with waxy filaments.	Conserve natural enemies that help in control. *See* 187.
Sticky honeydew and blackish sooty mold on leaves. Twigs and limbs may die back.	**Tuliptree scale**. Females ≤⅓ inch wide, irregularly hemispherical, and variably colored brown to gray with other-colored blotches.	Primarily a pest of *Liriodendron tulipifera* and deciduous magnolias. *See* 195.
Sticky honeydew, blackish sooty mold, whitish cast skins on leaves.	**Aphids**, including **Green peach aphid**. Pink to green insects in colonies on leaves and terminals.	Plants tolerate moderate populations. *See* 175.
Leaves stippled, bleached. Varnishlike specks on underside.	**Greenhouse thrips**. Tiny, slender, black adults or yellowish nymphs.	*See* 210.
Chewed leaves. Foliage may be rolled and tied together with silk.	**Omnivorous looper**. Yellow, green, or pink larvae, ≤1½ inches long, with green, yellow, or black stripes.	Larvae crawl in "looping" manner. *See* 150.
Brownish, grayish, tan, orange, or white bark encrustations. Rarely, dead or dying twigs or branches.	**California red scale, Greedy scale, Oleander scale**. Tiny, circular to oval individuals, often in colonies.	*See* 189–192.

WHAT THE PROBLEM LOOKS LIKE	PROBABLE CAUSE	COMMENTS
Galls or swellings on trunk and roots, usually near soil.	**Crown gall**. Bacterium persists in soil and infects plant via wounds.	*See* 108.
Bark stained brownish, exudes rancid fluid, often around crotches, wounds.	**Wetwood**, or **Slime flux**. Bacterial infections.	Usually do not cause serious harm to trees. *See* 110.

Mahonia, Berberis spp., Mahonia, Barberry, Oregon grape

Leaves stippled, bleached. Varnishlike specks on underside.	**Greenhouse thrips**. Tiny, slender, black adults or yellowish nymphs.	*See* 210.
Leaves chewed. Plants may be defoliated.	**Barberry looper**, *Coryphista meadii*. Green caterpillars, ≤1 inch long.	Vigorous plants tolerate moderate defoliation. Tolerate or apply *Bacillus thuringiensis* or another insecticide. *See* Loopers, 150.
Foliage discolored, reddening with irregular, brown to black dead spots. Orangish pustules or coating on leaves.	**Mahonia rusts**, *Cumminsiella mirabilissima, Puccinia* spp. Fungal diseases favored by wet foliage.	Avoid overhead watering. Vigorous plants tolerate moderate infection. *See* Rusts, 94.
Sticky honeydew and blackish sooty mold on plant. Tiny, mothlike insects (adult insects) on foliage. May be poor plant growth.	**Deer brush whitefly**, *Aleurothrixus interrogationis*. Oval, yellow, or tan nymphs and black pupae that suck sap on undersides of leaves.	Plants tolerate moderate densities. If severe, insecticidal soap or another insecticide may be applied. *See* Whiteflies, 182.
Sticky honeydew and blackish sooty mold on plant. Leaves may yellow and drop prematurely.	**Chinese wax scale**. Bulbous, irregular, brown, gray, or white waxy bodies (scales) on twigs.	*See* 196.
Sticky honeydew, blackish sooty mold, and whitish flocculent material on leaves or stems.	**Mealybug**, *Pseudococcus fragilis*. Oval, powdery, grayish or whitish insects with waxy filaments.	*See* 185.
Brownish, grayish, tan, or white encrustations on bark. Rarely, dead or declining twigs or branches.	**Greedy scale**. Nearly circular insects, <$^1/_{16}$ inch long.	*See* 190.

Malus spp., Crabapple

Foliage yellows and wilts. Branches or entire plant may die. May be white fungus beneath basal trunk bark.	**Armillaria root disease**. Fungus present in many soils. Favored by warm, wet soil. Persists for years in infected roots.	*See* 120.
Leaves discolor, stunt, wilt, and remain dead on tree or drop prematurely. Plants grow slowly and may die. Roots dark, decayed.	**Phytophthora root and crown rot**. Disease promoted by excess soil moisture and poor drainage.	*See* 122.
Foliage may yellow and wilt. Bark or wood may have bracketlike or fan-shaped fungal fruiting bodies. Limbs or entire plant may die.	**Wood decay**, or **Heart rot**, including *Ganoderma lucidum, Trametes versicolor*. Fungi that attack injured, old, or stressed trees.	*See* 66, 111.
Sudden wilting, then shriveling and blackening of shoots, blossoms, and fruit. Plant appears scorched.	**Fire blight**. Bacterium enters plants through blossoms.	Plant less-susceptible cultivars. *See* 84.
Shoots die back. Branches develop reddish brown lesions or cankers.	**Botryosphaeria canker and dieback; European canker**. Fungus persists in cankered wood and infects through leaf scars when plant is wet.	Prune out infected wood. *See* 98; Nectria Canker, 101.
Dead branches, limbs, or twigs. Plant declines, may die. Tiny, BB shot–sized holes in bark, which may ooze.	**Shothole borer**. Small, brown adult bark beetles, whitish larvae, tunnel beneath bark.	*See* 225.
Dieback of woody plant parts. Tunnels and larvae in wood.	**Pacific flatheaded borer**. Whitish larvae, ≤1 inch long, with enlarged head.	*See* 227.
Gray or whitish patches on foliage. Shoots may be stunted or distorted. Fruit may be mottled, russetted, or scabby.	**Powdery mildews**, including *Podosphaera leucotricha*. Fungal diseases favored by moderate temperatures, shade, and poor air circulation.	*See* 90.
Pale wax and gnarled, swollen bark on limbs, trunk, twigs, or roots. Growth may be stunted.	**Woolly apple aphid**. Small, gray or purplish, waxy insects, mostly on bark.	*See* Woolly Aphids, 177.

Malus spp. (continued)

WHAT THE PROBLEM LOOKS LIKE	PROBABLE CAUSE	COMMENTS
Sticky honeydew, blackish sooty mold, and whitish cast skins on foliage. Leaves may curl.	**Aphids**, including **Green apple aphid**, *Aphis pomi;* **Rosy apple aphid**, *Dysaphis plantaginea*. Small black, green, purplish, or waxy insects in groups on leaves and terminals.	*See* 175.
Sticky honeydew and blackish sooty mold on foliage.	**European fruit lecanium**. Yellow, orange, or brown, flattened or bulbous insect, often in groups.	*See* Lecanium Scales, 195.
Brown to gray encrustations on bark. May be declining or dead twigs or branches.	**Italian pear scale**, *Epidiaspis leperii;* **San Jose scale**. Tiny, circular to elongate insects.	Italian pear scale survives only under shelter such as lichens and moss on bark. The female's body is reddish purple beneath its cover. *See* Armored Scales, 189, 191.
Leaves stippled, bleached. May be dark excrement specks and sticky honeydew on fruit and leaves.	**Leafhoppers**, including **White apple leafhopper**, *Typhlocyba pomaria;* **Rose leafhopper**, *Edwardsiana rosae*.	Pale green, yellow, or white, wedgelike insects $\leq^1/_8$ inch long. Do not seriously threaten plant health. *See* 201.
Stippled, bleached, or reddened foliage. May be fine webbing on leaves or terminals. Leaves may drop prematurely. Fruit may be small and poorly colored.	**European red mite; McDaniel spider mite**, *Tetranychus mcdanieli;* **Pacific spider mite; Twospotted spider mite**. Brown, green, reddish, or yellow specks.	*See* Spider Mites, 246.
Pale blotches, spots, or winding tunnels in leaves. Leaves may drop prematurely.	**Leafminers**, *Phyllonorycter* spp. Pale larvae tunnel in leaves. Adults are brownish moths with silver or white.	Plants tolerate extensive mining. Conserve parasites, which usually control leafminers unless biological control is disrupted. *See* Foliage Miners, 216.
Foliage notched or clipped. Bark chewed or stripped around basal trunks and root crown.	**Cribrate weevil**, *Otiorhynchus cribricollis*. Adults are black or gray snout beetles, about $^3/_8$ inch long. Larvae are white grubs with brown head.	Bark girdling and leaf chewing of young trees are the problems. Root feeding and any injury to older trees are apparently not serious. *See* similar Black Vine Weevil, 160.
Leaves or blossoms chewed. Single branches or entire plant may be defoliated. Foliage may be rolled and tied together with silk. Fruit may have small gouges or brownish scars.	**Fall webworm; Fruittree leafroller; Obliquebanded leafroller**, *Choristoneura rosaceana;* **Omnivorous leafroller**, *Platynota stultana;* **Orange tortrix**, *Argyrotaenia citrana;* **Western tussock moth**.	Moth larvae $\leq 1^1/_2$ inches long. *See* Caterpillars, 149–153.
Maytenus boaria, Mayten		
Leaves discolor, wilt, stunt, and may drop prematurely. Branches or entire plant may die.	**Verticillium wilt**. Soil-dwelling fungus, infects through roots.	*See* 117.
Leaves discolor. Bark may canker or ooze. Limbs or entire plant may die.	**Borers**, including **Flatheaded appletree borer, Pacific flatheaded borer**.	Whitish larvae, ≤ 1 inch long, chew under bark. *See* 227.
Sticky honeydew and blackish sooty mold on foliage.	**Black scale, Nigra scale**. Yellowish to brown, flattened, oval insects or black and elongate or bulbous.	*See* 192.
Sticky honeydew and blackish sooty mold on plant. Bulbous, irregular, brown, gray, or white bodies (scales) on twigs.	**Chinese wax scale**. Bulbous to hemispherical waxy insects, suck sap.	*See* 196.
Sticky honeydew, blackish sooty mold, and whitish cast skins on plant.	**Aphids**. Small, greenish to black, pear-shaped insects often in groups on leaves and stems.	*See* 175.
Melaleuca spp., Melaleuca, Black tea tree, Granite bottlebrush, Myrtle, Paperbark		
Branches die back. Entire tree may die. May be white fungus beneath basal trunk bark.	**Armillaria root disease**. Fungus present in many soils. Favored by warm, wet soil. Persists for years in infected roots.	*See* 120.
Galls or swellings on roots. Plants may grow slowly.	**Root knot nematodes**. Tiny root-feeding roundworms.	*See* 293.

WHAT THE PROBLEM LOOKS LIKE	PROBABLE CAUSE	COMMENTS
Melia azedarach, Chinaberry, Texas umbrella tree		
Brownish, grayish, tan, or white encrustations on bark. Possibly decline or dieback of woody parts.	**Greedy scale**. Insects are circular to flattened, $<1/16$ inch long.	*See* 190.
Metasequoia glyptostroboides, Dawn redwood		
No serious invertebrate or pathogen pests have been reported in California. If plants are unhealthy, investigate whether they have been injured or lack good growing conditions or appropriate cultural care.		
Metrosideros spp., Metrosideros, Iron tree, New Zealand Christmas tree		
Leaves discolor, stunt, wilt, or drop prematurely. Plants grow slowly and may die. Roots dark, decayed.	**Phytophthora root and crown rot**. Disease promoted by excess soil moisture and poor drainage.	*See* 122.
Leaves discolor and wilt. Dead tree or dying limbs. Broad galleries beneath bark.	**Eucalyptus longhorned beetles**. Adults reddish brown with yellow on the back. Larvae whitish. Both ≤1 inch long.	Primarily pests of eucalyptus. *See* 233.
Brown, black, tan, or yellow spots or blotches on leaves. Spots may have dark or yellowish margins. Foliage may die and drop prematurely.	**Anthracnose, Blight**, and **Leaf spots**. Fungal diseases favored by cool, wet, conditions.	*See* 81, 83, 86.
Sticky honeydew and blackish sooty mold on foliage. Leaves and terminals pitted, distorted, and discolored.	**Eugenia psyllid**. Adults are tiny, leafhopperlike insects. Nymphs feed in pits on lower leaf surface.	Primarily a pest of eugenia. Young foliage of *Metrosideros excelsus* may be infested during winter. *See* 172.
Mimulus, Diplacus spp., Monkey flower		
Leaves have brown to yellow spots or blotches. Spots may have dark or yellowish margins. Foliage may shrivel and drop prematurely.	**Leaf spots**, including *Ramularia mimuli, Septoria mimuli*. Fungal diseases favored by prolonged wet conditions. Pathogens persist in plant debris.	*See* 86.
Leaves with dark or yellow spots and orangish pustules. Leaves may drop prematurely.	**Rusts**. Fungal pathogens favored by wet conditions.	Avoid overhead watering. Remove and dispose of infected leaves. *See* 94.
Leaves or shoots with powdery growth. Shoots or leaves may be stunted and distorted.	**Powdery mildews**, including *Erysiphe cichoracearum*. Fungi favored by moderate temperatures, shade, and poor air circulation.	*See* 90.
Leaves stippled or bleached. Leaves and shoots may die back.	**Seed bugs**, *Kleidocerys* spp. Small, brown, oval insects, suck plant juices.	No known controls. *See* True Bugs, 205.
Stippled or bleached leaves, varnishlike excrement specks on undersides.	**Thrips**. Tiny, slender, blackish or orangish insects.	*See* 209.
Flower buds mined or distorted. Flowering reduced. Insect pupal skins (exuviae) attached to flower buds.	**Gall midges**, *Asphondylia* spp. Larvae maggotlike, often several per bud. Adults tiny, delicate flies.	No known controls. Larvae reduce seeds, but established plants are not threatened. Several generations per year. *See* 214.
Sticky honeydew, blackish sooty mold, and grayish, flocculent material on plant. Foliage discolored. Plant growth slow.	**Golden mealybug**, *Nipaecoccus aurilanatus*. Females globular, $1/10$ inch long, purplish with felty, golden band marginally and on back.	Several generations occur each year. Natural enemies can help to control. *See* Mealybugs, 185.
Morus spp., Mulberry, Fruitless mulberry		
Angular, blackened areas on leaves. Young leaves and shoots distorted. Elongated lesions may occur on twigs.	**Bacterial blast, blight, and canker**. Disease favored by wet conditions.	*See* 83.
Leaves discolor, stunt, wilt, or drop prematurely. Basal trunk discolored. Minute white fungus growths may be visible beneath bark.	**Armillaria root disease**. Fungus present in many soils. Favored by warm, wet soil. Persists for years in infected roots.	*See* 120.
Leaves discolor and wilt. Bark may have dark, sunken cankers or coral-colored pustules.	**Nectria canker**. Fungus that primarily affects injured or stressed trees.	*See* 101.

Morus spp. (continued)

WHAT THE PROBLEM LOOKS LIKE	PROBABLE CAUSE	COMMENTS
Foliage yellows and wilts. May be bracketlike or fan-shaped fungal fruiting bodies on bark or wood. Branches or entire plant may die.	**Wood decay**, or **Heart rot**, including *Ganoderma applanatum, Schizophyllum commune*.	Fungi that infect injured or severely stressed trees. *See* 66, 111.
Sticky honeydew and blackish sooty mold on leaves. Blackish to gray, oval bodies with white waxy fringe (nymphs) on leaves.	**Whiteflies**, including **Giant whitefly; Mulberry whitefly**, *Tetraleurodes mori*. Small, mothlike adults and oval nymphs, suck sap.	Conserve natural enemies that help to control. *See* 182.
Sticky honeydew and blackish sooty mold on foliage. Cottony or waxy material on plant.	**Mealybugs**, including **Comstock mealybug**, *Pseudococcus comstocki*.	Powdery, grayish insects, $\leq^1/_4$ inch long. *See* 185.
Copious, misty, nonsticky liquid raining from plant. Surfaces covered with whitish residue.	**Glassy-winged sharpshooter**. Active, dark brown or gray leafhoppers, $\leq^1/_2$ inch long, suck xylem fluid.	Vectors *Xylella* pathogens. Report suspected glassy-winged sharpshooters to agricultural officials if found in areas where this pest is not known to occur. *See* 203.
Brownish, grayish, tan, or white encrustations on leaves, twigs, or branches. Rarely, declining or dead twigs or branches.	**California red scale, Oleander scale, San Jose scale**. Circular to oval insects, $<^1/_{16}$ inch long.	*See* 189–192.
Webbing or silk tents on ends of branches. Chewed leaves.	**Fall webworm**. Hairy, white to yellow larvae, $\leq$1 inch long, in colonies.	*See* 152.
Bark stained brownish, exudes rancid fluid, often around crotches, wounds.	**Wetwood**, or **Slime flux**. Bacterial infections.	Usually do not cause serious harm to trees. *See* 110.
Galls or swellings on roots.	**Root knot nematodes**. Tiny root-feeding roundworms.	*See* 293.
Myoporum spp., Myoporum, Lollipop tree[22]		
Leaves brown or yellow, wilted, and may drop prematurely. Branches or plant may die.	**Pythium root rot**, *Pythium* sp. Disease favored by moist, poorly drained soils.	*See* 123.
Leaves bleached or stippled with spots. Cast skins on underside of leaves.	**Blue-green sharpshooter**, *Graphocephala atropunctata*. Greenish, wedge-shaped insects, $\leq^1/_8$ inch long.	Plants tolerate moderate stippling. *See* 203.
Leaves curl, drop prematurely. Shoot tips severely distorted.	**Myoporum thrips**. Tiny, slender, brown to black adults and orange, pinkish, or white nymphs.	Ngaio tree (*M. laetum*) and *Myoporum* 'Pacificum' ground cover especially damaged. *See* 211.
Myrica spp., Wax myrtle, Bayberry, California wax myrtle		
Irregular, brown, tan, or white blotches or spots on leaves. Leaves may die and drop prematurely.	**Anthracnose**, Gnomonia myricae; **Leaf spots** and **Blight**, *Cronartium comptoniae, Lophodermium foliicola, Phyllosticta myricae*.	Fungi favored by wet conditions and persist in infected leaves and twigs. *See* 81, 86.
Sticky honeydew and blackish sooty mold on leaves and twigs. Stems may die back.	**European fruit lecanium**. Brown or yellowish, flattened to bulbous scale insects on twigs.	*See* Lecanium Scales, 195.
Sticky honeydew and blackish sooty mold on leaves. Blackish, oval bodies with white, waxy fringe (nymphs) on leaves.	**Mulberry whitefly**, *Tetraleurodes mori*. Tiny adults, white or yellowish and mothlike.	Conserve natural enemies that help to control. Difficult to control with sprays. *See* Whiteflies, 182.
Leaves discolored, wilted, and may drop prematurely. Branches or plant may die.	**Limb and trunk rot**, *Phellinus ferreus*. Decay fungus may produce bracketlike fruiting bodies on bark.	*See* Wood Decay, 111.
Nandina domestica, Nandina, Heavenly bamboo, Sacred bamboo		
Leaves turn yellow, then brown. Upper canopy dies back. Entire plant dies rapidly or slowly over several years.	**Bacterial leaf scorch**, or **Oleander leaf scorch**. Bacteria spread by certain leafhoppers, especially glassy-winged sharpshooter.	*See* 112.
Sticky honeydew and blackish sooty mold on foliage. Popcornlike bodies (egg sacs) on stems.	**Cottony cushion scale**. Orangish, flat immatures on leaves or cottony females on stems.	Normally controlled by natural enemies. *See* 130, 196.
Whitish, cottony material on bark or underside of leaves.	**Comstock mealybug**, *Pseudococcus comstocki*. Oblong, soft, powdery, waxy insects with filaments.	Control ants, reduce dust, avoid persistent pesticides that disrupt effective natural enemies. *See* Mealybugs, 185.

[22] Some species are invasive weeds. Other species may be better choices when planting.

WHAT THE PROBLEM LOOKS LIKE	PROBABLE CAUSE	COMMENTS
White, powdery growth on upper side of leaves. Leaves and terminals redden, curl, and twist.	**Powdery mildew**, *Erysiphe berberidis*. Pathogen favored by shade and poor air circulation.	*See 90.*
Foliage mottled, reddish.	**Nandina mosaic virus**. Disease spreads mechanically or by aphids.	Plants usually tolerate. *See 93.*

Nerium oleander, Oleander

Leaves turn brown or yellow, especially along margins and at tip. Leaves and shoots die. Entire plant dies rapidly or slowly over several years.	**Bacterial leaf scorch**, or **Oleander leaf scorch.** Bacteria spread by certain leafhoppers. Only molecular ELISA test can confidently diagnose this cause of damage.	Infected oleander and certain other hosts are severely damaged or killed. *See 112.*
Leaves brown, yellow, and wilt. Terminals may die back. Stems die back. Stems may have cankers, decay, or lesions, which can help distinguish this malady from oleander leaf scorch.	**Phoma stem and leaf blight**, *Phoma exigua*. Fungus infects through wounds, persists in infected tissue, and spreads during wet conditions.	Avoid wounding plants, except to prune off infected tissue only during prolonged dry weather, making cuts in healthy tissue. Provide a good growing environment and appropriate cultural care.
Leaves turn brown or yellow, especially along margins and at tip. Leaves may drop prematurely.	**Leaf burn**, or **Scorch**. Abiotic disorders with many potential causes, including frost, inappropriate irrigation, pesticide phytotoxicity, poor drainage, or saline soil or water.	Provide plants with a good growing environment and proper cultural care. *See* Abiotic Disorders, 43.
Leaves, blossoms, or stems with black to brown lesions, spots, or streaks. Wood may be cankered, dark, or die back.	**Bacterial blast, blight, and canker**. Disease favored by wet conditions.	*See 83.*
Leaves with brown, black, tan, or yellow spots or blotches. Spots may have dark or yellowish margins.	**Leaf spots**, including *Alternaria* sp., *Septoria oleandrina*. Favored by prolonged cool, wet, conditions. Persist in plant debris.	*See 86.*
Foliage may yellow and wilt. May be bracketlike or fan-shaped fungal fruiting bodies on bark or wood. Branches or entire plant may die.	**Wood decay**, or **Heart rot**, including *Schizophyllum commune*. Fungi that attack injured, old, or stressed plants.	*See 111.*
Sticky honeydew and blackish sooty mold on foliage.	**Black scale**. Orangish, flat immatures or brown to black, bulbous adults, may have raised H shape on back.	*See 192.*
Sticky honeydew and blackish sooty mold on foliage. May be cottony or waxy material on plant.	**Mealybugs**, including **Longtailed mealybug, Obscure mealybug**. Grayish, powdery, waxy insects.	*See 185.*
Sticky honeydew, blackish sooty mold, and whitish cast skins on leaves and terminals. New growth may be deformed.	**Oleander aphid**. Orangish or yellow insects with black on leaves, shoots, and flowers.	Water and prune less to reduce new growth that promotes aphids. Conserve natural enemies that help control. Hose forcefully with water. *See 175.*
Copious, misty, nonsticky liquid raining from plant. Surfaces covered with whitish residue.	**Glassy-winged sharpshooter**. Active, dark brown or gray leafhoppers, $\leq 1/2$ inch long, suck xylem fluid.	Vectors *Xylella* pathogens. Report suspected glassy-winged sharpshooters to agricultural officials if found in areas where this pest is not known to occur. *See 203.*
Brownish, grayish, tan, or white encrustations on leaves or twigs.	**Armored scales**, including **Greedy scale, Oleander scale**. Tiny, flattened, circular insects, $\leq 1/16$ inch long.	Rarely if ever cause serious damage to oleander. *See 189–192.*
Galls or knots on stems, bark, and occasionally on flower buds and leaves. Twigs may die back.	**Oleander gall**. Bacterium infects wet tissue through wounds.	*See* Olive Knot and Oleander Gall, 94.
Galls or swellings on trunk and roots, usually near soil.	**Crown gall**. Bacterium infects plant via wounds.	*See 108.*

WHAT THE PROBLEM LOOKS LIKE	PROBABLE CAUSE	COMMENTS
Notholithocarpus densiflorus, Tanoak, Tanbark oak		
Terminals bend downwards. Leaves and shoots brown, light green, yellow, or wilted. Trunks may ooze. Leaves, shoots, and eventually entire tree die.	**Sudden oak death**, *Phytophthora ramorum*. Pathogen spreads via airborne spores and contaminated plants and soil.	Primarily a problem in wildlands, killing many tanoak and true oaks there. *See* 105.
Foliage discolors, stunts, wilts, or drops prematurely. Plants grow slowly and may die. Roots dark, decayed.	**Phytophthora root and crown rot**. Disease favored by excess soil moisture and poor drainage.	*See* 122.
Foliage yellows and wilts. Branches or entire plant may die. Trees may fall over. May be fleshy or woody fungal growths on bark or around roots.	**Wood decay**, or **Heart rot**, including *Phellinus ferreus*, *Steccherinum* spp. Fungi that infect injured or severely stressed trees.	*See* 111.
Foliage discolors and wilts. Branches develop cankers and die back.	**Annulohypoxylon canker; Fusicoccum canker**, *Fusicoccum quercus*. Decay fungi that attack injured or stressed trees.	*Annulohypoxylon* produces dark hemispherical fruiting bodies on bark or wood. *See* 98.
Foliage or shoots turn yellow, then brown. Shoots may be bushy, short, or shriveled. Powdery, whitish growth on leaves or shoots.	**Powdery mildews**, including *Brasiliomyces trina*, *Microsphaera penicillata*, *Sphaerotheca lanestris*.	Fungi favored by moderate temperatures, shade, and poor air circulation. *See* 90.
Bleeding or frothy material bubbling from tiny holes in trunk or limbs. Fine boring dust may surround holes.	**Oak ambrosia and bark beetles**. Adults brown, $1/8$ inch long. White larvae tunnel beneath bark.	*See* 222.
Sticky honeydew and blackish sooty mold on foliage.	**Aphids**, including *Myzocallis* sp. Pear-shaped, greenish to yellow insects, usually in groups on leaves and terminals.	*See* 175.
Sticky honeydew and blackish sooty mold on foliage. Cottony or waxy material on plant.	**Mealybugs**, including *Pseudococcus* sp.	Powdery, grayish insects, $\leq 1/4$ inch long, with waxy filaments. *See* 185.
Sticky honeydew and blackish sooty mold on foliage. Black, brown, or yellow oval bodies (nymphs) often with white, waxy fringe, on leaves.	**Stanford whitefly**, *Tetraleurodes stanfordi*. Adults tiny, mothlike. Pupae black with white, waxy fringe.	Ignore insects, they apparently do not damage trees. *See* Whiteflies, 182.
Gray, pinkish, or whitish encrustations on bark, usually on the limb underside.	**Ehrhorn's oak scale**. Red insect, $1/25$ inch long, that occurs on bark beneath a gray to whitish fungus.	Heavy infestations may slow tree growth, but populations are often innocuous. *See* 197.
Brown, gray, tan, or white encrustations on bark or leaves. Rarely, declining or dead twigs or branches.	**Greedy scale; Tanoak scale**, *Aspidaspis densiflorae*. Tiny, oval to circular insects on bark.	Rarely if ever cause serious damage to plants. *See* 189–192.
Patches of dead leaves at end of branches.	**Oak twig girdler**. Adult cylindrical, metallic beetle, larvae whitish, in tunnels.	*See* 231.
Nyssa spp., Sour gum, Tupelo, Tupelo gum		
Chewed foliage. Leaves may be webbed with silk.	**Redhumped caterpillar, Tent caterpillars**.	Larvae hairy or colorful, ≤ 1 inch long. *See* 150–152.
Sticky honeydew and blackish sooty mold on foliage. Possible plant decline or dieback.	**European fruit lecanium**. Brown or yellow, flat or bulbous, immobile scale insects.	*See* Lecanium Scales, 195.
Olea europaea, Olive, Fruitless olive[23]		
Foliage fades, yellows, browns, or wilts, often scattered throughout canopy. Branches die. Entire plant may die.	**Verticillium wilt**. A soil-dwelling fungus that infects through roots.	Xylem discolors little or not at all in olive infected with Verticillium wilt. *See* 117.
Leaves turn yellow then brown, beginning at tips. Limb dieback scattered throughout the canopy. Entire plant dies, often slowly over several years.	**Bacterial leaf scorch**, or **Oleander leaf scorch**. Bacteria spread by certain leafhoppers, especially glassy-winged sharpshooter.	*See* 112.
Leaves discolor, stunt, wilt, or drop prematurely. Plants grow slowly and may die. Roots dark, decayed.	**Phytophthora root and crown rot**. Disease promoted by excess soil moisture and poor drainage.	*See* 122.

[23] For olive fruit pests, see ipm.ucanr.edu.

WHAT THE PROBLEM LOOKS LIKE	PROBABLE CAUSE	COMMENTS
Foliage may yellow and wilt. Bark or soil near trunk may have fungal fruiting bodies. Branches or plant may die.	**Wood decay**, or **Heart rot**, including *Abortiporus biennis,* which has variable fruiting bodies, often pale or pink, fan shaped to globular.	*See* 111.
Leaves discolor and wilt at branch tips. Dieback of some twigs. Tunnels under bark.	**Branch and twig borer**, *Melalgus (=Polycaon) confertus.* Adults black to brown beetles, ≤¹/₂ inch long. Larvae whitish.	Prune out affected parts. Eliminate nearby dying hardwoods in which beetles breed. Provide trees with proper cultural care. *See* Twig, Branch, and Trunk Boring Insects, 220.
Bark has boring dust or holes or oozes. Slow tree growth. Branches may wilt or die back.	**American plum borer, Ash borer**. Brown, green, pink, or white larvae, ≤1 inch long, bore beneath bark.	Distinguish between these differently managed species. *See* 235, 240.
Sticky honeydew and blackish sooty mold on foliage.	**Black scale**. Orangish, flat immatures or brown to black, bulbous adults, may have raised H shape on back.	*See* 192.
Sticky honeydew, blackish sooty mold, and flocculent white wax on leaves and twigs.	**Olive psyllid**, *Euphyllura olivina.* Pale green to tan insects, ≤¹/₁₀ inch long, suck phloem sap.	In areas with hot temperatures, prune off interior limbs to suppress psyllid populations due to heat exposure. *See* 172.
Brown, gray, or orangish encrustations on bark. Rarely, declining or dead twigs and branches.	**California red scale; Olive scale**, *Parlatoria oleae;* **Oleander scale**. Circular to oval, ≤¹/₁₆ inch long, on twigs and branches.	Rarely if ever cause serious damage to olive. Conserve natural enemies. *See* 189–192.
Copious, misty, nonsticky liquid raining from plant. Surfaces covered with whitish residue.	**Glassy-winged sharpshooter**. Active, dark brown or gray leafhoppers ≤¹/₂ inch long, suck xylem fluid.	Vectors *Xylella* pathogens. Report suspected glassy-winged sharpshooters to agricultural officials if found in areas where this pest is not known to occur. *See* 203.
Leaves develop blackish or green, circular blotches, ≤¹/₂ inch in diameter. Spots may have a faint yellow halo. Leaves fall prematurely and twigs may die due to leaf drop.	**Scab**, or **Olive peacock spot**, *Spilocaea oleaginea.* Fungus infects during wet winter weather and is most severe in lower canopy.	Usually harmless to trees. *See* 88.
Leaves turn yellow, brown, and drop prematurely. Limbs die back.	**Branch dieback**. Abiotic disorder with many potential causes, commonly caused by excess soil moisture and poor drainage, especially around the root crown.	Avoid wetting trunk or root crown. Avoid overhead irrigation around olive. Do not plant olive in frequently irrigated landscapes, such as lawns.
Decaying, dropped fruit and dark or oily stains beneath trees. May be pale maggots in fruit.	**Messy fruit**. Unharvested fruit naturally drop, which can create a slippery hazard beneath trees. Pale maggots of olive fruit fly, *Bactrocera oleae,* cause premature fruit drop.	Plant fruitless olive such as Majestic Beauty, Mother, and Swan Hill. Knock fruit onto tarps beneath trees. Prune off limbs overhanging pavement. Spray flowers with growth regulator before fruit set as directed on product label.
Galls or knots on stems and bark, and occasionally on flower buds and leaves. Twig dieback.	**Olive knot**, or **Bacterial gall**. Bacterium infects wet tissue through wounds.	*See* Olive Knot and Oleander Gall, 94.
Galls or swellings on roots. Plants may grow slowly.	**Root knot nematodes**. Tiny root-feeding roundworms.	*See* 293.
Parkinsonia =Cercidium spp., Palo Verde; *P. floridum,* Blue Palo Verde		
Foliage discolors and wilts and may drop prematurely. Decline or death of limbs. Bark discolored or bleeding.	**Flatheaded borers**, including **Flatheaded appletree borer**. Whitish larvae, ≤1 inch long, with enlarged head in tunnels.	Keep plants vigorous. Remove and dispose of damaged limbs. *See* 227.
Foliage discolors and wilts and may drop prematurely. Limbs or tree may decline and die. Basal roots have elliptical holes ≤³/₄ inch in diameter.	**Palo Verde root borer**, *Derobrachus geminatus.* Pale larvae, ≤5 inches long, bore in roots and basal trunk. Adults are dark longhorned beetles, ≤3¹/₂ inches long. Attacks unhealthy trees.	Protect trees from injury. Provide plants with good growing environment and appropriate cultural care. No direct control known to be effective. *See* Longhorned Beetles, or Roundheaded Wood Borers, 232.
Foliage yellows and wilts. Bark may have lesions that are dark, dry, crusty whitish, oily looking, or water-soaked. Branches or entire plant dies.	**Fusarium dieback**. Fungal disease spread by the polyphagous shothole borer, an ambrosia beetle, or bark beetle.	Occurs on *P. floridum* at least in southern California. *See* 115.

Parkinsonia =Cercidium spp. (continued)

WHAT THE PROBLEM LOOKS LIKE	PROBABLE CAUSE	COMMENTS
Leaves chewed and webbed with silk. Leaves may drop prematurely.	**Palo Verde webworm**, *Faculta inaequalis*. Adults are $1/4$-inch-long, tan moths. Larvae, $\leq 1/2$ inch long, feed mostly in silk tubes or webs.	Prefers foothills Palo Verde, *P. microphylla*. Otherwise healthy Palo Verde tolerate extensive leaf chewing. *See* Webworms, 152.
Leaves may dry up and drop prematurely. Shoots may be brown or yellow, bushy, or distorted into witches' brooms.	**Spider mites**. Tiny arthropods, often greenish or yellowish, suck sap.	*See* 246.
Green plants with smooth stems, thick, roundish leaves infesting branches. Shoots may distort into witches' brooms.	**Broadleaf mistletoe**. Parasitic, evergreen plant on host.	*See* 284.

Photinia spp., Photinia, Red tips

WHAT THE PROBLEM LOOKS LIKE	PROBABLE CAUSE	COMMENTS
Foliage yellows and wilts. Branches or entire plant dies. Rootlets decayed or sparse.	**Phytophthora root and crown rot**. Disease favored by wet soil and poor drainage.	*See* 122.
New foliage yellows, beginning along margins, but veins green. Leaves may develop brown blotches.	**Iron deficiency**. Abiotic disorder usually caused by poor soil conditions or unhealthy roots.	*See* 47.
Sudden wilting, then shriveling and blackening of shoots and blossoms. Plants appear scorched.	**Fire blight**. Bacterium infects plants through blossoms.	*See* 84.
Powdery, white growth and yellow blotches on leaves and sometimes on terminals and fruit. Shoots and leaves may be undersized or distorted.	**Powdery mildews**, *Podosphaera leucotricha, Sphaerotheca pannosa*. Fungal diseases favored by moderate temperatures, shade, and poor air circulation.	*See* 90.
Tiny brown, reddish, or purple leaf spots, may have yellow halos. Larger, dark areas on leaves. Leaves may drop prematurely.	**Entomosporium leaf spot**. A fungal disease promoted by wet foliage.	*See* 87.
Black to dark olive spots on leaves. Leaves yellow and may drop prematurely.	**Scab**, *Spilocaea* sp. A fungal disease spread by water from infected leaves and twigs.	Avoid overhead watering. Prune out infected twigs and leaves in fall. *See* 88.
Tiny brown, reddish, or purple leaf spots. Unlike Entomosporium leaf spot, spots lack dark centers. Leaves may drop prematurely.	**Physiological leaf spot**. Abiotic disorders with many potential causes. Cold stress is a common cause of damage, especially if plants are low-lying or shaded.	Apparently harmless to plants. Plant in full sun. Do not plant in low-lying spots, where cold air settles. Provide good growing conditions and proper cultural care.
Stippled, bleached, or reddened leaves with varnishlike specks on undersides.	**Greenhouse thrips**. Slender, black adults or yellow nymphs.	Look for insects to distinguish from similar damage caused by lace bugs. *See* 210.
Stippled or bleached leaves with varnishlike specks on undersides.	**Lace bug**, *Corythucha* spp. Adults $\leq 1/8$ inch long, wings lacy. Nymphs spiny.	Look for insects to distinguish from similar damage caused by thrips. *See* 207.
Blackish sooty mold, sticky honeydew, and whitish cast skins on plants.	**Aphids**. Small green, brown, black, or yellowish insects, often in groups.	*See* 175.
Leaf margins chewed, notched. Young plants may decline.	**Weevils**, including *Otiorhynchus* spp.; **Fuller rose beetle**, *Naupactus godmani*. Adults chew foliage. Larvae chew roots.	*See* 159.

Picea spp., Spruce[24]

WHAT THE PROBLEM LOOKS LIKE	PROBABLE CAUSE	COMMENTS
Needles mottled brown, light green, or yellow. Needles die and drop prematurely, often beginning with lower canopy. Plant growth slow.	**Needle blight and cast, Tar spots**, including *Lirula* sp., *Lophodermium piceae, Rhizosphaera kalkhoffii*. Fungi favored by cool, wet conditions in spring.	Remove nearby plants and weeds. Thin canopy and prune off lower branches to reduce humidity and improve air circulation. *See* 85.
Needles with dark or yellow spots. Shoots brown or yellow, bushy witches' brooms. Orangish pustules or pale growth on foliage. Needles may drop prematurely.	**Rusts**, including *Chrysomyxa arctostaphyli*. Fungi infect and develop when foliage is wet. Some infect alternate hosts, such as *C. arctostaphyli* on manzanita.	Avoid overhead watering. Plants tolerate moderate damage. Eliminating nearby alternate hosts may help in control. *See* 94.
Needles and terminals brown, yellow, and wilt. Roots or basal stem may be dark, decayed. Branches or entire plant dies.	**Phytophthora root and crown rot**. Disease favored by excess soil moisture and poor drainage.	*See* 122.

[24] For more information, see *Pests of the Native California Conifers*.

WHAT THE PROBLEM LOOKS LIKE	PROBABLE CAUSE	COMMENTS
Foliage may yellow and wilt. Bark or wood may have bracketlike or fan-shaped fungal fruiting bodies. Branches or entire plant may die.	**Wood decay**, or **Heart rot**, including *Ganoderma* spp., *Laetiporus gilbertsonii* (=*L. sulphureus*), *Pleurotus ostreatus*, *Schizophyllum commune*, *Stereum* sp., *Trametes* spp.	Fungi that attack injured, old, or stressed trees. *See* 66, 111.
Needles brown, yellow, or drop prematurely. Treetop or entire tree dies.	**Engraver beetles**. Adults are small, brown bark beetles. White larvae bore under bark.	*See* Pine Engravers, 225.
Needles turn brown or yellow on branch interiors and drop prematurely, leaving only young terminal needles. Tree may die.	**Spruce aphid**, *Elatobium abietinum*. Small, pear-shaped insects, dark to light green, in groups on older foliage.	Most common late winter to early spring. Can be a serious pest of spruce. *See* Aphids, 175.
Needles pale, mottled, or chlorotic.	**Pine needle scale**, *Chionaspis pinifoliae*. White, immobile armored scales, about $1/16$ inch long. Suck sap.	Vigorous plants tolerate moderate populations. Conserve beneficials. *See* Armored Scales, 189.
Needles bleached, stippled. Foliage color abnormally light green or yellowish.	**Spider mites**, including **Spruce spider mite**. Greenish specks, often in fine webbing at foliage base.	*See* Pine and Spruce Spider Mites, 247.
Needles chewed. Foliage browns. May be silk webbing on needles.	**Caterpillars**, including **Douglas-fir tussock moth**, *Orgyia pseudotsugata*; **Silverspotted tiger moth**, *Lophocampa argentata*. Caterpillars hairy, brownish with orange, red, or yellow.	Vigorous trees tolerate moderate damage. Apply *Bacillus thuringiensis* or another insecticide to control young larvae, if intolerable. *See* 144–152.
Needles chewed.	**Conifer sawflies**. Green larvae, ≤1 inch long, on needles.	*See* 153.
Terminals distorted, chewed, dead. Foliage may become busy, crooked. Roots or the basal trunk may be injured.	**Conifer twig weevils** or **Pine weevils**, including **White pine weevil**. Small black to brown weevils chew needles and shoots.	Grublike larvae chew on or within needles, shoots, or root crown; the most damaging species mine shoots. *See* 161.
Wet, white, frothy masses on needles or twigs.	**Spittlebugs**, including **Western pine spittlebug**, *Aphrophora permutata*. Green to black sucking insects, secrete spittle.	Tolerate; spittlebugs cause no apparent harm to trees. *See* 205.
Terminals galled, brown, light green, or purplish. Needles may have yellow spots. May be cottony or waxy material on bark or needles.	**Adelgids**, including **Cooley spruce gall adelgid**. Tiny, aphidlike insects feed on bark, at needle bases, and within galls.	*See* 181.
Sticky honeydew and blackish sooty mold on foliage. Brown to purplish insects clustered on foliage.	**Giant conifer aphids**, *Cinara* spp. Black, long-legged insects, larger than most aphid species.	Often infest only a single branch. In California, more common in south than in north. *See* 176.
Plants stunted, decline, may die. Powdery, waxy material may be visible on roots and around crown.	**Ground mealybugs**, *Rhizoecus* spp. Small, slender, pale insects, may be lightly covered with powdery wax, but lack marginal filaments.	*See* 187.
Pitchy masses 1 to 4 inches in diameter protruding from trunks and limbs. Limbs occasionally break.	**Douglas-fir pitch moth**. Dirty whitish larvae, ≤1 inch long, in pitch.	*See* Pitch Moths, 239.
Galls or swellings on trunk and roots, usually near soil, may be on branches.	**Crown gall**. Bacteria infect plant via wounds.	*See* 108.
Pieris spp., Pieris, Andromeda		
Foliage yellows and wilts. Branches or entire plant dies.	**Phytophthora root and crown rot.** Disease favored by wet soil and poor drainage.	*See* 122.
Brown, black, tan, or yellow spots or blotches on leaves. Spots may have dark or yellowish margins. Foliage may die and drop prematurely.	**Leaf spots**, *Alternaria* sp., *Phyllosticta* sp. Spread by air or splashing water. Favored by prolonged cool, wet conditions. Persist in plant debris.	*See* 86.
Leaves with brown, light green, or yellow lesions or dead spots, especially along leaf margins. Twigs and stems may canker or ooze.	**Ramorum blight**, *Phytophthora ramorum*. Pathogen spreads via airborne spores and contaminated plants and soil.	Primarily a problem in nurseries and on other hosts in wildlands, killing many oaks there. *See* Sudden Oak Death and Ramorum Blight, 105.

WHAT THE PROBLEM LOOKS LIKE	PROBABLE CAUSE	COMMENTS
Wilting or dead plants. Some roots stripped of bark or girdled near soil. Foliage may be notched around margins.	**Black vine weevil**. Adults are black or grayish snout beetles, about $3/8$ inch long. Larvae are white grubs with brown head.	Larvae feed on roots. Adults hide during day and feed at night. *See* 160.
Stippled or bleached leaves. Leaves may dry up and drop prematurely.	**Spider mites**, *Tetranychus* spp. Greenish specks, suck sap.	*See* 246.
Leaves stippled or bleached, with cast skins or varnishlike specks on underside.	**Andromeda lace bug**, *Stephanitis takeyai*. Black and clear insects, $\leq 1/8$ inch long, with lacy wings.	*See* Lace Bugs, 207.

Pinus spp., Pine[25]

WHAT THE PROBLEM LOOKS LIKE	PROBABLE CAUSE	COMMENTS
Needles discolor, stunt, wilt, or drop prematurely. Stems discolored, cankered, and dead. May be white fungus beneath basal trunk bark.	**Armillaria root disease**. Fungus present in many soils. Favored by warm, wet soil. Persists for years in infected roots.	*See* 120.
Needles discolor, stunt, wilt, or drop prematurely. Plants grow slowly and may die. Roots dark, decayed.	**Phytophthora root and crown rot**. Disease promoted by excess soil moisture and poor drainage.	*See* 122.
Needles yellow, wilt, and drop prematurely. Branches die back, and entire plant may die.	**Heterobasidion root disease**. Decay fungus that spreads through natural root grafts and airborne spores.	*See* 121.
Foliage yellows and wilts. Bark or wood may have bracketlike or fan-shaped fungal fruiting bodies. Branches or entire plant may die.	**Wood decay**, or **Heart rot**, including *Ganoderma* spp., *Laetiporus gilbertsonii* (=*L. sulphureus*), *Schizophyllum commune*, *Stereum* sp., *Trametes* spp.	Fungi that attack injured, old, or stressed trees. *See* 66, 111.
Slightly raised reddish or yellow spots on needles. Cankers, galls, or oozing on stems. Needles may drop prematurely. Terminals brown and die.	**White pine blister rust; Pine stem and cone rusts**. Fungi persist in infected bark. Disease is favored by cool, wet conditions.	White pine blister rust usually kills infected hosts. *See* 96.
Brown, reddish, or yellow bands or spots on needles. Needles drop prematurely, especially on lower branches.	**Pine needle rusts**, *Coleosporium* spp. Fungal diseases.	*See* 96.
Black, brown, reddish, or yellow lesions, spots, or streaks on needles or stems. Needles drop prematurely. Plants grow slowly. Branches may die back.	**Needle blight and cast, Leaf spots**. Fungal diseases favored by cool, wet conditions in spring.	*See* 85, 86.
Brown or yellow needles, especially at branch tips. Dead needles may persist on branches. Terminals or branches may die back.	**Blight**, or **Aleppo pine blight**. Abiotic disorders with many causes, including dry winds, excess light, fertilizer toxicity, low humidity, high temperatures, root injury, soil compaction, or too much or too little water.	Aleppo pine (*Pinus halepensis*) is commonly affected. Provide trees with a good growing environment and proper cultural care. *See* Abiotic Disorders, 43.
Needles wilt, turn yellow, then entirely brown in large groups. Twigs and treetop may die. Wood is stained black or brown.	**Diplodia canker**, *Sphaeropsis sapinea* =*Diplodia pinea*. A fungal disease that primarily affects injured or weakened pine.	*See* similar Oak Branch Canker and Dieback, 102.
Dead branches with clinging needles, mostly in upper canopy. Trunk cankers and branches exuding copious pitch.	**Pitch canker of pine**. A fungal disease.	Occurs in coastal areas of California. *See* 102.
Stunted, bushy foliage, possible dieback. Round swellings on branches, orangish (spore covered) in spring.	**Western gall rust**. Fungus that infects 2- and 3-needle pines.	*See* 96.
Pitchy masses 1 to 4 inches in diameter protruding from trunks and limbs. Limbs occasionally break.	**Douglas-fir pitch moth, Sequoia pitch moth**. Dirty, whitish larvae, ≤ 1 inch long, in pitch. Adults wasplike.	Avoid injuring pines, wounds attract moths. Plants tolerate these insects that feed shallowly beneath bark. *See* 239.
Sticky honeydew and blackish sooty mold on foliage. Whitish wax may cover needles. Possible yellowing of needles.	**Aphids**, including **Monterey pine aphid**, *Essigella californica*; **Woolly pine needle aphid**, *Schizolachnus piniradiatae*.	Tiny, green to gray insects. *See* 175.
Cottony white or grayish material on bark or needles. Slow pine growth.	**Pine bark adelgids**, *Pineus* spp. Tiny, purplish insects under cottony wax.	*See* 182.

[25] For more information, see *Pests of the Native California Conifers*.

WHAT THE PROBLEM LOOKS LIKE	PROBABLE CAUSE	COMMENTS
Sticky honeydew, blackish sooty mold, and waxy, whitish material on needles and twigs.	**McKenzie pine mealybug**, *Dysmicoccus piniculus*; **Obscure mealybug**. Oval, waxy-fringed insect, $\leq^1/_8$ inch long.	Conserve natural enemies. Plants tolerate moderate populations. Insecticidal soap can be applied. *See* Mealybugs, 185.
Sticky honeydew and blackish sooty mold on foliage. Brown to purplish insects clustered on foliage.	**Giant conifer aphids**. Dark, long-legged insects, $\leq^1/_5$ inch long.	Harmless to trees. *See* 176.
Sticky honeydew and blackish sooty mold on foliage. Possible yellowing of older needles. Male scales resemble rice grains on needles.	**Monterey pine scale**, *Physokermes insignicola*; **Irregular pine scale**. Females, $^1/_4$ inch long, resemble chips of marble or dark, shiny beads on twigs.	Vigorous trees tolerate moderate populations. *See* 194.
Interior needles turn brown or yellow and drop prematurely, leaving only young terminal needles. Tree may die.	**Spruce aphid**, *Elatobium abietinum*. Small, pear-shaped insects, dark to light green, in groups on older foliage.	Usually abundant only on spruce, where it can be a serious pest. *See* Aphids, 175.
Yellow mottling or dieback of needles.	**Pine needle scale**, *Chionaspis pinifoliae*; **Black pine leaf scale**, *Nuculaspis californica*. White, gray, or black armored scales, $^1/_{16}$ inch long, on needles.	Scales have several generations per year in warm areas, only one at cool sites. Plants tolerate moderate populations. Conserve natural enemies. *See* Armored scales, 189.
Stippled or bleached needles, more common on young pines.	**Spider mites**, *Oligonychus* spp. Green to pink specks on needles, suck sap.	*See* Pine and Spruce Spider Mites, 247.
Needles chewed, notched along length. Needles turn brown in late winter or spring. Damaged needles drop prematurely.	**Pine needle weevils**, *Scythropus* spp. Adults brownish snout beetles, $^1/_4$ inch long.	Adult damage to needles and larvae feeding on roots appear not to harm trees. No control known. *See* Weevils, 159.
Chewed needles.	**Conifer sawflies**, *Neodiprion* spp. Green larvae, ≤ 1 inch long, on needles.	*See* 153.
Chewed needles, may be webbed with silk.	**Silverspotted tiger moth**, *Lophocampa argentata*; **Tussock moths**. Dark, hairy larvae, $\leq 1^1/_4$ inches, may have colorful hairs or spots.	Adults brownish to tan moths and may have silvery spots. *See* 146, 152.
Mined buds and shoot tips. Killed tips give tree red or brown appearance. Foliage becomes bunchy-looking.	**Monterey pine tip moth, Nantucket pine tip moth, Ponderosa pine tip moth**. Orangish larvae in mines.	Adults, $^1/_3$ inch long, yellow, gray, to brown moths. Pines tolerate extensive tip mining. *See* 218.
Terminals distorted, chewed, dead. Foliage may become busy, crooked. Roots or the basal trunk may be injured.	**Conifer twig weevils**, or **Pine weevils**, including **White pine weevil**. Small black to brown weevils chew needles and shoots.	Larvae chew on or within needles, shoots, or the root crown. Tolerate or prune out damage. *See* 161.
Terminals brown, dead, or webbed with silk. Needles are clipped, mined, pale streaked, or wilted and drop prematurely. Foliage may be sparse. Plant growth stunted.	**Pine needle miners**, *Chionodes* spp., *Coleotechnites* spp. ; **Pine needle sheathminer**, *Zelleria haimbachi*. Tiny moths. Larvae feed in or on needles.	Where previously severe, a spring application to new growth of an insecticide registered for this use may provide control. *See* Foliage Miners, 216.
Needles distorted, short, twisted, yellow, and may drop prematurely. Plants may grow slowly.	**Pine bud mite**, *Trisetacus* sp. Tiny, elongate eriophyid mites that suck bud tissue.	*See* Gall Mites, or Eriophyids, 249.
Sections of shoot with greatly shortened needles with swollen bases.	**Monterey pine midge**, *Thecodiplosis piniradiatae*. White or orangish larvae in swollen needles.	Control generally not warranted. No management known. *See* Gall Midges, 214.
Wet, white, frothy masses of spittle on twigs or cones.	**Spittlebugs**, including **Western pine spittlebug**. Green to black insects in spittle.	Tolerate, spittlebugs cause no apparent harm to pines. *See* 205.
Tips of Monterey pine mined, but only for 1 or 2 inches. Tips die, often in crooked position.	**Monterey pine bud moth**, *Exoteleia burkei*. Larvae brownish yellow, $\leq^3/_{16}$ inch long, in mines.	Damage very localized and unlikely to harm tree. Prune out and dispose of affected tips, no other control known.
Small, leafless, orangish, upright plants on host stems. Distorted and slow plant growth. Branches die back.	**Dwarf mistletoe**, *Arceuthobium* spp. Host-specific parasitic plants that extract nutrients from host plant.	Prune out infected branches. Replace heavily infested plants with broadleaf species or conifers from other genera; dwarf mistletoes won't spread to unrelated species. *See* 284.

WHAT THE PROBLEM LOOKS LIKE	PROBABLE CAUSE	COMMENTS
Tree declining or dead. Boring dust or coarse granular material around tree base or on bark plates or branch crotches. Pitch tubes on bark.	**Bark beetles.** Brown to black, stout beetles ≤$^1/_4$ long. Larvae white grubs under bark.	*See 222.*
Foliage discolors. Tree or limbs declining or dead.	**Flatheaded borers, Longhorned beetles.** Whitish larvae, ≤1 inch long, tunneling beneath bark.	*See 227, 232.*
Plants stunted, decline, may die. Powdery, waxy material may be visible on roots and around crown.	**Ground mealybugs**, *Rhizoecus* spp. Small, slender, pale insects, may be lightly covered with powdery wax, but lack marginal filaments.	*See 187.*

Pistacia chinensis, Chinese pistache, Pistache

WHAT THE PROBLEM LOOKS LIKE	PROBABLE CAUSE	COMMENTS
Leaves discolor, stunt, wilt, or drop prematurely. Plants grow slowly and may die. Roots dark, decayed.	**Phytophthora root and crown rot.** Disease favored by excess soil moisture and poor drainage.	*See 122.*
Leaves discolor and wilt. Branches die back. Entire tree may die. May be white fungus beneath basal trunk bark.	**Armillaria root disease.** Fungus present in many soils. Favored by warm, wet soil. Persists for years in infected roots.	*See 120.*
Leaves brown, fade, yellow, or wilt, often scattered throughout canopy. Foliage may appear sparse, undersized. Plants may grow slowly. Branches die. Entire plant may die.	**Verticillium wilt**, *Verticillium albo-atrum*, *V. dahliae*. Soil-dwelling fungi that infect through roots.	*See 117.*
Leaves turn yellow or reddish and drop prematurely.	**Natural senescence.** Leaves often naturally drop sooner than on other species.	Providing trees with a good growing environment and appropriate cultural care may delay leaf drop.
Leaves, buds, or shoots blacken or brown and die. Branches or twigs may develop cankers or die.	**Botryosphaeria canker and dieback.** Pathogen persists in twigs and dead tissue and is favored by wet conditions.	*See 98.*
Leaves and shoots brown, yellow, and may die. Foliage may be distorted, undersized, or be covered with powdery, white growth.	**Powdery mildew**, *Oidium* sp. Fungal disease favored by moderate temperatures, shade, and poor air circulation.	*See 90.*
Leaves chewed. Foliage webbed with silk.	**Obliquebanded leafroller.** Greenish yellow larvae, ≤1$^1/_2$ inches long.	Adults are a small orangish moth with brown and white. *See Leafrollers, 149.*
Leaves chewed, may be tied with silk. Plant may be defoliated.	**Western tussock moth.** Dark, hairy larvae, ≤2 inches long, with red and yellow spots.	Adults are grayish moth with black and white. *See 152.*
Sticky honeydew and blackish sooty mold on foliage. Possible decline or dieback of twigs and branches.	**Black scale, Brown soft scale, European fruit lecanium, Frosted scale.**	Black, brown, orange, yellow, or waxy whitish, flattened or bulbous insects. *See 192–195.*
Sticky honeydew, blackish sooty mold, and whitish cast skins on foliage.	**Aphids.** Small insects, usually in groups, often pale yellowish, waxy, or white.	*See 175.*

Pittosporum spp., Pittosporum, Mock orange, Tobira, Victorian box

WHAT THE PROBLEM LOOKS LIKE	PROBABLE CAUSE	COMMENTS
Leaves discolor, stunt, wilt, and remain dead on plant or drop prematurely. Plants grow slowly and may die. Root crown dark, decayed.	**Phytophthora root and crown rot.** Disease favored by excess soil moisture and poor drainage.	*See 122.*
Flower buds, leaves, petals, or stems turn brown and die. Stems may ooze or be cankered.	**Bacterial blast, blight, and canker.** Disease favored by wet conditions.	*See 83.*
Leaves with brown, light green, or yellow spots or dead blotches.	**Ramorum blight**, *Phytophthora ramorum*. Pathogen spreads via airborne spores and contaminated plants and soil. Infects at least *Pittosporum undulatum*.	Primarily a problem in nurseries and on other hosts in wildlands, killing many oaks there. *See* Sudden Oak Death and Ramorum Blight, 105.
Twigs distorted, swollen, and pitted. Leaves dwarfed. Shoots may die back.	**Pit-making pittosporum scale**, *Planchonia* (=*Asterolecanium*) *arabidis*. Brown to white insects, ≤$^1/_8$ inch long, on twigs, often in pits.	In California, an occasional problem in the north. Management not investigated. *See* similar Oak Pit Scales, 198.

WHAT THE PROBLEM LOOKS LIKE	PROBABLE CAUSE	COMMENTS
Sticky honeydew, blackish sooty mold, and pale wax on plant.	**Pittosporum psyllid**, *Cacopsylla tobirae*. Flattened, greenish nymphs secrete wax strands.	On *Pittosporum tobira* in at least southern California. *See* Psyllids, 170.
Sticky honeydew, blackish sooty mold, and whitish cast skins on plant.	**Apple aphid**, *Aphis pomi;* **Melon aphid; Woolly apple aphid**. Bright green insects on terminals or leaves.	Conserve natural enemies that provide control. *See* Aphids, 175.
Sticky honeydew and blackish sooty mold on plant. May be cottony whitish material (egg sacs) on bark.	**Cottony cushion scale, Green shield scale**. Brownish, green, orange, red, or yellowish convex or flattened insects on bark or leaves.	*See* 130, 194, 196.
Sticky honeydew, blackish sooty mold, and cottony waxy material on plant.	**Longtailed mealybug, Obscure mealybug**. Powdery, grayish, segmented insects with fringe filaments, longer at tail.	*See* 187.
Brownish, grayish, tan, or white encrustations on bark. Rarely, declining or dead twigs or branches.	**Greedy scale**. Nearly circular insects, $^1/_{16}$ inch long, on bark.	*See* 190.
Leaves discolored, wilted, stunted, may drop prematurely. Branches or plant may die. Trees may fall over.	**Wood decay**, or **Heart rot**, including *Ganoderma applanatum*. Decay fungi that infect injured or severely stressed plants.	*Ganoderma* produces large, globular or bracketlike basidiocarps on lower trunks. *See* 66, 111.
Galls or swellings on roots.	**Root knot nematodes**. Tiny root-feeding roundworms.	*See* 293.
Platanus spp., Sycamore, London plane, Plane tree		
Foliage yellows and wilts. Branches or entire plant dies. Rootlets decayed or sparse.	**Phytophthora root and crown rot**. Disease favored by wet soil and poor drainage.	*See* 122.
Leaves smaller and fewer in number than normal. Rapid decline and death of entire tree.	**Sycamore canker stain**. Fungus enters wounds and spreads by contaminated equipment or tools or root grafts.	In California occurs at least in the San Joaquin Valley. *See* 107.
Foliage yellows and wilts. Bark may have lesions that are dark, dry, crusty whitish, oily looking, or water-soaked. Branches or entire plant dies.	**Fusarium dieback**. Fungal disease spread by the polyphagous shothole borer, an ambrosia beetle, or bark beetle.	On California sycamore in at least southern California. *See* 115.
Bark may ooze and have tiny holes. Foliage may yellow and wilts. Branches may die.	**Polyphagous shothole borer**. An ambrosia beetle, or bark beetle, which vectors *Fusarium*.	Occurs at least in southern California. *See* 222.
Foliage yellows and wilts. Bark or wood may have bracketlike or fan-shaped fungal fruiting bodies. Branches or entire plant may die.	**Wood decay**, or **Heart rot**, including *Ganoderma applanatum, Trametes hirsuta*. Fungi that attack injured, old, or stressed trees.	*See* 66, 111.
Leaves or shoots with white, powdery growth. Terminals may be distorted or stunted.	**Powdery mildew**, *Microsphaera alni*. A fungal disease favored by moderate temperatures, shade, and poor air circulation.	Disease most damaging on severely pruned trees. Plant resistant Columbia, Liberty, or Yarwood cultivars. *See* 90.
Leaves spotted, brown, or yellow, and may distort or drop prematurely. Cottony material in bark crevices overwinter.	**Sycamore scale**. Yellowish insects, $\leq^1/_{16}$ inch long, feed in center of yellow spot on lower leaf surface.	*See* 199.
Leaves, buds, and shoots distorted and discolored. Irregular, brown dead areas along leaf veins. Twigs die back.	**Anthracnose**, *Apiognomonia veneta* =*Discula veneta*. A fungal disease favored by wet conditions.	Branches may grow crookedly or be bent from regrowth after dieback. Grow resistant Bloodgood, Columbia, or Liberty cultivars. *See* 81.
Leaves with brown to tan spots. Infected tissue may drop, leaving leaf holes. Leaves may drop prematurely.	**Leaf spot**, *Stigmina platani-racemosae*. A fungal disease promoted by humid weather, wet leaves, and shading of leaves.	Can be severe during late summer and fall on sycamore exposed to ocean air. *See* 86.
Leaves with discolored spots, often between veins. Large spots chlorotic, then tan or brown.	**Summer leaf scorch**, or **Sunburn**. Noninfectious disorders, commonly appear during or after drought stress or high temperatures.	*See* 57.

Platanus spp. (continued)

WHAT THE PROBLEM LOOKS LIKE	PROBABLE CAUSE	COMMENTS
Leaves with irregular, brown or yellow blotches or spots. Leaves may become holey or tattered. Leaves may drop prematurely.	**Sycamore plant bug**, *Plagiognathus albatus*. Brown, greenish, or yellow bugs, ≤³/₁₆ inch long.	Dark, varnishlike bug excrement may be on undersides of leaves. *See* similar Ash Plant Bugs, 206.
Leaves stippled, bleached, and have cast skins and varnishlike specks. May be sticky honeydew on plant.	**Western sycamore lace bug**. *Corythucha confraterna*. Adults ≤¹/₈ inch long, wings lacy. Nymphs spiny.	*See* 207.
Leaves stippled and may become bleached.	**Sycamore spider mite**. Tiny, green arthropods, suck sap.	Problem in California mostly in interior valleys. *See* 246.
Sticky honeydew and blackish sooty mold on plant.	**European fruit lecanium, Frosted scale**. Hemispherical to flattened, brown, yellow, or whitish insects, suck sap.	*See* Lecanium Scales, 195.
Sticky honeydew, blackish sooty mold, and whitish cast skins on foliage.	**Aphids**, including **Melon aphid**. Small, greenish, blackish, or yellowish insects in groups.	*See* 175.
Webbing or silk tents on ends of branches. Chewed leaves.	**Fall webworm**. Hairy, white to yellow larvae, ≤1 inch long, in colonies.	*See* 152.
Young leaves skeletonized. Holes in leaves.	**Sycamore leaf skeletonizer**, *Gelechia desiliens*. Greenish larvae, ≤¹/₂ inch long, in tubular nest on leaves.	Plants tolerate extensive skeletonization. Tolerate or apply *Bacillus thuringiensis* or another insecticide if young moth larvae are abundant. *See* Foliage Miners, 216.
Brown to gray encrustations on bark. Possible stunting or dieback of woody parts.	**Oystershell scale**. Tiny, oyster-shaped insects, often in colonies.	*See* 191.
Decline or death of limbs or entire plant. Larvae, ≤1 inch long, tunneling under bark.	**Flatheaded appletree borer, Pacific flatheaded borer**. Whitish larvae with enlarged head in tunnels.	*See* 227.
Greatly roughened bark and boring dust on lower trunk and branch crotches. Slow tree growth. May be dark ooze from bark.	**American plum borer, Sycamore borer**. Brown to pink larvae that bore under bark.	Distinguish between these differently managed species. Trees are not seriously harmed by sycamore borer. *See* 236, 240.
Podocarpus spp., Podocarpus, African fern pine, Yew pine		
Brownish to orangish encrustations on bark.	**California red scale**. Circular to oval insects, <¹/₁₆ inch long.	Rarely if ever causes serious damage. *See* Armored Scales, 189.
Sticky honeydew, blackish sooty mold, and whitish cast skins on leaves. Bluish white bloom covering foliage.	**Podocarpus aphid**, *Neophyllaphis podocarpi*. Grayish insects, about ¹/₁₆ inch long, grouped on stems or leaves.	Plants tolerate extensive aphid feeding. *See* Aphids, 175.
Populus spp., Poplar, Cottonwood		
Leaves or terminals with powdery, white growth and yellow blotches. Shoots and leaves may be undersized or distorted.	**Powdery mildew**, *Uncinula adunca*. Fungal disease favored by moderate temperatures, shade, and poor air circulation.	*See* 90.
Leaves yellow, wilt, and may drop prematurely. Branches die back. Entire tree may die. May be white fungus beneath basal trunk bark.	**Armillaria root disease**. Fungus present in many soils. Favored by warm, wet soil. Persists for years in infected roots.	*See* 120.
Leaves wilted, discolored, may drop prematurely. Branches or entire plant may die. May be white fungal growth or dark crust on basal trunk, roots, or soil.	**Dematophora (Rosellinia) root rot**. Fungus favored by mild, wet conditions. Infects primarily through roots growing near infested plants.	Less common than *Armillaria*. *See* 121.
Black to brown lesions, spots, or streaks on blossoms, leaves, and stems. Cankers and brown streaks on wood. Limbs may die back.	**Bacterial blast, blight, and canker**. Disease favored by wet conditions.	*See* 83.
Leaves may discolor and wilt. Brownish, sunken lesions on trunk and large limbs. Small branches and twigs may be killed without apparent canker.	**Annulohypoxylon canker**, *Annulohypoxylon mediterraneum*; **Cytospora canker**. Fungal diseases most serious on low-vigor trees.	*See* 98, 100.

WHAT THE PROBLEM LOOKS LIKE	PROBABLE CAUSE	COMMENTS
Leaves yellow and wilt. Bark or wood may have bracketlike or fan-shaped fungal fruiting bodies. Branches or entire plant may die. Trees may fall over.	**Wood decay**, or **Heart rot**, including *Ganoderma* spp., *Laetiporus gilbertsonii* (=*L. sulphureus*), *Pleurotus ostreatus*, *Schizophyllum commune*, *Trametes* spp.	Fungi that infect injured, old, or stressed trees. *See* 66, 111.
Leaves with circular to irregular, tan or darker blotches or spots on leaves. Terminals may die back. Branches may canker on some hosts.	**Leaf spots** and **Shoot blights**, including *Mycosphaerella populorum* (on cottonwood), *Marssonina* spp. (poplar), *Septoria* sp. (cottonwood and poplar), *Venturia* spp. (on all).	Fungal diseases promoted and spread by water. *See* 86.
Leaves turn brown at the margins and tip. Leaves may be curled or puckered. Leaves may drop prematurely.	**Leaf burn**, or **Scorch**. Abiotic disorders. Common causes include drought, excess heat, injury to trunks or roots, overirrigation, and poor drainage.	Protect trees from injury. Provide good cultural care, especially appropriate irrigation. *See* Abiotic Disorders, 43.
Leaves with light to dark spots and orangish pustules. Leaves may drop prematurely.	**Rusts**, *Melampsora* spp. Fungi require moist conditions.	Avoid overhead watering. Vigorous plants tolerate moderate infection. *See* 94.
Leaves with pale blisters, spots, or felty masses of pale spores.	**Yellow leaf blister and spot**, *Taphrina* spp. Fungi favored by wet conditions during spring.	Rarely serious enough to threaten tree health or warrant control effort.
Leaves stippled or bleached, with cast skins and varnishlike specks.	**Lace bug**, *Corythucha* sp. Brown adults, ≤1/8 inch long, wings lacelike.	*See* 207.
Leaves chewed. Tree may be defoliated. Leaves may be tied together with silk. May be silken tents in tree.	**Caterpillars**, including **Fall webworm; Fruittree leafroller; Satin moth**, *Leucoma salicis*; **Silverspotted tussock moth**, *Lophocampa maculata*; **Tent caterpillars**.	*Leucoma* and *Lophocampa* larvae are ≤1 1/2 inches long, hairy, and black with yellow, white, or reddish. *See* 144–153.
Leaves chewed on scattered terminals. Caterpillars feeding in groups on shoots.	**Redhumped caterpillar; Spiny elm caterpillar; Western tiger swallowtail**, *Papilio rutulus*. Papilio larvae are bright green with eyespots and black and yellow markings. Adult *Papilio* are ≤2 inches long and yellow with black.	Spiny elm caterpillars and *Papilio* mature into attractive butterflies. Control not recommended. *See* 150, 151.
Leaves with elliptical blotches or holes, 1/8 to 1/4 inch long, or winding tunnels.	**Poplar shield bearer**, *Coptodisca* sp. Moth larvae mine foliage.	*See* 219.
Skeletonized leaf surfaces. No silk.	**Leaf beetles, Flea beetles**, *Altica* spp., *Chrysomela* spp., *Plagiodera* spp.	Adults dark or metallic, oval, ≤3/8 inch long. Larvae are dark, ≤1/2 inch long. *See* 155.
Sticky honeydew, blackish sooty mold, and whitish cast skins on leaves.	**Aphids**, including **Cloudywinged cottonwood aphid**, *Periphyllus populicola*; *Chaitophorus* spp. Brownish, gray, green, or yellowish, pear-shaped insects, suck sap.	Plants tolerate abundant aphids. *See* 175.
Sticky honeydew and blackish sooty mold on foliage. Possible decline or dieback of twigs or branches.	**Black scale; Brown soft scale; Cottony maple scale**, *Pulvinaria innumerabilis*; **European fruit lecanium scale**.	Oval, yellow to brown, and flattened; or black, brown, or cottony and bulbous. *See* 192–195.
Swellings (galls), often globular or purselike, on leaves and leaf petioles.	**Poplar gall aphids**. Tiny, grayish, waxy insects in galls.	*See* 177.
Brown to gray encrustations on bark. May be declining or dead twigs or branches.	**Oystershell scale, San Jose scale**. Tiny, oval to circular insects on bark.	*See* 189–192.
Dieback of branches or sometimes entire tree. Wet or dark spots on bark. Bark galled or gnarled.	**Carpenterworm; Clearwing moths**, including **American hornet moth** and **Western poplar clearwing; Flatheaded borers**, including the **Bronze birch borer** and **Bronze poplar borer; Longhorned beetles**, including *Saperda* spp.	Whitish larvae mine beneath bark or in wood. *See* 227–242.
Roughened twig bark. Possible twig dieback.	**Treehoppers**, including **Buffalo treehopper**, *Stictocephala bisonia*. Bright yellow to green insects, horny or spiny.	*See* 204.
Warty, woody swellings on twigs around buds. Terminals may be crooked.	**Cottonwood gall mite**, *Eriophyes parapopuli*. Tiny, elongate eriophyid mites in galls.	Plants tolerate extensive galling. *See* Gall Mites, or Eriophyids, 249.

Populus spp. (continued)

WHAT THE PROBLEM LOOKS LIKE	PROBABLE CAUSE	COMMENTS
Green plants with smooth stems, thick roundish leaves infesting branches.	**Broadleaf mistletoe**. Parasitic, evergreen plant on host.	*See* 284.
Bark stained brownish, exudes rancid fluid, often around crotches, wounds.	**Wetwood**, or **Slime flux**. Bacterial infections.	Usually do not cause serious harm to trees. *See* 110.
Prosopis spp., Mesquite		
Swollen growths on stems, may produce orange, powdery or sticky material. Foliage may distort, gall, or form stunted witches' brooms.	**Stem gall rust**, *Ravenelia holwayi*. Fungal disease infects through bark wounds.	Pruning and disposing of infected stems during dry conditions may provide some control. *See* similar Western Gall Rust, 96.
Black, brown, or tan encrustations on leaves or twigs. Rarely, declining or dead twigs or branches.	**Candidula scale**, *Hemiberlesia candidula;* **Mesquite scale**, *Xerophilaspis prosopidis*. Tiny, flattened, oval to circular insects.	Rarely if ever cause serious damage to plants. *See* Armored Scales, 189.
Sticky honeydew, blackish sooty mold, and cottony or waxy material on plant.	**Mealybugs**, including **Vine mealybug**. Powdery, grayish insects, $\leq 1/4$ inch long, with waxy filaments.	*See* 185.
Dead tree or dying limbs. Broad galleries beneath bark.	**Borers**, including **Flatheaded appletree borer; Mesquite girdler**, *Oncideres rhodosticta;* **Oldman longhorn**, *Schizax senex;* **Roundheaded mesquite borer**, *Megacyllene antennatus*.	Pale larvae that tunnel beneath bark. *See* Flatheaded Borers, or Metallic Wood Borers, 227; Longhorned Beetles, or Roundheaded Wood Borers, 232.
Prunus caroliniana, Carolina cherry laurel; *P. ilicifolia*, Holly-leaf cherry, Holly-leaved cherry, Evergreen cherry; *P. laurocerasus*, Cherry laurel, English laurel; *P. lyonii*, Catalina cherry		
Foliage yellows and wilts. Entire plant may die. May be white fungus beneath basal trunk bark.	**Armillaria root disease**. Fungus present in many soils. Favored by warm, wet soil. Persists for years in infected roots.	*See* 120.
Foliage yellows and wilts. Branches or entire plant dies.	**Phytophthora root and crown rot**. Disease favored by wet soil and poor drainage.	*See* 122.
Leaves with brown, reddish, or yellow discolored blotches. Infected blotches may drop, leaving holes in leaves. Leaves may drop prematurely.	**Leaf spots**, including the bacterium *Pseudomonas syringae* and fungi such as *Blumeriella* sp., *Cercospora* sp. Pathogens favored by wet conditions.	*See* 83, 86.
Small brownish, purplish, or reddish leaf spots, centers tan. Holes in leaf from dropped, infected tissue. Concentric lesions on branch.	**Shot hole**. A fungal disease favored by prolonged wet conditions.	*See* 89.
Leaves with necrotic blotches or spots. Spots may have reddish or yellow margins. Leaves may drop prematurely.	**Physiological leaf spot**. Abiotic disorders with many potential causes, including boron toxicity, direct injury to bark or roots, high temperatures, poor drainage, and too much or too little water.	Provide plants with good growing conditions and proper cultural care. *See* Abiotic Disorders, 43.
Powdery, white growth and brown or yellow blotches on leaves and sometimes on stems. Shoots and leaves may be undersized or distorted.	**Powdery mildews**, *Podosphaera* spp., *Sphaerotheca pannosa*. Fungal diseases favored by moderate temperatures, shade, and poor air circulation.	*See* 90.
Brownish, grayish, tan, or white encrustations on bark.	**Greedy scale, Oleander scale**. Insects are circular to flattened, $<1/16$ inch long.	Rarely if ever cause serious damage to plants. *See* 190, 191.
Dead twigs or branches. Plant declines, may die. Tiny shot holes in bark.	**Shothole borer**. Small, brown adult beetles, white larvae, tunnel beneath bark.	*See* 225.
Sticky honeydew, blackish sooty mold, and whitish cast skins on foliage.	**Aphids**. Small, blackish, brown, green, or yellowish insects, usually in groups.	*See* 175.
Sticky honeydew and blackish sooty mold on plant. Woody parts may decline and die back.	**Brown soft scale, Frosted scale**. Oval, flat or bulbous insects, may be brown, orangish, yellow, whitish, or waxy.	*See* 193; Lecanium Scales, 195.
Stippled or bleached leaves, varnishlike excrement specks on undersides.	**Greenhouse thrips**. Tiny, slender, blackish adults or yellowish nymphs.	*See* 210.

WHAT THE PROBLEM LOOKS LIKE	PROBABLE CAUSE	COMMENTS
Stippled, flecked, or bleached leaves. Leaves may drop prematurely.	**Mites**, including **Citrus red mite**. Tiny, sand-sized arthropods on leaves, often reddish, suck sap.	*See* Red Mites, 246.

Prunus cerasifera, Prunus hybrids, Plum, Flowering plum, Purple-leaf plum[26]

Leaves discolor, stunt, wilt, or drop prematurely. Stems discolored, cankered, and die. May be white fungus beneath basal trunk bark.	**Armillaria root disease**. Fungus present in many soils. Favored by warm, wet soil. Persists for years in infected roots.	*See* 120.
Leaves discolor, stunt, wilt, or drop prematurely. Plants grow slowly and may die. Roots dark, decayed.	**Phytophthora root and crown rot**. Disease favored by excess soil moisture and poor drainage.	*See* 122.
Leaves discolor, stunt, wilt, or drop prematurely, often on one side of plant. Stem xylem discolored. Stems or entire plant may die.	**Verticillium wilt**. Persists in soil, infects through roots.	*See* 117.
Leaves turn yellow then brown and drop prematurely. Entire plant dies rapidly or slowly over several years.	**Bacterial leaf scorch**, or **Oleander leaf scorch**. Bacteria spread by certain leafhoppers, especially glassy-winged sharpshooter.	*See* 112.
Leaves discolor and die. Stems, flowers, or fruit with dark lesions. Cankers possible on branches.	**Bacterial blast, blight, and canker**. Disease favored by wet conditions.	*See* 83.
Leaves reddened, curled, distorted in spring. Shoots thickened, distorted, may die. Leaves may drop prematurely.	**Plum pockets**, *Taphrina* spp. Fungal diseases promoted by moist spring weather or splashing irrigation water.	Apply Bordeaux, fixed copper, or synthetic fungicide in fall after leaf drop, repeat when buds swell.
Powdery, white growth and yellow blotches on leaves and sometimes on terminals and fruit. Shoots and leaves may be undersized or distorted.	**Powdery mildews**, *Podosphaera* spp., *Sphaerotheca pannosa*. Fungal diseases favored by moderate temperatures, shade, and poor air circulation.	*See* 90.
Sticky honeydew, blackish sooty mold, and whitish cast skins on plant. Leaves may curl or be covered with whitish wax.	**Aphids**, including **Green peach aphid; Leaf curl plum aphid**, *Brachycaudus helichrysi*; **Mealy plum aphid**, *Hyalopterus pruni*.	Small brownish, green, or yellowish, pear-shaped insects. *See* 175.
Sticky honeydew and blackish sooty mold on leaves and twigs. Possible dieback.	**Kuno scale**. Adult scales beadlike and dark shiny brown.	*See* 195.
Leaves chewed. Single branches or entire plant may be defoliated. Foliage may be rolled and tied together with silk.	**Fruittree leafroller, Omnivorous looper, Redhumped caterpillar, Tent caterpillars, Tussock moths**. Moth larvae, ≤1½ inches long.	*See* 149–152.
Galls or swellings may be on roots. Foliage may yellow or wilt. Roots may be darker or fewer than normal. Plants may grow slowly.	**Pin nematode, Ring nematode, Root lesion nematode, Root knot nematodes**. Tiny root-feeding roundworms.	*See* 293, 294.
Galls or swellings on trunk and roots, usually near soil.	**Crown gall**. Bacterium that infects plant via wounds.	*See* 108.
Dead twigs or branches. Plant declines, may die. Tiny, BB shot–size holes in bark.	**Shothole borer**. Small, brown bark beetles and whitish larvae tunnel beneath bark.	*See* 225.
Roughened bark, reddish brown granular material (frass) at base of trunk and main limbs. May be dark ooze from bark.	**American plum borer; Peachtree borer; Prune limb borer**, *Bondia comonana*. Brown to pink larvae that bore under bark.	Prevent injuries to bark. Provide proper cultural care. *See* 235, 240.

Pseudotsuga menziesii, Douglas-fir[27]

Foliage discolors and wilts. Branches die back. Entire tree may die. May be white fungus beneath basal trunk bark.	**Armillaria root disease**. Fungus present in many soils. Favored by warm, wet soil. Persists for years in infected roots.	*See* 120.
Foliage discolors, wilts, stunts, and may drop prematurely. Branches or plant may die. Plant base or roots dark or decayed.	**Phytophthora root and crown rot**. Disease favored by excess soil moisture and poor drainage.	*See* 122.

[26] For pests of fruiting plum, see *Pests of the Garden and Small Farm* and ipm.ucanr.edu.

[27] For more information, see *Pests of the Native California Conifers*.

Pseudotsuga menziesii (continued)

WHAT THE PROBLEM LOOKS LIKE	PROBABLE CAUSE	COMMENTS
Foliage discolors and wilts. Branches die back. Treetop or entire plant may die. Cankers on bark.	**Phomopsis canker**, *Phomopsis lokoyae*. A fungal disease that primarily affects young, stressed trees.	Provide appropriate soil moisture and cultural care. *See* Canker Diseases, 97.
Foliage browns, yellows, wilts, and drops prematurely. Branches die back. Entire plant may die.	**Heterobasidion root disease**. Decay fungus that spreads through natural root grafts and airborne spores.	*See* 121.
Soft, brown decay and/or gray, woolly growth on buds, flowers, leaves, and dead tissue.	**Botrytis blight**, or **Gray mold**. Fungal disease promoted by wet spring weather.	Improve airflow around plants. *See* 78.
Needles light green, yellow, or pale spotted. Orangish to yellow pustules on underside of needles. If severe, shoots develop reddish cankers and die back.	**Rusts**, including *Melampsora occidentalis*. Fungal diseases promoted by wet conditions.	*See* 94.
Needles brown, light green, yellow, or pale spotted, often beginning at tips. Tiny, black or orangish fruiting bodies on needles. Needles drop prematurely.	**Needle blight and cast**, including **Douglas-fir needle cast**, *Rhabdocline pseudotsugae, R. weirii*; **Swiss needle cast**, *Phaeocryptopus gaeumannii*.	*See* 85.
Needles discolor and die on branch terminals. Twigs or small branches cankered and may die.	**Ramorum blight**, *Phytophthora ramorum*. Pathogen spreads via airborne spores and contaminated plants and soil.	Primarily a problem in wildlands, killing many oaks there. *See* Sudden Oak Death and Ramorum Blight, 105.
Branch terminals turn brown and die. Tree deformed. Growth retarded.	**Douglas-fir twig weevil**, *Cylindrocopturus furnissi*. Adults black to gray. Pale larvae mine twigs. Primarily attacks small, stressed trees.	Provide appropriate irrigation and other cultural care. Prune and dispose of infested shoots during fall. *See* similar Conifer Twig Weevils, 161.
Yellow mottling or dieback of needles.	**Pine needle scale**, *Chionaspis pinifoliae*; **Black pineleaf scale**, *Dynaspidiotus californicus =Nuculaspis californica*. White, gray, or black armored scales, 1/16 inch long, on needles.	Scales have several generations per year in warm areas, only one at cool sites. Plants tolerate moderate populations. Conserve natural enemies. *See* Armored Scales, 189.
Interior needles turn brown or yellow and drop prematurely, leaving only young terminal needles. Tree may die.	**Spruce aphid**, *Elatobium abietinum*. Small, pear-shaped insects, dark to light green, in groups on older foliage.	Usually abundant only on spruce, where it can be a serious pest. *See* Aphids, 175.
Pale greenish or yellow needles. Foliage stippled. May be fine webbing at foliage base.	**Spider mites**, including **Spruce spider mite**. Tiny arthropods, often greenish, suck sap.	*See* Pine and Spruce Spider Mites, 247.
Sticky honeydew, blackish sooty mold, and whitish cast skins on needles. Brown to purplish insects clustered on foliage.	**Giant conifer aphids**. Dark, long-legged insects, ≤1/5 inch long.	Harmless to trees. *See* 176.
Sticky honeydew, blackish sooty mold, and whitish cast skins on needles.	**Aphids**, including **Monterey pine aphids**, *Essigella californica*.	Small, pear-shaped insects, usually green, may be in groups. *See* 175.
Cottony white tufts on needles. Needles have yellow spots.	**Adelgids**, including **Cooley spruce gall adelgid**. Tiny, purplish insects beneath cottony tufts.	*See* 181.
Wet, white, frothy masses on needles or twigs.	**Spittlebugs**, including **Western pine spittlebug**, *Aphrophora permutata*. Green to black sucking insects secrete spittle.	Tolerate; spittlebugs cause no apparent harm to trees. *See* 205.
Pitchy masses 1 to 4 inches in diameter protruding from trunks and limbs. Limbs occasionally break.	**Douglas-fir pitch moth**. Dirty whitish larvae, ≤1 inch long, in pitch.	*See* Pitch Moths, 239.
Globular growths 1/2 to 12 inches in diameter on stems. Shoots may die back.	**Bacterial gall**, possibly *Agrobacterium pseudotsugae*. Disease of crowded, stressed trees under wet conditions.	*See* similar Crown Gall, 108.
Needles galled or swollen. Possible needle drop and twig dieback.	**Needle and twig midges**, *Contarinia* spp. White larvae in swollen needles. Tiny, mosquitolike adults emerge from pupae in soil.	Apparently do not harm landscape trees. Prune out damaged shoots. *See* Gall Midges, 214.

WHAT THE PROBLEM LOOKS LIKE	PROBABLE CAUSE	COMMENTS
Buds or terminals curled, distorted, or swollen. Shoots may grow crookedly.	**Douglas-fir bud mite**, *Trisetacus* sp. Tiny, elongate eriophyid mites that suck bud tissue.	*See* Gall Mites, or Eriophyids, 249.
Needles are chewed. Roots or the basal trunk may be cankered or injured.	**Conifer twig weevils**. Small, black to brown weevils chew needles and shoots. Pale larvae bore in the root crown of dying or injured conifers.	Except for white pine weevil, most are secondary pests of minor importance. *See* 161.
Needles chewed or notched. Branches may yellow or die. Some roots stripped of bark or girdled near soil.	**Black vine weevil**. Adults are black or grayish snout beetles, about $^3/_8$ inch long. Larvae are white grubs with brown head.	*See* 160.
Needles chewed. Foliage browns. May be silk webbing on needles.	**Douglas-fir tussock moth**, *Orgyia pseudotsugata;* **Silverspotted tiger moth**, *Lophocampa argenta*. Caterpillars hairy, brownish with orange, red, or yellow.	Vigorous trees tolerate moderate damage. Apply *Bacillus thuringiensis* or another insecticide to control young larvae, if too abundant. *See* 146, 152.

Pyracantha spp., Pyracantha

Leaves yellow, wilt, and may drop prematurely. Branches die back. Entire tree may die. May be white fungus beneath basal trunk bark.	**Armillaria root disease**. Fungus present in many soils. Favored by warm, wet soil. Persists for years in infected roots.	*See* 120.
Leaves brown, yellow, and die, often beginning with older foliage. Browning of vascular tissue.	**Fusarium wilt**, *Fusarium oxysporum*. Fungus infects plant through roots.	Avoid injuring live tissue. Provide adequate drainage and appropriate irrigation. *See* 115.
Sudden wilting, shriveling, blackening of shoots, blossoms, fruits. Plant appears scorched.	**Fire blight**. Bacteria enter plant through blossoms.	*See* 84.
Black to dark olive spots on fruit and sometimes on leaves. Leaves yellow and may drop prematurely.	**Scab**, *Spilocaea pyracanthae*. A fungal disease spread by water from infected leaves and twigs.	Avoid overhead watering. Prune out infected twigs and leaves in fall. *See* 88.
Tiny, reddish to brown leaf spots, may have yellow halos. Larger, dark areas on leaves. Leaves may drop prematurely.	**Entomosporium leaf spot**. A fungal disease promoted by wet foliage.	*See* 87.
Powdery, white growth and yellow blotches on leaves and sometimes on terminals and fruit. Shoots and leaves may be undersized or distorted.	**Powdery mildews**, including *Podosphaera* sp. Fungal diseases favored by moderate temperatures, shade, and poor air circulation.	*See* 90.
Reddening or bronzing of foliage.	**Spider mites**, including **Southern red mite, Sycamore spider mite**.	Tiny brownish, green, or reddish arthropods on leaves. *See* 246.
Stippled or bleached leaves. Dark, varnishlike specks and cast skins on undersides of leaves.	**Hawthorn lace bug**, *Corythucha cydoniae*. Adults $\leq^1/_8$ inch long, wings lacy. Nymphs spiny.	*See* Lace Bugs, 207.
Sticky honeydew, blackish sooty mold, and whitish cast skins on plant.	**Aphids**, including **Apple aphid**, *Aphis pomi;* **Bean aphid**. Small green or black insects grouped on leaves or terminals.	Conserve natural enemies that can provide control. Hose with forceful water. *See* 175.
Sticky honeydew and blackish sooty mold on plant. Slow plant growth.	**European fruit lecanium, Kuno scale**. Brown or yellowish, beadlike, bulbous, or flattened insects on leaves or twigs.	*See* 195.
Sticky honeydew and blackish sooty mold on foliage. Popcornlike bodies (egg sacs) on bark.	**Cottony cushion scale**. Orangish, flat immatures or cottony females on bark.	Normally controlled by natural enemies. *See* 130, 196.
Brownish, grayish, tan, or white encrustations on bark. Rarely, declining or dead twigs or branches.	**Greedy scale, San Jose scale**. Tiny, circular to oval, flattened insects, often in colonies.	Rarely if ever cause serious damage to pyracantha. *See* 190, 191.
Chewed leaves. Plant may be defoliated.	**Tussock moths**. Hairy larvae, $\leq$1 inch long, may have colorful spots.	*See* 152.
Leaves chewed. Plant can be extensively defoliated.	**Indian walking stick**. Brown, twiglike insects $\leq$4 inches long, easily overlooked.	In California in at least the Central and South Coast. *See* 166.
Wood swellings (galls), cottony, waxy material on branches and roots.	**Woolly apple aphid**. Tiny, reddish, cottony or waxy insects.	Conserve natural enemies that help control. *See* Woolly Aphids, 177.

WHAT THE PROBLEM LOOKS LIKE	PROBABLE CAUSE	COMMENTS
Pyrus calleryana, Pear, Bradford pear, Callery pear, Flowering pear, Ornamental pear; *P. kawakamii*, Evergreen pear[28]		
Foliage yellows and wilts. Branches or entire plant may die. May be white fungus beneath basal trunk bark.	**Armillaria root disease**. Fungus present in many soils. Favored by warm, wet soil. Persists for years in infected roots.	*See* 120.
Sudden wilting, then shriveling and blackening of shoots, blossoms, and fruit. Plant appears scorched.	**Fire blight**. Bacterium enters plants through blossoms.	Plant less-susceptible cultivars. *See* 84.
Stems or flowers with dark lesions. Cankers possible on branches.	**Bacterial blast, blight, and canker**. Disease favored by wet conditions.	*See* 83.
Tiny, reddish to brown leaf spots, may have yellow halos. Larger, dark areas on leaves. Leaves may drop prematurely.	**Entomosporium leaf spot**. A fungal disease promoted by wet foliage.	*See* 87.
Black to dark olive spots on fruit, leaves, or stems. Leaves may distort, tear, yellow, or drop prematurely.	**Scab**, *Venturia pirina*. A fungal disease spread by water from infected leaves and twigs.	*See* 88.
Leaves discolor and wilt. Stems may have callus tissue, dark cankers, or coral-colored pustules.	**Cankers**, including *Nectria cinnabarina*, *N. galligena*, *Phomopsis* spp. Fungi that primarily affect injured or stressed trees.	*See* Nectria Canker, 101.
Sticky honeydew and blackish sooty mold on leaves and twigs. Dieback of twigs or branches possible.	**Black scale, Calico scale, European fruit lecanium**.	Black, brown, yellow, white, or mottled, flattened or bulbous insects ≤$^1/_5$ inch long. *See* 192–195.
Sticky honeydew and blackish sooty mold on foliage.	**Bean aphid, Green peach aphid, Melon aphid**. Small black, gray, green, or yellow insects on leaves and terminals.	*See* 176.
Sticky honeydew and blackish sooty mold on foliage. Cottony or waxy material on plant.	**Grape mealybug, Obscure mealybug**. Powdery, grayish insects, ≤$^1/_4$ inch long, with waxy filaments.	*See* 187.
Stippled, bleached, or reddened foliage. May be fine webbing on leaves or terminals. Leaves may drop prematurely.	**Brown mite**, *Bryobia rubrioculus*; **European red mite**; **McDaniel spider mite**, *Tetranychus mcdanieli*; **Pacific spider mite**; **Twospotted spider mite**.	Tiny brown, green, reddish, or yellow arthropods. *See* 246.
Brown to gray encrustations on bark. May be declining or dead twigs or branches.	**Italian pear scale**, *Epidiaspis leperii*; **San Jose scale**. Tiny, circular to elongate insects.	Italian pear scale survives only under shelter such as lichens and moss on bark. The female's body is reddish purple beneath its cover. *See* Armored Scales, 189–192.
Leaves or blossoms chewed. Single branches or entire plant may be defoliated. Foliage may be rolled and tied together with silk.	**Fruittree leafroller; Obliquebanded leafroller**, *Choristoneura rosaceana*; **Omnivorous leafroller**, *Platynota stultana*; **Orange tortrix**, *Argyrotaenia citrana*.	Moth larvae ≤1$^1/_2$ inches long. *See* 149.
Quercus spp., Oak[29]		
Leaves or shoots covered with white, powdery growth. Shoots brown or reddish, bushy, short, or shriveled.	**Powdery mildew; Witches' broom**, *Sphaerotheca lanestris*. Fungal diseases favored by moderate temperatures, shade, and poor air circulation.	Avoid cultural practices that stimulate excess growth, such as summer irrigation and heavy pruning. Prune off witches' brooms in winter. *See* 90.
Leaves discolor, stunt, wilt, or drop prematurely. Basal trunk discolored and may die. May be white fungus beneath basal trunk bark.	**Armillaria root disease**, or **Oak root fungus**. Fungus present in many soils. Favored by warm, wet soil. Persists for years in infected roots.	*See* 120.
Leaves discolor, stunt, wilt, or drop prematurely. Plants grow slowly and may die. Roots dark, decayed.	**Phytophthora root and crown rot**. Disease promoted by excess soil moisture and poor drainage.	*See* 122.

[28] For pests of fruiting pear, see *Pests of the Garden and Small Farm* and ipm.ucanr.edu.

[29] For more information, see *A Field Guide to Insects and Diseases of California Oaks* and *Oaks in the Urban Landscape*.

WHAT THE PROBLEM LOOKS LIKE	PROBABLE CAUSE	COMMENTS
Leaves brown, light green, yellow, or wilted. Trunks ooze, usually near the ground. Limbs and eventually entire tree dies. Trees may become heavily colonized by bark beetles.	**Sudden oak death**, *Phytophthora ramorum*. Pathogen spreads via airborne spores and contaminated plants and soil and infects through trunk bark. *Quercus* hosts include at least *Q. agrifolia, Q. chrysolepis, Q. kelloggii*, and *Q. parvula* var. *shrevei*.	Primarily a problem in wildlands, killing many oaks there. Do not move infected plants. Report to agricultural officials if found outside areas where pathogen is known to occur. *See* 105.
Foliage yellows and wilts. Bark may have lesions that are dark, dry, crusty whitish, oily looking, or water-soaked. Branches or entire plant dies.	**Fusarium dieback**. Fungal disease spread by the polyphagous shothole borer, an ambrosia beetle, or bark beetle.	Occurs on coast live oak in at least southern California. *See* 115.
Leaves discolor and wilt. Branches develop cankers and die back. Dark, hemispherical fruiting bodies may develop on bark.	**Cankers**, including **Annulohypoxylon canker**, *Annulohypoxylon thouarsianum*; **Nectria canker**, *Nectria peziza*.	Decay fungi that attack injured or stressed trees. *See* 98, 101.
Foliage may yellow and wilt. Tree grows slowly or declines. May be bracketlike or fan-shaped fungal fruiting bodies on bark or wood. Branches or entire plant may die.	**Wood decay**, or **Heart rot**, including *Ganoderma* spp., *Phellinus* spp., *Laetiporus gilbertsonii* (=*L. sulphureus*); **Canker rot**, *Inonotus* spp.	Fungi that attack injured, old, or stressed plants. *See* 66, 111.
Leaves yellowish, except along veins. Leaves small, drop prematurely. Branches die back.	**Iron deficiency**. Abiotic disorder usually caused by poor soil conditions or unhealthy roots.	Especially common in pin oak, *Quercus palustris. See* 47.
Leaves with discolored spots and brown, dead areas. Leaves on lower branches commonly are more severely affected. Can cause severe defoliation.	**Anthracnose**, *Apiognomonia errabunda* (=*Discula umbrinella*), *Septoria quercicola*. Fungal diseases infect new growth during wet, spring weather.	Difficult to manage and usually not serious enough to warrant control effort. *See* 81.
Leaves yellow, less abundant than normal. Bark bleeding, cracking, and staining on the main trunk.	**Bot canker**, *Diplodia corticola* (=*Botryosphaeria corticola*). A fungal disease; symptoms resemble those of goldspotted oak borer.	Occurs on coast live oak in southern California. *See* Oak Branch Canker and Dieback, 102.
Leaves wilt, turn yellow, then entirely brown in large groups. Wood is stained brown. Bark, branches, cambium, and sapwood die.	**Branch dieback**, *Diplodia quercina*. A fungal disease often associated with oak pit scales.	*See* Oak Branch Canker and Dieback, 102.
Leaves brown or white. Groups of entirely dead leaves remain on twigs, scattered throughout canopy. Death of current season's twigs.	**Twig blight**. Fungal diseases more severe if oak pit scale present.	*See* 86.
Patches of dead leaves at end of branches of live oaks.	**Oak twig girdlers**. Various beetles, larvae whitish and chew under bark.	*See* 231.
Leaves turn brown or yellow, especially along margins and at tip. Leaves may drop prematurely.	**Leaf burn**, or **Scorch**. Abiotic disorders with many potential causes, including watering near trunks, injuring trunks or roots, irrigating too frequently, planting irrigated landscapes beneath oaks, poor drainage, or salty soil or water.	Provide plants with a good growing environment and proper cultural care. *See* Abiotic Disorders, 43.
Leaves brown or scorched along margins and veins. Partially brown and entirely green leaves occur side-by-side on same twigs.	**Leaf scorch**, or **Twohorned oak gall wasp**. Galls cause portion of leaf to die terminal to where wasp larvae fed. Small, oblong galls or brown scars left by dropped galls occur on underside of leaf along vein.	Infests *Quercus agrifolia* and *Q. wislizenii*. Damage is apparently harmless to oaks. *See* 216.
Leaves, branches, catkins, or twigs with distorted growths or swellings (galls), which may be colorful.	**Cynipid gall wasps**. Adults are tiny wasps. Larvae whitish maggots in galls. Hundreds of species occur on oaks.	Most galls apparently do not harm trees. No control known for most oak galls, except pruning. *See* 215.
Twigs with black, brown, green, or red spherical swellings ≤4 inches in diameter.	**California gallfly**, or **Oak apple gall wasp**, *Andricus californicus*. Applelike plant growths caused by pale maggots feeding inside.	Galls do not harm trees. No control is recommended. Prune off galls if intolerable. *See* 215.

WHAT THE PROBLEM LOOKS LIKE	PROBABLE CAUSE	COMMENTS
Twigs with elongate swellings up to several inches long. Shoot terminals may die back.	**Spindle gall wasps**, including *Andricus, Callirhytis,* and *Disholcaspis* spp.	Trees generally tolerate. No control known, except pruning. *See 215.*
Leaves stippled or bleached. Severely infested leaves may brown and drop prematurely.	**Spider mites**, including **Sycamore spider mite**. Tiny arthropods, often green, suck sap.	Problem in California mostly in interior valleys, especially after insecticide spray during hot weather. *See 246.*
Leaves with many tiny, round, brown and yellow spots. Tiny, brown to yellow, round growths on underside of leaves.	**Jumping oak gall wasp**, *Neuroterus saltatorius.* Galls drop from leaves in summer and may be seen hopping on ground beneath oaks due to larva moving inside.	Occurs on various oaks in California, especially on *Quercus lobata* in the Central Valley. Damage is harmless to oaks. *See 215.*
Leaves with many tiny, irregular, brown and yellow spots. Young plants most susceptible.	**Oak leaf phylloxera**. Tiny, yellowish aphidlike insects on underside of leaves, suck sap.	Plants apparently not damaged. Tolerate or thoroughly apply soap to leaf underside or another insecticide when nymphs present, before severe spotting.
Bulges on upper leaf surface. Leaves galled, curled, may drop prematurely. Pale growth on either leaf surface.	**Oak leaf blister**, *Taphrina caerulescens.* Fungal disease promoted by wet foliage during leaf flush.	Provide oaks proper cultural care. No known control.
Raised blisters on upper leaf surface. Orange to pale felty depressions on leaf underside.	**Live oak erineum mite**. Tiny, elongate eriophyid mites.	No management known. *See 250.*
Sticky honeydew and blackish sooty mold on foliage. Globular gall-like bodies (mature scales) on leaves or twigs.	**Kermes scales**, including **Black-punctured kermes**, *Kermes nigropunctatus.* Spherical scales, ≤$^1/_4$ inch diameter. Most species are mostly brownish.	Usually not abundant enough to warrant control. *See* Scales, 188.
Sticky honeydew and blackish sooty mold on foliage.	**Oak lecanium scale**. Flattened, orangish nymphs to bulbous, brown adults on twigs.	Scale has one generation per year. Control usually not warranted. *See* Lecanium Scales, 195.
Sticky honeydew and blackish sooty mold on foliage. Dark, oval bodies (nymphs), about $^1/_{16}$ inch long on underside of leaves, often with white, waxy fringe.	**Stanford whitefly**, *Tetraleurodes stanfordi;* **Gelatinous whitefly**, *Aleuroplatus gelatinosus;* **Crown whitefly**. Adults tiny, mothlike insects that suck phloem sap.	Pupae black with white, waxy fringe. Ignore insects, they apparently do not damage trees. *See 182.*
Sticky honeydew and blackish sooty mold on foliage.	**Aphids**, including *Myzocallis* spp. Tiny, green to yellow insects, clustered on leaves.	Plants tolerate moderate aphid populations. Conserve natural enemies. *See 175.*
Sticky honeydew and blackish sooty mold on foliage. Leaves may curl or have rolled margins.	**Woolly oak aphids**, including *Stegophylla quercicola.* Small, greenish to bluish, cottony wax-covered insects.	Plants tolerate moderate aphid densities. *See* Woolly Aphids, 177.
Sticky honeydew, blackish sooty mold, and whitish wax on foliage.	**Obscure mealybug**. Powdery, gray insects with waxy filaments.	Conserve natural enemies that help in control. *See 187.*
Twig bark roughened. Terminals may die back. May be sticky honeydew and blackish sooty mold on foliage.	**Oak treehopper**. Adults green to brown, with red dots and horn on head, often on twigs with group of nymphs.	*See 204–205.*
White or pink, frothy material exudes from bark, often around wounds.	**Foamy bark canker**; **Foamy canker**, or **Alcoholic flux**.	*See 101.*
Brownish, sometimes rancid fluid exudes from bark, often around crotches or wounds. Bark stained.	**Wetwood**, or **Slime flux**. Bacterial infections.	Usually do not cause serious harm to trees. *See 110.*
Brownish to clear sticky material dripping from acorns, leaves, or twigs.	**Drippy nut disease**, or **Drippy acorn**. A bacterium that causes injured acorns to exude viscous liquid.	Associated with filbertworm, filbert weevils, gall wasps, and virtually any injury that allows bacteria to colonize. *See 80.*
Brownish, clear, frothy, or sticky liquid dripping from leaves, nuts, or twigs.	**Oak gall wasp nectar**. Certain *Andricus, Disholcaspis,* and *Dryocosmus* spp. cynipids induce oak galls that secrete nectar, which attracts ants that protect gall wasps from their natural enemies.	Control ants; over several years this can allow natural enemies to dramatically reduce populations of nectar-inducing gall wasps. *See 215.*
Copious, misty, nonsticky liquid raining from plant. Surfaces covered with whitish residue.	**Glassy-winged sharpshooter**. Active, dark brown or gray leafhoppers, ≤$^1/_2$ inch long, suck xylem fluid.	Vectors *Xylella* pathogens. Report suspected glassy-winged sharpshooters to agricultural officials if found in areas where this pest is not known to occur. *See 203.*

WHAT THE PROBLEM LOOKS LIKE	PROBABLE CAUSE	COMMENTS
Chewed leaves. Tree may be defoliated.	**California oakworm**. Dark to greenish larvae, ≤1¼ inches long, with yellow stripes.	In California, on oaks near coastal areas. *See* 147.
Leaves chewed, tied together with silk. Tree may be defoliated.	**Fruittree leafroller**. Green larvae, with black head and "shield" behind head. Larvae, ≤³/₄ inch long, wriggle vigorously when touched.	In California, common in warmer interior areas. *See* 149.
Chewed leaves. Silken mats, or "tents," sometimes seen in trees. Tree may be defoliated.	**Pacific tent caterpillar, Western tent caterpillar, Tussock moths**. Hairy, brownish to colorful caterpillars.	*See* 151, 152.
Leaves and shoot buds chewed. Trees can be defoliated.	**Oak leaf sawfly**, *Periclista* sp. Pale green larvae, ≤¹/₂ inch long.	Control generally not warranted. *See* Sawflies, 153.
Chewed leaves.	**Fuller rose beetle**, *Naupactus godmani*. Pale brown adult snout beetle, about ³/₈ inch long.	Adults hide during day and feed at night. Larvae feed on roots. *See* Weevils, 159.
Leaves chewed. Plant can be extensively defoliated.	**Indian walking stick**. Brown, twiglike insects ≤4 inches long, easily overlooked.	In California in at least the Central and South Coast. *See* 166.
Leaf surface etched. These "windows" may turn brown. White, ribbed, cigar-shaped cocoons on leaves or bark.	**Oak ribbed casemaker**. Larvae are ≤¹/₄ inch long.	*See* 218.
Leaves with elliptical blotches or holes, ¹/₈ to ¹/₄ inch long, or winding tunnels.	**Shield bearers**. Tiny larvae cut mined foliage from leaf.	*See* 219.
Gouged and etched leaves. Leaves may turn brown. Most new leaf damage appears from April to June.	**Live oak weevil**, *Deporaus glastinus*. Adults are dark, metallic-blue snout beetles, about ¹/₄ inch long.	Most common on *Quercus agrifolia*. Tolerate as damage does not threaten tree health. *See* Weevils, 159.
Gray, pinkish, or whitish encrustations on bark, usually on the limb underside of evergreen oaks. Leaves may drop prematurely.	**Ehrhorn's oak scale**. Reddish scale ¹/₂₅-inch that occurs in symbiotic colonies with the fungus *Septobasidium canescens*. Mostly in southern California.	Heavy infestations may slow oak growth or cause limb decline or dieback, but populations are often innocuous. *See* 197.
Gray encrustations on bark. May be declining or dead twigs or branches.	**Obscure scale**, *Melanaspis obscura*. Tiny, grayish, circular to oval insects on or under bark.	Occurs in California at least in Sacramento. Introduced parasite generally provides effective control. *See* Armored Scales, 189.
Rough bark, ring-shaped swellings. Dead twigs and branches. Dead leaves persist over winter on deciduous oak.	**Oak pit scales**. Pinhead-sized, brown to green insects, on bark in roundish swellings.	*See* 198.
Bark rough and flaking off of trunks and limbs, exposing discolored dark tissue underneath.	**Kuwana oak scale**. Red, roundish insects, ≤¹/₁₀ inch in diameter, covered with grayish or white wax.	Damaging only on blue oak, *Q. douglasii*. *See* 198.
Greatly roughened bark on lower trunk or major limb crotches. Slow growth.	**Sycamore borer**. Pink larvae, ≤³/₄ inch long, bore in bark or wood.	*See* 236.
Large holes, ≤¹/₂ inch diameter, in trunks and limbs. Gnarly bark. Slow tree growth. Limbs may fall.	**Carpenterworm**. Whitish larvae, ≤2¹/₂ inches long, with brown head, tunnel in wood.	*See* 241.
Bleeding or frothy material bubbling from tiny holes in trunk or limbs. Fine boring dust may surround holes. Limbs or entire tree may die.	**Oak ambrosia and bark beetles, Polyphagous shothole borer**. Adults brown bark beetles, ¹/₈ inch long. White larvae tunnel beneath bark.	*See* 222.
Trunk or limbs with roughened, wet, or oozing area. Cracked bark and dieback. Foliage may discolor and wilt. Tree may die.	**Flatheaded borers**, including **Flatheaded appletree borer, Goldspotted oak borer, Pacific flatheaded borer**. Whitish larvae under bark, may have enlarged head. Adults bullet shaped.	Many California black oak and coast live oak in southern California have been killed by goldspotted oak borer. *See* 227.
Holes in acorns. Tunnels inside acorns may contain insect larvae.	**Acorn moth, Filbert weevils, Filbertworm**. Yellow to whitish grubs or maggots that tunnel in nuts.	May reduce natural oak regeneration, but established trees are not damaged. *See* Filbert Weevils and Acorn Worms, 162.
Green plants with smooth stems, thick, roundish leaves infesting branches.	**Broadleaf mistletoe**. Parasitic, evergreen plant on host.	*See* 284.

WHAT THE PROBLEM LOOKS LIKE	PROBABLE CAUSE	COMMENTS
Rhamnus spp., Rhamnus, Buckthorn, Coffeeberry, Redberry		
Leaves discolor, stunt, wilt, or drop prematurely. Plants grow slowly and may die. Roots dark, decayed.	**Phytophthora root and crown rot.** Disease favored by excess soil moisture and poor drainage.	*See* 122.
Leaves with brown, light green, or yellow lesions or dead spots.	**Ramorum blight**, *Phytophthora ramorum*. Pathogen spreads via airborne spores and contaminated plants and soil. Hosts include at least *Rhamnus californica* and *R. purshiana*.	Primarily a problem in wildlands, killing many oaks there. *See* Sudden Oak Death and Ramorum Blight, 105.
Leaves or shoots turn yellow or brown and wilt. Leaves drop prematurely. Terminals die back. Branches may be cankered.	**Blight**, or **Twig dieback**, including *Diplodia frangulae, Phoma rhamnicola, Phomopsis communis*. Fungal diseases that primarily affect injured or stressed plants.	Protect plants from injury. Provide good cultural care, especially appropriate irrigation. Avoid overhead irrigation.
Leaves with brown to yellow blotches or scabby spots. Foliage may shrivel and drop prematurely.	**Leaf spots**, including *Cylindrosporium rhamni, Septoria blasdalei*. Fungal pathogens favored by prolonged cool, wet conditions.	Avoid wetting foliage. Use drip irrigation if feasible. Use good sanitation; promptly remove and dispose of debris and infected leaves. *See* 86.
Leaves or fruit with dark scabby or velvety spots.	**Scab**, *Venturia rhamni*. Fungal disease promoted by moist spring weather.	*See* 88.
Leaves with discolored spots and orangish pustules. Leaves may drop prematurely.	**Rusts**, including *Puccinia mesnieriana*. Pathogens favored by wet conditions. Fungal spores persist in plant debris.	Avoid overhead watering. Plants tolerate moderate populations. *See* 94.
Leaves or shoots with powdery, white growth or yellow blotches. Shoots and leaves may be undersized or distorted.	**Powdery mildews**, including *Stictis radiata*. Fungal diseases favored by moderate temperatures, shade, and poor air circulation.	*See* 90.
Sticky honeydew, blackish sooty mold, and whitish cast skins on plant.	**Aphids**, including *Aphis* sp., *Sitobion* sp. Colonies of small, pear-shaped insects, often greenish.	*See* 175.
Sticky honeydew and blackish sooty mold on plant. Tiny, white mothlike adults.	**Grape whitefly**, *Trialeurodes vittata*. Tiny, yellowish to translucent, oval nymphs. Mature nymphs (pupae) dark brown or mottled dark and yellowish.	Usually most severe on *R. californica* near grapes, especially in fall when insects move from senescent grape leaves. *See* Whiteflies, 182.
Rhaphiolepis spp., Rhaphiolepis, Indian hawthorn		
Tiny, reddish to brown leaf spots, may have yellow halos. Larger, dark areas on leaves. Leaves may drop prematurely.	**Entomosporium leaf spot.** A fungal disease promoted by wet foliage.	*See* 87.
Sudden wilting, then shriveling and blackening of shoots and blossoms. Plants appear scorched.	**Fire blight**. Bacterium infects plants through blossoms.	*See* 84.
Leaves discolor, stunt, wilt, or drop prematurely. Plants grow slowly and may die. Roots dark, decayed.	**Phytophthora root and crown rot**. Disease favored by excess soil moisture and poor drainage.	*See* 122.
Leaves wilt, turn brown, and die, often first on one side of plant. Dead leaves may remain on stems. Stem xylem discolored. Stems or entire plant may die.	**Verticillium wilt**, *Verticillium albo-atrum*. Fungus that persists in soil and infects through roots.	*See* 117.
Blackish sooty mold, sticky honeydew, and whitish cast skins on plant.	**Aphids**. Small, pear-shaped insects, often brownish, green, or yellowish, in groups.	*See* 175.
Leaf margins chewed, notched. Young plants may decline.	**Weevils**, including *Otiorhynchus* spp.; **Fuller rose beetle**, *Naupactus godmani*. Adults chew foliage. Larvae chew roots.	*See* 159.
Trunk or limbs with roughened, wet, or oozing area. Cracked bark and dieback.	**Borers**, including **Flatheaded borers**, *Chrysobothris* spp.	Pale beetle larvae bore under bark. *See* 227.

WHAT THE PROBLEM LOOKS LIKE	PROBABLE CAUSE	COMMENTS
Rhododendron spp., Azalea[30]		
New foliage yellows beginning along margins, but veins green. Leaves may develop brown blotches.	**Iron deficiency** or **Manganese deficiency**. Abiotic disorders usually caused by poor soil conditions or unhealthy roots.	*See* 47, 49.
Foliage yellows; older foliage may yellow first. Plants may grow slowly and produce fewer and smaller than normal leaves and shoots. Discolored foliage may drop prematurely.	**Nitrogen deficiency**. Abiotic disorder usually caused by poor soil conditions or unhealthy roots.	*See* 45.
Leaves turn brown or yellow, especially along margins and at tip. Leaves may drop prematurely.	**Leaf burn**, or **Scorch**. Abiotic disorders. Causes include drought, dry wind, excess fertilization, excess light, frost, saline irrigation water or soil, or transplant shock.	Provide plants with a good growing environment and proper cultural care. *See* Abiotic Disorders, 43.
Foliage yellows and wilts. Branches or entire plant dies.	**Root and crown rots**, *Phytophthora* spp., *Pythium* spp.	Diseases favored by excess soil moisture and poor drainage. *See* 122, 123.
Foliage may yellow, wilt, die back, or drop prematurely. Plant grows slowly. May be galls or swellings on roots.	**Root knot nematodes, Stubby root nematodes, Stunt nematode**. Tiny root-feeding roundworms.	*See* 293, 294.
Leaves with brown, light green, or yellow spots or dead blotches.	**Ramorum blight**, *Phytophthora ramorum*. Pathogen spreads via airborne spores and contaminated plants and soil.	Primarily a problem in nurseries and on other hosts in wildlands, killing many oaks there. *See* Sudden Oak Death and Ramorum Blight, 105.
Brownish to reddish spots and yellowing on leaves. Leaves may drop prematurely.	**Septoria leaf spot**, or **Leaf scorch**, *Septoria azaleae*. A fungal disease promoted and spread by wet conditions.	Keep foliage dry. Dispose of fallen leaves. *See* 88.
Leaves with pale green to yellow spots on upper surface and brown to purple spots on underside. Leaf undersides with blisters or orangish pustules. Leaves may drop prematurely.	**Rusts**, including *Chrysomyxa ledi, C. piperiana*. Pathogens favored by wet conditions. Fungal spores persist in plant debris.	Avoid overhead watering. Remove and dispose of infected leaves. Plants tolerate moderate populations. *See* 94.
Brown, purple, or yellow blotches on leaves. Powdery, whitish growth on leaves. Leaves may drop prematurely.	**Powdery mildew**, *Erysiphe polygoni*. Fungus develops on living tissue, favored by moderate temperatures, shade, and poor air circulation.	*See* 90.
Flowers with discolored blotches or round spots. Flowers collapse, become slimy and soft, and cling to leaves or stems.	**Ovulinia petal blight**. Fungus favored by cool, wet weather during flowering. Spores spread by splashing water.	*See* Azalea Petal Blight and Rhododendron Petal Blight, 78.
Soft, brown decay and/or gray, woolly growth on buds, flowers, leaves, and dead tissue.	**Botrytis blight**, or **Gray mold**. Fungus develops in plant debris or inactive tissue. Favored by high humidity and moderate temperatures. Spores airborne.	Unlike Ovulinia petal blight, infected flowers do not become slimy. *See* 78.
Leaves partly or all crisp, distorted, thickened, and brown or white. White or pinkish spores cover infected tissue.	**Leaf gall**, or **Kinnikinnick**, *Exobasidium vaccinii*. Fungus spreads by air only during wet weather.	Avoid overhead watering. Prune only when dry. Improve air circulation. Hand-pick or prune out galls. Rake and dispose of fallen leaves. Vigorous plants tolerate extensive leaf galling.
Sticky honeydew and blackish sooty mold on foliage. Tiny, whitish, mothlike insects (adult whiteflies) on foliage.	**Azalea whitefly**, *Pealius azaleae*; **Greenhouse whitefly; Rhododendron whitefly**, *Massilieurodes* (=*Dialeurodes*) *chittendeni*. Nymphs oval, flattened, yellowish to greenish, suck sap.	Cultivars vary greatly in susceptibility to these pests, which rarely cause serious damage to plants. *See* 182.
Sticky honeydew and blackish sooty mold on plant. Whitish cottony bodies on bark. Bark rough. Branches may deform. Possible dieback.	**Azalea bark scale**, or **Woolly azalea scale**, *Eriococcus azaleae*. Dark, reddish, oval insects underneath white wax, on bark crevices and in crotches.	Uncommon in California. If found in California, report to county agricultural commissioner. *See* similar European Elm Scale, 197.
Sticky honeydew and blackish sooty mold on leaves and twigs. Whitish cast skins on undersides of leaves.	**Aphids**, including *Macrosiphum* spp., *Masonaphis lambersi*. Tiny, pear-shaped insects, often brownish, green, or yellowish.	*See* 175.

[30] Also see Rhododendron.

Rhododendron spp. (continued)

WHAT THE PROBLEM LOOKS LIKE	PROBABLE CAUSE	COMMENTS
Leaves bleached, stippled, may turn brown and drop prematurely. Terminals may be distorted. Plant may have fine webbing.	**Mites**, including **Southern red mite, Twospotted spider mite**. Tiny, often green, pink, or red pests; may have two dark spots.	*See* 245.
Stippled or bleached leaves with dark, varnishlike excrement specks on undersides.	**Greenhouse thrips**. Tiny, slender, blackish adults or yellowish nymphs.	*See* 210.
Leaf edges notched or ragged, including on nearby hosts. Wilted or dying plants. Roots missing, debarked, girdled near soil surface.	**Black vine weevil; Obscure root weevil**, *Sciopithes obscurus*; **Woods weevil**, *Nemocestes incomptus*.	Adults dark brown to black or grayish snout weevils, ≤³/₈ inch long. Larvae root-feeding white grubs. *See* 160.
Leaves chewed. Plant can be extensively defoliated.	**Indian walking stick**. Brown, twiglike insects up to 4 inches long, easily overlooked.	In California in at least the Central and South Coast. *See* 166.
Browning of leaves or leaves tied together with silk. Leaves curled.	**Azalea leafminer**, *Caloptilia azaleella*. Greenish larvae, ≤¹/₂ inch long, secretive. Young larvae mine leaves, older larvae feed externally.	Vigorous plants tolerate extensive leaf damage. Control difficult. Clip and dispose of infested leaves. If needed, apply oil to mature larvae, or systemic insecticide can be applied for all larval stages.
Copious, misty, nonsticky liquid raining from plant. Surfaces covered with whitish residue.	**Glassy-winged sharpshooter**. Active, dark brown or gray leafhoppers ≤¹/₂ inch long, suck xylem fluid.	Vectors *Xylella* pathogens. Report suspected glassy-winged sharpshooters to agricultural officials if found in areas where this pest is not known to occur. *See* 203.
Rhododendron spp., Rhododendron		
New foliage yellows beginning along margins, but veins green. Leaves may develop brown blotches.	**Iron deficiency** or **Manganese deficiency**. Abiotic disorders usually caused by poor soil conditions or unhealthy roots.	*See* 47, 49.
Foliage yellows, older foliage may yellow first. Plants may grow slowly and produce fewer and smaller than normal leaves and shoots. Discolored foliage may drop prematurely.	**Nitrogen deficiency**. Abiotic disorder usually caused by poor soil conditions or unhealthy roots.	*See* 45.
Leaves turn brown or yellow, especially along margins and at tip. Leaves may drop prematurely.	**Leaf burn**, or **Scorch**. Abiotic disorders. Common causes include drought, dry wind, excess fertilization, excess light, frost, poor soil conditions, saline irrigation water or soil, too much or too little water, or transplant shock.	Provide plants with a good growing environment and proper cultural care. *See* Abiotic Disorders, 43.
Leaves discolor, stunt, wilt, or drop prematurely. Stems discolored, cankered, and die. May be white fungus beneath basal trunk bark.	**Armillaria root disease**. Fungus present in many soils. Favored by warm, wet soil. Persists for years in infected roots.	*See* 120.
Leaves discolored, wilted, stunted, may drop prematurely. Discolored bark may ooze sap. Branches or plant may die.	**Phytophthora root and crown rot, Pythium root rot**. Pathogens favored by excess soil moisture and poor drainage.	*See* 122, 123.
Leaves with pale green to yellow spots on upper surface and brown to purple spots on underside. Leaf undersides with blisters or orangish pustules. Leaves may drop prematurely.	**Rusts**, including *Chrysomyxa ledi, C. piperiana*. Pathogens favored by wet conditions. Fungal spores persist in plant debris.	Avoid overhead watering. Remove and dispose of infected leaves. Plants tolerate moderate populations. *See* 94.
Leaves with brown, red, tan, or yellow blotches, often with distinct, differently colored margins. Foliage may die and drop prematurely.	**Leaf spots**, *Phyllosticta* spp. Fungi spread by splashing water. Favored by prolonged cool, wet conditions.	Remove infected and fallen leaves. Provide good cultural care. *See* 86.
Leaves with brown, light green, or yellow spots or dead blotches. Twigs and stems with brown or black cankers, ooze, and die back.	**Ramorum blight; Phytophthora blight**, *Phytophthora syringae*. Spread via airborne or water-splashed spores and contaminated plants and soil.	*Phytophthora ramorum* primarily is a problem in nurseries and on other hosts in wildlands, killing many oaks there. *See* Sudden Oak Death and Ramorum Blight, 105.

WHAT THE PROBLEM LOOKS LIKE	PROBABLE CAUSE	COMMENTS
Brown, purple, or yellow blotches on leaves. Powdery, whitish growth on leaves. Leaves may drop prematurely.	**Powdery mildew**, including *Erysiphe polygoni*. Develops on living tissue, favored by moderate temperatures, shade, and poor air circulation.	*See* 90.
Leaves distorted, thickened, and brown or white. White or pinkish spores cover infected tissue. Flowers may be less abundant than normal.	**Leaf gall**, *Exobasidium vaccinii*. Fungus spreads by air only during wet weather.	Avoid overhead watering. Prune only when dry. Improve air circulation. Vigorous plants tolerate extensive leaf galling. Hand-pick or prune out galls. Rake and dispose of fallen leaves.
Flowers with discolored blotches or round spots. Flowers collapse, become slimy and soft, and cling to leaves or stems.	**Ovulinia petal blight**. Fungus favored by cool, wet weather during flowering. Spores spread by splashing water.	Remove and dispose of diseased blossoms. Do not overhead water. Improve air circulation. *See* Azalea Petal Blight and Rhododendron Petal Blight, 78.
Soft, brown decay and/or gray, woolly growth on buds, flowers, leaves, and dead tissue.	**Botrytis blight**, or **Gray mold**. Fungus develops in plant debris or inactive tissue. Favored by high humidity and moderate temperatures. Spores airborne.	Unlike Ovulinia petal blight, infected flowers do not become slimy. *See* 78.
Leaves stippled or bleached with varnishlike specks on undersides.	**Greenhouse thrips**. Tiny, slender, black adults or yellow nymphs.	*See* 210.
Leaves bleached, stippled, may turn brown and drop prematurely. Terminals may be distorted. Plant may have fine webbing.	**Mites**, including **Southern red mite, Twospotted spider mite**. Tiny, often green, pink, or red pests; may have two dark spots.	*See* 245.
Sticky honeydew and blackish sooty mold on foliage. Tiny, whitish, mothlike insects (adult whiteflies) on foliage.	**Whiteflies**, including **Azalea whitefly**, *Pealius azaleae;* **Ficus whitefly**, *Singhiella simplex;* **Greenhouse whitefly; Rhododendron whitefly**, *Massileurodes (=Dialeurodes) chittendeni*.	Nymphs oval, flattened, yellow to tan or greenish, suck sap. Cultivars vary greatly in susceptibility to these pests, which rarely cause serious damage to plants. *See* 182.
Sticky honeydew and blackish sooty mold on plant. Whitish cottony bodies on bark. Bark rough. Branches may deform. Possible dieback.	**Azalea bark scale**, or **Woolly azalea scale**, *Eriococcus azaleae*. Dark, reddish, oval insects underneath white wax, on bark crevices and in crotches.	Uncommon in California. If found in California, report to county agricultural commissioner. *See* similar European Elm Scale, 197.
Sticky honeydew and blackish sooty mold on leaves and twigs. Whitish cast skins on underside of leaves.	**Aphids**, including *Macrosiphum* spp., *Masonaphis lambersi*. Tiny, pear-shaped insects, often brownish, green, or yellowish.	*See* 175.
Wilted or dying plants. Roots missing, debarked, girdled near soil surface. Notched or ragged leaves, including on nearby hosts.	**Black vine weevil; Obscure root weevil**, *Sciopithes obscurus;* **Woods weevil**, *Nemocestes incomptus*.	Adults dark brown to black or grayish snout weevils, ≤$^3/_8$ inch long. Larvae white grubs in soil. *See* 160.
Foliage may yellow, wilt, die back, or drop prematurely. Plant grows slowly. May be galls or swellings on roots.	**Root knot nematodes, Stubby root nematodes, Stunt nematode**. Tiny root-feeding roundworms.	*See* 293, 294.
Rhus spp., Rhus, Lemonade berry, Sugarbush, Sumac, Wax tree		
Leaves discolor, stunt, wilt, or drop prematurely. Stems discolor, canker, and die. May be white fungus beneath basal trunk bark.	**Armillaria root disease**. Fungus present in many soils. Favored by warm, wet soil. Persists for years in infected roots.	*See* 120.
Leaves discolor and wilt. Stems may have dark, sunken cankers and dieback. May be small, coral-colored pustules on infected wood.	**Cankers**, including **Nectria canker**, *Nectria cinnabarina*. Fungi primarily attack injured or stressed trees.	*See* 101.
Leaves with black, brown, tan, or yellow spots or blotches. Foliage may die and drop prematurely. Twigs may die back.	**Leaf spots**, including *Cladosporium aromaticum, Gloeosporium toxicodendri, Phyllosticta* spp., *Pseudocercospora rhoina*.	Fungi spread in water and are favored by prolonged cool, wet conditions. *See* 86.
Leaves or shoots with powdery growth. Shoots or leaves may be stunted and distorted.	**Powdery mildews**, including *Erysiphe cichoracearum, Phyllactinia guttata*.	Fungi favored by moderate temperatures, shade, and poor air circulation. *See* 90.

Rhus spp. (continued)

WHAT THE PROBLEM LOOKS LIKE	PROBABLE CAUSE	COMMENTS
Sticky honeydew and blackish sooty mold on leaves or twigs. Whitish, waxy material on leaves and shoots.	**Sumac psyllids**, *Calophya* spp. Nymphs flattened, brown, orangish, or greenish. Eggs tiny, black.	Plants tolerate abundant psyllids. Conserve natural enemies, no other management recommended. *See Psyllids, 170.*
Sticky honeydew and blackish sooty mold on leaves or twigs. Terminals may die back and plants may decline.	**Black scale**. Brown to black, bulbous to flattened insects, $\leq^3/_{16}$ inch long. Raised H shape often on back.	*See 192.*
Sticky honeydew, blackish sooty mold, and whitish cast skins on plant.	**Aphids**. Small green, brownish, or yellowish insects, often in groups.	*See 175.*
Ribes spp., Ribes, Currant, Gooseberry		
Sticky honeydew, blackish sooty mold, and whitish cast skins on plant.	**Aphids**. Small green, brown, black, or yellowish insects, often in groups.	*See 175.*
Slightly raised, yellowish spots on underside of leaves and young stems. Leaves may drop prematurely.	**Rust**. Alternate host stage of white pine blister rust fungal disease that kills white (five-needle) pines.	*See 94, 96.*
Robinia spp., Locust, Black locust		
Leaves or shoots with powdery growth. Shoots or leaves may be stunted and distorted.	**Powdery mildews**, including *Erysiphe polygoni*. Fungi favored by moderate temperatures, shade, and poor air circulation.	*See 90.*
Leaves and branches wilt in spring. Stems may have callus tissue, dark or sunken cankers, or coral-colored pustules.	**Nectria canker**. Fungus primarily affects injured or stressed trees.	*See 101.*
Leaves may yellow and wilt. Bark may have bracketlike or fan-shaped fungal fruiting bodies. Branches or entire plant may die. Trees may fall over.	**Wood decay**, or **Heart rot**, including *Ganoderma* spp., *Laetiporus gilbertsonii* (=*L. sulphureus*). Fungi that infect injured, old, or stressed trees.	*See 66, 111.*
Limbs dying or dead. Large holes, $\leq^1/_2$ inch diameter, in trunks and limbs. Slow tree growth.	**Carpenterworm**. Whitish larvae, $\leq 2^1/_2$ inches long, with brown head, tunnel in wood.	*See 241.*
Limbs dying or dead. Bark has many tiny, round holes, which may ooze.	**Bark beetles**, including *Chramesus* spp. Adults $^1/_{10}$ inch long, dark, cylindrical. Whitish maggotlike larvae under bark.	Attack only severely stressed trees and dying limbs. Prune out infested wood. Provide trees proper care. *See 222.*
Leaves may yellow and wilt. Plants grow slowly or decline. Limbs or entire tree may die.	**Tenlined June beetle**, *Polyphylla decemlineata*. Cream-colored grubs, ≤ 2 inches long, chew roots.	More of a problem in sandy soils. *See* White Grubs and Scarab Beetles, 164.
Bark discolored, oozing, or swollen. Boring dust in crevices, at tree base. Dying limbs. Tree may decline.	**Black locust borer**, *Megacyllene robiniae*. Whitish larvae bore under bark. Adult is yellow and black longhorned beetle.	Attacks primarily stressed trees. Provide trees proper cultural care. *See* Longhorned Beetles, or Roundheaded Wood Borers, 232.
Roughened twig bark. Possible twig dieback.	**Treehoppers**, including **Buffalo treehopper**, *Stictocephala bisonia*.	Bright yellow to green insects, horny or spiny. *See 204.*
Sticky honeydew and blackish sooty mold on plant.	**Aphids**, including **Bean aphid, Melon aphid**, and other *Aphis* spp. Tiny, pear-shaped insects, often black, brownish, green, or yellow.	*See 175.*
Sticky honeydew and blackish sooty mold on plant.	**Frosted scale**. Hemispherical to flattened, brown, yellowish, or waxy insects, $\leq^1/_4$ inch long, suck sap.	*See Lecanium Scales, 195.*
Leaves chewed and may be tied with silk. Plant may be defoliated.	**Fruittree leafroller**. Green larvae, $\leq 1^1/_2$ inches long, with black head.	*See 149.*
Leaves stippled, bleached, or reddened.	**Spider mites**, including **European red mite**. Tiny, reddish to yellow arthropods, suck sap.	*See 246.*
Leaflets galled, margins rolled, or terminals podlike. Foliage may brown and drop prematurely.	**Pod gall midges**, including *Dasineura* sp., *Obolodiplosis robiniae*. Adults are tiny flies. Pale larvae occur in galled or rolled leaflets, then drop and pupate beneath plant.	Plants tolerate extensive galling. *See* similar Honeylocust Pod Gall Midge, 214.

WHAT THE PROBLEM LOOKS LIKE	PROBABLE CAUSE	COMMENTS
Rosa spp., Rose[31]		
White to gray growth, often powdery, on leaves, shoots, and buds. Leaves may become distorted and drop prematurely.	**Powdery mildew**. A fungal disease favored by moderate temperatures, shade, and poor air circulation.	Grow resistant cultivars. Reduce shadiness and improve air circulation. *See* 90.
Small, orange pustules, primarily on leaf undersides. Upper leaf surface may discolor. Leaves may drop prematurely.	**Rust**, *Phragmidium disciflorum*. Fungus favored by cool, moist weather. Spores airborne.	Avoid overhead watering and condensation. Fungicides help to prevent damage. *See* 94.
Dark spots with fringed margins on upper surface of leaves and succulent stems. Yellow areas develop around spots. Leaves may drop prematurely.	**Black spot**. Fungal spores spread by splashing water.	Grow resistant cultivars. Avoid wetting foliage. *See* 86.
Purplish, red, or dark brown spots on leaves. Leaf undersides covered with downy fungal growth. Leaves may yellow and drop prematurely.	**Downy mildew**, *Peronospora sparsa*. Spores produced only on living plants. Persistent spores carry fungus over unfavorable periods. Favored by moist, humid conditions.	Prune plants to improve air circulation. Avoid overhead watering. *See* 89.
Leaves yellow or with yellow blotches, lines, or intricate patterns. Leaves may distort. Plants may be stunted.	**Viruses**, including **Rose mosaic**, or **Prunus necrotic ringspot**. A complex of viruses can infect landscape roses.	Usually do not threaten plant health. Severity of symptoms varies greatly with rose variety and virus. *See* 92, 94.
Leaves curl downward and canes die. Leaves readily drop.	**Rose leaf curl**. Probably a virus.	Obtain virus-free stock. No known treatment. Tolerate or replace infected plants. *See* Mosaic and Mottle Viruses, 92.
Leaves emerging in spring are balled or curved on very short shoots with conspicuous vein clearing. Slow plant growth. Symptoms tend to disappear later in season.	**Rose spring dwarf**. Probably a virus.	Obtain virus-free plants. No known treatment. Tolerate or replace infected plants. *See* Mosaic and Mottle Viruses, 92.
Black to brown spots or streaks on leaves or stems. Blossoms, buds, or leaves darken and shrivel. Oozing lesions on twigs.	**Bacterial blast, blight, and canker**. Disease favored by wet conditions.	*See* 83.
Soft, brown decay and/or gray, woolly growth on buds, flowers, leaves, and dead tissue. Twig dieback and cane canker.	**Botrytis blight**, or **Gray mold.** Fungus develops in plant debris or inactive tissue. Favored by wet foliage, high humidity, and moderate temperatures.	Remove and dispose of fallen leaves and plant debris around plants. Improve air circulation and avoid wetting foliage. *See* 78.
Foliage discolors and wilts. Shoots die back. Brown cankers, sometimes with gray centers, on stems. Small, black, spore-producing structures (pycnidia) on dead tissue.	**Cankers** and **Dieback**, including *Botryosphaeria dothidea, Coniothyrium fuckelii, Cryptosporella umbrina, Nectria cinnabarina*. Fungi that primarily affect injured or stressed plants.	Provide proper cultural care to keep plants vigorous. *See* Botryosphaeria Canker and Dieback, 98, and Nectria Canker, 101.
Foliage fades, yellows, browns, or wilts, often scattered throughout canopy. Branches die. Entire plant may die.	**Verticillium wilt**. A soil-dwelling fungus that infects through roots.	*See* 117.
Leaves yellow, wilt, and may drop prematurely. Branches die back. Entire tree may die. May be white fungus beneath basal trunk bark.	**Armillaria root disease**. Fungus present in many soils. Favored by warm, wet soil. Persists for years in infected roots.	*See* 120.
Leaves discolor, stunt, wilt, or drop prematurely. Plants grow slowly and may die. Roots dark, decayed.	**Phytophthora root and crown rot**. Disease favored by excess soil moisture and poor drainage.	*See* 122.
Small, discrete spots to irregular, large blotches or holes over most of leaf surface. Leaves may drop prematurely.	**Leaf spots**, including *Mycosphaerella* spp., *Sphaceloma* spp. Fungi favored by wet conditions.	*See* 86.
Leaves bleached or stippled. Foliage may be finely webbed. Leaves may dry and drop prematurely.	**Spider mites**, including **Twospotted spider mite**. Tiny, greenish or yellowish arthropods, may have two dark spots.	Plants tolerate extensive leaf stippling, especially on older foliage. *See* 246.

[31] For more information, see the publications *Healthy Roses; Roses: Diseases and Abiotic Disorders Pest Notes;* and *Roses: Insect and Mite Pests and Beneficials Pest Notes.*

Rosa spp. (continued)

WHAT THE PROBLEM LOOKS LIKE	PROBABLE CAUSE	COMMENTS
Leaves bleached or stippled. Cast skins on underside of leaves.	**Rose leafhopper**, *Edwardsiana rosae*. Greenish to whitish, wedge-shaped insects, ≤1/8 inch long.	Plants tolerate extensive leaf stippling. Apply insecticidal soap or another insecticide if severe. *See* Leafhoppers, 201.
Sticky honeydew, blackish sooty mold, and whitish cast skins on plant. Blossoms may be distorted.	**Aphids**, including **Melon aphid; Rose aphid**; *Macrosiphum euphorbiae; Wahlgreniella nervata*.	Tiny, green to pink insects on terminals and buds. *See* 175.
Sticky honeydew and blackish sooty mold on foliage. May be whitish wax or cottony material on plant. Possible decline or dieback of canes.	**Black scale, Brown soft scale, Cottony cushion scale, European fruit lecanium, Frosted scale, Kuno scale**.	Flattened or bulbous and black, brownish, orangish, or yellow insects, may be cottony or waxy. *See* 192–195.
Sticky honeydew and blackish sooty mold on foliage. Tiny, powdery, white, mothlike insects.	**Greenhouse whitefly**. Flat, oval, tiny, yellow to green insects on leaf undersides.	*See* 184.
Sticky honeydew and blackish sooty mold on foliage. Cottony or waxy material on plant.	**Mealybugs**, including **Citrus mealybug**. Powdery, grayish insects, ≤1/4 inch long, may have waxy filaments.	Vigorous plants tolerate moderate populations. Conserve natural enemies that help in control. *See* 185.
Blossom petals or sepals streaked with brown.	**Thrips**, including **Madrone thrips**, *Thrips madroni*; **Western flower thrips**. Tiny, slender, yellow or black insects in blossoms.	Damage occurs before buds open. Thrips present in flowers may be pollen feeders, not petal-damaging species. *See* 209.
Blossoms develop green leaflike structures. Foliage is not affected.	**Rose phyllody**. Abiotic disorder usually caused by hot weather or drought stress when flower buds are forming.	Uncommon. Floribunda roses more often affected. *See* 60.
Chewed leaves or blossoms. Leaf margins notched, ragged.	**Fuller rose beetle**, *Naupactus godmani*. Pale brown adult snout beetle, about 3/8 inch long.	Adults hide during day and feed at night. Larvae feed on roots. *See* Weevils, 159.
Chewed blossoms, especially white and yellow flowers.	**Hoplia beetle**. Adult beetles about 1/4 inch long, mostly reddish brown with silver, black, or white.	Larvae feed on roots, but are not known to significantly damage rose roots. *See* 165.
Leaf undersides scraped, skeletonized. Large holes may be eaten in leaves.	**Roseslugs**, including **Bristly roseslug**. Green to yellowish sawfly larvae, ≤5/8 inch long, may be bristly.	*See* 154.
Holes punched in flowers and canes. Blossoms ragged.	**Rose curculios**. Red to black snout weevils, 1/4 inch long.	Small, whitish larvae in buds. *See* 163.
Chewed leaves. Buds may be mined or have holes. Leaves may be tied together with silk.	**Caterpillars**, including **Orange tortrix**, *Argyrotaenia citrana*; **Tussock moths; Fruittree leafroller; Tent caterpillars; Omnivorous looper**.	Hairy to naked larvae, ≤1 1/2 inches long. *See* 144–152.
Leaves chewed. Plant can be extensively defoliated.	**Indian walking stick**. Brown, twiglike insects up to 4 inches long, easily overlooked.	In California in at least the Central and South Coast. *See* 166.
Semicircular holes cut in margins of leaves or blossoms.	**Leafcutting bees**, *Megachile* spp. Robust bees about 1/2 inch long, line their nests with cut plant parts.	Not seriously harmful to otherwise healthy plants. Bees are beneficial pollinators and should not be killed.
Brownish, grayish, tan, orange, or white encrustations on canes. Rarely, canes may decline or die back.	**California red scale; Greedy scale; Latania scale; Rose scale**, *Aulacaspis rosae*; **San Jose scale**. Tiny, circular to oval insects, often in colonies.	Rarely if ever cause serous damage to rose in California. *See* 189–192.
Decline or death of canes or entire plant. Larvae, ≤1 inch long, tunneling in canes.	**Flatheaded appletree borer, Pacific flatheaded borer**. Whitish larvae with enlarged head in tunnels.	Keep plants vigorous. Prevent sunburn or sunscald. Remove and dispose of cane stubs from earlier pruning. *See* 227.
Tips of canes wilt in the spring and die back in summer. Spiral girdling (by larvae) in canes.	**Raspberry horntail**. Segmented, white larvae, ≤1 inch long. Adult sawflies black or black and yellow, 1/2 inch long, wasplike.	In California, mostly in interior valleys. *See* 243.

WHAT THE PROBLEM LOOKS LIKE	PROBABLE CAUSE	COMMENTS
Tips of canes bend, darken, distort, or wither and die. Flower buds darken and drop. Blossoms appear scorched.	**Rose midge**, *Dasineura rhodophaga*. Adults are tiny flies. Pale larvae occur between petals and sepals at flower bud base. Uncommon in California, reported only in Sonoma County.	Distinguish from similar-looking beneficial predaceous aphid midges. Prune and dispose of infested tissue. If severe, systemic insecticide can be applied in spring.
Buds, leaves, stems, or roots develop distorted swellings. Globular swellings may be spiny or resemble moss.	**Gall wasps**, including *Diplolepis* spp. *Diplolepis* wasp larvae cause a globular deformity that hardens and enlarges.	Do not threaten plant survival. Prune off undesired galls while green, before gall maker emerges as adult.
Galls or swellings on roots. Foliage may yellow and plants may grow slowly.	**Root knot nematodes; Dagger nematode**, *Xiphinema index*. Tiny root-feeding roundworms.	*See* 293, 294.
Roots smaller, darker than on healthy plants. Leaves may yellow. Growth may be stunted.	**Root lesion nematode**, *Pratylenchus* sp. Tiny root-feeding roundworms.	*See* 293.
Galls or enlarged, distorted tissue on stems and roots.	**Crown gall**. Bacteria persist in soil, spread in water, and infect through wounds.	*See* 108.
Swollen root tips with dense clusters of small rootlets. Foliage may yellow. Plants may grow slowly and die prematurely.	**Hairy root**, *Agrobacterium rhizogenes*. Bacteria persist in soil, spread in water, and infect through wounds.	Purchase high-quality plants. Avoid injuring plants during transplanting. Solarize soil before planting. *See* similar Crown Gall, 108.
Rosmarinus officinalis, Rosemary		
Brown, yellow, or wilted leaves. Roots or basal stem may be dark, decayed. Plant may die.	**Phytophthora root and crown rot**. Disease favored by excess soil moisture and poor drainage.	*See* 122.
Powdery, whitish growth on leaves or shoots. Foliage or shoots may distort or turn yellow or brown.	**Powdery mildew**, *Sphaerotheca* sp. Fungus favored by moderate temperatures, shade, and poor air circulation.	*See* 90.
Bleached, stippled leaves. Shoot tips may be distorted. Pale cast skins may be present on foliage.	**Ligurian leafhopper**, *Eupteryx decemnotata*. Colorful insect, $\leq 1/8$ inch long, hops away when disturbed.	Introduced leafhopper sucks foliage on various Lamiaceae plants. *See* Leafhoppers, 201.
Blackish sooty mold, sticky honeydew, and whitish cast skins on plant.	**Aphids**. Small green, brown, black, or yellowish insects, often in groups.	*See* 175.
Blackish sooty mold and sticky honeydew on plant. Tiny, white, mothlike insects on plant.	**Whiteflies**, including **Greenhouse whitefly**. Nymphs and pupae are flattened, oval, and pale.	*See* 182.
Wet, white, frothy masses of spittle on leaves or stems.	**Spittlebugs**. Insects covered in spittle, suck plant juices.	Tolerate, as they do not threaten plant health. Wash with forceful water. *See* 205.
Rubus spp., Salmonberry, Thimbleberry[32]		
Foliage fades, yellows, browns, or wilts, often scattered throughout canopy. Canes die. Entire plant may die.	**Verticillium wilt**. A soil-dwelling fungus that infects through roots. Symptoms sometimes appear first on only one side of the plant.	*See* 117.
Foliage discolors and wilts. Canes wilt and die back. Entire plant may die. May be white fungus beneath basal trunk bark.	**Armillaria root rot**. Fungus present in many soils. Favored by warm, wet soil. Persists for years in infected roots.	*See* 120.
Leaves with brown, light green, or yellow spots or dead blotches.	**Ramorum blight**, *Phytophthora ramorum*. Pathogen spreads via airborne spores and contaminated plants and soil.	Infects at least *Rubus spectabilis*. Primarily a problem in wildlands, killing many oaks there. *See* 105.
Soft, brown decay and/or gray, woolly growth on buds, flowers, fruit, leaves, and dead tissue.	**Botrytis blight**, or **Gray mold**. A fungal disease that thrives under moist conditions and cool to moderate temperatures.	*See* 78.
Purplish, red, or yellow blotches on leaves. Leaf undersides covered with downy fungal growth. Leaves may die and drop prematurely.	**Downy mildew**, *Peronospora sparsa*. Spores produced only on living plants. Resistant spores carry fungus over unfavorable periods. Favored by moist, humid conditions.	Prune plants to improve air circulation. Reduce humidity around plants, such as through drip or low-volume irrigation. Avoid overhead watering. *See* 89.

[32] Some species are invasive weeds. Other species may be better choices when planting.

Rubus spp. (continued)

WHAT THE PROBLEM LOOKS LIKE	PROBABLE CAUSE	COMMENTS
Powdery, white growth on underside and yellow blotches on upper side of leaves. Terminals and fruit may be covered with whitish growth. Shoots and leaves may be undersized or distorted.	**Powdery mildew**, *Sphaerotheca macularis*. Fungal disease favored by moderate temperatures, shade, and poor air circulation.	Thin canes and otherwise improve air circulation. Remove late-forming, infected suckers. Plant resistant caneberry varieties. *See* 90.
Orange or yellow pustules on canes or leaves. Foliage discolors and may drop prematurely.	**Rusts**, including *Arthuriomyces peckianus*, *Gymnoconia nitens*, and *Phragmidium rubi-idaei*. Fungal diseases favored by wet conditions.	Remove and dispose of infected canes and leaves. Avoid overhead irrigation. Provide good air circulation. *See* 94.
Blackish sooty mold, sticky honeydew, and whitish cast skins on plant.	**Aphids**. Small green, brown, black, or yellowish insects on leaves or fruit, especially under calyx.	*See* 175.
Leaves bleached. May be dark excrement specks on fruit.	**Leafhoppers**, including **White apple leafhopper**, *Typhlocyba pomaria*; **Rose leafhopper**, *Edwardsiana rosae*. Pale green, yellow, or white, wedgelike insects, ≤$^1/_8$ inch long.	Leafhoppers do not feed on fruit and do not seriously threaten plant health. *See* 201.
Tips of canes wilt in the spring and die back in summer. Spiral girdling (by larvae) in canes.	**Raspberry horntail**, *Hartigia cressoni*. Segmented white larvae, ≤1 inch long. Adult sawflies black or black and yellow, $^1/_2$ inch long, wasplike.	In California, mostly in interior valleys. *See* 243.
Wilting or dead plants. Some roots stripped of bark or girdled near soil. Foliage may be notched or clipped.	**Weevils**, including **Black vine weevil**; **Cribrate weevil**, *Otiorhynchus cribricollis*; **Fuller rose beetle**, *Naupactus godmani*.	Adults are black, brown, or gray snout beetles, ≤$^3/_8$ inch long. Larvae are root-feeding white grubs. *See* 160.
Grayish encrustations on bark. Rarely, dead or declining branches.	**Latania scale**. Individuals circular, <$^1/_{16}$ inch long, usually in groups.	*See* 190.

Salix spp., Willow		
Leaves with yellow spots and yellow to orangish, powdery pustules on lower surface. Leaves may drop prematurely.	**Rusts**, *Melampsora* spp. Fungal diseases that infect and develop when leaves are wet.	Damage usually not severe enough to warrant control action. *See* 94.
Leaves with oval or irregular, glossy black, thick, tarlike raised spots on upper leaf surface.	**Tar spot**, *Rhytisma salicinum*. Fungus most prevalent in moist environments.	Rarely if ever threatens plant health. Rake and dispose of leaves in the fall. *See* Leaf Spots, 86.
Leaves or twigs with irregular, small to large, black or brown blotches. Spots may have pale centers. Leaves may drop prematurely. Twigs may develop small cankers and dieback.	**Leaf and twig spots**, including *Marssonina apicalis*, *M. kriegeriana*; **Gray scab**, *Sphaceloma murrayae*. Fungi promoted and spread by wet conditions.	Rarely if ever threaten plant health. *See* Leaf Spots, 86; Scab, 88.
Leaves with black, brown, or dark olive spots in early spring. Leaves yellow and may drop prematurely. Twigs may develop small cankers or dieback.	**Scab**, or **Twig blight**, *Venturia chlorospora*, *V. saliciperda*. Fungal diseases spread by water from infected leaves and twigs.	Avoid overhead watering. Prune out infected twigs and leaves in fall. *See* 88.
Leaves brown, yellow, or drop prematurely. Terminals may die back. Branches or limbs may develop cankers. Infected wood is streaked brown.	**Bacterial blast, blight, and canker**. Disease favored by wet conditions.	*See* 83.
Leaves or shoots with powdery, white growth and yellow blotches. Shoots and leaves may be undersized or distorted.	**Powdery mildews**, including *Uncinula adunca*. Fungal diseases favored by moderate temperatures, shade, and poor air circulation.	*See* 90.
Leaves may yellow and wilt. Brownish, sunken lesions on limbs, trunk, or twigs. Small branches or twigs may die without any definite canker evident.	**Cankers, Twig blights**, including *Botryosphaeria ribis*, *Cytospora chrysosperma*, *Nectria* spp., *Valsa* spp. Fungal diseases most serious on injured or low-vigor trees.	*See* Botryosphaeria Canker and Dieback, 98; Cytospora Canker, 100; and Nectria Canker, 101.
Leaves turn brown or yellow, especially along margins and at tip. Leaves may drop prematurely.	**Leaf burn**, or **Scorch**. Abiotic disorders with many causes, including frost, inappropriate irrigation, poor drainage, or sunburn.	Provide plants with a good growing environment and proper cultural care. *See* Abiotic Disorders, 43.

WHAT THE PROBLEM LOOKS LIKE	PROBABLE CAUSE	COMMENTS
Leaves discolor and wilt. Decline or dieback of some branches or entire tree. Roughened bark may have dark or wet spots.	**Borers**, including **American hornet moth; Carpenterworm; Flatheaded appletree borer; Pacific flatheaded borer; Poplar borer,** *Saperda calcarata;* **Western poplar clearwing.**	Whitish larvae tunnel beneath bark or in wood. *See 227–242.*
Leaves may discolor, wilt, and drop prematurely. Bark may have bracketlike or fan-shaped fungal fruiting bodies. Branches or plant may die. Trees may fall over.	**Wood decay**, or **Heart rot**, including *Ganoderma* spp., *Pleurotus ostreatus, Schizophyllum commune, Stereum* sp., *Trametes* spp.	Fungi that attack injured, old, or stressed trees. *See 66, 111.*
Bark stained brownish, exudes rancid fluid, often around crotches, wounds.	**Wetwood**, or **Slime flux**. Bacterial infections.	Usually do not cause serious harm to trees. *See 110.*
Sticky honeydew and blackish sooty mold on plant. Twigs and branches may die.	**Giant bark aphid**, *Longistigma caryae.* Relatively large aphids, ≤¹/₄ inch long, grayish and black, in groups on bark.	Unlike most aphids, this bark-feeding species can be a serious pest, causing dieback. *See Aphids, 175.*
Sticky honeydew and blackish sooty mold on plant.	**Aphids**, including *Chaitophorus* spp.; **Giant willow aphid**, *Lachnus salignus;* **Melon aphid**. Green to brown insects, ≤¹/₈ inch long, clustered on leaves or twigs.	Plants tolerate moderate populations of leaf-feeding aphids. *See 175.*
Sticky honeydew and blackish sooty mold on foliage. May be pale waxiness on plants.	**Psyllids**, including *Psylla alba, P. americana.* Aphidlike insects, often brown, green, orangish, reddish, or white.	*See 170.*
Sticky honeydew and blackish sooty mold on foliage. Cottony or waxy material on plant.	**Mealybugs**, including **Obscure mealybug, Vine mealybug**. Powdery, grayish insects, ≤¹/₄ inch long, with waxy filaments.	Conserve natural enemies that help in control. *See 185.*
Sticky honeydew and blackish sooty mold on plant. Possible decline and dieback of woody parts. May be cottony material on plant.	**Brown soft scale; Cottony maple scale**, *Pulvinaria innumerabilis.* Yellow to brown, oval, flattened insects or cottony white, popcornlike bodies. Cottony maple scale has one generation per year.	Conserve natural enemies and tolerate moderate populations. *See 193;* similar Green Shield Scale, 194.
Leaves stippled or bleached, with cast skins and varnishlike specks.	**Lace bug**, *Corythucha* sp. Brown adults, ≤¹/₈ inch long, wings lacelike.	*See 207.*
Brownish, grayish, tan, or white encrustations on bark. May be decline or dieback of twigs or branches.	**Greedy scale, Latania scale, Oystershell scale, San Jose scale**. Tiny, circular to elongate insects on twigs and branches.	*See 189–192.*
Leaves skeletonized.	**Leaf beetles**, including *Altica bimarginata; Chrysomela aeneicollis; Syneta albida;* **California willow beetle**, *Melasomida californica;* **Cottonwood leaf beetle**. Brown to black adults and larvae.	Vigorous plants tolerate moderate leaf damage. Provide proper cultural care, including adequate water for willows, which are adapted to moist soils. *See 155.*
Leaves chewed. May be silken tents in tree.	**Fall webworm; Fruittree leafroller; Omnivorous looper; Redhumped caterpillar; Tent caterpillars; Tussock moths; Satin moth**, *Leucoma salicis.* Naked, spiny or hairy larvae, ≤2 inches long.	Vigorous plants tolerate moderate defoliation. Prune out colonies of caterpillars confined to a few branches. Apply *Bacillus thuringiensis* or another insecticide when caterpillars are young if intolerable. *See 149–152.*
Leaves chewed.	**Cerisyi's sphinx, Eyed sphinx, or One-eyed sphinx**, *Smerinthus cerisyi.* Adult mostly brown, tan, and reddish moth. Caterpillar green with pink anal horn, ≤2 inches long.	*See Caterpillars, 144.*
Leaves chewed, often on scattered branches. Larvae present may be feeding in groups.	**Spiny elm caterpillar**, or **Mourningcloak butterfly; Western tiger swallowtail**, *Papilio rutulus. Papilio* larvae are bright green with eyespots, black and yellow markings; adult is ≤2 inches long, yellow with black.	Caterpillars mature into highly attractive butterflies. Control not recommended. *Bacillus thuringiensis* kills young larvae. *See 151.*
Prominent, red, globular or elongate swellings (galls) on leaves, terminals, or twigs.	**Willow gall sawflies**. Tiny, whitish larvae feed in galls. Adults are small, stout wasps.	Galls do not harm plant. *See 216.*

Salix spp. (continued)

WHAT THE PROBLEM LOOKS LIKE	PROBABLE CAUSE	COMMENTS
Roundish brown, reddish, or green galls with pointed end on shoots.	**Willow beaked-gall midge**, *Mayetiola* (=*Rabdophaga*) *rigidae*. Larva feeding in new buds causes distortion.	Galls do not harm plant. *See* Gall Midges, 214.
Roundish, grayish to red blisters, beadlike swellings, or fuzzy growths on leaves.	**Blister gall mites**, including *Vasates laevigatae*. Microscopic eriophyid mites living in groups.	*See* Gall Mites, or Eriophyids, 249.
Galls or swellings on trunk and roots, usually near soil, may be on branches.	**Crown gall**. Bacteria infect plant via wounds.	*See* 108.
Salvia spp., Salvia, Sage		
Leaves or shoots with powdery growth. Shoots or leaves may be distorted or stunted.	**Powdery mildews**, including *Erysiphe cichoracearum, Phyllactinia guttata*.	Fungi favored by moderate temperatures, shade, and poor air circulation. *See* 90.
Leaves with brown or yellow spots and orangish pustules. Leaves may drop prematurely.	**Rusts**, including *Puccinia mellifera*. Fungi favored by wet conditions.	Avoid overhead watering. Plants tolerate moderate populations. *See* 94.
Leaves with brown, gray, reddish, yellow, or whitish blotches or spots. Leaves may drop prematurely.	**Leaf spots**, including *Mycosphaerella audibertiae, Septoria rhabdocarpa*.	Fungal diseases that infect and develop when foliage is wet. *See* 86.
Blackish sooty mold, sticky honeydew, and whitish cast skins on plant.	**Aphids**. Small green, brown, black, or yellowish insects, often in groups.	*See* 175.
Stippled or bleached leaves, varnishlike excrement specks on undersides.	**Thrips**. Tiny, slender, blackish or yellowish insects.	*See* 209.
Sambucus spp., Elderberry		
Foliage yellows and wilts. Stems may have dark, sunken cankers. Limbs or entire tree may die.	**Cankers**, including **Botryosphaeria canker and dieback; Diaporthe stem canker and dieback**, *Diaporthe sociabilis*; **Nectria canker**; *Sphaeropsis* spp. Fungi primarily affect injured or stressed trees.	Avoid wounding plants, except to prune off infected limbs. Provide a good growing environment and appropriate cultural care. *See* 98, 101.
Foliage yellows and wilts. Bark or wood may have bracketlike or fan-shaped fungal fruiting bodies. Branches or entire plant may die.	**Wood decay**, or **Heart rot**, including *Ganoderma* spp., *Phellinus igniarius, Schizophyllum commune*. Fungi that attack injured, old, or stressed trees.	*See* 66, 111.
Foliage and stems have brown to yellow spots or blotches. Spots may have dark or yellowish margins. Foliage may shrivel and drop prematurely.	**Leaf and stem spots**, including *Cercospora* sp., *Ramularia glauca, R. sambucina, Septoria sambucina, Stigmina* spp. Fungi spread by splashing water. Diseases favored by prolonged wet conditions.	Avoid wetting foliage. Use drip irrigation where feasible. Promptly remove and dispose of plant debris and infected leaves. *See* Leaf Spots, 86.
Bark with wet spots or sawdustlike boring material. Limbs may decline and die.	**Elder borers**, *Desmocerus* spp. Adult longhorned beetles bluish, greenish, or blackish with gold or orange and feed on elderberry flowers. Larvae bore in living trees, which usually are not killed.	Attack stressed trees and dying limbs. Valley elderberry longhorn beetle (*D. californicus*) and its habitat (dying elderberry) may be protected by law. *See* Longhorned Beetles, or Roundheaded Wood Borers, 232.
Sticky honeydew, blackish sooty mold, and whitish cast skins on plant.	**Aphids**, including **Bean aphid**. Small, green or black insects in groups.	*See* 175.
Sticky honeydew and blackish sooty mold on leaves and twigs. May be dieback of twigs or branches.	**European fruit lecanium**. Black, brown, or yellow, bulbous or flattened insects, $\leq 1/5$ inch long.	*See* 195.
Sapium spp., Chinese tallow tree, Japanese tallow tree[33]		
Leaves discolor and wilt. Branches die back. Entire tree may die. May be white fungus beneath basal trunk bark.	**Armillaria root disease**. Fungus present in many soils. Favored by warm, wet soil. Persists for years in infected roots.	*See* 120.
Twig terminals die back in fall. Foliage curls, turns brown or black, and dies.	**Leaf burn**, or **Scorch**. Abiotic disorders with many potential causes, including freeze or cold damage.	Avoid excess fertilization, irrigation, and pruning that stimulate excess growth, especially in fall. Provide a good growing environment and proper cultural care.

[33] Some species are invasive weeds. Other species may be better choices when planting.

WHAT THE PROBLEM LOOKS LIKE	PROBABLE CAUSE	COMMENTS
Sticky honeydew and blackish sooty mold on plant. Waxy, cottony material on leaves or twigs.	**Obscure mealybug**. Powdery, whitish insects, ≤$1/_8$ inch long, with filaments, longest at tail. Suck sap.	*See* 187.
Schinus molle, Pepper tree, California pepper tree, Peruvian pepper tree[34]		
Leaves discolor, stunt, wilt, or drop prematurely. Stems discolored, cankered, and die. May be white fungus beneath basal trunk bark.	**Armillaria root disease**. Fungus present in many soils. Favored by warm, wet soil. Persists for years in infected roots.	*See* 120.
Leaves discolor, stunt, wilt, or drop prematurely, often on one side of plant. Stem xylem discolored. Stems or entire plant may die.	**Verticillium wilt**. Fungus persists in soil, infects through roots.	*See* 117.
Foliage yellows and wilts. Bark or wood may have bracketlike or fan-shaped fungal fruiting bodies. Branches or entire plant may die.	**Wood decay**, or **Heart rot**, including *Ganoderma* spp., *Laetiporus gilbertsonii* (=*L. sulphureus*), *Schizophyllum commune*, *Trametes* spp.	Fungi that attack injured, old, or stressed trees. *See* 66, 111.
Leaves turn brown or yellow, especially along margins and at tip. Leaves may drop prematurely.	**Leaf burn**, or **Scorch**. Abiotic disorders with many potential causes, including frost, inappropriate irrigation, or poor drainage.	Provide plants with a good growing environment and proper cultural care. *See* Abiotic Disorders, 43.
Chewed leaves. Foliage may be rolled and tied together with silk.	**Omnivorous looper**. Yellow, green, or pink larvae, ≤$1^1/_2$ inches long, with green, yellow, or black stripes.	*See* 150.
Roundish pits in leaflets, petioles, and twigs. Sparse foliage.	**Peppertree psyllid**. Tiny, green adults, $1/_{16}$ inch long. Nymphs flattened and in pits, suck sap.	*See* 174.
Stippled, bleached foliage.	**Citrus thrips**. Tiny, slender, yellow.	Rarely if ever seriously damages pepper tree. *See* 209.
Sticky honeydew and blackish sooty mold on foliage. May be declining or dead twigs or branches.	**Barnacle scale**, *Ceroplastes cirripediformis*; **Black scale; Green shield scale; Hemispherical scale**, *Saissetia coffeae*; **Wax scales**.	Small black, brown, green, yellowish, waxy, or whitish, flattened or bulbous insects. *See* 192, 194, 196.
Brownish, grayish, tan, or white encrustations on bark. Rarely, declining or dead twigs or branches.	**Greedy scale, Oleander scale**. Tiny, oval to circular insects on bark.	Rarely if ever cause serious damage to plants. *See* 189–192.
Holes in parallel lines on trunk or large branches.	**Sapsuckers**. Various species of sap-feeding woodpeckers, make holes in bark of many woody plant species.	Damage is primarily cosmetic if trees are otherwise healthy. *See* 220; *Woodpeckers Pest Notes* at ipm.ucanr.edu.
Sequoia sempervirens, Coast redwood, Redwood, Sequoia[35]		
Foliage discolors, wilts, and drops prematurely. Branches die back. Entire tree may die. May be white fungus beneath basal trunk bark.	**Armillaria root disease**. Fungus present in many soils. Favored by warm, wet soil. Persists for years in infected roots.	*See* 120.
Foliage discolors. Plants grow slowly and may die. Roots or basal stem may be dark or decayed.	**Phytophthora root and crown rot**, *Phytophthora cinnamomi*.	Pathogen promoted by excess soil moisture and poor drainage. *See* 122.
Needles and shoots turn brown, tan, or yellow and may die. Dead, dying, and falling branchlets.	**Needle blight**, or **Scorch**. Abiotic disorders. Causes include alkaline water, drought stress, dry wind, high temperatures, and salty spray onto foliage.	Some needles and branchlets are shed naturally. Provide trees with a good growing environment and appropriate cultural care. *See* Abiotic Disorders, 43.
Needles turn brown, tan, or yellow, beginning in inner, lower canopy. Affected needles and shoots drop, leaving only green terminals.	**Cercospora needle blight**, *Cercospora sequoiae*. Fungal disease favored by warm, wet conditions.	Avoid sprinkling foliage. Improve air circulation. *See* Needle Blight and Cast, 85.
Needles discolor and die on branch terminals, primarily on sprouts on lower trunk. Twigs with cankers and dieback.	**Ramorum blight**, *Phytophthora ramorum*. Pathogen spreads via airborne spores and contaminated plants and soil.	Primarily a problem in wildlands, killing many oaks there. *See* Sudden Oak Death and Ramorum Blight, 105.

[34] Brazilian pepper tree (*Schinus terebinthifolius*) is an invasive weed. Other species may be better choices when planting.

[35] For more information, see *Pests of the Native California Conifers*.

Sequoia sempervirens (continued)

WHAT THE PROBLEM LOOKS LIKE	PROBABLE CAUSE	COMMENTS
Dead or dying branches. Cankers on trunk or limbs. Dying branches with resinous lesions.	**Botryosphaeria canker and dieback, Cytospora canker.** Fungal diseases, often affecting drought-stressed trees.	Provide good growing conditions and appropriate cultural care, especially adequate irrigation. *See* 98, 100.
Resinous lesions or cankers on bark. Branches or treetop die back.	**Redwood canker**, *Coryneum* sp. Fungal disease, usually at warm, dry sites.	Prune out and dispose of diseased branches. Irrigate adequately. *See* similar Cypress Canker, 99.
Terminals die back. Twigs may develop brown, sunken cankers. Soft, brown decay and/or gray, woolly growth on buds, leaves, and dead tissue.	**Botrytis blight**, or **Gray mold.** Fungus favored by wet conditions. Spreads by airborne spores.	*See* 78.
Black, dark brown, or gray encrustations on shoots or needles. Needles may yellow.	**Black araucaria scale**, *Lindingaspis rossi;* **Redwood scale**, *Aonidia shastae.* Tiny, circular to oval armored scales.	More a problem in California in the south. Vigorous trees tolerate moderate populations. Conserve natural enemies. *See* Armored Scales, 189.
Browning of tips, beginning in fall, worst in late winter and spring.	**Cypress tip miner.** Silvery tan moths, green larvae, about $1/4$ inch long.	*See* 217.
Terminals brown and die several inches back from tip.	**Redwood bark beetle**, *Phloeosinus sequoiae.* Tiny, dark reddish bark beetle adults bore in twigs. Pale larvae bore in broken or recently killed limbs.	Terminal feeding does not seriously harm trees. *See* Cedar, Cypress, and Redwood Bark Beetles, 224.
Older foliage dark, stippled, may contain fine webbing.	**Spruce spider mite.** Sand-sized specks on needles.	Mites feed during cool weather. *See* Pine and Spruce Spider Mites, 247.
Globular galls, up to several inches in diameter on branches.	**Unidentified cause.** Believed to be a genetic disorder. Tree vigor not seriously affected.	No control except to eliminate tree. Tolerate, does not spread to other trees.
Sequoiadendron giganteum, Giant sequoia, Big tree, Sierra redwood[36]		
Treetop, branches dying. Some branches reddish, pitchy, or grayish, bare.	**Botryosphaeria canker and dieback.** Fungus primarily affecting injured or stressed trees.	Avoid planting sequoia in hot areas outside of native range. Affects drought-stressed trees. Provide proper irrigation. *See* 98.
Foliage may yellow and wilt. Bark or wood may have bracketlike or fan-shaped fungal fruiting bodies. Limbs or entire plant may die.	**Wood decay**, or **Heart rot**, including *Schizophyllum commune, Stereum* sp. Fungi that attack injured, old, or stressed trees.	*See* 111.
Stippled foliage. Foliage color abnormally light green or yellowish.	**Spruce spider mite.** Greenish specks, often in fine webbing at foliage base.	Highest populations occur during spring and fall. *See* Pine and Spruce Spider Mites, 247.
Terminals die back. Twigs may develop brown sunken cankers. Soft, brown decay and/or gray, woolly growth on buds, leaves, and dead tissue.	**Botrytis blight**, or **Gray mold.** Fungus favored by wet conditions. Spreads by airborne spores.	*See* 78.
Needles turn brown, tan, or yellow, beginning in inner, lower canopy. Needles may drop prematurely, leaving only green terminals.	**Needle blight and cast**, *Cercospora sequoiae.* Fungal disease favored by warm, wet conditions.	Avoid wetting foliage. Improve air circulation. *See* 85.
Brownish or grayish encrustations on foliage. Foliage may yellow.	**Redwood scale**, *Aonidia shastae.* Oval to circular insects, each $\leq 1/16$ inch long, often in colonies.	Conserve natural enemies that provide control. Vigorous plants tolerate moderate populations. *See* Armored Scales, 189.
Sophora japonica, Japanese pagoda tree		
Leaves discolor and wilt. Branches die back. Entire tree may die. May be white fungus beneath basal trunk bark.	**Armillaria root disease.** Fungus present in many soils. Favored by warm, wet soil. Persists for years in infected roots.	*See* 120.
Chewed leaves. Plants may be defoliated.	**Genista caterpillar**, *Uresiphita reversalis.* Caterpillars, $\leq 1 1/4$ inches long, green to orange with black and white hairs.	Introduced to control brooms, which are often considered weeds. Apply *Bacillus thuringiensis* or another insecticide. *See* Caterpillars, 144.

[36] For more information, see *Pests of the Native California Conifers.*

WHAT THE PROBLEM LOOKS LIKE	PROBABLE CAUSE	COMMENTS
Sorbus spp., Mountain ash		
Sudden wilting, then shriveling and blackening of blossoms or shoots. Plant appears scorched. Shoots may develop cankers, which may ooze.	**Fire blight**. Bacterium enters plants through blossoms.	Plant less-susceptible cultivars. *See 84.*
Brownish, sunken lesions on trunk and large limbs. Small branches and twigs may be killed without any apparent canker.	**Cankers**, including **Cytospora canker, Nectria canker**. Fungal diseases primarily affecting injured or severely stressed trees.	*See 97–101.*
Leaves with orangish pustules or light to dark spots. Leaves may drop prematurely. Terminals may die back.	**Rusts**, *Gymnosporangium* spp. Fungi that infect and develop during moist conditions.	Avoid overhead watering. Vigorous plants tolerate moderate infection. *See Cedar, Cypress, and Juniper Rusts, 95.*
Leaves may discolor and wilt. Bark may be holey, roughened, or ooze dark liquid. May be reddish brown, granular material (frass) on trunk or limbs.	**Borers**, including **American plum borer; Flatheaded appletree borer; Longhorned beetle**, *Saperda* sp. Brown, pink, or whitish larvae, ≤2 inches long, that bore under bark.	*See 240;* Flatheaded borers, or Metallic Wood Borers, 227; Longhorned Beetles, or Roundheaded Wood Borers, 232.
Stippled, bleached, or reddened foliage.	**Spider mites**, including **European red mite**; *Oligonychus* sp. Tiny, reddish to yellow arthropods, suck sap.	*See Red Mites, 246.*
Sticky honeydew, blackish sooty mold, and whitish cast skins on plant. May be whitish wax on plant.	**Apple aphid**, *Aphis pomi;* **Melon aphid; Woolly apple aphid**. Bright green to grayish insects on terminals or leaves.	Conserve natural enemies that can provide control. *See Aphids, 175.*
Sticky honeydew and blackish sooty mold on foliage. Possible decline or dieback of twigs or branches.	**Frosted scale**. Brown, yellow, or waxy insects, bulbous or flattened, on leaves or twigs.	*See Lecanium Scales, 195.*
Grayish or brown encrustations on bark. May be declining or dead twigs or branches.	**Oystershell scale; San Jose scale; Walnut scale**, *Diaspidiotus* (=*Quadraspidiotus*) *juglansregiae*. Tiny, oval to circular insects on bark.	*See 189–192.*
Leaves or blossoms chewed. Single branches or entire plant may be defoliated. Foliage may be rolled and tied together with silk.	**Caterpillars**, including **Obliquebanded leafroller**, *Choristoneura rosaceana*. Moth larvae, ≤1¹/₂ inches long.	*See Leafrollers, 149.*
Leaves scraped, skeletonized, and webbed together with silk.	**Leafminers**, including **Apple-and-thorn skeletonizer**, *Choreutis pariana*. Adult *Choreutis* is dark brown moth, <¹/₂ inch long. Larvae are ≤1 inch long and greenish or yellow.	Plants usually tolerate damage. If intolerable prune off infested foliage. *See Foliage Miners, 216.*
Spiraea spp., Spirea		
Sudden wilting, then shriveling and blackening of shoots and blossoms. Plants appear scorched.	**Fire blight**. Bacterium infects plants through blossoms.	*See 84.*
Powdery, white growth and yellow blotches on leaves and sometimes on terminals. Shoots and leaves may be undersized or distorted.	**Powdery mildews**, including *Podosphaera* spp., *Stictis radiata*. Fungal diseases favored by moderate temperatures, shade, and poor air circulation.	*See 90.*
Leaves with brown, black, tan, or yellow spots or blotches. Foliage may die and drop prematurely.	**Leaf spots**, including *Cercospora rubigo*. Spread by air or splashing water. Favored by prolonged cool, wet conditions.	*See 86.*
Sticky honeydew, blackish sooty mold, and whitish cast skins on leaves. Leaves curled.	**Spirea aphid**, *Aphis spiraecola*. Green insects, <¹/₈ inch long, clustered on growing leaves and tips.	Vigorous plants tolerate moderate populations. *See Aphids, 175.*
Leaves bleached, stippled, may turn brown and drop prematurely. Terminals may be distorted. Plant may have fine webbing.	**Spider mites**, including *Eotetranychus* sp. Tiny arthropods, often green, pink, red, or yellowish.	*See 246.*

WHAT THE PROBLEM LOOKS LIKE	PROBABLE CAUSE	COMMENTS
Strelitzia spp., Bird of paradise, Giant bird of paradise		
Leaves discolor, wilt, stunt, or drop prematurely. Plants stunted. Root system reduced, small roots rotted.	**Pythium root rot**. Soilborne pathogen favored by excess soil moisture and poor drainage.	*See* 123.
Leaves discolor, wilt, stunt, or drop prematurely. Stems discolored, cankered, may ooze sap and die. May be white fungus beneath basal trunk bark.	**Armillaria root disease**. Fungus present in many soils. Favored by warm, wet soil. Persists for years in infected roots.	*See* 120.
Soft, brown decay and/or gray, woolly growth on buds, flowers, leaves, and dead tissue. Stem may be girdled.	**Botrytis blight**, or **Gray mold**. Fungus develops in plant debris or inactive tissue. Favored by wet conditions and moderate temperatures. Spores airborne.	*See* 78.
Sticky honeydew and blackish sooty mold on foliage. Foliage may yellow. Plant may have cottony material (egg sacs).	**Citrus mealybug; Longtailed mealybug; Obscure mealybug**. Powdery, gray insects with waxy filaments.	*See* 186, 187.
Sticky honeydew and blackish sooty mold on foliage. Copious, white, waxy material may be on plant. Leaves may yellow and wither.	**Giant whitefly; Greenhouse whitefly; Iris whitefly**, *Aleyrodes spiraeoides*. Nymphs and pupae are flattened, oval, and pale. Adults are tiny, mothlike insects.	*See* 182.
Sticky honeydew, blackish sooty mold, and pale wax on fronds.	**Coconut mealybug**, *Nipaecoccus nipae*. Adult females are oval shaped and orangish with marginal wax filaments.	*See* 186.
Sticky honeydew, blackish sooty mold, and whitish cast skins on plant. Foliage may yellow.	**Aphids**. Tiny, pear-shaped insects, often green, yellowish, or blackish.	*See* 175.
Sticky honeydew and blackish sooty mold on foliage. Foliage may yellow and plant may die back.	**Brown soft scale; Nigra scale**, *Parasaissetia nigra*. Orange, black, brown, or yellow, flattened to bulbous, oval insects.	*See* 193.
Stems or leaves have brown, gray, tan, reddish, orange, or white encrustations. Foliage may yellow. Rarely, plant may decline or die back.	**California red scale; Cycad scale; Dictyospermum scale**, *Chrysomphalus dictyospermi*; **Greedy scale; Latania scale; Oleander scale**.	Oval to round, flattened insects, $\leq^1/_{16}$ inch long. *See* Armored Scales, 189.
Plants may stunt, decline, or, rarely, die. Powdery, waxy material may be visible on roots and around crown.	**Ground mealybugs**. Small, slender, pale insects, may be lightly covered with powdery wax, but lack marginal filaments.	*See* 187.
Foliage yellows, wilts, and dies. Basal stem has holes or decay.	**Crown borer**, *Opogona omoscopa*. Dark, brownish moth. Grayish larva, $\leq^3/_4$ inch long, head dark, bores in plant.	Probably a secondary pest attracted to decaying tissue. Avoid wounding plants. Provide good cultural care. Avoid excess irrigation. Use good sanitation; remove debris and dying plants.
Syringa spp., Syringa, Japanese tree lilac, Lilac		
Leaves or stems with black to brown spots and streaks. Blossoms, buds, or leaves darken and shrivel. Oozing lesions on twigs.	**Bacterial blast, blight, and canker**. Disease favored by wet conditions.	Grow *Syringa josikaea, S. komarowii, S. microphylla, S. pekinensis*, and *S. reflexa*, which are less susceptible than many *S. vulgaris* cultivars. *See* 83.
Leaves yellow and wilt. Branches die back. Entire plant may die. May be white fungus beneath basal trunk bark.	**Armillaria root disease**. Fungus present in many soils. Favored by warm, wet soil. Persists for years in infected roots.	*See* 120.
Leaves fade, yellow, brown, wilt, often scattered throughout canopy or first on one side of plant. Branches die. Entire plant may die.	**Verticillium wilt**. A soil-dwelling fungus that infects through roots.	*See* 117.
Foliage yellows and wilts. Bark or wood may have bracketlike or fan-shaped fungal fruiting bodies. Branches or entire plant may die.	**Wood decay**, or **Heart rot**, including *Trametes versicolor*. Fungi that attack injured, old, or stressed trees.	*See* 111.

WHAT THE PROBLEM LOOKS LIKE	PROBABLE CAUSE	COMMENTS
Leaves fade and wilt, often scattered throughout canopy. Branches die. Boring dust, ooze, or holes on trunk or branches.	**Lilac borer**, or **Ash borer**. White larvae, ≤1 inch long, bore in wood. Adults are wasplike moths.	*See* 235.
Leaves turn brown or yellow, especially along margins and at tip. Leaves may drop prematurely.	**Leaf burn**, or **Scorch**. Abiotic disorders with many potential causes, including extreme temperatures, excess light, inappropriate irrigation, or poor drainage.	Provide plants with a good growing environment and proper cultural care. *See* Abiotic Disorders, 43.
Leaves brown, mostly along edges and at tips. Leaf buds and shoot tips die.	**Ramorum blight**, *Phytophthora ramorum*. Pathogen spreads via airborne spores and contaminated plants and soil.	Primarily a problem in wildlands, killing many oaks there. *See* Sudden Oak Death and Ramorum Blight, 105.
Leaves or shoots with gray or white, powdery growth. Leaves or terminals may distort.	**Powdery mildew**, *Microsphaera* sp. A fungal disease favored by moderate temperatures, shade, and poor air circulation.	Usually does no serious harm to lilac. *See* 90.
Soft, brown decay and/or gray, woolly growth on buds, flowers, leaves, and dead tissue.	**Botrytis blight**, or **Gray mold**. Fungus develops in plant debris or inactive tissue. Spores airborne.	Favored by high humidity and moderate temperatures. *See* 78.
Leaves with brown, black, tan, or yellow spots or blotches. Spots may have dark or yellowish margins. Foliage may die and drop prematurely.	**Leaf spots**, including *Alternaria* sp. Spread by air or splashing water. Favored by prolonged cool, wet conditions. Persist in plant debris.	*See* 86.
Leaves with brown, yellow, or whitish blotches. Leaves may be rolled at edges or tip.	**Leafminers**, including **Lilac leafminer**, *Caloptilia syringella*. Greenish larvae tunnel in leaves. Adults are small, brownish moths.	Plants tolerate extensive mining. If problem has been severe, systemic insecticide can be applied during spring. *See* Foliage Miners, 216.
Sticky honeydew and blackish sooty mold on foliage. Possible plant decline or dieback.	**European fruit lecanium**. Brown to yellow, flat or bulbous, immobile scale insects.	*See* Lecanium Scales, 195.
Brown to gray encrustations on twigs or branches. May be stunted or dying woody parts.	**Oystershell scale**. Individuals resemble miniature oysters, 1/16 inch long.	*See* 191.
Chewed leaves or blossoms.	**Fuller rose beetle**, *Naupactus godmani*. Pale brown adult snout beetle, about 3/8 inch long.	Adults hide during day and feed at night. Larvae feed on roots. *See* Weevils, 159.
Plants stunted, decline, may die. Powdery, waxy material may be visible on roots and around crown.	**Ground mealybugs**, *Rhizoecus* spp. Small, slender, pale insects, may be lightly covered with powdery wax, but lack marginal filaments.	*See* 187.
Syzygium paniculatum, Eugenia, Australian brush cherry, Brush cherry		
Foliage or flowers curl, turn brown or black, and die. Branches may die back.	**Leaf burn**, or **Scorch**. Common causes include freeze or cold damage, poor drainage, and too much or too little water.	*See* Abiotic Disorders, 43.
Sticky honeydew and blackish sooty mold on foliage. Leaves and terminals pitted, distorted, and discolored.	**Eugenia psyllid**. Adults are tiny, leafhopperlike insects. Nymphs feed in pits on lower leaf surface.	*See* 172.
Sticky honeydew and blackish sooty mold on foliage. Tiny, whitish, mothlike adult insects.	**Woolly whitefly**. Nymphs are oval, waxy, or cottony insects.	Usually under good biological control unless natural enemies are disrupted. *See* 184.
Sticky honeydew and blackish sooty mold on foliage. May be cottony material (egg sacs) on plant.	**Green shield scale, Pyriform scale**. Brownish, green, or yellowish convex or flattened insects on bark or leaves.	*See* 192, 194.
Copious, misty, nonsticky liquid raining from plant. Surfaces covered with whitish residue.	**Glassy-winged sharpshooter**. Active, dark brown or gray leafhoppers, ≤1/2 inch long, suck xylem fluid.	Vectors *Xylella* pathogens. Report suspected glassy-winged sharpshooters to agricultural officials if found in areas where this pest is not known to occur. *See* 203.
Grayish, orange, or brownish encrustations on bark or foliage. Rarely, stunted, declining, or dead branches.	**California red scale**. Tiny, circular, flattened insects on stems or leaves.	Rarely if ever causes significant damage to eugenia. *See* Armored Scales, 189.

WHAT THE PROBLEM LOOKS LIKE	PROBABLE CAUSE	COMMENTS
Taxus spp., Yew		
Foliage discolored, wilted, stunted, drops prematurely. Discolored bark may ooze sap. Branches or plant may die.	**Phytophthora root and crown rot.** Disease favored by excess soil moisture and poor drainage.	_See_ 122.
Leaves yellow and wilt. Bark or wood may have bracketlike or fan-shaped fungal fruiting bodies. Branches or entire plant may die. Trees may fall over.	**Wood decay**, or **Heart rot**, including _Laetiporus gilbertsonii_ (=_L. sulphureus_). Fungi that infect injured, old, or stressed trees.	_See_ 111.
Sticky honeydew and blackish sooty mold on foliage. Cottony, waxy material on plant.	**Obscure mealybug.** Powdery, grayish, waxy insects with fringe filaments.	_See_ 187.
Brown, gray, or purplish encrustations on bark. Rarely, declining or dead twigs or branches.	**Oleander scale; Purple scale**, _Lepidosaphes beckii_. Tiny, circular to elongate insects, often in colonies.	Rarely if ever cause serious damage to yew. _See_ 189–192.
Needles notched or clipped. Foliage may yellow or wilt. General decline of plant may occur.	**Black vine weevil.** Adults $^3/_8$ inch long, black or grayish snout beetles, active at night.	Larvae feed on roots. _See_ 160.
Tecomaria capensis, Cape honeysuckle		
Foliage or flowers curl, turn brown or black, may die.	**Leaf burn**, or **Scorch**. Abiotic disorders with many potential causes, including freeze or cold damage.	_See_ Abiotic Disorders, 43.
Teucrium spp., Germander		
No serious invertebrate or pathogen pests have been reported in California. If plants are unhealthy, investigate whether they have been injured or lack good growing conditions or appropriate cultural care.		
Thuja spp., _Thuja occidentalis_, _Thuja orientalis_ =_Platycladus orientalis_, Arborvitae; _Thuja plicata_, Western red cedar		
Foliage turns brown at base of branches, evenly scattered throughout canopy, during summer or fall. Needles may drop prematurely.	**Flagging.** Abiotic disorders with many potential causes, commonly caused by root injury, drought stress, and hot weather.	Protect trees from injury. Avoid disturbing soil around roots. Provide good cultural care, such as appropriate irrigation during prolonged drought.
Brown or yellow dying foliage on branches. Cankers and resinous exudate on limbs or trunk.	**Cypress canker.** A fungal disease.	Arborvitae is occasionally infected. Primarily affects _Cupressus_ spp. _See_ 99.
Foliage yellows and wilts. Branches or entire plant dies. Rootlets decayed or sparse.	**Phytophthora root and crown rot.** Disease favored by wet soil and poor drainage.	Commonly affects _Thuja plicata_ near turfgrass and irrigated landscape. _See_ 122.
Discolored or dying foliage. Tiny, oval to circular bodies on foliage.	**Juniper scale**, _Carulaspis juniperi_; **Minute cypress scale**. Tiny armored scales.	Populations in California seldom warrant control, except on Italian cypress or in nurseries. _See_ 190.
Browning of tips beginning in fall, becoming worse in late winter to spring.	**Cypress tip miner.** Greenish larvae, $\leq^1/_8$ inch long, tunnel in foliage.	_See_ 217.
Terminals distorted, wilting, or dead.	**Leaffooted bugs.** Mostly brown bug with yellow or bright colors and enlarged, flattened hind legs. Suck plant juices.	Plants probably tolerate. _See_ 205; _Leaffooted Bug Pest Notes_ at ipm.ucanr.edu.
Dead twigs on tree. Lateral twigs killed about 6 inches from tip.	**Cedar** and **Cypress bark beetles.** Small, dark adults bore in twigs. Pale larvae bore in broken or recently killed limbs.	Terminal feeding does not seriously harm trees. _See_ Cedar, Cypress, and Redwood Bark Beetles, 224.
Branches killed, sometimes to trunk. Coarse boring dust at trunk wounds and branch crotches.	**Cypress bark moth**, _Laspeyresia cupressana_. Larvae, $\leq^1/_2$ inch long, under bark of trunk and limbs.	Provide plants with proper cultural care. Avoid excess water and fertilizer that promote rapid growth and thin bark. Avoid wounding bark, as insects colonize wounds.
Needles are chewed. Roots or the basal trunk may be cankered or injured.	**Conifer twig weevils.** Small, black to brown weevils that chew needles and shoots. Pale larvae bore in the root crown of dying or injured conifers.	Except for white pine weevil, most are secondary pests of minor importance. _See_ 161.
Chewed needles.	**Conifer sawflies, Cypress sawfly.** Greenish larvae, ≤ 1 inch long, on needles.	_See_ 153, 154.

WHAT THE PROBLEM LOOKS LIKE	PROBABLE CAUSE	COMMENTS
Sticky honeydew and blackish sooty mold on foliage.	**Arborvitae aphid**, *Dilachnus tujafilinus*. Brown to gray insects, $^1/_8$ inch long, on leaves and twigs.	Vigorous plants tolerate moderate populations. *See* Aphids, 175.
Tilia spp., Linden		
Sticky honeydew, blackish sooty mold, and whitish cast skins on leaves. *Tilia cordata* and *T. europaea* are more susceptible lindens.	**Linden aphid**, *Eucallipterus tiliae*. Small, yellowish insects, with black.	Conserve *Trioxys curvicaudus* parasites and other natural enemies. Plant *Tilia platyphyllos* or other species with more hairy leaf undersides. *See* Aphids, 175.
Sticky honeydew and blackish sooty mold on leaves. Twigs and limbs may die back.	**Tuliptree scale**. Females ≤$^1/_3$ inch, irregularly hemispherical, and variably colored brown to gray with other-colored blotches.	Primarily a pest of tulip tree and deciduous magnolias. *See* 195.
Leaves with elongate or pointed growths on upper surface.	**Linden gall mite**, *Eriophyes tiliae*. Microscopic, wormlike eriophyid mites that induce leaf galls.	These mites do not seriously harm trees. No control is recommended. *See* Gall Mites, or Eriophyids, 249.
Leaves may discolor and wilt. Bark may have boring frass, elliptical holes, or oozing liquid. Branches or entire tree may die.	**Borers**, including **Longhorned beetles**, *Saperda* spp. White larvae, ≤$1^1/_2$ inches long, that bore beneath bark.	Usually secondary pests that attack injured or severely stressed trees. *See* 232.
Foliage or stems yellow and wilt. Bracketlike or fan-shaped fungal fruiting bodies may be on bark or wood. Branches or entire plant may die.	**Wood decay**, or **Heart rot**, including *Pleurotus ostreatus, Trametes versicolor*. Fungi that attack injured, old, or stressed plants.	*See* 111.
Tipuana tipu, Tipu, Rosewood tree		
Leaves stippled, distorted, and drop prematurely. Sticky honeydew, dark sooty mold, and waxy pellets on plant.	**Tipu psyllid**, *Platycorypha nigrivirga*. Dark brown, green, and orangish insects.	*See* Psyllids, 170.
Ulmus spp., Elm		
Foliage yellows, then wilts, usually first in one part of canopy. Curled, dead, brown leaves remain on tree.	**Dutch elm disease**. Fungal disease spreads by bark beetles and root grafts.	Do not confuse with elm leaf beetle feeding that causes skeletonized leaves. *See* 113.
Foliage fades, yellows, browns, or wilts, often scattered throughout canopy. Branches die. Entire plant may die.	**Verticillium wilt**. A soil-dwelling fungus that infects through roots.	*See* 117.
Foliage discolors, wilts, and drops prematurely. Branches die back. Entire tree may die. May be white fungus beneath basal trunk bark.	**Armillaria root disease**. Fungus present in many soils. Favored by warm, wet soil. Persists for years in infected roots.	*See* 120.
Foliage with irregular, black, tarlike spots. Premature leaf drop. Perennial cankers on limbs and trunk. Dieback.	**Chinese elm anthracnose**. A fungal disease affecting only Chinese (evergreen) elm (*Ulmus parvifolia*).	In California usually a problem only near the coast. Plant resistant cultivars or species. *See* 81, 99.
Foliage wilts and branches die back. Stems may have dark cankers, callus tissue, or coral-colored pustules.	**Nectria canker**, *Nectria cinnabarina*. Fungus primarily affects injured or stressed trees.	On elm most common on Chinese elm (*Ulmus parvifolia*). *See* 101.
Foliage may yellow and wilt. Bark or wood may have bracketlike or fan-shaped fungal fruiting bodies. Limbs or entire plant may die.	**Wood decay**, or **Heart rot**, including *Ganoderma* spp., *Laetiporus gilbertsonii* (=*L. sulphureus*), *Pleurotus ostreatus*, *Schizophyllum commune, Stereum* sp., *Trametes* spp.	Fungi that attack injured, old, or stressed trees. *See* 66, 111.
Foliage brown or yellow, especially along leaf margins and tips. Leaves may drop prematurely.	**Leaf burn**, or **Scorch**. Abiotic disorders with many potential causes. Too little water may be most common cause when summer rainfall–adapted elms are planted in arid California.	Provide plants with a good growing environment and proper cultural care, especially appropriate irrigation. *See* Abiotic Disorders, 43.

WHAT THE PROBLEM LOOKS LIKE	PROBABLE CAUSE	COMMENTS
Leaves discolored with irregular, yellowish pattern. Abnormal leaf size.	**Viruses**, including **Elm mosaic virus**. Spread mechanically in sap or seed. Certain elm viruses are apparently spread by nematodes.	May slow growth, but otherwise harmless to elms. Provide proper cultural care, especially proper water. No other treatment. *See* Mosaic and Mottle Viruses, 92.
Leaves with powdery, white growth. May be tiny, black, overwintering bodies later.	**Powdery mildew**, *Phyllactinia guttata*. A fungal disease favored by moderate temperatures, shade, and poor air circulation.	Generally not severe enough to warrant control. *See* 90.
Sticky honeydew, blackish sooty mold, and whitish cast skins on leaves. Leaves may curl.	**Aphids**, including **Elm leaf aphid**, *Tinocallis ulmifolii*; **Woolly aphids**, *Eriosoma* spp.; **Chinese elm aphid**, *Tinocallis ulmiparvifoliae*. Tiny, green insects clustered on leaves.	Trees are not damaged by aphids. *See* 175.
Sticky honeydew and blackish sooty mold on plant. Possible dieback.	**European elm scale**. Dark, reddish, oval insects with white, waxy fringe.	Commonly at twig crotches or on undersides of limbs, especially on Chinese elm (*Ulmus parvifolia*). *See* 197.
Sticky honeydew and blackish sooty mold on leaves and twigs. Possible plant dieback.	**European fruit lecanium scale, Frosted scale**. Brown, gray, yellow, white, or waxy, flattened to bulbous insects on twigs or leaves.	*See* Lecanium Scales, 195.
Stippled, bleached leaves and whitish cast skins. Sticky honeydew and blackish sooty mold may be present.	**Leafhoppers**, including *Empoasca* sp.; **Rose leafhopper**. Pale green to white, $\leq^1/_8$ inch long, wedge-shaped insects.	Tolerate, apparently do not harm elms. *See* 201.
Leaves skeletonized, some small holes. Leaves turn yellow, brown, and fall. Yellowish pupae may be found around tree base.	**Elm leaf beetle**. Adults greenish with black, longitudinal stripes. Larvae black to green, $\leq^1/_4$ inch long.	*See* 155.
Leaves chewed. Often, only a single branch is defoliated.	**Spiny elm caterpillar**, or **Mourningcloak butterfly; Western tiger swallowtail**, *Papilio rutulus*. Dark, hairy caterpillars, $\leq 1^1/_2$ inches long.	Ignore, they do not harm tree, or prune out infested branches. *Bacillus thuringiensis* controls young larvae. *See* 151.
Chewed leaves. Foliage may be webbed or contain silken tents.	**Fall webworm, Fruittree leafroller, Omnivorous looper**. Larvae ≤ 1 inch long. May be in webbed foliage.	*See* 149–153.
Woody swellings (galls), or cottony, waxy material on branches or roots.	**Woolly apple aphid**. Tiny, reddish, cottony, or waxy insects.	Conserve natural enemies that help control. *See* Woolly Aphids, 177.
Decline of branches or entire tree. Canopy yellowing but leaves not chewed. Tiny shot holes in bark.	**Elm bark beetles**. Small, dark, stout adults. Whitish larvae tunnel beneath bark.	Beetles can transmit Dutch elm disease fungi. *See* 113, 224.
Large holes, $\leq^1/_2$ inch in diameter, in trunks and limbs. Limbs may die back or drop. Slow tree growth.	**Carpenterworm**. Whitish larvae, $\leq 2^1/_2$ inches long, with brown head, tunnel in wood.	*See* 241.
Bark exudes white, frothy material, often around wounds, has pleasant odor.	**Foamy canker**. Unidentified cause, possibly a bacterium.	Foamy material appears for only short time during warm weather. *See* 101.
Bark stained brownish, exudes rancid fluid, often around crotches, wounds.	**Wetwood**, or **Slime flux**. Bacterial infections.	Usually do not cause serious harm to trees. *See* 110.
Umbellularia californica, California bay, Bay, California bay laurel, Oregon myrtle, Pepperwood		
Leaves discolor and wilt, usually in spring. Stems may have dark, sunken cankers. May be small, coral-colored pustules on infected wood.	**Cankers**, including **Nectria canker**, *Nectria cinnabarina, N. coccinea, N. peziza*. Fungi primarily attack injured or stressed trees.	*See* 101.
Foliage yellows and wilts. Bark or wood may have bracketlike or fan-shaped fungal fruiting bodies. Branches or entire plant may die.	**Wood decay**, or **Heart rot**, including *Ganoderma* spp., *Phellinus* spp., *Schizophyllum commune*. Fungi that attack injured, old, or stressed trees.	*See* 66, 111.

WHAT THE PROBLEM LOOKS LIKE	PROBABLE CAUSE	COMMENTS
Small, black, angular spots and large, irregular, brown spots on leaves.	**Anthracnose, Leaf blights**, *Pseudomonas* sp., *Kabatiella* sp. Bacterial and fungal pathogens favored by prolonged rainy springs.	Usually not serious enough to threaten tree health or warrant control effort. *See* 81, 86.
Leaves with brown or dark lesions or dead spots often bordered by yellow, usually on leaf tips or edges.	**Ramorum blight**, *Phytophthora ramorum*. Pathogen spreads via airborne spores and contaminated plants and soil.	Primarily a problem in wildlands, killing many oaks there. *See* Sudden Oak Death and Ramorum Blight, 105.
Sticky honeydew and blackish sooty mold on leaves or twigs.	**California laurel aphid**, *Euthoracaphis umbellulariae*. Grayish insects, $^1/_{16}$ inch long, resembling immature whiteflies or scales on undersides of leaves.	Ignore, even heavy populations apparently do not harm tree. *See* 176.
Sticky honeydew and blackish sooty mold on leaves or twigs. Plants may decline.	**Black scale**. Brown to black, bulbous to flattened insects, $\leq^3/_{16}$ inch long. Raised H shape often on back.	*See* 192.
Plant has sticky honeydew and blackish sooty mold. Plant grows slowly. Foliage may yellow.	**Pyriform scale**, *Protopulvinaria pyriformis*. Triangular insects, $^1/_8$ inch long and brown, yellow, or mottled red.	*See* 192.
Sticky honeydew and blackish sooty mold on plant. Bulbous, irregular, brown, gray, or white bodies (scales) on twigs.	**Wax scales**. Bulbous to hemispherical, waxy insects, suck sap.	*See* 196.
Brownish, grayish, tan, or white encrustations on twigs. Rarely, twig or branch dieback.	**Greedy scale, Oleander scale**. Oval to circular insects, $\leq^1/_{16}$ inch long.	Rarely if ever cause serious damage to plants. *See* 189–192.
Dieback of occasional twigs.	**Branch and twig borers**, including **Leadcable borer**, *Scobicia declivis; Melalgus* (=*Polycaon*) *confertus*. Adults $^1/_4$ to $^1/_2$ inch long, tunnel in twigs.	Keep plants vigorous, provide proper cultural care. Prune out affected parts. Eliminate nearby dead hardwood where beetles breed. *See* Twig, Branch, and Trunk Boring Insects, 220.
Viburnum spp., Viburnum		
Foliage yellows and wilts. Branches or entire plant dies.	**Phytophthora root and crown rot**. Disease favored by wet soil and poor drainage.	*See* 122.
Leaves wilted, discolored, may drop prematurely. Branches or entire plant may die. May be white fungal growth or dark crust on basal trunk, roots, or soil.	**Dematophora (Rosellinia) root rot**. Fungus favored by mild, wet conditions. Infects primarily through roots growing near infested plants.	Less common than *Phytophthora. See* 121.
Leaves and stems with brown, water-soaked lesions. Shoots and stems may darken and die back.	**Bacterial blast, blight, and canker.** Disease favored by wet conditions.	*See* 83.
Leaves with dark or discolored blotches. Leaves may drop prematurely.	**Leaf spots**, *Cercospora* spp., *Phyllosticta* sp. Fungi favored by wet conditions.	*See* 86.
Leaves discolor and wilt. Stems may die back.	**Ramorum blight**, *Phytophthora ramorum*. Pathogen spreads via airborne spores and contaminated plants and soil.	Primarily a problem in nurseries and on other hosts in wildlands, killing many oaks there. *See* Sudden Oak Death and Ramorum Blight, 105.
Powdery, whitish growth on leaves and shoots. Terminals may distort.	**Powdery mildews**, including *Microsphaera sparsa, Microsphaera penicillata*.	*See* 90.
Stippled, bleached leaves with varnishlike specks on undersides.	**Greenhouse thrips**. Tiny, slender, black adults or yellow immatures.	*See* 210.
Leaves bleached, stippled, may turn brown and drop prematurely. Terminals may be distorted. Plant may have fine webbing.	**Mites**, including **Southern red mite**. Tiny arthropods, often green, pink, or red.	*See* Red Mites, 246.
Sticky honeydew and blackish sooty mold on foliage. May be distorted terminals.	**Aphids**, including **Bean aphid**. Dull black or green insects, $<^1/_8$ inch long.	*See* 175.
Brown encrustations on bark. May be stunting or dieback of woody parts.	**Oystershell scale**. Tiny, immobile insects, resembling miniature oysters.	*See* 191.

Viburnum spp. (continued)

WHAT THE PROBLEM LOOKS LIKE	PROBABLE CAUSE	COMMENTS
Foliage chewed or notched around margins. Some roots stripped of bark or girdled near soil. Young plants may wilt or die.	**Weevils**, including **Black vine weevil; Woods weevil**, *Nemocestes incomptus*. Adults are dark snout beetles, ≤³/₈ inch long.	Larvae are root-feeding white grubs with brown head. *See* 159.
Vitex agnus-castus, Chaste tree		
Leaves discolor, stunt, wilt, or drop prematurely. Plants grow slowly and may die. Roots dark, decayed.	**Phytophthora root and crown rot.** Disease favored by excess soil moisture and poor drainage.	*See* 122.
Leaves yellow and wilt. Entire plant may die. May be white fungus beneath basal trunk bark.	**Armillaria root disease.** Fungus present in many soils. Favored by warm, wet soil. Persists for years in infected roots.	*See* 120.
Leaves turn brown or yellow, especially along margins and at tip. Leaves may drop prematurely.	**Leaf burn**, or **Scorch**. Abiotic disorders with many causes, including salinity, poor drainage, and too much or too little water.	Provide plants with a good growing environment and proper cultural care. *See* Abiotic Disorders, 43.
Twigs distorted, swollen, and pitted. Leaves dwarfed. Shoots may die back.	**Pit-making pittosporum scale**, *Planchonia* (=*Asterolecanium*) *arabidis*. Brown to white insects, ≤¹/₈ inch long, on twigs, often in pits.	In California, an occasional problem in the north. Management not investigated. *See* similar Oak Pit Scales, 198.
Brownish, grayish, tan, or white encrustations on twigs.	**Oleander scale**. Tiny, flattened, circular insects, ≤¹/₁₆ inch long.	Rarely if ever causes serious damage to plants. *See* 191.
Sticky honeydew, blackish sooty mold, and whitish cast skins on plant.	**Aphids**. Small, pear-shaped insects, commonly black, brown, green, or yellowish, often in groups.	*See* 175.
Weigela spp., Weigela		
Whitish, cottony material on bark or underside of leaves.	**Comstock mealybug**, *Pseudococcus comstocki*. Oblong, soft, powdery, waxy insects with filaments.	Control ants, reduce dust, avoid persistent pesticides that disrupt effective natural enemies. *See* Mealybugs, 185.
Wisteria spp., Wisteria		
Foliage yellows and wilts. Branches or entire plant dies.	**Phytophthora root and crown rot.** Disease favored by excess soil moisture and poor drainage.	*See* 122.
Leaves brown, yellow or die. Branches may die back. Wood may canker or ooze.	**Cankers**, including **Botryosphaeria canker and dieback, Phomopsis canker**, *Phomopsis wistariae*.	Fungi primarily affect injured and stressed plants. *See* 98.
Leaves have pale blotches, mottling, spots, or line patterns.	**Viruses**, including **Wisteria vein mosaic virus**. Spreads mechanically, e.g., during propagation.	Rarely if ever threatens plant health. Infected landscape plants cannot be cured. *See* 92–94.
Sticky honeydew, blackish sooty mold, and whitish cast skins on plant.	**Aphids**. Small green, black, brown, or yellowish pear-shaped insects, often in groups.	*See* 175.
Sticky honeydew and blackish sooty mold on foliage.	**Calico scale**. Adults globular, black with white or yellow spots.	Rarely if ever threatens plant health. *See* 193.
White encrustations on woody parts and leaves.	**Wisteria scale**, *Chionaspis wistariae*. Elongate, <¹/₁₆ inch long.	Effect of scale on plant unknown. *See* Armored Scales, 189.
Leaves may yellow and wilt. Decline of plant. Dying branches.	**Spotted tree borer**, *Synaphaeta guexi*. Larvae are whitish grubs, ≤³/₄ inch long, tunnel in woody parts. Adults are mostly grayish with black, orange, and white.	Provide proper cultural care to keep plants vigorous. Prune out and dispose of damaged plant parts. *See* Longhorned Beetles, or Roundheaded Wood Borers, 232.
Leaves may yellow and wilt. Plants grow slowly or decline. Limbs or entire tree may die.	**Tenlined June beetle**, *Polyphylla decemlineata*. Cream-colored grubs, ≤2 inches long, chew roots.	More of a problem in sandy soils. *See* White Grubs and Scarab Beetles, 164.

WHAT THE PROBLEM LOOKS LIKE	PROBABLE CAUSE	COMMENTS
Xylosma congestum, Xylosma		
Leaves with dark brown, circular blotches, ≤³/₄ inch in diameter. Orangish pustules on leaf underside. Leaves may drop prematurely.	**Xylosma rust**, *Melampsora medusae*. Fungal disease favored by wet foliage.	Avoid overhead watering. Vigorous plants can tolerate a moderate infection. *See* Rusts, 94.
Branches die back. Entire plant may die. May be white fungus beneath basal trunk bark.	**Armillaria root disease**. Fungus present in many soils. Favored by warm, wet soil. Persists for years in infected roots.	*See* 120.
Foliage has sticky honeydew, blackish sooty mold, and copious, white, waxy material. Leaves may yellow, wither, and drop prematurely.	**Whiteflies**, including **Giant whitefly**. Adults are tiny, whitish, mothlike insects. Nymphs and pupae are flattened, oval, translucent, and greenish or yellow.	*See* 182.
Leaves bleached, stippled, may turn brown and drop prematurely. Terminals may be distorted. Plant may have fine webbing.	**Spider mites**, *Tetranychus* spp. Tiny, often green, pink, or red pests; may have two dark spots.	*See* 246.
Zelkova serrata, Zelkova, Japanese zelkova		
Foliage yellows, then wilts, usually first in one part of canopy. Curled, dead, brown leaves remain on tree.	**Dutch elm disease**. Fungal disease spread by bark beetles and root grafts.	Zelkova is moderately resistant to this disease. *See* 113.
Leaves discolor, stunt, wilt, or drop prematurely. Stems discolored, cankered, and dead. May be white fungus beneath basal trunk bark.	**Armillaria root disease**. Fungus present in many soils. Favored by warm, wet soil. Persists for years in infected roots.	*See* 120.
Leaves and branches wilt in spring. Stems may have dark cankers, callus tissue, or coral-colored pustules.	**Nectria canker**, *Nectria cinnabarina*. Fungus primarily affects injured or stressed trees.	*See* 101.
Pale blotches or irregular, black, tarlike spots on leaves. Leaves may drop prematurely.	**Anthracnose**, or **Chinese elm anthracnose canker**. A fungal disease.	In California usually severe only on Chinese elm, *Ulmus parvifolia*, near the coast. *See* 81, 99.
Sticky honeydew and blackish sooty mold on leaves and twigs.	**Calico scale**. Black, brown, and white, mottled, flattened to bulbous insects on twigs.	Rarely if ever threatens plant health. *See* 193.
Leaves skeletonized, some small holes. Leaves turn yellow, brown, and drop prematurely.	**Elm leaf beetle**. Adults greenish with black, longitudinal stripes. Larvae black to green, ≤¹/₄ inch long. Pest is unable to complete development on zelkova.	A problem only when beetles move from nearby favored elms (*Ulmus* spp.), where control should be targeted. *See* 155.
Galls or swellings on trunk and roots, usually near soil.	**Crown gall**. Bacterium that infects plant via wounds.	*See* 108.

Names of Pests and Plants

Both common and scientific names are used to identify organisms. Because different humans (*Homo sapiens*) may use different names for the same organism, names are often a source of confusion.

Scientists use a unique two-word combination for each animal, plant, and microorganism. This scientific name provides the surest identification because scientific names are used according to agreed-upon rules. Although scientific names are sometimes changed based on new information, each organism has only one valid scientific name, which is used throughout the world. If a scientific name has recently been changed, both names may be printed, *Cotesia* (=*Apanteles*), with the currently correct name listed first followed by an equal sign and the synonym or former name (sometimes in parentheses).

The first word of a scientific name, the genus or generic name, is capitalized. The second word, the specific name, or specific epithet, is not. Both words are italicized and are Latinlike or in Latin so scientists can understand what organism others are referring to regardless of nationality and native language. After its first use in the text, the genus name is often abbreviated; for example, *Eucalyptus globulus* is shortened to *E. globulus*. When multiple species within the same genus are discussed together, species may be abbreviated as "spp." (e.g., *Eucalyptus* spp.). When referring to only one species, "sp." is used.

Some species exist in multiple forms that differ significantly from each other and are called subspecies (abbreviated ssp.) or various other terms. For example, the insect-pathogenic bacterium *Bacillus thuringiensis* includes *Bacillus thuringiensis* ssp. *kurstaki*, which is toxic to caterpillars (larvae of moths and butterflies) and *B. thuringiensis* ssp. *israelensis*, which is toxic to fly larvae (e.g., mosquito larvae). Other subspecies-type categories used by plant pathologists include special forms, or *forma specialis* (f. sp.), and *pathovar* (pv.). For example, *Ceratocystis fimbriata* f. sp. *platani* is a wilt fungus that causes canker stain of sycamore. *Pseudomonas savastanoi* pv. *nerii* is a leaf- and stem-galling bacterium that infects oleander and olive.

Botanists and horticulturalists use various terms for two forms of the same plant species, including subspecies (ssp. or subsp.), varieties (var.), and cultivars, or cultivated varieties, (cv.). Named hybrids between two plant species have the cross symbol, ✕, between their genus and specific epithet, such as *Platanus* ✕ *acerifolia* for the London plane sycamore tree.

Scientific names are used in a hierarchical organization that includes the order and family names. These hierarchical names show relationships among organisms, as illustrated here for the common convergent lady beetle (*Hippodamia convergens*):

Kingdom: Animalia (animals)
Phylum: Arthropoda (arthropods)
Class: Insecta (insects)
Order: Coleoptera (beetles)
Family: Coccinellidae (lady beetles)
Genus: *Hippodamia*
Specific epithet: *convergens*

Besides the two-part scientific name, many plants, insects, and diseases also have common names. Common names are familiar to more people than are scientific names, and they are often easier to pronounce and remember. However, there are serious problems with common names. There are no clear rules for deciding what is the correct common name of most organisms. A single common name is often used to refer to several distinctly different organisms. The same organism can have several common names, some of which may be known and used only by people in certain locations. Common names may also be inaccurate; *pineapple* refers to a plant that is very unlike pines and apples. *Ladybug* refers to certain beetles, which are very different from the insects that scientists call true bugs. The sequoia pitch moth (*Synanthedon sequoiae*) infests only pines; it never attacks giant sequoia (*Sequoiadendron giganteum*) or coast redwood (*Sequoia sempervirens*). Many important organisms, including most species of beneficial predators, parasites, and pathogens, have no common name, often because they are tiny or known mostly only to scientists.

Both common and scientific names are used in this book. Pest scientific and common names are given in the index, and both names are used together in the major section discussing that pest as listed in the table of contents. Scientific names and common names are also used in the "Tree and Shrub Pest Tables" for pests not detailed elsewhere.

Plant scientific names are generally avoided, except in the index and tables. Common names are more widely known for most plants and can be found in the problem-solving tables (Chapter 9) and other references, such as *The New Sunset Western Garden Book* (Brenzel 2012) and *Trees and Shrubs of California* (Stuart and Sawyer 2001). Some plants are mentioned so often that using their scientific names in this book would consume too much space and appear awkward. For many plants, such as camellia, citrus, and rhododendron, the genus name and common name are the same, except that the common name is not capitalized or italicized. Exceptions include plants named after people, such as Douglas-fir, which is capitalized because the tree is named for the nineteenth-century botanist David Douglas.

Sources for names used in this book include: *An Annotated Checklist of Woody Ornamental Plants of California, Oregon, & Washington* (McClintock and Leiser 1979), *Common Names of Arachnids* (Breene 1995), *Common Names of Insects & Related Organisms* 1997 (Bosik 1997), *Composite List of Weeds* (Alex et al. 1989), *Fungi on Plants and Plant Products in the United States* (Farr et al. 1989), *The Jepson Manual: Higher Plants of California* (Hickman 1993), *Jepson eFlora* (Jepson Flora Project 2013), and *Weeds of California and Other Western States* (DiTomaso and Healy 2007).

List of Figures and Tables

List of Figures and Tables

Suggested Reading[1]

These publications are sources of further information. For publications cited in figures and tables as the sources of information and illustrations, see "Literature Cited."

Abiotic Disorders of Landscape Plants: A Diagnostic Guide. 2003. L. R. Costello, E. J. Perry, N. P. Matheny, J. M. Henry, and P. M. Geisel. Oakland: UC ANR Publication 3420.

Almonds: UC IPM Pest Management Guidelines: Insects and Mites. 2012. F. G. Zalom, C. Pickel, W. J. Bentley, D. R. Haviland, and R. A. Van Steenwyk. University of California Statewide Integrated Pest Management Program. Oakland: UC ANR Publication 3431.

An Annotated Checklist of Woody Ornamental Plants of California, Oregon, & Washington. 1979. E. McClintock and T. Leiser. Oakland: UC ANR Publication 4091.

Annual Bluegrass Pest Notes. 2012. M. LeStrange, P. M. Geisel, D. W. Cudney, C. L. Elmore, and V. A. Gibeault. University of California Statewide Integrated Pest Management Program. Oakland: UC ANR Publication 7464.

Anthracnose Pest Notes. 2009. A. Crump. University of California Statewide Integrated Pest Management Program. Oakland: UC ANR Publication 7420.

Ants Pest Notes. 2012. M. K. Rust and D.-H. Choe. University of California Statewide Integrated Pest Management Program. Oakland: UC ANR Publication 7411.

Aphids Pest Notes. 2013. M. L. Flint. University of California Statewide Integrated Pest Management Program. Oakland: UC ANR Publication 7404.

Apple and Pear Scab Pest Notes. 2011. D. D. Giraud, R. B. Elkins, and W. D. Gubler. University of California Statewide Integrated Pest Management Program. Oakland: UC ANR Publication 7413.

Arboriculture: Integrated Management of Landscape Trees, Shrubs, and Vines. 4th ed. 2003. R. W. Harris, J. R. Clark, and N. P. Matheny. Prentice-Hall. Englewood Cliffs, NJ.

Asian Citrus Psyllid. 2006. E. E. Grafton-Cardwell, K. E. Godfrey, M. E. Rogers, C. C. Childers, and P. A. Stansly. Oakland: UC ANR Publication 8205.

Asian Citrus Psyllid and Huanglongbing Disease Pest Notes. 2013. E. E. Grafton-Cardwell and M. P. Daugherty. University of California Statewide Integrated Pest Management Program. Oakland: UC ANR Publication 74155.

Avocado Lace Bug Pest Notes. 2007. G. S. Bender, J. G. Morse, M. S. Hoddle, and S. H. Dreistadt. University of California Statewide Integrated Pest Management Program. Oakland: UC ANR Publication 74134.

Bark Beetles Pest Notes. 2008. S. J. Seybold, T. D. Paine, and S. H. Dreistadt. University of California Statewide Integrated Pest Management Program. Oakland: UC ANR Publication 7421.

Bee Alert: Africanized Honey Bee Facts. 2002. V. Lazaneo. Oakland: UC ANR Publication 8068.

Bee and Wasp Stings Pest Notes. 2011. E. C. Mussen. University of California Statewide Integrated Pest Management Program. Oakland: UC ANR Publication 7449.

Bermudagrass Pest Notes. 2007. D. W. Cudney, C. L. Elmore, and C. E. Bell. University of California Statewide Integrated Pest Management Program. Oakland: UC ANR Publication 7453.

Biological Control and Natural Enemies Pest Notes. 2014. S. H. Dreistadt. University of California Statewide Integrated Pest Management Program. Oakland: UC ANR Publication 74140.

Biological Control of Insect Pests and Weeds. 1964. P. DeBach and E. I. Schlinger, eds. New York: Reinhold Publishing Corp.[2]

The Biology and Management of Landscape Palms. 2012. D. R. Hodel. Porterville, CA: Western Chapter International Society of Arboriculture.

Black Scale Pest Notes. 2012. E. J. Fichtner and M. W. Johnson. University of California Statewide Integrated Pest Management Program. Oakland: UC ANR Publication 74160.

Black Widow and Other Widow Spiders Pest Notes. 2009. R. S. Vetter. University of California Statewide Integrated Pest Management Program. Oakland: UC ANR Publication 74149.

Bordeaux Mixture Pest Notes. 2010. J. C. Broom and D. R. Donaldson. University of California Statewide Integrated Pest Management Program. Oakland: UC ANR Publication 7481.

Boxelder Bug Pest Notes. 2014. E. J. Perry and K. Windbiel-Rojas. University of California Statewide Integrated Pest Management Program. Oakland: UC ANR Publication 74114.

Brown Marmorated Stink Bug Pest Notes. 2014. C. Ingels and L. Varela. University of California Statewide Integrated Pest Management Program. Oakland: UC ANR Publication 74169.

Brown Recluse and Other Recluse Spiders Pest Notes. 2008. R. S. Vetter. University of California Statewide Integrated Pest Management Program. Oakland: UC ANR Publication 7468.

1. University of California publications are available for free download (in certain instances) or can be purchased by visiting the anrcatalog.ucanr.edu or ipm.ucanr.edu World Wide Web sites.
2. Publications out of print. Copies may be available for reference at libraries.

California Ground Squirrel Pest Notes. 2010. T. P. Salmon and W. P. Gorenzel. University of California Statewide Integrated Pest Management Program. Oakland: UC ANR Publication 7438.

California Insects. 1979. J. A. Powell and C. L. Hogue. Berkeley: UC Press.

California Master Gardener Handbook. 2nd ed. 2014. D. R. Pittenger, ed. Oakland: UC ANR Publication 3382.

California Oakworm Pest Notes. 2009. S. Swain, S. A. Tjosvold, and S. H. Dreistadt. University of California Statewide Integrated Pest Management Program. Oakland: UC ANR Publication 7422.

Carpenter Ants Pest Notes. 2009. J. H. Klotz, M. K. Rust, and L. D. Hansen. University of California Statewide Integrated Pest Management Program. Oakland: UC ANR Publication 7416.

Carpenter Bees Pest Notes. 2004. M. L. Flint, ed. University of California Statewide Integrated Pest Management Program. Oakland: UC ANR Publication 7407.

Carpenterworm Pest Notes. 2010. P. M. Geisel. University of California Statewide Integrated Pest Management Program. Oakland: UC ANR Publication 74105.

Citrus: UC IPM Pest Management Guidelines: Insects, Mites, and Snails. 2013. E. E. Grafton-Cardwell, J. G. Morse, N. V. O'Connell, P. A. Phillips, C. E. Kallsen, and D. R. Haviland. University of California Statewide Integrated Pest Management Program. Oakland: UC ANR Publication 3441.

Clearwing Moths Pest Notes. 2013. J. F. Karlik, S. A. Tjosvold, and S. H. Dreistadt. University of California Statewide Integrated Pest Management Program. Oakland: UC ANR Publication 7477.

Cliff Swallows Pest Notes. 2005. T. P. Salmon and W. P. Gorenzel. University of California Statewide Integrated Pest Management Program. Oakland: UC ANR Publication 7482.

Clovers Pest Notes. 2007. R. Smith, D. W. Cudney, and C. L. Elmore. University of California Statewide Integrated Pest Management Program. Oakland: UC ANR Publication 7490.

Codling Moth Pest Notes. 2011. J. L. Caprile and P. M. Vossen. University of California Statewide Integrated Pest Management Program. Oakland: UC ANR Publication 7412.

Color-Photo and Host Keys to the Armored Scales of California. 1982. R. J. Gill. Scale and Whitefly Key #5. Sacramento: California Department of Food and Agriculture.[2]

Color-Photo and Host Keys to California Whiteflies. 1982. R. J. Gill. Scale and Whitefly Key #2. Sacramento: California Department of Food and Agriculture.[2]

Color-Photo and Host Keys to the Mealybugs of California. 1982. R. J. Gill. Scale and Whitefly Key #3. Sacramento: California Department of Food and Agriculture.[2]

Color-Photo and Host Keys to the Soft Scales of California. 1982. R. J. Gill. Scale and Whitefly Key #4. Sacramento: California Department of Food and Agriculture.[2]

Common Groundsel Pest Notes. 2006. C. A. Wilen. University of California Statewide Integrated Pest Management Program. Oakland: UC ANR Publication 74130.

Common Names of Arachnids. 1995. R. G. Breene. South Padre Island, TX: American Tarantula Society.

Common Names of Insects & Related Organisms 1997. J. J. Bosik, ed. Lanham, MD: Entomological Society of America.

Common Purslane Pest Notes. 2007. D. W. Cudney, C. L. Elmore, and R. H. Molinar. University of California Statewide Integrated Pest Management Program. Oakland: UC ANR Publication 7461.

Common-Sense Pest Control. 1991. W. Olkowski, S. Daar, and H. Olkowski. Newton, CT: The Taunton Press.

Compatible Plants Under and Around Oaks. 1991. B. W. Hagen, B. D. Coate, and G. Keater. Oakland: California Oak Foundation.

Compendium of Rose Diseases and Pests. 2nd ed. 2007. R. K. Horst and R. Cloyd. St. Paul, MN: American Phytopathological Society.

Composite List of Weeds. 1989. J. F. Alex, G. A. Bozarth, C. T. Bryson, J. W. Everest, E. P. Flint, F. Forcella, D. W. Hall, H. F. Harrison, Jr., L. W. Hendrick, L. G. Holm, D. E. Seaman, V. Sorensen, H. V. Strek, R. H. Walker, and D. T. Patterson. Champaign, IL: Weed Science Society of America.

Conenose Bugs Pest Notes. 2013. L. Greenberg, J. O. Schmidt, S. A. Klotz, and J. H. Klotz. University of California Statewide Integrated Pest Management Program. Oakland: UC ANR Publication 7455.

Controlling Bark Beetles in Wood Residue and Firewood. 1996. S. R. Sanborn. Sacramento: California Department of Forestry and Fire Protection. Tree Notes 3. Online at ceres.ca.gov/foreststeward/pdf/treenote3.pdf.

Cottony Cushion Scale Pest Notes. 2012. E. E. Grafton-Cardwell. University of California Statewide Integrated Pest Management Program. Oakland: UC ANR Publication 7410.

Crabgrass Pest Notes. 2010. R. H. Molinar and C. L. Elmore. University of California Statewide Integrated Pest Management Program. Oakland: UC ANR Publication 7456.

2. Publications out of print. Copies may be available for reference at libraries.

Creeping Woodsorrel and Bermuda Buttercup Pest Notes. 2010. M. LeStrange, C. L. Elmore, and D. W. Cudney. University of California Statewide Integrated Pest Management Program. Oakland: UC ANR Publication 7444.

Dandelions Pest Notes. 2006. D. W. Cudney and C. L. Elmore. University of California Statewide Integrated Pest Management Program. Oakland: UC ANR Publication 7469.

Destructive and Useful Insects. 5th ed. 1993. R. L. Metcalf and R. A. Metcalf. New York: McGraw-Hill.

Determining Daily Reference Evapotranspiration (Eto). 1987. R. L. Snyder, W. O. Pruitt, and D. A. Shaw. Oakland: UC ANR Publication 21426.

Diaprepes Root Weevil. 2004. E. E. Grafton-Cardwell, K. E. Godfrey, J. E. Peña, C. W. McCoy, and R. F. Luck. Oakland: UC ANR Publication 8131.

Diseases of Forest and Shade Trees of the United States. 1971. G. H. Hepting. Washington, DC: USDA Agricultural Handbook 386.

Diseases of Pacific Coast Conifers. 1993. R. F. Scharpf, ed. Washington, DC: USDA Agricultural Handbook 521.

Diseases of Trees and Shrubs. 1987. W. A. Sinclair, H. H. Lyon, and W. T. Johnson. Ithaca, NY: Cornell University Press.

Dodder Pest Notes. 2010. W. T. Lanini, D. W. Cudney, G. Miyao, and K. J. Hembree. University of California Statewide Integrated Pest Management Program. Oakland: UC ANR Publication 7496.

Drip Irrigation in the Home Landscape. 1999. L. Schwankl and T. Prichard. Oakland: UC ANR Publication 21579.

Drywood Termites Pest Notes. 2014. V. R. Lewis, A. M. Sutherland, and M. I. Haverty. University of California Statewide Integrated Pest Management Program. Oakland: UC ANR Publication 7440.

Easy On-Site Tests for Fungi and Viruses in Nurseries and Greenhouses. 1997. J. N. Kabashima, J. D. MacDonald, S. H. Dreistadt, and D. E. Ullman. Oakland: UC ANR Publication 8002.

Elm Leaf Beetle Pest Notes. 2014. S. H. Dreistadt and A. B. Lawson. University of California Statewide Integrated Pest Management Program. Oakland: UC ANR Publication 7403.

Eucalyptus Longhorned Borers Pest Notes. 2009. T. D. Paine, S. H. Dreistadt, and J. G. Millar. University of California Statewide Integrated Pest Management Program. Oakland: UC ANR Publication 7425.

Eucalyptus Redgum Lerp Psyllid Pest Notes. 2006. T. D. Paine, S. H. Dreistadt, R. W. Garrison, and R. J. Gill. University of California Statewide Integrated Pest Management Program. Oakland: UC ANR Publication 7460.

Eucalyptus Tortoise Beetle Pest Notes. 2009. J. G. Millar, T. D. Paine, J. A. Bethke, R. W. Garrison, K. A. Campbell, and S. H. Dreistadt. University of California Statewide Integrated Pest Management Program. Oakland: UC ANR Publication 74104.

Evaluation of Hazard Trees in Urban Areas. 1991. N. P. Matheny and J. R. Clark. Urbana, IL: International Society of Arboriculture.

Evapotranspiration and Irrigation Water Requirements. 1990. M. E. Jensen, R. D. Burman, and R. G. Allen, eds. New York: American Society of Civil Engineers.

Fertilizing Landscape Trees. 2001. E. Perry and G. W. Hickman. Oakland: UC ANR Publication 8045.

Field Bindweed Pest Notes. 2011. S. D. Wright, C. L. Elmore, and D. W. Cudney. University of California Statewide Integrated Pest Management Program. Oakland: UC ANR Publication 7462.

A Field Guide to Insects and Diseases of California Oaks. 2006. T. J. Swiecki and E. A. Bernhardt. Berkeley: USDA Forest Service General Technical Report PSW-GTR-197. Online at www.treesearch.fs.fed.us/pubs/25928.

Field Identification Guide for Light Brown Apple Moth in California Nurseries. 2014. S. A. Tjosvold, N. B. Murray, M. Epstein, O. Sage, and T. Gilligan. Davis: University of California Statewide Integrated Pest Management Program. Online at www.ipm.ucanr.edu/PDF/PMG/LBAMinCAnurseries.pdf.

Fire Blight Pest Notes. 2011. B. L. Teviotdale. University of California Statewide Integrated Pest Management Program. Oakland: UC ANR Publication 7414.

Floriculture and Ornamental Nurseries: UC IPM Pest Management Guidelines. 2013. S. T. Koike, C. A. Wilen (Diseases); J. A. Bethke (Insects and Mites); C. A. Wilen (Mollusks); J. J. Stapleton, M. V. McKenry, and A. T. Ploeg (Nematodes); and C. A. Wilen (Weeds). University of California Statewide Integrated Pest Management Program. Oakland: UC ANR Publication 3392.

Flower Flies (Syrphidae) and Other Biological Control Agents for Aphids in Vegetable Crops. 2008. R. L. Bugg, R. G. Colfer, W. E. Chaney, H. A. Smith, and J. Cannon. Oakland: UC ANR Publication 8285.

Fungi on Plants and Plant Products in the United States. 1989. D. F. Farr, G. F. Bills, G. P. Chamuris, and A. Y. Rossman. 1989. St. Paul, MN: American Phytopathological Society.

Giant Whitefly Pest Notes. 2006. T. S. Bellows, J. N. Kabashima, and K. L. Robb. University of California Statewide Integrated Pest Management Program. Oakland: UC ANR Publication 7400.

Glassy-Winged Sharpshooter Pest Notes. 2007. L. G. Varela, J. M. Hashim-Buckey, C. A. Wilen, and P. A. Phillips. University of California Statewide Integrated Pest Management Program. Oakland: UC ANR Publication 7492.

Goldspotted Oak Borer Field Identification Guide. S. Hishinuma, T. W. Coleman, M. L. Flint, and S. J. Seybold. Davis: University of California Statewide Integrated Pest Management Program. Online at www.ipm.ucdavis.edu/PDF/MISC/GSOB_field-identification-guide.pdf.

Goldspotted Oak Borer Pest Notes. 2013. M. L. Flint, M. I. Jones, T. W. Coleman, and S. J. Seybold. University of California Statewide Integrated Pest Management Program. Oakland: UC ANR Publication 74163.

Grasshoppers Pest Notes. 2013. M. L. Flint. University of California Statewide Integrated Pest Management Program. Oakland: UC ANR Publication 74103.

Green Kyllinga Pest Notes. 2011. D. A. Shaw, C. A. Wilen, D. W. Cudney, and C. L. Elmore. University of California Statewide Integrated Pest Management Program. Oakland: UC ANR Publication 7459.

Guideline Specifications for Nursery Tree Quality. 2009. D. Burger, B. Coate, L. Costello, R. Crudup, J. Geiger, B. Hagen, R. Harris, B. Kempf, J. Koch, B. Ludekens, G. McPherson, M. Ozonoff, E. Perry, and M. Robert. Visalia, CA: Urban Tree Foundation. Online at http://ccuh.ucdavis.edu/industry/files/GuidelineSpecificationsforNurseryTreeQuality.pdf/at_download/file.

A Guide to Estimating Irrigation Water Needs of Landscape Plantings in California. 2000. Sacramento: California Department of Water Resources. Online at www.water.ca.gov/wateruseefficiency/docs/wucols00.pdf.

Hackberry Woolly Aphid Pest Notes. 2014. A. B. Lawson and S. H. Dreistadt. University of California Statewide Integrated Pest Management Program. Oakland: UC ANR Publication 74111.

Handbook of Turfgrass Insect Pests. 2nd ed. 2012. R. L. Brandenburg and C. Freeman, eds. Lanham, MD: Entomological Society of America.

Healthy Roses: Environmentally Friendly Ways to Manage Pests and Disorders in Your Garden and Landscape. 2nd ed. J. F. Karlik, M. L. Flint, and D. Golino. 2009. Oakland: UC ANR Publication 21589.

Herbicide Resistance: Definition and Management Strategies. 2000. T. S. Prather, J. M. DiTomaso, and J. S. Holt. Oakland: UC ANR Publication 8012.

Hiring a Pest Control Company Pest Notes. 2006. C. A. Wilen, D. L. Haver, M. L. Flint, P. M. Geisel, and C. L. Unruh. University of California Statewide Integrated Pest Management Program. Oakland: UC ANR Publication 74125.

Hobo Spider Pest Notes. 2006. R. S. Vetter. University of California Statewide Integrated Pest Management Program. Oakland: UC ANR Publication 7488.

Home Landscaping for Fire. 2007. G. Nader, G. Nakamura, M. De Lasaux, S. Quarles, and Y. Valachovic. Oakland: UC ANR Publication 8228.

Home Orchard: Growing Your Own Deciduous Fruit and Nut Trees. 2007. C. A. Ingels, P. M. Geisel, and M. V. Norton. Oakland: UC ANR Publication 3485.

Hoplia Beetle Pest Notes. 2010. E. J. Perry. University of California Statewide Integrated Pest Management Program. Oakland: UC ANR Publication 7499.

House Mouse Pest Notes. 2011. R. M. Timm. University of California Statewide Integrated Pest Management Program. Oakland: UC ANR Publication 7483.

Indian Walking Stick Pest Notes. 2011. D. H. Headrick and C. A. Wilen. University of California Statewide Integrated Pest Management Program. Oakland: UC ANR Publication 74157.

Insect Pest Management Guidelines for California Landscape Ornamentals. 1987. C. S. Koehler. Oakland: UC ANR Publication 3317.[2]

Insects Affecting Ornamental Conifers in Southern California. 1967. L. R. Brown and C. O. Eads. Berkeley: UC Agricultural Experiment Station Bulletin 834.[2]

Insects and Diseases of Woody Plants of the Central Rockies. 2000. W. Cranshaw, D. Leatherman, B. Kondratieff, R. Stevens, and R. Wawrzynski. Ft. Collins: Colorado State University.

Insects and Mites of Western North America. 1958. E. O. Essig. New York: MacMillan.[2]

Insects of the Los Angeles Basin. C. L. Hogue. 1993. Los Angeles: Natural History Museum of Los Angeles County.

Insects That Feed on Trees and Shrubs. 2nd ed. 1988. W. J. Johnson, and H. H. Lyon. Ithaca, NY: Cornell University Press.

Integrated Pest Management for Almonds. 2nd ed. 2002. L. L. Strand. University of California Statewide Integrated Pest Management Program. Oakland: UC ANR Publication 3303.

Integrated Pest Management for Apples and Pears. 2nd ed. 1999. B. L. P. Ohlendorf. University of California Statewide Integrated Pest Management Program. Oakland: UC ANR Publication 3340.

2. Publications out of print. Copies may be available for reference at libraries.

Integrated Pest Management for Citrus. 3rd ed. 2012. S. H. Dreistadt. University of California Statewide Integrated Pest Management Program. Oakland: UC ANR Publication 3303.

Integrated Pest Management for Floriculture and Nurseries. 2001. S. H. Dreistadt. University of California Statewide Integrated Pest Management Program. Oakland: UC ANR Publication 3402.

Integrated Pest Management for Stone Fruits. 1999. L. L. Strand. University of California Statewide Integrated Pest Management Program. Oakland: UC ANR Publication 3389.

Integrated Pest Management for Walnuts. 3rd ed. 2003. L. L. Strand. University of California Statewide Integrated Pest Management Program. Oakland: UC ANR Publication 3270.

Invasive Plants Pest Notes. 2007. C. E. Bell, J. M. DiTomaso, and C. A. Wilen. University of California Statewide Integrated Pest Management Program. Oakland: UC ANR Publication 74139.

IPM in Practice: Principles and Methods of Integrated Pest Management. 2012. M. L. Flint. University of California Statewide Integrated Pest Management Program. Oakland: UC ANR Publication 3418.

The IPM Practitioner. Berkeley: Bio-Integral Resource Center.

Jepson eFlora. Jepson Flora Project, eds. 2013. Online at http://ucjeps. berkeley.edu/IJM.html.

The Jepson Manual: Vascular Plants of California. 2012. B. G. Baldwin, D. Goldman, D. J. Keil, R. Patterson, T. J. Rosatti, and D. Wilken, eds. Berkeley: UC Press.

A Key to the Most Common and/or Economically Important Ants of California With Color Photographs. 1983. P. Haney, P. A. Philips, and R. Wagner. Oakland: UC ANR Publication 21433.

Key to Identifying Common Household Ants. 2005. C. A. Reynolds, M. L. Flint, M. K. Rust, P. S. Ward, R. L. Coviello, and J. H. Klotz. Davis: University of California Statewide Integrated Pest Management Program. Online at www.ipm.ucanr. edu/TOOLS/ANTKEY.

Kikuyugrass Pest Notes. 2011. C. A. Wilen, D. W. Cudney, C. L. Elmore, and V. A. Gibeault. University of California Statewide Integrated Pest Management Program. Oakland: UC ANR Publication 7458.

Lace Bugs Pest Notes. 2014. S. H. Dreistadt. University of California Statewide Integrated Pest Management Program. Oakland: UC ANR Publication 7428.

Landscape Maintenance Pest Control. 2006. P. J. O'Connor-Marer. University of California Statewide Integrated Pest Management Program. Oakland: UC ANR Publication 3493.

Landscape Plant Problems: A Pictorial Diagnostic Manual. 2000. R. S. Byther, C. R. Foss, A. L. Antonelli, R. R. Maleike, and V. M. Bobbitt. Pullman: Washington State University.

Lawn and Residential Pest Control: A Guide for Maintenance Gardeners. 2009. S. Cohen, M. L. Flint, and N. Hines. University of California Statewide Integrated Pest Management Program. Oakland: UC ANR Publication 3510.

Leaffooted Bug Pest Notes. 2014. C. Ingels and D. Haviland. University of California Statewide Integrated Pest Management Program. Oakland: UC ANR Publication 74168.

Leafrollers on Ornamental and Fruit Trees Pest Notes. 2010. W. J. Bentley. University of California Statewide Integrated Pest Management Program. Oakland: UC ANR Publication 7473. *Life Stages of California Red Scale and Its Parasitoids.* 1995. L. D. Forster, R. F. Luck, and E. E. Grafton-Cardwell. Oakland: UC ANR Publication 21529.

Living Among the Oaks: A Management Guide for Homeowners. Undated. S. G. Johnson. Oakland: UC ANR Publication 21538.

Managing Insects and Mites with Spray Oils. 1991. N. J. Davidson, J. E. Dibble, M. L. Flint, P. J. Marer, and A. Guye. Oakland: UC ANR Publication 3347.

Mealybugs in California Vineyards. 2002. K. E. Godfrey, K. M. Daane, W. J. Bentley, R. J. Gill, and R. Malakar-Kuenen. Oakland: UC ANR Publication 21612.

Mistletoe Pest Notes. 2006. E. J. Perry and C. L. Elmore. University of California Statewide Integrated Pest Management Program. Oakland: UC ANR Publication 7437.

Mushrooms and Other Nuisance Fungi in Lawns Pest Notes. 2012. M. LeStrange, C. A. Frate, and R. M. Davis. University of California Statewide Integrated Pest Management Program. Oakland: UC ANR Publication 74100.

Myoporum Thrips Pest Notes. 2013. J. A. Bethke and L. Bates. University of California Statewide Integrated Pest Management Program. Oakland: UC ANR Publication 74165.

Natural Enemies Handbook: The Illustrated Guide to Biological Pest Control. 1998. M. L. Flint and S. H. Dreistadt. Oakland: UC ANR Publication 3386.

Nematodes Pest Notes. 2010. E. J. Perry and A. T. Ploeg. University of California Statewide Integrated Pest Management Program. Oakland: UC ANR Publication 7489.

The New Sunset Western Garden Book. 2012. 9th ed. K. N. Brenzel, ed. Birmingham, AL: Oxmoor House.

Nursery Guide for Diseases of Phytophthora ramorum on Ornamentals: Diagnosis and Management. 2005. S. A. Tjosvold, K. R. Beumeyer, C. Blomquist, and S. Frankel. Oakland: UC ANR Publication 8156.

Nutsedge Pest Notes. 2010. C. A. Wilen, M. E. McGiffen, and C. L. Elmore. University of California Statewide Integrated Pest Management Program. Oakland: UC ANR Publication 7432.

Oak Pit Scales Pest Notes. 2013. P. M. Geisel and E. J. Perry. University of California Statewide Integrated Pest Management Program. Oakland: UC ANR Publication 7470.

Oaks in the Urban Landscape: Selection, Care, and Preservation. 2011. L. R. Costello, B. W. Hagen, and K. S. Jones. Oakland: UC ANR Publication 3518.

Oleander Leaf Scorch Pest Notes. 2008. C. A. Wilen, J. S. Hartin, M. J. Henry, M. Blua, and A. H. Purcell. University of California Statewide Integrated Pest Management Program. Oakland: UC ANR Publication 7480.

Olive Fruit Fly Pest Notes. 2009. F. G. Zalom, R. Van Steenwyk, H. J. Burrack, and M. W. Johnson. University of California Statewide Integrated Pest Management Program. Oakland: UC ANR Publication 74112.

Olive Knot Pest Notes. 2011. E. J. Fitchner. University of California Statewide Integrated Pest Management Program. Oakland: UC ANR Publication 74156.

Pacific Northwest Landscape Integrated Pest Management (IPM) Manual: Culture of Key Trees & Shrubs, Problem Diagnosis and Management Options. 2005. V. M. Bobbitt, A. L. Antonelli, C. R. Foss, R. M. Davidson Jr., R. S. Byther, and R. R. Maleike. Pullman: Washington State University.

Palm Diseases in the Landscape Pest Notes. 2009. D. R. Hodel. University of California Statewide Integrated Pest Management Program. Oakland: UC ANR Publication 74148.

Peach Leaf Curl Pest Notes. 2012. J. C. Broom and C. A. Ingels. University of California Statewide Integrated Pest Management Program. Oakland: UC ANR Publication 7426.

Pesticides: Safe and Effective Use in the Home and Landscape Pest Notes. 2006. C. A. Wilen, D. L. Haver, M. L. Flint, P. M. Geisel, and C. L. Unruh. University of California Statewide Integrated Pest Management Program. Oakland: UC ANR Publication 74126.

Pests of the Garden and Small Farm. 2nd ed. 1998. M. L. Flint. University of California Statewide Integrated Pest Management Program. Oakland: UC ANR Publication 3332.

Pests of the Native California Conifers. 2003. D. L. Wood, T. W. Koerber, R. F. Scharpf, and A. J. Storer. Berkeley: UC Press.

Pests of the West. 2nd ed. 1998. W. Cranshaw. Fort Collins: Colorado State University.

Pitch Canker Disease in California. 2013. K. S. Camilli, J. Marshall, D. Owen, T. Gordon, and D. Wood. Sacramento: California Department of Forestry and Fire Protection. Tree Notes 32.

Pitch Canker Diseases of Pines: A Technical Review. 2003. B. Aegerter, T. Gordon, A. Storer, and D. Wood. Oakland: UC ANR Publication 21616.

Pitch Canker Pest Notes. 2013. C. L. Swett and T. R. Gordon. University of California Statewide Integrated Pest Management Program. Oakland: UC ANR Publication 74107.

Pitch Moths Pest Notes. 2013. S. V. Swain and S. H. Dreistadt. University of California Statewide Integrated Pest Management Program. Oakland: UC ANR Publication 7479.

Plant Health Care for Woody Ornamentals. 1997. J. E. Lloyd, ed. Champaign, IL: International Society of Arboriculture.

Planting Landscape Trees. 2001. G. W. Hickman and P. Svihra. Oakland: UC ANR Publication 8046.

Plants Resistant or Susceptible to Verticillium Wilt. 1981. A. H. McCain, R. D. Raabe, and S. Wilhelm. Oakland: UC ANR Publication 2703.

Pocket Gophers Pest Notes. 2009. T. P. Salmon and R. A. Baldwin. University of California Statewide Integrated Pest Management Program. Oakland: UC ANR Publication 7433.

Poison Oak Pest Notes. 2009. J. M. DiTomaso and W. T. Lanini. University of California Statewide Integrated Pest Management Program. Oakland: UC ANR Publication 7431.

Powdery Mildew on Ornamentals Pest Notes. 2009. W. D. Gubler and S. T. Koike. University of California Statewide Integrated Pest Management Program. Oakland: UC ANR Publication 7493.

A Property Owner's Guide to Reducing Wildfire Threat. D. S. Farnham. 1995. Oakland: UC ANR Publication 21539.

Protecting Trees When Building on Forested Land. 1983. C. S. Koehler, R. H. Hunt, D. F. Lobel, and J. Geiger. Oakland: UC ANR Publication 21348.

Psyllids Pest Notes. 2014. J. N. Kabashima, T. D. Paine, K. M. Daane, and S. H. Dreistadt. University of California Statewide Integrated Pest Management Program. Oakland: UC ANR Publication 7423.

Puncturevine Pest Notes. 2006. C. A. Wilen. University of California Statewide Integrated Pest Management Program. Oakland: UC ANR Publication 74128.

Questions and Answers About Tensiometers. 1981. A. W. Marsh. Oakland: UC ANR Publication 2264.

Rabbits Pest Notes. 2010. T. P. Salmon and W. P. Gorenzel. University of California Statewide Integrated Pest Management Program. Oakland: UC ANR Publication 7447.

Rebugging Your Home and Garden: A Step By Step Guide to Modern Pest Control. 1996. R. Troetschler, A. Woodworth, S. Wilcomer, J. Hoffmann, and M. Allen. Palo Alto, CA: PTF Press.

Recognizing Tree Hazards: A Photographic Guide for Homeowners. 1999. L. R. Costello, B. Hagen, and K. S. Jones. Oakland: UC ANR Publication 21584.

Redhumped Caterpillar Pest Notes. 2010. M. L. Flint, ed. University of California Statewide Integrated Pest Management Program. Oakland: UC ANR Publication 7474.

Red Imported Fire Ant Pest Notes. 2007. L. Greenberg, J. H. Klotz, and J. N. Kabashima. University of California Statewide Integrated Pest Management Program. Oakland: UC ANR Publication 7487.

Removing Honey Bee Swarms and Established Hives Pest Notes. 2012. E. C. Mussen. University of California Statewide Integrated Pest Management Program. Oakland: UC ANR Publication 74159.

Residential, Industrial, and Institutional Pest Control. 2nd ed. 2006. P. J. O'Connor-Marer. University of California Statewide Integrated Pest Management Program. Oakland: UC ANR Publication 3334.

Roses: Diseases and Abiotic Disorders Pest Notes. 2009. J. F. Karlik and M. L. Flint. University of California Statewide Integrated Pest Management Program. Oakland: UC ANR Publication 7463.

Roses: Insect and Mite Pests and Beneficials Pest Notes. 2008. M. L. Flint and J. F. Karlik. University of California Statewide Integrated Pest Management Program. Oakland: UC ANR Publication 7466.

The Safe and Effective Use of Pesticides. 2nd ed. 2000. P. J. O'Connor-Marer. University of California Statewide Integrated Pest Management Program. Oakland: UC ANR Publication 3324.

Sago Palms in the Landscape. 2001. P. M. Geisel, C. L. Unruh, and P. M. Lawson. Oakland: UC ANR Publication 7466.

The Scale Insects of California Part 1: The Soft Scales. R. J. Gill. 1988. Sacramento: California Department of Food and Agriculture. Online at http://www.cdfa.ca.gov/phpps/PPD/PDF/Technical_Series_01.pdf.

Scale Insects of California Part 2: The Minor Families. R. J. Gill. 1993. Sacramento: California Department of Food and Agriculture. Online at http://www.cdfa.ca.gov/phpps/PPD/PDF/Technical_Series_02.pdf.

Scale Insects of California Part 3: The Armored Scales. R. J. Gill. 1997. Sacramento: California Department of Food and Agriculture. Online at http://www.cdfa.ca.gov/phpps/PPD/PDF/Technical_Series_03.pdf.

Scales Pest Notes. 2014. J. N. Kabashima and S. H. Dreistadt. University of California Statewide Integrated Pest Management Program. Oakland: UC ANR Publication 7408.

SelecTree: A Tree Selection Guide. 2001. J. L. Reimer and W. Mark. San Luis Obispo: California State University. Online at http://selectree.calpoly.edu.

Slugs: A Guide to the Invasive and Native Fauna of California. 2009. R. J. McDonnell, T. D. Paine, and M. J. Gormally. Oakland: UC ANR Publication 8336.

Snails and Slugs Pest Notes. 2009. M. L. Flint and C. A. Wilen. University of California Statewide Integrated Pest Management Program. Oakland: UC ANR Publication 7427.

Soil and Fertilizer Management. 2014. B. Faber, J. Walworth, D. D. Giraud, and D. Silva. In California Master Gardener Handbook, 2nd ed., D. R. Pittenger, ed. Oakland: UC ANR Publication 3382.

Soil Solarization: A Natural Mechanism of Integrated Pest Management. 1995. J. J. Stapleton and J. E. DeVay. In Novel Approaches to Integrated Pest Management, R. Reuveni, ed. Boca Raton, FL: Lewis Publishers.

Soil Solarization for Gardens & Landscapes Pest Notes. 2008. J. J. Stapleton, C. A. Wilen, and R. H. Molinar. University of California Statewide Integrated Pest Management Program. Oakland: UC ANR Publication 74145.

Sooty Mold Pest Notes. 2011. F. F. Laemmlen. University of California Statewide Integrated Pest Management Program. Oakland: UC ANR Publication 74108.

Spider Mites Pest Notes. 2011. L. D. Godfrey. University of California Statewide Integrated Pest Management Program. Oakland: UC ANR Publication 7405.

Spiders Pest Notes. 2007. R. S. Vetter. University of California Statewide Integrated Pest Management Program. Oakland: UC ANR Publication 7442.

Spotted Spurge and other Spurges Pest Notes. 2009. R. H. Molinar, D. W. Cudney, C. L. Elmore, and A. Sanders. University of California Statewide Integrated Pest Management Program. Oakland: UC ANR Publication 7445.

Spotted Wing Drosophila Pest Notes. 2011. J. L. Caprile, M. L. Flint, M. P. Bolda, J. A. Grant, R. Van Steenwyk, and D. R. Haviland. University of California Statewide Integrated Pest Management Program. Oakland: UC ANR Publication 74158.

Stages of the Cottony Cushion Scale (Icerya purchasi) *and its Natural Enemy, the Vedalia Beetle* (Rodolia cardinalis). 2002. B. Grafton-Cardwell. Oakland: UC ANR Publication 8051.

Sticky Trap Monitoring of Insect Pests. 1998. S. H. Dreistadt, J. P. Newman, and K. L. Robb. Oakland: UC ANR Publication 21572.

Subterranean and Other Termites Pest Notes. 2014. V. R. Lewis, A. M. Sutherland, and M. I. Haverty. University of California Statewide Integrated Pest Management Program. Oakland: UC ANR Publication 7415.

Sudden Oak Death Pest Notes. 2010. J. M. Alexander and S. V. Swain. University of California Statewide Integrated Pest Management Program. Oakland: UC ANR Publication 7498.

Sustainable Landscaping in California. 2014. J. Hartin, P. Geisel. A. Harivandi, and R. Elkins. Oakland: UC ANR Publication 8504.

Sycamore Scale Pest Notes. 2010. S. H. Dreistadt. University of California Statewide Integrated Pest Management Program. Oakland: UC ANR Publication 7409.

A Technical Study of Insects Affecting the Elm Tree in Southern California. 1966. L. R. Brown and C. O. Eads. Berkeley: UC Agricultural Experiment Station Bulletin 821.[2]

A Technical Study of Insects Affecting the Oak Tree in Southern California. 1965. L. R. Brown and C. O. Eads. Berkeley: UC Agricultural Experiment Station Bulletin 810.[2]

A Technical Study of Insects Affecting the Sycamore Tree in Southern California. 1965. L. R. Brown and E. O. Eads. Berkeley: UC Agricultural Experiment Station Bulletin 818.[2]

Thrips Pest Notes. 2014. J. A. Bethke, S. H. Dreistadt, and L. G. Varela. University of California Statewide Integrated Pest Management Program. Oakland: UC ANR Publication 7429.

Training Young Trees for Structure and Form. 1999. L. R. Costello. Oakland: UC ANR Publication 6580D (DVD).

Tree Guidelines for Coastal Southern California Communities. 2000. G. McPherson, J. R. Simpson, P. J. Peper, Q. Xiao, and K. I. Scott. Sacramento: Local Government Commission. Online at www.lgc.org/freepub/energy/guides.

Tree Guidelines for Inland Empire Communities. 2001. G. McPherson, J. R. Simpson, P. J. Peper, Q. Xiao, D. R. Pittenger, and D. R. Hodel. Sacramento: Local Government Commission. Online at www.lgc.org/freepub/energy/guides.

Tree Guidelines for San Joaquin Valley Communities. 1999. G. McPherson, J. R. Simpson, P. J. Peper, and Q. Xiao. Sacramento: Local Government Commission. Online at www.lgc.org/freepub/energy/guides.

Trees and Development: A Technical Guide to Preservation of Trees During Land Development. 1998. N. Matheny and J. Clark. Champaign, IL: International Society of Arboriculture.

Trees and Shrubs of California. 2001. J. D. Stuart and J. O. Sawyer. Berkeley: UC Press.

Trees for Saving Energy. 1991. R. Thayer. Oakland: UC ANR Publication 21485.

Trees Under Power Lines: A Homeowner's Guide. 1989. L. R. Costello, A. M. Berry, F. J. Chan, and R. R. Novembri. Oakland: UC ANR Publication 21470.

Turfgrass Insects of the United States and Canada. 2nd ed. 1999. P. J. Vittum, M. G. Villani, and H. Tashiro. Ithaca, NY: Cornell University Press.

Turfgrass Pests. 1989. A. D. Ali and C. L. Elmore, eds. Oakland: UC ANR Publication 4053.

Twig Blight and Branch Dieback of Oaks in California. 1989. L. R. Costello, E. I. Hecht-Poinar, and J. R. Parmeter. Oakland: UC ANR Publication 21462.

Urban Entomology. 1978. W. Ebeling. Berkeley: UC Division of Agricultural Science.[2]

Urban Pest Management of Ants in California. 2010. J. Klotz, L. Hansen, H. Field, M. Rust, D. Oi, and K. Kupfer. Oakland: UC ANR Publication 3524.

Urban Trees and Ozone Formation: A Consideration for Large-Scale Plantings. 2012. J. Karlik and D. Pittenger. Oakland: UC ANR Publication 8484.

Voles (Meadow Mice) Pest Notes. 2010. T. P. Salmon and W. P. Gorenzel. University of California Statewide Integrated Pest Management Program. Oakland: UC ANR Publication 7439.

Walnut Husk Fly Pest Notes. 2009. C. Pickel. University of California Statewide Integrated Pest Management Program. Oakland: UC ANR Publication 7430.

Walnut: UC IPM Pest Management Guidelines. 2013. C. Pickel, J. A. Grant, W. J. Bentley, J. K. Hasey, W. W. Coates, R. A. Van Steenwyk (Insects and Mites); J. E. Adaskaveg, R. P. Buchner, G. T. Browne, W. D. Gubler (Diseases); M. V. McKenry, B. B. Westerdahl (Nematodes); A. Shrestha, R. B. Elkins, K. Hasey, and K. K. Anderson. University of California Statewide Integrated Pest Management Program. Oakland: UC ANR Publication 3471.

Water Conservation Tips for the Home Lawn and Garden. 2001. P. M. Geisel and C. L. Unruh. Oakland: UC ANR Publication 8036.

Water Management. 2014. J. Hartin and B. Faber. In California Master Gardener Handbook, 2nd ed., D. R. Pittenger, ed. Oakland: UC ANR Publication 3382.

Water Quality: Its Effects on Ornamental Plants. 1985. D. S. Farnham, R. F. Hasek, and J. L. Paul. Oakland: UC ANR Publication 2995.

2. Publications out of print. Copies may be available for reference at libraries.

Weed Management in Landscapes Pest Notes. 2007. C. A. Wilen and C. L. Elmore. University of California Statewide Integrated Pest Management Program. Oakland: UC ANR Publication 7441.

Weed Pest Identification and Monitoring Cards. 2013. J. M. DiTomaso. Oakland: UC ANR Publication 3541.

Weeds of California and Other Western States. 2007. J. M. DiTomaso and E. A. Healy. Oakland: UC ANR Publication 3488.

Weeds of the West. 1991. T. D. Whitson, L. C. Burrill, S. A. Dewey, D. W. Cudney, B. E. Nelson, R. D. Lee, and R. Parker. Wyoming Agric. Extension. Jackson, WY. Available as Oakland: UC ANR Publication 3350.

Western Forest Insects. 1977. R. L. Furniss and V. M. Carolin. Washington, DC: USDA Miscellaneous Publication 1339.[2] Online at www.fs.fed.us/r6/nr/fid/wid.shtml.

Whiteflies Pest Notes. 2002. M. L. Flint. University of California Statewide Integrated Pest Management Program. Oakland: UC ANR Publication 7401.

Wild Blackberries Pest Notes. 2010. J. M. DiTomaso. University of California Statewide Integrated Pest Management Program. Oakland: UC ANR Publication 7434.

Wildlife Pest Control Around Gardens and Homes. 2nd ed. 2006. T. P. Salmon, D. A. Whisson, and R. E. Marsh. Oakland: UC ANR Publication 21385.

Wood-Boring Beetles in Homes Pest Notes. 2010. V. R. Lewis and S. J. Seybold. University of California Statewide Integrated Pest Management Program. Oakland: UC ANR Publication 7418.

Wood Decay Fungi in Landscape Trees Pest Notes. 2011. G. W. Hickman, E. J. Perry, and R. M. Davis. University of California Statewide Integrated Pest Management Program. Oakland: UC ANR Publication 74109.

Woodpeckers Pest Notes. 2005. T. P. Salmon, D. A. Whisson, and R. E. Marsh. University of California Statewide Integrated Pest Management Program. Oakland: UC ANR Publication 74124.

Wood Preservation. 1992. P. J. Marer and M. Grimes. University of California Statewide Integrated Pest Management Program. Oakland: UC ANR Publication 3335.

Wood Wasps and Horntails Pest Notes. 2010. E. C. Mussen. University of California Statewide Integrated Pest Management Program. Oakland: UC ANR Publication 7407.

Woody Landscape Plants. 2014. D. R. Hodel and D. R. Pittenger. In California Master Gardener Handbook, 2nd ed., D. R. Pittenger, ed. Oakland: UC ANR Publication 3382.

Woody Weed Invaders Pest Notes. 2008. J. M. DiTomaso and G. B. Kyser. University of California Statewide Integrated Pest Management Program. Oakland: UC ANR Publication 74142.

Yellow Starthistle Pest Notes. 2007. J. M. DiTomaso, G. B. Kyser, W. T. Lanini, C. D. Thomsen, and T. S. Prather. University of California Statewide Integrated Pest Management Program. Oakland: UC ANR Publication 7402.

2. Publications out of print. Copies may be available for reference at libraries.

Literature Cited[1]

These publications are cited in figures and tables as the sources of information and illustrations. For publications referred to in text as sources of further information, see the "Suggested Reading."

Adams, D. Undated. *Pitch Canker—An Introduced Disease.* Davis: California Department of Forestry and Fire Protection. Unpublished.

Agrios, G. N. 1997. *Plant Pathology.* 4th ed. San Diego: Academic Press.

Ahrens, W. H., ed. 1994. *Herbicide Handbook.* 7th ed. Champaign, IL: Weed Science Society of America.

Antonelli, A. L., and R. L. Campbell. 1984. *Root Weevil Control on Rhododendrons.* Pullman: Washington State University Extension Bulletin 0970.

Bosik, J. J., ed. 1997. *Common Names of Insects & Related Organisms 1997.* Lanham, MD: Entomological Society of America. Online at www.entsoc.org/common-names.

Brennan, E. B., G. F. Hrusa, S. A. Weinbaum, and W. Levison. 2001. Resistance of *Eucalyptus* species to *Glycaspis brimblecombei* (Homoptera: Psyllidae) in the San Francisco Bay Area. *Pan-Pacific Entomologist* 77:249–253.

Burger, D., B. Coate, L. Costello, R. Crudup, J. Geiger, B. Hagen, R. Harris, B. Kempf, J. Koch, B. Ludekens, G. McPherson, M. Ozonoff, E. Perry, and M. Robert. 2009. *Guideline Specifications for Nursery Tree Quality.* Visalia, CA: Urban Tree Foundation. Online at http://ccuh.ucdavis.edu/industry/files/GuidelineSpecificationsforNurseryTreeQuality.pdf/at_download/file.

Costello, L. R., C. S. Koehler, and W. W. Allen. 1987. *Fuchsia Gall Mite.* Oakland: UC ANR Publication 7179.

Crump, A. 2009. *Anthracnose Pest Notes.* University of California Statewide Integrated Pest Management Program. Oakland: UC ANR Publication 7420.

Dahlsten, D. L., S. M. Tait, D. L. Rowney, and B. J. Gingg. 1993. A monitoring system and developing ecologically sound treatments for elm leaf beetle. *Journal of Arboriculture* 19(4):181–186. Online at http://joa.isa-arbor.com.

Dallara, P. L., A. J. Storer, T. R. Gordon, and D. L. Wood. 1995. *Current Status of Pitch Canker Disease in California.* Sacramento: California Department of Forestry and Fire Protection. Tree Notes 20.

Farr, D. F., G. F. Bills, G. P. Chamuris, and A. Y. Rossman. 1989. *Fungi on Plants and Plant Products in the United States.* St. Paul, MN: American Phytopathological Society.

Flint, M. L., M. I. Jones, T. W. Coleman, and S. J. Seybold. 2013. *Goldspotted Oak Borer Pest Notes.* University of California Statewide Integrated Pest Management Program. Oakland: UC ANR Publication 74163.

Frankie, G. W., J. B. Fraser, and J. F. Barthell. 1986. Geographic distribution of *Synanthedon sequoia* and host plant susceptibility on Monterey pine in adventive and native stands in California (Lepidoptera: Sesiidae). *Pan-Pacific Entomologist* 62:29–40.

Hanks, L. M., T. D. Paine, J. G. Millar, and J. L. Hom. 1995. Variation among *Eucalyptus* species in resistance to eucalyptus longhorned borer in southern California. *Entomologia Experimentalis et Applicata* 74:185–194.

Harris, R. W., J. R. Clark, and N. P. Matheny. 1999. *Arboriculture: Integrated Management of Landscape Trees, Shrubs, and Vines.* 3rd ed. Englewood Cliffs, NJ: Prentice-Hall.

Hickman, G. W. and P. Svihra. 2001. *Planting Landscape Trees.* Oakland: UC ANR Publication 8046.

Hickman, J. C., ed. 1993. *The Jepson Manual: Higher Plants of California.* Berkeley: UC Press.

Kaya, H. K. 1993. Contemporary issues in biological control with entomopathogenic nematodes. Taipei City, Taiwan: *Food and Fertilizer Technology Center Extension Bulletin* 375.

Koehler, C. S., W. W. Allen, and L. R. Costello. 1985. Fuchsia gall mite management. *California Agriculture* 39(7, 8):10–12.

Koehler, C. S., and W. W. Moore. 1983. Resistance of several members of the Cupressaceae to the cypress tip miner, *Argyresthia cupressella. Journal of Environmental Horticulture* 1:87–88.

Koehler, C. S., W. S. Moore, and B. Coate. 1983. Resistance of *Acacia* to the acacia psyllid, *Psylla uncatoides. Journal of Environmental Horticulture* 1:65–67.

Marer, P. J. 1991. *Residential, Industrial, and Institutional Pest Control.* University of California Statewide Integrated Pest Management Program. Oakland: UC ANR Publication 3334.

McCain, A. H., R. D. Raabe, and S. Wilhelm. 1981. *Plants Resistant or Susceptible to Verticillium Wilt.* Oakland: UC ANR Publication 2703.

McPherson, G., L. Costello, J. Harding, S. Dreistadt, M. L. Flint, and S. Mezger. 2009. National Elm Trial: Initial Report from Northern California. *Western Arborist.* Fall 2009:32–36.

Munro, J. A. 1963. Biology of the ceanothus stem-gall moth, *Periploca ceanothiella,* with consideration of its control. *Journal of Research on the Lepidoptera* 1:183–190.

1. University of California publications are available for free download (in certain instances) or can be purchased by visiting the anrcatalog.ucanr.edu or ipm.ucanr.edu World Wide Web sites.

410 • LITERATURE CITED

O'Connor-Marer, P. J. 2000. *The Safe and Effective Use of Pesticides*. 2nd ed. University of California Statewide Integrated Pest Management Program. Oakland: UC ANR Publication 3324.

Ohr, H. D., G. A. Zentmyer, E. C. Pond, and L. J. Klure. 1980. *Plants in California Susceptible to Phytophthora cinnamomi*. Oakland: UC ANR Publication 21178.

Paine, T. D. and S. H. Dreistadt. 2007. *Psyllids Pest Notes*. University of California Statewide Integrated Pest Management Program. Oakland: UC ANR Publication 7423.

Peterson, A. 1956. *Larvae of Insects*. Part 1. Ann Arbor, MI: Edwards Brothers.

——. 1960. *Larvae of Insects*. Part 2. Ann Arbor, MI: Edwards Brothers.

Retzinger, E. J. and C. Mallory-Smith. 1997. Classification of herbicides by site of action for weed resistance management strategies. *Weed Technology* 11:384–393.

Scriven, G. T. and R. F. Luck. 1980. Susceptibility of pines to attack by the Nantucket pine tip moth in southern California. *Journal of Economic Entomology* 73:318–320.

Smith, R. F. and K. S. Hagen. 1956. Enemies of spotted alfalfa aphid. *California Agriculture* 10(4):8–10.

Southern California Research Learning Center. 2012. *Mediterranean Ecosystem*. Thousand Oaks, CA. Online at www.mednscience.org.

Svihra, P. 1994. Principles of eradicative pruning. *Journal of Arboriculture* 20:262–272.

Svihra, P. and B. Duckles. 1999. Bronze Birch Borer. Novato, CA: UC Cooperative Extension Marin County. Pest Alert 2.

Teviotdale, B. L. 1991. Further tests on effectiveness of disinfectants for preventing transmission of fireblight. Kearney: *UC Plant Protection Quarterly* 2(1):5–6.

——. 1992. Efficacy of trisodium phosphate and sodium hypochlorite as disinfectants to prevent transmission of fire blight. Kearney: *UC Plant Protection Quarterly* 2(2):9–10.

——. 2011. *Fire Blight Pest Notes*. University of California Statewide Integrated Pest Management Program. Oakland: UC ANR Publication 7414.

Teviotdale, B. L., M. F. Wiley, and D. H. Harper. 1991. How disinfectants compare in preventing transmission of fire blight. *California Agriculture* 45(4):21–23.

USDA APHIS. 2012. *List of Regulated Hosts and Plants Proven or Associated with Phytophthora ramorum*. Online at www.aphis.usda.gov/plant_health/plant_pest_info/pram.

Weed Science Society of America. 2010. *Summary of Herbicide Mechanism of Action According to the Herbicide Resistance Action Committee (HRAC) and Weed Science Society of America (WSSA) Classification*. Online at http://wssa.net/weed/resistance.

Wilen, C. 2012. Should you be worried about herbicide resistance? Davis: University of California Statewide Integrated Pest Management Program. *Green Bulletin* 2(3):1–2. Online at http://ipm.ucanr.edu/greenbulletin.

Glossary

abdomen. The posterior body division of an arthropod.

abiotic disorder. A disease caused by factors other than a pathogen, such as adverse environmental conditions or inappropriate cultural practices.

alkaline. Basic, having a high pH or pH greater than 7.

annual. A plant that normally completes its life cycle of seed germination, vegetative growth, reproduction, and death in a single year.

antenna (pl., **antennae**). The paired, segmented sensory organs on each side of the head.

auricle. A small, earlike projection from the base of a leaf or petal.

available water. The amount of water held in the soil that can be extracted by plants.

Bacillus thuringiensis. A bacterium that causes disease in certain insects, most commonly caterpillars; formulations of the several subspecies of B.t. are used as insecticides.

bacterium (pl., **bacteria**). A single-celled, microscopic organism that lacks a nucleus. Some bacteria cause plant or animal diseases.

biological control. The action of parasites, predators, or pathogens in maintaining another organism's population density at a lower average level than would occur in their absence. Biological control may occur naturally in the field or result from manipulation or introduction of biological control agents (natural enemies) by people.

biotic disease. Disease caused by a pathogen, such as a bacterium, fungus, phytoplasma, or virus.

botanical. Derived from plants or plant parts, as with pyrethrin insecticides.

broad-spectrum pesticide. A pesticide that kills a large number of unrelated species.

B.t. or **Bt.** Abbreviations for *Bacillus thuringiensis*.

buffering capacity. The ability of soil to resist change in pH.

cambium. Thin layer of undifferentiated, actively growing plant tissue between the phloem and the xylem.

canker. A dead, discolored, often sunken area (lesion) on a branch, root, stem, or trunk.

canopy. The leafy parts of vines or trees.

caterpillar. The immature stage of butterflies and moths.

chlorophyll. The green pigment of plants that captures the energy from sunlight necessary for photosynthesis.

chlorosis. Yellowing or bleaching of normally green plant tissue.

cornicle. Two tubular structures projecting from the rear of an aphid's abdomen.

cotyledon. A leaf formed within the seed and present on a seedling at germination; seed leaf. Cotyledons typically have a different appearance than true leaves, which develop entirely after a seed germinates.

crawler. The active first instar of certain types of insects, such as scales and mealybugs.

crochets. Tiny hooks on the prolegs of caterpillars.

crown. The plant part near the soil surface where the main stem (trunk) and roots join. Also used in forestry to refer to the topmost limbs on a tree or shrub.

cultivar. An identifiable strain within a plant species that is specifically bred for particular properties; sometimes used synonymously with *variety*.

degree-day. A unit combining temperature and time that is used in monitoring growth and development of organisms. Also called heat unit.

delayed dormant. The treatment period beginning when buds begin to swell and ending with the beginning of green tip development or just before the emergence of new leaves.

developmental threshold. The lowest temperature at which growth occurs in a given species.

disease. An unhealthy condition of a plant that interferes with its normal structure, function, or economic value.

dormant. Inactive during periods of adverse environmental conditions, commonly during winter or cold weather.

drift. The aerial dispersal of a substance such as a pesticide beyond the intended application area.

dwarfing. Stunting of normal growth characterized in plants by smaller-than-normal leaves and stems.

entomopathogenic nematodes. Nematodes that, in combination with symbiotic bacteria, kill insects.

evapotranspiration. The loss of soil moisture due to evaporation from the soil surface and transpiration by plants.

field capacity. The amount of water held in soil after saturation and the drainage and runoff of excess water.

frass. Solid fecal material produced by certain insects.

fruiting bodies. In fungi, reproductive structures containing spores.

fungicide. A pesticide used to control fungi.

fungus (pl., **fungi**). A multicellular, generally microscopic organism lacking chlorophyll, such as mildew, mold, rust, or smut. The fungus body normally consists of filamentous strands called mycelium and reproduces through dispersal of spores.

gall. Localized swelling or outgrowth of plant tissue, often formed by the plant in response to the action of an insect, pathogen, or other pest.

girdle. Damage that completely encircles a stem or root, often resulting in death of plant parts above or below the girdle.

ground cover. Any of various low, dense-growing plants, such as ivy and pachysandra, used for covering the ground, as in places where it is difficult to grow grass.

hazard. The risk of danger from pesticides. Hazard, or risk, is a function of two factors—toxicity and potential exposure to the toxic substance.

heat unit. *See* **degree-day**.

herbicide. A pesticide used to control weeds.

honeydew. An excretion from insects, such as aphids, mealybugs, whiteflies, and soft scales, consisting of modified plant sap and composed mostly of water and sugars.

horticultural oils. Highly refined petroleum oils that are manufactured specifically to control pests on plants.

host. A plant or animal that provides sustenance for another organism.

hypha. (pl., **hyphae**). One of the filaments forming the body (vegetative, nonreproductive structure), or mycelium, of a fungus.

inorganic. Containing no carbon; generally used to indicate materials (e.g., fertilizers) that are of mineral origin.

instar. The larval or nymphal stage of an immature insect between successive molts.

integrated pest management (IPM). A pest management strategy that focuses on long-term prevention or suppression of pest problems through a combination of techniques such as encouraging biological control, using resistant varieties, and adopting alternate cultural practices such as modifying irrigation or pruning to make the habitat less conducive to pest development. Pesticides are used only when careful monitoring indicates they are needed according to preestablished guidelines or treatment thresholds, or to prevent pests from significantly interfering with the purposes for which plants are being grown.

invertebrate. An animal having no internal skeleton, such as an earthworm or insect.

juvenile. Immature form of a nematode that hatches from an egg and molts several times before becoming an adult.

larva (pl., **larvae**). The immature form of insects that develop through complete metamorphosis including egg, several larval stages, pupa, and adult. In mites, the first-stage immature is also called a larva.

lesion. Localized area of diseased or discolored tissue.

ligule. In many grasses, a short membranous projection on the inner side of the leaf blade at the point where the leaf blade and leaf sheath meet.

mandibles. Jaws; the forward-most pair of mouthparts of an insect.

metamorphosis. The change in form that takes place as insects grow from an immature to adult.

microbial pesticides. Bacteria, fungi, viruses, or other microorganisms that are commercially produced for control of invertebrates, plant pathogens, or weeds.

microorganism. An organism of microscopic size, such as a bacterium, fungus, phytoplasma, or virus.

molt. In insects and other arthropods, the forming of a new cuticle (skin) that precedes shedding of the old skin (ecdysis), part of the process of development into a larger and older instar or metamorphosis into the next life stage.

monitoring. Carefully watching and recording information on the abundance, activities, development, and growth of organisms or other factors on a regular basis over a period of time, often utilizing very specific procedures.

mulch. A layer of material placed on the soil surface to prevent weed growth and improve plant health.

mummy. The crusty skin of an aphid whose inside has been consumed by a parasite. An unharvested nut remaining on the tree.

mycelium. A mass of branching or interwoven hyphae that are the vegetative (nonreproductive) structure on most true fungi.

mycoplasma. Living organisms smaller than bacteria, now called phytoplasmas; they have a unit membrane but no cell wall as do bacteria.

mycorrhizae. Beneficial associations between plant roots and fungi.

narrow-range oil. A highly refined petroleum oil that is manufactured specifically to control pests on plants; also called superior, supreme, or horticultural oil.

natural enemies. Predators, parasites, or pathogens that are considered beneficial because they attack and kill pests. The organisms are used in the biological control of pests.

necrosis. Death of tissue accompanied by dark discoloration, usually occurring in a well-defined part of a plant, such as the portion of a leaf between leaf veins or the xylem or phloem in a stem or tuber.

nymph. Immature stage of insects that develop through incomplete metamorphosis, such as aphids, grasshoppers, leafhoppers, mealybugs, and true bugs that hatch from eggs and gradually acquire adult form through a series of molts without passing through a pupal stage.

organic. A material (e.g., a pesticide) whose molecules contain carbon and hydrogen atoms. Also refers to plants or animals that are grown without the use of synthetic fertilizers or pesticides.

oviposit. To lay or deposit eggs.

parasite. An organism that lives and feeds in or on a larger organism (the host) without killing the host directly. Also, an insect that spends its immature stages in or on the body of a host that dies just before the parasite emerges (this type is more accurately called a parasitoid). Unlike many other parasites, adult parasitoids are free-living.

pathogen. A disease-causing organism.

perennial. A plant that can live 3 or more years and flower at least twice.

pesticide. Any substance or mixture intended for preventing, destroying, repelling, killing, or mitigating problems caused by fungi, insects, nematodes, rodents, weeds, or other pests; and any other substance or mixture intended for use as a plant growth regulator, defoliant, or desiccant.

pesticide resistance. The result of genetic selection, when a pest population is no longer controlled by a pesticide that previously provided control.

pest resurgence. Rapid rebound of a pest population after it has been controlled.

petiole. The stalk connecting a leaf to a stem.

pH. A value used to express the relative degree of acidic or basic conditions; the hydrogen ion concentration as expressed in a negative logarithmic scale ranging from 0 to 14.

pheromone. A substance secreted by an organism to affect the behavior or development of other members of the same species. Sex pheromones that attract the opposite sex for mating are used in monitoring or management of certain insects.

phloem. The food-conducting tissue of a plant, made up of sieve tubes, companion cells, phloem parenchyma, and fibers.

photosynthesis. The process by which plants use sunlight to convert carbon dioxide and water into sugars.

phytoplasma. Living organisms smaller than bacteria, formerly called mycoplasmas or mycoplasma-like organisms; they have a unit membrane but no cell wall, as do bacteria.

phytotoxicity. Ability of a material such as a pesticide or fertilizer to cause injury to plants.

postemergence herbicide. Herbicide applied after the emergence of weeds.

predator. Organism (including certain insects and mites) that attacks, kills, and feeds on several or many other individuals (its prey) during its lifetime.

preemergence herbicide. Herbicide applied before emergence of weeds.

presence-absence sampling. A sampling method that involves recording only whether members of the population being sampled (such as an insect pest) are present or absent on a sample unit (such as a leaf), rather than counting the numbers of individuals. Also called binomial sampling.

proleg. A fleshy, unsegmented leglike appendage, as on the abdomen of caterpillars; the presence, location, and number of these appendages help to distinguish among larvae of different types of insects, such as larvae of moths in comparison with those of beetles or sawflies.

propagules. Any part of an organism from which progeny can grow, including bulbs, tubers, and seeds of plants, and sclerotia and spores of fungi.

prothorax. The anterior of the three thoracic segments of an insect.

pupa. Nonfeeding stage between larva and adult in insects with complete metamorphosis.

pupate. To develop from the larval stage to the pupa.

resistant. Able to tolerate conditions (such as pesticide sprays or pest damage) harmful to other species or other strains of the same species, but not immune to such conditions.

rhizome. A horizontal, underground shoot, especially one that forms roots at the nodes to produce new plants.

sanitation. Activity that reduces the spread of pathogen inoculum, such as removal and destruction of infected plant parts or cleaning of contaminated tools and equipment.

sclerotium (pl., **sclerotia**). A compact mass of hardened mycelium that serves as a dormant stage in some fungi.

secondary outbreak. The increase in the number of a nontarget species to harmful (pest) levels following a pesticide application to control a different species that was the target pest; caused by destruction of natural enemies that normally control the nontarget species.

sedges. Group of grasslike herbaceous plants that, unlike grasses, have unjointed stems. Stems are usually solid and often triangular in cross-section.

seed leaf. Leaf formed in a seed and present on a seedling at germination; cotyledon.

selective pesticide. Pesticides that are toxic primarily to the target pest (and often related species), leaving most other organisms, including natural enemies, unharmed.

sign. Presence of pathogen reproductive structures (e.g., mushrooms or rust pustules) or vegetative structures (hyphae, mycelium) that indicate that plants may be diseased.

solarization. Heating soil to temperatures that are lethal to pests by applying clear plastic to the soil surface for 4 to 6 weeks to capture solar energy during warm, sunny weather.

sooty mold. Dark coating on foliage or fruit formed by the mycelia of fungi that live on honeydew secreted by certain insects.

sp. Abbreviation for a single species.

species. A group of individual organisms that in nature can interbreed to produce fertile offspring.

spore. Seedlike reproductive structure produced by certain fungi and other organisms that is capable of growing into a new individual under proper conditions.

spp. Abbreviation for multiple species.

stolon. Trailing aboveground stem or shoot, often rooting at the nodes and forming new plants.

symptom. Outward expression by a host that it is unhealthy, such as chlorosis, necrosis, or wilting that occurs in diseased plants.

synthetic organic pesticides. Manufactured pesticides produced from petroleum and containing largely carbon and hydrogen atoms in their basic structure.

systemic. Capable of moving throughout a plant or other organism, usually in the vascular system.

target pest. A pest species that a control action is intended to manage.

tensiometer. Device for measuring soil moisture, such as a mechanical device consisting of a buried tube of water that develops a partial vacuum as surrounding soil dries out.

terminal. The growing tip of a stem, especially the main stem.

thorax. The second of three major divisions in the body of an insect, and the one bearing the legs and (if present) wings.

tolerance. Inherent lack of susceptibility to a pesticide. Also, the ability of a plant to grow in spite of infection by a pathogen.

translocated herbicide. Herbicide that is able to move throughout a plant, such as to roots after being applied to leaf surfaces.

transpiration. Evaporation of water vapor from plants, mostly through tiny leaf openings (stomata).

tuber. An enlarged, fleshy, underground stem with buds capable of producing new plants.

vascular system. The system of plant tissues that conducts water, mineral nutrients, and products of photosynthesis through the plant, consisting of the xylem and phloem.

vector. Organism able to transport and transmit a pathogen to a host.

virus. Noncellular, submicroscopic pathogen that can multiply only within living cells of other organisms and is capable of producing disease symptoms in some plants and animals.

xylem. Plant tissue that conducts water and nutrients from the roots up through the plant.

Index

An *italic "f"* after a page number, such as 45f, refers to a photo or an illustration; an italic *"t"*, as in 76t, refers to a table. **Bold** page numbers refer to the main discussion for topics with multiple page referrals. Pests and disorders listed in Chapter 9, "Tree and Shrub Pest Tables," are not indexed here; consult those tables to see what pests are reported for each landscape plant, and for relevant page numbers in the text.

A

Biologicals (pesticides), 75, 76t, 140–141
Birch (*Betula*), 312–313
Bird of paradise (*Strelitzia*), 388
Bird of paradise bush (*Caesalpinia*), 315
Birds, as predators, 136
Blackberry (*Rubus*), 287–288
Black leaf spot (*Stegophora ulmea*), 81, 82t
Black locust (*Robinia*), 378
Black medic (*Medicago lupulina*), 275
Black scale (*Saissetia oleae*), 130, 192–193
Black spot (*Diplocarpon rosae*), 86–87
Black tea tree (*Melaleuca*), 352
Black vine weevil (*Otiorhynchus sulcatus*), 129f, 160–161
Black walnut (*Juglans*), 111f, 343–344
Blastobasis (acorn moth), 162
Bleach, 73
Blights, 83–86
Blossom and fruit diseases, 77–80
Bluegrass (*Poa annua*), 277–278
Blue-green sharpshooter (*Graphocephala atropunctata*), 203
Blue gum, 212f
Bluegum psyllid (*Ctenarytaina eucalypti*), 171
Blue oak, 198f
Blue Palo Verde (*Parkinsonia floridum*), 357–358
Boisea rubrolineata (western boxelder bug), 206–207
Bordeaux mixture, 75, 76t
Bordered plant bug (*Largus cinctus*), 207
Boric acid ant baits, 169–170
Boring insects. *See specific insects, such as ambrosia beetles, bark beetles*
Boron excess, 52–53
Botanicals (pesticides), 75, 133t, 141, 268–269
Bot canker (*Diplodia corticola*), 102
Bot, or ficus, canker (*Neofusicoccum mangiferae*), 100–101
Botryosphaeria (ficus canker), 100–101
Botryosphaeria canker and dieback (*Botryosphaeria, Fusicoccum*), 98–99
Botryosphaeria corticola (bot canker), 102
Botryosphaeria dothidea (Botryosphaeria canker and dieback), 98–99, 102
Botryosphaeria stevensii (Raywood ash canker and decline), 104–105
Botrytis blight (*Botrytis, Botryotinia*), 78, 79f
Bottlebrush (*Callistemon*), 315
Bottle tree (*Brachychiton*), 133f, 314

Bougainvillea, 26, 313–314
Bougainvillea looper (*Disclisioprocta stellata*), 150
Bowlegged fir aphid (*Cinara curvipes*), 177f
Box (*Buxus*), 314–315
Box elder (*Acer negundo*), 193f, 301
Boxelder bug, 206–207
Boxwood (*Buxus*), 314–315
Brachychiton (bottle tree), 314
Brachymeria ovata (parasitic wasp), 148f
Bradford pear (*Pyrus calleryana*), 370
Brahea (palm), 307–309
Branch beat sampling, 128
Branch symptoms and causes, 296–297
Brassica residue (mustard meal), 292
Brazilian butterfly tree (*Bauhinia*), 312
Brenneria quercina (drippy nut disease, drippy acorn), 80
Brisbane box (*Lophostemon conferta*), 350
Bristly roseslug (*Cladius difformis*), 154
Broadleaf herbicides, 54, 268
Broad mite (*Polyphagotarsonemus latus*), 249
Broad-spectrum pesticides
  defined, 10
  phytotoxicity, 54
  selective use, 10, 268
  toxicity to natural enemies, 10, 131–133, 142
Bronze birch borer (*Agrilus anxius*), 227f
Bronze poplar borer (*Agrilus liragus*), 227
Brooms (weeds), 259
Brown felt blight (*Neopeckia coulteri, Herpotrichia juniperi*), 85
Brown garden snail (*Cornu aspersum*), 251
Brown lacewing, 137
Brown marmorated stink bug (*Halyomorpha halys*), 208, 209f
Brown soft scale (*Coccus hesperidum*), 193
Brush cherry (*Syzygium paniculatum*), 389
Bt (*Bacillus thuringiensis*), 133t, 140, 147
Bucculatrix albertiella (oak ribbed casemaker), 218
Buckeye (*Aesculus*), 303
Buckthorn (*Rhamnus*), 374
Buckwheat (*Eriogonum*), 133f, 330
Buddleia, 314
Buffalo treehopper (*Stictocephala bisonia*), 204
Buffering capacity, 53–54
Bulk density of soils, 44
Bunya-Bunya tree (*Araucaria*), 306

Burclover (*Medicago polymorpha*), 275
Bush anemone (*Carpenteria californica*), 317
Bush morning glory (*Convolvulus cneorum*), 324
Butia (palm), 307–309
Buttercup oxalis (*Oxalis pes-caprae*), 286
Butterflies, 126
Buxus (boxwood), 314–315

C

Cactus, 317
Caesalpinia, 315
Calcium carbonate, 53–54
Cales noacki (parasitic wasp), 184
Calico scale (*Eulecanium cerasorum*), 193
California bay (*Umbellularia californica*), 107f, 175f, 392–393
California bay laurel (*Umbellularia californica*), 107f, 175f, 392–393
California black oak, 106f, 198f, 215f
California black walnut (*Juglans*), 36f, 343–344
California buckeye (*Aesculus californica*), 303
California burclover (*Medicago polymorpha*), 275
California Christmas berry tingid (*Corythucha incurvata*), 208f
California fan palm (*Washingtonia filifera*), 87, 109f, 307–309
California fivespined ips (*Ips paraconfusus*), 224f
California gallfly (*Andricus californicus*), 215f
California hazel (*Corylus*), 325–326
California honeysuckle (*Lonicera*), 349–350
California Irrigation Management Information System (CIMIS), 33
California laurel aphid (*Euthoracaphis umbellulariae*), 175f, 176f
California laurel borer (*Rosalia funebris*), 232f
California Nursery Stock Nematode Certification program, 292
California oakworm (*Phryganidia californica*), 126f, 145f, 147
California pepper tree (*Schinus molle*), 174f, 282f, 385
California red scale, 188f, 189f
California wax myrtle (*Myrica*), 354
Caliroa cerasi (pear sawfly), 154
Callery pear (*Pyrus calleryana*), 370
Callirhytis perdens (oak gall wasp), 215f

Cycad scale (*Furchadaspis zamiae*), 189, 190f

Cycas (sago palm, cycad), 190f, 329

Cyclamen mite (*Phytonemus pallidus*), 249

Cydia latiferreana (filberworm), 162

Cynipids (gall wasps), 215–216

Cynodon dactylon (bermudagrass), 271, 272f, 281

Cyperus (nutsedges), 279–280

Cypress (*Cupressus*), 99–100, 328–329

Cypress bark beetle (*Phloeosinus*), 224

Cypress bark mealybug (*Ehrhornia cupressi*), 186–187, 188

Cypress bark moths, 35, 99

Cypress canker (*Seiridium cardinale*), 99–100

Cypress rust (*Gymnosporangium*), 95

Cypress sawflies (*Susana*), 154

Cypress scale (*Carulaspis minima*), 190–191

Cypress tip miner (*Argyresthia cupressella*), 217–218

Cyst nematode (*Heterodera*), 292f

Cystotheca (powdery mildew), 91

Cytospora canker (*Cytospora*), 100

D

Dactylifera (date palm), 110, 115–116, 162, 307–309

Dactylopius (cochineal scale), 196

Dagger nematode (*Xiphinema*), 294

DANGER label, 10

Daphne, 329

Dasineura gleditchiae (honeylocust pod gall midge), 214

Dasineura rhodophaga (rose midge), 163

Dawn redwood (*Metasequoia glyptostroboides*), 353

Deciduous trees and shrubs
  energy conservation and, 24
  growth cycle, 18f
  pruning, 37, 38

Decollate snail (*Rumina decollata*), 252

DED (Dutch elm disease), 113–115

Degree-day monitoring, 129
  citricola scale, 194
  elm leaf beetle, 156
  European elm scale, 197
  pine tip moth, 218

Deicing salts, 51

Delphastus pusillus (lady beetle), 185f

Dematophora root rot (*Rosellinia necatrix*), 121

Dendroctonus brevicomis (western pine beetle), 226

Dendroctonus valens (red turpentine beetle), 222f, 223, 225, 226

Deodar cedar (*Cedrus*), 320

Desert catalpa, desert willow (*Chilopsis linearis*), 322

Designing landscapes. *See* Landscape design

Diagnosing problems. *See also specific plants and plant problems*
  overview, 5–6
  common problems, by plant species, 299–395
  diseases, 66–68
  invertebrates, 127–129
  problem-solving guide, 296–298

Diamond scale (*Phaeochoropsis neowashingtoniae*), 66f, 87

Diaphorina citri (Asian citrus psyllid), 116–117, 170f

Diaprepes abbreviatus (diaprepes root weevil), 136f, 161–162

Diaspidiotus juglansregiae (walnut scale), 192

Diaspidiotus perniciosus (San Jose scale), 191–192

Diatomaceous earth, 141

Dicamba, 268

Digitaria (crabgrass), 278

Dinapate wrighti (giant palm borer), 163

Diomus pumilio (lady beetle), 170, 171f

Diplacus (monkey flower), 353

Diplocarpon mespili (Entomosporium leaf spot), 87–88

Diplocarpon rosae (black spot), 86–87

Diplodia (oak branch canker and dieback), 102

Disclisioprocta stellata (bougainvillea looper), 150

Discula (anthracnose), 81–83

Discula fraxinea (ash anthracnose), 81f, 82f

Discula platani (sycamore anthracnose), 81f

Disease
  defined, 65
  monitoring and diagnosis, 66–68
  symptoms, causes, 67–68, 296–298
  transmission, 14, 38f, 65–66, 69, 73

Disease-free plants. *See* Pathogen-free nursery stock

Disease management. *See also* Pesticides
  biological control, 74
  excluding foreign pests, 5, 69
  fertilization, 35, 71

irrigation, 29–30, 70–71

microorganisms, beneficial, 20, 74

mulch pathogens, 70

plant stress, 29–30, 68

prevention of disease, 68

pruning, 36–37, 72t, 73

resistant plants, 22–24, 68–69, 82t, 91t, 118t

sanitation, 14, 37, 38f, 73

selecting plants, 68–69, 70

site and design, 70

soil solarization, 74, 256–257

weed and insect control, 73–74

Disholcaspis washingtonensis (gall wasp), 215f

Disinfecting tools. *See* Sanitizing tools

Disposal of pesticides, 12f, 14

Dodder (*Cuscuta*), 282–283

Dodonaea, 329

Dog vomit fungus (*Fuligo septica*), 119f

Dogwood (*Cornus*), 82t, 324–325

Dolichovespula (yellowjackets), 244

Domesticated animals, to control weeds, 265

Dormant-season oils, 10, 142. *See also* Narrow-range oils

Douglas-fir (*Pseudotsuga menziesii*), 367–369

Douglas-fir needle cast (*Rhabdocline*), 85

Douglas-fir pitch moth (*Synanthedon novaroensis*), 237f, 239

Douglas-fir tussock moth (*Orgyia pseudotsugata*), 152

Downy mildew (*Peronospora, Plasmopara*), 89–90

Drab pit scale (*Asterodiaspis quercicola*), 198

Dracaena, Cordyline, 324

Dragon tree (*Cordyline, Dracaena*), 324

Drainage
  assessing, 25
  cause of disease, 71
  changes to, 40
  planting depth and, 26

Drip irrigation, 28, 30–31, 33. *See also* Irrigation

Drippy nut disease, drippy acorn (*Brenneria quercina*), 80

Drop crotch pruning, 37

Drosophila suzukii (spotted wing drosophila), 245

Drought-adapted plantings, 22, 29–30, 34, 41

Locust (*Robinia*), 378
Locust clearwing, or western poplar clearwing (*Paranthrene robiniae*), 236, 237*f*, 238, 239
Lollipop tree (*Myoporum*), 211*f*, 354
London plane (*Platanus*), 363–364. *See also* Sycamore
Longhorned beetles, 232–234
Longtailed mealybug (*Planococcus longispinus*), 187
Loopers, 150
Lophocampa argentata (silverspotted tiger moth), 146*f*
Lophostemon conferta (Tristania), 350
Lygaeus kalmii (small milkweed bug), 207
Lygus bug, 136*f*
Lyonothamnus floribundus (ironwood), 35*f*, 350
Lysiphlebus testaceipes (parasitic wasp), 177
Lysol, 73*t*

M

Maconellicoccus hirsutus (pink hibiscus mealybug), 185*f*, 186*f*
Madrone (*Arbutus menziesii*), 99*f*, 306–307
Madrone leaf miner (*Marmara arbutiella*), 217*f*
Madrone shield bearer (*Coptodisca arbutiella*), 219
Magnesium deficiency, 50
Magnolia, 350–351
Maidenhair tree (*Ginkgo biloba*), 53*f*, 337–338
Malacosoma (tent caterpillars), 151–152
Mallow (*Malva parviflora*), 276
Malus (crabapple), 351–352
Malva parviflora (little mallow), 276
Mammillaria (cactus), 317
Management methods, 8, 129–130. *See also* Biological control; Pesticides; *specific pests*
Manganese deficiency, 45*t*, 49
Manure, 47, 52
Manzanita (*Arctostaphylos*), 175*f*, 219*f*, 309–310
Maple (*Acer*), 64*f*, 117*f*, 301–303
Mare's tail (*Conyza canadensis*), 274
Marigolds, nematode suppressing, 292
Marmara arbutiella (madrone leaf miner), 217*f*
Marmara gulosa (citrus peelminer), 217*f*
Marssonina (anthracnose), 81–83
Material Safety Data Sheet (MSDS), 10

Mayten (*Maytenus boaria*), 352
Meadow spittlebug (*Philaenus spumarius*), 205
Mealybug destroyer lady beetle (*Cryptolaemus montrouzieri*), 134, 188
Mealybugs, 185–188
Measuringworms, or loopers, 150
Mechanical controls, 8, 129–130
Mechanical injuries, 60–61
Medicago lupulina (black medic), 275
Medicago polymorpha (California burclover), 275
Mediterranean climate, 21
Mediterranean fruit flies, 245
Megachile (leafcutter bee), 244*f*
Megaphragma mymaripenne (parasitic wasp), 210–211
Melaleuca, 54*f*, 352
Melia azedarach (Chinaberry), 353
Meloidogyne (root knot nematode), 291, 293
Melon aphid (*Aphis gossypii*), 176
Merhynchites (rose curculios), 163
Mesocriconema xenoplax (ring nematode), 294
Mesquite (*Prosopis*), 366
Metallic wood borers, 227–231
Metamorphosis of insects, mites, 126
Metaphycus (parasitic wasp), 132*t*, 193, 200
Metasequoia glyptostroboides (dawn redwood), 353
Metrosideros (iron tree), 353
Mexican orange (*Choisya ternata*), 322
Microbial insecticides, 133*t*, 140–141
Microlarinus (weevils), 265, 266*f*
Microorganisms, beneficial, 65, 74, 136
Microsphaera (powdery mildew), 90–91
Mildews, 89–91
Milk snail (*Otala lactea*), 251
Mimosa webworm (*Homadaula anisocentra*), 147, 152–153
Mimulus, Diplacus (monkey flower), 353
Mineral deficiencies. *See* Nutrient deficiencies
Minute cypress scale (*Carulaspis minima*), 190–191
Minute pirate bugs (*Orius, Anthocoris*), 127*f*, 138, 170, 212*f*
Mistletoes, 284–285
Mites
   overview, 245–246
   broad mite, 249

cyclamen mite, 249
false spider mites, 246*f*, 248–249
   life stages, 126, 246*f*
   management, 130, 131*f*, 133–134, **248–249**
   predaceous mites, 139, 212*f*, 246*f*, 248*f*, **249**
spider mites
      persea mite, 247
      pine and spruce spider mites, 247
      red mite, 246, 247*f*
      twospotted spider mite, 245*f*, 246, 248*f*, 249*f*
Mixing pesticides, 14, 269
Mock orange (*Choisya ternata*), 322
Mock orange (*Pittosporum*), 362–363
Modes of action, 11
Molds, 89–91
Molluscicide baits, 252
Monarthrum (oak ambrosia beetles), 222
Monitoring
   overview of techniques, 6–7
   disease, 67–68
   ET (evapotranspiration), 32–33, 126–127
   hazardous trees, 39
   invertebrates, 127–129
   soil moisture, 32–33
   weeds, 254–255
Monkey flower (*Mimulus, Diplacus*), 353
Monkey puzzle tree (*Araucaria*), 306
Mononchus (predaceous nematodes), 289, 290*f*
Monterey cypress (*Cupressus macrocarpa*), 21, 35, 187*f*, **328–329**
Monterey cypress scale (*Xylococculus macrocarpae*), 198
Monterey pine, 59, 103*f*, 161, 194*f*, **360–361**
Monterey pine scale (*Physokermes insignicola*), 194
Monterey pine tip moth (*Rhyacionia pasadenana*), 218
Morus (mulberry), 353–354
Mosaic viruses, 92–94
Mosses, 288
Moths, 126, 126*f*, 127*t*, 144–153
Mottle viruses, 92–94
Mounds, berms, or raised beds, 24–25, 30*f*, 31*f*, 33, 71

X